Lecture Notes in Computer Science 14074

Founding Editors

Gerhard Goos
Juris Hartmanis

Editorial Board Members

The series Lecture Notes in Computer Science (LNCS), including its subseries Lecture Notes in Artificial Intelligence (LNAI) and Lecture Notes in Bioinformatics (LNBI), has established itself as a medium for the publication of new developments in computer science and information technology research, teaching, and education.

LNCS enjoys close cooperation with the computer science R & D community, the series counts many renowned academics among its volume editors and paper authors, and collaborates with prestigious societies. Its mission is to serve this international community by providing an invaluable service, mainly focused on the publication of conference and workshop proceedings and postproceedings. LNCS commenced publication in 1973.

Jiří Mikyška · Clélia de Mulatier ·
Maciej Paszynski · Valeria V. Krzhizhanovskaya ·
Jack J. Dongarra · Peter M. A. Sloot
Editors

Computational Science – ICCS 2023

23rd International Conference
Prague, Czech Republic, July 3–5, 2023
Proceedings, Part II

Springer

Editors
Jiří Mikyška (ID)
Czech Technical University in Prague
Prague, Czech Republic

Clélia de Mulatier (ID)
University of Amsterdam
Amsterdam, The Netherlands

Maciej Paszynski (ID)
AGH University of Science and Technology
Krakow, Poland

Valeria V. Krzhizhanovskaya (ID)
University of Amsterdam
Amsterdam, The Netherlands

Jack J. Dongarra (ID)
University of Tennessee at Knoxville
Knoxville, TN, USA

Peter M. A. Sloot (ID)
University of Amsterdam
Amsterdam, The Netherlands

ISSN 0302-9743   ISSN 1611-3349 (electronic)
Lecture Notes in Computer Science
ISBN 978-3-031-36020-6   ISBN 978-3-031-36021-3 (eBook)
https://doi.org/10.1007/978-3-031-36021-3

This Springer imprint is published by the registered company Springer Nature Switzerland AG
The registered company address is: Gewerbestrasse 11, 6330 Cham, Switzerland

Preface

Welcome to the 23rd annual International Conference on Computational Science (ICCS - https://www.iccs-meeting.org/iccs2023/), held on July 3–5, 2023 at the Czech Technical University in Prague, Czechia.

In keeping with the new normal of our times, ICCS featured both in-person and online sessions. Although the challenges of such a hybrid format are manifold, we have always tried our best to keep the ICCS community as dynamic, creative, and productive as possible. We are proud to present the proceedings you are reading as a result.

ICCS 2023 was jointly organized by the Czech Technical University in Prague, the University of Amsterdam, NTU Singapore, and the University of Tennessee.

Standing on the Vltava River, Prague is central Europe's political, cultural, and economic hub.

The Czech Technical University in Prague (CTU) is one of Europe's largest and oldest technical universities and the highest-rated in the group of Czech technical universities. CTU offers 350 accredited study programs, 100 of which are taught in a foreign language. Close to 19,000 students are studying at CTU in 2022/2023. The Faculty of Nuclear Sciences and Physical Engineering (FNSPE), located along the river bank in Prague's beautiful Old Town (Staré Mesto) and host to ICCS 2023, is the only one in Czechia to offer studies in a broad range of fields related to Nuclear Physics and Engineering. The Faculty operates both fission (VR-1) and fusion (GOLEM Tokamak) reactors and hosts several cutting-edge research projects, collaborating with a number of international research centers (CERN, ITER, BNL-STAR, ELI).

The International Conference on Computational Science is an annual conference that brings together researchers and scientists from mathematics and computer science as basic computing disciplines, as well as researchers from various application areas who are pioneering computational methods in sciences such as physics, chemistry, life sciences, engineering, arts, and humanitarian fields, to discuss problems and solutions in the area, identify new issues, and shape future directions for research.

Since its inception in 2001, ICCS has attracted increasingly higher-quality attendees and papers, and this year is not an exception, with over 300 participants. The proceedings series have become a primary intellectual resource for computational science researchers, defining and advancing the state of the art in this field.

The theme for 2023, **"Computation at the Cutting Edge of Science"**, highlights the role of Computational Science in assisting multidisciplinary research. This conference was a unique event focusing on recent developments in scalable scientific algorithms; advanced software tools; computational grids; advanced numerical methods; and novel application areas. These innovative novel models, algorithms, and tools drive new science through efficient application in physical systems, computational and systems biology, environmental systems, finance, and others.

ICCS is well known for its excellent lineup of keynote speakers. The keynotes for 2023 were:

- **Helen Brooks**, United Kingdom Atomic Energy Authority (UKAEA), UK
- **Jack Dongarra**, University of Tennessee, USA
- **Derek Groen**, Brunel University London, UK
- **Anders Dam Jensen**, European High Performance Computing Joint Undertaking (EuroHPC JU), Luxembourg
- **Jakub Šístek**, Institute of Mathematics of the Czech Academy of Sciences & Czech Technical University in Prague, Czechia

This year we had 531 submissions (176 to the main track and 355 to the thematic tracks). In the main track, 54 full papers were accepted (30.7%); in the thematic tracks, 134 full papers (37.7%). A higher acceptance rate in the thematic tracks is explained by the nature of these, where track organizers personally invite many experts in a particular field to participate in their sessions. Each submission received at least 2 single-blind reviews (2.9 reviews per paper on average).

ICCS relies strongly on our thematic track organizers' vital contributions to attract high-quality papers in many subject areas. We would like to thank all committee members from the main and thematic tracks for their contribution to ensuring a high standard for the accepted papers. We would also like to thank *Springer, Elsevier,* and *Intellegibilis* for their support. Finally, we appreciate all the local organizing committee members for their hard work in preparing for this conference.

We are proud to note that ICCS is an A-rank conference in the CORE classification.

We hope you enjoyed the conference, whether virtually or in person.

July 2023

Jiří Mikyška
Clélia de Mulatier
Maciej Paszynski
Valeria V. Krzhizhanovskaya
Jack J. Dongarra
Peter M. A. Sloot

Organization

The Conference Chairs

General Chair

Valeria Krzhizhanovskaya   University of Amsterdam, The Netherlands

Main Track Chair

Clélia de Mulatier   University of Amsterdam, The Netherlands

Thematic Tracks Chair

Maciej Paszynski   AGH University of Science and Technology, Poland

Scientific Chairs

Peter M. A. Sloot   University of Amsterdam, The Netherlands | Complexity Institute NTU, Singapore

Jack Dongarra   University of Tennessee, USA

Local Organizing Committee

LOC Chair

Jiří Mikyška   Czech Technical University in Prague, Czechia

LOC Members

Pavel Eichler   Czech Technical University in Prague, Czechia
Radek Fučík   Czech Technical University in Prague, Czechia
Jakub Klinkovský   Czech Technical University in Prague, Czechia
Tomáš Oberhuber   Czech Technical University in Prague, Czechia
Pavel Strachota   Czech Technical University in Prague, Czechia

Thematic Tracks and Organizers

Advances in High-Performance Computational Earth Sciences: Applications and Frameworks – IHPCES

Takashi Shimokawabe, Kohei Fujita, Dominik Bartuschat

Artificial Intelligence and High-Performance Computing for Advanced Simulations – AIHPC4AS

Maciej Paszynski, Robert Schaefer, Victor Calo, David Pardo, Quanling Deng

Biomedical and Bioinformatics Challenges for Computer Science – BBC

Mario Cannataro, Giuseppe Agapito, Mauro Castelli, Riccardo Dondi, Rodrigo Weber dos Santos, Italo Zoppis

Computational Collective Intelligence – CCI

Marcin Maleszka, Ngoc Thanh Nguyen

Computational Diplomacy and Policy – CoDiP

Michael Lees, Brian Castellani, Bastien Chopard

Computational Health – CompHealth

Sergey Kovalchuk, Georgiy Bobashev, Anastasia Angelopoulou, Jude Hemanth

Computational Modelling of Cellular Mechanics – CMCM

Gabor Zavodszky, Igor Pivkin

Computational Optimization, Modelling, and Simulation – COMS

Xin-She Yang, Slawomir Koziel, Leifur Leifsson

Computational Social Complexity – CSCx

Vítor V. Vasconcelos. Debraj Roy, Elisabeth Krüger, Flávio Pinheiro, Alexander J. Stewart, Victoria Garibay, Andreia Sofia Teixeira, Yan Leng, Gabor Zavodszky

Computer Graphics, Image Processing, and Artificial Intelligence – CGIPAI

Andres Iglesias, Lihua You, Akemi Galvez-Tomida

Machine Learning and Data Assimilation for Dynamical Systems – MLDADS

Rossella Arcucci, Cesar Quilodran-Casas

MeshFree Methods and Radial Basis Functions in Computational Sciences – MESHFREE

Vaclav Skala, Samsul Ariffin Abdul Karim

Multiscale Modelling and Simulation – MMS

Derek Groen, Diana Suleimenova

Network Models and Analysis: From Foundations to Complex Systems – NMA

Marianna Milano, Pietro Cinaglia, Giuseppe Agapito

Quantum Computing – QCW

Katarzyna Rycerz, Marian Bubak

Simulations of Flow and Transport: Modeling, Algorithms, and Computation – SOFTMAC

Shuyu Sun, Jingfa Li, James Liu

Smart Systems: Bringing Together Computer Vision, Sensor Networks and Machine Learning – SmartSys

Pedro Cardoso, Roberto Lam, Jânio Monteiro, João Rodrigues

Solving Problems with Uncertainties – SPU

Vassil Alexandrov, Aneta Karaivanova

Teaching Computational Science – WTCS

Angela Shiflet, Nia Alexandrov

Reviewers

Zeeshan Abbas
Samsul Ariffin Abdul Karim
Tesfamariam Mulugeta Abuhay
Giuseppe Agapito
Elisabete Alberdi
Vassil Alexandrov
Nia Alexandrov
Alexander Alexeev
Nuno Alpalhão
Julen Alvarez-Aramberri
Domingos Alves
Sergey Alyaev
Anastasia Anagnostou
Anastasia Angelopoulou
Fabio Anselmi
Hideo Aochi
Rossella Arcucci
Konstantinos Asteriou
Emanouil Atanassov
Costin Badica
Daniel Balouek-Thomert
Krzysztof Banaś
Dariusz Barbucha
Luca Barillaro
João Barroso
Dominik Bartuschat
Pouria Behnodfaur
Jörn Behrens
Adrian Bekasiewicz
Gebrail Bekdas
Mehmet Belen
Stefano Beretta
Benjamin Berkels
Daniel Berrar
Piotr Biskupski
Georgiy Bobashev
Tomasz Boiński
Alessandra Bonfanti

Carlos Bordons
Bartosz Bosak
Lorella Bottino
Roland Bouffanais
Lars Braubach
Marian Bubak
Jérémy Buisson
Aleksander Byrski
Cristiano Cabrita
Xing Cai
Barbara Calabrese
Nurullah Çalık
Victor Calo
Jesús Cámara
Almudena Campuzano
Cristian Candia
Mario Cannataro
Pedro Cardoso
Eddy Caron
Alberto Carrassi
Alfonso Carriazo
Stefano Casarin
Manuel Castañón-Puga
Brian Castellani
Mauro Castelli
Nicholas Chancellor
Ehtzaz Chaudhry
Théophile Chaumont-Frelet
Thierry Chaussalet
Sibo Cheng
Siew Ann Cheong
Lock-Yue Chew
Su-Fong Chien
Marta Chinnici
Bastien Chopard
Svetlana Chuprina
Ivan Cimrak
Pietro Cinaglia

Noélia Correia
Adriano Cortes
Ana Cortes
Anna Cortes
Enrique Costa-Montenegro
David Coster
Carlos Cotta
Peter Coveney
Daan Crommelin
Attila Csikasz-Nagy
Javier Cuenca
António Cunha
Luigi D'Alfonso
Alberto d'Onofrio
Lisandro Dalcin
Ming Dao
Bhaskar Dasgupta
Clélia de Mulatier
Pasquale Deluca
Yusuf Demiroglu
Quanling Deng
Eric Dignum
Abhijnan Dikshit
Tiziana Di Matteo
Jacck Długopolski
Anh Khoa Doan
Sagar Dolas
Riccardo Dondi
Rafal Drezewski
Hans du Buf
Vitor Duarte
Rob E. Loke
Amir Ebrahimi Fard
Wouter Edeling
Nadaniela Egidi
Kareem Elsafty
Nahid Emad
Christian Engelmann
August Ernstsson
Roberto R. Expósito
Fangxin Fang
Giuseppe Fedele
Antonino Fiannaca
Christos Filelis-Papadopoulos
Piotr Frąckiewicz

Alberto Freitas
Ruy Freitas Reis
Zhuojia Fu
Kohei Fujita
Takeshi Fukaya
Wlodzimierz Funika
Takashi Furumura
Ernst Fusch
Marco Gallieri
Teresa Galvão Dias
Akemi Galvez-Tomida
Luis Garcia-Castillo
Bartłomiej Gardas
Victoria Garibay
Frédéric Gava
Piotr Gawron
Bernhard Geiger
Alex Gerbessiotis
Josephin Giacomini
Konstantinos Giannoutakis
Alfonso Gijón
Nigel Gilbert
Adam Glos
Alexandrino Gonçalves
Jorge González-Domínguez
Yuriy Gorbachev
Pawel Gorecki
Markus Götz
Michael Gowanlock
George Gravvanis
Derek Groen
Lutz Gross
Tobias Guggemos
Serge Guillas
Xiaohu Guo
Manish Gupta
Piotr Gurgul
Zulfiqar Habib
Yue Hao
Habibollah Haron
Mohammad Khatim Hasan
Ali Hashemian
Claire Heaney
Alexander Heinecke
Jude Hemanth

Marcin Hernes
Bogumila Hnatkowska
Maximilian Höb
Rolf Hoffmann
Tzung-Pei Hong
Muhammad Hussain
Dosam Hwang
Mauro Iacono
Andres Iglesias
Mirjana Ivanovic
Alireza Jahani
Peter Janků
Jiri Jaros
Agnieszka Jastrzebska
Piotr Jedrzejowicz
Gordan Jezic
Zhong Jin
Cedric John
David Johnson
Eleda Johnson
Guido Juckeland
Gokberk Kabacaoglu
Piotr Kalita
Aneta Karaivanova
Takahiro Katagiri
Mari Kawakatsu
Christoph Kessler
Faheem Khan
Camilo Khatchikian
Petr Knobloch
Harald Koestler
Ivana Kolingerova
Georgy Kopanitsa
Pavankumar Koratikere
Sotiris Kotsiantis
Sergey Kovalchuk
Slawomir Koziel
Dariusz Król
Elisabeth Krüger
Valeria Krzhizhanovskaya
Sebastian Kuckuk
Eileen Kuehn
Michael Kuhn
Tomasz Kulpa
Julian Martin Kunkel

Krzysztof Kurowski
Marcin Kuta
Roberto Lam
Rubin Landau
Johannes Langguth
Marco Lapegna
Ilaria Lazzaro
Paola Lecca
Michael Lees
Leifur Leifsson
Kenneth Leiter
Yan Leng
Florin Leon
Vasiliy Leonenko
Jean-Hugues Lestang
Xuejin Li
Qian Li
Siyi Li
Jingfa Li
Che Liu
Zhao Liu
James Liu
Marcellino Livia
Marcelo Lobosco
Doina Logafatu
Chu Kiong Loo
Marcin Łoś
Carlos Loucera
Stephane Louise
Frederic Loulergue
Thomas Ludwig
George Lykotrafitis
Lukasz Madej
Luca Magri
Peyman Mahouti
Marcin Maleszka
Alexander Malyshev
Tomas Margalef
Osni Marques
Stefano Marrone
Maria Chiara Martinis
Jaime A. Martins
Paula Martins
Pawel Matuszyk
Valerie Maxville

Pedro Medeiros
Wen Mei
Wagner Meira Jr.
Roderick Melnik
Pedro Mendes Guerreiro
Yan Meng
Isaak Mengesha
Ivan Merelli
Tomasz Michalak
Lyudmila Mihaylova
Marianna Milano
Jaroslaw Miszczak
Dhruv Mittal
Miguel Molina-Solana
Fernando Monteiro
Jânio Monteiro
Andrew Moore
Anabela Moreira Bernardino
Eugénia Moreira Bernardino
Peter Mueller
Khan Muhammad
Daichi Mukunoki
Judit Munoz-Matute
Hiromichi Nagao
Kengo Nakajima
Grzegorz J. Nalepa
I. Michael Navon
Vittorio Nespeca
Philipp Neumann
James Nevin
Ngoc-Thanh Nguyen
Nancy Nichols
Marcin Niemiec
Sinan Melih Nigdeli
Hitoshi Nishizawa
Algirdas Noreika
Manuel Núñez
Joe O'Connor
Frederike Oetker
Lidia Ogiela
Ángel Javier Omella
Kenji Ono
Eneko Osaba
Rongjiang Pan
Nikela Papadopoulou

Marcin Paprzycki
David Pardo
Anna Paszynska
Maciej Paszynski
Łukasz Pawela
Giulia Pederzani
Ebo Peerbooms
Alberto Pérez de Alba Ortíz
Sara Perez-Carabaza
Dana Petcu
Serge Petiton
Beata Petrovski
Toby Phillips
Frank Phillipson
Eugenio Piasini
Juan C. Pichel
Anna Pietrenko-Dabrowska
Gustavo Pilatti
Flávio Pinheiro
Armando Pinho
Catalina Pino Muñoz
Pietro Pinoli
Yuri Pirola
Igor Pivkin
Robert Platt
Dirk Pleiter
Marcin Płodzień
Cristina Portales
Simon Portegies Zwart
Roland Potthast
Małgorzata Przybyła-Kasperek
Ela Pustulka-Hunt
Vladimir Puzyrev
Ubaid Qadri
Rick Quax
Cesar Quilodran-Casas
Issam Rais
Andrianirina Rakotoharisoa
Célia Ramos
Vishwas H. V. S. Rao
Robin Richardson
Heike Riel
Sophie Robert
João Rodrigues
Daniel Rodriguez

Marcin Rogowski

Sergio Rojas

Diego Romano

Albert Romkes

Debraj Roy

Adam Rycerz

Katarzyna Rycerz

Mahdi Saeedipour

Arindam Saha

Ozlem Salehi

Alberto Sanchez

Ayşin Sancı

Gabriele Santin

Vinicius Santos Silva

Allah Bux Sargano

Azali Saudi

Ileana Scarpino

Robert Schaefer

Ulf D. Schiller

Bertil Schmidt

Martin Schreiber

Gabriela Schütz

Jan Šembera

Paulina Sepúlveda-Salas

Ovidiu Serban

Franciszek Seredynski

Marzia Settino

Mostafa Shahriari

Vivek Sheraton

Angela Shiflet

Takashi Shimokawabe

Alexander Shukhman

Marcin Sieniek

Joaquim Silva

Mateusz Sitko

Haozhen Situ

Leszek Siwik

Vaclav Skala

Renata Słota

Oskar Slowik

Grażyna Ślusarczyk

Sucha Smanchat

Alexander Smirnovsky

Maciej Smołka

Thiago Sobral

Isabel Sofia

Piotr Sowiński

Christian Spieker

Michał Staniszewski

Robert Staszewski

Alexander J. Stewart

Magdalena Stobinska

Tomasz Stopa

Achim Streit

Barbara Strug

Dante Suarez

Patricia Suarez

Diana Suleimenova

Shuyu Sun

Martin Swain

Edward Szczerbicki

Tadeusz Szuba

Ryszard Tadeusiewicz

Daisuke Takahashi

Osamu Tatebe

Carlos Tavares Calafate

Andrey Tchernykh

Andreia Sofia Teixeira

Kasim Terzic

Jannis Teunissen

Sue Thorne

Ed Threlfall

Alfredo Tirado-Ramos

Pawel Topa

Paolo Trunfio

Hassan Ugail

Carlos Uriarte

Rosarina Vallelunga

Eirik Valseth

Tom van den Bosch

Ana Varbanescu

Vítor V. Vasconcelos

Alexandra Vatyan

Patrick Vega

Francesc Verdugo

Gytis Vilutis

Jackel Chew Vui Lung

Shuangbu Wang

Jianwu Wang

Peng Wang

Katarzyna Wasielewska
Jarosław Wątróbski
Rodrigo Weber dos Santos
Marie Weiel
Didier Wernli
Lars Wienbrandt
Iza Wierzbowska
Maciej Woźniak
Dunhui Xiao
Huilin Xing
Yani Xue
Abuzer Yakaryilmaz
Alexey Yakovlev
Xin-She Yang
Dongwei Ye
Vehpi Yildirim
Lihua You

Drago Žagar
Sebastian Zając
Constantin-Bala Zamfirescu
Gabor Zavodszky
Justyna Zawalska
Pavel Zemcik
Wenbin Zhang
Yao Zhang
Helen Zhang
Jian-Jun Zhang
Jinghui Zhong
Sotirios Ziavras
Zoltan Zimboras
Italo Zoppis
Chiara Zucco
Pavel Zun
Karol Życzkowski

Contents – Part II

Advances in High-Performance Computational Earth Sciences: Applications and Frameworks

Artificial Intelligence and High-Performance Computing for Advanced Simulations

Biomedical and Bioinformatics Challenges for Computer Science

Hierarchical Relative Expression Analysis in Multi-omics Data

Marcin Czajkowski, Krzysztof Jurczuk, and Marek Kretowski

ICCS 2023 Main Track Full Papers (Continued)

Automated Identification and Location of Three Dimensional Atmospheric Frontal Systems

Stefan Niebler[1]([✉]), Bertil Schmidt[1][iD], Holger Tost[2][iD], and Peter Spichtinger[2][iD]

[1] Institut für Informatik, Johannes Gutenberg-Universität, 55099 Mainz, Germany
stnieble@uni-mainz.de
[2] Institut für Physik der Atmosphäre, Johannes Gutenberg-Universität,
55099 Mainz, Germany

Abstract. We present a novel method to identify and locate weather fronts at various pressure levels to create a three dimensional structure using weather data located at the North Atlantic. It provides statistical evaluations regarding the slope and weather phenomena correlated to the identified three dimensional structure. Our approach is based on a deep neural network to locate 2D surface fronts first, which are then used as an initialization to extend them to various height levels. We show that our method is able to detect frontal locations between 500 hPa and 1000 hPa.

Keywords: Weather Prediction · Atmospheric Physics · Deep Learning · Climate Change

1 Introduction

Weather fronts are important synoptic scale phenomena influencing the atmospheric environment and weather conditions surrounding it. They can be correlated to extreme weather events such as extreme precipitation [4,14] or cyclones [16]. Fronts are usually depicted on a 2D-spatial grid, showing the global position of a front at the surface level [8,14]. While this depiction can be sufficient to satisfy the information about the current location of a front, it neglects the fact that a front is actually a three dimensional, inclined structure instead of just a line in 2D-space. Previous methods for detecting fronts relied on data with low spatial resolution, which makes it hard to analyze frontal inclination due to horizontal coarseness. With recent high resolution data sets such as ERA5 [6] we can now provide spatially more accurate locations of fronts which allows us to determine improved representations of the 3D-structure of fronts.

Classical computational approaches for the detection of spatial 2D fronts rely on derivations of a thermal field such as wet bulb temperature, or equivalent potential temperature, respectively [1,7,8]. Many of these methods need coarser resolutions or a smoothed thermal field, since they were developed for synoptic scale features on low resolutions. The field of frontal detection is still highly

J. Mikyška et al. (Eds.): ICCS 2023, LNCS 14074, pp. 3–17, 2023.
https://doi.org/10.1007/978-3-031-36021-3_1

active and tools are still refined regularly; e.g. [17]. Deep learning (DL) based methods have recently been successfully applied to several different problems in atmospheric physics from cyclone tracking [5] to forecasting [11,15]. These methods have shown to be able to exceed the performance of traditional numerical methods for example in the field of numerical weather prediction (NWP). Machine Learning (ML) has also been applied to the task of front detection. Previous work to detect 2D fronts used random forests [3] and neural networks [2,10,12]. In contrast to traditional algorithms these methods can be trained on high resolution grids and may therefore be directly applied to data sets at this resolution. In recent work [14], we presented a deep neural network for detecting and classifying frontal lines directly on ERA5 renalysis data. The used data is provided as a latitude-longitude (lat-lon) grid at a resolution of 0.25° along both horizontal axes. In contrast to other DL-based front detection methods this model directly outputs the identified fronts as thin lines allowing for high spatial accuracy. Additionally, the model classifies the detected fronts into 4 types (*warm, cold, stationary, occluded*). While the evaluation was restricted to areas covering parts of the northern Hemisphere extending from North America over the Atlantic to Europe, we also showed qualitatively plausible results for other parts of the globe.

None of these methods is able to detect the actual three dimensional shape of a front. So far there is only one approach for the detection of three dimensional frontal fields [9]. It applies Hewson's method to several height levels. However, the approach performs several iterations of smoothing before frontal detection, which removes fine scale features, reducing the gained information from high resolution data. This is necessary due to the applied frontal detection algorithm, which relies on a smoothed temperature field for good results, since it was developed for coarse resolution data sets.

A typical problem for DL-based methods is the need for labeled training data. While the ECMWF provides atmospheric grids for several decades there exists - to the best of our knowledge - no data set regarding three dimensional structure of weather fronts. As such supervised training of a network to detect three dimensional fronts is not directly possible and other methods are needed to complement DL-based front detection.

In this work we propose a new method based on our recently described DL model [14]. We use the 2D fronts inferred by the network and add a new post processing pipeline, which extends them to frontal surfaces within the three dimensional ERA5 grid. To avoid processing a complete 3D grid, we also use the information obtained from the network output such as the frontal location and classification for more efficient computation. Furthermore our model does not rely on fixed thermal thresholds to filter out wrong results such as the method proposed by [9]. Instead we use an adaptive threshold to filter false positives per frontal object. As such our method is – to the best of our knowledge – the first to enable statistical analysis of three dimensional fronts over multiple timestamps. It is able to provide qualitatively good results, which are in agreement with theories regarding the shape and characteristics of a front. As we are not restricted to singular cases, our method can be a valuable tool in research of front related weather phenomena. In comparison to previous methods our algorithm can be directly applied onto the ERA5 grid, without any smoothing, making full use of high resolution data.

2 Data

2.1 Atmospheric Data

Processing and evaluation is performed on ERA5 reanalysis data using pressure levels extending from 1000 hPa (surface) to 500 hPa (approx. 5 km above sea level). For our algorithm we use data covering the area $[-90, 50]° E$ and $[90, 0]° N$. As we want to extract wide cross sections we only take surface front pixels into account, that are located within the region spanning $[-65, 25]° E$ and $[65, 25]° N$. This region mostly covers the North Atlantic, minimizing influences of orography such as mountains. The area also contains the northern mid latitudes, where frontal systems are commonly located [18]. The used ERA5 data does not mask invalid grid points which would be located within the terrain (e.g. data at 1000 hPa in Greenland), but rather uses interpolation to provide values for these areas. We do not alter this data except for the calculation of the equivalent potential temperature θ_e, which is a conserved quantity of reversible phase transitions, and thus well suited for investigations of fronts. It can be calculated from the atmospheric variables (pressure p, temperature T, water vapor mixing ratio q_v). We used the approximation provided by MetPy v1.3.1 [13].

2.2 Frontal Types

Fronts are distributed into the 4 categories: **warm, cold, stationary, occluded**. Warm and cold fronts have a clear structure consisting of a warm and a cold side. Stationary fronts are mainly defined by their travelling speed and not by their gradient direction. Occluded fronts on the other hand are the result of a cold and a warm front overlapping. Thus gradient strength is reduced at lower levels, while they are split into two parts at higher altitudes, where the two fronts have not overlapped yet. Thus, we mainly focus on determining the three dimensional structure of warm and cold fronts, even though the network is able to provide surface fronts for all 4 categories [14].

3 Methods

Our pipeline consists of 4 stages processing the detected 2D surface fronts provided by the deep neural network. For each pixel in the surface front we calculate the normal direction and extract a cross section of the underlying thermal field along this vector. We evaluate each point of a cross section and perform local optimization to find the positions of the front at each processed pressure level within each cross section. The workflow is illustrated in Fig. 1 and explained in more detail in the following subsections.

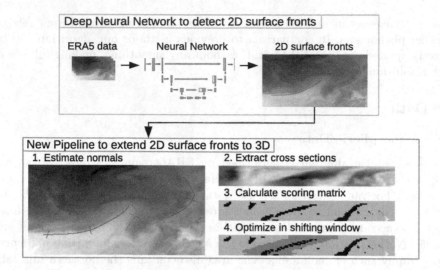

Fig. 1. Top Box: 2D surface fronts are detected from ERA5 data using a deep neural network. Bottom Box: Steps performed in our pipeline for one normal per front: (1) Normal directions of surface fronts are estimated; (2) cross sections along these normal directions are extracted; (3) each pixel is evaluated, creating a scoring matrix; (4) the minimum within a shifting window (turquoise area) is calculated and returned as front (red dots). (Color figure online)

3.1   Surface Front Detection

Prior to our new pipeline we locate and classify surface fronts within the ERA5 data grid. For this task we use our neural network described in [14]. The network was trained using the frontal labels provided by two different weather services (German Weather Service, North American Weather Service), both of which covering parts of the Northern Atlantic region in their analysis. Our evaluation also shows that the network performs well over the sea covered surface - such as the Northern Atlantic, which we use in this work. The network output consists of a multi channel 2D grid where each channel denotes the locations of the different frontal classes (*warm, cold, stationary, occluded*) on the ERA5 grid.

3.2   Cross Sections

Given the locations of the surface fronts for each class we identify frontal objects by their connectivity and ignore small objects in the following steps. For each front remaining we extract a cross section along an estimated normal for each pixel in each determined surface front. Each cross section contains values for the equivalent potential temperature θ_e, at each pressure level in our data set for 100 sample locations before as well as 100 sample locations behind the location of the surface fronts. Sample locations are determined by estimating a normal and then sampling the temperature field at 20 km intervals along the normal direction. As the earth can be approximated by a sphere we estimate the normal on the tangential plane of each pixel on a sphere rather than on the flat lat-lon grid. Variable values for sampling locations not located directly on a grid point are interpolated. Points where the cross section exceeds the limits of our input region, as well as cross sections containing invalid data (e.g. due to an error in the files or interpolation with values outside the data region) are removed. The orientation of each cross section is chosen such that the sign of the dot product with the wind is positive.

Oversampling. The determined surface fronts may be smaller - in terms of pixel count - than the same front at higher altitudes. For example a front may extend in a radial fashion where the radius increases with altitude, leading to a larger circumference of the circular sector. This may result in comparatively sparse samples of higher altitudes, even though the lower levels may be densely sampled. To compensate for this we create multiple samples from each surface front pixel by additionally creating cross sections with slightly deviated positions and/or angles. For angular deviations of each point we additionally sample cross sections where the normal is rotated by $\pm 30°$, resulting in 3 different rotations including the non-rotated cross sections. Additionally we sample each of those 3 normal directions at shifted positions. We evaluate a total of 9 shifted positions. These shifted sample points are calculated as the points at a distance of $[-4, -3, ..., 3, 4]$ from the surface front pixel in the direction perpendicular to the non rotated normal direction. This results in a total sampling of 27 cross sections per point.

Orientation Correction. It is important that cross sections are correctly oriented, because the algorithm may not correctly process wrongly oriented samples, as they exhibit contrary features such as an inverted sign of the gradient. However, in some cases the wind may be oriented (near) perpendicular to the front-normal. In these cases the orientation based on the wind direction may be unreliable, leading to wrongly oriented cross sections. Oversampling - mainly caused by the rotation - may further add some wrongly oriented samples. As such we adjust the orientation of samples by taking batches of each frontal objects pixels and reorient each sample within a batch based on their sign of the dot product with the mean direction of the batch.

3.3   Extension to Other Pressure Levels

Once we obtain the 2D cross-section slices for each surface front pixel we can calculate the front location for each height level within that slice.

In a first step we flip the orientation of our cold fronts such that the cold region is located right to the center, as it should already be the case for warm fronts. This way we can calculate the positions regardless of the frontal type and potential differences in the sign of the gradients. A front typically bends over its cold region, meaning that the purely warm side has little to no information. We can therefore cut large parts of the warm region located left of the center of each cross section to reduce computation, without a loss of information. We cut the first 60 pixels of each cross section, keeping a buffer of 40 pixels left of the center.

To find the frontal location in each cross section we first score each pixel of the cross section to create a score matrix. This score matrix is created evaluating the scoring functions as described in Subsect. 3.4. Once the scoring matrix is calculated we can determine the frontal structure. Starting from the surface front which we initially locate at the center of the bottom of our cross section, we crop a smaller region of 40 pixels width centered at the x position of our surface front position and a height (extent in y-coordinate) of $h = 1$ pixel. We call this the current window. The optimal front location is then determined by minimizing a scoring function within the current window. Once we determined h separation points (1 point per pressure level) we shift our window such that it is now centered at the mean x position determined as our front location and covering the next set of h pressure levels. With this windowing approach we restrict the front detection to a smaller region where the simple optimization is applicable, as we assume that only one front is located within the window. The shifting of the window with increasing altitude allows us to potentially follow the fronts inclination across the whole width of each cross section, without the need to increase the window size. This method may fail if the difference between two succeeding height levels is larger than the distance covered by the window. Overall our approach essentially reduces the problem of locating a front in a three dimensional space to a set of 2D local optimization problems. As there is no dependency between cross section - even of the same object - the method can be easily adapted for parallel processing using multiple threads on GPUs or CPUs.

3.4   Scoring Function

The presented algorithm finds the optimal position as the position that optimizes a scoring function at each pressure level. Our scoring function consists of three parts each being connected to a typical characteristic of weather fronts.

$$L_g = \nabla_x \theta_e \tag{1}$$

$$L_e = \left|\left|\left(\mathrm{var}[\theta_e^{i,j+1}, ..., \theta_e^{i,j+k}], \mathrm{var}[\theta_e^{i,j-k}, ..., \theta_e^{i,j-1}]\right)\right|\right|_2, i, j \in \mathbf{N} \tag{2}$$

$$L_d = \mathrm{mean}[\theta_e^{i,j+1}, ..., \theta_e^{i,j+k}] - \mathrm{mean}[\theta_e^{i,j-k}, ..., \theta_e^{i,j-1}], i, j \in \mathbf{N} \tag{3}$$

where index i indicate the vertical level, j denotes the horizontal grid position, and we set $k = 20$ as half the running window size of 40 respectively. Equation 1 describes the horizontal gradient of the temperature field. As we orient each sample such that the cold side is located at higher x indices the location of the strongest negative gradient is considered to be the ideal position for our front. This is consistent with the classic definition where the front is located at the zero pass of the thermal front parameter, which in our case simply becomes the 2nd derivative of θ_e. We can use an optimization approach for this equation instead of a commonly seen threshold based decision, as

a) we only look at a locally restricted area, where we can assume that the frontal gradient is the most dominant effect

b) due to our initialization we already know that a front is located here, so we have a lower risk of false positives.

However, we need to respect that fronts may not extend all the way up to 500 hPa. We therefore added a variable threshold that only points where Eq. 1 is among the lowest 10% of all gradients within a front at each pressure level are considered valid locations for a front. As a result, points with very weak gradients are omitted and not seen as fronts. The latter threshold is adaptable to the general surroundings meaning that it works for both very strong fronts with a strong gradient as well as weaker fronts where the gradient may be not as dominant.

Equation 2 describes our assumption that the air masses before and after a front should be locally consistent, i.e., we do not expect high variation in the temperature within one air mass. As a result regions with very high temperature variance in any air mass are considered less ideal locations for a front.

Finally, Eq. 3 describes the fact that the cold air mass is considered to have a lower mean equivalent potential temperature than the warm air mass. Our final scoring function therefore can be described as

$$L = w_g L_g + w_e L_e + w_d L_d \tag{4}$$

with positive weights $w_g, w_e, w_d \in \mathbf{R}$. For our evaluations we set all weights to 1.

3.5   Final Processing

Once we calculated the separation point for each height level, we have obtained a 1D array consisting of the optimal frontal position for each level. If we combine all these result vectors for all cross section of a frontal object, we can create a point cloud sampling of the 3D frontal object. If desired one could use post processing techniques to create a 2D-surface instead. However, for our purposes point clouds are sufficient as we are interested in statistical evaluation of the resulting fronts. As our initial surface fronts are already classified we also have a classification for the three dimensional fronts by simply copying the initial frontal type for all levels.

4   Results

We present several applications of our method and show results for the evaluated dataset consisting of hourly reanalysis data for 2016. Due to a lack of a ground truth dataset of annotated 3D fronts we can only provide qualitative evaluations. These evaluations show that the characteristics of our three dimensional fronts are in accordance with classical theoretical knowledge surrounding fronts and their expected behaviour. Further we can see that aggregated results as well as the visual presentation are plausible. Figure 2 shows warm and cold fronts as identified by our proposed method at 1000 hPa, 850 hPa and 650 hPa for 2016-09-23 00:00 UTC over a background of θ_e (interval: $[270, 360]$ K). We can clearly see an inclination of the central meridional warm front, as it "moves" eastward with increasing altitude. While the central cold front stays roughly at the same location, indicating a much steeper inclination instead. Note that occluded and stationary fronts are not shown.

For all our tests we first calculated surface front positions and saved them to disk. This way the rest of the pipeline could be executed without the need for a GPU. In these evaluations we removed results where the optimal separation point was found to be at 0 location within the reduced cross section grid. In addition, all locations where no valid separation location was found were also set to 0 before the filter was applied, removing those as well from the evaluation. This results in approximately 7% to 20% of cross sections being filtered per level. Here we can observe that the lower and upper levels contain more filtered samples compared to the levels around 850 hPa, which indicates that frontal characteristics are more pronounced at this level instead of high altitudes or near surface. Potential causes why no valid separation could be found include falsely identified surface fronts, weakly expressed frontal characteristics at the edges of a front, bad samples caused by the additional sampling or that the front is not present at this level at all.

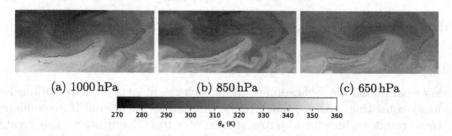

(a) 1000 hPa   (b) 850 hPa   (c) 650 hPa

270 280 290 300 310 320 330 340 350 360
θ_e (K)

Fig. 2. Locations of cold (blue) and warm (red) fronts at 1000 hPa (a), 850 hPa (b) and 650 hPa (c) for 2016-09-23 00:00 UTC. Shown Snippet Coordinates: $[70, 30]°$N, $[-80, 10]°$E. Background: θ_e (interval: $[270, 360]$ K). (Color figure online)

4.1   Probability Density Distribution of Temperature Difference

A typical warm or cold front is classified by a strong temperature gradient between two air masses. During the passing of a warm front the temperature rises, while for a cold front it falls. For our cross sections this means that a warm front should have a cold side to the right side of the center (i.e. the side the front has not yet passed) and a warm side at the left of the center (i.e. the side where the front already passed through). For a cold front the location of the warm and cold side are swapped.

To highlight the effect of a passing front we plotted the distribution of the difference of the mean temperature in a region 200 km behind the identified front and the mean temperature in the region 200 km ahead of the front at 500 hPa, 850 hPa and 1000 hPa for the results of our algorithm. We further plot the results of a baseline at 850 hPa, where the locations at which we evaluate the temperature difference within a cross section were randomly sampled instead.

Algorithm Results. Figure 3 shows the histogram of temperature differences for fronts identified by our algorithm. As expected we can see for all presented levels that the temperature does indeed rise (fall) with the passing of warm (cold) fronts. We can further see that the strongest temperature difference can be found at 850 hPa, a level commonly used for detecting fronts. Further the plot shows that the temperature difference at 1000 hPa is higher than at 500 hPa, indicating that the temperature gradient at higher levels is less expressed than at (near) surface level. For both cold and warm fronts we can see that a small percentage of identified front locations express a wrong sign in the temperature difference, which is most likely caused by false positive frontal positions or wrongly oriented samples.

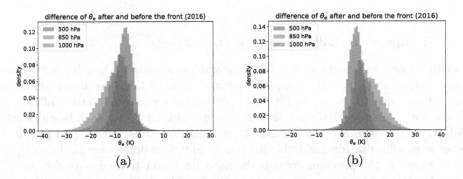

(a)   (b)

Fig. 3. Histograms showing the distribution of temperature differences across (a) cold and (b) warm fronts as calculated by our method at different height levels.

Randomly Sampled Baseline. As a comparison we plotted the temperature differences at 850 hPa if the frontal separation was uniformly randomly sampled instead of using our algorithm in Fig. 4. The data for the random samples consists of the same cross sections used for the cold (warm) fronts. To create these plots we aggregated the results every 168 h (1 week) within 2016. In addition to the histogram we also fitted both a normal and a Laplace distribution to the data. As can be seen the Laplace distribution provides a good fit for the temperature difference of the random samples. Further the choice of warm or cold front cross sections barely changes the distribution of the random samples. In both cases the mean μ and the peak is located near 0. We realize that in this case the position of the standard deviation σ of both distributions almost coincide, thus it makes sense to use σ as a measure for testing the quality of the derived probability densities.

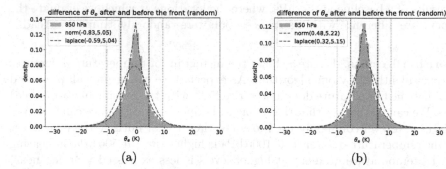

(a)   (b)

Fig. 4. Histograms of temperature differences for randomly separated cross sections. Additionally displays a fitted laplace and normal probability density function as well as mean ± 1 or 2 standard deviations.

4.2   Temperature Difference at Various Levels

In this section we present results for all height levels, showing how the algorithm fairs against the randomly sampled case and showing that the algorithm does provide meaningful results. In Fig. 5, we plotted the mean temperature difference of the random separation case, as well as $\mu \pm \sigma$ and $\mu \pm 2\sigma$ for each height level. Additionally boxplots of the results of our algorithm at each evaluated pressure level were added to highlight the differences to the randomly sampled case.

As seen in the previous section the most dominant temperature difference, even for the random case, lies around 850 hPa. This is in agreement with various non-DL-based front detection algorithms, that rely on the thermal field at 850 hPa to determine the location of fronts. Our results confirm that this commonly used but empirically chosen pressure level is indeed a good choice.

In comparison to the random case, the plots show that for most cases our results are at least at a σ distance from μ, with the more extreme samples even exceeding the 2σ distance. Mean (green triangle) and median (orange line) also tend to be located closely to the 2σ line, sometimes even exceeding it.

As mentioned in Subsect. 4.1 and Fig. 4, we can see that the random fronts temperature difference distribution can be approximated using a laplace distribution $L(\mu, b)$ with mean $\mu \in \mathbf{R}$ and scale $b \in \mathbf{R}$. We know that the standard deviation can be calculated as $\sigma = \sqrt{2}b$ and from this we can calculate that a point randomly sampled from a laplace distribution has a probability of approx. 75% (resp. 94%) to lie within the interval $[\mu - \sigma, \mu + \sigma]$ (resp. $[\mu - 2\sigma, \mu + 2\sigma]$). This further enhances our point, that the extracted 3D fronts are meaningful.

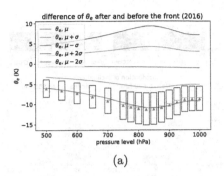
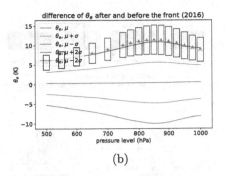

(a)   (b)

Fig. 5. Temperature difference across the detected fronts for different pressure levels for (a) cold and (b) warm fronts as box plots including their mean (green triangle) and median (orange line). Additionally the mean (μ) as well as the one and two σ intervals for the randomly separated fronts is inserted as orientation. (Color figure online)

4.3   Evaluation of Frontal Inclination

In a third evaluation we determine the inclination of weather fronts and provide a histogram of how the inclination is distributed across fronts of each type. We determine the inclination in degrees using the following formula:

$$\text{inclination} = \arctan\left(\frac{|\text{pressure\_difference (hPa)}|}{\text{horizontal\_distance (km)}}\right) \tag{5}$$

While the formula uses different units, it has the advantage of better showing the inclination, as opposed to describing the vertical difference in terms of km, since the atmosphere's length scales are very different and thus the aspect ratio (vertical vs. horizontal scale) is very small. For instance, the horizontal extent of a pixel is 20 km while the vertical distance from 1000 hPa to 500 hPa is only approximately 5 km. Determining the inclination on this scale would make it hard to see differences in the angles and results in mostly near zero inclinations. In Fig. 6 we show the mean cold (a) and warm (b) front as well as the average inclination of the frontal objects from 1000 hPa to 600 hPa (left) and from 1000 hPa to 850 hPa (right). To calculate the inclination of a frontal object we first determine the median position of the frontal separation points for each pressure level. The inclination between two pressure levels is then described by the line connecting the resulting positions for the corresponding pressure levels (e.g. 1000 hPa and 850 hPa or 600 hPa). The median position is chosen as it is more robust towards outliers, where the algorithm may have failed to determine one of the separation points correctly. The mean inclination of frontal objects as depicted in Fig. 6 is then obtained by connecting the median separation positions averaged over all fronts. In case of the warm fronts (Fig. 6b) we can clearly see the expected flat inclination, while for the cold fronts (Fig. 6a) we can see a rather steep inclination of near 90° and a slight backward tilt for the inclination. Both of these observations are in accordance with the typical theory surrounding the shape and inclination of warm and cold fronts.

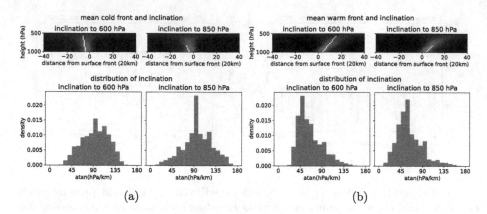

Fig. 6. (a): Average frontal separation for cold fronts in red. Light blue line indicates the average inclination from surface ($1000hPa$) to $600hPa$ (left) or $850hPa$ (right). Histograms show the distribution of inclinations accordingly. (b): as (a), but for warm fronts. (Color figure online)

4.4   Runtime

Tests were run on a compute node with an Intel 2630v4 CPU with 10 cores and 128GB RAM. For the pipeline we used precalculated fronts from the mentioned neural network. Thus, frontal locations can be directly forwarded to the pipeline. The pipeline itself was implemented in Python.

Parallelization is performed by splitting the dataset into multiple batches of 6 days at hourly time resolution resulting in 144 timestamps per batch. For each batch all types of fronts were calculated, with two calculations for occluded fronts. Since intermediate cross sections are stored in RAM instead of main memory we requested 10GB for each batch. Processing a batch took less than 150 minutes, varying with the amount of frontal pixels present at each times-tamp. Per sample timestamp this results in a runtime of approximately 1 minute. As we did not store the cross section results our evaluations also had to run the complete pipeline, resulting in similar runtimes. Reducing the size of cross sections as well as the oversampling rate could further reduce runtime and memory consumption, however we decided to keep those values high, to allow a more thorough evaluation of the results, as runtime optimization was not a main focus of this work.

4.5   Potential Extensions

ERA5 data contains theoretically invalid regions such as high pressure levels in Greenland. In addition to these errors there is also a strong influence at the ground level temperature based on the orography. As a result, some fronts, located in the vicinity of land surface, may exhibit dominating ground gradients based not on the frontal activity but rather the orography. These may lead to wrongly extracted frontal locations on the lower levels, where this effect is more dominant.

Our algorithm is only able to extend identified surface fronts. It is not designed to identify new fronts that may not extend all the way to the surface. Further, occluded fronts are currently not investigated, as the algorithm does not support the case, where a front splits into several parts at higher altitudes. Especially as in the case of an occluded front these parts have differently signed gradients.

Further, we only evaluated our algorithm within the North Atlantic region. Performance on other regions needs to be further evaluated, especially regarding orographic features such as high mountains, which are less represented in our data. However, it is well known that fronts as synoptic weather features are heavily perturbed by mountainous regions on continents, thus loosing their clear structure. Therefore, it is even not clear that extending our algorithm to such regions is really meaningful.

5   Conclusions and Future Work

In this paper we presented a novel pipeline for finding three dimensional structures of frontal lines based on a surface front detection neural network. Our algorithm uses the prior knowledge of frontal characteristics and surface front positions to expand frontal lines from a two dimensional grid to three dimensional structures. Our model is the first that enables statistical analysis of three dimensional fronts. Therefore, we can provide a statistical evaluation of characteristics of automatically detected three dimensional fronts, which previous methods were not suited to do. Our method may thus provide important insights, deepening the understanding of fronts and their three dimensional structure and how these are connected to other weather phenomena. The code of the new pipeline will be available at https://github.com/stnie.

Possible enhancements of the proposed method include extensions to other regions and providing more statistical evaluations regarding similarities and differences in the frontal structure over the globe. This would enable further investigation on how the structure of a front influences phenomena often correlated to frontal activity such as extreme precipitation. Our current pipeline focuses on the warm and the cold weather fronts. While occluded fronts can possibly be detected and found by our algorithm, it is currently not suited to describe the nature of such fronts as two overlapping fronts. It may therefore be of interest to extend this approach to cover the more complex shapes of occluded fronts.

Interesting computational extensions would include the generalization of the proposed method to enable more complex structures including the detection of frontal three dimensional volumes instead of surfaces. Storing the cross sections on hard disk would require several TB of memory due to the oversampling of each pixel and the large horizontal extent. To circumvent this we process each timestamp individually. Even with multiple cores running in parallel and by splitting the dataset into multiple batches, computation still takes several hours.

For a better applicability and faster experimentation we therefore plan to provide a parallelized version of the pipeline on GPU systems.

Another interesting research direction would be the design of a neural network similar to our previous work, to directly detect three dimensional fronts. However, as currently there is no valid ground truth 3D data regarding weather fronts, such an approach is not feasible at the moment. We do however believe that creation of such a network would allow us to leverage the deep neural networks' ability to model complex shapes and phenomena. This approach could, as proven in the 2D case, create clearer results and reduce computation time. In addition it may be able to better generalize, i.e. frontal activity over challenging terrain or detecting the more complex shaped frontal types.

Acknowledgement. The study is supported by the project "Big Data in Atmospheric Physics (BINARY)", funded by the Carl Zeiss Foundation (grant P2018-02-003). We acknowledge the ECMWF for providing access to the ERA5 reanalysis data and the ZDV of JGU for providing access to Mogon II. We further acknowledge Daniel Kunkel for supporting us with data management and thank Michael Wand for fruitful discussions.

References

1. Berry, G., Reeder, M.J., Jakob, C.: A global climatology of atmospheric fronts. Geophys. Res. Lett. **38**(4), 1–5 (2011)
2. Biard, J., Kunkel, K.: Automated detection of weather fronts using a deep learning neural network. Adv. Statist. Climatol. Meteorol. Oceanography **5**, 147–160 (2019)
3. Bochenek, B., Ustrnul, Z., Wypych, A., Kubacka, D.: Machine learning-based front detection in central Europe. Atmosphere **12**(10), 1312 (2021)
4. Catto, J.L., Pfahl, S.: The importance of fronts for extreme precipitation. J. Geophys. Res. Atmospheres **118**(19), 10791–10801 (2013)
5. Giffard-Roisin, S., Yang, M., Charpiat, G., Kumler Bonfanti, C., Kégl, B., Monteleoni, C.: Tropical cyclone track forecasting using fused deep learning from aligned reanalysis data. Front. Big Data **3**, 1 (2020)
6. Hersbach, H., et al.: The era5 global reanalysis. Q. J. R. Meteorol. Soc. **146**(730), 1999–2049 (2020)
7. Hewson, T.D.: Objective fronts. Meteorol. Appl. **5**(1), 37–65 (1998)
8. Jenkner, J., Sprenger, M., Schwenk, I., Schwierz, C., Dierer, S., Leuenberger, D.: Detection and climatology of fronts in a high-resolution model reanalysis over the alps. Meteorol. Appl. **17**(1), 1–18 (2010)
9. Kern, M., Hewson, T., Schätler, A., Westermann, R., Rautenhaus, M.: Interactive 3D visual analysis of atmospheric fronts. IEEE Trans. Visual Comput. Graphics **25**(1), 1080–1090 (2019)
10. Lagerquist, R., McGovern, A., II, D.J.G.: Deep learning for spatially explicit prediction of synoptic-scale fronts. Weather Forecast. **34**(4), 1137–1160 (2019)
11. Lam, R., et al.: GraphCast: learning skillful medium-range global weather forecasting (2022). https://doi.org/10.48550/ARXIV.2212.12794
12. Matsuoka, D., et al.: Automatic detection of stationary fronts around Japan using a deep convolutional neural network. SOLA **15**, 154–159 (2019)

13. May, R.M., et al.: MetpPy: a meteorological python library for data analysis and visualization. Bullet. Am. Meteorol. Soc. **103**(10), E2273–E2284 (2022)
14. Niebler, S., Miltenberger, A., Schmidt, B., Spichtinger, P.: Automated detection and classification of synoptic-scale fronts from atmospheric data grids. Weather Climate Dyn. **3**(1), 113–137 (2022)
15. Pathak, J., et al.: FourCastNet: a global data-driven high-resolution weather model using adaptive Fourier neural operators (2022). https://doi.org/10.48550/ARXIV. 2202.11214
16. Pfahl, S., Sprenger, M.: On the relationship between extratropical cyclone precipitation and intensity. Geophys. Res. Lett. **43**(4), 1752–1758 (2016)
17. Sansom, P.G., Catto, J.L.: Improved objective identification of meteorological fronts: a case study with era-interim. Geoscientific Model Develop. Discussions **2022**, 1–19 (2022)
18. Schemm, S., Sprenger, M., Wernli, H.: When during their life cycle are extratropical cyclones attended by fronts? Bullet. Am. Meteorol. Soc. **99**(1), 149–166 (2018). https://doi.org/10.1175/BAMS-D-16-0261.1

Digital Twin Simulation Development and Execution on HPC Infrastructures

Marek Kasztelnik[1]([⊠])(iD), Piotr Nowakowski[1,2](iD), Jan Meizner[1,2](iD),
Maciej Malawski[1,2](iD), Adam Nowak[2](iD), Krzysztof Gadek[2](iD), Karol Zajac[2](iD),
Antonino Amedeo La Mattina[3,4](iD), and Marian Bubak[2](iD)

[1] ACC Cyfronet AGH University, Kraków, Poland
m.kasztelnik@cyfronet.pl
[2] Sano Centre for Computational Medicine, Kraków, Poland
[3] Department of Industrial Engineering,
Alma Mater Studiorum - University of Bologna, Bologna, Italy
[4] Medical Technology Lab, IRCCS Istituto Ortopedico Rizzoli, Bologna, Italy

Abstract. The Digital Twin paradigm in medical care has recently
gained popularity among proponents of translational medicine, to enable
clinicians to make informed choices regarding treatment on the basis of
digital simulations. In this paper we present an overview of functional
and non-functional requirements related to specific IT solutions which
enable such simulations - including the need to ensure repeatability and
traceability of results - and propose an architecture that satisfies these
requirements. We then describe a computational platform that facilitates
digital twin simulations, and validate our approach in the context of a
real-life medical use case: the BoneStrength application.

Keywords: Personalized medicine · Digital shadow · Digital execution
environment · HPC

1 Motivation

Applications of the Digital Twin paradigm in medical care have recently gained
popularity among proponents of translational medicine. A digital twin is typically described as a virtual representation of a real-life process. In the context of

This work was supported by the EDITH, a coordination and support action funded
by the Digital Europe program of the European Commission under grant agreement
No. 101083771. This work was also supported by the European Union's Horizon 2020
research and innovation program under grant agreement Sano No. 857533 as well as
the Sano project carried out within the International Research Agendas program of
the Foundation for Polish Science, co-financed by the European Union under the European Regional Development Fund. This work was (partly) supported by the European
Union's Horizon 2020 research and innovation program under grant agreement ISW
No. 101016503. We also gratefully acknowledge Poland's high-performance computing
infrastructure PLGrid (HPC Centers: ACK Cyfronet AGH) for providing computer
facilities and support within computational grant No. PLG/2022/015850.

J. Mikyška et al. (Eds.): ICCS 2023, LNCS 14074, pp. 18–32, 2023.
https://doi.org/10.1007/978-3-031-36021-3_2

medical interventions, it refers to a computational model which represents a specific patient, usually focusing on a specific ailment or pathological process, and enables a medical care professional to make informed choices regarding treatment on the basis of digital simulations. The model can fetch patient data from databases or from the IoT devices worn by the patient [30], in order to simulate the given ailment/organ/etc.

Based on our expertise connected with building patient digital twins on HPC we discovered a set of patterns (as well as antipatterns) for the calculations executed and data managed on supercomputers by researchers. (1) Researchers tend to try to run models as fast as possible and focus on results, often forgetting about model versioning and traceability. (2) Data is usually transferred to HPC from a local computer into the user's personal directory, which is not accessible by other team members. (3) When a calculation for a given use case is finished, other calculations are sometimes executed, which often leads to overwriting result data. As a result, it is difficult or even impossible for other researchers to recreate a specific experimental setup. Moreover, the problem of research sustainability is not new (see e.g. [33] and [27]) and there are even dedicated institutes that assist in performing sustainable research (e.g. [17]). The Cyfronet DICE team [5] has been involved in such initiatives for a long time [32]. We build tools that simplify the way data is managed, versioned, and made accessible to others.

The objective of this paper is to present an overview of functional and non-functional requirements related to specific IT solutions which enable simulations for Digital Twins in medicine - including the need to ensure repeatability and traceability of results. We propose an architecture that satisfies these requirements and promotes the principles of 3R[1]. Subsequently, we present the Model Execution Environment [20] – a reference implementation of our concept. We validate our approach in the context of a real-life medical use case: BoneStrength – an in silico trial solution for efficacy evaluation of treatments preventing proximal femur fracture. Given that we do not focus on interactions between the model and the real patient (instead relying on data fetched from external databases or IoT devices), the described solution can be treated as a digital user shadow, or digital execution environment.

2   Typical Requirements of Digital Twins Simulation

Digital twins are intended to stand in for the patient when medical simulations need to be carried out. Accordingly, the concept of a digital twin should be understood as a set of data which represents the specific patient in relation to a specific condition or treatment process, in particular circumstances. The data in question is not typically restricted in any manner - indeed, it may include unstructured textual data (such as measurement results), binary data (images and scans), structured repositories (databases storing patient information), or even free-text descriptions, such as those provided by medical practitioners who interact with the patient.

---

[1] Repeatability, Replicability, Reproducibility.

A system capable of performing digital twin simulations, must – unless geared for a specific procedure or disease type – be agnostic as to the supported classes and formats of data items. This is a primary requirement facing such systems, which, in turn, translates into the need to support (by means of integration) a comprehensive data repository, where various data items may be queried, retrieved and fed into the computational models which constitute the given simulation workflow.

In addition to the above, the platform must provide access to a computational infrastructure – given that, naturally, the very concept of digital twin simulations implies that data needs to be processed in some manner. Depending on the scope and specific aims of the simulation, a variety of specific requirements may emerge. First of all, the scale of the simulation must be taken into account. The following types of computational resources may be required:

- standalone servers (for small-scale simulations),
- cloud computing infrastructures (for simulations in which a moderately sized set of data is processed using complex algorithms),
- classical HPC (High Performance Computing) solutions such as computing clusters (for scale-out studies which involve processing large amounts of data and "parameter study" types of computations).

Additional functional requirements associated with digital twin simulations refer to the properties of the underlying computational infrastructure - both in terms of hardware and software. The following notable requirements are often encountered:

- availability of specific hardware components, mainly in the context of GPGPU processing,
- availability of specific software packages and libraries - which are often commercial and sometimes costly to use,
- ensuring security of data and computational models themselves - as digital twin simulations frequently process sensitive medical data, special arrangements must be made to guarantee that such data is protected against unauthorized access.

3   A Critical Review of Platforms for Digital Twins

One of the important aspects that we have focused on was the thorough State of the Art analysis, to identify potential candidates for a digital twin platform. We focused on infrastructural as well as software aspects.

Relevant medical simulations may be run on a wide range of computational platforms, ranging from dedicated workstations, through local clusters to large-scale Cloud [31] and HPC Systems [26]. Each of those systems provides distinct features and challenges. As local workstations lack sufficient power for serious computations, the preferred option is usually to utilize more sophisticated systems – which, due to their multi-tenant and shared nature, require appropriate integration with an execution platform.

The most straightforward type of infrastructure is a computational cloud, as while the platform is usually multi-tenant in nature (even for private clouds), the user usually obtains unrestricted access to a set of instances. However, not every model can be run in the cloud. Even though modern cloud providers offer dedicated solutions for batch or HPC-like use cases such as Azure Batch [4], virtualization overhead and the need for flexibility may nevertheless hinder performance when compared to purpose-built HPC systems.

The aforementioned HPC systems provide great performance in terms of computational power, storage and interconnect; however, they carry drawbacks in other areas that need to be addressed by the platform. Those are mostly related to ensuring secure access, which, while standardized to some degree, may vary from system to system (GSI-SSH [8], plain password-based SSH, key pairs, native API), as well as the multi-tenant nature of the system, which makes it more difficult to install additional end-user software, and also imposes the need to strictly control access to data.

Clearly, the infrastructure by itself is not sufficient to solve the problem of digital twin simulations. Another important aspect relates to the software components required to build an integrated simulation platform. Due to the complex nature of such applications, the most manageable way to express and run simulations is via a specialized workflow system [22].

An example of a workflow platform is the Arvados Workflow System [3]. It is based on the well-developed Common Workflow Language (CWL) [21] and provides a mechanism for integration with a wide range of aforementioned infrastructures both cloud- and HPC-based; however there are two significant drawbacks. First and foremost, integration with HPC requires significant modification of the HPC cluster itself, including deployment of tools and services both on the login and compute nodes. For large-scale clusters, this may be infeasible for operational and/or security reasons. Secondly, even though Arvados supports Singularity (now referred to as Apptainer) containers [2] commonly used for HPC, it is clearly stated that this support is still experimental and Docker should be used instead. However, Docker is not commonly encountered on multi-tenant HPC systems for security reasons. It is also important to mention that Arvados requires some form of containers (Docker or Singularity/Apptainer) to deploy workflows on the HPC cluster, and cannot operate on standalone jobs.

Another large and well-known workflow system is Pegasus [23]. The system is robust and offers the ability to run computations on a wide range of infrastructures including grids, clusters and clouds. It also provides advanced features such as provenance and error recovery. However, while the above features are desirable, we have decided that at the current stage the overall complexity of this solution is too great given the requirements of the Digital Twin Platform. On top of that, we would still need to implement domain-specific features that are not present in this generic platform.

Additionally we have analyzed the High-Performance Computing and Cloud Computing with Unified Big-Data Workflows platform created in the scope of the LEXIS Project [25]. The platform integrates HPC and Cloud computing enabling multi-system and multi-site deployment of workloads as well as provides required utilities such as orchestration, billing, AAI, API and Portal. HPC access is enabled by custom middleware called HEAppE [9] that utilizes SSH and SCP/Rsync protocols for running jobs on clusters. The Cloud component utilizes Alien4Cloud [1] with extended TOSCA [19] templates for deployment to platforms such as OpenStack [13]. Regarding data, iRODS and EUDAT/B2SAFE [6] are used. Access is secured by the Keycloak [11] based AAI system where accounts are mapped to cluster accounts, as well as appropriate custom integration with iRODS [24]. While the overall work done in the scope of the LEXIS Project is impressive and we believe that the platform may be useful for large multi-site deployments involving both HPC and Clouds the significant deployment overhead, which involves installation of custom middleware (both on HPC and cloud compute sites, as well as on storage sites), along with a dedicated AAI system, is too great given the stated needs of our research. Moreover, as in the case of the Pegasus system described above, the platform is not dedicated to the medical domain and would require extensions to be fit for such purpose. Finally, we have found that one of the crucial components, Alen4Cloud, is no longer maintained according to its official website (since July 2022); thus, the user would need to provide their own maintenance services, or find a substitute - which might require significant effort as this component lies at the core of the platform.

4   The Concept of a Universal Platform for Digital Twins

Our aim was to create a platform for digital twin simulations that may be used by scientists to submit their computations to e-infrastructures such as HPC clusters in an easy and accessible way. This concept is presented in Fig. 1. Users interact with the platform through a web browser after being authenticated and authorized by an appropriate IDP, which also can generate required credentials (such as a user grid proxy certificate, to be used for access to the e-infrastructure). When the simulation is started by the user, all required code, along with input data, is automatically transferred to the infrastructure (such as an HPC cluster). The platform monitors the status of the simulation and notifies the user when it is finished. It also provides a way to download and compare results online.

Our conceptual work on the presented platform was guided by the following principles:

- **Model versioning** - previous versions of the model are stored and may be referred to if needed
- **Repeatable runs (3R vision)**
  - **Repeatability** - a researcher can reliably repeat their own computation provided the conditions are the same (same team, same experimental setup)

- **Replicability** - an independent group can obtain the same results using the authors' own artifacts (different team, same experimental setup)
- **Reproducibility** - an independent group can obtain the same result using artifacts that they develop completely independently (different team, different experimental setup)

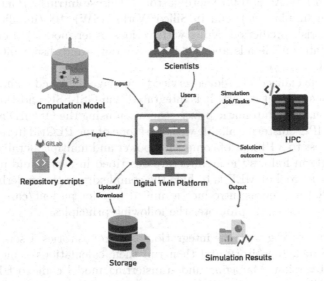

Fig. 1. General concept of the platform.

Following up on the above vision, we also adhered to the key requirement to enable a group of scientists with various backgrounds (such as medicine, physics, chemistry and computer science) to take part in digital simulation-driven experiments via a coherent and manageable system. As mentioned above, the system should enable execution of computational models controlled by a set of scripts with a versioning system, on the one hand enabling collaborative editing, while on the other hand tagging specific versions that may be later selected to suit the researchers' needs. Those models need to operate on data stored in a storage backend appropriate for the compute infrastructure. Another important goal is to streamline access to said compute infrastructures, which involves abstraction of the underlying APIs such as remote access mechanisms and system tools (including queuing tools). In addition to the above, the platform needs to provide a straightforward way to display, download and analyze simulation results.

The platform should also provide a mechanism to reuse models while supporting custom artifacts, thus realizing our 3R vision by producing consistent results regardless of conditions (teams, artifacts, setup).

5   Overview of Implementation

The Model Execution Environment (MEE) is a prototype implementation of the architecture described in the previous section. It was deployed [12] and it is used to validate core digital twin creation and modeling concepts in the scope of applications from a range of projects, including EurValve [7] (human heart simulations), PRIMAGE [14] (Neuroblastoma, Diffuse Intrinsic Pontine Glioma tumor growth simulations), and In Silico World (ISW) [10] (in silico trials for bone fracture risk prediction, along with various other models). Below we discuss implementation details and list the advantages and disadvantages of the proposed tool.

MEE is a specialized high-level service to manage data and computations in the context of a patient cohort. It is integrated (via GSI-SSH and Proxy Certificate delegation, with automatic proxy generation using the OpenID mechanism) with several HPC clusters available with the scope of the PLGrid infrastructure[2], delivering access to 8 PFlops of computing power and multiple petabytes of storage. The platform follows the architecture described in Sect. 4 and presented in Fig. 1. The main goal of MEE is to hide the complexity of the underlying infrastructure (HPC) as well as introduce a unified way for patient/case data to be stored and maintained. It promotes the following principles:

- **Model versioning** with the integration of git repositories. Users can simply push code to a repository and then run their calculations using the specified model version. Managing and transferring model code to HPC is done automatically.
- **Repeatable runs** achieved through integration with git repositories. Each run record model stores a version (git SHA) which can be used to rerun the same calculation. As a result, we can meet 3R (repeatability, replicability, reproducibility) criteria.
- **Integration with HPC.** MEE is integrated with the PLGrid infrastructure which allows us to delegate user rights from MEE to the HPC supercomputers.
- **Organized way to store patient data and the calculation results.** The main building block of the MEE environment is a single patient. Each patient has a dedicated storage space on a storage resource (e.g. HPC, S3) where patient data is stored. Inside, we also have a space for calculation-specific inputs and results. Patients can be grouped into cohorts upon which simulation campaigns can be executed (a campaign involves running the same pipeline for each member of a given cohort). This unification represents another step towards realizing the 3R concept.

The user interacts with MEE by using a local web browser (no further dependencies need to be installed on the user's machine). The first step is for the MEE administrator to define a set of simulations - which we call **steps** (see Fig. 2), by

---

[2] The PLGrid infrastructure is a joint effort of the largest HPC centers in Poland. It offers coherent management of users, groups and computational grants, as well as unified access to the integrated HPC clusters [28].

providing credentials enabling access to the simulation's code repository. Each step can define a list of required inputs needed to start the calculation. Steps can be grouped into **flows**. A flow is a template used to build a pipeline for a specific patient/case. The patient data structure defines the location where the personalized input data is stored. It contains a list of pipelines executed on the patient data. Each pipeline defines the output location and a list of computations executed on HPC. Each computation needs to be configured before it can be started (a specific version of the simulation, as well as the required input parameters, need to be selected).

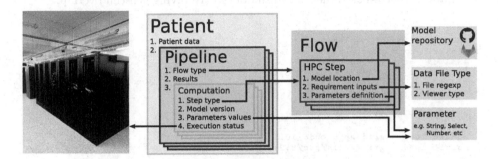

Fig. 2. Data structure used in MEE.

Pipelines consist of separate computations which are called pipeline steps. In most cases, these are pieces of software (e.g. Matlab scripts, CFD simulations, etc.) executed on the HPC cluster. Each pipeline step can be configured for integration with a collaborative source control project. For this purpose, MEE is integrated with the Git versioning systems (e.g. http://github.com/, http://gitlab.com/ services can be used). There, MEE users are able to apply the typical features of a sophisticated source control tool to collaboratively develop, share and test their code (e.g. a simulation). Inside the repository, the template of the HPC queuing configuration script, as well as the rest of the simulation source code, should be stored. In order to launch a computation on HPC, the user needs to select which model version should be used. The model version is taken from the Git repository (MEE shows all repository branches and versions). When this value is selected, the queuing configuration template file is downloaded from the repository and converted to a final queuing system configuration file (in our case, a Slurm [16] startup script) dedicated for the selected case. To enable customization, MEE delivers a set of helpers which can be applied in the queuing configuration script in order to customize it. Notable helpers are briefly listed below:

- **clone_repo** injects code responsible for cloning the simulation repository in the version specified by the user.
- **stage_in input-file-type** searches for the simulation input file in the results of past computations, pipeline input, and patient input directories.

- **stage_out file-path** uploads simulation results to the pipeline output directory.
- **value_of parameter-key** injects the value specified by the user while starting the calculation.

Below we present an example queuing configuration template script taken from the PRIMAGE project. It runs an agent-based simulation which is a hybrid model pertinent to the tissue level: patches of the whole tumor. The simulation comprises a continuous automaton representing the microenvironment, discrete agents representing neuroblasts and Schwann cells occupying the microenvironment, and a centre-based mechanical model for resolving cell-cell overlap.

```
 1 #!/bin/bash -l
 2 #SBATCH -N 2
 3 #SBATCH --ntasks-per-node=2
 4 #SBATCH --mem-per-cpu=3GB
 5 #SBATCH --time=04:00:00
 6 #SBATCH -A {% value_of grant_id %}
 7 #SBATCH -p plgrid-gpu
 8 #SBATCH --gres=gpu:2
 9
10 {% clone_repo %}
11
12 module load plgrid/apps/cuda/10.1
13 module load plgrid/tools/gcc/8.2.0
14
15 nvidia-smi
16
17 {% stage_in amb-input amb-input.json %}
18
19 ./amb-input Models/Prototype_v2.0/ABM13.4/FGPU_NB --in amb-input.json --
      primage amb-output.json
20
21 {% stage_out amb-output.json %}
```

Listing 1.1. Example queuing configuration template script.

Lines 2 to 8 define the required resources for the simulation. In line 6 we can see how to inject the value of the parameter selected by the user while starting the simulation. The `clone_repo` directive (line 10) injects code to clone the simulation repository in the version selected by the user. The certificate used to clone the repository is registered in the Gitlab/Github as a deploy key. It has only read capability, thus the computation cannot push any modification the the repository. In line 17 we request input for the simulation which should be downloaded from the platform. Next, the simulation is started, and in line 21 we request upload of simulation results to the pipeline output directory. Once the simulation begins, the template is converted to a customized script, specific for the selected patient and pipeline:

```
 1 #!/bin/bash -l
 2 #SBATCH -N 2
 3 #SBATCH --ntasks-per-node=2
 4 #SBATCH --mem-per-cpu=3GB
 5 #SBATCH --time=04:00:00
 6 #SBATCH -A plgprimage4
 7 #SBATCH -p plgrid-gpu
 8 #SBATCH --gres=gpu:2
 9
10 export SSH_DOWNLOAD_KEY="-----BEGIN RSA PRIVATE KEY-----
```

```
11 xxxxxxxxxxxxxxxxxxxxxxxxxxxxxxxxxxxxxxxxxxxxxxxxxxxxxx
12 -----END RSA PRIVATE KEY-----
13 "
14 ssh-agent bash -c '
15   ssh-add <(echo "$SSH_DOWNLOAD_KEY");
16   git clone git@gitlab.com:primageproject/Models
17   cd 'basename cyfronet/Models .git'
18   git reset --hard c0719f1c7e2990554821181a505d056a837eb236'
19
20 module load plgrid/apps/cuda/10.1
21 module load plgrid/tools/gcc/8.2.0
22
23 nvidia-smi
24
25 curl -o "Models/Prototype_v2.0/ABM13.4/amb-input.json" "https://mee.s3p.
       cloud.cyfronet.pl/production/patients/case2341/inputs/amb-input.json?X
       -Auth-Secrets=xxxxxx"
26
27 ./amb-input Models/Prototype_v2.0/ABM13.4/FGPU_NB --in amb-input.json --
       primage amb-output.json
28
29 curl -T amb-output.json "https://mee.s3p.cloud.cyfronet.pl/production/
       patients/case2341/pipelines/32/outputs/amb-output.json?X-Auth-Secrets=
       xxxxx"
```

Listing 1.2. Customized for the patient and pipeline queue system starting script.

To run the pipeline on the HPC cluster the user simply needs to log in to the system and click the run button. Underneath, during the login process, MEE asks the PLGrid IDP[3] about the user's grid proxy certificate, which is later on saved in an encrypted form in the MEE database. The Grid proxy certificate enables user rights delegation and it is used to submit Slurm jobs to HPC by using the Rimrock service [15]. MEE monitors job execution and notifies the user about the result. Once the calculation is finished, the user can click a link to download results. Whenever a new pipeline is launched, a separate storage space is created. As a result, we can be sure that outputs are never overwritten.

In order to perform the same calculations on multiple patients, MEE supports cohorts and campaigns (see Fig. 3). A cohort provides a way to group patients. Once it is created, the user can schedule a campaign: this creates a pipeline with a shared configuration for each cohort patient. The difference between these pipelines is that each one has inputs dedicated to a specified patient and produces outputs in a dedicated storage space. Once the campaign ends, the user can inspect or download results.

---

[3] PLGrid identity provider, which is capable of generating user proxy certificates that delegate user rights to the HPC infrastructure.

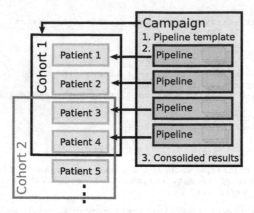

Fig. 3. The cohort groups patients and allows execution of the same set of calculations for each cohort member in the scope of the campaign.

6 Example of Usage - The BoneStrength Application

We will demonstrate the usage of the MEE platform on the basis of BoneStrength - an in Silico Trial solution for efficacy evaluation of treatments preventing proximal femur fracture. It is one of the two Fast Application Track solutions in the ISW project and consists of two main models.

- A Finite Element (FE) model, which predicts bone strength with a patient-specific model in a side fall condition as a function of the direction of impact.
- A stochastic patient-specific model of the side fall, which predicts the probability distribution of the impact force that a large number of random falls would cause in a certain patient.

The standard approach is patient-driven and it was born as a Digital Twin solution. The simulation computes 28 falls (by varying the falling load direction in the antero-posterior and medio-lateral directions, in order to explore all possible fall configurations) per year of a single patient and then estimate the average risk of fracture. A validated strain-based failure criterion allow predicting the femoral strength, which means the force causing bone failure. By calculating the ratio between the number of simulated falls that caused a fracture and the total number of simulated falls, and considering the falling annual rate, the Absolute Risk of Fracture at time zero (e.g., at the time the CT scan was performed) (ARF0) can be calculated [18]. Another approach, called "Markov" version of BoneStrength, is a simulation that aims to estimate the number of fractured patients over certain observation time. The very first phase is to generate series of falling along 10 years and randomly assign their occurrence to the patients. Each single case simulation will mark the patient as already fractured if the falling lead exceeds the femur fracture load. This indicates that the next simulations will not be executed for this patient. As it is a stochastic process, a bunch of realizations – named campaign – are required to obtain an average result.

FE model simulations use ANSYS Mechanical APDL as solver, with MPI parallelism to speed-up the computations. Each execution requires around 5-6GB of disk space, 20GB of RAM and 4-8 cores, and needs 1-2 core-hours to be completed. Since multiple configurations for many patients need to be simulated, approximately 100 000 executions in total are intended to be processed including both BoneStrength model versions. In this case, array jobs are used for submitting multiple runs at the same time. In general, patient-driven simulations are characterized by the association with a specific subject or case described with its patient-related inputs. When it comes to large-scale realizations, where a set of patients – named cohort – is going to be simulated under the same scenario, but with different parameters, researchers must be able to identify the single runs in order to detect most interesting results (e.g., find patient and parameters that lead to bone fracture). The concept of the BoneStrength workflow is presented in Fig. 4.

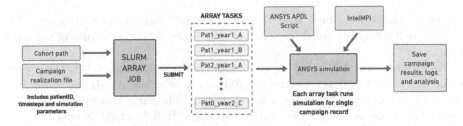

Fig. 4. Simulation workflow of the Markov BoneStrength version.

7   Evaluation - Assessment of Proof of Concept

The presented architecture and MEE technology were used to simulate two Markov versions of the BoneStrength application on a cohort of 1080 patients, running a total of 7500 simulations as a campaign per model version. Each submitted campaign task allocated 28 GB of RAM and 4 CPUs, used ANSYS software and parallelized simulations with IntelMPI. The campaign was run on Prometheus Cluster at Cyfronet. The resource usage presented in Fig. 5 indicates that each thread used in the ANSYS solver used an average of 813 CPU-seconds and 6.3 GB of RAM. The Slurm mechanism reports over 90% of job efficiency, which is the ratio of the total useful runtime of a job to the wall-clock time requested by the user, expressed as a percentage value. This shows that almost the entire CPU time for all allocated cores was utilized during the job's elapsed time – which, in turn, indicates that HPC resource wastage is kept to a minimum.

The observed distribution of campaign CPU time allocation reflects the usual distribution resulting from different inputs/meshes of patients' bones or the difficulty of finding a solution by the ANSYS software. Some reported outliers are a consequence of using an external ANSYS license server which sometimes forced

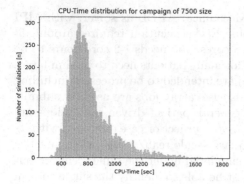

CPU-Time distribution for campaign of 7500 size

	CPU-Time	Memory used	Efficiency
Mean	813.2 sec	6448.9 MB	94.5 %
Median	769.8 sec	6447.0 MB	92.9 %

Fig. 5. HPC infrastructure resource usage for campaign run of 7500 simulations of BoneStrength. The figure presents distribution of simulation's CPU time, average memory usage and HPC job efficiency regarding the provided allocation.

a reconnection during high load times. Thus, the course of the campaign was limited to 400 simultaneously running jobs in order to avoid overuse of software licenses.

As a part of the ISW project, the conducted research allowed us to prepare initial sensitivity analyses for the BoneStrength solution, obtaining result files from bone analysis and fracture risk prediction. The same concept of a Digital Twin simulation is going to be applied in other BoneStrength model versions and workflows, as well as other patient-specific medical solutions.

8   Conclusions and Future Work

The presented architecture and its reference implementation (the MEE platform) have been validated in the course of three projects (EurValve, PRIMAGE and InSilocoWorld) implementing medical applications from different fields of the VPH domain. The feedback received from applications running on the platform enabled us to identify platform elements which should be improved, as well as new features which should be implemented in MEE. Additionally, based on the Model Execution Environment, a separate platform was created for the purpose of the PROCESS Project [29] called the Interactive Execution Environment (IEE). Given its reliance on an advanced version of MEE, it provides a comprehensive set of infrastructural features. Moreover, the IEE features a mechanism for running Singularity containers which can be useful for provisioning of models. In the near future we intend to change the way patient, pipeline, and computation data is stored to support easy usage traceability (who used the data, who produced the results, etc.) We will also investigate how to introduce a more generic abstraction of the computational process, which will enable support for additional computational platforms such as native Kubernetes containers.

References

1. Alien4cloud. https://alien4cloud.github.io. Accessed 18 Apr 2023
2. Apptainer - the container system for secure high-performance computing. https://apptainer.org. Accessed 11 Apr 2023
3. Arvados workflow system. https://arvados.org. Accessed 11 Apr 2023
4. Azure batch. https://azure.microsoft.com/en-us/products/batch. Accessed 11 Apr 2023
5. Distributed computing environments (dice) team. https://dice.cyfronet.pl. Accessed 11 Apr 2023
6. Eudat collaborative data infrastructure. https://www.eudat.eu. Accessed 18 Apr 2023
7. Eurvalve: Personalized decision support for heart valve disease. https://eurvalve.sites.sheffield.ac.uk. Accessed 11 Apr 2023
8. Gsi-ssh. https://grid.ncsa.illinois.edu/ssh. Accessed 11 Apr 2023
9. Heappe middleware. https://heappe.eu/web/. Accessed 18 Apr 2023
10. In silico world: Lowering the barriers to a universal adoption of in silico trials. https://insilico.world. Accessed 11 Apr 2023
11. Keycloak open source identity and access management. https://www.keycloak.org. Accessed 18 Apr 2023
12. Model execution environment. https://mee.cyfronet.pl. Accessed 11 Apr 2023
13. Openstack. https://www.openstack.org. Accessed 18 Apr 2023
14. Primage, medical imaging, artificial intelligence, childhood cancer research. https://www.primageproject.eu. Accessed 11 Apr 2023
15. Rimrock: Robust remote process controller controller. https://rimrock.plgrid.pl. Accessed 11 Apr 2023
16. Slurm workload manager. https://slurm.schedmd.com. Accessed 11 Apr 2023
17. Software sustainability institute. https://software.ac.uk. Accessed 11 Apr 2023
18. Bhattacharya, P., Altai, Z., Qasim, M., Viceconti, M.: A multiscale model to predict current absolute risk of femoral fracture in a postmenopausal population. Biomech. Model. Mechanobiol. **18**(2), 301–318 (2019)
19. Brogi, A., Soldani, J., Wang, P.W.: TOSCA in a nutshell: promises and perspectives. In: Villari, M., Zimmermann, W., Lau, K.-K. (eds.) ESOCC 2014. LNCS, vol. 8745, pp. 171–186. Springer, Heidelberg (2014). https://doi.org/10.1007/978-3-662-44879-3_13
20. Bubak, M., et al.: The EurValve model execution environment. Interface Focus **11**(1), 20200006 (2021)
21. Crusoe, M.R., et al.: Methods included: Standardizing computational reuse and portability with the common workflow language. Commun. ACM **65**(6), 54–63 (2022). https://doi.org/10.1145/3486897
22. Deelman, E., Gannon, D., Shields, M., Taylor, I.: Workflows and e-science: an overview of workflow system features and capabilities. Future Gener. Comput. Syst. **25**(5), 528–540 (2009)
23. Deelman, E., et al.: The evolution of the Pegasus workflow management software. Comput. Sci. Eng. **21**(4), 22–36 (2019). https://doi.org/10.1109/MCSE.2019.2919690
24. García-Hernández, R.J., Golasowski, M.: Supporting keycloak in iRODS systems with OpenID authentication. presented at cs3-workshop on cloud storage synchronization and sharing services. https://indico.cern.ch/event/854707/contributions/3681126. Accessed 18 Apr 2023

25. Hachinger, S., et al.: Leveraging High-Performance Computing and Cloud Computing with Unified Big-Data Workflows: the LEXIS Project. In: Curry, E., Auer, S., Berre, A.J., Metzger, A., Perez, M.S., Zillner, S. (eds.) Technologies and Applications for Big Data Value. Springer, Cham (2022). https://doi.org/10.1007/978-3-030-78307-5_8

26. Jadczyk, T., Malawski, M., Bubak, M., Roterman, I.: Examining protein folding process simulation and searching for common structure motifs in a protein family as experiments in the gridspace2 virtual laboratory. In: Bubak, M., Szepieniec, T., Wiatr, K. (eds.) Building a National Distributed e-Infrastructure–PL-Grid. LNCS, vol. 7136, pp. 252–264. Springer, Heidelberg (2012). https://doi.org/10.1007/978-3-642-28267-6_20

27. Katz, D.S.: Fundamentals of software sustainability (2018). https://danielskatz blog.wordpress.com/2018/09/26/fundamentals-of-software-sustainability/

28. Kitowski, J., Wiatr, K., Dutka, Ł, Szepieniec, T., Sterzel, M., Pająk, R.: Domain-specific services in polish e-infrastructure. In: Bubak, M., Kitowski, J., Wiatr, K. (eds.) eScience on Distributed Computing Infrastructure. LNCS, vol. 8500, pp. 1–15. Springer, Cham (2014). https://doi.org/10.1007/978-3-319-10894-0_1

29. Meizner, J., et al.: Towards exascale computing architecture and its prototype: Services Infrastruct. **39**, 860–880 (2021). https://www.cai.sk/ojs/index.php/cai/article/view/2020_4_860

30. Minerva, R., Lee, G.M., Crespi, N.: Digital twin in the IoT context: a survey on technical features, scenarios, and architectural models. Proc. IEEE **108**(10), 1785–1824 (2020). https://doi.org/10.1109/JPROC.2020.2998530

31. Nowakowski, P., et al.: Cloud computing infrastructure for the VPH community. J. Comput. Sci. **24**, 169–179 (2018)

32. Nowakowski, P., et al.: The collage authoring environment. Procedia Comput. Sci. **4**, 608–617 (2011)

33. Venters, C., et al.: Software sustainability: The modern tower of babel. CEUR Workshop Proceed. **1216**, 7–12 (2014)

Numerical Simulation of the Octorotor Flying Car in Sudden Rotor Stop

Naoya Takahashi[1(✉)], Ritsuka Gomi[1], Ayato Takii[1,2], Masashi Yamakawa[1], Shinichi Asao[3], and Seiichi Takeuchi[3]

[1] Kyoto Institute of Technology, Matsugasaki, Sakyo-ku, Kyoto 606-8585, Japan
m2623017@edu.kit.ac.jp
[2] RIKEN Center for Computational Science, 7-1-26 Minatojima-minami-machi, Chuo-ku, Kobe 650-0047, Hyogo, Japan
[3] College of Industrial Technology, 1-27-1, Amagasaki 661-0047, Hyogo, Japan

Abstract. Currently, manned drones, also known as flying cars, are attracting attention, but due to legal restrictions and fear of accidents, it is difficult to fly them. We therefore present a numerical simulation of the sudden stop of the rotor of an octrotor flying car. In this paper, we consider the interaction between fluid and rigid-body in a 6-degrees of freedom flight simulation of a flying car. For the purpose, the attitude of the aircraft is determined based on the force generated from the flow field around the aircraft due to the rotation of the rotor. The motion of the aircraft is obtained from the equations of motion of translation and rotation, and Newton's equation of motion and Euler's equation of rotation are used. A multi-axis sliding mesh is adopted for the rotation of the rotor, and calculations with multiple rotating bodies in the computational grid are performed. In addition, we use the motion computational domain (MCD) method to represent the free motion of the octrotor flying car by the motion computational domain itself. Using the above method, we will show the appropriate rotation method from various rotor stop patterns, demonstrate the safety of the octrotor flying car, and clarify the behavior of the aircraft and the surrounding flow field.

Keywords: CFD · moving grids · coupled computation · flying car

1 Introduction

In recent years, the development of flying cars that can travel in airspace has been the focus of attention as a revolution in mobility. Flying cars, especially VTOL (Vertical Take Off and Landing), are new vehicles that fly using multiple rotors. In addition, the shape of the airframe is not predetermined, and various mechanisms are being evaluated and compared [1]. Most VTOL are electric driven nowadays, called eVTOL, which are environmentally friendly and can travel between cities. However, the safety of flying cars must be improved for inter-city travel. Aerodynamic performance is an important factor in aircraft, and many studies on drag and lift forces on aircraft bodies have been reported [2]. Aerodynamic around the rotor have been studied by Seokkwan Yoon et al.

© The Author(s), under exclusive license to Springer Nature Switzerland AG 2023
J. Mikyška et al. (Eds.): ICCS 2023, LNCS 14074, pp. 33–46, 2023.
https://doi.org/10.1007/978-3-031-36021-3_3

[3] using numerical simulations. However, most of these studies are based on steady-state flight tests and wind tunnel tests. In actual flight, steady-state conditions are rare and the phenomena are more complex. Although flight tests of actual flying vehicles are being conducted around the world, legal regulations and the fear of accidents prevent easy flight. In addition, many tests with actual aircraft are required to improve the safety of the vehicle, and the time and cost for development are likely to increase. Therefore, it is necessary to establish a more practical flight prediction method using numerical method.

Unmanned drones have often become inoperable and crashed due to damage to the propeller parts or external disturbances such as wind [4]. Unmanned drones rarely cause damage if the location of the crash is taken into account, but manned drones must be able to land without losing their posture in an emergency because a person is aboard the drone. Therefore, even if one rotor stops, the flying car needs to maintain hovering with the lift force of the other rotors. However, it leads to crash accidents because it is difficult to control the attitude when the rotor suddenly stops. In order to solve this problem, it is first necessary to clarify how the flying car is affected by the air flow when the rotor suddenly stops, and how the flight behavior of the car is affected by that. However, in the flight simulation of the flying car, it is necessary to couple the interaction between the flight dynamics, which deals with the movement of the flying car, and the fluid dynamics, which handles the fluid around the flying car. This numerical simulation is very difficult to perform because it involves such a very complicated moving boundary problem. In contrast, the authors have proposed a method that combines the MCD (Moving Computational Domain) method [5] based on the unstructured mesh finite volume method and the multi-axis sliding mesh method [6]. The MCD method applies the finite volume method to a 4-dimensional inspection volume and discretizes it, enabling calculations while strictly satisfying geometric conservation laws. In addition, the multi-axis sliding mesh method can represent motions such as rotation by forming computational domains divided by regions with different states of motion. Yamakawa et al. [7] used this method to analyze the submarine flow and clarify the effect of the shape of the free water surface.

It is thought that the flight behavior of the flying car is able to be simulated by applying the multi-axis sliding mesh method to multiple rotor sections that are independent of each other. In this study, flight simulations of the octorotor flying car with fluid and rigid-body interaction are performed to investigate the behavior of the flying car in the case of sudden stop of rotors. In addition, as a method to maintain stable hovering when the rotor suddenly stops, we examine the effectiveness of the method of rotating other rotors in reverse.

2 Numerical Approach

2.1 Governing Equation

In present work, since the flow around the propeller is included, the Reynolds number is approximately 3,000,000 and the maximum Mach number is 0.55. Therefore, it was treated as a compressible fluid and the Euler equations were used without considering viscosity. In addition, the ideal gas equation of state was used as the governing equation. The governing equations are shown in Equations. (1)–(3).

$$\frac{\partial q}{\partial t} + \frac{\partial E}{\partial x} + \frac{\partial F}{\partial y} + \frac{\partial G}{\partial z} = 0,$$

(1)

$$q = \begin{bmatrix} \rho \\ \rho u \\ \rho v \\ \rho w \\ e \end{bmatrix}, E = \begin{bmatrix} \rho u \\ \rho u^2 + p \\ \rho uv \\ \rho uw \\ u(e+p) \end{bmatrix}, F = \begin{bmatrix} \rho v \\ \rho uv \\ \rho v^2 + p \\ \rho vw \\ v(e+p) \end{bmatrix}, G = \begin{bmatrix} \rho w \\ \rho uw \\ \rho vw \\ \rho w^2 + p \\ w(e+p) \end{bmatrix},$$

(2)

$$p = (\gamma - 1)\left\{e - \tfrac{1}{2}\rho\left(u^2 + v^2 + w^2\right)\right\},$$

(3)

where q is the conserved quantity vector, and E, F, and G are the inviscid flux vectors in the x, y, and z directions, respectively. In addition, ρ is density, u, v, and w are velocities in the x, y, and z directions, respectively, p is pressure, and e is total energy per unit volume. The inviscid flux vectors are evaluated using Roe's velocity-difference separation [8] and MUSCL methods. Green-Gauss method and Hishida's van Leer-like limiter are used for variable reconstruction [9]. The specific heat ratio γ is assumed to be 1.4 in this study. The proposed model is simulated by means of a two-step rational Runge-Kutta numerical method.

2.2 Unstructured Moving-Grid Finite-Volume Method

This study investigates the behavior of the flying car when its rotor comes to a sudden stop. Conventionally, the flow field around the aircraft has been calculated by placing the object to be calculated at the center of the calculation and applying uniform flow. However, for this case, a uniform flow cannot reproduce free motion because the aircraft's attitude is expected to change significantly due to the sudden stop of the rotor. Therefore, a moving-grid finite volume method is used, in which the computational domain can be moved by the motion of the aircraft. The unstructured moving-grid finite volume method applies the finite volume method to a four-dimensional inspection volume, including time and space, and discretizes it. Therefore, even if the lattice is moved, the calculation can be performed while strictly satisfying the geometric conservation law. As shown in Fig. 1, this method can reproduce free motion without restrictions.

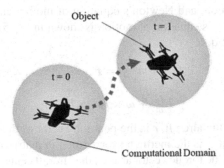

Fig. 1. Moving Computational Domain Method

2.3 Sliding Mesh Approach

The rotor section of the flying car requires a separate computational domain, unlike the kinematic state of the airframe. Therefore, the sliding mesh method, which enables the transfer of physical quantities across multiple computational domains, is used to calculate the fluid flow. This method slides the mesh itself at a specific interface, and the volume does not change due to deformation of the mesh. As shown in Fig. 2, one element is in contact with one or more other elements on the sliding surface, and the interpolation of physical quantities between regions is performed using Eq. (4).

$$q_{bi} = \frac{1}{S_i} \sum_{j \in i} q_j S_{ij}, \tag{4}$$

where q_{bi} is the physical quantity possessed by the virtual cell in cell i, q_j is the physical quantity possessed by cell j, S_{ij} is the overlapped area of cell i and cell j, and S_i is the area of cell i in the sliding plane. In this study, all computational domains consist of nine regions, one around the aircraft and eight in the rotor section, and the rotor rotation is reproduced using the sliding mesh method.

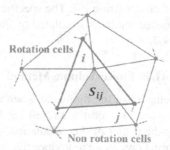

Fig. 2. Sliding Mesh Method

2.4 Coupled Computation

In this study, the position and attitude of the flying car are determined by weakly coupled calculations that take into account the interaction between the fluid and the rigid-body of the aircraft. The motion of the aircraft can be obtained from the equations of motion for translation and rotation, and Newton's equation of motion and Euler's equation of rotation are used. Newton's equation of motion is shown in Eq. 5, and Euler's equation of rotation is shown in Eq. 6.

$$m\frac{d^2 r}{dt^2} = F, \tag{5}$$

$$I\frac{d\omega}{dt} + \omega \times I\omega = T, \tag{6}$$

where m is the mass of the aircraft, r is the position vector of the center of the aircraft, and F is the force vector. I is the inertia tensor written in matrix form, ω is the angular velocity vector, and T is the torque vector around the aircraft center. The force and torque are calculated from the pressure applied to the surface of the aircraft.

3 Flight Simulation of Flying Car

3.1 Computational Model

The computational model (Fig. 3) is based on SkyDrive's SD-03 [10], for which actual flight experiments were conducted. Table 1 shows the specifications of the computational model. The rotor, arranged on the same axis, is a contra-rotating type. The model was created with the aircraft size of 4[m] as the representative length and the computational model as 1[–]. The computational model is an unstructured mesh and was created using MEGG3D [11, 12]. The total number of elements is about 3,000,000, and the rotor part is about 120,000 by itself. In this study, a sliding mesh method was used to reproduce the rotation of the rotor. Therefore, as shown in Fig. 4, the entire computational domain was constructed by fitting the rotor domain to the airframe domain.

Table 1. Specifications of the computational model

Aircraft weight	400 [kg]
Aircraft Sizes(x,y,z)	4, 2, 4 [m]
Number of rotors	8
Number of rotor blades	3
Hovering rotor speed	1930 [rpm]

Fig. 3. .Computational surface grid

Figure 5 shows a cross-sectional view of the computational domain and a mesh cross section around the airframe of the flying car. The computational domain is a sphere with the computational target in the center. The size of the sphere is 30L, which is 30 times the size of the flying car.

Fig. 4. Sliding Mesh region and mesh around rotors

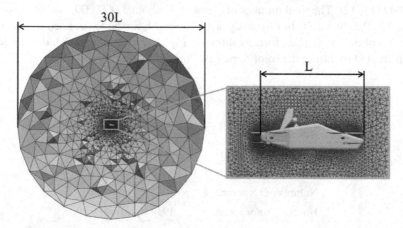

Fig. 5. Computational Domain

3.2 Computational Conditions

Table 2 shows the initial values of variables and boundary conditions. The characteristic velocity and density are set to 340.29 [m/s] and 1.247 [kg/m^3], respectively. The variables are nondimensionalized by these characteristic values.

Table 2. Initial condition

ρ	1.0
p	$1.0 / \gamma$
u, v, w	0.0, 0.0, 0.0
Aircraft surface	Slip wall condition
Outer boundary	Riemann invariant boundary condition

3.3 Control Method

The rotor speed is controlled by PD control, which is widely used in the world. In this study, we conduct a numerical experiment in which the rotor is suddenly stopped and the flight attitude is forcibly broken on the simulation. In that case, it works to maintain the correct flight attitude by automatically controlling the rotation speed of other rotors. The rotation of the rotor is determined by the superposition of four manipulated variables: the throttle that controls altitude, the aileron that controls roll, the rudder that controls yaw, and the elevator that controls pitch. In order to hover, the target angles of roll, pitch, and yaw angle are all set to zero. The target altitude is set to 0.5[m]. After confirming that the roll, pitch, yaw angle, and altitude have reached their target values and that the motion is stable, the rotor is brought to the sudden stop. Figure 6 shows the axes of roll, pitch, and yaw angle and the direction of rotation in the computational model. Equations (7)–(10) show the four transfer functions used in control.

$$\frac{p(s)}{u_{Roll}} = \frac{LK_T}{I_{xx}s},\tag{7}$$

$$\frac{q(s)}{u_{Pitch}} = \frac{LK_T}{I_{zz}s},\tag{8}$$

$$\frac{r(s)}{u_{Yaw}} = \frac{K_A}{I_{yy}s},\tag{9}$$

$$\frac{v_{lift}(s)}{u_{Throttle}} = \frac{K_T}{ms},\tag{10}$$

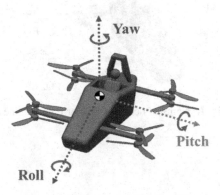

Fig. 6. 3-axis angle of Flying Car

where u_{Roll}, u_{Pitch}, u_{Yaw}, $u_{Throttle}$ are the control inputs for the 3-axis attitude and thrust, p, q, and r are the angular velocities of roll, pitch, and yaw angle, and v_{lift} is the speed in the upward direction. L is the distance of the rotor from the aircraft center of gravity, I is the moment of inertia of the aircraft, K is a coefficient obtained from the lift force and torque [13], and m is the weight of aircraft. Equations (11)–(13) show the relevant mixing equations. The mixing equation is the ratio of the rotation speed assigned to each rotor from the control input obtained in Equations. (7)–(10).

$$\bar{M} = \begin{bmatrix} 1 & 1 & \cdots & 1 \\ -\bar{r}_{Z1} & -\bar{r}_{Z2} & \cdots & -\bar{r}_{ZN} \\ \bar{r}_{X1} & \bar{r}_{X2} & \cdots & \bar{r}_{XN} \\ e_{p1} & e_{p2} & \cdots & e_{pN} \end{bmatrix},\tag{11}$$

$$M = \overline{M}^T \left(\overline{M}\,\overline{M}^T \right)^{-1}, \tag{12}$$

$$D = M \begin{bmatrix} u_{Throttle} \\ u_{Roll} \\ u_{Pitch} \\ u_{Yaw} \end{bmatrix}, \tag{13}$$

where $\bar{r}_X$ and $\bar{r}_Z$ are the dimensionless distances from the aircraft center of gravity to the rotor axis of rotation, and e_p is a scalar that defines the direction of rotor rotation, where forward pitch propellers are set to 1 and reverse pitch propellers to −1. D is the number of rotations assigned to each rotor.

4   Calculation Results

As shown in Fig. 7, each rotor is named FLU, FLL, FRU, FRL, RLU, RLL, RRU, and RRL. The direction of rotor rotation is distinguished by color. The initial letter of each name, such as FLU (Front Left Upper), is used here.

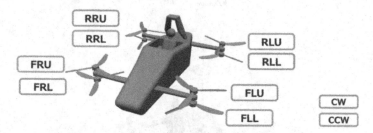

Fig. 7. Name of octorotor (Color figure online)

4.1   Stopping FLU

Figure 8 shows the results of the sudden stop of one of the FLU rotors. The graphs of eight rotor speed versus time and 3-axis attitude versus time are shown. Before the rotor stops, eight rotors are hovering while maintaining their normal motion and attitude. After the rotor stops, control functions to maintain the attitude with the remaining seven rotors. It can be confirmed that the respective rotor speeds were affected before and after the stop. This is because the FLL, which is positioned below the FLU, is trying to maintain its attitude by increasing its rotation speed through control due to the lack of thrust from the FLU. In order to maintain the 3-axis attitude of the roll-pitch-yaw angle, the rotor speeds of FRL and RLL increased, and the rotor speed of RRL decreased. However, there is a 4[deg] tilt of the pitch angle in the attitude angle, confirming the effect on the aircraft's attitude. The large effect on the roll angle is due to the control gain. The yaw angle was also affected by the torque to the aircraft due to one of the rotors stopping.

Therefore, when the rotor stops, the danger of a crash can be avoided by sacrificing one angle because it is difficult to restore all angles to normal.

Figure 9 shows a contour plot of the velocity before and after the rotor stopped suddenly. From Fig. 9, the velocity magnitude after the rotor stops increases and the rotor tries to maintain its attitude. It is also confirmed that asymmetric flow in the left/right and front/rear rotors causes instability in the fluid flow on the underside of the flying car. Therefore, it is expected that the flying car will continue to slide sideways while maintaining its current attitude.

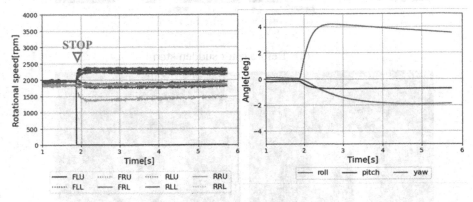

Fig. 8. Time histories of rotational speed(rpm) and angle(deg) in FLU stop (Color figure online)

4.2 Stopping FLU-FLL

This is the result of two contra-rotating rotors, FLU-FLL, stopping simultaneously. Figure 10 shows the trajectory of the flying car at various times due to FLU and FLL sudden stops. Figure 11 shows the isosurface of Q-criterion (Q = 0.01). From Fig. 10, the altitude could not be maintained during the FLU-FLL sudden stop, and the flying car crashed while rotating. At this time, the lift force suddenly is lost from one rotor, causing the rotor to lose its attitude, and the other six rotors could not recover.

From Fig. 11, it can be confirmed that the entire body is caught in the vortex after the rotor suddenly stops. As a result, the aircraft could not maintain its attitude by control and crashed. Considering the balance of forces on the aircraft, it is necessary to operate the rotors of the RRU-RRL, which are located diagonally opposite to the stopped rotor, in a new way.

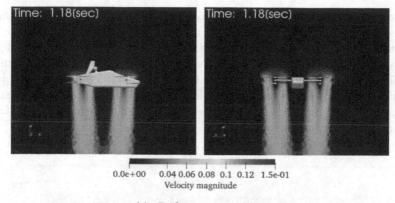

(a) Before rotor sudden stop

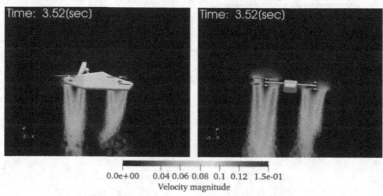

(b) After rotor sudden stop

Fig. 9. Velocity distribution in the plane across the propeller on the left and flont sides at the FLU stop

4.3 No Rotation Speed Limit in Stopping FLU-FLL

When the FLU-FLL was stopped, the aircraft's attitude significantly collapsed and the aircraft crashed. This was considered to be due to a decrease in lift force from one side. Therefore, to balance the forces applied to the aircraft, the control method of the rotor located at the opposite angle was changed. The direction of rotation of the rotor is determined by its shape. However, we thought that it would be possible to obtain a negative lift force by turning the rotor in the opposite direction. Therefore, we made it possible to rotate in the opposite direction by eliminating the lower limit of the rotation speed control. Two of the FLU-FLL are stopped, but the rotation speed limit for each rotor is eliminated to allow reverse rotation. Figure 12 shows graphs of 8 rotor speed versus time and 3-axis attitude versus time. The graphs in Fig. 8 and Fig. 12 compare the results when the rotor speed limit is 0–4000 [rpm] and when the rotor speed is not limited as shown in Fig. 8 and Fig. 12, respectively. From Fig. 12, it can be confirmed that the rotation of RRU-RRL (yellow line), which is located diagonally, shifts to the

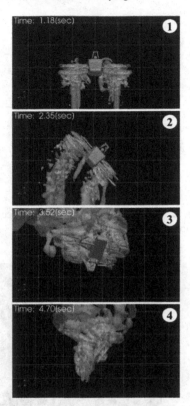

Fig. 10. Time histories of position y-z in FLU-FLL stop

Fig. 11. Isosurfaces of Q-criterion (Q = 0.01) at FLU-FLL stop

reverse direction as soon as the rotation of FLU-FLL stops. The reverse rotation of the rotors generates thrust in the reverse direction of the normal rotation, which enables the flying car to maintain its 3-axis attitude angle and maintain its altitude without crashing. By enabling reverse rotation of the rotor and obtaining reverse thrust, it is possible to approach stable flight.

Figure 13 shows the isosurfaces of Q-criterion (Q = 0.01) with FLU-FLL sudden stops and no speed limit. As shown in Fig. 13, the reverse rotation confirms that the fluid flows in the opposite direction to the normal direction and produces the opposite thrust.

Figure 14 shows the difference in lift force between forward and reverse rotation depending on the rotor geometry used in this study. The vertical axis is the lift force, and the horizontal axis is the rotation speed. The results show that reverse rotation results in a negative lift force value compared to normal rotation. These results indicate that reverse rotation at the appropriate timing can reduce the hazardous behavior of the rotor when it comes to a sudden stop.

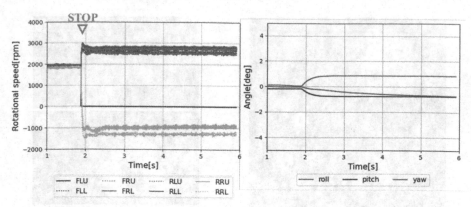

Fig. 12. Time histories of rotational speed(rpm) and angle in FLU-FLL stop (No limit) (Color figure online)

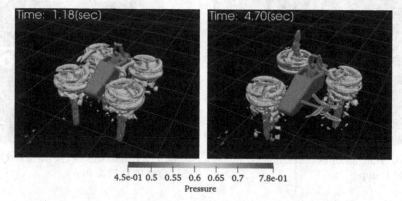

Fig. 13. Isosurfaces of Q-criterion (Q = 0.01) at FLU-FLL stop (No Limit)

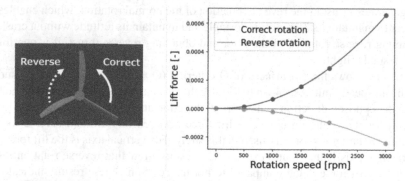

Fig. 14. Difference in lift force between correct and reverse rotation speeds

5 Conclusions

A 6-degrees of freedom flight simulation of an octorotor flying car with a sudden rotor stop was performed by calculating the coupling of fluid and rigid-body interactions. The MCD method based on the unstructured moving mesh finite volume method and the multi-axis sliding mesh method were combined as the computational method. In the simulation of the rotor sudden stop, one and two of the eight rotors were stopped, and the rotation speed assignment was evaluated based on the obtained 3-axis attitude of the flying car. It was confirmed that a one-rotor stop affects the attitude of the flying car, but is unlikely to cause a dangerous situation such as a crash. However, the two-rotor stop may cause a crash depending on the arrangement of the rotors, and it was found that countermeasures were necessary. Therefore, reverse rotation is possible in a rotor that has not stopped. The reverse rotation was confirmed to stabilize the attitude of the flying car by obtaining thrust in the opposite direction of the normal rotation. In the future, we would like to be able to investigate aircraft behavior with other propeller stop and evaluate safety for many different aircraft models.

Acknowledgments. This paper is based on results obtained from a project, JPNP14004, subsidized by the New Energy and Industrial Technology Development Organization (NEDO).

References

1. Alessandro, B., et al.: Electric VTOL configurations comparison. Aerospace **6**(3) (2019)
2. Barcelos, D., et al.: Experimental study of the aerodynamic loads on the airframe of a multirotor UAV. In: 65th Aeronautics Conference, CASI AERO 21 (2021)
3. Seokkwan, Y., et al.: Computational Study of Flow Interactions in Coaxial Rotors (2016)
4. Alberto, S., et al.: A Technocritical Review of Drones Crash Risk Probabilistic Consequences and its Societal Acceptance (2015)
5. Watanabe, K., et al.: Moving computational domain method and its application to fow around a highspeed car passing through a hairpin curve. J. Comput. Sci. Technol. **3**(2), 449–459 (2009)
6. Takii, A., Yamakawa, M., Asao, S., Tajiri, K.: Six degrees of freedom numerical simulation of tilt-rotor plane. In: Rodrigues, J.M.F., et al. (eds.) ICCS 2019. LNCS, vol. 11536, pp. 506–519. Springer, Cham (2019). https://doi.org/10.1007/978-3-030-22734-0_37
7. Yamakawa, M., Yoshioka, K., Asao, S., Takeuchi, S., Kitagawa, A., Tajiri, K.: Numerical simulation of free surface affected by submarine with a rotating screw moving underwater. In: Paszynski, M., Kranzlmüller, D., Krzhizhanovskaya, V.V., Dongarra, J.J., Sloot, P.M.A. (eds.) ICCS 2021. LNCS, vol. 12747, pp. 268–281. Springer, Cham (2021). https://doi.org/10.1007/978-3-030-77980-1_21
8. Roe, P.L.: Approximate Riemann solvers, parameter vectors, and diference schemes. J. Comput. Phys. **43**(2), 357–372 (1981)
9. Hishida, M., et al.: A new slope limiter for fast unstructured CFD solver FaSTAR. In: Proceedings of 42nd Fluid Dynamics Conference/Aerospace Numerical Simulation Symposium. Japan Aerospace Exploration Agency, JAXA-SP-10-012, pp. 85–90 (in Japanese) (2010)

10. SkyDrive Inc Homepage. https://en.skydrive2020.com/. Accessed 23 Jan 2023
11. Ito, Y., et al.: Surface triangulation for polygonal models based on CAD data. Int. J. Numer. Methods Fluids **39**(1), 75–96 (2002)
12. Ito, Y.: Challenges in unstructured mesh generation for practical and efficient computational fluid dynamics simulations. Comput. Fluids **85**(1), 47–52 (2013)
13. Gomi, R., et al.: Flight simulation from take of to yawing of eVTOL airplane with coaxial propellers by fluid-rigid body interaction (2023)

Experimental Study of a Parallel Iterative Solver for Markov Chain Modeling

Valerio Besozzi, Matteo Della Bartola, and Luca Gemignani$^{(\boxtimes)}$ (iD)

Dipartimento di Informatica, Università di Pisa, 56127 Pisa, Italy
luca.gemignani@unipi.it

Abstract. This paper presents the results of a preliminary experimental investigation of the performance of a stationary iterative method based on a block staircase splitting for solving singular systems of linear equations arising in Markov chain modelling. From the experiments presented, we can deduce that the method is well suited for solving block banded or more generally localized systems in a parallel computing environment. The parallel implementation has been benchmarked using several Markovian models.

Keywords: Iterative methods · parallel algorithms · Markov chains

1 Introduction

The solving of linear algebraic systems lies at the core of many scientific and engineering simulations. Discrete-state models are widely employed for modeling and analysis of large networks and systems such as communication networks, allocation schemes, computer systems and population processes. If the future evolution of the system depends only on the current state of the system and not on the past history, the system may be represented by a Markov chain. For a homogeneous, irreducible, continuous time Markov chain with N states, the long-term behavior of the system is determined by the stationary probability vector $\boldsymbol{\pi} \in \mathbb{R}^N$ such that

$$Q^T \boldsymbol{\pi} = \mathbf{0}, \quad \boldsymbol{\pi} \geq \mathbf{0}, \quad e^T \boldsymbol{\pi} = 1, \tag{1}$$

where $Q \in \mathbb{R}^{N \times N}$ is the transition rate matrix, or the infinitesimal generator of the Markov chain, and $e = [1, \ldots, 1]^T$. Since Q is irreducible, by the Perron-Frobenius Theorem [17] we find that Q has rank $N - 1$ and, therefore, $\boldsymbol{\pi}$ spans the kernel of Q^T. The computation of $\boldsymbol{\pi}$ amounts to solve the homogeneous

This work has been supported by the project PRA_2020_61 of the University of Pisa and by the Spoke 1 "FutureHPC & BigData" of the Italian Research Center on High-Performance Computing, Big Data and Quantum Computing (ICSC) funded by MUR Missione 4 Componente 2 Investimento 1.4: Potenziamento strutture di ricerca e creazione di "campioni nazionali di R&S (M4C2-19)" - Next Generation EU (NGEU).

J. Mikyška et al. (Eds.): ICCS 2023, LNCS 14074, pp. 47–61, 2023.
https://doi.org/10.1007/978-3-031-36021-3_4

linear system (1). A review of numerical methods for solving (1) can be found in [20, 26]. Active research in this area is focused on the development of techniques, methods and data structures, which minimize the computational (space and time) requirements for solving the linear system (1) when Q is large and sparse. One of such techniques is parallelization.

Iterative methods are generally preferred for solving large linear systems of equations because they are insensitive to fill-in and accuracy issues [20]. Stationary iterative methods like Gauss-Seidel (GS), Jacobi, and Successive Over-Relaxation (SOR) are interesting on their own and have further applications as preconditioners for projection methods like CG and GMRES. Experimental studies demonstrated that for Markov chain problems (1) block methods based on matrix splittings such as block Jacobi and block Gauss-Seidel give better convergence than other projection methods (see [27] and the references given therein).

Among classical iterative methods, the Gauss-Seidel method has several interesting features. It is a classical result that on a nonsingular M-matrix the Gauss-Seidel method converges faster than the Jacobi method [5, Corollary 5.22]. Moreover it can be implemented just using one iteration vector which is an important feature for huge systems. The SOR method with the optimal relaxation parameter can be better yet, but, however, choosing an optimal SOR relaxation parameter is difficult for many problems. Therefore, the Gauss–Seidel method is very attractive in practice and it is also used as preconditioner in combination with other iterative schemes. A classical example is the multigrid method for partial differential equations, where using Gauss–Seidel or SOR as a smoother typically yields good convergence properties [28].

Parallel implementations of Gauss-Seidel method have been designed for certain regular problems, for example, the solution of Laplace's equations by finite differences, by relying upon red-black coloring or more generally multi-coloring schemes to provide some parallelism [19]. In most cases, constructing efficient parallel true Gauss-Seidel algorithms is challenging and Processor-Block (or localized) Gauss-Seidel is often used [25]. Recent examples with applications to Markov chain modeling are the methods proposed in [6] and [1]. Here, each processor performs Gauss-Seidel as a subdomain solver for a block Jacobi method. While Processor-Block Gauss-Seidel methods are easy to parallelize, the overall convergence of the resulting iterative scheme can suffer.

In order to cope with the parallelization of Gauss-Seidel type methods while retaining the same convergence rate, in [14] staircase splittings were introduced by proving that, for consistently ordered matrices [21], the iterative scheme based on such partitionings splits into independent computations and at the same time exhibits the same convergence rate as the classical Gauss-Seidel iteration. A specialization of this result for block tridiagonal matrices had already appeared in [2]. More recently, in [10] the computational interest of staircase splittings has been broadened by showing that for a nonsingular M-matrix A in block lower Hessenberg form the asymptotic rate of convergence of the block staircase method is better than the asymptotic rate of convergence of the block Gauss-

Seidel method applied to A. A further extension with applications to accelerating certain fixed point iterations for Markov chain modeling is given in [9]. These results are quite surprising since the matrix M in the block staircase partitioning of $A = M - N$ is much more sparse than the corresponding block lower triangular matrix M of the Gauss-Seidel splitting.

The contribution of this paper is twofold. The matrix $A = Q$ in (1) is singular and the comparison theorems proved in [10] do not extend to the singular case, while classical results for singular systems [15,16] do not apply to our methods. The first aim is to gain an understanding of how (block) staircase and (block) Gauss-Seidel type methods compare when applied for solving large and sparse Markov Chain problems. In particular, we are interested in the case where A is banded or localized around the main diagonals. The second goal is to perform this comparison in a parallel computing environment. To do this we have implemented a block staircase iterative solver for the parallel computation of the vector π. The properties of this method are examined experimentally. In particular, numerical experiments are performed to compare our method with an implementation of the composite solver proposed in [6] in terms of traditional efficiency measures for parallel algorithms. Our experimental evidence indicates that the block staircase partitioning generally works quite well when compared to block Gauss-Seidel for block banded or more generally localized matrices [4] with entries decaying away from the main diagonals. A discussion of the results is presented together with some conclusions and insights for future work.

2 Mathematical Background

Let $P = (p_{i,j}) \in \mathbb{R}^{N \times N}$ be a transition probability matrix of a homogeneous ergodic Markov Chain with N states. Then P is irreducible and row-stochastic, that is, $p_{i,j} \geq 0$, $1 \leq i,j \leq N$, and $Pe = e$ with $e^T = [1,\ldots,1]$. The matrix $Q^T = I_N - P^T$ is a singular M-matrix. Observe that $e^T Q^T = 0^T$. Since Q^T is also irreducible, by the Perron-Frobenius Theorem it follows that the kernel of Q^T is spanned by a vector π such that $\pi > 0$ and $e^T \pi = 1$. This vector is called the stationary probability distribution vector of the Markov Chain.

The computation of π amounts to solve the homogeneous linear system $Q^T \pi = 0$ under the normalization $e^T \pi = 1$. Iterative methods based on the power iteration can be used [20]. The computational efficiency and the convergence properties of these algorithms can benefit of a block partitioning of the matrix Q^T. Let us assume that

$$Q^T = \begin{bmatrix} Q_{1,1} & \cdots & Q_{1,n} \\ \vdots & \vdots & \vdots \\ Q_{n,1} & \cdots & Q_{n,n} \end{bmatrix},$$

where $Q_{i,j} \in \mathbb{R}^{n_i \times n_j}$, $1 \leq i,j \leq n$, $\sum_{i=1}^{n} n_i = N$. A regular splitting of the matrix Q^T is a partitioning $Q^T = M - N$ with $M^{-1} \geq 0$ and $N \geq 0$. Since $Q^T \pi = 0$ we find that $M\pi = N\pi$ which gives $M^{-1}N\pi = \pi$. It is well known that the spectral radius of $M^{-1}N$ is equal to 1 and $\lambda = 1$ is a simple eigenvalue of $M^{-1}N$ [24]. This does not immediately imply that $\lambda = 1$ is the dominant eigenvalue of $M^{-1}N$, that is, that for the remaining eigenvalues λ of $M^{-1}N$ it holds $|\lambda| < 1$.

Example 1. Let $Q^T = \begin{bmatrix} 1 & -1 & 0 & 0 \\ -1/2 & 1 & -1/2 & 0 \\ 0 & -1/2 & 1 & -1/2 \\ 0 & 0 & -1 & 1 \end{bmatrix}$. The Jacobi splitting with

$M = I_4$ is a regular splitting but the iteration matrix $M^{-1}N$ has eigenvalues $\{-1, 1, -1/2, 1/2\}$. The Gauss-Seidel splitting is a regular splitting and the corresponding iteration matrix has eigenvalues $\{1, 1/4, 0, 0\}$.

By graph-theoretic arguments [24] it follows that for a regular splitting the matrix $M^{-1}N$ is permutationally similar to a block matrix $T = \begin{bmatrix} 0 & T_{1,2} \\ 0 & T_{2,2} \end{bmatrix}$ where $T_{2,2}$ is square, irreducible and non-negative and every row of the possibly nonempty matrix $T_{1,2}$ is nonzero. A non-negative square matrix A is primitive if there is $m \geq 1$ such that $A^m > 0$. By the Perron-Frobenius Theorem we obtain that $\lambda = 1$ is the dominant eigenvalue of $M^{-1}N$ if $T_{2,2}$ is primitive. Hereafter, this condition is always assumed. Under this assumption the classical power iteration is eligible for determining a numerical approximation of the vector $\boldsymbol{\pi}$.

The method based on the (block) Jacobi splitting is very convenient to vectorize and to parallelize. As shown in the simple example above it can suffer from convergence problems. In this respect, the (block) Gauss-Seidel iteration generally outperforms the Jacobi algorithm. Processor-Block (or localized) Gauss-Seidel schemes provide a reliable compromise between parallelization and convergence issues. One such hybrid adaptation is described in [6]. Suppose that the matrix Q^T is partitioned as

$$Q^T = \begin{bmatrix} Q^{(1)^T} \\ \hline \vdots \\ \hline Q^{(p)^T} \end{bmatrix}, \quad Q^{(j)^T} = \begin{bmatrix} Q_{m_j,1} & \cdots & Q_{m_j,n} \\ \vdots & \vdots & \vdots \\ Q_{m_j+r_j-1,1} & \cdots & Q_{m_j+r_j-1,n} \end{bmatrix},$$

with $m_1 = 1$, $m_j = \sum_{i=1}^{j-1} r_i + 1$, $2 \leq j \leq p$, $\sum_i^p r_i = n$. The iterative scheme in [6] exploits the regular splitting where

$$M = \text{diag}\left[M_1, \ldots, M_p\right], \quad M_j = \begin{bmatrix} Q_{m_j,m_j} & & & \\ Q_{m_j+1,m_j} & Q_{m_j+1,m_j+1} & & \\ \vdots & & \ddots & \\ Q_{m_{j+1}-1,m_j} & \cdots & & Q_{m_{j+1}-1,m_{j+1}-1} \end{bmatrix}$$

and, hence, M is a block diagonal matrix with block lower triangular blocks. The resulting scheme proceeds as follows:

$$\begin{cases} M\boldsymbol{x}^{(k+1)} = N\boldsymbol{x}^{(k)} \\ \boldsymbol{x}^{(k+1)} = \dfrac{\boldsymbol{x}^{(k+1)}}{e^T\boldsymbol{x}^{(k+1)}} \end{cases}, \quad k \geq 1. \tag{2}$$

If p is the number of processors, then Algorithm 1 is a possible implementation of this scheme starting from the skeleton proposed in [6]. A different approach to

---

Algorithm 1. This algorithm approximates the vector π by means of the method in [6]

1: Initialization
2: **while** $err \geq tol$ & $it \leq maxit$ **do**
3:     **parfor** $j = 1,\ldots,p$ **do**
4:         **for** $k = m_j,\ldots,m_j + r_j - 1$ **do**
5:             $\mathcal{J}_k = \{s \in \mathbb{N}\colon m_j \leq s < k\}$
6:             $z_k \leftarrow -Q_{k,k}^{-1}\left(\sum_{s \in \mathcal{J}_k} Q_{k,s} z_s + \sum_{s \in \{1,\ldots,n\} \setminus \mathcal{J}_k} Q_{k,s} x_s\right)$
7:         **end for**
8:     **end parfor**
9:     $z \leftarrow \frac{z}{e^T z}$
10:    $err \leftarrow \| z - x \|_1$; $x \leftarrow z$; $it \leftarrow it + 1$
11: **end while**
12: **return** x

---

parallelizing stationary iterative solvers was taken in [14]. The approach is based on the exploitation of a suitable partitioning of the matrix Q^T where M has a "zig-zag" pattern around the main diagonal referred to as a staircase profile. More specifically, we can associate with Q^T the stair matrices $M_1 \in \mathbb{R}^{N \times N}$ and $M_2 \in \mathbb{R}^{N \times N}$ defined by

$$M_1 = \begin{bmatrix} Q_{1,1} & & & & & \\ Q_{2,1} & Q_{2,2} & Q_{2,3} & & & \\ & & Q_{3,3} & & & \\ & & Q_{4,3} & Q_{4,4} & Q_{4,5} & \\ & & & & Q_{5,5} & \\ & & & & \times & \times \times \\ & & & & & \times \end{bmatrix}$$

and

$$M_2 = \begin{bmatrix} Q_{1,1} & Q_{1,2} & & & & \\ & Q_{2,2} & & & & \\ & Q_{3,2} & Q_{3,3} & Q_{3,4} & & \\ & & & Q_{4,4} & & \\ & & & Q_{5,4} & Q_{5,5} & Q_{5,6} \\ & & & & \times & \\ & & & & \times & \times \times \end{bmatrix}.$$

These matrices are called stair matrices of type 1 and 2, respectively. Staircase splittings of the form $Q^T = M - N$ where M is a stair matrix have two remarkable features:

1. The solution of a linear system $M\boldsymbol{x} = \boldsymbol{b}$ can be carried out in two parallel steps since all even and all odd components of $\boldsymbol{x}$ can be computed concurrently.
2. In terms of convergence these splittings inherit some advantages of the block Gauss-Seidel method. If Q^T is block tridiagonal, then it can be easily proved that the iteration matrices associated with block Gauss-Seidel and block staircase splittings have the same eigenvalues and therefore the same convergence rate. In [10] it is shown that the spectral radius of the iteration matrix generated by the staircase splitting of an invertible M-matrix in block lower Hessenberg form is not greater than the spectral radius of the corresponding iteration matrix in the block Gauss-Seidel method. For singular matrices the convergence of the scheme (2) depends on the spectral gap between the dominant eigenvalue equal to 1 and the second eigenvalue $\gamma = \max\{|\lambda| : \lambda \text{ eigenvalue of } M^{-1}N \text{ and } \lambda \neq 1\}$. Experimentally (see Example 3 below) the block staircase iteration still performs similarly to the block Gauss-Seidel method when applied for solving linear systems with singular M-matrices in block Hessenberg form. The same behavior is observed for matrices that are localized around the main diagonals. Examples are the covariance matrices with application to the spatial kriging problem (compare with [11]). Other non artificial examples are discussed in Sect. 3.

Example 2. For the matrix of Example 1 the staircase splitting of the first type gives an iteration matrix having the same eigenvalues $\{1, 1/4, 0, 0\}$ of the Gauss-Seidel scheme.

Example 3. We have performed several numerical experiments with randomly generated singular M-matrices in banded block lower Hessenberg form having the profile depicted in Fig. 1.

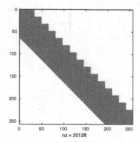

Fig. 1. Spy plot

Table 1. Convergence comparisons

kp \ k		16	32	64
16	ρ_1			1.6
	ρ_2			1.5
32	ρ_1			1.6
	ρ_2			1.4
64	ρ_1	∞	33.6	2.0
	ρ_2	1.0	1.0	1.4

The size of the blocks is k, the block size of the matrix is n and the lower bandwidth is $4k$. We compare the performance of the block Jacobi (BJ), block Gauss-Seidel (BGS) and block staircase (BS) methods for different sizes of the block partitioning denoted as kp. In the following Table 1 we show the maximum value of $\rho_1 = \log(1/\gamma_{BGS})/\log(1/\gamma_{BJ})$ and of $\rho_2 = \log(1/\gamma_{BGS})/\log(1/\gamma_{BS})$ – where γ_M is the second eigenvalue of the iteration matrix of size $N = k \times n$

generated by the method $M \in \{BGS, BJ, BS\}$ over 1000 experiments with $k = 64$ and $n = 16$. The value ∞ indicates that in some trials the block Jacobi method does not converge due to the occurrence of two or more eigenvalues equal to 1 in magnitude. The results demonstrate that the spectral gap of the iteration matrix in the block staircase method remains close to that one of BGS.

The following Algorithm 2 provides an implementation of (2) using $M = M_1$. For $i = 1, \ldots, n$ let

$$\mathcal{J}_i = \begin{cases} \{i-1, i+1\} \cap \{1, \ldots, n\}, & i \text{ even} \\ i, & i \text{ odd.} \end{cases}$$

Algorithm 2. This algorithm approximates the vector π by means of the method (2) with $M = M_1$

1: Initialization
2: **while** $err \geq tol$ & $it \leq maxit$ **do**
3:    **parfor** $i = 1, \ldots, n$ **do**
4:       $z_i \leftarrow -\sum_{k \notin \mathcal{J}_i} Q_{i,k} x_k$
5:    **end parfor**
6:    **parfor** $i = 1, 3, \ldots, n$ **do**
7:       $z_i \leftarrow Q_{i,i}^{-1} z_i$
8:    **end parfor**
9:    **parfor** $i = 2, 4, \ldots, n$ **do**
10:       $z_i \leftarrow Q_{i,i}^{-1} \left(z_i - \sum_{k \in \mathcal{J}_i} Q_{i,k} z_k \right)$
11:    **end parfor**
12:    $z \leftarrow \frac{z}{e^T z}$
13:    $err \leftarrow \| z - x \|_1$; $x \leftarrow z$; $it \leftarrow it + 1$
14: **end while**
15: **return** x

In the next section numerical experiments are performed to compare the performance of Algorithm 1 and Algorithm 2 for solving singular systems of linear equations arising in Markov chain modeling.

3 Numerical Experiments

The experiments have been run on a server with two Intel Xeon E5-2650v4 CPUs with 12 cores and 24 threads each, running at 2.20 GHz. Hyper-threading was disabled so that the number of logical processors is equal to physical processors (cores). To reduce the variance of execution times, the number of threads is set to the number of physical cores and each thread is mapped statically to one core. The parallel implementations of Algorithm 1 –referred to as JGS algorithm– and Algorithm 2 –referred to as STAIR1 or STAIR2 algorithm depending on the staircase splitting– are based on OpenMP. Specifically, we have used C++20 with the help of Armadillo [22,23] which also provides integration with LAPACK [3] and OpenBLAS [13].

Our test suite consists of the following transition matrices:

1. The transition rate matrix associated with the queuing model described in
   [7]. This is a complex queuing model, a BMAP/PHF/1/N model with retrial
   system with finite buffer and non-persistent customers. We do not describe
   in detail the construction of this matrix, as it would take some space, but
   refer the reader to [6, Sects. 4.3 and 4.5]. The buffer size is denoted as n.
   The only change with respect to the paper is that we fix the orbit size to a
   finite capacity k (when the orbit is full, customers leave the queue forever).
   We set $n = 15$, which results in a block upper Hessenberg matrix Q of size
   $N = 110400$ with $k \times k$ blocks of size 138.
2. The transition rate matrix for the model described in Example 1 of [20]. The
   model describes a time-sharing system with n terminals which share the same
   computing resource. The matrices are nearly completely decomposable (NCD)
   so that classical stationary iterative methods do not perform satisfactorily as
   the spectral gap is pathologically close to unity. We set $n = 80$ which gives a
   matrix of size $N = 91881$.
3. The transition rate matrix for the model described in Example 3 of [20]. The
   model describes a multi-class, finite buffer, priority system. The buffer size is
   denoted as n. This model can be applied to telecommunications modeling, and
   has been used to model ATM queuing networks as discussed in [20, 26]. We
   note that the model parameters can be selected so that the resulting Markov
   chain is nearly completely decomposable. We set $n = 100$ which results in a
   matrix of size $N = 79220$.
4. The transition rate matrix generated by the set of mutual-exclusion problems considered in [8]. In these problems, n distinguishable processes share a
   certain resource. Each of these processes alternates between a sleeping state
   and a resource using state. However, the number of processes that may concurrently use the resource is limited to r, where $1 \leq r \leq n$, so that when a
   process wishing to move from the sleeping state to the resource using state
   finds r processes already using the resource, that process fails to access the
   resource and returns to the sleeping state. We set $n = 16$ and $r = 12$ so that
   the transition matrix has size $N = 64839$.

All the considered transition matrices are sparse matrices. In Fig. 2 we show the
spy plots of the matrices generated in tests 1-4. The matrices are stored using
the compressed sparse column format. With this method, only nonzero entries
are kept in memory. However, despite evident merits this solution has also some
drawbacks. In particular, we notice that certain operations such as submat calls
become relatively expensive because they could be creating temporary matrices.

Clearly, the size of the block partitioning of the matrix seriously affects the
performance of iterative methods. There is an extensive literature on this topic
(see for instance [12, 18] and the references given therein). We have not tried the
partition algorithms described in [12, 18] and in this paper we explore the use of
blocks of equal size $n_i = \ell$, $1 \leq i \leq n-1$. A test for the change of two consecutive
iterates to be less than a prescribed tolerance or the number of iterations to be

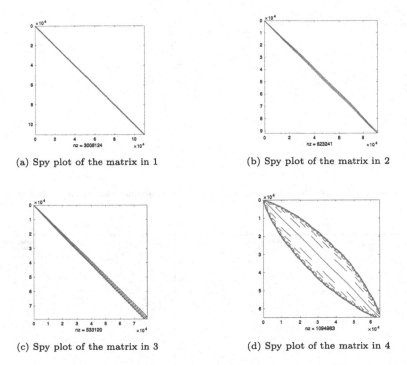

(a) Spy plot of the matrix in 1

(b) Spy plot of the matrix in 2

(c) Spy plot of the matrix in 3

(d) Spy plot of the matrix in 4

Fig. 2. Illustration of the sparsity pattern of the matrices in our test suite

greater than a given bound is used as stopping criterion. In other words, we stop
the iteration if

$$\| \; x^{(k)} - x^{(k+1)} \; \|_1 \leq \epsilon \; \vee \; k \geq \text{maxit}.$$

In all the experiments reported below we have used $\epsilon = -1.0e - 9$ and
maxit=$1.0e + 4$. A larger number of iterations is carried out in some cases to
check for divergence of the iterative process. For each algorithm and experiment
we measure the sequential completion time T_{seq}, the parallel completion time
on m threads $T_{par}(m)$, the speedup $S_p(m) = T_{seq}/T_{par}(m)$ and the efficiency
$E(m) = S_p(m)/m$. We also report a plot of the residual $\| \; x^{(k)} - x^{(k+1)} \; \|_1$ to
analyze the convergence of the iterative scheme.

In the first experiment we consider the transition rate matrix generated in
1. with $n = 15$ and $k = 800$. The matrix has size $N = 110400$ and the block
partitioning is determined by setting $\ell = 512$. The matrix is block lower Hessenberg and block banded. In Fig. 3 we show the plots of completion time, residual,
speedup and efficiency generated for this matrix. The convergence of block staircase methods is better than the convergence of the block Gauss-Seidel method.
The superlinear speedup is a cache effect. The best sustained computation rate
is about 6.5 Gflops on 24 cores. We think that the attained level of performance
is quite low with respect to the theoretical peak performance due to the use of
sparse linear algebra functionalities.

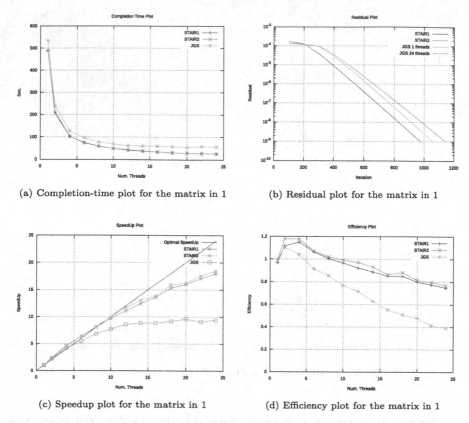

(a) Completion-time plot for the matrix in 1

(b) Residual plot for the matrix in 1

(c) Speedup plot for the matrix in 1

(d) Efficiency plot for the matrix in 1

Fig. 3. Illustration of the performance of algorithms JGS, STAIR1 and STAIR2 for the matrix in 1.

In Fig. 4 we show the plots generated for the matrix in 2. with $n = 80$. The size is $N = 91881$. The matrix is symmetric with bandwidth 1135. We set $\ell = 2048$ so that the matrix is block tridiagonal. The crazy behavior of JGS in Fig. 4 and in Fig. 5 depends on the fact that the block partitioning varies with the number of threads. In particular, for the nearly completely decomposable example this makes possible divergence or false convergence cases. In the course of experimenting with the JGS method, we have tried various policies for the distribution of blocks to different processors. The chosen policy also goes to affect the behavior and convergence of the algorithm as the number of threads varies. It was chosen in the end to use as fair a distribution policy as possible, trying in general to allocate in each thread the same number of blocks and eventually, in the case of any remaining blocks, evenly distributing the remainder.

In Fig. 5 we show the plots generated for the matrix in 3. with $n = 100$. The matrix has size $N = 79220$, lower bandwidth 2370 and upper bandwidth 1585. We set $\ell = 2048$ so that the matrix is block banded in block lower Hessenberg form.

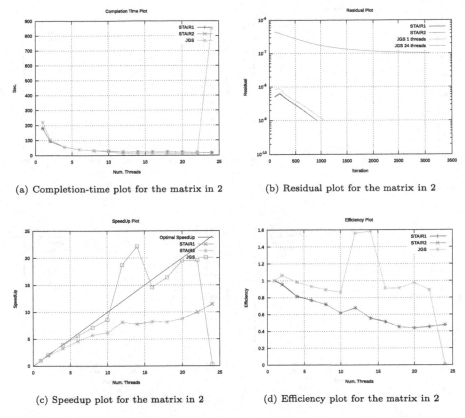

(a) Completion-time plot for the matrix in 2   (b) Residual plot for the matrix in 2

(c) Speedup plot for the matrix in 2   (d) Efficiency plot for the matrix in 2

Fig. 4. Illustration of the performance of algorithms JGS, STAIR1 and STAIR2 for the matrix in 2.

The results in Figs. 4, 5 clearly highlight that the convergence of the JGS method may deteriorate as the number of threads increase since the iteration becomes close to a pure block Jacobi method. In particular for $m = 24$ threads in both examples the method does not converge and the process stops as the maximum number of iterations has been reached. Finally, in Fig. 6 we illustrate the plots of completion time and speedup for the matrix generated by 4. with $n = 16$, and $r = 12$. The matrix has size $N = 64839$ and bandwidth 13495. We set $\ell = 256$ so that the matrix is block banded. Since the entries are very rapidly decaying away from the main diagonal all methods perform quite well and the number of iterations in the JGS method is quite insensitive to the number of threads.

Concerning the parallel performance, we recall that the server has only 24 combined physical cores, and going above 12 required communication between the different CPUs, which inevitably reduces the efficiency of the parallelization. When the number of threads is quite small it is generally observed that the bigger the block size, the shorter is the execution time. Differently, as the number

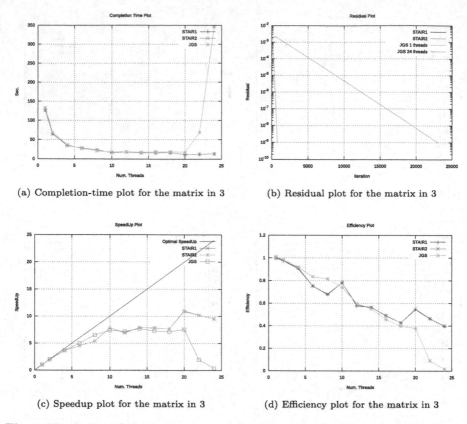

(a) Completion-time plot for the matrix in 3

(b) Residual plot for the matrix in 3

(c) Speedup plot for the matrix in 3

(d) Efficiency plot for the matrix in 3

Fig. 5. Illustration of the performance of algorithms JGS, STAIR1 and STAIR2 for the matrix in 3.

of threads increases small blocks promote the parallelism. The reason can be attributed to two main factors: load balancing and communication overhead. In Fig. 7 we show the the speedup plot for the test 3. with $\ell = 512$ and $\ell = 1024$, respectively. The comparison with the results reported in Fig. 5 with $\ell = 2048$ indicates some improvements. These effects are highlighted in all the conducted experiments. Considering that most consumer hardware has between 2 and 8 or 16 cores, this shows that the proposed method is generally well tuned for the currently available architectures.

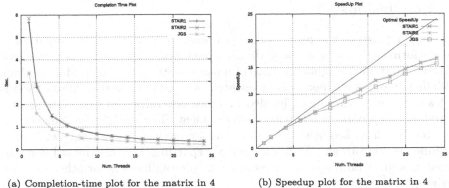

(a) Completion-time plot for the matrix in 4

(b) Speedup plot for the matrix in 4

Fig. 6. Illustration of the performance of algorithms JGS, STAIR1 and STAIR2 for the matrix in 4.

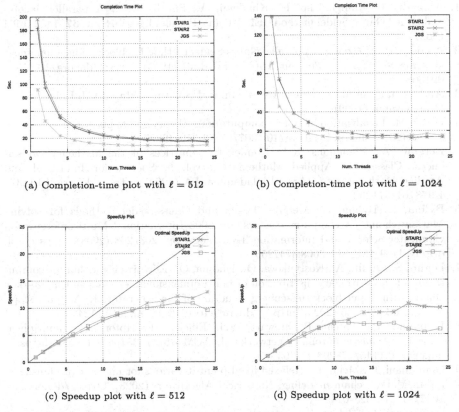

(a) Completion-time plot with $\ell = 512$

(b) Completion-time plot with $\ell = 1024$

(c) Speedup plot with $\ell = 512$

(d) Speedup plot with $\ell = 1024$

Fig. 7. Illustration of the completion time and speedup plot for the matrix 3 with different block sizes

4   Conclusions

This paper presents the results of a preliminary experimental investigation of the performance of a stationary iterative method based on a block staircase splitting for solving singular systems of linear equations arising in Markov chain modeling. From the experiments presented, we can deduce that the method is well suited for solving block banded or more generally localized systems in a shared-memory parallel computing environment. The parallel implementation has been benchmarked using several Markovian models. In the future we plan to examine the performance of block staircase splittings in a distributed computing environment and, moreover, their use as preconditioners for other iterative methods.

References

1. Ahmadi, A., Manganiello, F., Khademi, A., Smith, M.C.: A parallel Jacobi-embedded Gauss-Seidel method. IEEE Trans. Parallel Distrib. Syst. **32**, 1452–1464 (2021)
2. Amodio, P., Mazzia, F.: A parallel Gauss-Seidel method for block tridiagonal linear systems. SIAM J. Sci. Comput. **16**(6), 1451–1461 (1995). https://doi.org/10.1137/0916084
3. Anderson, E., et al.: LAPACK Users' Guide. USA, third edn, SIAM, Philadelphia, Pennsylvania (1999)
4. Benzi, M.: Localization in matrix computations: theory and applications. Presented at the (2016). https://doi.org/10.1007/978-3-319-49887-4_4
5. Berman, A., Plemmons, R.J.: Nonnegative matrices in the mathematical sciences, Classics in Applied Mathematics, vol. 9. Society for Industrial and Applied Mathematics (SIAM), Philadelphia, PA (1994). https://doi.org/10.1137/1.9781611971262
6. Bylina, J., Bylina, B.: Merging Jacobi and Gauss-Seidel methods for solving Markov chains on computer clusters. In: 2008 International Multiconference on Computer Science and Information Technology, pp. 263–268 (2008). https://doi.org/10.1109/IMCSIT.2008.4747250
7. Dudin, S., Dudin, A., Kostyukova, O., Dudina, O.: Effective algorithm for computation of the stationary distribution of multi-dimensional level-dependent Markov chains with upper block-Hessenberg structure of the generator. J. Comput. Appl. Math. **366**, 112425 (2020). https://doi.org/10.1016/j.cam.2019.112425
8. Fernandes, P., Plateau, B., Stewart, W.J.: Efficient descriptor-vector multiplications in stochastic automata networks. J. ACM **45**(3), 381–414 (1998). https://doi.org/10.1145/278298.278303
9. Gemignani, L., Meini, B.: Relaxed fixed point iterations for matrix equations arising in Markov chain modeling. Numerical Algorithms (2023). https://doi.org/10.1007/s11075-023-01496-y
10. Gemignani, L., Poloni, F.: Comparison theorems for splittings of M-matrices in (block) Hessenberg form. BIT **62**(3), 849–867 (2022). https://doi.org/10.1007/s10543-021-00899-4
11. Ghadiyali, H.S.: Partial gauss-seidel approach to solve large scale linear systems, Master's thesis, Florida State University (2016). https://purl.flvc.org/fsu/fd/FSU_2016SP_Ghadiyali_fsu_0071N_13280

12. Klevans, R.L., Stewart, W.J.: From queueing networks to Markov chains: the XMARCA interface. Perform. Eval. **24**(1), 23–45 (1995). https://doi.org/10.1016/0166-5316(95)00007-K
13. Lawson, C.L., Hanson, R.J., Kincaid, D.R., Krogh, F.T.: Basic linear algebra subprograms for Fortran usage. ACM Trans. Math. Softw. **5**(3), 308–323 (1979). https://doi.org/10.1145/355841.355847
14. Lu, H.: Stair matrices and their generalizations with applications to iterative methods. I. A generalization of the successive overrelaxation method. SIAM J. Numer. Anal. **37**(1), 1–17 (1999). https://doi.org/10.1137/S0036142998343294
15. Marek, I., Szyld, D.B.: Iterative and semi-iterative methods for computing stationary probability vectors of Markov operators. Math. Comp. **61**(204), 719–731 (1993). https://doi.org/10.2307/2153249
16. Marek, I., Szyld, D.B.: Comparison of convergence of general stationary iterative methods for singular matrices. SIAM J. Matrix Anal. Appl. **24**(1), 68–77 (2002). https://doi.org/10.1137/S0895479800375989
17. Meyer, C.: Matrix analysis and applied linear algebra. Society for Industrial and Applied Mathematics (SIAM), Philadelphia, PA (2000). https://doi.org/10.1137/1.9780898719512
18. O'Neil, J., Szyld, D.B.: A block ordering method for sparse matrices. SIAM J. Sci. Statist. Comput. **11**(5), 811–823 (1990). https://doi.org/10.1137/0911048
19. Ortega, J.M., Voigt, R.G.: Solution of partial differential equations on vector and parallel computers. SIAM Rev. **27**(2), 149–240 (1985). https://doi.org/10.1137/1027055
20. Philippe, B., Saad, Y., Stewart, W.J.: Numerical methods in Markov chain modeling. Oper. Res. **40**(6), 1156–1179 (1992). https://www.jstor.org/stable/171728
21. Saad, Y.: Iterative methods for sparse linear systems. Society for Industrial and Applied Mathematics, Philadelphia, PA, second edn. (2003). https://doi.org/10.1137/1.9780898718003
22. Sanderson, C., Curtin, R.: Armadillo: a template-based C++ library for linear algebra. J. Open Source Softw. **1**, 26 (2016)
23. Sanderson, C., Curtin, R.: A user-friendly hybrid sparse matrix class in C++. In: Davenport, J.H., Kauers, M., Labahn, G., Urban, J. (eds.) Mathematical Software - ICMS 2018, pp. 422–430. Springer International Publishing, Cham (2018)
24. Schneider, H.: Theorems on M-splittings of a singular M-matrix which depend on graph structure. Linear Algebra Appl. **58**, 407–424 (1984). https://doi.org/10.1016/0024-3795(84)90222--2
25. Shang, Y.: A distributed memory parallel Gauss-Seidel algorithm for linear algebraic systems. Comput. Math. Appl. **57**(8), 1369–1376 (2009). https://doi.org/10.1016/j.camwa.2009.01.034
26. Stewart, W.J.: Introduction to the numerical solution of Markov chains. Princeton University Press, Princeton, NJ (1994)
27. Touzene, A.: A new parallel algorithm for solving large-scale Markov chains. J. Supercomput. **67**(1), 239–253 (2014)
28. Wallin, D., Löf, H., Hagersten, E., Holmgren, S.: Multigrid and Gauss-Seidel smoothers revisited: parallelization on chip multiprocessors. In: Egan, G.K., Muraoka, Y. (eds.) Proceedings of the 20th Annual International Conference on Supercomputing, ICS 2006, Cairns, Queensland, Australia, 28 June - 01 July 2006, pp. 145–155. ACM (2006). https://doi.org/10.1145/1183401.1183423

Impact of Mixed-Precision: A Way to Accelerate Data-Driven Forest Fire Spread Systems

Carlos Carrillo$^{(\boxtimes)}$ ⓘ, Tomás Margalef ⓘ, Antonio Espinosa ⓘ,
and Ana Cortés ⓘ

Computer Architecture and Operating Systems Department, Universitat Autónoma
de Barcelona, Barcelona, Spain
{carles.carrillo,tomas.margalef,antoniomiguel.espinosa,
ana.cortes}@uab.cat

Abstract. Every year, forest fires burn thousands of hectares of forest around the world and cause significant damage to the economy and people from the affected zone. For that reason, computational fire spread models arise as useful tools to minimize the impact of wildfires. It is well known that part of the forest fire forecast error comes from the uncertainty in the input data required by the models. Different strategies have been developed to reduce the impact of this input-data uncertainty during the last few years. One of these strategies consists of introducing a data-driven calibration stage where the input parameters are adjusted according to the actual evolution of the fire using an evolutionary scheme. In particular, the approach described in this work consists of a Genetic Algorithm (GA). This calibration strategy is computationally intensive and time-consuming. In the case of natural hazards, it is necessary to maintain a balance between accuracy and time needed to calibrate the input parameters. Most scientific codes have over-engineered the numerical precision required to obtain reliable results. In this paper, we propose to use a mixed-precision methodology to accelerate the calibration of the input parameters involved in forest fire spread prediction without sacrificing the accuracy of the forecast. The proposed scheme has been tested on a real fire. The results have led us to conclude that using the mixed-precision approach does not compromise the quality of the result provided by the forest fire spread simulator and can also speed up the whole evolutionary prediction system.

Keywords: Mixed-precision · Forest Fire Spread Simulator · Input
data Uncertainty

1 Introduction

The need to anticipate potential fire behavior and its resultant impacts has led to the development of the field of fire modeling. In recent decades, various physical and mathematical models have been developed to provide reliable forecasting of

J. Mikyška et al. (Eds.): ICCS 2023, LNCS 14074, pp. 62–76, 2023.
https://doi.org/10.1007/978-3-031-36021-3_5

fire behavior and optimize firefighting resource management during outbreaks. Simulators implementing forest fire spread models need several input parameters to describe the characteristics of the circumstances where the fire occurs to evaluate its future propagation. However, there are serious difficulties in acquiring precise values of certain parameters at the same places where the fire is taking place, often because the hazard itself distorts the measurements. To minimize the input data uncertainty when dealing with forest fire spread prediction, we focus on a Two-stage prediction scheme [1]. The fundamental aim of this approach is to introduce an intermediate adjustment stage to improve the estimation of certain input parameters according to the observed behavior of the forest fire. In particular, we focus on a well-known Two-stage Strategy based on Genetic Algorithms (GA) [7]. The GA generates an initial population where each individual consists of a particular configuration of the input parameters that will be fed into the underlying forest fire spread simulator. The population will evolve using the so-called genetic operators (elitism, mutation, selection, and crossover) to obtain an improved set of parameters that reproduce the observed past behavior of the fire best. To evaluate the simulation quality of each individual, the landscape where the fires are taking place is represented as a grid of cells, and each cell will have assigned a state according to its belonging to either a real burnt area or a simulated burnt area. There are four possible states for a given cell: cells that were burnt in both fires, the real and the simulated fire, *(Hits)*, cells burnt in the simulated fire but not in reality *(False Alarms)*, cells burnt in reality but not in the simulated fire *(Misses)* and cells that were not burnt in any case *(Correct negatives)* [8] [10]. These four possibilities are used to construct a 2×2 contingency table, [4]. A perfect simulation system would have data only on the main diagonal. In the particular case of forest fire, the difference between real and simulated fire is computed using Eq. 1, [16]. The resulting error is used to rank the individuals in a list from lower to higher error.

$$\in = \frac{Misses + FA}{Hits + Misses} \tag{1}$$

Most scientific codes use double-precision by default for all their floating point variables. It is taken for granted that all real number variables must be implemented using double-precision types. But in many cases, the computational cost of this decision is not well weighted and other alternatives could be considered. In many cases, not all variables in a program need this double-precision. Different studies have demonstrated the potential benefits that mixed-precision approaches can provide to many different kinds of scientific codes since it is possible to achieve substantial speed-ups for both compute and memory-bound algorithms requiring little code effort and acceptable impact on functionality, [3,15]. The main principle is to review programs from the observation that, in many cases, a single-precision solution to a problem can be refined to the degree where double-precision accuracy is achieved. However, an excessive precision reduction of a set of relevant variables can lead to accuracy losses that may finally produce unreliable results.

In this paper, we present a straightforward methodology to review the need for double-precision variables to improve the global performance of the applications. We apply the methodology to the code of the forest fire spread simulator. We have modified it to select the set of variables that can use single-precision without disturbing the general evolution of the forest fire spread simulation and, even more, to determine the performance improvements. We will use well-known double-precision fire behavior results to compare and verify the quality of each mixed-precision approach generated. In particular, as a forest fire spread simulator, we used FARSITE (*Fire Area Simulator*) [9], which is a widely accepted and validated wildfire spread simulator [2] based on the Rothermel's model [17]. We have concentrated on the principal algorithms defining FARSITE floating number processing performance. Most complex modules include the fire perimeter expansion and the perimeter reconstruction that account for the 67% of the total execution time, 7% for the perimeter expansion, and 60% for the perimeter reconstruction modules. We apply the *Two-stage Strategy* to calibrate the input values required for the prediction during this work. Simulation is based on the execution of GA to evaluate hundreds of simulations, in our case, 10 generations and 128 individuals per generation. If the complexity of running these individuals can be reduced, even if the improvement is not very noticeable, it will be cumulative for all generations and the final prediction. To preserve the accuracy of the obtained prediction, we only modify the precision of those variables that do not significantly impact the fire evolution.

This paper is organized as follows. Section 2 exposes the proposed mixed-precision methodology applied to a forest fire spread prediction system. Subsequently, in Sect. 3, the experimental study and the obtained results are reported. Finally, in Sect. 4, the main conclusions and open lines of this work are summarized.

2   Mixed-Precision Methodology

Most computational science codes have overestimated the needed numerical precision of a model leading to a situation where simulators are using more precision than required without considering whether this precision is really needed. By using a more appropriate choice of 32-bit and 64-bit floating point arithmetic, the performance of some scientific applications was significantly enhanced without affecting the accuracy of the results, [5,11,13]. A careful way of managing floating-point precision that outstrips double-precision performance motivated us to study if using the mixed-precision variables allows reducing the execution time of the forest fire spread simulations without losing accuracy. To effectively compare the proposed mixed-precision model against a reference implementation (double-precision scheme) we need to ensure not to affect the quality of the results.

One Key point of our methodology is to evaluate the simulation quality when changing the variables of FARSITE that can use single precision. First, we need to define some metrics to quantify the quality of the mixed-precision simulation. The Forest area is divided into cells, and each cell will have a meaning. The cells around the map that have been burnt by neither the reference simulation nor the mixed-precision simulation are considered *Correct Negatives* (CN). Those cells that have been burnt in both simulated fires are called *Hits*. The cells that are only burnt in the Reference simulation and are not burnt in the mixed-precision simulation are called *Misses*. The cells burnt in the mixed-precision simulation, but the Reference simulation does not reach them, are called *False Alarms* (FA), see Fig. 1.

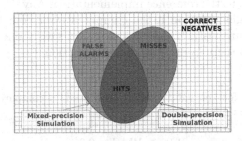

Fig. 1. Events involved in metrics related to model verification.

Three different quality metrics are used to verify the proposed mixed-precision implementation, [4]. The first metric is the *Bias score or frequency Bias* (BIAS), which represents the ratio of the number of correct forecasts over the number of correct observations, Eq. 2. It represents the normalized symmetric difference between the real and simulated maps.

$$BIAS = \frac{Hits + FA}{Hits + Misses} \tag{2}$$

The second metric is the *False Alarm Rate* (FAR). The *FAR* measures the proportion of the wrong events forecast, see Eq. 3. The *FAR* is sensitive to *False Alarms* but ignores the misses. A perfect comparison has a *FAR* value equal to 0.

$$FAR = \frac{FA}{(Hits + FA)} \tag{3}$$

The last metric is the *Probability of Detection of hits rate* (POD), see Eq. 4. *POD* considers the observed and positively estimated events. Thus, it represents the probability of an event being detected. The *POD* is sensitive to hits but ignores the false alarms. Its ideal value for the *POD* is 1.

$$POD = \frac{Hits}{Hits + Misses} \tag{4}$$

In order to evaluate the impact of the variables of FARSITE that can use single-precision, we have to estimate how the use of mixed-precision affects the general prediction of forest fire behavior. We can summarize our methodology into three steps:

1. Define a reference execution without modifying the precision of any variable. The obtained simulation will be used as a reference against which we will validate any mixed-precision approach taken later on.
2. Define an accuracy threshold. Table 1 displays the thresholds used for the different metrics when the mixed-precision simulations are compared against the reference.
3. Individual precision reduction of each variable and evaluating the impact of this reduction versus the reference implementation. Any variable analyzed can be defined in single-precision if the computed error complies with all three thresholds simultaneously. If the error does not satisfy one of the thresholds, the variable must keep its original double-precision.

Table 1. Thresholds for the validation metrics used to compare the reference and mixed-precision simulations.

1.05 > BIAS > 0.95
FAR < 0.05
POD > 0.95

In our case, an individual accuracy test was done for each variable to analyze the global impact of the mixed-precision implementation. For each test, the precision of a single variable was reduced, while the rest of the variables of the program kept their original form, and a new fire spread simulation was done. The obtained simulation was compared with the reference, and a new *BIAS*, *FAR*, and *POD* were calculated. If the simulation does not exceed the thresholds, the precision reduction for this variable was accepted. The variable maintained its original double-precision definition if the validation values were beyond the threshold limits. Then, the methodology analyzes the next variable and reduces its precision, and the validation process starts again. A new mixed-precision forest fire simulation model was obtained with the resulting final list of variables.

At the end of this analysis process, 266 different variables were tested, 221 for the point expansion algorithm and 45 for the perimeter reconstruction algorithm. We found that nearly 74% of the point expansion algorithm variables could safely use single-precision. For the perimeter reconstruction algorithm, only 15% of the variables could use single-precision without compromising the accuracy of the prediction.

This methodology is expensive in terms of execution time and resources because it is necessary to perform one simulation for each variable. If simulations consume 60 s, we need around 4.5 h to apply the methodology completely. However, because this mixed-precision model is going to be used for calculating hundreds of simulations, we expect the benefits of overcoming this initial cost. Moreover, the mixed-precision fire spread model is not dependent on the forest fire scenario, which means we can use it in any future wildfire to predict its behavior.

3 Experimental Study and Results

To analyze the impact of the mixed-precision in the propagation of the fire front and into the forest fire spread simulator performance, we have selected an event belonging to the database of EFFIS (*European Forest Fire Information System*) [6]. The study case took place in 2009 in the region of Nuñomoral, Spain. The forest fire began on July 25th, and the total burnt area was 3,314 ha. In Fig. 2, we show the fire perimeters at three different time instants: t_0 (July 26th at 11:27 am), t_1 (July 27th at 10:32am) and t_2 (July 28th at 11:15 am). The grid size of this fire is $100m$ X $100m$. All calculations reported here were performed using a cluster with 128 cores. The execution platforms are CPU Intel(R) Xeon(R) CPU E5-2620 v3 @ 2.40GHz, with 6 cores. This research used a population size of 128 individuals, and the number of iterations (generations) was set to 10. For this analysis, we assess the accuracy of the output when the mixed-precision is utilized compared to the reference simulation.

The Two-stage Strategy has two different steps: the calibration and prediction stages. In the calibration step, we have to perform hundreds of simulations to evaluate the input parameters that reproduce the real fire spread better. In order to improve the analysis in the prediction stage, the 5 individuals with the lowest error value after the calibration stage are used to perform fire predictions. The error of each simulation is calculated using Eq. 1, see Fig. 3. To effectively compare the double and mixed-precision implementations, we force both implementations to use the same individuals per generation during the calibration process. This ensures that the performance difference between both implementations is not a consequence of using a different combination of input parameters. Therefore, the same best individuals are selected for both implementations at the end of the calibration process.

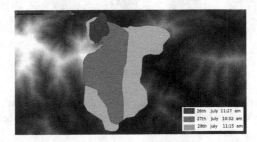

Fig. 2. Digital Elevation map of Nuñomoral fire area and the three different perimeters of the forest fire. The *Perimeter 1 - red* was used as initial perimeter (ignition Perimeter), *Perimeter 2 - green* was used in the calibration stage, and the perimeter to be predicted is *Perimeter 3 - blue*, [6]. (Color figure online)

In Fig. 3(a), we can see the evolution of the computed error of the simulation through the generations of the calibration stage. The progression of the computed error reveals that the simulated perimeter of the fire becomes closer to the real fire perimeter. Moreover, we can see that the error is lower for the double-precision simulation than for the mixed-precision. Nevertheless, at the end of the calibration process, the error of the tenth generation is very close for both implementations.

Figure 3(b) reveals the computed error of the predictions when the same five best individuals are used in both implementations. In all predictions, the error for both implementations is very close. In the example, we have used the third simulation, as it has the lowest error prediction.

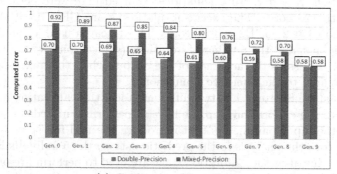

(a) Calibration Stage Error

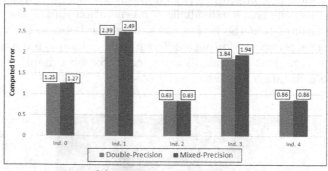

(b) Prediction Stage Error

Fig. 3. Computed error per individual using Eq. 1 of the Nuñomoral's fire when double-precision (red) and mixed-precision (blue) are used. (Color figure online)

Figure 4 details the difference in the fire evolution between double (green line) and mixed-precision (red line) when the second individual simulation is used. We can see that the difference between the perimeter evolution of both implementations increases as the fire extension widens. As we said, the precision reduction produces a small difference when Rothermel's model is applied. This difference is accumulative and increases with the velocity of the propagation of the fire front. We notice that in some parts of the maps, the perimeter of the fire spreads very fast.

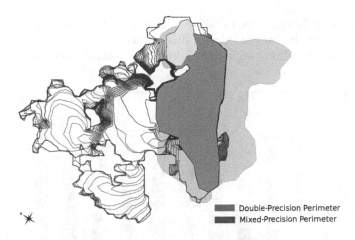

Fig. 4. Obtained simulations of the Nuñomoral's fire after the prediction stage when the third simulation is used. The initial perimeter is drawn in *green* area. The *dark red* line indicates the fire propagation when mixed-precision is used, and the *dark green* line indicates the fire propagation when we use double-precision. The *light blue* area represents the final perimeter of the fire. (Color figure online)

Table 2 compares the best five fire predictions when the two implementations are used considering the difference between sets. It shows that the number of cells burnt by both simulations is very close. In this case, the computed validation values for the best individual prediction are $BIAS= 1.012$, $FAR= 0.018$, and $POD= 0.993$. A BIAS value higher than one means that the mixed-precision implementation burns more area than the reference simulation. The mixed-precision implementation produces the same prediction of the fire behavior when the third individual prediction is used. Its validation values are $BIAS= 1.000$, $FAR= 0.000$ and $POD= 1.000$. The Computed validation confirms that both simulations are similar and well below the defined thresholds.

In any case, a consequence of the mixed-precision implementation is that, due to its lower accuracy, the fire perimeter shape is simplified in some circumstances. This outcome has an indirect effect in reducing the execution time. When the shape is smoother, see Fig. 4, we need less number of points to represent the fire front; therefore, the time consumed by the point expansion and the perimeter reconstruction algorithms decreases, which generates an execution performance improvement. The penalty for paying is an error increment versus the reference but kept within the acceptable threshold values.

Table 2. Obtained results of comparing the prediction individuals between the double-precisionand when mixed-precision using the *Two-stage Strategy*.

	1st Pred.	2nd Pred.	3th Pred.	4th Pred.	5th Pred.
BIAS	1.012	1.002	1.000	0.998	1.002
FAR	0.018	0.032	0.000	0.017	0.005
POD	0.993	0.970	1.000	0.980	0.997

As mentioned above, when a wildfire is simulated, the most basic information is related to the evolution tendency of fire behavior, which is the most relevant information the firefighter can use to tackle these hazards efficiently.

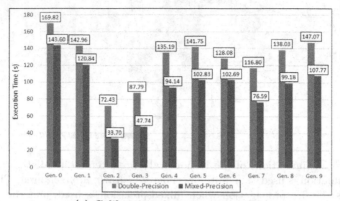

(a) Calibration time per generation.

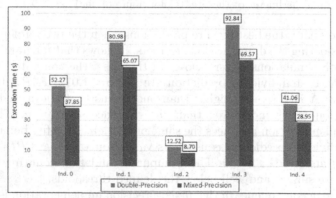

(b) Prediction time for best simulations.

Fig. 5. Execution time for two different implementations, the reference (red) and the mixed-precision (blue) in Nuñomoral's fire. (Color figure online)

In Fig. 5 we illustrate the execution time for predicting the forest fire behavior. Figure 5(a) displays the execution time per generation in the calibration stage. In our particular case, all the individuals of the same generation are executed in parallel; therefore, the execution time of a generation is determined by the slowest individual.

This figure shows a considerable reduction in the execution time when the mixed-precision is used. The highest performance improvement is obtained in the third generation, where the execution time is reduced by 53.5% when the mixed-precision is applied, representing a speed up of 2.15. The worst performance improvement is obtained in the first and second generations, where the computed speed up is around 1.18 in both cases. Figure 5(b) illustrates the execution time of each individual from the prediction stage. As in the calibration stage, the usage of mixed-precision shows a significant reduction in the execution time. The maximum speed up, 1.44, is obtained by the third best individual. If we focus only on the best individual, the speed up calculated is 1.38, which represents a substantial performance improvement. Therefore, we can confirm that using mixed-precision methodology reduces the execution time.

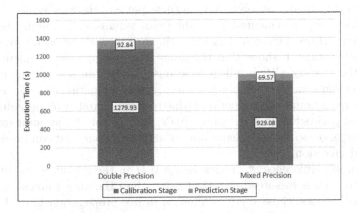

Fig. 6. Execution time of the *Two-stage Strategy* for the different scenarios: double and mixed-precision in Nuñomoral's fire.

Figure 6 details the execution time invested in completing a forest fire spread prediction. The green column represents the execution time of the *calibration Stage* and the blue column the *prediction stage* for the two different implementations: double and mixed-precision. We can see that, in this fire scenario, the reduction of the execution time is notable, 374.12 seconds, 350.85 seconds for the Calibration Stage, and 23.27 for the Prediction Stage, which represents an improvement around the 27.3% of the execution time, 27,4% for the Calibration Stage and 25.1% for the Prediction Stage. The computed speeds up are 1.40 for Calibration Stage and 1.33 for the Prediction Stage, respectively. The speed up of the whole execution is 1.37. We note that for the case of Nuñomoral's fire, the execution time reduction is significant.

In order to better understand the computational reasons for the performance improvements of the mixed-precision in the *Two-stage Strategy*, we used a profiling tool, *Linux prof*, to get the basic performance metrics of the experiments and compare the results of the execution. We will use CPU Cycles Per machine code Instruction (CPI) as a metric for comparing the computational cost of both implementations. The CPI is the ratio between the number of CPU cycles over the number of instructions executed. It reflects the average number of CPU cycles needed to complete an instruction. Thus, CPI indicates how much latency is in the system and can be a useful measure of how an application is performing. A low CPI indicates that a few cycles are needed to execute an instruction.

However, if the CPI is high, the execution of a single instruction requires a large number of CPU cycles, which indicates that the application has poor performance. Because CPI is affected by changes in the number of CPU cycles an application takes or changes in the number of instructions executed, it is best used for comparison when only one part of the ratio changes. The goal is to decrease the CPI in certain parts of the code as well as the whole application. In our particular case, we want to understand the impact of mixed precision in the complexity of calculating specific functionality. CPI helps to determine the reduction of the processing complexity in those parts of the code where the mixed precision has been implemented, as in the point expansion and perimeter reconstruction algorithms. Figure 7(a) displays the average CPI for each generation in the calibration stage. It shows that the usage of mixed-precision implementation improves the average CPI in all generations. As we can see, the highest CPI reduction is found in the fourth generation, with a reduction of 28%. This substantial computational cost reduction is due to a high number of individuals with an extended perimeter. In these individuals, the weight of the point expansion algorithm is large, so we have many potential instructions that can be simplified with mixed precision.

This notable difference is a consequence that in the reference execution, the fire perimeter tends to spread slightly further than the mixed precision simulations. In some cases, these small differences in fire propagation have the consequence of a sudden acceleration of the fire propagation in areas of the terrain with the worst possible combination of the factors that drive the fire evolution. These factors are the slope, aspect, fuel, and wind. When these four factors are combined in a specific adversarial way, the acceleration of the propagation of the fire front increases exponentially when the fire arrives at this area. In this case, double-precision simulation implementation introduces artificially complicated cases that significantly increase the computational cost of the simulations. When the mixed-precision implementation is used, some complex calculations are not executed as we avoid some particularly complex simulation cases: high error simulations with extra perimeter points and more work to simulate an extensive area propagation. On top of this, it is important to consider how these special cases affect the final predicted burned area. Ultimately, most of these expensive cases are discarded as they produce high error rates well above the quality threshold limits. In summary, mixed-precision avoids the execution of long, complex high-error cases whose results are not useful for the final prediction.

Another consideration that justifies this computational cost reduction is due to the *Distance Resolution*, see [9]. When the fire front propagation is accelerated, the perimeter points will propagate forward a long distance. As explained above, the Distance Resolution limits the maximum expansion of a single perimeter point in a single time iteration. For that reason, the propagation of these perimeter points is divided into a long set of shorter propagation sub-steps, which implies a greater consumption of resources by the perimeter expansion algorithm. When mixed-precision is applied, the propagation is much simpler as

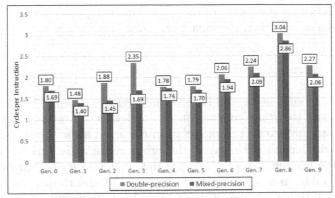

(a) Calibration CPI per generation (avg.)

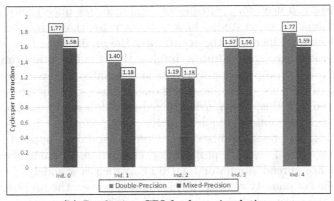

(b) Prediction CPI for best simulations

Fig. 7. CPI of the *Two-stage Strategy* for two different implementations, reference and mixed-precision in Nuñomoral's fire.

it needs a few steps. If we evaluate the whole calibration stage, the obtained CPI is 1.91 for the reference simulation and 1.67 for the mixed-precision implementation, representing a reduction of 12.60% of the computational cost. So, we can conclude that the utilization of the mixed-precision produces a notable reduction of the computational cost maintaining the quality of the simulation. In Fig. 7(b), we can observe the CPI of the prediction stage. It shows that, as in the calibration stage, the utilization of the mixed-precision provides a fair reduction of the computational cost.

In this scenario, the highest CPI decrease is found in the second-best case, where the computational cost decreases from 1.40 CPI in the reference simulation to 1.18 when the mixed-precision implementation is used. This represents a reduction of the 15.5% of the computational cost. In a real forest fire emergency, the velocity to predict future fire behavior is crucial; for that reason, only the best individual is used to predict the fire spread. If we focus on the best individual, we see a reduction of the 10.5% of the CPI. This computational cost reduction improves performance when the mixed-precision approach is employed.

4 Conclusions

Forest fire is a significant natural hazard that every year causes important damages. Models and their implementation in simulators can estimate their behavior, but they are not exempt from a certain degree of error. The quality results of these models depend not only on the propagation equations describing the behavior of the fire but also on the input data required to initialize the model. Typically, this data is subjected to a high degree of uncertainty and variability during the evolution of the event due to the dynamic nature of some of them, such as meteorological conditions or moisture contents in the vegetation. Consequently, there is a need for strategies to minimize this uncertainty to provide better predictions. Previous studies have demonstrated that using the *Two-stage Strategy* undoubtedly increases the accuracy of the predictions. However, due to its characteristics, *Two-stage Strategy* implies that the execution time invested in predicting the fire propagation increases significantly. Different works expose that the generalized use of double-precision in most scientific codes could be notably reduced. For this reason, using mixed-precision would pay back in terms of performance improvement, requiring little coding effort.

In this work, we present a mixed-precision methodology applied to improve the performance of a forest fire propagation simulation. In particular, we focus on the forest fire spread simulator FARSITE, which is a widely accepted tool in the scientific community related to this field. However, the proposed methodology is simulator independent and could be applied to any other. A relevant outcome from the experimental study is that around 74% of the variables used to calculate the point expansion algorithm can be defined in single-precision. On the other hand, only 15% of the variables used in the perimeter reconstruction module could be used with single-precision because that part represents, in some cases, around 60% of the total simulation time. For that reason, the performance improvement using the mixed-precision is limited in this particular case.

The experimental study performed analyzes the impact of the mixed-precision on the *Two-stage Strategy* where a genetic algorithm is used in a data-driven way to determine the most suitable values of the input simulator parameters based on the current evolution of the forest fire. The proposed mixed-precision methodology has been tested using a real fire that took place in 2009 in the region of Nuñomoral, Spain. The results suggest that using mixed-precision can provide remarkable performance improvements without compromising the simulation results in terms of quality. The implementation of the mixed prediction reduces the execution time by around the 30% of the whole process. We have to take into account that the improvement is accumulative. One way to improve the prediction quality of the fire spread is to increase the number of generations and the number of individuals per generation. In this case, the potential improvement provided by the mixed-precision implementation will be higher.

Moreover, all future simulations will benefit from this implementation without additional effort. In terms of quality, for short fire front propagation, the difference between both obtained perimeters is negligible. However, this difference is accumulative; therefore, when the fire propagation increases, this difference is more discernible. As we saw, in some cases, the double-precision implementation introduces an artificial complexity in the simulations that significantly increases the computational cost of those executions and, therefore, the time invested into the *Two-stage Strategy*. Finally, although the evolutionary mixed-precision data-driven forest fire spread system described in this work has been focused on accelerating the forecast of wildfire's evolution, it should be considered that the complete methodology can be generalized for any other natural hazard forecast systems.

This study is a starting point for utilizing the mixed-precision methodology in the forest fire spread simulation field. Future works include the utilization of different tools, see [12,14,18], to evaluate a different set of variables that can be used in single-precision, to check out the best combination to optimize the performance benefit with the lowest accuracy reduction. In addition, using accelerators like GPUs could increment the impact of the methodology on the performance of the simulator, as the savings in mixed-precision could be higher.

Acknowledgement. This work has been granted by the Spanish Ministry of Science and Innovation MCIN AEI/10.13039/501100011033 under contract PID2020-113614RB-C21 and by the Catalan government under grant 2021 SGR-00574.

References

1. Abdalhaq, B., Cortés, A., Margalef, T., Luque, E.: Enhancing wildland fire prediction on cluster systems applying evolutionary optimization techniques. Future Gener. Comp. Syst. **21**(1), 61–67 (2005). https://dx.doi.org/10.1016/j.future.2004.09.013
2. Arca, B., Laconib, M., Maccionib, A., Pellizzaroa, G., Salisb, M.: P1. 1 validation of Farsite model in Mediterranean area (2014)
3. Baboulin, M., et al.: Accelerating scientific computations with mixed precision algorithms. Comput. Phys. Commun. **180**(12), 2526–2533 (2009). https://doi.org/10.1016/j.cpc.2008.11.005
4. Bennett, N.D., et al.: Characterising performance of environmental models. Environ. Modell. Softw. **40**, 1–20 (2013). https://doi.org/10.1016/j.envsoft.2012.09.011
5. Buttari, A., Dongarra, J., Langou, J., Langou, J., Luszczek, P., Kurzak, J.: Mixed precision iterative refinement techniques for the solution of dense linear systems. Int. J. High Performance Comput. Appl. **21**(4), 457–466 (2007). https://doi.org/10.1177/1094342007084026
6. Centre, J.R.: European forest fire information system. https://forest.jrc.ec.europa.eu/effis/ (2011)
7. Denham, M., Wendt, K., Bianchini, G., Cortés, A., Margalef, T.: Dynamic data-driven genetic algorithm for forest fire spread prediction. J. Comput. Sci. **3**(5), 398–404 (2012). https://dx.doi.org/10.1016/j.jocs.2012.06.002
8. Ebert, B.: Forecast verification: issues, methods and FAQ. https://www.cawcr.gov.au/projects/verification (2012)

9. Finney, M.A.: Farsite: fire area simulator-model development and evaluation. Research Paper RMRS-RP-4 Revised 236 (1998)
10. Gjertsen U, O.V.: The water phase of precipitation: a comparison between observed, estimated and predicted values. https://dx.doi.org/10.1016/j.atmosres. 2004.10.030 (2005)
11. Jia, X., et al.: Highly scalable deep learning training system with mixed-precision: Training imagenet in four minutes. CoRR abs/1807.11205 (2018). https://arxiv. org/abs/1807.11205
12. Lam, M.O., de Supinksi, B.R., LeGendre, M.P., Hollingsworth, J.K.: Poster: Automatically adapting programs for mixed-precision floating-point computation. In: 2012 SC Companion: High Performance Computing, Networking Storage and Analysis, p. 1424 (2012). https://doi.org/10.1109/SC.Companion.2012.232
13. Maynard, C., Walters, D.: Mixed-precision arithmetic in the endgame dynamical core of the unified model, a numerical weather prediction and climate model code. Comput. Phys. Commun. **244**, 69–75 (2019). https://www.sciencedirect. com/science/article/pii/S0010465519302127
14. Menon, H., et al.: Adapt: Algorithmic differentiation applied to floating-point precision tuning. In: Proceedings of the International Conference for High Performance Computing, Networking, Storage, and Analysis. SC 2018, IEEE Press (2019). https://doi.org/10.1109/SC.2018.00051
15. Prikopa, K.E., Gansterer, W.N.: On mixed precision iterative refinement for eigenvalue problems. In: Proceedings of the International Conference on Computational Science, ICCS 2013, Barcelona, Spain, 5–7 June 2013. Procedia Computer Science, vol. 18, pp. 2647–2650. Elsevier (2013). https://doi.org/10.1016/j.procs.2013.06. 002
16. Rodríguez, R., Cortés, A., Margalef, T.: Injecting dynamic real-time data into a DDDAS for forest fire behavior prediction. In: Allen, G., Nabrzyski, J., Seidel, E., van Albada, G.D., Dongarra, J., Sloot, P.M.A. (eds.) ICCS 2009. LNCS, vol. 5545, pp. 489–499. Springer, Heidelberg (2009). https://doi.org/10.1007/978-3-642-01973-9_55
17. Rothermel, R.: A mathematical model for predicting fire spread in wildland fuels (US Department of Agriculture, Forest Service, Inter- mountain Forest and Range Experiment Station Ogden, UT, USA 1972)
18. Rubio-González, C., et al.: Precimonious: tuning assistant for floating-point precision. In: Proceedings of the International Conference on High Performance Computing, Networking, Storage and Analysis. SC 2013, Association for Computing Machinery, New York, NY, USA (2013). https://doi.org/10.1145/2503210.2503296

RL-MAGE: Strengthening Malware Detectors Against Smart Adversaries

Adarsh Nandanwar, Hemant Rathore$^{(\boxtimes)}$, Sanjay K. Sahay, and Mohit Sewak

Department of CS & IS, BITS Pilani, Goa Campus, Goa, India
{f20180396,hemantr,ssahay,p20150023}@goa.bits-pilani.ac.in

Abstract. Today, android dominates the smartphone operating systems market. As per Google, there are over 3 billion active android users. With such a large population depending on the platform for their daily activities, a strong need exists to protect android from adversaries. Historically, techniques like signature and behavior were used in malware detectors. However, machine learning and deep learning models have now started becoming a core part of next-generation android malware detectors. In this paper, we step into malware developers/adversary shoes and ask: Are machine learning based android detectors resilient to reinforcement learning based adversarial attacks? Therefore, we propose the *RL-MAGE* framework to investigate the adversarial robustness of android malware detectors. The RL-MAGE framework assumes the grey-box scenario and aims to improve the adversarial robustness of malware detectors. We designed three reinforcement learning based evasion attacks *A2C-MEA*, *TRPO-MEA*, and *PPO-MEA*, against malware detectors. We investigated the robustness of 30 malware detection models based on 2 features (android permission and intent) and 15 distinct classifiers from 4 different families (machine learning classifiers, bagging based classifiers, boosting based classifiers, and deep learning classifiers). The designed evasion attacks generate adversarial applications by adding perturbations into the malware so that they force misclassifications and can evade malware detectors. The attack agent ensures that the adversarial applications' structural, syntactical, and behavioral integrity is preserved, and the attack's cost is minimized by adding minimum perturbations. The proposed TRPO-MEA evasion attack achieved a mean evasion rate of 93.27% while reducing the mean accuracy of 30 malware detectors from 85.81% to 50.29%. We also propose the *ARShield* defense strategy to improve the malware detectors' classification performance and robustness. The TRPO-MEA ARShield models achieved 4.10% higher mean accuracy and reduced the mean evasion rate of re-attack from 93.27% to 1.05%. Finally, we conclude that the RL-MAGE framework improved the detection performance and adversarial robustness of malware detectors.

Keywords: Android Malware · Deep Learning · Evasion Attack · Machine Learning · Greybox Environment · Reinforcement Learning

© The Author(s), under exclusive license to Springer Nature Switzerland AG 2023
J. Mikyška et al. (Eds.): ICCS 2023, LNCS 14074, pp. 77–92, 2023.
https://doi.org/10.1007/978-3-031-36021-3_6

1 Introduction

Since the launch of the android operating system in 2008, it has rapidly gained popularity. During Google I/O 2021, Google announced that the number of active android users had surpassed 3 billion [4]. However, due to newer versions of android releasing frequently, OEMs have a hard time catching up, resulting in version fragmentation. They usually stop updating older smartphones due to the high cost of maintenance. According to a study, over 40% of android users no longer receive security updates, making them vulnerable to malware attacks [1]. This leads users to lean towards anti-malware software/malware detectors to detect malware or prevent malware attacks.

Android occupies over 70% market share of smartphone operating systems [2]. Due to this over-dependence on one platform, its security is critical. In the past, signature, heuristic, specification, sandbox, etc. techniques have been used to detect malware [13,20]. With the rapid developments in machine learning, it has also found its use in malware detection and has shown a good success rate in detecting old and new malware. However, researchers have demonstrated that machine learning and deep learning classification models are susceptible to adversarial attacks [9,21]. This generated doubts about the reliability of the machine learning based malware detectors against adversarial attacks and if they are safe for real-world deployment. It is essential to be aware of the weaknesses of the malware detectors before it gets exploited by malware developers/adversaries.

The adversarial attack scenarios can be grouped into three broad categories based on the attacker's knowledge of the target system. First, is *white-box scenario*, where the attacker has full knowledge, like, the dataset used to train/test classifiers, features used, classifiers used, and other parameters, etc. It is far from reality, where the attacker generally has partial information about the malware detection ecosystem called the *grey-box scenario*. Finally, the third category is *black-box scenario*, where the attacker has no information about the malware detector. In recent years, many researchers have proposed adversarial attacks on malware detectors in a white-box scenario with good evasion rates [10,14,23]. However, still limited work has been done in the grey-box scenario. Therefore, in this work, we focus on the grey-box scenario where the attacker knows about the dataset and features used but has zero information about the classifiers and other parameters used in the malware detector.

Using an adversarial evasion attack, the attacker can bypass the malware detector and install malware on the target system, putting the user's data at risk. Research on adversarial attacks and adversarial defenses complement each other, both aiming to improve the adversarial robustness of machine learning based malware detectors. In this work, we propose a framework called the *Reinforcement Learning based Malware Attack in Greybox Environment (RL-MAGE)* to investigate and improve the adversarial robustness of the machine learning based malware detectors. Reinforcement learning has recently become more popular because of rapid advancements in computational and storage capabilities. In this work, we conduct an in-depth comparative study on the performance of three different *on-policy* reinforcement learning algorithms in malware evasion and adversarial defense tasks. We designed three *Malware Evasion Attacks (MEA)* namely *A2C-MEA*, *TRPO-MEA*, and *PPO-MEA*, in the RL-MAGE framework and compared their performance. The *MEA* perform untargeted evasion attacks on malware detectors that aim to introduce *type-II false-negative errors*. The attacks are designed to achieve high evasion rates while minimizing the execution cost by limiting the perturbations allowed by the agent. The attack agent also ensures the adversarial applications' structural, syntactical, and behavioral integrity. We tested the performance of 30 distinct malware detectors based on 4 families of classifiers against the MEA attacks. Later, we propose an adversarial retraining based defense strategy called the *ARShield* to improve the adversarial robustness of malware detectors and defend against adversarial attacks. With this work, we make the following contributions:

1. We proposed *Reinforcement Learning based Malware Attack in Greybox Environment (RL-MAGE)* to investigate the detection performance and robustness of malware detectors.
2. We developed 30 malware detectors based on 2 features (*android permissions* and *android intents*) and 15 distinct classifiers from 4 families, namely: *machine learning classifiers*, *bagging based classifiers*, *boosting based classifiers*, and *deep learning classifiers*. The 30 malware detector models achieved a mean accuracy and AUC of 85.81% and 0.86, respectively.
3. We designed 3 *Malware Evasion Attacks (MEA)* to expose vulnerabilities in the above 30 malware detectors models. The *A2C-MEA* achieved a mean evasion rate of 91.40% with just 1.74 mean perturbations that reduced the mean accuracy from 85.81% to 51.22% in 30 malware detectors. The *TRPO-MEA* achieved a mean evasion rate of 93.27% with just 1.95 mean perturbations and reduced the mean accuracy from 85.81% to 50.29% in the same 30 malware detectors. Finally, *PPO-MEA* accomplished a 92.17% mean evasion rate with 2.05 mean perturbations, reducing the mean accuracy from 85.81% to 50.79% in 30 malware detectors. We also list the most pertubated android permissions and android intents during the above attacks.
4. We developed *ARShield defense* to improve the robustness and detection performance of malware detectors. The mean evasion rate of *A2C-MEA*, *TRPO-MEA*, and *PPO-MEA* reattacks against *ARShield models* was drastically dropped from 91.40% to 1.32%, 93.27% to 1.05% and 92.17% to 1.40%, respectively. The mean accuracy of *A2C-ARShield*, *TRPO-ARShield*, and *PPO-ARShield* models also improved from 85.81% to 89.89%, 89.91%, and 89.83%, respectively.

The rest of the paper is organized as follows. The proposed RL-MAGE framework, Malware Evasion Attack (A2C-MEA, TPRO-MEA, and PPO-MEA), and ARShield defense are explained in Sect. 2. Experimental setup is described in Sect. 3. Experimental results of baseline malware detectors, MEA attack on baseline detectors, ARShield detectors, and MEA reattack on ARShield detectors are discussed in Sect. 4. Related work is explained in Sect. 5, and the conclusion is presented in Sect. 6.

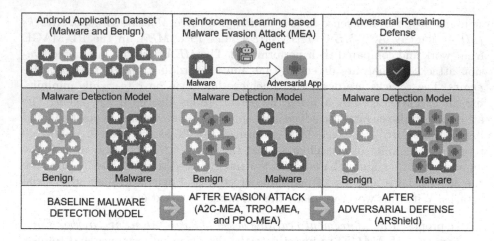

Fig. 1. Proposed RL-MAGE framework to improve detection performance and robustness of malware detectors.

2 Adversarial Attack and Defense

2.1 Proposed RL-MAGE Framework

Figure 1 summarises the *Reinforcement Learning based Malware Evasion Grey box Environment (RL-MAGE)* framework visually. The RL-MAGE framework aims to improve the malware detectors to defend against *Malware Evasion Attacks (MEA)*. The framework consists of 3 steps. The **Step-1** is dataset gathering, followed by constructing baseline malware detection models. The **Step-2** aims to develop malware evasion attacks using reinforcement learning against baseline malware detection models. These attacks generate adversarial applications to evade the malware detectors. The **Step-3** is to develop an adversarial defense *(ARShield)* to counter the evasion attacks and to improve the robustness of the malware detectors.

2.2 Malware Evasion Attack (MEA)

A reinforcement learning setup consists of an *environment* and an *agent*. The environment defines the observation space, action-space and returns a reward for an action performed by the agent. The agent interacts with the environment and adjusts its internal weight based on the *reward* received for the *action* performed. In our case, the state of the environment is a multi-binary representation of the android application in terms of its features (android permission/android intent). The environment calculates the reward by taking the difference between the probability of the state being *malware* for the current state and the following state (after taking action) as given by the underlying malware detector.

Steps involved in the MEA attack on malware detectors are described in Fig. 2. The attack process can be divided into 2 phases: *training phase* and *attack phase*.

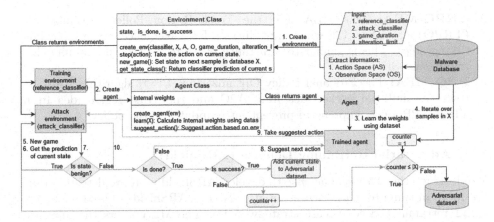

Fig. 2. Overview of the steps involved in the RL-MAGE based adversarial attack to create an adversarial dataset.

The *training phase* is performed once for the reinforcement learning agent to learn the environment and devise a strategy to navigate it. *Exploration* is done in a training environment which is constructed using reference malware detectors. The environment is designed to sequentially go through all the available malware applications and expose the agent to all types of applications for it to learn. The agent is restricted by the maximum number of perturbations/alterations that it is allowed to make. The limit is 5 for android permissions and 4 for android intents. Once the agent learns from the training environment using the rewards received for its actions, we move to the next phase.

In the *attack phase*, we create attack environments for each malware detection model being attacked and release the agent in them. If the environment's state is classified as *benign*, we reset the environment and move to the next malware. Whenever the state of the environment is classified as *malware*, the agent performs actions from the environment's action space until we encounter the stop condition. The stop condition occurs either when the malware application is misclassified as benign or if the number of permissible actions has crossed the limit. In case of misclassification, the final state of the environment is used to rebuild the android application and added to the adversarial dataset.

Reinforcement learning agents may use different algorithms to learn the strategy to navigate the environment. These are used to adjust the internal weights and parameters using the states, actions, and rewards while interacting with the training environment. In this work, we explore the *on-policy algorithms* where the policy used to take action is the same policy being evaluated and updated. We train the following Malware Evasion Agents (MEAs) for our attacks:

1. **A2C-MEA:** The Advantage Actor Critic (A2C) MEA uses the A2C algorithm [11], which is a synchronous and deterministic variant of Asynchronous Advantage Actor Critic (A3C). It avoids the use of a replay buffer by using multiple workers.

2. **TRPO-MEA:** This MEA uses the Trust Region Policy Optimization (TRPO) algorithm [17], which introduces a KL-Divergence constraint while taking steps. This is similar to a distance measure between probability distributions.
3. **PPO-MEA:** The Proximal Policy Optimization (PPO) algorithm [18] borrows the multiple worker idea from A2C and the Trust region idea from TRPO. PPO uses clipping to prevent huge updates to the policy.

2.3   Adversarial Retraining Defense (ARShield)

We designed the *ARShield defense* to develop ARShield malware detectors that are more immune to malware evasion attacks. The ARShield is based on adversarial retraining based defense strategy [9,13]. The MEA alters the malware applications to force the malware detectors to misclassify. When there is a misclassification, the final state of the environment is saved after converting it back into a feature vector, in the same format as the original dataset used to train the malware classifiers. A new dataset is constructed by including the original dataset with their original labels used to train baseline models and the new adversarial samples with their labels set to malware. We then retrain all the malware detectors using this new adversarial dataset resulting in more robust ARShield models that are more immune to malware evasion attacks.

3   Experimental Setup

The dataset used for the experiments contains 5721 benign applications downloaded from the *Google Play Store* [5] and screened using *VirusTotal* [6]. The dataset also contains 5553 malicious applications from the Drebin dataset [7]. We then use the reverse engineering tool *Apktool* [3] to develop a parser that extracts *android permission* and *android intent* usage in each android application. This parser recorded 195 permissions and 273 intents to create 2 feature vectors (android permission and android intent). We then train 30 different malware detection models using 15 distinct classification algorithms belonging to 4 families (basic machine learning classifiers, bagging based classifiers, boosting based classifiers, and deep learning classifiers). Table 1 shows classifies and their corresponding categories. The detailed discussion on the design and parameters of these malware detection models is explained in other papers [15,19]. We used a train test split of 70:30. All the experiments were conducted on the *Google Colab Pro* platform using the *Python* programming language using *scikit-learn*, *TensorFlow*, and *OpenAI Gym*.

3.1   Performance Metrics

In this work, we use the following performance metrics to evaluate different phases of the RL-MAGE framework.

Table 1. Classifiers and their families used in RL-MAGE framework.

Machine Learning (ML) Classifier(s)	Logistic Regression (LR) Support Vector Machine (SVM) Kernel Support Vector Machine (KSVM) Decision Tree (DT)
Bagging based Classifier(s)	Random Forest (RF) Bootstrap Aggregation Logistic Regression (BALR) Bootstrap Aggregation Kernel Support Vector Machine (BAKSVM)
Boosting based Classifier(s)	Adaptive Boosting (AB) Gradient Boosting (GB) eXtreme Gradient Boosting (XGB) Light Gradient Boosting Machine (LGBM)
Deep Learning (DL) Classifier(s)	Deep Neural Network 0 Hidden Layer (DNN0L) Deep Neural Network 1 Hidden Layer (DNN1L) Deep Neural Network 2 Hidden Layer (DNN2L) Deep Neural Network 4 Hidden Layer (DNN4L)

- **Accuracy:** is used to measure the performance of a malware detector. It is the ratio of the number of samples correctly classified by the total number of samples. Higher accuracy is better.
- **Mean Accuracy (MA):** is the average of the accuracies of many malware detectors.
- **Accuracy Reduction:** is the difference between the accuracy of a malware detector before and after the attack. It is used to evaluate the robustness of a malware detector. Lower accuracy reduction indicates that a malware detector is more immune to attacks.
- **Evasion Rate (ER):** is the percentage of adversarial samples (generated by an attack agent) to the number of malware samples in a dataset. A high evasion rate indicates a strong adversarial attack and a weak malware detector.
- **Mean Evasion Rate (MER):** is the average evasion rate of a group of adversarial attacks against a group of malware detectors.
- **Perturbation Application Percentage (PAP)** is the number of adversarial applications in which a particular feature was modified (perturbated) by an attack agent to the total number of adversarial applications.
- **Perturbation Frequency Percentage (PFP):** of a feature is the percentage ratio of the number of times that feature was modified (perturbated) to the total number of perturbations during an attack.

4   Experimental Results

This section will discuss the results of all the conducted experiments. It includes the baseline models, MEA attack, ARShield defense, and MEA re-attack on ARShield models.

4.1   Baseline Android Malware Detection Models

We trained 30 unique baseline malware detection models based on 15 classifies (from 4 families: Machine Learning Classifiers, Bagging based Classifiers, Boosting based Classifiers, Deep Learning Classifiers) and two static features (permission and intent). We measure the performance of these models using Mean Accuracy (MA) and Area under the ROC Curve (AUC). The Mean Accuracy (MA) of 93.28% was achieved by the 15 permission based malware detection models. The maximum MA of 94.27% was obtained by 4 DNN models and the minimum MA of 92.26% by 4 Boosting Models. The average AUC for 15 permission based models was 0.93. On the other hand, the MA of 78.34% was obtained by the 15 intent based malware detection models. The highest and lowest MA of 78.73% and 78.07% were obtained by 4 DNN and 4 Boosting models, respectively. We obtained a mean AUC of 0.78 for the 15 intent based malware detection models. We observe that the permission based malware detection models perform much better than the intent based detection models. The mean accuracy and AUC of all 30 malware detection models were 85.81% and 0.86, respectively.

4.2   Adversarial Attack on Malware Detection Models

The MEA agents (A2C-MEA, TRPO-MEA and PPO-MEA) modify the malware samples (by adding perturbations) such that the generated adversarial samples can evade the malware detection models. The alterations made in the attack add perturbations (permissions and intents) while preserving the syntactic and semantic integrity of the modified application. The attack is designed to minimize the alterations, while their count is limited to 5 for permission models and 4 for intent models. We use metrics such as the Evasion Rate (ER), Mean Evasion Rate (MER), and change in Mean Accuracy (MA) to evaluate the attack strategies.

4.2.1   Evasion Rate @ MEA Attack

Evasion Rate (ER) is defined as the percentage of malware applications (which are classified as malware by the baseline models) that are successfully converted to adversarial applications by introducing alterations/perturbations such that they are forcefully misclassified as benign by the same baseline models. MER is the mean evasion rate across classifier families or features. Figure 3 plots the MER over the number of alterations by the 3 MEAs (A2C-MEA, TRPO-MEA, and PPO-MEA) against different families of malware detection models.

During the *A2C-MEA attack*, we observed an MER of 84.57% against the 15 permission based models while making just 2.20 perturbations on average. While against the 15 intent based models, the MER obtained was 98.22% with 1.27 mean perturbations. We record the maximum MER in deep learning models, 99.68% on permission based models, and 99.62% on intent based models. The *TRPO-MEA attack* obtained an MER of 89.44% using 2.52 mean perturbations against 15 permission based models. The highest MER of 99.65% was seen in 4 DL models. The MER obtained against 15 intent models was 97.10% using average 1.38 perturbations. Boosting models showed the maximum MER of 98.65%.

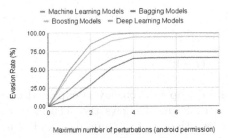

(a) **Mean Evasion Rate** of **Permission** based Malware Detection Models with **A2C-MEA**

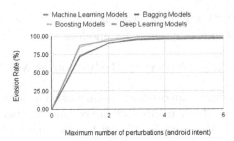

(b) **Mean Evasion Rate** of **Intent** based Malware Detection Models with **A2C-MEA**

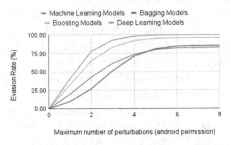

(c) **Mean Evasion Rate** of **Permission** based Malware Detection Models with **TRPO-MEA**

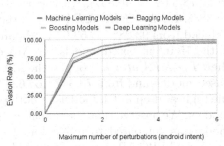

(d) **Mean Evasion Rate** of **Intent** based Malware Detection Models with **TRPO-MEA**

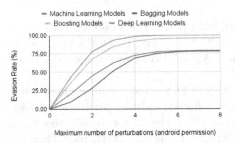

(e) **Mean Evasion Rate** of **Permission** based Malware Detection Models with **PPO-MEA**

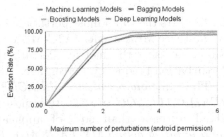

(f) **Mean Evasion Rate** of **Intent** based Malware Detection Models with **PPO-MEA**

Fig. 3. Performance of **Malware Evasion Attack** (A2C-MEA, TRPO-MEA and PPO-MEA) against different android malware detection models.

Finally, in the case of *PPO-MEA attack*, we achieved an MER of 87.36% and 96.99% against 15 permission and 15 intent based models with just 2.42 and 1.69 perturbations, respectively. We again observe the highest MER on deep learning models. We can see that the MER in intent based models is much higher compared to the permission based models, showing us that intent based models are more vulnerable to MEA attacks. Finally, the A2C-MEA, TRPO-MEA, and

PPO-MEA achieved an MER of 91.40%, 93.27%, 92.17%, respectively, against 30 malware detection models.

4.2.2 Accuracy Reduction @ MEA Attack

After a successful malware evasion attack, a drop in accuracy is expected due to a high number of forced misclassifications. Figure 4 shows the accuracies of the different families of baseline malware detection models (blue bar) and after the MEA attack on baseline models (red bars). The *A2C-MEA attack* on 15 permission based models dropped the MA from 93.28% to 55.89%. DL models show the highest drop of 44.93%. The MA for 15 intent based models dropped from 78.34% to 46.55%, with 4 boosting models showing the highest drop of 32.38%. In the *TRPO-MEA attack*, the MA of the 15 permission based models reduced from 93.28% to 53.66%. Maximum drop in DL models (44.91%). While in the 15 intent based models, the MA dropped to 46.92% from 78.34%. Finally, the *PPO-MEA attack* dropped the MA from 93.28% to 54.62% in 15 permission based models and from 78.34% to 46.96% in 15 intent based models. We observe the highest drop in DL models in both permissions (44.89%) and intent (32.07%) models. We observe that the reduction in accuracy is greater in permission based models as compared to intent based models. Finally, the A2C-MEA, TRPO-MEA, and PPO-MEA reduced the MA from 85.81% to 51.22%, 85.81% to 50.29% and 85.81% to 50.79%, respectively, in 30 malware detection models.

4.2.3 Vulnerability List

Table 2 lists the top 5 most pertubated android permissions and android intents by the three MEA agents (A2C-MEA, TRPO-MEA, and PPO-MEA). They are evaluated using the metrics Perturbation Application Percentage (PAP) and Perturbation Frequency Percentage (PFP).

The *android.permission.USE_CREDENTIALS* and *android.permission. READ_CALL_LOG* are the two most pertubated android permissions, being added in more than 60% of applications. On the other hand, the intent *android.intent.action.MY_PACKAGE_REPLACED* is modified in more than 70% of applications making it the most pertubated android intent.

Out of all the perturbations made during the attacks on permission based classifiers, *android.permission.USE_CREDENTIALS* and *android.permission. READ_CALL_LOG* were around 50% of them. For A2C-MEA, these 2 had a combined PFP of 63.41%. Whereas, out of all intents, *android.intent. action.MY_PACKAGE_REPLACED* had the highest PFP. It contributed to more than 70% in A2C-MEA and TRPO-MEA.

4.3 ARShield Defense Strategy

The final stage of the proposed framework is the ARShield defense strategy. We measure the improvements in the detection and robustness performance of the ARShield models over the baseline models in terms of accuracy, evasion rate, and change in accuracy of the models after MEA attacks.

4.3.1   Detection Performance

Figure 4 shows the accuracies of the 4 families of malware detection models at various stages of the RL-MAGE framework. On applying ARShield defense using the A2C-MEA adversarial samples, the Mean Accuracy (MA) of permission based models increased from 93.28% to 95.39%, an improvement of 2.11%, and the MA of intent based models jumped from 78.34% to 84.38%, which is an improvement of 6.04%. We see maximum improvement in MA for boosting models, of 2.41% and 6.22% in permission and intent based models respectively.

Table 2. List of top android permissions and android intents that are most perturbed during A2C-MEA, TRPO-MEA and PPO-MEA attacks on malware detection models.

Perturbation	Name	Frequency Percentage	Sample Percentage
A2C-MEA			
Android Permission (Maximum 5 perturbations)	android.permission.USE_CREDENTIALS	32.41	73.69
	android.permission.READ_CALL_LOG	31.00	70.48
	android.permission.READ_PROFILE	19.28	43.84
	android.permission.CAMERA	15.65	35.58
	android.permission.GET_ACCOUNTS	0.52	1.18
Android Intent (Maximum 4 perturbations)	android.intent.action.MY_PACKAGE_REPLACED	55.80	70.67
	android.intent.action.MEDIA_BUTTON	33.00	41.79
	android.intent.action.SEND	5.98	7.57
	android.intent.action.TIMEZONE_CHANGED	2.46	3.12
	android.intent.category.DEFAULT	1.38	1.74
TRPO-MEA			
Android Permission (Maximum 5 perturbations)	android.permission.USE_CREDENTIALS	26.22	68.48
	android.permission.READ_CALL_LOG	23.19	60.58
	android.permission.READ_PROFILE	19.04	49.73
	android.permission.CAMERA	11.56	30.19
	android.permission.GET_ACCOUNTS	8.28	21.63
Android Intent (Maximum 4 perturbations)	android.intent.action.MY_PACKAGE_REPLACED	56.78	78.58
	android.intent.action.MEDIA_BUTTON	15.59	21.57
	android.intent.action.SEND	8.65	11.98
	android.intent.action.TIMEZONE_CHANGED	4.23	5.85
	android.intent.category.MONKEY	2.84	3.94
PPO-MEA			
Android Permission (Maximum 5 perturbations)	android.permission.USE_CREDENTIALS	27.37	67.98
	android.permission.READ_CALL_LOG	25.40	63.07
	android.permission.READ_PROFILE	20.96	52.06
	android.permission.GET_ACCOUNTS	10.51	26.10
	android.permission.CAMERA	9.53	23.66
Android Intent (Maximum 4 perturbations)	android.intent.action.MY_PACKAGE_REPLACED	40.76	68.24
	android.intent.category.BROWSABLE	12.48	20.89
	android.intent.action.PACKAGE_REMOVED	10.77	18.02
	android.intent.action.TIMEZONE_CHANGED	9.24	15.47
	android.intent.action.MEDIA_BUTTON	8.58	14.36

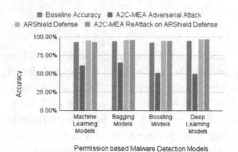

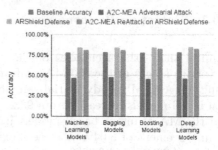

(a) **Performance** of **Permission** based Malware Detection Models against **A2C-MEA**

(b) **Performance** of **Intent** based Malware Detection Models against **A2C-MEA**

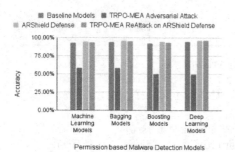

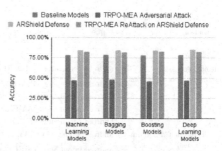

(c) **Performance** of **Permission** based Malware Detection Models against **TRPO-MEA**

(d) **Performance** of **Intent** based Malware Detection Models against **TRPO-MEA**

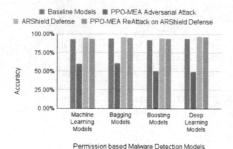

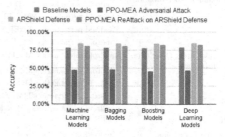

(e) **Performance** of **Permission** based Malware Detection Models against **PPO-MEA**

(f) **Performance** of **Intent** based Malware Detection Models against **PPO-MEA**

Fig. 4. Performance of different android malware detection models against A2C-MEA, TRPO-MEA and PPO-MEA

On applying the TRPO-ARShield, we observe a boost in the mean accuracy of the permission models by 2.00%, from 93.28% to 95.28%. Maximum improvement of 2.23% is shown by bagging Classifiers. In the case of intent based models, we observe a more significant increase of 6.19%, from 78.34% to 84.53%. Deep learning classifiers showed the most improvement of 6.37%.

Similarly, the PPO-ARShield improved the MA of permission and intent models from 93.28% to 95.44% and 78.34% to 84.22%, which is an improvement of 2.16% and 5.88%, respectively. Here too, we notice boosting models improving the most. MA of permission and intent based models improved by 2.33% and 6.03% respectively. From the data, we observe that the ARShield defense positively impacts the malware detection capabilities of the models, leading to an overall accuracy improvement of 4.06% across all models. Out of the four families, boosting classifiers benefit the most with ARShield defence.

4.3.2 Robustness Performance

Apart from improving the malware detection performance of the malware detection models in normal circumstances, the ARShield defense also improves the adversarial robustness of the models to help them defend against adversarial attacks.

In the case of A2C-MEA reattack on 15 permission ARShield models, we observe an MER of 1.27% with a reduction in MA from 95.39% to 94.69% (0.7% drop). Whereas in 15 intent ARShield models, we record a drop in MA of 2.28%, from 84.38% to 82.10%, and an MER of 1.38%. Performing the TRPO-MEA attack on the 15 permission ARShield models leads to a drop in MA from 95.28% to 94.48%, which is a dip of just 0.80%. The MER obtained is 1.25%. On the other hand, in 15 intent based ARShield models, we get an MER of 0.84% with a 2.40% drop in MA from 84.53% to 82.14%. In the case of the PPO-MEA attack on permission ARShield models, we get an MER of 1.14%, reducing the MA from 95.44% to 94.72% (0.72% drop). The same attack on intent ARShield models resulted in an MA drop from 84.22% to 81.72% (2.50% drop) and an MER of 1.66%.

Permisison based ARShield models of basic machine learning category show the highest MER during reattack. We get 2.34%, 1.20%, and 1.74% MER with A2C-MEA, TRPO-MEA and PPO-MEA respectively. On the other hand, intent based ARShield models of bagging category show the highest MER during reattack. We observe 3.14%, 3.00% and 3.67% with the above 3 MEAs respectively. The above results show that the newly enforced ARShield models are much more effective and robust against adversarial attacks.

5   Related Work

Since Papernot et al. in 2016 highlighted that deep neural networks are vulnerable to adversarial attacks, there has been a lot of focus on adversarial attacks on neural networks [12]. However, attacks on applications of traditional machine learning have not received similar attention as image classification problems. With billions of users using android daily, the platform's security against malware is of high priority. Researchers like Taheri et al., Rathore et al., and Grosse et al. came up with novel adversarial attacks like generative adversarial network (GAN), single policy attack (SPA), and adversarial sample crafting on android malware detectors and achieved good results [10,16,20,23]. But these attacks were proposed for white box scenarios which are not representative of the real world. Greybox and black box scenarios more closely represent the real world.

Researchers have extensively studied adversarial attacks in the image classification application on grey and black box environments. In 2021, Sinha et al. fooled DNNs by modifying just 1 pixel in sample images with a mean evasion rate of 48.33% using the differential evolution attack in a grey box scenario [22]. However, there is still a lack of research on adversarial attacks on android malware detection in a partial or no-knowledge environment. Rathore et al. proposed a multi-policy attack that attacked android permission based malware detectors in a grey box scenario but achieved an evasion rate of just 53.20% across 8 machine learning models [14]. In 2021, Rathore et al. used GAAN to achieve a evasion rate of 94.69% but with 10 bit alterations [16]. When it comes to black box scenarios, Bostani et al. developed EvadeDroid achieved 81.07% evasion rate but used 8 android features and 52.48 ± 29.45 alterations [8]. Using ShadowDroid, Zhang et al. claimed a 100% evasion rate using 5.86 mean alterations [24]. But they used 8 different types of android features and studied just one classifier (SVM) without focusing on the generality of the attack which is very important in a partial knowledge scenario where the nature of the classifier is unknown.

6   Conclusion

In this work, we propose a novel RL-MAGE framework to improve the classification performance and robustness of android malware detectors. We designed three reinforcement learning agents, A2C-MEA, TRPO-MEA, and PPO-MEA, for evasion attacks and the ARShield defense strategy to improve malware detectors. The 30 baseline malware detectors achieved mean accuracy of 85.81%. Out of the three on-policy reinforcement learning algorithms, we observe that TRPO is the best algorithm for evasion as well as defense task. TRPO-MEA achieved the highest evasion rate of 93.27% with 1.95 mean alterations in 30 malware detectors. The TRPO-ARShield models also achieved the highest mean accuracy of 89.91% while displaying the lowest evasion rate of just 1.05% during re-attack.

Acknowledgement. One of the authors Dr. Sanjay K. Sahay is thankful to Data Security Council of India for financial support to work on the Android malware detection system.

References

1. More than one billion Android devices at risk of malware threats. https://www.which.co.uk/news/article/more-than-one-billion-android-devices-at-risk-of-malware-threats-aXtug2P0ET0d
2. Android Statistics (2023). https://www.businessofapps.com/data/android-statistics/
3. Apktool (2023). https://ibotpeaches.github.io/Apktool/
4. Google I/O (2023). https://io.google/2021/program/content/?lng=en
5. Google Play Store (2023). https://play.google.com/store/
6. VirusTotal (2023). https://www.virustotal.com/gui/home/upload
7. Arp, D., Spreitzenbarth, M., Hubner, M., Gascon, H., Rieck, K.: Drebin: effective and explainable detection of android malware in your pocket. In: Network and Distributed System Security Symposium (NDSS), vol. 14, pp. 23–26 (2014)
8. Bostani, H., Moonsamy, V.: Evadedroid: a practical evasion attack on ML for black-box android malware detection (2021). arXiv preprint arXiv:2110.03301
9. Demetrio, L., Coull, S.E., Biggio, B., Lagorio, G., Armando, A., Roli, F.: Adversarial exemples: a survey and experimental evaluation of practical attacks on machine learning for windows malware detection. ACM Trans. Privacy Secur. (TOPS) **24**(4), 1–31 (2021)
10. Grosse, K., Papernot, N., Manoharan, P., Backes, M., McDaniel, P.: Adversarial perturbations against deep neural networks for malware classification. arXiv preprint arXiv:1606.04435 (2016)
11. Mnih, V., et al.: Asynchronous methods for deep reinforcement learning. In: International Conference on Machine Learning (ICML), pp. 1928–1937 (2016)
12. Papernot, N., McDaniel, P., Jha, S., Fredrikson, M., Celik, Z.B., Swami, A.: The limitations of deep learning in adversarial settings. In: IEEE European Symposium on Security and Privacy (IEEE EuroS&P), pp. 372–387. IEEE (2016)
13. Qiu, J., Zhang, J., Luo, W., Pan, L., Nepal, S., Xiang, Y.: A survey of android malware detection with deep neural models. ACM Comput. Surv. (CSUR) **53**(6), 1–36 (2020)
14. Rathore, H., Sahay, S.K., Nikam, P., Sewak, M.: Robust android malware detection system against adversarial attacks using q-learning. Inf. Syst. Front. 1–16 (2021)
15. Rathore, H., Sahay, S.K., Rajvanshi, R., Sewak, M.: Identification of significant permissions for efficient android malware detection. In: Gao, H., J. Durán Barroso, R., Shanchen, P., Li, R. (eds.) BROADNETS 2020. LNICST, vol. 355, pp. 33–52. Springer, Cham (2021). https://doi.org/10.1007/978-3-030-68737-3_3
16. Rathore, H., Samavedhi, A., Sahay, S.K., Sewak, M.: Robust malware detection models: learning from adversarial attacks and defenses. Forensic Sci. Int. Digit. Investig. **37**, 301183 (2021)
17. Schulman, J., Levine, S., Abbeel, P., Jordan, M., Moritz, P.: Trust region policy optimization. In: ICML, pp. 1889–1897 (2015)
18. Schulman, J., Wolski, F., Dhariwal, P., Radford, A., Klimov, O.: Proximal policy optimization algorithms. arXiv preprint arXiv:1707.06347 (2017)
19. Sewak, M., Sahay, S.K., Rathore, H.: Deepintent: implicitintent based android ids with e2e deep learning architecture. In: IEEE 31st PIMRC, pp. 1–6. IEEE (2020)
20. Sewak, M., Sahay, S.K., Rathore, H.: Value-approximation based deep reinforcement learning techniques: an overview. In: 2020 IEEE 5th International Conference on Computing Communication and Automation (ICCCA), pp. 379–384. IEEE (2020)

21. Sewak, M., Sahay, S.K., Rathore, H.: DRLDO: a novel DRL based de-obfuscation system for defence against metamorphic malware. Def. Sci. J. **71**(1), 55–65 (2021)
22. Sinha, S., Saranya, S.: One pixel attack analysis using activation maps. Ann. Roman. Soc. Cell Biol. 8397–8404 (2021)
23. Taheri, R., Javidan, R., Shojafar, M., Vinod, P., Conti, M.: Can machine learning model with static features be fooled: an adversarial machine learning approach. Clust. Comput. **23**(4), 3233–3253 (2020). https://doi.org/10.1007/s10586-020-03083-5
24. Zhang, J., Zhang, C., Liu, X., Wang, Y., Diao, W., Guo, S.: ShadowDroid: practical black-box attack against ML-based android malware detection. In: International Conference on Parallel and Distributed Systems, pp. 629–636. IEEE (2021)

Online Runtime Environment Prediction for Complex Colocation Interference in Distributed Streaming Processing

Fan Liu[1,2], Weilin Zhu[1,2], Weimin Mu[1(✉)], Yun Zhang[1], Mingyang Li[1], Can Ma[1], and Weiping Wang[1]

[1] Institute of Information Engineering, Chinese Academy of Sciences, Beijing, China
{liufan,zhuweilin,muweimin,zhangyun,limingyang,macan,
wangweiping}@iie.ac.cn
[2] School of Cyber Security, University of Chinese Academy of Sciences,
Beijing, China

Abstract. To improve system resource utilization, multiple operators are co-located in the distributed stream processing systems. In the colocation scenarios, the node runtime environment and co-located operators affect each other. The existing methods mainly study the impact of the runtime environment on operator performance. However, there is still a lack of in-depth research on the interference of operator colocation to the runtime environment. It will lead to inaccurate prediction of the performance of the co-located operators, and further affect the effect of operator placement. To solve these problems, we propose an online runtime environment prediction method based on the operator portraits for complex colocation interference. The experimental results show that compared with the existing works, our method can not only accurately predict the runtime environment online, but also has strong scalability and continuous learning ability. It is worth noting that our method exhibits excellent online prediction performance for runtime environments in large-scale colocation scenarios.

Keywords: Complex Colocation · Interference · Runtime Environment Prediction · Fused Deep Convolutional Neural Networks · Distributed Stream Processing

1 Introduction

The distributed stream processing systems (DSPSs) offer an effective means to analyze and mine the real-time value of data. They support many data stream processing applications (DSPAs), which are composed of various operators that implement specific computing logic. To improve system resource utilization, multiple operators are deployed and run in colocation. In the colocation scenario, the runtime environment and co-located operators affect each other [1–4]. Specifically, the competition for shared resources by co-located operators will interfere

J. Mikyška et al. (Eds.): ICCS 2023, LNCS 14074, pp. 93–107, 2023.
https://doi.org/10.1007/978-3-031-36021-3_7

with the runtime environment of nodes. In turn, the runtime environment will lead to operator performance fluctuations.

Now many studies focus on the impact of the runtime environment on operator performance [5,6]. Zhao et al. [7] proposes an incremental learning method to predict the performance of serverless functions under a partial interference environment. Patel et al. [8] studies whether the runtime environment can meet the performance requirements of multiple jobs when the latency-critical and background jobs are co-located. Few studies analyze the interference of operator colocation to the runtime environment [9,10]. Xu et al. [11] weights each job's interference to the runtime environment and designs a linear summation model to predict the runtime environment. Li et al. [12] computes the mean and variance of the interference of all co-located games to represent the runtime environment. However, the above methods are too simple and rough, and lack in-depth research on the interference of operator colocation to the runtime environment. It will lead to inaccurate prediction of the performance of the co-located operators, and further affect the effect of operator placement. The overall interference of co-located operators is complex and time-varying. So it is impossible to accurately predict the runtime environment with simple statistical methods. Besides, these methods are offline methods based on various benchmarks, and cannot continuously learn online.

In this paper, we propose an online runtime environment prediction method based on the operator portraits for complex colocation interference. Our method can accurately quantify and predict the key runtime environment metrics. The contributions of our work are as follows:

- To the best of our knowledge, our work is the first to deeply study the interference of operator colocation to the runtime environment. We use the operator portraits and fused deep convolution neural networks to model the colocation interference. On this basis, we predict the runtime environment in complex colocation scenario online.
- We design a runtime environment prediction model that fuses the deep convolutional neural networks and spatial pyramid pooling strategy [13]. Compared with the traditional padding models, our model can theoretically adapt to the learning of any scale colocation. Therefore, our model has strong scalability in the context of the rapid development of hardware technology.
- The experimental results show that our method can accurately predict the runtime environment in complex colocation scenarios online. In addition, our method has strong scalability and online learning ability to continuously improve prediction performance.

We organize the rest of this paper as follows. Section 2 elaborates the motivation of our work. Section 3 presents the design and implementation of our method. Section 4 describes the experimental results. Section 5 concludes our work.

2   Motivation

2.1   Interference Characteristics of Operator Colocation to Runtime Environment

Non-additivity of Colocation Interference. In the DSPSs, due to different resource contention of operators, multiple operator instances, and time-varying input load, the interference of co-located operators to the runtime environment is more complicated. The overall interference of co-located operators is not the sum of each operator's interference. We illustrate with the numerical calculation operator *calculator* and the sorting operator *sorter* as an example to illustrate. As shown in Fig. 1, the average CPU utilization of operator *calculator* is 15.19%, and that of operator *sorter* is 17.57% from Epoch t to $t + 850$. The actual measured average CPU utilization is 35.63% when they are co-located. It is greater than the sum of each CPU utilization (i.e. 32.76%). This is because when two operators are co-located, the CPU resources are time-division multiplexed. In addition to the overhead of operator solo-run, the overhead of CPU resource contention, inter-process switching, and site reservation is also increased. Moreover, for runtime environment metrics such as the cache hit rate, CPU interrupt times, and context switch times, the colocation interference cannot be superimposed.

Instability of Interference. The interference of operators to the runtime environment is time-varying and unstable. We analyze the interference when eleven different types of operators solo or co-located run. Specifically, we describe the runtime environment characteristics with nine key metrics, including the load, CPU utilization, cache hit rate, and context switching times, etc. We collect the runtime environment metrics at regular intervals to form multivariate time series. Then, we use the *Unit Root Test* method [14] to test the stationarity of the time series. We find that for some combinations of co-located operators and some runtime environment metrics, the *ADF test statistic* value is greater than the *Test critical value under 10% level*, and the *P-value* is also large, as shown in Fig. 2. Experimental results show that the interference of operators to the runtime environment is time-varying and unstable. Therefore, it is unreasonable and inaccurate to describe the runtime environment with a fixed value or mean and variance.

2.2   Importance of Interference of Operator Colocation to Runtime Environment

SLA Guarantee. In the DSPSs, the runtime environment analysis is the premise of operator performance prediction and operator placement. Besides, to process the input load in real time, it is necessary to predict the runtime environment before the operators are actually co-located. In addition, when the colocation changes, it is necessary to predict the runtime environment instantaneously. Therefore, if we can predict the runtime environment online, we can better meet the end-to-end latency requirements.

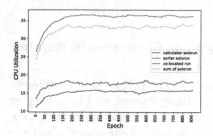

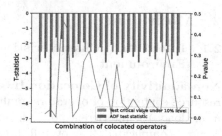

Fig. 1. The CPU utilization of the operator *calculator* and *sorter* when they solo or co-located run.

Fig. 2. The Stationarity Analysis of Interference of Different Operator Colocation to the CPU Utilization.

System Stability. If we can accurately predict the runtime environment online, we can effectively support operator placement decisions, avoid or reduce operator migration and adjust jitter, and thus improve system stability.

3   Design and Implementation

3.1   Overview

Figure 3 describes the design architecture of our method, which contains three core modules: the operator interference profiler (OIProfiler), the colocation interference learner (CILearner), and the runtime environment predictor (REPredictor). We adopt the OIProfiler to build the portraits of single operators offline, and the CILearner to continuously learn the portraits of co-located operators online. Based on the portraits of single and co-located operators, we use the REPredictor to predict the runtime environment.

The work process consists of two main stages: the offline solo-run and the online colocation-run. In the offline solo-run stage, a single operator runs on an empty node, and the Metric Collector collects the key runtime environment metrics over some time at regular intervals. Then the OIProfiler analyzes the metrics and builds the portrait of the operator.

In the online colocation-run stage, multiple operators run on an empty node simultaneously, and the Metric Collector collects the key runtime environment metrics at regular intervals continuously. Then the CILearner builds or updates the portraits of the co-located operators. At last, the REPredictor refers to the portraits of single operators and co-located operators to predict the runtime environment interfered by arbitrary co-located operators. Besides, as the collected samples increase, our CILearner dynamically updates the built portraits and creates the portraits of new co-located operators. Our REPredictor also updates adaptively to realize online prediction.

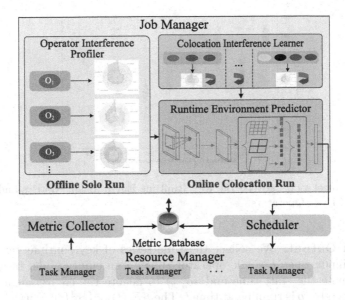

Fig. 3. Architecture.

3.2 OIProfiler: Operator Interference Profiler

Key Runtime Environment Metrics. By summarizing previous work [7, 15, 16], we can collect the runtime environment metrics through the system performance analysis tools and benchmarks. The system performance analysis tools directly monitor the various runtime environment metrics of nodes. The benchmark method customizes benchmarks sensitive to different resources and indirectly quantifies the runtime environment by the performance degradation of benchmarks. The benchmarks occupy more node resources by comparison, which themselves affect the runtime environment. And the benchmark method cannot collect the runtime environment metrics online. Therefore, we adopt the system performance analysis tools to collect the runtime environment metrics.

In this paper, we use the *top, perf* [17], and *vmstat* tools to monitor 42 runtime environment metrics, including CPU, memory, bandwidth, L1 instruction and data cache, LLC cache, context switch, and branch prediction, etc. However, some metrics have no or low correlation with the operator performance. To avoid overfitting and improve prediction efficiency, we use the *Pearson Correlation Coefficient* [18] to measure the correlations between the runtime environment metrics and the performance of eleven operators with different resource contention characteristics. Finally, we select nine key metrics, including *load_average_1min, cpu_system, cpu_user, l1_dcache_load_misses_rate, llc_load_misses_rate, llc_store_misses_rate, branch_misses_rate, ipc,* and *system_cs*. The correlation between the key runtime environment metrics and the performance of eleven operators is shown in Fig. 4.

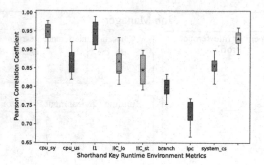

Fig. 4. The correlation between the Key Runtime Environment Metrics and the Performance of Eleven Operators.

Input and Output. We collect the time series of key runtime environment metrics as input when a single operator runs on an empty node. In detail, we use t to denote the start time and $se_o^i(t)$ to denote the environment metric i when the operator o is running at time t. The $SE_o(t) = \{se_o^1(t), se_o^2(t), ..., se_o^n(t)\}$ represents the key runtime environment metrics at time t. So we use the dataset $S_o = \{SE_o(t), SE_o(t+1), SE_o(t+2), ..., SE_o(t+w)\}$ as the input, in which w represents the collection duration.

We build the operator portrait as output, denoted as $SP_o = \{sp_o^1, sp_o^2, ..., sp_o^n\}$. The sp_o^i represents the interference of the operator o to the runtime environment metric i, in which $i \in \{1, 2, ..n\}$.

Profiling Module. As explained in Sect. 2, the interference of a single operator to the runtime environment is time-varying and unstable. We utilize the *first-order difference* method [19] to process the time series of key runtime environment metrics. We find that the time series of all metrics after the difference are stable. As shown in Fig. 5, we take the metrics of *load_average_1min* and *cpu_user* as an example to illustrate. The *ADF test statistic* value of each operator is less than the *Test critical value under* 1% *level*. The *P-value* is much less than the significance level of 0.01 and very close to zero. These show that the original hypothesis that the time series after differencing is not stationary can be strictly rejected. Therefore, we use the mean values of the time series after differencing to image the single operators.

3.3 CILearner: Colocation Interference Learner

Input and Output. We collect the time series of key runtime environment metrics as input when multiple operators co-located run on an empty node. In detail, we use O to denote the co-located operator set and $ce_O^i(t)$ to denote the environment metric i when the co-located operator set O is running at time t. The $CE_O(t) = \{ce_O^1(t), ce_O^2(t), ..., ce_O^n(t)\}$ represents the key runtime environment metrics at time t. So we use the dataset $C_O = $

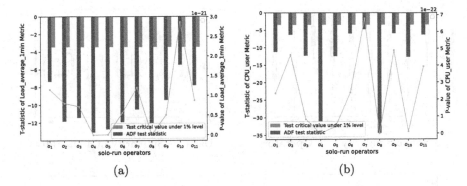

Fig. 5. The Stationarity Analysis of Time Series of *load_average_1min* and *cpu_user* Metrics after Difference.

$\{CE_O(t), CE_O(t+1), CE_O(t+2), ..., CE_O(t+w)\}$ as the input, in which w represents the collection duration.

We build the portrait of the co-located operators as output, denoted as $CP_O = \{cp_O^1, cp_O^2, ..., cp_O^n\}$. The cp_O^i represents the interference of the co-located operator set O to the runtime environment metric i, in which $i \in \{1, 2, ..n\}$.

Learning Module.

Observation. Co-located operators started in different orders interfere equally with the runtime environment.

We start $2, 4, 8, 16, 24$ and 32 operators in different orders respectively. The startup interval between adjacent operators is randomly between one second and ten minutes. As the co-located operators run, we get the time series of key runtime environment metrics. We adopt the *Pearson Correlation Coefficient* [18] to measure the shape similarity between pairwise time series, and the *Dynamic Time Warping (DTW)* [20] to measure the distance similarity.

We take the colocation of four and twenty-four operators as an example to illustrate. We compare the correlation and distance between pairwise time series. As shown in Fig. 6, the minimum correlation of any metrics are is greater than 0.9 and the distances are less than 0.1. Therefore, we get that the runtime environment is independent of the operator startup order. Furthermore, we find that although the time series of key runtime environment metrics are unstable, those after the difference are stable. Therefore, we use the mean values of the time series after the difference to image the co-located operators. Moreover, after collecting new existing co-located samples, we update the mean value to realize continuous learning.

3.4 REPredictor: Runtime Environment Predictor

Input Processing. Our REPredictor predicts the runtime environment based on the portraits of each co-located operator. The colocation scale is variable.

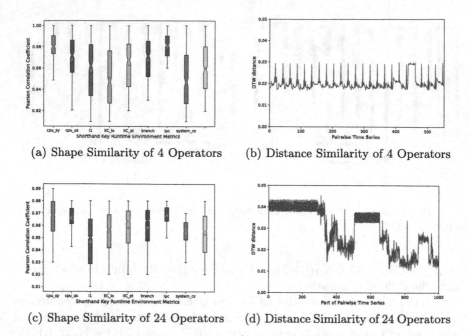

(a) Shape Similarity of 4 Operators

(b) Distance Similarity of 4 Operators

(c) Shape Similarity of 24 Operators

(d) Distance Similarity of 24 Operators

Fig. 6. The Shape Similarity and Distance Similarity Analysis of Pairwise Time Series at Different Colocation Scales.

But the existing prediction models require the input size to be fixed. Therefore, the traditional method is to unify the input based on the maximum colocation capacity of nodes. If the colocation scale is less than the maximum colocation capacity, the method completes the input with zero padding. Our method organizes the input into scalable images, which can scale vertically with the change of colocation scale.

In Fig. 7, we vividly describe the input of traditional and our methods. We assume that the operator portrait contains three metrics, which are denoted by the gold, blue and green squares in the figure. The maximum colocation capacity is assumed to be ten. Therefore, the traditional method unifies the input to 30 dimensions, which is obtained by multiplying the operator portrait dimension by the maximum colocation capacity. The part less than the maximum capacity is padded zeros, as shown in the purple dotted box in the figure. Our method flexibly scales the input image vertically according to the colocation scale.

Input and Output. As mentioned above, we organize the portraits of each co-located operator into image format as input. We use o_i to denote the ith co-located operator, m to denote the colocation scale, and $O(o_i \in O, i = \{1, 2, ..., m\})$ to denote the co-located operator set. The input of our REPredictor is expressed as:

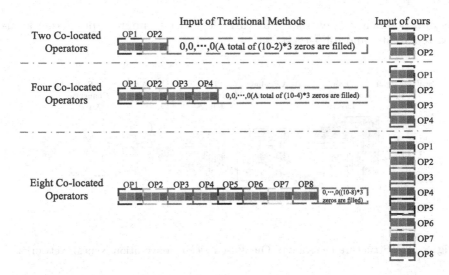

Fig. 7. The Input of Traditional Methods and Ours. (Color figure online)

$$
P_O = \begin{bmatrix} sp_{o_1}^1 & sp_{o_1}^2 & \cdots & sp_{o_1}^n \\ sp_{o_2}^1 & sp_{o_2}^2 & \cdots & sp_{o_2}^n \\ \vdots & \vdots & \ddots & \vdots \\ sp_{o_m}^1 & sp_{o_m}^2 & \cdots & sp_{o_m}^n \end{bmatrix}
$$

Our REPredictor predicts the final runtime environment as output. The output is expressed as: $CP_O = \{cp_O^1, cp_O^2, ..., cp_O^n\}$.

Predicting Module. As mentioned above, our REPredictor organizes the input into a scalable image. Because the colocation scale is variable, the image size is also not fixed. In this paper, we introduce the spatial pyramid pooling (SPP) Strategy [13] to remove the fixed-size constraint. Specifically, we design fused deep convolutional neural networks by adding an SPP layer between the last convolutional layer and the fully connected layer, as shown in Fig. 8. Firstly, the convolutional part runs in a sliding-window manner to learn features for co-located operator images of arbitrary size. The convolutional part consists of two convolutional layers, a pooling layer, and two convolutional layers. Their outputs are *feature maps*. After the last convolutional layer, the SPP layer pools the features and generates fixed-length outputs. We use the blocks of three different sizes to extract features, and put these three grids on the *feature maps* to get different spatial bins. Then we extract a feature from each Spatial bin to obtain fixed-dimensional feature vectors. At last, we feed the feature vectors to the fully connected layer to predict the final runtime environment.

Our predicting networks can generate a fixed-length runtime environment output regardless of the colocation scale. The networks use multi-level spatial bins and pool features extracted at variable scales to improve prediction accuracy, scalability, and robustness.

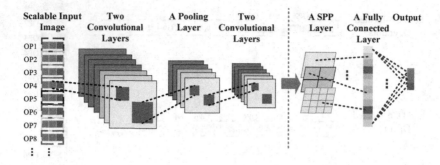

Fig. 8. The Structure Diagram of Our Fused Deep Convolution Neural Networks.

4   Experiments

4.1   Settings and Datasets

Settings. We experiment on a cluster of twelve servers. There are two kinds of servers in the cluster: two GPU servers and ten CPU servers. Each GPU server contains 36 cores Intel Xeon CPU E5-2697 v4 2.30 GHz, 256 GB memory, two NVIDIA GeForce GTX 1060ti cards, and 500 GB disks. We use one GPU server to run Job Manager, Scheduler, and MetricDatabase, and another GPU server to train and evaluate our proposed model. Each CPU server contains 10 cores Intel Xeon CPU Gold 5115 2.4 GHz, 256 GB memory, 480G SSD, and 2.4T SAS. We use the CPU servers to run Task Manager to execute operators. Besides, we train and evaluate our REPredictor with python 3.7 and tensorflow 1.15.0.

Datasets. We experiment on DataDock [21], our distributed stream processing system. We collect the key runtime environment metrics when the operators solo or co-located run, and build four databases, as shown in Table 1. The samples in the SRE database are collected at every second interval for 20 min offline. The SIP database is built by our OIProfiler with samples from the SRE database. The samples in the CRE database are continuously collected online at every second interval. The CIP database is built by our CILearner with samples from the CRE database.

We design eleven operators with different resource contention characteristics. These operators are implemented in C language to eliminate the impact of factors such as garbage collection on the runtime environment. In the experiment, the operators run at full load, that is, the input rate of the operators is greater than their processing capacity. Besides, each operator can be copied into multiple instances, so that we can simulate the complex colocation scenarios.

Table 1. The Datasets of Our Experiments.

ID	Name	Description
1	SRE	The key runtime environment metrics collected when the operators run offline and solo
2	SIP	The portraits of interference of single operators to the runtime environment
3	CRE	The key runtime environment metrics when the operators run online and co-located
4	CIP	The portraits of interference of co-located operators to the runtime environment

4.2   Evaluation

We evaluate the performance of runtime environment prediction from three aspects: accuracy, scalability, and continuous learning ability. We use the *Root Mean Square Errors(RMSE)* and *Mean Absolute Errors(MAE)* as the evaluation metric. $RMSE = \sqrt{\frac{1}{n}\sum_{i=1}^{n}(y-\hat{y})^2}$, $MAE = \frac{1}{n}\sum_{i=1}^{n}|y-\hat{y}|$, where y is the actual runtime environment, and $\hat{y}$ is the predicted value. Besides, we repeat the experiments 100 times and compute the average results to eliminate outliers and reduce random errors.

Prediction Accuracy. To evaluate the accuracy of runtime environment prediction, we compare with the GAugur [12], PYTHIA [11], and Paragon [22]. For a set of co-located operators, the GAugur method computes the mean and variance of the portraits of all co-located games to characterize the runtime environment. The PYTHIA method computes the linear sum of the portraits of all co-located games to characterize the runtime environment. The Paragon method calculates the sum of the portraits of all co-located games to characterize the runtime environment. We compare the accuracy of four methods for runtime environment prediction when 2, 4, 8, 16, 24, and 32 operators are co-located.

As is shown in Fig. 9, Our method outperforms other methods. Especially when the colocation scale increases, the advantages of our method are more significant. This is because we can learn more complex hidden features of colocation interference by using the fused deep convolutional neural networks. Besides, with the development of hardware technology, the colocation scale is increasing. Our method shows better prediction performance in large-scale colocation scenarios. In addition, our method can directly predict the runtime environment. It is particularly suitable for online prediction and solves the cold start problem well.

Prediction Scalability. The colocation scale is variable. To evaluate the scalability of runtime environment prediction, we compare our method with the MLP and CNN models. These models use the traditional padding method to unify the input.

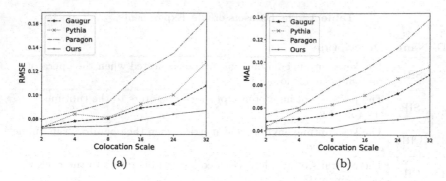

Fig. 9. The Prediction Accuracy of Different Methods.

The main parameters of the models are as follows. For the MLP and CNN networks, we set the maximum number of co-located operators to 32 and unify the input to 288 dimensions. In the MLP networks, we use *relu* as the activation function, *adam* as the optimizer, and *mse* as the loss, and set five hidden layers with sizes 512, 256, 128, 64, and 32. The CNN networks consist of two convolutional layers, a pooling layer, two convolutional layers, and a fully connected layer. And we use *relu* as the activation function, *adam* as the optimizer, *mse* as the loss, and set the six layers with sizes 288, 128, $(2,2)$, 64, 32, 9. The composition of our model is the same with the CNN networks, only adding an SPP layer between the last convolutional layer and the fully connected layer. And we set the SPP layer with block sizes of $(1*1, 2*2, 4*4)$ and the other parameters are the same as the CNN networks.

As is shown in Table 2, the prediction performance of MLP model is the worst, and the CNN model is comparable to ours. Our method solves the scalability problem while ensuring good prediction performance.

Continuous Learning Ability. To verify the continuous learning ability of our method, we compare the prediction performance at different running times and sample capacities. As shown in Fig. 10, as the running time or sample capacity increases, the prediction accuracy of our methods continues to improve. This is because our method can collect samples online and learn new operator colocation scenarios.

Table 2. The Prediction Scalability of Different Methods.

Co-located Numbers	MLP		CNN		**Ours**	
	RMSE	MAE	RMSE	MAE	**RMSE**	**MAE**
2	0.0793	0.0473	0.0729	0.0437	**0.0723**	**0.0416**
4	0.0836	0.0502	0.0748	0.0429	**0.0737**	**0.0431**
8	0.0891	0.0543	0.0759	0.0441	**0.0742**	**0.0425**
16	0.0928	0.0612	0.0784	0.0473	**0.0795**	**0.0483**
24	0.1003	0.0685	0.0847	0.0538	**0.0843**	**0.0501**
32	0.1172	0.0783	0.0837	0.0574	**0.0825**	**0.0529**

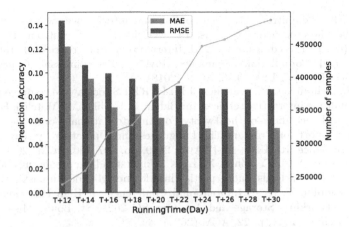

Fig. 10. The Continuous Learning Ability of Our Method.

5   Conclusion

In this paper, we propose an online runtime environment prediction method for complex colocation interference. It contains three core modules: the OIProfiler, the CILearner, and the REPredictor. Firstly, the OIProfiler builds the portraits of single operators in the form of offline solo-run. Then, the CILearner builds the portraits of co-located operators in the form of online colocation-run. At last, we use the REPredictor to model the portraits of single operators and co-located operators to predict the runtime environment with our fused deep convolutional neural networks. The experiments on the real-world datasets demonstrate that our method is better than the state-of-the-art methods. Our method can not only accurately predict the runtime environment online for complex colocation interference, but also has strong scalability and continuous learning ability.

References

1. Zhang, H., Geng, X., Ma, H.: Learning-driven interference-aware workload parallelization for streaming applications in heterogeneous cluster. IEEE Trans. Parallel Distrib. Syst. **32**(1), 1–15 (2021)
2. Gan, Y., et al.: Seer: leveraging big data to navigate the complexity of performance debugging in cloud microservices. In: Bahar, I., Herlihy, M., Witchel, E., Lebeck, A.R. (eds.) Proceedings of the Twenty-Fourth International Conference on Architectural Support for Programming Languages and Operating Systems. ASPLOS 2019, Providence, RI, USA, 13–17 April 2019, pp. 19–33, ACM (2019)
3. HoseinyFarahabady, M., Zomaya, A.Y., Tari, Z.: Qos- and contention- aware resource provisioning in a stream processing engine. In: 2017 IEEE International Conference on Cluster Computing. CLUSTER 2017, Honolulu, HI, USA, 5–8 September 2017, pp. 137–146. IEEE Computer Society (2017)
4. Buddhika, T., Stern, R., Lindburg, K., Ericson, K., Pallickara, S.: Online scheduling and interference alleviation for low-latency, high-throughput processing of data streams. IEEE Trans. Parallel Distrib. Syst. **28**(12), 3553–3569 (2017)

5. Romero, F., Delimitrou, C.: Mage: online and interference-aware scheduling for multi-scale heterogeneous systems. In: Evripidou, S., Stenström, P., O'Boyle, M.F.P. (eds.) Proceedings of the 27th International Conference on Parallel Architectures and Compilation Techniques. PACT 2018, Limassol, Cyprus, 01–04 November 2018, pp. 19:1–19:13. ACM (2018)

6. Chen, S., Delimitrou, C., Martínez, J.F.: PARTIES: qos-aware resource partitioning for multiple interactive services. In: Bahar, I., Herlihy, M., Witchel, E., Lebeck, A.R. (eds.) Proceedings of the Twenty-Fourth International Conference on Architectural Support for Programming Languages and Operating Systems. ASPLOS 2019, Providence, RI, USA, 13–17 April 2019, pp. 107–120. ACM (2019)

7. Zhao, L., Yang, Y., Li, Y., Zhou, X., Li, K.: Understanding, predicting and scheduling serverless workloads under partial interference. In: de Supinski, B.R., Hall, M.W., Gamblin, T. (eds.) International Conference for High Performance Computing, Networking, Storage and Analysis, SC 2021, St. Louis, Missouri, USA, 14–19 November 2021, p. 22. ACM (2021)

8. Patel, T., Tiwari, D.: CLITE: efficient and qos-aware co-location of multiple latency-critical jobs for warehouse scale computers. In: IEEE International Symposium on High Performance Computer Architecture. HPCA 2020, San Diego, CA, USA, 22–26 February 2020, pp. 193–206. IEEE (2020)

9. Zhang, Y., Laurenzano, M.A., Mars, J., Tang, L.: Smite: precise qos prediction on real-system SMT processors to improve utilization in warehouse scale computers. In: 47th Annual IEEE/ACM International Symposium on Microarchitecture. MICRO 2014, Cambridge, United Kingdom, 13–17 December 2014, pp. 406–418. IEEE Computer Society (2014)

10. Delimitrou, C., Kozyrakis, C.: Quasar: resource-efficient and qos-aware cluster management. In: Balasubramonian, R., Davis, A., Adve, S.V. (eds.) Architectural Support for Programming Languages and Operating Systems. ASPLOS 2014, Salt Lake City, UT, USA, 1–5 March 2014, pp. 127–144. ACM (2014)

11. Xu, R., Mitra, S., Rahman, J., Bai, P., Zhou, B., Bronevetsky, G., Bagchi, S.: Pythia: improving datacenter utilization via precise contention prediction for multiple co-located workloads. In: Ferreira, P., Shrira, L. (eds.) Proceedings of the 19th International Middleware Conference, Middleware 2018, Rennes, France, 10–14 December 2018, pp. 146–160. ACM (2018)

12. Li, Y., et al.: Gaugur: quantifying performance interference of colocated games for improving resource utilization in cloud gaming. In: Weissman, J.B., Butt, A.R., Smirni, E. (eds.) Proceedings of the 28th International Symposium on High-Performance Parallel and Distributed Computing. HPDC 2019, Phoenix, AZ, USA, 22–29 June 2019, pp. 231–242. ACM (2019)

13. He, K., Zhang, X., Ren, S., Sun, J.: Spatial pyramid pooling in deep convolutional networks for visual recognition. IEEE Trans. Pattern Anal. Mach. Intell. **37**(9), 1904–1916 (2015)

14. Dickey, D.A.: Dickey-fuller tests. In: Lovric, M. (ed.) International Encyclopedia of Statistical Science, pp. 385–388, Springer, Heidelberg (2011). https://doi.org/10.1007/978-3-642-04898-2_210

15. Courtaud, C., Sopena, J., Müller, G., Pérez, D.G.: Improving prediction accuracy of memory interferences for multicore platforms. In: IEEE Real-Time Systems Symposium. RTSS 2019, Hong Kong, SAR, China, 3–6 December 2019, pp. 246–259. IEEE (2019)

16. Chen, Q., Yang, H., Guo, M., Kannan, R.S., Mars, J., Tang, L.: Prophet: precise qos prediction on non-preemptive accelerators to improve utilization in warehouse-scale computers. In: Chen, Y., Temam, O., Carter, J. (eds.) Proceedings of the

Twenty-Second International Conference on Architectural Support for Programming Languages and Operating Systems. ASPLOS 2017, Xi'an, China, 8–12 April 2017, pp. 17–32. ACM (2017)

17. perf. https://perf.wiki.kernel.org/index.php/Tutorial
18. Amannejad, Y., Krishnamurthy, D., Far, B.H.: Predicting web service response time percentiles. In: 12th International Conference on Network and Service Management. CNSM 2016, Montreal, QC, Canada, 31 October–4 November 2016, pp. 73–81. IEEE (2016)
19. Lan, Y., Neagu, D.: Applications of the moving average of nth-order difference algorithm for time series prediction. In: Alhajj, R., Gao, H., Li, J., Li, X., Zaïane, O.R. (eds.) ADMA 2007. LNCS (LNAI), vol. 4632, pp. 264–275. Springer, Heidelberg (2007). https://doi.org/10.1007/978-3-540-73871-8_25
20. Geler, Z., Kurbalija, V., Ivanovic, M., Radovanovic, M., Dai, W.: Dynamic time warping: Itakura vs sakoe-chiba. In: Koprinkova-Hristova, P.D., Yildirim, T., Piuri, V., Iliadis, L.S., Camacho, D. (eds.) IEEE International Symposium on INnovations in Intelligent SysTems and Applications. INISTA 2019, Sofia, Bulgaria, 3–5 July 2019, pp. 1–6. IEEE (2019)
21. Mu, W., Jin, Z., Wang, J., Zhu, W., Wang, W.: BGElasor: elastic-scaling framework for distributed streaming processing with deep neural network. In: Tang, X., Chen, Q., Bose, P., Zheng, W., Gaudiot, J.-L. (eds.) NPC 2019. LNCS, vol. 11783, pp. 120–131. Springer, Cham (2019). https://doi.org/10.1007/978-3-030-30709-7_10
22. Delimitrou, C., Kozyrakis, C.: Paragon: Qos-aware scheduling for heterogeneous datacenters. In: Sarkar, V., Bodík, R. (eds.) Architectural Support for Programming Languages and Operating Systems. ASPLOS 2013, Houston, TX, USA, 16–20 March 2013, pp. 77–88. ACM (2013)

ICCS 2023 Main Track Short Papers

On the Impact of Noisy Labels
on Supervised Classification Models

Rafał Dubel[1], Agata M. Wijata[2,3]([✉]) [ID], and Jakub Nalepa[1,3]([✉]) [ID]

[1] Faculty of Automatic Control, Electronics and Computer Science,
Department of Algorithmics and Software, Silesian University of Technology,
Akademicka 16, 44-100 Gliwice, Poland
jnalepa@ieee.org
[2] Faculty of Biomedical Engineering, Silesian University of Technology,
Roosevelta 40, 41-800 Zabrze, Poland
awijata@ieee.org
[3] KP Labs, Konarskiego 18C, 44-100 Gliwice, Poland
{awijata,jnalepa}@kplabs.pl

Abstract. The amount of data generated daily grows tremendously in
virtually all domains of science and industry, and its efficient storage,
processing and analysis pose significant practical challenges nowadays.
To automate the process of extracting useful insights from raw data,
numerous supervised machine learning algorithms have been researched
so far. They benefit from annotated training sets which are fed to the
training routine which elaborates a model that is further deployed for a
specific task. The process of capturing real-world data may lead to acqur-
ing noisy observations, ultimately affecting the models trained from such
data. The impact of the label noise is, however, under-researched, and
the robustness of classic learners against such noise remains unclear. We
tackle this research gap and not only thoroughly investigate the classification capabilities of an array of widely-adopted machine learning models
over a variety of contamination scenarios, but also suggest new metrics
that could be utilized to quantify such models' robustness. Our extensive
computational experiments shed more light on the impact of training set
contamination on the operational behavior of supervised learners.

Keywords: Supervised machine learning · binary classification · label
noise · robustness

AMW was supported by the Silesian University of Technology, Faculty of Biomedical
Engineering grant (07/010/BK_23/1023). JN was supported by the Silesian University
of Technology Rector's grant (02/080/RGJ22/0026).

Supplementary Information The online version contains supplementary material
available at https://doi.org/10.1007/978-3-031-36021-3_8.

1   Introduction

The amount of acquired data grows tremendously in virtually all domains, spanning across medical imaging [15,22], text analysis and categorization [7], speech recognition [20], predictive maintenance [8], and many others. Gathering such enormous amounts of data of different modalities, however, poses new practical challenges concerned with its automated analysis and exploitation using data-driven techniques. In *supervised machine learning* (ML), we benefit from the acquired training data coupled with ground-truth labels to build models that are deployed to process incoming observations in a plethora of classification and regression tasks. Although deep learning—which benefits from the automated representation learning paradigm—established the state of the art in a multitude of fields, classic ML techniques are still widely used and researched due to their simplicity, resource frugality (which is especially important while deploying them on e.g., edge devices [6]), enhanced interpretability [13], and reduced requirements on the amount of training data necessary to elaborate well-generalizing models, effectively dealing with the unseen data.

Independently of the type of an ML model, we need to face the problem of noise which may easily affect the training data (also, training sets may be weakly-labeled [12]). Such noise may have different sources—it can be a result of a human or a sensor error, incorrectly designed data acquisition process, wrongly interpreted data or even hostile actions [10]. In general, we distinguish two types of the data noise, being random or systematic, and influencing supervised learners: the (i) attribute (feature) noise, and (ii) the label noise [2]. In the former case, the features, corresponding to the observed objects in the training set are contaminated, whereas in the latter scenario, the class labels are mistakenly assigned to training examples, due to e.g., an incorrect data annotation process or human errors/bias. Here, the label noise commonly leads to more severe consequences, as it can directly mislead the learning process [21] resulting in random predictions [9], and it cannot be compensated by other (not noisy) features if others are contaminated with noise. Also, noisy training data commonly leads to overly complex models.

Understanding its impact on the capabilities of ML techniques, thus robustifying them against such unexpected data-level contaminations is of paramount practical importance. The vast majority of works concerning this issue focus on developing more comprehensive and computationally-intensive processing pipelines [14], coping with noisy ML datasets, through the identification or reduction of noise, and pruning such contaminated training samples [1,2]. On the other hand, there are significantly less studies investigating the influence of noise on the "vanilla" versions of ML algorithms. Capturing the empirical behavior of supervised learners and understanding their intrinsic robustness against data-level noise can, however, influence the selection of an ML model for implementation, given the characteristics of the data acquisition and operation environment [17].

In this work, we address the above-discussed research gap, and thoroughly investigate an array of widely-adopted supervised ML classifiers trained from the datasets contaminated with different levels of label noise, in a variety of contamination scenarios (Sect. 2). Our extensive computational experiments (Sect. 3), performed over artificially-generated and benchmark datasets shed more light on the impact of the training set imbalance, cardinality and characteristics on the overall performance of ML models trained from noisy labels. We belive that the results reported in our study can constitute an interesting point of departure for further research on robustifying classic ML learning models elaborated from large, imbalanced and (potentially) noisy training sets, hence on enhancing their practical utility in real-life data acquisition environments.

2 Materials and Methods

In this study, we aim at quantifying the impact of the class-label noise which contaminates the training set T (the contaminated training set is denoted as T') on the generalization capabilities of supervised learners, calculated over the unseen test set Ψ that is *not* affected by the label noise (Fig. 1). As we target the binary classification problems, we flip the class labels of randomly selected training examples to the opposite one (i.e., the positive-class label is swapped to the negative-class label, or vice versa) in our simulation process. To capture various real-world scenarios, we investigate the following simulations:

- **Uniform label noise**, in which the same percentage η of class labels are flipped in both classes (positive and negative).
- **Positive-class label noise**, in which a given percentage η of training vectors originally belonging to the positive class are swapped and become the negative-class examples. For simplicity, we assume that the positive class corresponds to the majority class.
- **Negative-class label noise**, in which a given percentage η of training vectors originally belonging to the negative class are swapped and become the positive-class examples. For simplicity, we assume that the negative class corresponds to the minority class.
- **Random label noise**, in which the class labels of a given percentage η of all training vectors are swapped. Here, the contaminated vectors are randomly drawn, without considering their original class labels.

The following levels of the label noise contamination are considered in this study: $\eta \in \{0\%, 5\%, 10\%, 20\%, 30\%, 40\%, 50\%, 70\%, 90\%\}$ (note that we target uncontaminated T's, as well as extremely noisy training sets). To understand the robustness of the most popular learners, we deploy the following models in our pipeline (which is independent of the ML model): k-Nearest Neighbors ($k = 5$), linear support vector machines ($C = 0.025$), Gaussian Process classifiers with the radial basis function kernel, Decision Trees (with the Gini impurity measuring the quality of the split, max. depth of the tree: 7), Random Forests (max. depth of the tree: 7, max. number of trees: 50), Multi-Layer Perceptron (MLP) classifier (with rectified linear unit activations), AdaBoost classifier (max. number of estimators: 100), and the Quadratic Discriminant Analysis-based models [18].

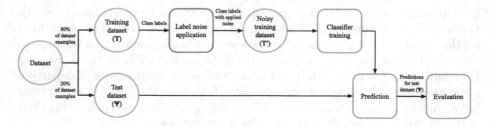

Fig. 1. A high-level flowchart of our computational experiments, in which a dataset is split into training and test subsets (T and Ψ, respectively, containing e.g., 80% and 20% of all training examples). The ML models are trained using a contaminated training set (T'), and their performance is quantified over the uncontaminated Ψ.

To investigate the classification capabilities of the learners, we quantify their performance over the unseen (uncontaminated) Ψ's using classic metrics such as accuracy (Acc), sensitivity (Sen), specificity (Spe), F1-score [19] and the Matthews correlation coefficient (MCC), with MCC commonly used for imbalanced classification, as the most robust quality metric [5]. The MCC values range from -1 (very strong negative relationship between ground-truth labels and prediction) to 1 (very strong positive relationship between them), whereas all other metrics range from zero to one, with one corresponding to the perfect classification performance. Additionally, we propose two auxiliary metrics:

- **D1**—it quantifies the stability of the ML model trained from contaminated T's ($\eta = 0\%-40\%$). Here, the "robustness" is calculated as the mean standard deviation of the obtained MCC scores (the smaller, the better).
- **D2**—it is calculated as the mean MCC score obtained by the ML model trained from contaminated T's with $\eta = 0\% - 40\%$ (the larger, the better).

The models were trained and validated on (i) synthesized and (ii) almost 40 benchmark datasets, the latter acquired from the KEEL and sklearn repositories, and manifesting different imbalance ratio across positive- and negative-class examples (Table 1). The synthetic datasets were generated using the `make_moons`, `make_circles`, `make_blobs` and `make_classification` sklearn functions with various parameterizations, concerning the number of training examples (ranging from 100 up to 5000), and the number of features (up to 60 features, with and without redundant ones). The majority of the KEEL/sklearn benchmark datasets include below 1000 training examples (with a few having above 5000 of them), with the mean and median of 1976 and 611 training vectors, respectively. The majority of benchmarks have up to 50 features (attributes). For full details of the investigated datasets, see the supplement available at https://gitlab.com/agatawijata/impact-of-noisy-labels.

Table 1. The aggregated imbalance ratio of all benchmarks (KEEL and sklearn).

Minimum	Mean	Median	Maximum
1.00	1.37	1.22	1.88

3 Experimental Results

The experiments were split into those focused on investigating all models over the (i) synthetic and (ii) benchmark (KEEL/sklearn) datasets (all experiments, for all ML models and datasets were executed in 10 independent runs, and the results were aggregated). The results obtained over the simulated datasets were consisted across all ML models, and showed that their robustness increases with the larger number of training vectors, as the uncontaminated examples were able to effectively compensate those affected by the label noise (see Fig. 2; for brevity, we present the MCC scores—all other metrics are available in the supplementary material). Similar, albeit not as obvious observations may be drawn for the models trained over T'''s with varying numbers of features. Here, the robustness of the models tends to increase for larger dimensionalities of the dataset, but—especially for smaller T'''s, the generalization capabilities can drop, due to the inherent curse of dimensionality issues (as the effective number of "correct" training examples is further decreased once they are contaminated with the class-label noise). We can hypothesize, however, that increasing the number of training examples could make the models more robust against feature-level noise here—this requires further investigation. Overall, the experiments over the synthetic data showed that Gaussian Process, AdaBoost, MLP and SVM classifiers offer the best generalization while trained over contaminated T'''s.

The results obtained over all benchmark datasets are rendered in Fig. 3, presenting the D1 and D2 metrics quantifying the "robustness" of the models against different noise contamination scenarios (D1 should be minimized, whereas D2—maximized; note the reversed D1 axis). We can observe that the MLP classifier elaborates the best aggregated prediction quality for the uniformly and randomly (across classes) applied class-label noise (the largest D2), with the linear support vector machines offering the best stability of predictions (reflected in the lowest D1 values). It is of note that the label noise applied to separate classes (either the positive or negative, corresponding to the majority and minority ones) led to consistent behavior of the investigated machine learning models. However, for the datasets with a larger imbalance ratio (commonly larger than 1.8), contaminating the minority classes triggered a visible drop in the MCC scores. This can be attributed to the fact that contaminating the minority class with noise makes the learning process much more challenging, as the intrinsic properties of the minority-class examples may be lost or severely modified through the noise injection. On the other hand, contaminating the majority class with noise may indeed act as an additional regularizer—there are techniques which exploit the noise injection to synthetically generate training examples, e.g., in the context of augmenting training sets for deep learning models targeting the hyperspec-

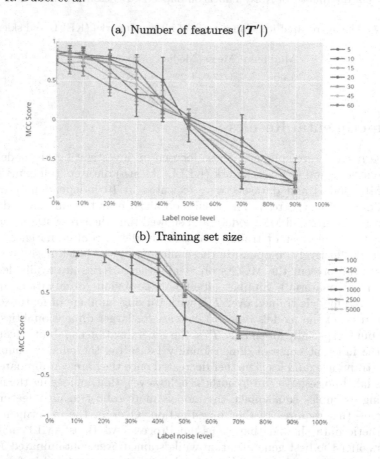

Fig. 2. The MCC scores obtained by the Gaussian Process classifier once the training set was contaminated by (a) the uniform label noise (for different numbers of features simulated using the `make_classification` dataset, and by (b) the positive-class label noise for different numbers of training set examples generated using the `make_circles` function. The number of features and training set examples is given in the legend.

tral image classification and segmentation [16]. In their recent work, Beinecke and Heider showed that deploying Gaussian Noise Up-Sampling, which effectively selects the minority-class examples and adds noise to the data points in order to smooth the class boundary can indeed reduce overfitting in some clinical decision making tasks as well [3]. Also, it worked on par with well-established the Synthetic Minority Over-sampling [4] and Adaptive Synthetic Sampling [11] approaches, and even outperformed those algorithms on selected datasets, showing the potential of utilizing noise simulations in training set augmentation routines.

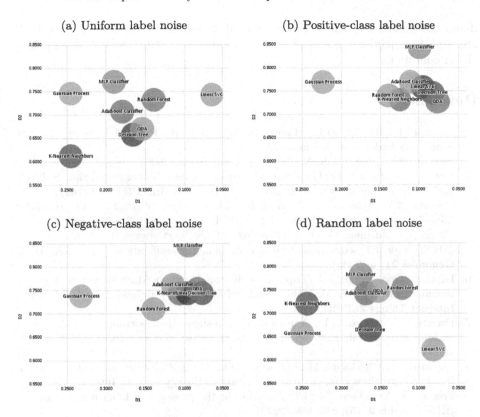

Fig. 3. The scatter plots of our D1 and D2 metrics quantifying the robustness of all investigated models over all benchmark datasets and noise contamination scenarios (D1 should be minimized, whereas D2—maximized).

4   Conclusions

Training supervised learners from noisy data has become an important practical issue, given the amount of data acquired on a daily basis. Such data may be contaminated with feature and class label noise due to various reasons, ranging from the operator bias, incorrect acquisition process or malfunctioning of the sensory system. Such noisy data, however, directly affects supervised learners trained from such data. Understanding the robustness of ML models against class label noise remains under-researched—we tackled this research gap, and thoroughly investigated an array of established models, following a variety of noise contamination scenarios. On top of that, we proposed new metrics that can be utilized to quantify the robustness of ML models against various levels of noise. Our extensive experiments, performed over synthetic and benchmark datasets revealed that there are indeed ML models which are more robust against label noise (the robustness directly depends on the size of the training set). We believe that the results reported in this work can constitute an exciting departure point for further research focused on developing the noise-robust models, and on objectively quantifying their robustness against real-life data acquisition conditions.

References

1. Awasthi, P., Balcan, M.F., Haghtalab, N., Urner, R.: Efficient learning of linear separators under bounded noise (2015)
2. Balcan, M.F., Haghtalab, N.: Noise in classification (2020)
3. Beinecke, J., Heider, D.: Gaussian noise up-sampling is better suited than SMOTE and ADASYN for clinical decision making. BioData Min. **14**(1), 49 (2021)
4. Chawla, N.V., Bowyer, K.W., Hall, L.O., Kegelmeyer, W.P.: SMOTE: synthetic minority over-sampling technique. J. Artif. Int. Res. **16**(1), 321–357 (2002)
5. Chicco, D., Jurman, G.: The advantages of the Matthews correlation coefficient (MCC) over F1 score and accuracy in binary classification evaluation. BMC Genomics **21**, 6 (2020)
6. Dhar, S., Guo, J., Liu, J.J., Tripathi, S., Kurup, U., Shah, M.: A survey of on-device machine learning: an algorithms and learning theory perspective. ACM Trans. Internet Things **2**(3), 3450494 (2021)
7. Duarte, J.M., Berton, L.: A review of semi-supervised learning for text classification. Artif. Intell. Rev. **56**, 1–69 (2023). https://doi.org/10.1007/s10462-023-10393-8
8. Es-sakali, N., Cherkaoui, M., Mghazli, M.O., Naimi, Z.: Review of predictive maintenance algorithms applied to HVAC systems. Energy Rep. **8**, 1003–1012 (2022)
9. Frenay, B., Verleysen, M.: Classification in the presence of label noise: a survey. IEEE TNNLS **25**(5), 845–869 (2014)
10. Gupta, S., Gupta, A.: Dealing with noise problem in machine learning data-sets: a systematic review. Procedia Comput. Sci. **161**, 466–474 (2019)
11. He, H., Bai, Y., Garcia, E.A., Li, S.: ADASYN: adaptive synthetic sampling approach for imbalanced learning. In: Proceedings of IEEE WCCI, pp. 1322–1328 (2008)
12. Kawulok, M., Nalepa, J.: Towards robust SVM training from weakly labeled large data sets. In: Proceedings of IAPR ACPR, pp. 464–468 (2015)
13. Kotowski, K., Kucharski, D., et al.: Detecting liver cirrhosis in computed tomography scans using clinically-inspired and radiomic features. Comput. Biol. Med. **152**, 106378 (2023)
14. Leung, T., Song, Y., Zhang, J.: Handling label noise in video classification via multiple instance learning. In: Proceedings of IEEE ICCV, pp. 2056–2063 (2011)
15. Nalepa, J., Kotowski, K., et al.: Deep learning automates bidimensional and volumetric tumor burden measurement from MRI in pre- and post-operative glioblastoma patients. Comput. Biol. Med. **154**, 106603 (2023)
16. Nalepa, J., Myller, M., Kawulok, M.: Training- and test-time data augmentation for hyperspectral image segmentation. IEEE Geosci. Remote Sens. Lett. **17**(2), 292–296 (2020)
17. Nettleton, D.F., Orriols-Puig, A., Fornells, A.: A study of the effect of different types of noise on the precision of supervised learning techniques. Artif. Intell. Rev. **33**(4), 275–306 (2010)
18. Pedregosa, F., et al.: Scikit-learn: machine learning in Python. J. Mach. Learn. Res. **12**(85), 2825–2830 (2011)

19. Powers, D.: Evaluation: from precision, recall and F-measure to ROC, informedness, markedness and correlation (2020)
20. Pradana, W.A., Adiwijaya, K., Wisesty, U.N.: Implementation of support vector machine for classification of speech marked Hijaiyah letters based on Mel frequency cepstrum coefficient feature extraction. J. Phys. Conf. Ser. **971**(1), 012050 (2018)
21. Sáez, J.A., Galar, M., Luengo, J., Herrera, F.: Analyzing the presence of noise in multi-class problems: alleviating its influence with the One-vs-One decomposition. Knowl. Inf. Syst. **38**(1), 179–206 (2012). https://doi.org/10.1007/s10115-012-0570-1
22. Wijata, A.M., Nalepa, J.: Unbiased validation of the algorithms for automatic needle localization in ultrasound-guided breast biopsies. In: Proceedings of IEEE ICIP, pp. 3571–3575 (2022)

Improving Patients' Length of Stay Prediction Using Clinical and Demographics Features Enrichment

Hamzah Osop[1(✉)], Basem Suleiman[2,3], Muhammad Johan Alibasa[4],
Drew Wrigley[2], Alexandra Helsham[2], and Anne Asmaro[2]

[1] Nanyang Technological University, Singapore, Singapore
hamzah.osop@ntu.edu.sg
[2] The University of Sydney, Sydney, Australia
basem.suleiman@sydney.edu.au,
{dwri5016,ahel9827,aasm3350}@uni.sydney.edu.au
[3] The University of New South Wales, Sydney, Australia
b.suleiman@unsw.edu.au
[4] Telkom University, Kabupaten Bandung, Indonesia
alibasa@telkomuniversity.ac.id

Abstract. Predicting patients' length of stay (LOS) is crucial for efficient scheduling of treatment and strategic future planning, in turn reduce hospitalisation costs. However, this is a complex problem requiring careful selection of optimal set of essential factors that significantly impact the accuracy and performance of LOS prediction. Using an inpatient dataset of 285k of records from 14 general care hospitals in Vermont, USA from 2013–2017, we presented our novel approach to incorporate features to improve the accuracy of LOS prediction. Our empirical experiment and analysis showed considerable improvement in LOS prediction with an XGBoost model RMSE score of 6.98 and R2 score of 38.24%. Based on several experiments, we provided empirical analysis of the importance of different feature sets and its impact on predicting patients' LOS.

Keywords: machine learning · length of stay · electronic health records

1 Introduction

The global cost of healthcare is rising faster than any increase in provisions for its funding. Thus, with an aging population, there is an added pressure to reduce the costs associated with patient treatment in hospitals [6]. Patients' length of stay (LOS) and hospital readmissions are factors that make up the true cost of hospitalisation [10]. LOS remains one of the biggest drivers of costs and a determinant in patients' lives saved within healthcare. Predicting the LOS allows for more effective and efficient planning within the hospital. It also improves the scheduling of elective surgeries, while also supporting the long-term

J. Mikyška et al. (Eds.): ICCS 2023, LNCS 14074, pp. 120–128, 2023.
https://doi.org/10.1007/978-3-031-36021-3_9

strategic planning of the hospital [11]. A prime example of the significance of LOS, besides the availability of health equipment, has been COVID-19 which unprecedentedly tested the effectiveness of healthcare systems across the globe.

The goal of this paper is to predict LOS using machine learning methodologies to support hospitals in LOS management. We enrich our data by contributing to novel feature extraction techniques utilising International Classification of Disease (ICD) 9 and ICD10 codes, and admission information. We provide an empirical evaluation by training and testing five machine learning models to evaluate and to present the best performing model, to identify opportunities for future research in this domain.

In this paper, we present a preliminary study on a methodological approach for predicting patients' LOS based on the Vermont dataset, a real-world hospitalisation dataset. Predicting LOS is very challenging as it requires employing an optimal set of diverse features that are often not available in many datasets [12]. Hence, we present our empirical approach for enriching the Vermont dataset by incorporating features including Charlson Comorbidity Index (CCI), rank procedure severity, categorise procedures, patient median income and median LOS of the prior year. Besides these features, the Vermont dataset includes a large number of other features which might be necessary for prediction LOS [3]. Thus, we also present a machine learning method to identify the optimal and essential features that are crucial for predicting LOS. We construct several regression models to predict LOS using the enriched and selected features of Vermont dataset. We present our empirical experiments and show the best-performing regression model that significantly outperforms other benchmarks in terms of RMSE and R2 score. Our empirical results further highlight the importance of three categories of features and their impact on the accuracy of LOS prediction.

2   Related Work

There are a number of factors that influence and contribute to a patient's LOS and in turn, the ability to predict LOS. Buttigieg et al. [3] summarised factors that impacted LOS from 46 research papers, with relationships identified between emergency department crowding, early patient transfer to specialists, date and time of admission, access to early imaging and patient income to name a few. The relationship between patient demographics and LOS is prevalent in prior studies. Particularly, age was identified as one of the main drivers, explaining for 87% of LOS variability [10]. This can also be seen when predicting patient discharge [1], LOS and readmission after colorectal resection [8] with age being one of the most predictive features of LOS. Further, patients' payment type has also emerged as a predictor of LOS, for example, in predicting LOS amongst cardiac patients [7]. The socio-economic position of a patient also influences LOS, with low income associated with inadequate housing increasing LOS by 24% [3].

Past studies have used machine learning to predict LOS from inpatient data. Daghistani et al. [7] used four machine learning algorithms, including Random Forest (RF), Bayesian Network (BN), Support Vector Machine (SVM) and Artificial Neural Networks (ANN). They found that RF algorithm performed best on

70 features derived from demographic information, cardiovascular risk factors, vital signs on admission and admission criteria. Further developments on LOS prediction include extending it to predict the time of patient discharge. Barnes et al. [1] conducted a study to classify if a patient will be discharged at 2 pm or midnight that day, using data available at 7 am daily. While this use case is a slight variation of LOS prediction, it highlights the range of applications and value of predicting LOS. The key feature used were the elapsed length of stay, observation status, age, reason for visit and day of the week. However, these past studies did not include various rich features such as comorbidity measures, and other feature extraction techniques as they mostly used demographic and hospital admission information.

3 Methodology

3.1 Dataset and Exploratory Analysis of Vermont Dataset

We utilised the Vermont Hospital Dataset, consisting of inpatient discharge data (260k records), outpatient procedures and services data (8.3M records), and emergency department data (1.3M records) collected across Vermont's 14 general care hospitals using the Vermont Uniform Hospital Discharge Data System (VUHDDS). The dataset spanned between 2013 to 2017, included attributes such as (i) diagnostic discharge data, (ii) patient socio-demographic characteristics, (iii) ICD9 and ICD10 codes for diagnosis and reasons for admission, (iv) patient treatment and services provided, (v) length of hospital stay, and (vi) financial data such as billing and charges.

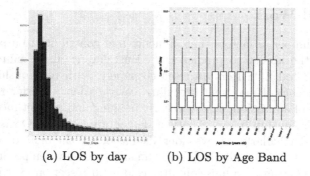

(a) LOS by day   (b) LOS by Age Band

Fig. 1. Distribution of Length of Stay (LOS)

Initial analysis of the dataset revealed a strong skew in the number of stay days, with about 93% of LOS being ten days or less (refer to Fig. 1(a)). Given the age banding and a disproportion towards low stay days, an analysis of patients' age group with LOS showed a general association of older patients with a longer period of stay as shown in Fig. 1(b). The primary diagnostic group (ccsdx) mapped patients' diagnosis to a grouped category of diagnoses, was analysed with the distribution of LOS. The analysis of the top 15 ccsdx codes showed a significant amount of variance between these categories, suggesting a strong relationship between diagnosis and patient LOS.

3.2 Data Processing and Feature Engineering

To accommodate missing values, data variables were approximated, turned to factors or removed completely. The dataset was anonymised and without features like medical history, readmission visits, medication, or clinical data such as laboratory testing, the dataset was further enriched using secondary data sources. The ICD9 and 10 codes consisted of unique string values and were difficult to consume by machine learning models. Hence, the ICD codes were mapped to (1) replace 'NA' values, (2) map ICD9 to ICD10 codes, (3) calculate the Charlson Comorbidity Index, (4) rank procedure severity and, (5) categorise procedures, making it more meaningful.

Handling Missing Values. There were several data fields that had high counts of 'NA' values: the diagnosis codes fields (DX1-DX20) and procedure codes fields (PX1-PC20). ICD mappings were used in place of the missing values.

Aligning ICD9 and ICD10 Data. The dataset contained a mix of ICD9 and ICD10 codes depending on the year the codes were used. To create consistency in the use of ICD codes, all ICD9 codes were mapped to ICD10, based on mappings provided by the Bureau of Economic Research (NBER).

Calculating Charlson Comorbidity Index (CCI). The CCI predicts the 10-year mortality of a patient with a range of comorbid procedures [5]. The CCI calculation was performed using the ICD library in R, where each ICD9/10 codes correspond to a certain score.

ICD Procedure Classes. Two conversions were performed on the procedure data fields. First, procedures were classified according to the first character in the ICD codes which specified the classes of procedures performed. This classified each ICD code into categories with the corresponding nature of procedures. The second conversion categorised the procedures into four classes of Minor Diagnostic, Major Diagnostic, Minor Therapeutic and Major Therapeutic. The categorisation was determined based upon whether a procedure was (1) diagnostic or therapeutic, and (2) performed in an operating room.

Median Income. Patients with lower socioeconomic status have been shown to stay in hospitals longer as their external environment is not adequate for proper care requirements. Given the details of the 14 hospitals in the dataset, the zip codes of those hospitals were extracted. The zip codes were then mapped to publicly available median income data sources [4].

Median LOS for Prior Year. Given the correlation between LOS and age, a new feature called Median LOS of Prior Year was introduced into the dataset. Based on the patient's age group, the median LOS from the previous year was calculated and labelled in the dataset as median_intage.

Y Variable Log Transformation. Due to the dataset being highly skewed, a log transformation on the LOS data field was performed to convert it towards being normally distributed. This is to ensure that models such as linear regression could deliver the best model performance.

4   Experiments

4.1   Constructing Model to Predict LOS

We selected four regression models to empirically evaluate the LOS prediction, namely Linear Regression, SVM, RF and XGBoost. The dataset was split into training and testing data, where training data contained 2014 to 2016 records, and testing data 2017 records. Altogether, the training data had 157k records and the testing data, 53k records.

We only selected optimal features for the model to achieve balance between model performance and computational effort. The feature selection was performed using RF and Recursive Feature Elimination (RFE). Utilising RF approach allowed for the features information gain to be isolated while taking into consideration the impact of multivariate nature of the problem. RFE was a wrapper type feature selection algorithm, by fitting a model using all features, and then ranking the features by importance. It recursively discarded the least important features and re-fitted the model. This continued until the optimal features were selected for the model. As such, we did not consider the correlation analysis of the data features.

Optimisation of model performance was implemented using Random Search hyperparameter tuning. This method required input for the tuneLength parameter, which defined the total number of parameter combinations to be evaluated. In this case, the parameter was set to ten. This method of tuning was selected over alternates such as Grid Search due to the computational challenges that came with exhaustive search. Studies such as [2] have also shown than Random Search is far more efficient that Grid Search for hyperparameter tuning.

4.2   Results

We based the benchmark for the performance of LOS regression on the study by Liu et al. [9], given the similarities to our dataset. However, we were not able to obtain the benchmark dataset to replicate and compare the model result with our dataset. Both our dataset and the benchmark dataset consisted of administrative data, such as demographics and admission diagnosis. The similarities end there with the benchmark study using two additional features of Laboratory Acute

Physiology Score (LAPS) and Comorbidity Point Score (COPS) derived from its dataset. Similarly, we adopted the CCI and Prior-Year LOS fea-tures. The benchmark employed a linear regression with an R2 of 0.124 and RMSE of 173.4.

Predicting LOS (Regression). Table 1 summarised the LOS prediction results using different regression models. The best regression model, XGBoost, significantly outperformed the benchmark in both RMSE and R2. Although the underlying datasets were different, the massive lift in model performance suggested the methodologies adopted in this study had the potential to improve the performance of models trained on similar datasets. The XGBoost model could predict the patients' LOS based on information available at admission. However, the model might not be strong enough for implementation, with the RMSE above the median LOS of the population at 3 days. As shown in the table, the SVM model performed far worse compared to the other models. Considering a big number of data that we used, the possible reason is that the SVM model was unable to find a clear decision boundary based on the provided features.

Table 1. Regression model performance for all diagnosis

Model	RMSE	R2
Baseline	8.2	N/A
Linear Regression	7.82	34.33%
SVM	399.7	8.87%
Random Forest	7.15	37.42%
XGBoost	6.98	38.24%
Liu et al. [9]	173.5	12.4%

Feature Importance. There is significant value in understanding the drivers of LOS, including the ability to generalise the insights. Our results showed that the optimal number of features based on collection of dataset are 175 and 18 for all diagnosis and heart & Circulatory diagnosis, respectively. In this study, the data features were categorised into three groups, including features that originated from the base dataset and features from secondary data sources. The key features for the models were summarised as follows:

- **Features originated from base data**: Among features from base data, attributes related to primary diagnosis are most important. Fields such as ccsdx (primary diagnosis) and MDC (major diagnostic category), ccspx (primary medical procedures), are in the top 10 most important features of most models.
- **Features derived from base data**: This study created over 1,000 new features from base data. The prior year median LOS features turn out to

be very predictive of LOS, seemingly more predictive than the features they were created from. This reflects prior-year LOS could be a good indicator for future studies.

- **Features derived from secondary sources**: Although adding new features from secondary data sources has improved the model performance, not many features in this category are identified are as the top 10 features for regression or classification models.

5   Discussion

The initial motivation of the feature analysis was to assist hospitals in identifying key improvement areas in LOS management. As most fields in the base data were out of control for hospitals, the key features identified might not provide good guidance for hospital LOS management in practice. Future research might need to analyse this dataset together with hospital operational data to provide actionable insights for hospitals. Interestingly, the median income of hospitals was predictive of LOS in the classification model. Typically, the literature used the median income based on patient location, rather than hospital location. Perhaps the socio-economic region that a hospital is based in can be used as a proxy in the absence of patient location.

One of the most successful experiments in this study was the creation of prior median LOS features, where new features were created by calculating the median LOS of the prior year for each attribute of the column. This approach applied the methodologies of time series analysis on based data generated over 1000 new features, over 200 of which were kept in the final model. These new features significantly improved the model performance and many of the new features were considered important for many models in the predictive features' analysis. This was a way to make better use of a dataset with time series information, not limited to a similar clinical dataset, for regression and even classification problems. It enabled studies to generate meaningful features without sourcing external data. The only drawback of the approach was the first-year records could not be used in model training and testing, which reduced the data size for model development.

Like many other LOS datasets used in prior studies, the Vermont dataset was highly skewed. This limited the performance of many models which assumed the training data was normally distributed. The Y variable log transformation was a good solution to this problem. Log transformation converted the dataset towards normal distribution and made the dataset better suited for model training. While this approach was simple, it could have been overlooked. This study reaffirmed the effectiveness of the method and would recommend it for consideration in future studies that utilises skewed data. Despite outperforming the benchmark, this prediction might not have been accurate enough to support a decision in practice. It should be highlighted that despite the significant enrichment made in this study, predicting the exact LOS continues to remain a challenge.

6   Conclusion and Future Work

We propose a methodological approach for predicting patients' LOS using a real-world hospital dataset. We empirically enriched the data by incorporating relevant features that contributed to improving LOS prediction. Our empirical experiments showed that our prediction approach outperformed the identified benchmarks that used regression models. Compared to the benchmark research, we introduced an empirical approach that uniquely improved the regression-based LOS prediction. Through the feature selection process, the optimal number of features was selected for each model type. This variance in the number of features highlighted that different models would require different levels of data and different covariates. While our approach seems to only be applicable to the Vermont dataset, most electronic health records would contain similar data fields as our dataset. In our future works, replicating the prediction model using similarly typed electronic health records could provide a meaningful comparison of model accuracy. The findings above, therefore, add towards academic studies and medical research. The key features could provide research teams with possible directions in LOS reduction-related research.

Acknowledgements. We would like to thank the State of Vermont for providing us with the Vermont Uniform Hospital Discharge Dataset in agreement with its Public Use policy. We also thank Jirong Liu and Young Rang Choi for helping out with the visual data analysis needed for this work.

References

1. Barnes, S., Hamrock, E., Toerper, M., Siddiqui, S., Levin, S.: Real-time prediction of inpatient length of stay for discharge prioritization. J. Am. Med. Inform. Assoc. JAMIA **23**, e2–e10 (2015)
2. Bergstra, J., Bengio, Y.: Random search for hyper-parameter optimization. J. Mach. Learn. Res. **13**, 281–305 (2012)
3. Buttigieg, S.A., Abela, L., Pace, A.: Variables affecting hospital length of stay: a scoping review. J. Health Organ. Manage. **32**, 463–493 (2018)
4. Center, M.P.S.: Zip code characteristics: mean and median household income (2020). https://www.psc.isr.umich.edu/dis/census/Features/tract2zip/
5. Charlson, M.: Charlson comorbidity index (CCI) (2020). https://www.mdcalc.com/charlson-comorbidity-index-cci
6. Clarke, A.: Why are we trying to reduce length of stay? evaluation of the costs and benefits of reducing time in hospital must start from the objectives that govern change. BMJ Qual. Saf. **5**(3), 172–179 (1996)
7. Daghistani, T., El Shawi, R., Sakr, S., Ahmed, A., Thwayee, A., Al-Mallah, M.: Predictors of in-hospital length of stay among cardiac patients: a machine learning approach. Int. J. Cardiol. **288**, 140–147 (2019)
8. Kelly, M., Sharp, L., Dwane, F., Kelleher, T., Comber, H.: Factors predicting hospital length-of-stay and readmission after colorectal resection. BMC health Serv. Res. **12**, 77 (2012)

 9. Liu, V., Kipnis, P., Gould, M.K., Escobar, G.J.: Length of stay predictions: improvements through the use of automated laboratory and comorbidity variables. Med. Care **48**, 739–744 (2010)
10. Masip, J.: The relationship between age & hospital length of stay: a quantitative correlational study, Ph. D. thesis, University of Phoenix (2019)
11. Pendharkar, P., Khurana, H.: Machine learning techniques for predicting hospital length of stay in pennsylvania federal and specialty hospitals. Int. J. Comput. Sci. Appl. **11**, 45–56 (2014)
12. Turgeman, L., May, J., Sciulli, R.: Insights from a machine learning model for predicting the hospital length of stay (LOS) at the time of admission. Expert Syst. Appl. **78**, 376–385 (2017)

Compiling Tensor Expressions into Einsum

Julien Klaus$^{(\boxtimes)}$, Mark Blacher, and Joachim Giesen

Friedrich Schiller University Jena, Jena, Germany
{julien.klaus,mark.blacher,joachim.giesen}@uni-jena.de

Abstract. Tensors are a widely used representations of multidimensional data in scientific and engineering applications. However, efficiently evaluating tensor expressions is still a challenging problem, as it requires a deep understanding of the underlying mathematical operations. While many linear algebra libraries provide an Einsum function for tensor computations, it is rarely used, because Einsum is not yet common knowledge. Furthermore, tensor expressions in textbooks and scientific articles are often given in a form that can be implemented directly by using nested for-loops. As a result, many tensor expressions are evaluated using inefficient implementations. For making the direct evaluation of tensor expressions multiple orders of magnitude faster, we present a tool that automatically maps tensor expressions to highly tuned linear algebra libraries by leveraging the power of Einsum. Our tool is designed to simplify the process of implementing efficient tensor expressions, and thus making it easier to work with complex multidimensional data.

Keywords: tensor expressions · einsum · domain specific languages · mathematics of computing

1 Introduction

Tensors are higher-dimensional generalizations of vectors and matrices. We can think of tensors as multidimensional arrays. Tensors are used in various applications. For example, an RGB-image can be represented by a tensor A_{ijk} with three dimensions, where the RGB-values of the first pixel can be accessed through A_{00k}. In general, computations over such tensors are mostly written in a form that uses summation symbols and access the tensors by indices. Such computations are typically implemented by nesting for-loops. This, however, can be quite inefficient, when compared to highly-tuned libraries for working on tensors, like NumPy [8], PyTorch [13] and TensorFlow [1]. These libraries use a function called *Einsum*. Although Einsum is quite powerful, it requires some knowledge to use it efficiently and correctly. Therefore, we present a transformation of tensor expressions into Einsum expressions, that can be mapped directly to multiple Einsum libraries or backends. For the transformation, tensor expressions are specified in a simple formal language that is close to the language used in textbooks and scientific articles.

In this article, all steps are described on the example of the Tucker decomposition [17]. This decomposition is used to decompose a high-dimensional tensor

J. Mikyška et al. (Eds.): ICCS 2023, LNCS 14074, pp. 129–136, 2023.
https://doi.org/10.1007/978-3-031-36021-3_10

into a tensor with fewer entries than the original tensor, and a set of matrices. Although this decomposition works with tensors of arbitrary order, we will focus, for simplicity, on tensors of order three, that is decomposing a tensor $A \in \mathbb{R}^{I \times J \times K}$ into a tensor $Z \in \mathbb{R}^{L \times M \times N}$ and matrices $B \in \mathbb{R}^{L \times I}, C \in \mathbb{R}^{M \times J}, D \in \mathbb{R}^{N \times K}$, where $L \ll I, M \ll J$, and $N \ll L$. To obtain such a decomposition, we have to solve the following optimization problem:

$$\min_{Z,B,C,D} \sqrt{\sum_{i=1}^{I} \sum_{j=1}^{J} \sum_{k=1}^{K} \left(A_{ijk} - \sum_{l=1}^{L} \sum_{m=1}^{M} \sum_{n=1}^{N} Z_{lmn} B_{li} C_{mj} D_{nk} \right)^2}.$$

Evaluating the objective function of this problem entails six nested loops. By using highly-tuned backends, the objective function can be computed orders of magnitude more efficiently than the naive implementation.

Related Work. There are already approaches that map tensor expressions into various backends [7,14,15]. But most of them are either only usable for linear algebra expressions [2,11], need additional information about the expressions' variables and parameters [2,16], do not map into backends, but optimize the loops directly [3,4,12], or do not allow unary operations [9,18]. We present a solution that can handle tensors of arbitrary orders, support unary and binary operations, and does not need any additional information about variables and parameters, like symmetries.

2   Understanding Einsum Notation

Einsum notation is a generalization of the Einstein summation convention, introduced by Einstein in 1916 [6]. The Einstein summation convention simply means summing over shared indices. For example, a matrix-vector product $Ab = \sum_j A_{ij} b_j$ is written as $A_{ij} b_j$, when using the Einstein summation convention. Implicitly, the summation is performed over the shared index, in this case j. In contrast to the Einstein summation convention, which sums over shared indices, the output of the operation is explicitly defined in Einsum notation. Since the tensor names are not relevant for the description of the operation, only the indices of the expression are retained. Therefore, in Einsum notation our matrix-vector product can be written simply as `ij,j->i`. It is important to note, that this makes Einsum notation more general than the Einstein summation convention. In Einsum notation we could also write `ij,j->ij`, which describes the operation of elementwise multiplying the vector represented by the second operand with each column of the first operand, resulting in a matrix. As we can use as many indices as we like, describing tensors of arbitrary dimension is possible, which makes Einsum notation a powerful tool. A slightly more complex example, is the partial expression $\sum_l Z_{nml} D_{lk}$ from the Tucker decomposition. This expression calculates the tensor $E \in \mathbb{R}^{N \times M \times K}$ as

$$E_{nmk} = \sum_{l=1}^{L} Z_{nml} D_{lk}.$$

In Einsum notation, this operation is written as `nml,lk->nmk`, where `nml` are the indices of Z (left operand), `lk` are the indices of D (right operand), and `nmk` are the indices of E (result).

Users are not accustomed to directly write Einsum expressions, but are used to a language typically used in textbooks. In textbooks, tensor expressions are almost exclusively written in a form that makes sums and multiplications explicit by using indices. Some examples of explicit expressions and their translation into Einsum are shown in Table 1.

Table 1. Example tensor expressions in their Einsum Notation and our Tensor Expression language.

Operation	Explicit expression	Einsum notation
Scalar times vector	`s*a[j]`	`,j->j`
Vector times vector	`a[i]*b[i]`	`i,i->i`
Vector outer product	`a[i]*b[j]`	`i,j->ij`
Matrix times vector	`A[i,j]*b[j]`	`ij,j->i`
Inner product	`sum[i](a[i]*b[i])`	`i,i->`
Batch matrix multiplication	`sum[k](A[b,i,k]*B[b,k,j])`	`bik,bkj->bij`
Marginalization (sum over axes)	`sum[i,l,n,o](A[i,l,m,n,o])`	`ilmno->m`
Mahalanobis distance	`sum[i,j](a[i]*A[i,j]*b[j])`	`i,ij,j->`

Although Einsum is quite flexible, it lacks frequently used element-wise functions like exp and log, and binary operations such as $+, -$ between Einsum expressions. For instance, our example of the Tucker decomposition, also uses the square-root function and the difference of two terms. To overcome these issues, we need a language to describe tensor expressions with the additional operations.

3   A Language for Tensor Expressions

In this section, we describe a simple formal language for explicit tensor expressions, which is close to standard textbook form. We extend a language for linear algebra expressions [11] to tensors, by allowing multiple indices for variables and sums. Thereby, the language becomes powerful enough to cover arbitrary tensor expressions and most classical machine learning problems, even problems not contained in standard libraries like scikit-learn [5]. An EBNF grammar for the language is shown in Fig. 1.

The formal language supports the combination of arbitrary tensors with unary and binary operators, as well as numbers. A special operation is the summation operation *sum*, which includes a list of non-optional indices. The list of indices describes the dimensions over which of the underlying tensors are contracted.

$$\langle expr \rangle \quad ::= \langle term \rangle \: \{ ('+' \mid '-') \: \langle term \rangle \}$$

$$\langle term \rangle \quad ::= ['-'] \: \langle factor \rangle \: \{ ('*' \mid '/') \: ['-'] \: \langle factor \rangle \}$$

$$\langle factor \rangle \quad ::= \langle atom \rangle \: [\, '^\wedge{}' \: \langle factor \rangle \,]$$

$$\langle atom \rangle \quad ::= number \mid \langle function \rangle \: '(' \: \langle expr \rangle \: ')' \mid \langle variable \rangle$$

$$\langle function \rangle \quad ::= \text{'sin'} \mid \text{'cos'} \mid \text{'exp'} \mid \text{'log'} \mid \text{'sign'} \mid \text{'sqrt'} \mid \text{'abs'} \mid \text{'sum'} \: '[' \: \langle indices \rangle \: ']'$$

$$\langle variable \rangle \quad ::= alpha+ \: [\, '[' \: \langle indices \rangle \: ']' \,]$$

$$\langle indices \rangle \quad ::= \langle index \rangle \: \{ ',' \: \langle index \rangle \}$$

$$\langle index \rangle \quad ::= alpha$$

Fig. 1. EBNF grammar for tensor expressions. In this grammar, *number* is a place-holder for an arbitrary floating point number and *alpha* for Latin characters.

In the formal language, the Tucker decomposition reads as

$$\mathrm{sqrt}(\mathrm{sum}[i,j,k]((A[i,j,k] - \mathrm{sum}[n,m,l](Z[n,m,l] * B[n,i] * C[m,j] * D[k,l]))^\wedge 2))$$

A point worth emphasizing is that indices always select scalar entries of tensors, which makes every operation an operation between scalars. Expressions that conform to the grammar from Fig. 1 are parsed into an expression tree. An expression tree $G = (V, E)$ is a directed tree, where every node $v \in V$ has a specific label, which can be either an operation, a tensor name, a number, or a list of indices.

Furthermore, we assign to each node its dimension, that is, the order of the tensor, after evaluating the node, described by its indices. For example, a node of dimension i, j, k has the order three. The dimension *dim* is computed recursively for each node $n \in V$ as

$$\mathrm{dim}(n) = \begin{cases} \text{indices of } n & \text{, if n is a variable} \\ \emptyset & \text{, if n is a number} \\ \bigcup_{\text{children of } n} \mathrm{dim}(c) & \text{, if n is an operation node} \\ \bigcup_{\text{children of } n} \mathrm{dim}(c) \setminus \{\text{indices of the sum}\} & \text{, if n is a } sum \text{ node} \end{cases}$$

For leaf nodes, the dimension is the index list, for all other nodes, except for the special *sum* nodes, the dimension is the union of the dimension of their children.

Since the *sum* operation removes indices, the dimension of a *sum* node is the union of their childrens' dimension, minus the indices of the *sum* node. For example, the inner summation of the Tucker decomposition sum[n,m,l](Z[n,m,l] * B[n,i] * C[m,j] * D[k,l]), has the dimension

$$\underbrace{\{i,j,k,l,m,n\}}_{\text{union of its children dimensions}} \setminus \overbrace{\{l,m,n\}}^{\text{summation indices}} = \{i,j,k\}.$$

Additionally, for each index, we determine its size (number of entries in this dimension), through the tensors, that refer to the index. For example, the tensor Z_{lmn}, has the indices lmn, and so we know that l has the size of Z's first dimension, and analogously for m and n.

An expression tree for the Tucker decomposition is shown in Fig. 2. Our task, and the contribution of this paper, becomes to compile expressions that conform to the grammar in Fig. 1 into Einsum expressions that can be evaluated efficiently.

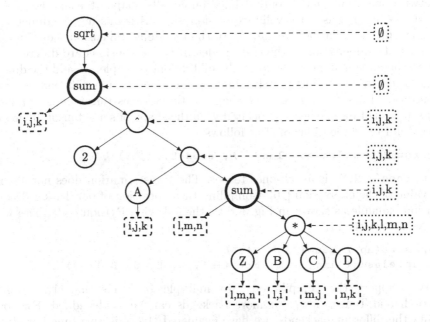

Fig. 2. Expression tree for the Tucker decomposition problem with different node types. Bold nodes indicate sum operations, dashed rectangle nodes denote indices, and all other nodes are either common operations or tensors. For operations and sum nodes, we show the indices (dimension) after performing the operation in dotted rectangles on the right side.

4   Transformation and Compilation into Einsum

The information that we need for the compilation is available in the expression tree. The compilation is performed in a recursive manner, mapping the label of a node to the corresponding library function and then continuing with the children of the node. There are, however, two special cases in the compilation process that cannot be dealt with by simple mappings.

First, at *sum* nodes, we check if the child node is a multiplication node. If this is the case, we combine the summation and multiplication in one operation. For example, during the compilation of our Tucker decomposition when we arrive at the node sum[l,m,n](Z[l,m,n]*B[l,i]*C[m,j]*D[n,k]), we can combine the summation and multiplication into

```
einsum('lmn,li,mj,nk->ijk', Z, B, C, D),
```

which is much shorter and easier to read than compiling the sum and multiplication individually

```
einsum('ijklmn->ijk',
  einsum('lmn,li,mj,nk->ijklmn', Z, B, C, D)).
```

Second, at binary operation nodes, the dimensions of the two operand tensors have to be equal, because as already mentioned, each binary operation is a pointwise operation, that is only defined on equally shaped tensors. For example, $A_{ijk} - E_{ijklmn}$ is not a valid expression, since A is missing the dimensions lmn. During compilation, we can verify the matching dimensions condition by checking the dimensions of the child nodes. If the dimensions are different, we extend the missing dimensions for each child. In our example, we add the dimensions lmn to node A. Fortunately, this is possible in Einsum notation by using an additional all-ones tensor. The shape of the all-ones tensor is determined, because each of its indices is associated with the shape of some tensor. Thus, we can extend the dimensions of A as follows

```
einsum('ijk,lmn->ijklmn', A, ones(L, M, N)),
```

where `ones(L,M,N)` is an all-ones tensor. The transformation does not change the value of the expression [10]. Compiling the expression of our Tucker decomposition example as shown in Fig. 3 gives the following Python code (here with the NumPy backend):

```
np.sqrt(np.einsum('ijk->',(A -
  np.einsum('lmn,li,mj,nk->ijk', Z, B, C, D))**2)).
```

We support the compilation into multiple backends, namely, NumPy, PyTorch and TensorFlow, but more backends can be easily added. For comparing the different backends, we have evaluated the objective function of our Tucker decomposition example on random tensors $A \in \mathbb{R}^{s,s,s}, Z \in \mathbb{R}^{10,10,10}, B \in \mathbb{R}^{10,s}, C \in \mathbb{R}^{10,s}, D \in \mathbb{R}^{10,s}$, with $s \in \{25, 50, 100, 150, 200\}$.

For the measurements, we use a machine with an Intel i9-10980XE 18-core processor (36 hyperthreads) running Ubuntu 20.04.5 LTS with 128 GB of RAM. Each core has a base frequency of 3.0 GHz and a max turbo frequency of 4.6 GHz, and supports the AVX-512 vector instruction set. Table 2 shows the relative speed-up of our compiled code using NumPy, TensorFlow and PyTorch, with respect to a baseline implementation with simple nested for-loops. This shows that, our approach can speed up the evaluation of tensor expressions up to four orders of magnitude.

Table 2. Relative speed-up of evaluating the objective function of the Tucker decomposition. The speed-up is computed as $S_s^{\text{method}} = T_s^{\text{baseline}} / T_s^{\text{method}}$, where *method* can be NumPy, TensorFlow or PyTorch and T is the runtime of the *method*.

Backend	$s = 25$	$s = 50$	$s = 100$	$s = 150$	$s = 200$
NumPy	221	248	283	268	278
TensorFlow	31	255	2 083	6 601	16 005
PyTorch	5 777	70 941	191 796	370 973	371 042

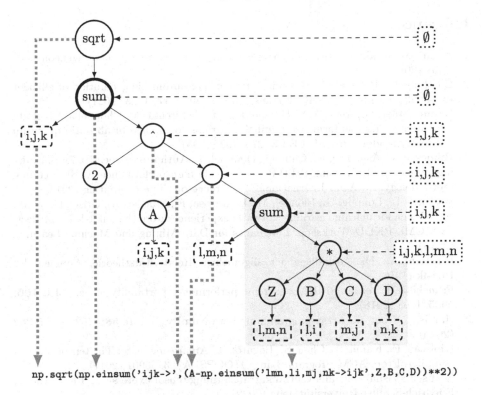

```
np.sqrt(np.einsum('ijk->',(A-np.einsum('lmn,li,mj,nk->ijk',Z,B,C,D))**2))
```

Fig. 3. Expression tree with compiled NumPy code for the Tucker decomposition problem. For each operation node, a dotted gray arrow points to the corresponding part in the code. Note, that the *sum* and child multiplication nodes are merged into one Einsum operation. The Einsum notation for such operations can be read-off from the tree.

5 Conclusion

We have presented a recursive algorithm for compiling tensor expressions into multiple Einsum backends. The compiled expressions evaluate orders of magnitude faster than their straightforward Python implementations using for-loops. To make our approach and its implementation easily accessible, the implementation and the code for the experiments are available at http://github.com/julien-klaus/tec.

Acknowledgements. This work was supported by the Carl Zeiss Foundation within the project *Interactive Inference* and from the Ministry for Economics, Sciences and Digital Society of Thuringia (TMWWDG), under the framework of the Landesprogramm ProDigital (DigLeben-5575/10-9).

References

1. Abadi, M., et al.: TensorFlow: Large-scale machine learning on heterogeneous systems (2015)
2. Barthels, H., Psarras, C., Bientinesi, P.: Linnea: automatic generation of efficient linear algebra programs. ACM Trans. Math. Softw. **47**, 1–26 (2021)
3. Baumgartner, G., Auer, A.A., Bernholdt, D.E., Bibireata, A., Choppella, V., et al.: Synthesis of high-performance parallel programs for a class of ab initio quantum chemistry models. Proc. IEEE **93**, 276–292 (2005)
4. Bilmes, J.A., Asanovic, K., Chin, C., Demmel, J.: Author retrospective for optimizing matrix multiply using PHiPAC: a portable high-performance ANSI C coding methodology. In: ACM International Conference on Supercomputing (2014)
5. Buitinck, L., Louppe, G., Blondel, M., Pedregosa, F., Mueller, A., Grisel, O., et al.: API design for machine learning software: experiences from the scikit-learn project. In: ECML PKDD Workshop: Languages for Data Mining and Machine Learning (2013)
6. Einstein, A.: Die Grundlage der allgemeinen Relativitätstheorie. Annalen der Physik (1916)
7. Franchetti, F., et al.: SPIRAL: extreme performance portability. Proc. IEEE **106**, 1935–1968 (2018)
8. Harris, C.R., et al.: Array programming with NumPy. Nature **585**(7825), 357–362 (2020)
9. Kjolstad, F., Kamil, S., Chou, S., Lugato, D., Amarasinghe, S.: The tensor algebra compiler. Proc. ACM Program. Lang. **1**(OOPSLA), 3133901 (2017)
10. Klaus, J.: Visualizing, analyzing and transforming tensor expressions, Ph. D. thesis, Friedrich-Schiller-Universität Jena (2022)
11. Klaus, J., Blacher, M., Giesen, J., Rump, P.G., Wiedom, K.: Compiling linear algebra expressions into efficient code. In: Computational Science - ICCS 2022–22nd International Conference (2022)
12. Nuzman, D., et al.: Vapor SIMD: auto-vectorize once, run everywhere. In: Proceedings of the CGO 2011 (2011)
13. Paszke, A., et al.: PyTorch: an imperative style, high-performance deep learning library. In: Advances in Neural Information Processing Systems, vol. 32, pp. 8024–8035 (2019)
14. Psarras, C., Barthels, H., Bientinesi, P.: The linear algebra mapping problem. arXiv preprint arXiv:1911.09421 (2019)
15. Sethi, R., Ullman, J.D.: The generation of optimal code for arithmetic expressions. J. ACM **17**, 715–728 (1970)
16. Spampinato, D.G., Fabregat-Traver, D., Bientinesi, P., Püschel, M.: Program generation for small-scale linear algebra applications. In: Proceedings of the 2018 International Symposium on Code Generation and Optimization (2018)
17. Tucker, L.R.: Some mathematical notes on three-mode factor analysis. Psychometrika **31**, 279–311 (1966). https://doi.org/10.1007/BF02289464
18. Vasilache, N., et al.: Tensor comprehensions: framework-agnostic high-performance machine learning abstractions. CoRR (2018)

r-softmax: Generalized Softmax with Controllable Sparsity Rate

Klaudia Bałazy[✉], Łukasz Struski, Marek Śmieja, and Jacek Tabor

Jagiellonian University, Kraków, Poland
klaudia.balazy@doctoral.uj.edu.pl

Abstract. Nowadays artificial neural network models achieve remarkable results in many disciplines. Functions mapping the representation provided by the model to the probability distribution are the inseparable aspect of deep learning solutions. Although softmax is a commonly accepted probability mapping function in the machine learning community, it cannot return sparse outputs and always spreads the positive probability to all positions. In this paper, we propose r-softmax, a modification of the softmax, outputting sparse probability distribution with controllable sparsity rate. In contrast to the existing sparse probability mapping functions, we provide an intuitive mechanism for controlling the output sparsity level. We show on several multi-label datasets that r-softmax outperforms other sparse alternatives to softmax and is highly competitive with the original softmax. We also apply r-softmax to the self-attention module of a pre-trained transformer language model and demonstrate that it leads to improved performance when fine-tuning the model on different natural language processing tasks.

Keywords: Sparse probability function · Controlling sparsity level · Softmax alternative

1 Introduction

Deep learning models excel in various domains, including computer vision, natural language processing (NLP), among others. Mapping the numerical output of a neural network into a probability distribution on a discrete set is crucial for many machine learning models. In classification, it describes the probability over classes, while in the attention mechanism for NLP, it indicates which words in a text are contextually relevant to other words. Softmax [4,11] is the well accepted standard for probability mapping, as it is easily evaluated and differentiated.

Although softmax [4,11] is the most widely applied probability mapping function in machine learning it cannot return sparse outputs. This means it assigns a non-zero probability to every component, making it difficult to identify insignificant elements. As a result, it does not return the number of relevant labels, making it necessary to define a threshold below which the label is considered negative. This threshold often necessitates a hyperparameter selection process, which adds computational overhead, especially in multi-label classification.

© The Author(s), under exclusive license to Springer Nature Switzerland AG 2023
J. Mikyška et al. (Eds.): ICCS 2023, LNCS 14074, pp. 137–145, 2023.
https://doi.org/10.1007/978-3-031-36021-3_11

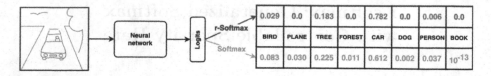

0.029	0.0	0.183	0.0	0.782	0.0	0.006	0.0
BIRD	PLANE	TREE	FOREST	CAR	DOG	PERSON	BOOK
0.083	0.030	0.225	0.011	0.612	0.002	0.037	10^{-13}

Fig. 1. Comparison of softmax and r-softmax for multi-label classification. Our r-softmax can produce zero probabilities indicating negative classes, making it more intuitive and interpretable than softmax.

Some of the noteworthy alternatives to softmax include the spherical softmax [3], the multinomial probit [1], softmax approximations [2] and Gumbel-Softmax [8]. As this paper introduces a novel sparse alternative to softmax, we focus on existing sparse probability mapping functions. Sparsemax [12] projects an input vector onto the probability simplex, producing sparse outputs. Unfortunately, it generally performs worse than softmax. In [9], a general family of probability mapping functions, including softmax and sparsemax, was defined, and a strategy for designing convex loss functions was proposed, including an alternative loss for sparsemax that improved its experimental performance.

We introduce r-softmax, a sparse alternative to the softmax function, that solves the issue of non-zero probabilities and provides intuitive control of the sparsity rate. Users can specify the sparsity rate r, representing the desired fraction of zero values, or train the model to determine its value using gradient descent. This eliminates the need for an additional mechanism like a threshold to identify positive labels in multi-label classification, as shown in Fig. 1.

We evaluate r-softmax as a function determining probabilities of classes in a multi-label classification problem and as a function determining the significance probability of elements in the attention mechanism. In multi-label classification, we benchmark r-softmax on real and synthetic datasets and find it outperforms other sparse alternatives to softmax, like sparsemax [12] and sparsehourglass [9], and competes with the original softmax using an optimal threshold for positive labels. For the attention mechanism, we replace the pre-trained transformer language model's softmax with r-softmax and show that our modification improves the fine-tuned model's performance on various NLP tasks.

2 Sparse Version of Softmax

We introduce r-softmax, a sparse probability mapping with controllable sparsity.

Problem Motivation. Probability mapping functions transform a real-valued response $x = (x_1, \ldots, x_n) \in \mathbb{R}^n$ of the neural network into a probability vector $p = (p_1, \ldots, p_n)$, where $p_i \geq 0$ and $\sum_{i=1}^{n} p_i = 1$. The most commonly used function to parameterize this probability is softmax:

$$\text{softmax}(x) = \left(\frac{\exp(x_1)}{\sum\limits_{i=1}^{n} \exp(x_i)}, \ldots, \frac{\exp(x_n)}{\sum\limits_{i=1}^{n} \exp(x_i)} \right).$$

The limitation of softmax is its inability to return sparse outputs with zero probabilities. Sparse outputs are very useful for example in (i) multi-label classification, where zero-probabilities indicate absence of labels, or (ii) self-attention layers, where they allow to ignore irrelevant keys.

The Weighted Softmax. With the motivation of constructing a probability mapping function capable of producing sparse output vectors, we introduce the weighted softmax as a generalization of the traditional softmax. By appropriately parameterizing its weights, the weighted softmax can reduce to a typical softmax or binary one-hot vector form where one coordinate contains a value of 1 and the rest are set to 0. It can also enable sparse probability mapping functions that fall between the two extremes.

Let $x = (x_1, \ldots, x_n) \in \mathbb{R}^n$ be a point, associated with vector of weights $w = (w_1, \ldots, w_n) \in \mathbb{R}^n_+$, where $\sum_{i=1}^{n} w_i > 0$. We define a weighted softmax by the following formula:

$$\text{softmax}(x, w) = \left(\frac{w_1 \exp(x_1)}{\sum\limits_{i=1}^{n} w_i \exp(x_i)}, \ldots, \frac{w_n \exp(x_n)}{\sum\limits_{i=1}^{n} w_i \exp(x_i)} \right).$$

The weighted softmax is a proper probability distribution (all components are non-negative and sum to 1) that reduces to classical softmax for a constant weight vector. Unlike softmax, it can return zero probability at some coordinates by setting the corresponding weight to zero, and produce one-hot vectors by setting exactly one non-zero weight.

To achieve a smooth transition between softmax and binary one-hot vectors, we construct t-softmax, in which all weights depend on a single parameter $t > 0$:

$$\text{t-softmax}(x, t) = \text{softmax}(x, w_t), \tag{1}$$

where $w_t = (w_t^1, \ldots, w_t^n)$ and $w_t^i = \text{ReLU}(x_i + t - \max(x))$. All weights w_i are nonnegative with at least one positive weight, satisfying the definition of weighted softmax. We can observe that w_i equals zero when the absolute difference between x_i and the maximum value $\max(x)$ is greater than or equal to t.

The following examines how t-softmax changes with varying values of t:

Theorem 1. *Let $x \in \mathbb{R}^n$ be a data point and let $t \in (0, \infty)$. Then*

- *the limit of t-softmax(x, t) is softmax(x) as t approaches infinity,*
- *if x reaches unique max at index k, then*

$$\text{t-softmax}(x, t) = \text{onehot}(\arg\max_i(x)), \tag{2}$$

for $t \in (0, x_k - \max_{i \neq k}(x)]$, where onehot$(i) \in \mathbb{R}^n$ is a vector consisting of zeros everywhere except k-th position where 1 is located.

Proof. The first property is a consequence of t-softmax$(x, t) = $ softmax$(x, \frac{w_t}{t})$, and if t approaches infinity then $\frac{w_t}{t}$ goes to 1, leading to softmax$(x, 1) = $ softmax(x). The last property follows directly from the definition of t-softmax.

Controlling the Number of Non-zero Values Using r-softmax. Instead of learning the optimal value of t as discussed above, there are situations in which we would like to have the ability to explicitly decide how many components returned by t-softmax should be zero. For this purpose, we introduce a parameter $r \in [0, 1]$ that we call a *sparsity rate*. The sparsity rate r represents the fraction of zero components we would like to obtain in the output probability distribution.

Recall that $w_i^t = 0$ for $i = 1, \ldots, n$ if $|x_i - \max(x)| \geq t$, as defined in Eq. (1). To control the number of non-zero weights, we can inspect the range $[\min(x), \max(x)]$ and select t such that $x_i < t < x_j$, where x_i and x_j are two distinct elements in $x_1, \ldots, x_n$, in increasing order. This will zero out the i-th component while keeping the j-th component non-zero. We can use the quantile of the set of x's coordinates $x_1, \ldots, x_n$ to implement this rule. The q-quantile quantile(x, q) outputs the value v in $[\min(x), \max(x)]$ such that the probability of $x_i : x_i \leq v$ equals q. If the quantile lies between x_i and x_j with indices i and j in the sorted order, we use linear interpolation to compute the result as $x_i + \alpha \cdot (x_j - x_i)$, where α is the fractional part of the computed quantile index. Setting $q = 0$ or $q = 1$ in quantile(x, q) will return the lowest or highest value of x, respectively.

We define r-softmax as a probability mapping function with a fixed sparsity rate $r \in [0, 1]$ as shown in Eq. (3), where $t_r = -$quantile$(x, r) + \max(x)$.

$$\text{r-softmax}\,(x, r) = \text{t-softmax}(x, t_r). \tag{3}$$

The parameterization of t_r ensures that a fraction of r components will be zero. When r-softmax(x, r) is applied to $x = (x_1, \ldots, x_n) \in \mathbb{R}^n$ and $r = \frac{k}{n}$, for $k \leq n$, the output will be a probability distribution with k zero coordinates. The r-softmax reduces model complexity by eliminating less probable components. Our experiments demonstrate the benefits of this approach, particularly in the self-attention mechanism for NLP tasks.

3   Experiments

We benchmarked r-softmax against basic softmax and other sparse probability mapping functions (such as sparsemax and sparsehouglass) in multi-label classification and self-attention blocks of pre-trained language models. Our results demonstrate that r-softmax outperforms the other functions in most cases.[1]

---

[1] Code with r-softmax is available at https://github.com/gmum/rsoftmax.

3.1   Alternative to Softmax in Multi-label Classification

Multi-label classification is a crucial problem in various domains, including image classification, where a single class description may not suffice due to multiple object classes in an image. In these models, the final element is a function that maps network output to a probability vector representing class membership probabilities. Softmax is often used, but other functions, such as those introducing sparse probability distributions, have also been investigated [9,12].

r-softmax for Multi-label Classification. To use r-softmax in multi-label classification, we need a proper loss function. We cannot directly apply cross-entropy loss as r-softmax can return zeros, which makes the logarithm undefined. We follow the approach used in [9] to overcome this. Given input x, logits z and a probability distribution $\eta = y/|y|_1$ over labels y, we define our loss function:

$$\mathcal{L}(z, y) = \|y \cdot (\text{r-softmax}(z, r) - \eta)\|_2^2 + \sum_{y_i=1, y_j=0} \max\left(0, \eta_i - (z_i - z_j)\right), \quad (4)$$

where y_i is i-th coordinate of the vector y (similarly for z and η). The first term approximates the positive label probability given by r-softmax$(z, r)_i$, while the second term pushes negative label logits away from positive ones by η_i.

Datasets. We tested different probability mapping functions on multi-label classification using synthetic data (similarly to [9]), with various possible output classes, average number of labels per sample and document length. We also evaluated considered functions on two real datasets (VOC 2007 [6] and COCO [10]).

Experimental Setting. We compare our method with sparsemax, sparsehourglass, and softmax. To handle the issue of obtaining zero values with softmax, we use various thresholds p_0 to consider a class as negative. For softmax we use cross-entropy loss, for sparsehourglass we use function proposed by [9] and for sparsemax we test functions: sparsemax+huber [12] and sparsemax+hinge [9].

We use a two-layer neural network for synthetic datasets and pre-trained ResNet models [7] with an additional linear layer for classification for real datasets. The models are trained with different learning rates. Our r-softmax is parameterized by the sparsity rate r that is found using an additional classification layer. The multi-label classification cost function for r-softmax includes cross-entropy loss to evaluate the correctness of the number of labels.

We report the best validation score after the models reach stability. The F1 score is used as quality metric for multi-label classification models, as it is based on the returned classes rather than target scores (e.g., mean average precision).

Results on Synthetic Datasets. Figure 2 compares using r-softmax and other functions for multi-label classification on the synthetic data validation set. In Fig. 2a, we show that r-softmax consistently outperforms other functions,

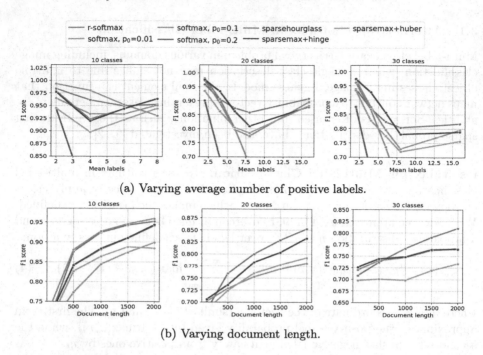

(a) Varying average number of positive labels.

(b) Varying document length.

Fig. 2. Different probability mapping functions for multi-label classification on synthetic datasets with varying output class numbers. For large output class numbers, r-softmax seems to be the most beneficial choice.

particularly when the dataset has a large number of possible output classes. In Fig. 2b, we demonstrate the impact of the average document length on model performance and show that r-softmax achieves the best results for most configurations, especially for a larger number of possible output classes. Our method, r-softmax, is the preferred choice as it provides the most benefits in the investigated scenarios.

Results on Real Datasets. We also evaluate r-softmax on real multi-label datasets VOC and COCO, see Table 1. Our r-softmax outperforms other sparse softmax alternatives and is competitive with the original softmax.

3.2   Alternative to Softmax in the Self-attention Block in Transformer-Based Model

Transformer-based [13] models are widely used in NLP and rely on attention mechanisms to identify important information. Self-attention modules in each layer apply softmax to generate weights for all tokens, but this assigns non-zero weight to even insignificant tokens. We propose replacing softmax with r-softmax function to obtain sparse probability distributions that allow the model to ignore irrelevant tokens, which we show to be beneficial in the following section.

Table 1. Performance of different probability mappings for multi-label classification on VOC and COCO datasets. Our r-softmax outperforms other sparse mapping functions and is competitive with softmax (with threshold selection).

Experimental setup	VOC (F1)	COCO (F1)
Softmax (p_0=0.05)	75.05	71.38
Softmax (p_0=0.10)	78.87	72.29
Softmax (p_0=0.15)	**79.43**	69.22
Softmax (p_0=0.20)	79.07	64.88
Softmax (p_0=0.30)	75.88	54.76
Sparsemax+huber	66.84	52.30
Sparsemax+hinge	71.91	65.67
Sparsehourglass	71.35	64.85
r-softmax	77.90	**72.56**

Table 2. Comparing different probability mapping functions in pretrained BERT$_{BASE}$ self-attention blocks for finetuning on GLUE benchmarks. Our r-softmax with a specific sparsity level outperforms other approaches.

Experiment setup	MRPC (Acc)	RTE (Acc)	SST-2 (Acc)	QNLI (Acc)	QQP (Acc)
Softmax	84.56	68.95	92.32	**91.76**	91.12
Sparsemax	68.38	52.71	79.82	55.57	77.18
Sparsehourglass	68.38	52.71	79.24	70.99	76.04
r-softmax	**85.54**	**71.84**	**92.89**	91.73	**91.13**

Experimental Setting. We experiment with BERT$_{BASE}$ [5] language model and focus on probability mapping function in each self-attention block during fine-tuning. We compare the performance of baseline softmax with sparsemax, sparsehourglass, and r-softmax replacements on GLUE benchmark classification tasks [14]. The final best validation score is reported after fine-tuning for 5 epochs for MRPC and 3 epochs for others, with different learning rates. We linearly increase hyperparameter r for r-softmax from 0 to desired sparsity $r \in 0.05, 0.1, 0.15, 0.2, 0.5$ during training.

Results. The results for GLUE tasks obtained by the best run in the grid search are summarized in Table 2. Using r-softmax instead of softmax generally improves the performance of the fine-tuned transformer-based model. Sparsemax and sparsehourglass do not perform very well in this application. We found that introducing a small sparsity rate produced the best results for r-softmax. The optimal sparsity rates for the QQP, MRPC, QNLI, RTE, and SST-2 tasks were $r = 0.1, 0.15, 0.15, 0.2, 0.2$ respectively. These results suggest that eliminating irrelevant elements is beneficial, but excluding too many can cause the model to lose important context or hinder gradient flow during learning.

4   Conclusions

We proposed r-softmax, a generalization of softmax that produces sparse probability distributions with a controllable sparsity rate. We applied r-softmax to multi-label classification and self-attention tasks and showed that it outperforms or is highly competitive with baseline models.

Acknowledgements. The work of Klaudia Bałazy and Łukasz Struski was supported by the National Centre of Science (Poland) Grant No. 2020/39/D/ST6/01332. The research of Jacek Tabor was carried out within the research project "Bio-inspired artificial neural network" (grant no. POIR.04.04.00-00-14DE/18-00) within the Team-Net program of the Foundation for Polish Science co-financed by the European Union under the European Regional Development Fund. The work of Marek Śmieja was supported by the National Centre of Science (Poland) Grant No. 2022/45/B/ST6/01117. Klaudia Balazy is affiliated with Doctoral School of Exact and Natural Sciences at the Jagiellonian University.

References

1. Albert, J.H., Chib, S.: Bayesian analysis of binary and polychotomous response data. J. Am. Stat. Assoc. **88**(422), 669–679 (1993)
2. Bouchard, G.: Efficient bounds for the softmax function and applications to approximate inference in hybrid models. In: NIPS Workshop for Approximate Bayesian Inference in Continuous/Hybrid Systems (2007)
3. de Brébisson, A., Vincent, P.: An exploration of softmax alternatives belonging to the spherical loss family. arXiv preprint arXiv:1511.05042 (2015)
4. Bridle, J.S.: Probabilistic interpretation of feedforward classification network outputs, with relationships to statistical pattern recognition. In: Soulié, F.F., Hérault, J. (eds.) Neurocomputing. NATO ASI Series, vol. 68, pp. pp. 227–236. Springer, Heidelberg (1990). https://doi.org/10.1007/978-3-642-76153-9_28
5. Devlin, J., et al.: BERT: pre-training of deep bidirectional transformers for language understanding, vol. 1, pp. 4171–4186. ACL (2019)
6. Everingham, M., et al.: The PASCAL Visual Object Classes Challenge 2007 (VOC2007) Results (2007)
7. He, K., et al.: Deep residual learning for image recognition. In: Proceedings of the IEEE Conference on Computer Vision and Pattern Recognition, pp. 770–778 (2016)
8. Jang, E., et al.: Categorical reparameterization with Gumbel-softmax. arXiv preprint arXiv:1611.01144 (2016)
9. Laha, A., et al.: On controllable sparse alternatives to softmax. NeurIPS (2018)
10. Lin, T.-Y., et al.: Microsoft COCO: common objects in context. In: Fleet, D., Pajdla, T., Schiele, B., Tuytelaars, T. (eds.) ECCV 2014. LNCS, vol. 8693, pp. 740–755. Springer, Cham (2014). https://doi.org/10.1007/978-3-319-10602-1_48
11. Luce, R.D.: Individual choice behavior: a theoretical analysis. Courier Corporation (2012)

12. Martins, A., et al.: From softmax to sparsemax: a sparse model of attention and multi-label classification. In: ICML2016, vol. 48, pp. 1614–1623. JMLR.org (2016)
13. Vaswani, A., et al. s: Attention is all you need. In: NeurIPS, pp. 5998–6008 (2017)
14. Wang, A., et al.: GLUE: A multi-task benchmark and analysis platform for natural language understanding (2018)

Solving Uncertainly Defined Curvilinear Potential 2D BVPs by the IFPIES

Andrzej Kużelewski$^{(\boxtimes)}$ ⓘ, Eugeniusz Zieniuk ⓘ, and Marta Czupryna ⓘ

Institute of Computer Science, University of Bialystok, Ciolkowskiego 1M,
15-245 Bialystok, Poland
{a.kuzelewski,e.zieniuk,m.czupryna}@uwb.edu.pl

Abstract. The paper presents the interval fast parametric integral equations system (IFPIES) applied to model and solve uncertainly defined curvilinear potential 2D boundary value problems with complex shapes. Contrary to previous research, the IFPIES is used to model the uncertainty of both boundary shape and boundary conditions. The IFPIES uses interval numbers and directed interval arithmetic with some modifications previously developed by the authors. Curvilinear segments in the form of Bézier curves of the third degree are used to model the boundary shape. However, the curves also required some modifications connected with applied directed interval arithmetic. It should be noted that simultaneous modelling of boundary shape and boundary conditions allows for a comprehensive approach to considered problems. The reliability and efficiency of the IFPIES solutions are verified on 2D complex potential problems with curvilinear domains. The solutions were compared with the interval solutions obtained by the interval PIES. All performed tests indicated the high efficiency of the IFPIES method.

Keywords: Interval fast parametric integral equations system ·
Interval numbers · Directed interval arithmetic · Uncertainty

1 Introduction

The interval fast parametric integral equations system (IFPIES) [1] is a robust numerical tool for solving uncertainly defined boundary value problems (BVPs). It is based on successors of the original parametric integral equations system (PIES) such as the interval parametric integral equations system (IPIES) [2] and the fast parametric integral equations system (FPIES) [3].

The IPIES was developed to solve uncertainly defined problems. In traditional modelling and solving BVPs, the shape of the boundary, boundary conditions and some other parameters of the considered domain (i.e. material properties) are defined precisely using real numbers. In practice, we should measure some physical quantities to obtain these data. However, the accuracy of determining the physical quantity is affected by, e.g. gauge reading error, inaccuracy of measurement instruments or approximations of the models used in the analysis

© The Author(s), under exclusive license to Springer Nature Switzerland AG 2023
J. Mikyška et al. (Eds.): ICCS 2023, LNCS 14074, pp. 146–153, 2023.
https://doi.org/10.1007/978-3-031-36021-3_12

of measurements. Therefore, we should consider the uncertainty of the domain description in modelling and solving BVPs.

Classical mathematical models require exact values of the data. Therefore, the direct consideration of uncertainty is not possible. However, many known methods were modified to consider uncertainty (e.g. [4–6]). Some of them applied interval numbers and interval arithmetic to the methods of modelling and solving BVPs. Therefore, the interval finite element method (IFEM) [7], the interval boundary element method (IBEM) [8], and the IPIES were obtained. However, either the IFEM or the IBEM considered only the uncertainty of material parameters or boundary conditions. Only in a few papers some parameters of the boundary shape (e.g. radius or beam length) were uncertainly defined. Therefore, the possibility of simultaneous consideration of all uncertainties mentioned above in the IPIES [2] becomes a significant advantage.

Although the IPIES has other advantages (e.g., defining the boundary by curves widely used in computer graphics that uses a small number of interval control points) inherited from the PIES, there are also some disadvantages. The main is connected with dense non-symmetric coefficient matrices and Gaussian elimination applied to solve the final system of algebraic equations. Unfortunately, the application of interval arithmetic and interval numbers also significantly slows the computational speed and utilizes more memory (RAM) than in the PIES. Usually, to accelerate computations, parallel computing methods (e.g. MPI or OpenMP) and graphics processing unit (GPU) for numerical calculations (such as CUDA or OpenCL) are commonly used. In our previous papers, we also proposed parallelization of the PIES by OpenMP [9] and CUDA [10] to reduce the time of computations. However, the use of these methods did not affect reducing RAM consumption. Therefore, to solve complex (large-scale) uncertainly defined problems using a standard personal computer (PC), we had to apply the fast multipole method (FMM) [11] to the IPIES in a similar way as in the FPIES. The FMM allows to significant reduction the RAM utilization [12]. It also reduces computation time.

The main goal of this paper is to present the IFPIES applied for numerical solving of 2D potential complex BVPs with uncertainly defined boundary shapes and boundary conditions. Simultaneous consideration of both uncertainties in describing the domain becomes a comprehensive approach to solving practical BVPs. The efficiency and accuracy of the IFPIES are tested on the potential problems with curvilinear domains.

2   Modelling Uncertainties in the IFPIES

In previous papers (e.g. [2]), we described some problems during the application of either classical [13] or directed [14] interval arithmetic for modelling boundary problems with uncertainties. Hence, we also proposed some modifications during the application of the directed interval arithmetic, such as mapping arithmetic operators to the positive semi-axis into the IPIES (clearly described in [2]). The same strategy was applied in the IFPIES.

The general form of the IFPIES formula was presented in [1], has the following form:

$$\frac{1}{2}u_l(\widehat{s}) = \sum_{j=1}^{n} \mathbb{R}\left\{ \int_{s_{j-1}}^{s_j} \widehat{U}_{lj}^{*(c)}(\widehat{s},s)p_j(s)J_j^{(c)}(s)ds \right\} -$$

$$\sum_{j=1}^{n} \mathbb{R}\left\{ \int_{s_{j-1}}^{s_j} \widehat{P}_{lj}^{*(c)}(\widehat{s},s)u_j(s)J_j^{(c)}(s)ds \right\}, \qquad (1)$$

$$l = 1, 2, ..., n, \ s_{l-1} \le \widehat{s} \le s_l, \ s_{j-1} \le s \le s_j,$$

where: $\widehat{s}$ and s are defined in the parametric coordinate system (as real values), s_{j-1} (s_{l-1}) correspond to the beginning while s_j (s_l) to the end of interval segment S_j (S_l), n is the number of parametric segments that creates a boundary of the domain in 2D, $\widehat{U}_{lj}^{*(c)}(\widehat{s},s)$ and $\widehat{P}_{lj}^{*(c)}(\widehat{s},s)$ are modified interval kernels (complex function), $J_j^{(c)}(s)$ is the interval Jacobian, $u_j(s)$ and $p_j(s)$ are interval parametric boundary functions on individual segments S_j of the interval boundary, $\mathbb{R}$ is the real part of complex function.

In this paper, for modelling uncertainly defined boundary shapes, curvilinear segments in the form of interval Bézier curves of the third degree are used:

$$S_j(s) = a_j s^3 + b_j s^2 + c_j s + d_j, \quad 0 \le s \le 1, \qquad (2)$$

where vector $S_j(s) = [S_j^{(1)}(s), S_j^{(2)}(s)]^T$ is composed of two interval components connected with the direction of coordinates in 2D Cartesian reference system: $S_j^{(1)} = [\underline{S}_j^{(1)}, \overline{S}_j^{(1)}], S_j^{(2)} = [\underline{S}_j^{(2)}, \overline{S}_j^{(2)}]$. The $j = \{1, 2, ..., n\}$ is the number of segment created boundary, and s is a variable in the parametric reference system. Coefficients a_j, b_j, c_j, d_j have also form of vectors composed of two intervals (similarly to $S_j(s)$). They are computed using interval points describing particular segments of the boundary as presented in Fig. 1:

$$a_j = P_{e(j)} - 3P_{i2(j)} + 3P_{i1(j)} - P_{b(j)}, \quad b_j = 3(P_{i2(j)} - 2P_{i1(j)} + P_{b(j)}),$$

$$c_j = 3(P_{i1(j)} - P_{b(j)}), \quad d_j = P_{b(j)},$$

where coordinates of all points P, regardless of their subscript, have the form of a vector of intervals:

$$P = [P^{(1)}, P^{(2)}]^T = \left[[\underline{P}^{(1)}, \overline{P}^{(1)}], [\underline{P}^{(2)}, \overline{P}^{(2)}] \right]^T.$$

Boundary conditions are uncertainly defined using interval boundary functions $u_j(s)$ and $p_j(s)$ which are approximated by the following series:

$$u_j(s) = \sum_{k=0}^{N} u_j^{(k)} L_j^{(k)}(s), \quad p_j(s) = \sum_{k=0}^{N} p_j^{(k)} L_j^{(k)}(s), \qquad (3)$$

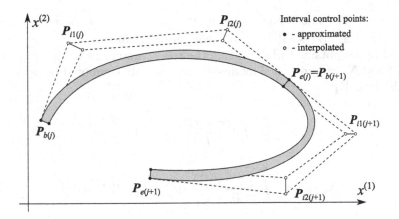

Fig. 1. The interval Bézier curve of the third degree used to define a segment of the boundary

where $\boldsymbol{u}_j^{(k)} = [\underline{u}_j^{(k)}, \overline{u}_j^{(k)}]$ and $\boldsymbol{p}_j^{(k)} = [\underline{p}_j^{(k)}, \overline{p}_j^{(k)}]$ are unknown or given interval values of boundary functions in defined points of the segment j, N - is the number of terms in approximating series (4), which approximated boundary functions on the segment j and $L_j^{(k)}(s)$ – the base functions (Lagrange polynomials) on segment j.

3   Solving the IFPIES

The process of solving the IFPIES is connected with the application of the FMM into the PIES. The FMM uses the tree structure to transform interactions between segments describing boundary into interactions between the cells (groups of segments). Also, the Taylor expansion is used to approximate the PIES's modified kernels. The process of applying the FMM into the PIES is clearly described in [3].

At last, integrals in (1) have the following form [1]:

$$\int_{s_{j-1}}^{s_j} \widehat{U}_{lj}^{*(c)}(\widehat{s}, s)\, p_j(s)\, \boldsymbol{J}_j^{(c)}(s)ds = \frac{1}{2\pi}\sum_{l=0}^{N_T}(-1)^l\cdot$$

$$\left\{\sum_{k=0}^{N_T}\sum_{m=l}^{N_T}\frac{(k+m-1)!\cdot \boldsymbol{M}_k(\boldsymbol{\tau}_c)}{(\boldsymbol{\tau}_{el}-\boldsymbol{\tau}_c)^{k+m}}\cdot\frac{(\boldsymbol{\tau}_{el}'-\boldsymbol{\tau}_{el})^{m-l}}{(m-l)!}\right\}\frac{(\widehat{\boldsymbol{\tau}}-\boldsymbol{\tau}_{el}')^l}{l!},$$

$$\int_{s_{j-1}}^{s_j} \widehat{P}_{lj}^{*(c)}(\widehat{s}, s)\, u_j(s)\, \boldsymbol{J}_j^{(c)}(s)ds = \frac{1}{2\pi}\sum_{l=0}^{N_T}(-1)^l\cdot \qquad (4)$$

$$\left\{\sum_{k=1}^{N_T}\sum_{m=l}^{N_T}\frac{(k+m-1)!\cdot \boldsymbol{N}_k(\boldsymbol{\tau}_c)}{(\boldsymbol{\tau}_{el}-\boldsymbol{\tau}_c)^{k+m}}\cdot\frac{(\boldsymbol{\tau}_{el}'-\boldsymbol{\tau}_{el})^{m-l}}{(m-l)!}\right\}\frac{(\widehat{\boldsymbol{\tau}}-\boldsymbol{\tau}_{el}')^l}{l!}.$$

where: N_T is the number of terms in the Taylor expansion, $\hat{\tau} = S_l^{(1)}(\hat{s}) + iS_l^{(2)}(\hat{s})$, $\tau = S_j^{(1)}(s) + iS_j^{(2)}(s)$, complex interval points τ_c, τ_{el}, τ_c', τ_{el}' are midpoints of leaves obtained while tracing the tree structure (see [15]). Expressions $M_k(\tau_c)$ and $N_k(\tau_c)$ are called moments (and they are computed twice only) and have the form [1]:

$$M_k(\tau_c) = \int_{s_{j-1}}^{s_j} \frac{(\tau - \tau_c)^k}{k!} p_j(s) J_j^{(c)}(s) ds,$$

$$N_k(\tau_c) = \int_{s_{j-1}}^{s_j} \frac{(\tau - \tau_c)^{k-1}}{(k-1)!} n_j^{(c)} u_j(s) J_j^{(c)}(s) ds.$$

(5)

where $n_j^{(c)} = n_j^{(1)} + in_j^{(2)}$ the complex interval normal vector to the curve created segment j.

The IFPIES, similarly to the original PIES, is written at collocation points whose number corresponds to the number of unknowns. However, in the IFPIES, the system of algebraic equations $A \cdot x = b$ is produced implicitly, i.e. only the result of multiplication of the matrix A by the vector of unknowns x is obtained, contrary to the explicit form in the PIES. Therefore, an iterative GMRES solver [16] modified by the application of directed interval arithmetic directly integrated with the FMM was applied in the IFPIES. Also, the GMRES solver was applied to the IPIES to obtain a more reliable comparison.

4   Numerical Results

The example is the gear-shaped plate presented in Fig. 2. The problem is described by Laplace's equation. The boundary contains 2 048 curvilinear interval segments. Interval boundary conditions are also presented in Fig. 2 (where u - Dirichlet and p - Neumann boundary conditions). Tests are performed on a PC based on Intel Core i5-4590S with 16 GB RAM. Application of the IPIES and the IFPIES are compiled by g++ 7.5.0 (-O2 optimization) on 64-bit Ubuntu Linux OS (kernel 5.4.0).

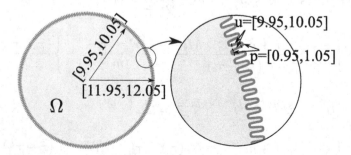

Fig. 2. Considered the gear-shaped modelled by curvilinear segments

The first research focused on finding the optimal number of tree levels in the IFPIES from the speed of computations and RAM utilization point of view. Approximation of the IFPIES kernels uses 25 terms in the Taylor series, and the GMRES tolerance is equal to 10^{-8}.

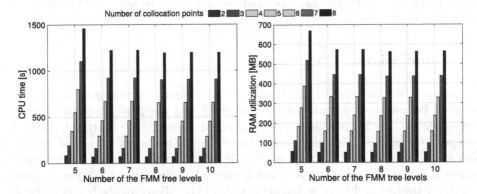

Fig. 3. Comparison of computation time and RAM utilization of the IFPIES for different tree levels

As can be seen from Fig. 3, the shortest time of computations and the smallest number of utilized memory for all numbers of collocation points is obtained for 8 tree levels. Therefore, that number is used in other research.

The subsequent research focused on the CPU time, RAM utilization and accuracy of the IFPIES compared to the IPIES only due to the lack of methods of solving problems with uncertainly defined boundary shape and boundary conditions. The same number of terms in the Taylor series and the value of GMRES tolerance as in the previous example are used. The number of collocation points is the same in each segment, which is changed from 2 to 8. Therefore, we should solve the system of 4 096 to 16 384 equations, respectively.

Table 1. Comparison between the IFPIES and the IPIES

Number of		CPU time [s]		RAM utilization [MB]		MSE	
col. pts	eqs	*IFPIES*	*IPIES*	*IFPIES*	*IPIES*	inf	sup
2	4 096	70.73	131.24	51.31	390	0.0	0.0
3	6 144	159.22	310.25	97	896	$4.41 \cdot 10^{-15}$	$1.43 \cdot 10^{-15}$
4	8 192	284.84	565.46	157	1 578	$6.29 \cdot 10^{-15}$	$5.96 \cdot 10^{-15}$
5	10 240	449.05	910.28	236	2 455	$6.23 \cdot 10^{-14}$	$2.55 \cdot 10^{-14}$
6	12 288	653.36	1 349.4	327	3 523	$1.17 \cdot 10^{-11}$	$2.39 \cdot 10^{-13}$
7	14 336	900.19	1 892.83	437	4 787	$3.33 \cdot 10^{-13}$	$6.74 \cdot 10^{-11}$
8	16 384	1 192.09	2 559.22	561	6 243	$6.95 \cdot 10^{-10}$	$6.14 \cdot 10^{-11}$

As can be seen from Table 1, the IFPIES is about 2 times faster and uses up to 10 times less RAM than the IPIES. To prove the accuracy of the proposed method, the mean square error (MSE) between the lower and upper bound (infimum and supremum) of the IFPIES and the IPIES solutions are computed. The IFPIES is as accurate as the IPIES. The mean square error (MSE) between both methods is very low and does not exceed 10^{-9}. Hence, the IFPIES is as accurate as the IPIES.

5 Conclusions

The paper presents the IFPIES in solving 2D potential curvilinear uncertainly defined boundary value problems. The IFPIES was previously applied in modelling and solving 2D polygonal potential problems with the uncertainly defined shape of the boundary. Applied interval modelling of boundary shape and boundary conditions allows for including the uncertainty of measurement data (measurement errors) in calculations, which is ignored in classic practical design. Also, applying the fast multipole technique in the IFPIES allows for the highly efficient solving of complex engineering problems on a standard PC in a reasonable time. However, the real power of the IFPIES is connected with low RAM utilization. The IPIES for solving the problems with a system of 16 384 equations uses over 6 GB RAM, while the IFPIES requires about 0.5 GB of RAM.

Obtained results suggest that the direction of research should be continued. Our further research should cover problems modelled by other than Laplace's equations.

References

1. Kużelewski, A., Zieniuk, E., Czupryna, M.: Interval modification of the Fast PIES in solving 2D potential BVPs with Uncertainly defined polygonal boundary shape. In: Groen, D., de Mulatier, C., Paszynski, M., Krzhizhanovskaya, V.V., Dongarra, J.J., Sloot, P.M.A. (eds.) Computational Science – ICCS 2022. ICCS 2022. Lecture Notes in Computer Science, vol. 13351. Springer, Cham (2022). https://doi.org/10.1007/978-3-031-08754-7_3
2. Zieniuk, E., Kapturczak, M., Kużelewski, A.: Modification of interval arithmetic for modelling and solving uncertainly defined problems by interval parametric integral equations system. In: Shi, Y., et al. (eds.) ICCS 2018. LNCS, vol. 10862, pp. 231–240. Springer, Cham (2018). https://doi.org/10.1007/978-3-319-93713-7_19
3. Kużelewski, A., Zieniuk, E.: The fast parametric integral equations system in an acceleration of solving polygonal potential boundary value problems. Adv. Eng. Softw. **141**, 102770 (2020)
4. Fu, C., Zhan, Q., Liu, W.: Evidential reasoning based ensemble classifier for uncertain imbalanced data. Inf. Sci. **578**, 378–400 (2021)
5. Wang, C., Matthies, H.G.: Dual-stage uncertainty modeling and evaluation for transient temperature effect on structural vibration property. Computat. Mech. **63**(2), 323–333 (2019)

6. Gouyandeh, Z., Allahviranloo, T., Abbasbandy, S., Armand, A.: A fuzzy solution of heat equation under generalized Hukuhara differentiability by fuzzy Fourier transform. Fuzzy Sets Syst. **309**, 81–97 (2017)
7. Ni, B.Y., Jiang, C.: Interval field model and interval finite element analysis. Comput. Methods Appl. Mech. Eng. **360**, 112713 (2020)
8. Zalewski, B., Mullen, R., Muhanna, R.: Interval boundary element method in the presence of uncertain boundary conditions, integration errors, and truncation errors. Eng. Anal. Boundary Elem. **33**(4), 508–513 (2009)
9. Kużelewski, A., Zieniuk, E.: OpenMP for 3D potential boundary value problems solved by PIES. In: Simos, T.E., et al. (eds.) 13th International Conference of Numerical Analysis and Applied Mathematics ICNAAM 2015, AIP Conference Proceedings, vol. 1738, 480098. AIP Publishing LLC., Melville (2016). https://doi.org/10.1063/1.4952334
10. Kuzelewski, A., Zieniuk, E., Boltuc, A.: Application of CUDA for acceleration of calculations in boundary value problems solving using PIES. In: Wyrzykowski, R., Dongarra, J., Karczewski, K., Waśniewski, J. (eds.) PPAM 2013. LNCS, vol. 8385, pp. 322–331. Springer, Heidelberg (2014). https://doi.org/10.1007/978-3-642-55195-6_30
11. Greengard, L.F., Rokhlin, V.: A fast algorithm for particle simulations. J. Comput. Phys. **73**(2), 325–348 (1987)
12. Liu, Y.J., Nishimura, N.: The fast multipole boundary element method for potential problems: a tutorial. Eng. Anal. Boundary Elem. **30**(5), 371–381 (2006)
13. Moore, R.E.: Interval Analysis. Prentice-Hall, Englewood Cliffs, New York (1966)
14. Markov, S.M.: On directed interval arithmetic and its applications. J. Univ. Comput. Sci. **1**(7), 514–526 (1995)
15. Kużelewski, A., Zieniuk, E.: Solving of multi-connected curvilinear boundary value problems by the fast PIES. Comput. Methods Appl. Mech. Eng. **391**, 114618 (2022)
16. Saad, Y., Schultz, M.H.: GMRES: a generalized minimal residual algorithm for solving non-symmetric linear systems. SIAM J. Sci. Stat. Comput. **7**, 856–869 (1986)

Improving LocalMaxs Multiword Expression Statistical Extractor

Joaquim F. Silva[✉][iD] and Jose C. Cunha[iD]

NOVA LINCS, NOVA School of Science and Technology, Costa da Caparica, Portugal
{jfs,jcc}@fct.unl.pt

Abstract. LocalMaxs algorithm extracts relevant Multiword Expressions from text *corpora* based on a statistical approach. However, statistical extractors face an increased challenge of obtaining good practical results, compared to linguistic approaches which benefit from language-specific, syntactic and/or semantic, knowledge. First, this paper contributes to an improvement to the LocalMaxs algorithm, based on a more selective evaluation of the cohesion of each Multiword Expressions candidate with respect to its neighbourhood, and a filtering criterion guided by the location of stopwords within each candidate. Secondly, a new language-independent method is presented for the automatic self-identification of stopwords in *corpora*, requiring no external stopwords lists or linguistic tools. The obtained results for LocalMaxs reach Precision values of about 80 % for English, French, German and Portuguese, showing an increase of around $12-13$ % compared to the previous Local-Maxs version. The performance of the self-identification of stopwords reaches high Precision for top-ranked stopword candidates.

Keywords: Multiword Expressions · Statistical Extractor · LocalMaxs algorithm · Stopwords

1 Introduction

Multiword Expressions (MWEs) are sequences of consecutive words in text *corpora*, that is n-grams, having a meaningful content, semantically more or less strong, e.g., the 2-gram "financial crisis", the 3-gram "world population growth", the 5-gram "International Conference on Computer Science". MWEs are useful: *i*) for unsupervised clustering and classification of documents; *ii*) as document keywords; *iii*) for indexing; *iv*) in Statistical Machine Translation. The extraction of MWEs tries to identify semantically relevant n-grams occurring in a *corpus*, by symbolic (morphosyntactic or semantic) or statistical methods [1,2]. The latter have the advantage of language-independence. The evaluation of MWE relevance is subjective, always made with reference to some context, which may include thematic terms, e.g. "global warming", specific to subject fields, or more

This work is supported by NOVA LINCS (UIDB/04516/2020) with the financial support of FCT.IP.

general expressions in a language, occurring across multiple domains, still characterised as semantic units. So, the evaluation of the quality of the output of an automatic extractor should be done by a human jury. LocalMaxs multiword statistical extractor [2] is based on two main aspects: *i*) the cohesion between the words within each *n*-gram; *ii*) a criterion to evaluate the relative cohesion of the *n*-gram with respect to its neighbourhood. In fact, if the words of an *n*-gram are cohesive among themselves, then the *n*-gram is probably semantically strong, and therefore relevant. We propose to improve the Precision of LocalMaxs extractor [2], through a more selective evaluation of the cohesion of each MWE candidate. Apart from that, we present an automatic language-independent stopwords identification method, for general purpose application. In the following we present the improvement to LocalMaxs, the new method for stopwords identification and conclusions. A guide for model reproducibility is in https://github.com/OurName1234/OurFiles/releases/tag/v1.

2   Improvement to LocalMaxs Statistical Extractor

2.1   Background on MWE Statistical Extraction

The foundations of automatic term extraction have been developed for several decades [1, 3–5]. Statistical regularities in natural language texts have been identified, leading to several statistical association/cohesion measures, e.g. MI [5], χ^2 [6], *Dice* [7], *SCP* [2], *Loglike* [4], c-value, among others, and their application in text processing tasks. Alternatively, language-dependent linguistic or combined linguistic-statistical approaches, e.g. [8], may achieve better quality results. Some methods use Machine Learning approaches. For example, [9] used C4.5 algorithm to classify candidates, but it strongly depends on the high quality of the training data. Concerning the statistical approaches, Xtract [1] identifies collocations reporting around 80 % Precision, using MI measure [5]. In [10], lexical collocations were extracted with *t-score*, reporting about 60 % Precision. Another approach, *mwetoolkit* [11] for MWE extraction provides integration with web search engines and with a machine learning tool for the creation of supervised MWE extraction models if annotated data is available. *Mwetoolkit* uses *t-score*, MI [5], *Dice* [7] and *Log-likelihood* [4] measures, and the highest F1[1] value reported was 30.57 % (56.83 % Precision and 20.91% Recall).

MWE extraction methods may use stopwords in some step of the extraction pipeline, namely either associated to the *corpora* preprocessing or to the candidate selection. However, unlike the proposal in Sect. 3, most proposals rely on predefined stopwords lists, which may not be available for some languages.

The Cohesion Measures. The degree of relevance of an MWE tends to be reflected in the degree of cohesion among its component words. Some widely used cohesion measures, such as $MI(.)$ [5], $\chi^2(.)$ [6], $Dice(.)$ [7] and $SCP(.)$ [2], were originally designed to measure the cohesion between just two consecutive

---

[1] $F1 = \frac{2 \cdot Precision \times Recall}{Precision + Recall}$.

words. The improvement we propose applies to n-grams with more than two words, $(w_1 \ldots w_n)$, with $n \geq 2$, following [2] to generalise the cohesion measures.

$$MI\_f((w_1 \ldots w_n)) = \log\left(\frac{p(w_1 \ldots w_n)}{\frac{1}{n-1} \sum\limits_{i=1}^{i=n-1} p(w_1 \ldots w_i)\, p(w_{i+1} \ldots w_n)}\right) \tag{1}$$

$$\chi^2\_f((w_1 \ldots w_n)) = \frac{\left(N\, f(w_1 \ldots w_n) - Avp\right)^2}{Avp\,(N - Avx)\,(N - Avy)} \tag{2}$$

where

$$Avp = \frac{1}{n-1} \sum\limits_{i=1}^{i=n-1} f(w_1 \ldots w_i)\, f(w_{i+1} \ldots w_n) \tag{3}$$

$$Avx = \frac{1}{n-1} \sum\limits_{i=1}^{i=n-1} f(w_1 \ldots w_i) \qquad Avy = \frac{1}{n-1} \sum\limits_{i=2}^{i=n} f(w_i \ldots w_n) \tag{4}$$

$$Dice\_f((w_1 \ldots w_n)) = \frac{2\, f(w_1 \ldots w_n)}{\frac{1}{n-1} \sum\limits_{i=1}^{i=n-1} f(w_1 \ldots w_i) + f(w_{i+1} \ldots w_n)} \tag{5}$$

$$SCP\_f((w_1 \ldots w_n)) = \frac{f(w_1 \ldots w_n)^2}{\frac{1}{n-1} \sum\limits_{i=1}^{i=n-1} f(w_1 \ldots w_i)\, f(w_{i+1} \ldots w_n)} \tag{6}$$

2.2   The Previous Version

LocalMaxs previous version [2] is reviewed in Definition 1.

Definition 1. *Let* $W = w_1, \ldots w_n$ *be an n-gram,* $g(.)$ *a generic cohesion function and* $frq(W)$ *the absolute frequency of occurrence of W in a corpus. Let:* $\Omega_{n-1}(W)$ *be the set of* $g(.)$ *values of all contiguous (n−1)-grams contained in W;* $\Omega_{n+1}(W)$ *be the set of* $g(.)$ *values of all contiguous (n+1)-grams containing W;* $len(W)$ *be the length (number of words) of W. W is an MWE if and only if,*

$$frq(W) > 1 \,\wedge\, \big(for \forall x \in \Omega_{n-1}(W), \forall y \in \Omega_{n+1}(W)$$

$$(len(W) = 2 \wedge g(W) > y) \,\vee\, (len(W) > 2 \wedge g(W) > \frac{x+y}{2})\big)$$

Thus, previous version of LocalMaxs can be seen as a function $LocalMaxs(g(.))$ parameterised by a cohesion function $g(.)$. The generic function $g(.)$ in Definition 1 can be instantiated with any cohesion measure as long as it is extended for n-grams, with $n \geq 2$. This previous version, besides extracting semantically strong MWE, e.g. "climate change" and "inflation rate in eurozone", it also extracts some n-grams, e.g. "even though", "having established", which, despite frequently co-occurring in *corpora*, are semantically irrelevant. These false MWEs prevent the previous extractor from reaching higher Precision values.

2.3 The Improved Version of LocalMaxs

Two modifications to the criterion for selecting MWEs in LocalMaxs are proposed, as described in Definition 2: a) using a generalised mean for the evaluation of the relative cohesion; b) using a filtering criterion based on stopwords.

Definition 2. *Let: $W = w_1, \ldots w_n$ be an n-gram in a corpus C; $frq(W)$ the absolute frequency of occurrence of W in C. Let function $LocalMaxs(g(.), p, S)$ have three parameters: $g(.)$, a generic cohesion function; an integer $p \geq 1$; and S, the set of stopwords in corpus C. Let: $\Omega_{n-1}(W)$ be the set of $g(.)$ values of the two contiguous $(n-1)$-grams contained in W in C; $\Omega_{n+1}(W)$ be the set of $g(.)$ values of all contiguous $(n+1)$-grams that contain W in the same corpus; $len(W)$ be the length (number of words) of W. W is an MWE if and only if,*

$$\Big(\big(len(W) = 2 \wedge g(W) \geq \max(\Omega_{n+1}(W))\big) \vee$$

$$\big(len(W) > 2 \wedge g(W) \geq (\frac{\max(\Omega_{n-1}(W))^p + \max(\Omega_{n+1}(W))^p}{2})^{\frac{1}{p}}\big)\Big)$$

$$\wedge \ frq(W) > 1 \ \wedge w_1 \notin S \wedge w_n \notin S$$

The interpretation of the two modifications proposed is as follows.

a) Using a generalised mean: instead of the arithmetic mean in the condition $g(W) > \frac{\max(\Omega_{n-1}(W)) + \max(\Omega_{n+1}(W))}{2}$, implicit in Definition 1, we propose the generalised mean in the condition $g(W) \geq (\frac{\max(\Omega_{n-1}(W))^p + \max(\Omega_{n+1}(W))^p}{2})^{\frac{1}{p}}$. In this last condition, p is an integer parameter of the extractor. Thus, If $p > 1$, it implies that the cohesion of the n-gram W necessary to consider W as an MWE, tends to be greater than in the case of the arithmetic mean, since for $p > 1$, $g(W)$ will have to be closer to the largest value between $\max(\Omega_{n-1}(W))$ and $\max(\Omega_{n+1}(W))$. In fact, for the arithmetic mean, it suffices that $g(W)$ is superior by an infinitesimal to the value that is at the same distance between $\max(\Omega_{n-1}(W))$ and $\max(\Omega_{n+1}(W))$. Therefore, being more demanding with regard to the value of $g(W)$, this new condition tends to produce fewer False Positives, which is reflected in a higher Precision value of the MWE selected by the extractor. However, this new condition can decrease the Recall value, since some true MWE may not have a sufficient $g(W)$ value to be selected as such.

b) Using a filtering criterion based on stopwords: in a language, usually, a subset of words — called stopwords — is identified, having low semantic content and occurring very frequently in *corpora*, e.g.: "the", "in", "of", in English. This improvement to LocalMaxs considers the judicious location of the stopwords within the MWE candidates to be selected by the extractor. Thus, Definition 2 includes the additional requirement that, for W to be an MWE, its leftmost and rightmost words must not be in the *corpus* stopwords set, that is $w_1 \notin S \wedge w_n \notin S$. This helps rejecting MWE candidates as "even though", "regarded as", "having established" .... In Sect. 2.4, a comparison of results shows the effectiveness of these modifications, LocalMaxs impicled version being seen as a function $LocalMaxs(g(.), p, S)$ parameterised by a cohesion function $g(.)$, an integer p representing the exponent of the generalised mean, and the set of stopwords S, used in the selection criterion of the Definition 2.

2.4   Experimental Evaluation of the Improved LocalMaxs Version

The Corpora. We used two English *corpora* (EN6.0 Mw and EN0.5 Mw) with 6 019 951 and 500 721 words; and one *corpus* for each of the following languages: French (FR6.1 Mw) with 6 079 056 words, German (DE6.0 Mw) with 6 036 023 words, and Portuguese (PT6.1 Mw) with 6 061 118 words. These were collected from https://linguatools.org/tools/corpora/wikipedia-monolingual-corpora/.

The Evaluation Criterion. The evaluation was made by a jury of three persons. Concerning the Precision of the whole set of n-grams extracted as MWEs by each of the LocalMaxs versions, a large enough random sample, Q, was taken. An n-gram in Q is considered a True Positive MWE, if and only if the majority of evaluators agree. The ratio given by the size of TP (the set of True Positive MWEs) over the size of Q, estimates the Precision. Concerning Recall, a large enough random sample of true MWEs (the *TrREs* set), is obtained by human evaluation from each *corpus*. Let R be the set given by the intersection of *TrREs* with the full set extracted by each LocalMaxs version. The ratio of the size of R over the size of *TrREs* estimates the Recall.

Discussion of Experimental Results. Lines under $LocalMaxs(g(.), p, S)$ correspond to Definition 2 with parameters instantiated as described in Table 1 caption. Results show the obtained improvements, where all four cohesion measures present better Precision and Recall than the previous version of the extractor. For all five *corpora*, χ^2_f presents the best values for the combined F1 score metric. The results show that the Precision obtained by the improved algorithm reaches about 80 %, consistently across all considered language *corpora*. Besides the language-independence nature of the algorithm (definitions 1 and 2), the results encourage using this improvement in other languages.

The results show that the replacement of the arithmetic mean with the generalised mean in Definition 2 introduces improvements, as the best result was obtained for $p = 2$ for every cohesion measure. When exponent $p = 1$, the arithmetic mean in Definition 1 is equivalent to the generalised mean in Definition 2. So, the difference between $LocalMaxs(g(.), 1, S)$ and $LocalMaxs(g(.))$ is only due to the restriction that the leftmost and the rightmost word of an MWE can not be a stopword. Then, we conclude that both the aforementioned stopwords restriction and the use of the generalised mean (with $p = 2$) have important contributions to the overall Precision improvement to LocalMaxs, and their orthogonal individual effects add together. In fact, for the example of EN6.0 Mw *corpus* and χ^2_f, we have an increase from 68.5 % (for $LocalMaxs(\chi^2\_f)$) to 74.2 % (for $LocalMaxs(\chi^2\_f), 1, S)$), that is 5.7 % due to stopwords restriction, and another increase from 74.2 % to 80.5 % (for $LocalMaxs(\chi^2\_f, 2, S)$), that is 6.3 % due to $p = 2$. Similar contributions happen to the other language *corpora* as shown in Table 1. Overall, the modifications from Definition 2 lead to a significant improvement in Precision, around $12-13\,\%$.

Table 1. Precision and Recall results for each version of LocalMaxs algorithm considering its parameters, for the *corpora* in Sect. 2.4. Lines under the $LocalMaxs(g(.), p, S)$ header present results of the improved version (Definition 2), using four different $g(.)$ cohesion functions, different values for the p exponent, and S instantiated, for each *corpus*, by method in Sect. 3. Results under the $LocalMaxs(g(.))$ header refer to the previous version (Definition 1) for each of the same four cohesion functions.

		$LocalMaxs(g(.), p, S)$									
		EN 6.0Mw		EN 0.5Mw		FR 6.1Mw		DE 6.0Mw		PT 6.1Mw	
$g(.)$	p	Prc	Rec	Prc	Rec	Prc	Rec	Prc	Rec	Prc	Rec
χ^2_f	1	74.2	77.0	73.3	78.5	72.8	75.3	74.3	75.8	73.0	75.0
	2	**80.5**	75.0	**79.3**	75.8	**79.5**	73.8	**80.0**	74.3	**79.8**	73.8
	3	75.0	70.5	74.8	71.0	73.8	69.8	75.0	69.3	75.3	70.0
$Dice\_f$	1	63.3	81.5	61.8	82.3	63.5	82.0	62.0	80.3	62.8	82.3
	2	69.3	80.0	68.0	80.8	72.0	81.0	68.8	78.8	69.5	80.3
	3	64.0	76.5	62.8	77.5	65.3	77.3	64.0	78.3	65.0	76.0
MI_f	1	69.3	50.5	68.3	52.0	70.3	49.3	69.0	50.8	69.8	50.3
	2	76.8	48.0	75.3	49.5	77.0	47.3	76.0	48.3	75.3	47.0
	3	69.8	44.5	68.3	45.3	69.5	45.5	68.3	46.0	70.3	45.0
SCP_f	1	70.0	55.5	70.0	56.8	71.3	54.8	72.0	56.3	71.3	54.0
	2	78.0	55.0	77.5	56.0	78.8	54.0	79.3	53.8	77.8	55.3
	3	71.3	51.0	72.3	51.5	73.0	51.8	71.8	50.8	72.0	50.3
		$LocalMaxs(g(.))$									
		EN 6.0Mw		EN 0.5Mw		FR 6.1Mw		DE 6.0Mw		PT 6.1Mw	
		Prc	Rec	Prc	Rec	Prc	Rec	Prc	Rec	Prc	Rec
χ^2_f	–	**68.5**	77.0	**67.3**	78.3	**66.8**	76.8	**67.0**	75.8	**66.5**	76.0
$Dice\_f$	–	60.5	79.5	59.3	80.8	58.8	80.8	58.5	79.8	60.0	79.0
MI_f	–	63.8	50.5	62.5	50.8	64.0	48.3	62.0	48.8	63.3	49.3
SCP_f	–	67.0	55.5	65.8	56.8	65.5	56.0	66.3	55.3	64.8	54.8

3   Identifying the Set of Stopwords in *Corpora*

We propose a new general purpose method for the automatic self-identification of stopwords from each *corpus*. This can be also used to instantiate the set S in Definition 2, preserving its language independence, unlike almost all proposals, which depend on predefined stopwords lists, not always available.

Background on Stopwords Identification. To identify the stopwords in a *corpus*, morphosyntactic approaches are language-dependent and pose difficulties in handling multilingual *corpora*, unlike statistical approaches. According to [12], the stopword lists based on Zipf's law are reliable but very expensive to carry out. In [13], to identify stopwords in *corpora*, the authors use the Rocchio classifier and the *IDF* (Inverse Document Frequency) measure, which requires the frequency of each word per document, an information that may not be available. Another proposal [14] identifies stopwords in *corpora* by using Term Frequency (TF). Stopwords are not all equally meaningless and, in [14], are ordered by meaningless according to the used criterion. Using the stopwords lists from http://snowball.tartarus.org/algorithms/LANG/stop for several languages, they measure the Precision for several cases of the top n stopwords selected.

A New Stopwords Identification Criterion. The proposed stopwords identification criterion combines two factors: the *number of neighbours each word* has in the *corpus*; and the *number of syllables of each word*. Concerning the first factor, number of neighbours of each word, by analysing the occurrence of each word in a *corpus*, it is observed that the higher the number of its distinct left or right neighbouring n-grams, the less meaningful the word is. When considering the number of distinct 2-gram neighbours of each word, this produced better results, by empirical analysis, than any other n-gram size, for the purpose of word meaningfulness. Indeed, large numbers of neighbours are associated with meaningless words; small number of neighbours suggest non stopwords. The second factor, the number of syllables in a word, reflects the effort to pronounce the word. In fact, words with a higher number of syllables tend to occur less often.

The method relies on the combination of the two factors, defining a $NeigSyl(w)$ function, given by dividing the number of 2-gram neighbours each distinct word w has in a *corpus*, by the number of syllables of w. We verified that the higher the value of this quotient, the less meaningful the word. By ordering the words in ranks (r) according to decreasing $NeigSyl(w)$ values, this method allows separating stopwords from content words. Because the list sorted by decreasing $NeigSyl(.)$ values from a *corpus* does not show a clear boundary between stopwords and content words, we need to define a cutoff to automatically separate these two word groups, corresponding to the rank b such that $b = \text{argmax}_r \left(|f(r+\Delta_k) - f(r)| \geq \Delta_k \right)$, where: $f(r) = NeigSyl(word(r))$; $word(r)$ stands for the word corresponding to rank r; Δ_k is a fixed integer distance in the rank ordering. To ignore irregular sharp variations in $NeigSyl(.)$ for neighbouring ranks, we set Δ_k greater than 1; the value leading to the best results was $\Delta_k = 4$.

Experimental Evaluation. Using the same reference stopwords lists as in [14], and considering four *corpora*, each one having around 6 Million words (presented in Sect. 2.4), when comparing the Precision values for our criterion *vs* the approach in [14], we obtained the following pairs of values, for the lists of top ranked 50 & 100 stopword candidates: (1.0 & .87) *vs* (0.82 & 0.74) for English; (.96 & .83) *vs* (.74 & .61) for French; (.96 & .82) *vs* (.84 & .74) for German; (.96 & .84) *vs* (.78 & .62) for Portuguese. The proposed criterion achieves a better Precision than using Term Frequency alone [14], showing values around 10 % higher.

4   Conclusions

An improvement is proposed to the LocalMaxs statistical extractor that significantly increases its Precision values, as shown for four natural languages. The proposed method introduces the requirement that an MWE has no stopwords in its leftmost or rightmost words, as well as the replacement of the arithmetic mean with the generalised mean in the criterion for evaluating the relative importance of the cohesion of n-grams in the context of their neighbourhoods. Language independence is maintained and the Precision of the LocalMaxs extractor improves by around 12–13%, reaching about 80 % for the language *corpora* tested. Also, a novel method for automatically identifying the stopwords of each working *corpus* is proposed, which considers both the number of syllables and the number of neighbouring n-grams of each word, leading to improved Precision in the identification of top-n ranked stopwords by language-independent statistical methods, when compared to using Term Frequency only. The method does not depend on predefined stopwords lists, possibly unavailable for some languages. It can also be used as a general tool for monolingual or multilingual *corpora*.

References

1. Smadja, F.: Retrieving collocations from text: Xtract. Comput. Linguist. **19**, 143–177 (1993)
2. Silva, J., Mexia, J., Coelho, C., Lopes, G.: A statistical approach for multilingual document clustering and topic extraction form clusters. Pliska Studia Mathematica Bulgarica **16**, 207–228 (2004)
3. Justeson, J., Katz, S.: Technical terminology: some linguistic properties and an algorithm for identification in text. Nat. Lang. Eng. **1**(1), (1995)
4. Dunning, T.: Accurate methods for the statistics of surprise and coincidence. Comput. Linguist. **19**(1), 61–74 (1993)
5. Church, K., Hanks, P.: Word association norms, mutual information, and lexicography. Comput. Linguist. **16**(1), 22–29 (1990)
6. Gale, W., Church, K.W.: In concordance for parallel texts. In: Proceedings of the Seventh Annual Conference of the UW Centre of the new OED and Text Research, Using Corpora, pp. 40–62. Oxford (1991)
7. Dice, L.R.: Measures of the amount of ecologic association between species. Ecology **26**(3), 297–302 (1945)
8. Pazienza, M.T., Pennacchiotti, M., Zanzotto, F.M.: Terminology extraction: an analysis of linguistic and statistical approaches. In: Knowledge Mining, pp. 255–279. Springer, Berlin Heidelberg (2005). https://doi.org/10.1007/3-540-32394-5_20
9. Witten, I., Paynter, G., Frank, E., Gutwin, C., Nevill-Manning, C.: KEA: practical automatic keyphrase extraction. CoRR, cs.DL/9902007 (1999)
10. Evert, S., Krenn, B.: Using small random samples for the manual evaluation of statistical association measures. Comput. Speech Lang. **19**(4), 450–466 (2005). Special issue on Multiword Expression
11. Ramisch, C., Villavicencio, A., Boitet, C.: mwetoolkit: a framework for multiword expression identification. In: Proceedings of the 7th International Conference on Language Resources and Evaluation (LREC'10). ELRA (2010)

12. Tsz-Wai Lo, R., He, B., Ounis, I.: Automatically building a stopword list for an information retrieval system. J. Digit. Inf. Manag. **3**, 3–8 (2005)
13. Makrehchi, M., Kamel, M.S.: Automatic extraction of domain-specific stopwords from labeled documents. In: Macdonald, C., Ounis, I., Plachouras, V., Ruthven, I., White, R.W. (eds.) ECIR 2008. LNCS, vol. 4956, pp. 222–233. Springer, Heidelberg (2008). https://doi.org/10.1007/978-3-540-78646-7_22
14. Ferilli, S., Izzi, G.L., Franza, T.: Automatic stopwords identification from very small corpora. In: Stettinger, M., Leitner, G., Felfernig, A., Ras, Z.W. (eds.) ISMIS 2020. SCI, vol. 949, pp. 31–46. Springer, Cham (2021). https://doi.org/10.1007/978-3-030-67148-8_3

PIES in Multi-region Elastic Problems Including Body Forces

Agnieszka Bołtuć[(✉)] and Eugeniusz Zieniuk

Institute of Computer Science, University of Bialystok, Bialystok, Poland
{a.boltuc,e.zieniuk}@uwb.edu.pl

Abstract. The paper presents the formulation of a parametric integral equation system (PIES) for boundary problems with piecewise homogeneous media and body forces. The multi-region approach is used, in which each region is treated separately and modeled globally by a Bezier surface. Each subarea can have different material properties, and different body loads can act on it. Finally, they are connected by dedicated conditions. Two examples are solved to confirm the effectiveness of the proposed approach. The results are compared with analytical solutions and those received from other numerical methods (FEM, BEM).

Keywords: PIES · Multi-region analysis · Body forces · Bezier surfaces

1 Introduction

In many elasticity problems, the considered body is made up of different materials or mechanical material properties (e.g. Young's modulus, Poisson's ratio) vary in some piecewise fashion. The approach to this type of problem differs significantly in the best-known numerical methods for solving them. The oldest and most popular finite element method (FEM) [1,2] is characterized by the general strategy regardless of the problem. Therefore, the whole body is always divided into finite elements, for which the same or various material properties can be posed. It can be said that different materials are taken into account automatically. On the other hand, the number of required elements and nodes is the largest here. The boundary element method (BEM) [3,4] reduces the problem size, because modeling is limited to the boundary only. Bodies in which material properties vary piecewise are approximated here by a system of homogeneous bodies. Such an approach is called multi-region formulation. However, the methods based on the boundary integral generate the dense resulting matrix, while in FEM, it is sparse. The method developed by the authors, called the parametric integral equation system (PIES), also applies to bodies made up of subdomains with different material properties [5]. PIES significantly reduces the number of input data necessary for modeling the shape, because only the boundary is created using parametric curves. Thus, discretization into elements is eliminated.

© The Author(s), under exclusive license to Springer Nature Switzerland AG 2023
J. Mikyška et al. (Eds.): ICCS 2023, LNCS 14074, pp. 163–170, 2023.
https://doi.org/10.1007/978-3-031-36021-3_14

But what about problems with piecewise constant material properties in which body forces also appear? The approach in FEM does not change, because various forces can be posed on different finite elements. In BEM, the domain should be created. This process is technically very similar to discretization in FEM, but the used elements are called cells. There are some body forces for which only the boundary can be defined, because the domain integral is transformed into the boundary. However, the general approach requires the application of cells for each region separately. PIES has been used to solve problems with body forces, but only in homogeneous domains [6,7]. It significantly simplifies the way of modeling, because the whole area on which the body forces act is created using a single Bezier surface of any degree. This, in turn, is reduced to just setting control points.

This paper presents PIES for multi-region elastic problems, but also including body forces. Each considered region with different material properties and various body forces is modeled globally using the Bezier surface and its control points. Then they are connected by the compatibility and equilibrium conditions at the common interface. PIES formula for such problems is developed together with a numerical solution scheme. Two examples are solved, confirming the approach's effectiveness and accuracy.

2   PIES for Elasticity with Body Forces

The isotropic linear elastic solids with body forces are considered. The governing equations, known as Navier's equations, are expressed by

$$\mu u_{i,jj} + \frac{\mu}{1 - 2\nu} u_{j,ji} + b_i = 0, \tag{1}$$

where μ is the shear modulus, ν is the Poisson's ratio, u_i is the displacement, b_i is the body force and commas represent differentiation with respect to spatial coordinates ($i, j = 1, 2$ for 2D).

The Eq. (1) can be transformed into the corresponding integral equation using the strategy described in [8]

$$0.5u_l(\bar{s}) = \sum_{j=1}^{n} \int_{s_{j-1}}^{s_j} \left\{ \mathbf{U}_{lj}^*(\bar{s}, s)\mathbf{p}_j(s) - \mathbf{P}_{lj}^*(\bar{s}, s)\mathbf{u}_j(s) \right\} J_j(s) ds$$
$$+ \sum_{k=1}^{m} \int_{\Omega_k} \bar{U}_l^*(\bar{s}, \mathbf{y})\mathbf{b}_k(\mathbf{y}) d\Omega(\mathbf{y}), \tag{2}$$

where $J_j(s)$ is the Jacobian of transformation to the parametric reference system, $l, j = 1..n$, $s_{l-1} \le \bar{s} \le s_l$, $s_{j-1} \le s \le s_j$ and n, m are the number of segments and subregions.

Functions $\mathbf{u}_j(s)$, $\mathbf{p}_j(s)$ describe the distribution of displacements and tractions on the boundary, respectively. On each segment, one is prescribed, and the other is to be solved. The function $\mathbf{b}(\mathbf{y})$ represents the vector of body forces.

The fundamental solutions for displacement $\mathbf{U}^*_{lj}(\bar{s}, s)$ and traction $\mathbf{P}^*_{lj}(\bar{s}, s)$ are presented explicitly in [8]. The solution $\bar{U}^*_l(\bar{s}, \mathbf{y})$ can be found in [6].

PIES' boundary and domain are defined in a parametric reference system using well-known computer graphics tools like curves and surfaces [9]. They are analytically incorporated into the formalism of PIES by functions η [8], which represent the distance between two boundary/domain points.

The collocation method [10] is used for the PIES solution. Unknown boundary functions are approximated by series with arbitrary basis functions, e.g., Legendre or Chebyshev polynomials [6–8]. The number of expressions in the series affects the accuracy of the solutions. Only this parameter should be increased to reduce the error, without interfering with the shape and re-discretization. After substituting approximation series to (2) and writing the resulting equation for all collocation points, the PIES matrix form is obtained

$$\mathbf{H}\mathbf{u} = \mathbf{G}\mathbf{p} + \mathbf{b}, \tag{3}$$

where $\mathbf{H}$, $\mathbf{G}$ are square matrices of boundary integrals from (2), while $\mathbf{b}$ is the vector of integrals over the domain.

After solving Eq. (3) only the boundary solutions are obtained. To calculate results within the domain, the integral identities for displacements and stresses are used. They are presented in [5,8] (without body forces).

3   Multi-region Formulation

As mentioned in previous sections, for bodies with piecewise homogeneous materials, it is necessary to consider more than one region. Then such a body can be approximated by a system of homogeneous bodies with different material constants. To illustrate the problem, consider a region Ω consisting of two subregions Ω_1 and Ω_2. They are separated by an interface boundary Γ_I and surrounded by respectively Γ_1 and Γ_2. Each region has different mechanical material properties and different body forces can act on it (Fig. 1). The analysis of such a problem consists in considering each region separately [3], which results in the following matrix equations for Ω_1 and Ω_2

$$\begin{bmatrix} \mathbf{H}_1 & \mathbf{H}_1^I \end{bmatrix} \begin{bmatrix} \mathbf{u}_1 \\ \mathbf{u}_1^I \end{bmatrix} = \begin{bmatrix} \mathbf{G}_1 & \mathbf{G}_1^I \end{bmatrix} \begin{bmatrix} \mathbf{p}_1 \\ \mathbf{p}_1^I \end{bmatrix} + \{\mathbf{b}_1\}, \tag{4}$$

$$\begin{bmatrix} \mathbf{H}_2 & \mathbf{H}_2^I \end{bmatrix} \begin{bmatrix} \mathbf{u}_2 \\ \mathbf{u}_2^I \end{bmatrix} = \begin{bmatrix} \mathbf{G}_2 & \mathbf{G}_2^I \end{bmatrix} \begin{bmatrix} \mathbf{p}_2 \\ \mathbf{p}_2^I \end{bmatrix} + \{\mathbf{b}_2\}, \tag{5}$$

where $\mathbf{H}_1$, $\mathbf{G}_1$ contains the boundary integrals over Γ_1, $\mathbf{H}_2$, $\mathbf{G}_2$ over Γ_2, while $\mathbf{H}_1^I$, $\mathbf{G}_1^I$ over the interface boundary Γ_I from Ω_1 and $\mathbf{H}_2^I$, $\mathbf{G}_2^I$ over Γ_I from Ω_2.

The Eqs. (4) and (5) are connected by compatibility and equilibrium conditions at the interface boundary Γ_I

$$\mathbf{u}_1^I = \mathbf{u}_2^I = \mathbf{u}^I, \mathbf{p}_1^I = \mathbf{p}_2^I = \mathbf{p}^I. \tag{6}$$

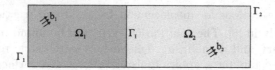

Fig. 1. Multiregion body.

The inclusion of (6) in (4) and (5) results in the following matrix equation

$$
\begin{bmatrix} H_1 & H_1^I & -G_1^I & 0 \\ 0 & H_2^I & G_2^I & H_2 \end{bmatrix} \begin{bmatrix} u_1 \\ u^I \\ p^I \\ u_2 \end{bmatrix} = \begin{bmatrix} G_1 & 0 \\ 0 & G_2 \end{bmatrix} \begin{bmatrix} p_1 \\ p_2 \end{bmatrix} + \begin{Bmatrix} b_1 \\ b_2 \end{Bmatrix}. \tag{7}
$$

Matrices in (7) are block-banded, and each block corresponds to one region. The matrix on the left contains overlaps between blocks at the common interface.

When calculating solutions in the domain, the separate identity is required for each region. Which one is used depends on the region of the examined point.

4   Modeling of Regions

The way of modeling problems with various material properties strongly depends on the method used. In FEM [1,2], the procedure is general and does not differ from the case where the material properties are constant over the whole area. It comes from the fact that the domain is divided into finite elements on which various properties can be applied (Fig. 2a). BEM [3,4] proposes two approaches: modeling only the boundary of separate regions (if the integral over the domain is transformed to the boundary) by boundary elements (Fig. 2b) or modeling the boundary and the domains of the regions using cells (Fig. 2c). Both approaches could be implemented in PIES, but the way of modeling the shape is entirely different, because no division into elements or cells is required.

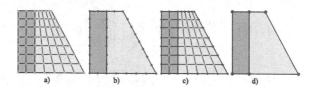

Fig. 2. Modeling in a) FEM, b) BEM (only the boundary), c) BEM (the boundary and the domain), d) PIES.

As shown in Fig. 2d, in PIES, each region is modeled by a separate Bezier surface [9]. They can be of various degrees, e.g., the bilinear surface, used for polygonal shapes, requires only 4 corner points to be given. For curved region shapes, surfaces of higher degrees are used. However, the third-degree patch has sufficient design flexibility, and higher degrees require longer processing time. It consists of 12 control points for modeling the boundary and 4 responsible for the shape in 3D (in 2D problems, they are not essential). However, the formalism of PIES allows the use of surfaces other than Bezier.

Comparing the approaches presented in Fig. 2 on the sample shape and discretization schemas, it can be seen that PIES significantly reduces the set of required nodes (from many elements/cells in FEM/BEM to a few corner points). In PIES, the accuracy depends on the number of expressions in the approximation series, not the number of data used for modeling. For the same shape, always minimal data set is required. It comes from the fact that the shape approximation is separated from the approximation of the solutions.

5 Tested Examples

5.1 Example 1

The first problem concerns elastic analysis with a centrifugal load. A square plate (Fig. 3) rotates about the x-axis with angular velocity $\omega = 100rad/s$. There is a discontinuity in the density distribution and material properties: $\rho_1 = 1 \times 10^{-6}$, $E = 210GPa$, $v = 0.2$ for $0 \le y \le 50$ and $\rho_2 = 2 \times 10^{-6}$, $E = 160GPa$, $v = 0.3$ for $50 < y \le 100$.

As seen in Fig. 3, the problem in PIES is modeled using two bilinear Bezier surfaces, one for each region. They have been defined by 6 control (corner) points. Boundary functions (u and p) are approximated by series with Chebyshev polynomials of the first kind with 7 expressions for each boundary segment.

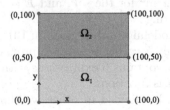

Fig. 3. Square plate with two materials rotating about the x-axis.

The exact solution for the one-dimensional problem is given in [11] by

$$
\begin{aligned}
\sigma_y &= \frac{\rho_1\omega^2}{8}\left[2L(L-2y) - (L-2y)^2\right] + \rho_2\omega^2\frac{3L^2}{8}, 0 \le y \le 50, \\
\sigma_y &= \rho_2\omega^2\left[L(L-y) - \frac{(L-y)^2}{2}\right], 50 < y \le 100.
\end{aligned}
\tag{8}
$$

The considered geometry in other numerical methods requires posing elements, e.g., in [11] 80 constant boundary elements. In this paper, the FEM model is used for comparison purposes. Two meshes are applied, with 400 and 1600 8-noded quadrilateral finite elements. Finally, the number of solved equations is 112 in PIES and 2562/9922 in FEM (depending on the mesh).

Figure 4 shows the comparison of stress distribution at $x = 50$. Only FEM results with a finer mesh are presented, as the average relative error obtained at the considered cross-section equals 1.54% (for a coarser mesh it is 1.6%). For the PIES method, the error equals 1.02%. Additionally, the computational times are compared. PIES solved the problem in 0.772 s, while FEM in 2.05 s.

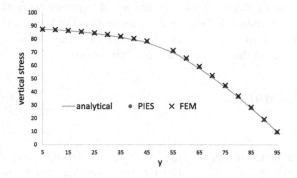

Fig. 4. Stress σ_y distribution at $x = 50$.

5.2   Example 2

The second example concerns a footing on a horizontal layer of soil (Fig. 5a) under a uniform compressive load with a magnitude of 1 ton/m and a self-weight (of the footing and soil). The considered material properties are: $E = 1 \times 10^4$ ton/m^2, $v = 0.4$, $\gamma = 2$ ton/m^3 for the soil, and $E = 2 \times 10^6$ ton/m^2, $v = 0.2$, $\gamma = 2.4$ ton/m^3 for the footing.

As shown in Fig. 5b, modeling both regions in PIES using 2 bilinear Bezier surfaces requires posing only 8 corner points. Approximation series for u and p for each boundary segment contains 6 expressions with Chebyshev polynomials of the first kind used as basis functions.

The same problem can be defined in other well-known methods like FEM or BEM. However, the number of elements and nodes is much higher than the number of corner points applied in PIES. For example, in the model created in BEM, 75 quadratic boundary elements are used, while FEM requires 243 8-noded finite elements.

The first test concerns the analysis of displacements u_y at the top boundary of the region Ω_2. Table 1 presents the results obtained by PIES after solving the system of 144 equations, by BEM with 362 equations and FEM with 1612 equations. The vertical stress values for both regions at $x = 0$ are also obtained and compared with BEM and FEM (Fig. 6).

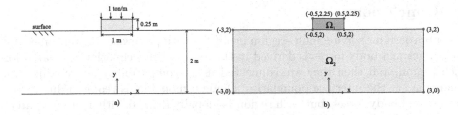

Fig. 5. Footing on a horizontal layer: a) definition, b) modeling.

Table 1. Vertical displacements along the boundary $y = 2$.

x	FEM	BEM	PIES
3	−0.003278	−0.003277	−0.003277
2.75	−0.003279	−0.003279	−0.003278
2.5	−0.003283	−0.003281	−0.003283
2.25	−0.003292	−0.003290	−0.003291
2	−0.003305	−0.003306	−0.003305
1.75	−0.003326	−0.003327	−0.003326
1.5	−0.003359	−0.003357	−0.003360
1.25	−0.003409	−0.003402	−0.003411
1	−0.003489	−0.003480	−0.003490
0.75	−0.003623	−0.003622	−0.003621
0.5	−0.003964	−0.003876	−0.003915

The PIES solutions presented in Table 1 and Fig. 6 are very close to FEM and BEM results. In element methods, however, a much larger number of input data necessary for modeling and the number of equations in the solved system are observed. The evaluated computational times are 1.43 s for FEM and 0.954 s for PIES. For BEM, it is less than 1 s, but accurate reading is impossible.

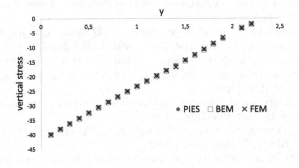

Fig. 6. Vertical stresses in the cross-section $x = 0$.

6 Conclusions

The approach for solving multi-region problems with piecewise constant material properties and body forces is derived in this paper. PIES equation is created for each region, and then they are connected using compatibility and equilibrium conditions at the interface boundaries. Domain modeling is required due to the presence of body forces, but each region is globally defined with a surface patch. Therefore, dividing them into elements or cells is eliminated.

The proposed formulation is tested on two examples: centrifugal and gravitational loads. The results are compared with exact and numerical solutions (FEM, BEM). They agree (or even are more accurate) with a significantly smaller number of data for modeling and solving the problem. Moreover, the computational time in the examined cases also proves in favor of PIES.

The limitation of the method can be a very complex shape that cannot be modeled with a single Bezier patch. Then different types of surfaces that are more flexible and can define domains other than quadrangular should be applied. This is one of the future research goals. It is also planned to use the method to solve elastoplastic problems with piecewise constant material properties and then also with body forces. In addition, in this paper, authors did not analyze singularities that occur in problems with multiple domains due to, e.g., reentrant corners. This issue should also be the subject of a detailed examination.

References

1. Ameen, M.: Computational elasticity. Alpha Science International Ltd., Harrow UK (2005)
2. Zienkiewicz, O.C.: The Finite Element Methods. McGraw-Hill, London (1977)
3. Aliabadi, M.H.: The boundary element method, vol. 2. Applications in Solids and Structures. John Wiley and Sons Ltd, Chichester (2002)
4. Gao, X., Davies, T.G.: Boundary Element Programming in Mechanics. Cambridge University Press, Cambridge UK (2002)
5. Zieniuk, E., Bołtuć, A.: A non-element method of solving the two-dimensional Navier-Lame equation in problems with non-homogeneous polygonal subregions. TASK QUARTERLY 12(1), 1001–1014 (2008)
6. Bołtuć, A., Zieniuk, E.: Modeling domains using Bézier surfaces in plane boundary problems defined by the Navier-Lame equation with body forces. Eng. Anal. Bound. Elem. 35, 1116–1122 (2011)
7. Bołtuć, A., Zieniuk, E.: PIES in problems of 2D elasticity with body forces on polygonal domains. J. Theor. Appl. Mech. 49(2), 369–384 (2011)
8. Zieniuk, E., Bołtuć, A.: Non-element method of solving 2D boundary problems defined on polygonal domains modeled by Navier equation. Int. J. Solids Struct. 43(25–26), 7939–7958 (2006)
9. Salomon, D.: Curves and Surfaces for Computer Graphics. Springer, USA (2006). https://doi.org/10.1007/0-387-28452-4
10. Gottlieb, D., Orszag, S.: Numerical Analysis of Spectral Methods: Theory and Applications. SIAM, Philadelphia (1977)
11. Ochiai, Y.: Stress analysis with centrifugal load in non-homogeneous materials by triple-reciprocity boundary element method. Int. J. Numer. Meth. Bio. 26, 1331–1342 (2010)

Fine-Tuning Large Language Models for Answering Programming Questions with Code Snippets

Vadim Lomshakov[1]([✉])[iD], Sergey Kovalchuk[1][iD], Maxim Omelchenko[1],
Sergey Nikolenko[2,3][iD], and Artem Aliev[1]

[1] Huawei, St. Petersburg, Russia
{vadim.lomshakov,sergey.kovalchuk,artem.aliev}@huawei.com,
maxim.omelchenko@huawei-partners.com
[2] AI Center, National University of Science and Technology MISIS, Moscow, Russia
sergey@logic.pdmi.ras.ru
[3] St. Petersburg Department of the Steklov Institute of Mathematics,
St. Petersburg, Russia

Abstract. We study the ability of pretrained large language models (LLM) to answer questions from online question answering fora such as *Stack Overflow*. We consider question-answer pairs where the main part of the answer consists of source code. On two benchmark datasets—CoNaLa and a newly collected dataset based on *Stack Overflow*—we investigate how a closed-book question answering system can be improved by fine-tuning the LLM for the downstream task, prompt engineering, and data preprocessing. We use publicly available autoregressive language models such as GPT-Neo, CodeGen, and PanGu-Coder, and after the proposed fine-tuning achieve a BLEU score of 0.4432 on the CoNaLa test set, significantly exceeding previous state of the art for this task.

Keywords: Program synthesis · Question answering · Large language models

1 Introduction

Modern natural language processing (NLP) widely employs large language models (LLMs) that extract knowledge from text implicitly, via pretraining, and then can perform, e.g., open domain question answering (QA) without access to any external context or knowledge [13,18,20]. These LLMs provide an alternative way to the design of QA systems, without an external knowledge base and explicit retrieval components; fine-tuning a pretrained LLM with (question, answer) pairs is usually sufficient to train it for a given domain. For very large models such as GPT-3 [4], this approach may even lead to results competitive with retrieval-based methods on open domain QA without any fine-tuning. The pretraining procedure is always the most important part, and the data and process of pretraining directly affects the quality shown on downstream tasks [25].

J. Mikyška et al. (Eds.): ICCS 2023, LNCS 14074, pp. 171–179, 2023.
https://doi.org/10.1007/978-3-031-36021-3_15

Recent success of *Codex* [5] in program synthesis has demonstrated that pre-trained LLMs can be successfully adapted from NLP to the domain of source code. *Codex* is a GPT language model fine-tuned on a large corpus of publicly available code from *GitHub*; it powers *GitHub CoPilot* [5] that allows programmers to generate code in an IDE from a natural language query. Previously, the TranX model attempted to learn a neural semantic parser from scratch [27], but latest works have shown that using pretrained large language models (LLM) as part of the architecture works better. This includes the recently developed BERT+TAE [17], TranX+BERT [1], and MarianCG [21] models, the latter is a pretrained neural machine translation model (MarianMT) fine-tuned on code.

Program synthesis focuses on correctly implementing some functionality defined with a natural language query; researchers often use competitive programming problems to evaluate program synthesis models [5,11,14]. However, empirical studies of programming-related QA websites such as *Stack Overflow*[1] have shown that real life questions of programmers are not limited to defining a functionality. For instance, the work [2] classifies all posts into different categories such as "Conceptual" (*Why...? Is it possible...? Why something works?*), "API usage" (*How to implement something? Way of using something?*), "Discrepancy" (*Does not work, What is the problem...?*), and others, proposing regular expressions to define these categories. This gap leads to our research question: how effective are pretrained LLMs in answering the questions of real life programmers, even if we focus on questions answered with code snippets?

In this work, we use several publicly available GPT-based LLMs pretrained on code as backbones for solving closed-book QA in the *Stack Overflow* domain, evaluating on the *CoNaLa* dataset [26] and our own QA dataset collected from *Stack Overflow*. We introduce several approaches for fine-tuning, prompt engineering, and data preprocessing, achieving new state of the art results on *CoNaLa*. Below, Sect. 2 introduces the data, Sect. 3 outlines our approach, Sect. 4 discusses the evaluation study, and Sect. 5 concludes the paper.

2   Dataset

We consider programming-related QA with short code snippets generated as answers to real world problems. Thus, we focus on data with the following properties: (i) "API usage" questions according to the taxonomy shown in [2]; (ii) questions that consist of a short textual description (≤ 200 characters) without explicit source code in them; (iii) answers with explicit code snippets giving a solution to the proposed problem; (iv) to be more focused, we have limited our study to *Python* as one of the most popular programming languages.

First, we use the existing publicly available dataset *CoNaLa*[2] that satisfies our requirements [26]. The dataset consists of 2 879 examples, 2 379 in the training set and 500 in the test set, crawled from *Stack Overflow* and then manually curated by human annotators. Second, we have prepared our own dataset

---

[1] https://stackoverflow.com/.

[2] https://conala-corpus.github.io/.

based on original publicly available *Stack Overflow* data[3]. We have selected questions with the tag *"Python"* and used only the title text as the question. As ground truth, we have selected answers that earned maximum scores according to *Stack Overflow* data. To filter code-containing questions we search the text for <pre><code> in HTML and select questions with no explicit code paired with answers with a single code snippet. Finally, we have used the regular expressions proposed in [2] to select the "API usage" category of questions. We have cleaned the code by removing comments and selecting only snippets with correct syntax (no parsing errors). After these steps, we obtained a dataset with 10 522 question-answer pairs. We set aside 1 000 entries for the test set, using only questions from 2021 and later to avoid possible data leaks since we use LLMs trained on publicly available data. Since we use only the titles of *Stack Overflow* posts, the questions are short: 90% of the questions in the final dataset have between 5 and 17 words.

3   Methods

3.1   Large Language Models

As backbones, we have selected several LLMs for our study that are (i) publicly available, (ii) computationally inexpensive, and (iii) pretrained on code. Specifically, we have used: (i) CoPilot [5], an industrial solution based on *Codex*; (ii) GPT-Neo-2.7B (GPT-Neo below) [3] that shows high performance compared to *Codex* [24]; (iii) CodeGen-mono-2B (CodeGen) [16] that was trained on the *Pile* dataset [9] and separately on the code from *BigQuery* and *BigPython*; (iv) PanGu-Coder-2.6B (PanGu-Coder) [6] that was pretrained on *GitHub* code with a combination of autoregressive (causal) and masked language modeling losses in two stages, with the second stage using paired natural language and source code data. Both PanGu-Coder and CodeGen have equivalent or better performance on the *HumanEval* dataset than similarly sized Codex models [5].

3.2   Fine-Tuning, Prompt Engineering, and Data Preprocessing

We propose several techniques for solving our particular QA problem for short text questions answered by code snippets. First, we fine-tune the selected pretrained models on training sets from both *CoNaLa* (denoted as FT:C in Table 1) and *StackOverflow* (FT:SO in Table 1). For fine-tuning, we used the AdamW optimizer with $\beta_1 = 0.9$, $\beta_2 = 0.999$, $\epsilon = 10^{-8}$, with no weight decay, learning rate 5e-06 with linear decay, batch size 40, and half-precision (fp16).

Second, we experimented with various prompt engineering techniques to wrap the question into a context better suited for a specific LLM. For CodeGen and GPT-Neo models, the best prompt has turned out to be simply wrapping the question into a multiline comment: """question""" answer during training and

---

[3] https://data.stackexchange.com/.

`"""question"""` on inference. For PanGu-Coder, we have used the prompt format used in its pretraining: `<comments> question <python> answer <eoc>`. We have also experimented with prefix tuning from the *OpenPrompt* library [7] but it has not led to improvements on our datasets.

Third, we have found that adding domain-specific knowledge such as the names of code libraries can dramatically improve the quality of generated answers. For that purpose, we have trained a classifier on *StackOverflow* data to predict the usage of *Python* libraries. Specifically, we selected 317 300 posts (this set does not intersect with either *CoNaLa* or *StackOverflow* datasets) with only one `import` statement in the best answer, selected top 200 most used libraries as classifier labels, used a pretrained MPNet-base model [22] to encode the titles of these posts, and trained a support vector machine (SVM) for classification with these embeddings as input and the corresponding libraries as output. The resulting classifier obtains 0.47 Prec@1, 0.76 Prec@5, and 0.48 Recall@1 on a held-out test set. Then we automatically annotate the prompts by adding top 5 predicted import statements before the question; this is shown as +I in Table 1.

Fourth, we have tried variable name substitution similar to [1], i.e., replacing variable names and string literals in both question and answer with special tokens (`var_i`, `lst_i` etc.); all new tokens were added to the LLM vocabulary. We used it only for the *CoNaLa* dataset since it contains special labeling for variable names and constants in the question body (+R in Table 1). Finally, for the *StackOverflow* dataset we also have post body fields in addition to the title; we have tried to add the post bodies to the question, concatenating them with the titles and truncating the result to 270 tokens (+B in Table 1).

3.3   Evaluation Procedure

For model evaluation on test sets, we have used both general-purpose NLP metrics—BLEU, BERTScore [28] and Rouge [15]—and metrics developed for program synthesis: CodeBLEU [19] and Ruby [23]. Note that while the *CoNaLa* leaderboard uses BLEU[4], recent studies show that direct application of BLEU and other automated metrics may lead to issues in code generation evaluation [8]. Unfortunately, there is still no good alternative, although recent studies suggest that Ruby is better aligned with human evaluation [12]; in our experiments, all metrics seem to agree on what the best models are. For *CoNaLa*, we used the test set provided by the authors; for the *StackOverflow* dataset, the test dataset is a random sample of 1 000 questions dated 2021 and later, while questions dated 2020 and earlier were used for training. In both cases, we randomly split the training set in the 90:10 ratio into train and validation parts.

4   Results

Table 1 shows our evaluation results; for *CoNaLa*, we show the best BLEU scores from the leaderboard by recently developed BERT+TAE [17], TranX+BERT [1],

Table 1. Evaluation study. Best results are shown in bold, results that differ from them statistically insignificantly are underlined; FT:C – fine-tuned on CoNaLa; FT:SO – on SO; +I – import classifier; +R – variable replacement; +B – using the question body.

	Model	Options	BertScore	Rouge	CodeBLEU	Ruby	BLEU
CoNaLa dataset	BERT+TAE [17]	—	—	—	—	—	0.3341
	TranX+BERT [1]	—	0.8998	0.4616	0.4742	0.5478	0.3420
	MarianCG [21]	—	0.8982	0.5000	0.5001	0.5121	0.3443
	CoPilot [5]	—	0.8468	0.3376	0.2602	0.2595	0.1642
	CoPilot	+I	0.8761	0.4169	0.3531	0.3447	0.2230
	GPT-Neo [3]	—	0.7345	0.0425	0.0723	0.1524	0.0104
	GPT-Neo	FT:C	0.8688	0.4304	0.3841	0.4319	0.1633
	CodeGen [16]	—	0.8064	0.3093	0.2767	0.2749	0.1120
	CodeGen	FT:C	0.9015	0.5522	0.4850	0.5599	0.3171
	CodeGen	FT:C+R	0.8685	0.3711	0.4242	0.4712	0.2085
	CodeGen	FT:C+I	0.9115	0.5998	0.5696	0.6325	0.4319
	PanGu-Coder [6]	—	0.8825	0.4599	0.4421	0.4774	0.2121
	PanGu-Coder	FT:C	0.9217	0.5981	0.6032	0.6375	0.4098
	PanGu-Coder	FT:C+I	**0.9235**	**0.6061**	**0.6122**	**0.6511**	**0.4432**
StackOverflow dataset	TranX+BERT [1]	—	0.8286	0.1001	0.3829	0.2390	0.0515
	CoPilot [5]	—	0.7939	0.0965	0.0827	0.0864	0.0234
	GPT-Neo [3]	—	0.7552	0.0472	0.1187	0.1165	0.0107
	GPT-Neo	FT:SO	0.8052	0.1130	0.1362	0.0956	0.0464
	GPT-Neo	FT:C	0.7956	0.1237	0.2919	0.1676	0.0477
	CodeGen [16]	—	0.7438	0.0688	0.1469	0.1131	0.0205
	CodeGen	FT:SO	0.8021	0.1242	0.1562	0.0994	0.0508
	CodeGen	FT:C	0.8217	0.1490	0.3310	0.2025	0.0719
	PanGu-Coder [6]	—	0.8305	0.1497	0.2149	0.1387	0.0717
	PanGu-Coder	FT:SO	0.8448	0.1825	0.3280	0.2033	0.1087
	PanGu-Coder	FT:C	0.8386	0.1607	**0.4323**	**0.2467**	0.0555
	PanGu-Coder	FT:SO+I	0.8445	0.1843	0.3325	0.2041	0.1099
	PanGu-Coder	FT:SO+B	**0.8540**	**0.2452**	0.3217	0.2418	**0.1519**

and MarianCG [21] models; we have reproduced the results of TranX+BERT and used the MarianCG test set results to supplement BLEU with other metrics. Here, the best results were obtained by PanGu-Coder FT:C+I, followed by CodeGen FT:C+I; both set a new state of the art. PanGu-Coder was also best among the original pretrained models, but our proposed classification-based procedure (+I) gives significant improvements. GPT-Neo loses significantly, perhaps because it was originally pretrained only on the *Pile* dataset, while CodeGen and PanGu-Coder have comparable results; both were pretrained on code, but CodeGen also used the *Pile*, so we see that pretraining on a natural language corpus does not affect quality as much as suggested in [5]. Note how our proposed pretraining and fine-tuning heuristics make CodeGen results very close to PanGu-Coder, although pretrained versions show very different performance.

Table 2. Sample answers of PanGu-Coder FT:C+I (Ex. 1), FT:SO+B (Ex. 2, 3) and FT:SO+I (Ex. 2.1, 3.1).

Example 1. **Q**: *download a file 'url' over HTTP and save to "10MB"*

A: `urllib.request.urlretrieve(url, '10MB')`

Example 2. **Q**: *delete all specific element from Python list*

I have a list like this. a = ['\n', 'a', 'b', '\n', 'c', '\n'] As far as I know, the remove, pop, delete method only removes one by one. If I want to remove all '\ n' from this list, the only way to do

A: a = ['\n', 'a', 'b', '\n', 'c', '\n']

```
a = [x for x in a if x!= '\n']
```

Example 2.1. **Q**: *delete all specific element from Python list*

I: re json itertools sys random

A: a = [1,2,3,4,5,6,7,8,9,10]

```
a[4:] = a[4:] + a[4:][4:]
```

Example 3. **Q**: *iterate odd indexes and add them back to original list*

I want to add "..." after every odd index in my list (or after every second word), but I can't quite figure out how to do it.

A: def add_odds(lst):

```
    return [lst[i] + '...' if i\%2 else lst[i] for i in range(len(lst))]
```

Example 3.1. **Q**: *iterate odd indexes and add them back to original list*

I: `itertools collections random scipy os`

A: a = [1,2,3,4,5,6,7,8,9,10]

```
for i in range(len(a)-2):

    if a[i] % 2 == 0:

        a.append(a.pop(i)+a.pop(i+1))

print(a)
```

On the *StackOverflow* dataset (Table 1), the best results were obtained by PanGu-Coder FT:SO+B, i.e., fine-tuned on *StackOverflow* and using concatenated question bodies. In general, results for *StackOverflow* are worse than for *CoNaLa*: real questions from Q&A web fora are more challenging for models than specially rephrased questions. In particular, a question's title often does not contain complete information, or the wording of the question is not familiar to the model (i.e., they do not represent similar doc-strings). On the other hand, the +I heuristic here helps much less than for *CoNaLa*, probably because titles of real questions often already contain library names while the *CoNaLa* dataset removes them to achieve "pure" program synthesis. We have run the bootstrap test for difference between means [10] with 100 000 samples and confidence levels 0.055 (*CoNaLa*) and 10^{-4} (*StackOverflow*). In Table 1, best results are shown in bold, and results that do *not* significantly differ from them are underlined.

Table 2 shows sample PanGu-Coder answers that illustrate our techniques. In Example 1, the classifier has produced `urllib` as a suggestion, and the model has used it successfully. In Examples 2 and 3, the titles (italicized first line) are not informative enough, as shown by the answers of FT:SO+I (Examples 2.1 and 3.1 in Table 2), but adding the question body leads to correct answers.

5   Conclusion

In this work, we have demonstrated that simple fine-tuning on downstream tasks such as closed-book QA dramatically improves the quality of LLMs applied to code generation. In particular, we have achieved a new state of the art result on the CoNaLa dataset with BLEU score 0.4432. We have presented several fine-tuning and prompt engineering techniques that can improve the performance of a variety of LLMs, and we believe that similar approaches can work in other scenarios; e.g., fine-tuning also helps significantly on our newly collected *Stack-Overflow* dataset (very different from *CoNaLa*). Understanding real-life noisy natural language queries remains a challenging task for LLMs, evidenced by much better performance on *CoNaLa* where the questions are manually curated, and relatively low results overall (despite them being state of the art). Another problem is that common metrics, even code-specific ones [8], often lead to incorrect evaluation due to the diversity of correct code. We view these problems as important directions for further work.

Acknowledgements. The work of Sergey Nikolenko was prepared in the framework of the strategic project "Digital Business" within the Strategic Academic Leadership Program "Priority 2030" at NUST MISiS.

References

1. Beau, N., Crabbé, B.: The impact of lexical and grammatical processing on generating code from natural language. In: Findings of the Association for Computational Linguistics: ACL 2022, pp. 2204–2214. Association for Computational Linguistics, Dublin, Ireland (2022). https://doi.org/10.18653/v1/2022.findings-acl.173
2. Beyer, S., Macho, C., Di Penta, M., Pinzger, M.: What kind of questions do developers ask on Stack Overflow? A comparison of automated approaches to classify posts into question categories. Empir. Softw. Eng. **25**(3), 2258–2301 (2019). https://doi.org/10.1007/s10664-019-09758-x
3. Black, S., Gao, L., Wang, P., Leahy, C., Biderman, S.: GPT-Neo: large Scale autoregressive language modeling with mesh-tensorflow (2021). https://doi.org/10.5281/zenodo.5297715
4. Brown, T.B. et al.: Language models are few-shot learners (2020). https://doi.org/10.48550/ARXIV.2005.14165

5. Chen, M. et al.: Evaluating large language models trained on code. CoRR abs/2107.03374 (2021), https://arxiv.org/abs/2107.03374

6. Christopoulou, F. et al.: PanGu-Coder: program synthesis with function-level language modeling (2022). https://doi.org/10.48550/ARXIV.2207.11280

7. Ding, N. et al.: Openprompt: an open-source framework for prompt-learning. arXiv preprint arXiv:2111.01998 (2021)

8. Evtikhiev, M., Bogomolov, E., Sokolov, Y., Bryksin, T.: Out of the bleu: how should we assess quality of the code generation models? (2022). https://doi.org/10.48550/ARXIV.2208.03133

9. Gao, L., et al.: The pile: An 800 GB dataset of diverse text for language modeling. arXiv preprint arXiv:2101.00027 (2020)

10. Hall, P., Hart, J.D.: Bootstrap test for difference between means in nonparametric regression. J. Am. Statist. Assoc. **85**(412), 1039–1049 (1990). https://doi.org/10.1080/01621459.1990.10474974

11. Hendrycks, D., Basart, S., Kadavath, S., Mazeika, M., Arora, A., Guo, E., Burns, C., Puranik, S., He, H., Song, D., Steinhardt, J.: Measuring coding challenge competence with apps (2021). https://doi.org/10.48550/ARXIV.2105.09938

12. Kovalchuk, S.V., Lomshakov, V., Aliev, A.: Human perceiving behavior modeling in evaluation of code generation models. In: Proceedings of the 2nd Workshop on Natural Language Generation, Evaluation, and Metrics (GEM), pp. 287–294. ACL, Abu Dhabi, UAE (2022). https://aclanthology.org/2022.gem-1.24

13. Lee, N., Li, B.Z., Wang, S., Yih, W.T., Ma, H., Khabsa, M.: Language models as fact checkers? (2020). https://doi.org/10.48550/ARXIV.2006.04102

14. Li, Y. et al.: Competition-level code generation with alphacode (2022). https://doi.org/10.48550/ARXIV.2203.07814

15. Lin, C.Y.: ROUGE: a package for automatic evaluation of summaries. In: Text Summarization Branches Out, pp. 74–81. Association for Computational Linguistics, Barcelona, Spain (2004). https://aclanthology.org/W04-1013

16. Nijkamp, E. et al.: CodeGen: an open large language model for code with multi-turn program synthesis (2022). https://doi.org/10.48550/ARXIV.2203.13474

17. Norouzi, S., Cao, Y.: Semantic parsing with less prior and more monolingual data. CoRR abs/2101.00259 (2021). https://arxiv.org/abs/2101.00259

18. Petroni, F. et al.: Language models as knowledge bases? (2019). https://doi.org/10.48550/ARXIV.1909.01066

19. Ren, S. et al.: CodeBLEU: a method for automatic evaluation of code synthesis (2020). https://doi.org/10.48550/ARXIV.2009.10297

20. Roberts, A., Raffel, C., Shazeer, N.: How much knowledge can you pack into the parameters of a language model? (2020). https://doi.org/10.48550/ARXIV.2002.08910

21. Soliman, A.S., Hadhoud, M.M., Shaheen, S.I.: MarianCG: a code generation transformer model inspired by machine translation. J. Eng. Appl. Sci. **69**(1), 1–23 (2022)

22. Song, K., Tan, X., Qin, T., Lu, J., Liu, T.Y.: MPNet: masked and permuted pretraining for language understanding. In: Proceedings of the 34th International Conference on Neural Information Processing Systems. NIPS2020, Curran Associates Inc., Red Hook, NY, USA (2020)

23. Tran, N., Tran, H., Nguyen, S., Nguyen, H., Nguyen, T.: Does BLEU score work for code migration? In: 2019 IEEE/ACM 27th International Conference on Program Comprehension (ICPC), pp. 165–176 (2019)

24. Xu, F.F., Alon, U., Neubig, G., Hellendoorn, V.J.: A systematic evaluation of large language models of code (2022). https://doi.org/10.48550/ARXIV.2202.13169

25. Ye, Q. et al.: Studying strategically: learning to mask for closed-book QA (2020). https://doi.org/10.48550/ARXIV.2012.15856
26. Yin, P., Deng, B., Chen, E., Vasilescu, B., Neubig, G.: Learning to mine aligned code and natural language pairs from stack overflow. In: 2018 IEEE/ACM 15th Intl. Conf. on Mining Software Repositories (MSR), pp. 476–486 (2018)
27. Yin, P., Neubig, G.: TRANX: A transition-based neural abstract syntax parser for semantic parsing and code generation. In: Proceedings of the 2018 Conference on Empirical Methods in Natural Language Processing: System Demonstrations, pp. 7–12. ACL, Brussels, Belgium (2018). https://doi.org/10.18653/v1/D18-2002
28. Zhang, T., Kishore, V., Wu, F., Weinberger, K.Q., Artzi, Y.: Bertscore: Evaluating text generation with BERT (2019). https://doi.org/10.48550/ARXIV.1904.09675

Introducing a Computational Method to Retrofit Damaged Buildings under Seismic Mainshock-Aftershock Sequence

Farzin Kazemi[1(✉)], Neda Asgarkhani[1], Ahmed Manguri[1,2], Natalia Lasowicz[1], and Robert Jankowski[1]

[1] Faculty of Civil and Environmental Engineering,
Gdańsk University of Technology, ul. Narutowicza 11/12, 80-233 Gdansk, Poland
{farzin.kazemi,neda.asgarkhani,ahmed.manguri,natalia.lasowicz,
jankowr}@pg.edu.pl
[2] Civil Engineering Department, University of Raparin, Rania, Kurdistan Region, Iraq

Abstract. Retrofitting damaged buildings is a challenge for engineers, since commercial software does not have the ability to consider the local damages and deformed shape of a building resulting from the mainshock record of an earthquake before applying the aftershock record. In this research, a computational method for retrofitting of damaged buildings under seismic mainshock-aftershock sequences is proposed, and proposed computational strategy is developed using Tcl programming code in OpenSees and MATLAB. Since the developed programming code has the ability of conducting nonlinear dynamic analysis (e.g. Incremental Dynamic Analysis (IDA)), different types of steel and reinforced concrete structures, assuming different intensity measures and engineering demands, can be on the benefit of this study. To present the ability of method, the 4-Story and 6-Story damaged steel structures were selected. Then, the linear Viscous Dampers (VDs) are used for retrofitting of the damaged structures, and IDAs were performed under aftershock records. The results showed that the proposed method and computational program could improve the seismic performance level of damaged structures subjected to the mainshock-aftershock sequences. In addition, the damaged floor level of the building is recognized by programming code and can be effectively considered for local retrofit schemes.

Keywords: Computational Method · Damaged-Building · Retrofitting of Buildings · Mainshock-Aftershock Sequence

1 Introduction

Nowadays, seismic activity is known as an external threat to buildings due to its unpredictable external loads that may impose sudden force on the story levels where the weight of the building is concentrated. Investigations have been carried out to propose procedures for assuming the effects of external loads such as pounding phenomenon [1] and impact force [2], which can cause local damages of structures [3, 4]. Then, retrofitting strategies for controlling the lateral loads effects [5], controlling the pounding force [6], and improving the seismic performance level of structures using buckling-restrained brace [7], knee brace [8, 9] and semi-rigid connections having shape memory alloys [10, 11] were proposed. Although it is possible to use dissipative devices such as linear Viscous Dampers (VDs), it is not beneficial to construct buildings with these expensive devices. Therefore, most of buildings have not been equipped with VDs and may be exposed to damages of mainshock or seismic sequences [12].

Studies have been conducted to provide the information of using linear and nonlinear VDs having different types of floor level distribution, and their influences on the Residual Drift (RD) and Interstory Drift (ID) of the steel structures [13]. Deringöl and Güneyisi [14] investigated the effectiveness of the using VDs in the seismically isolated steel buildings. Hareen and Mohan [15] introduced an energy-based method for retrofitting of reinforced concrete buildings using VDs, which improved the seismic performance. Pouya et al. [16] investigated the failure mechanism of conventional bracing system under mainshock-aftershock sequences. To overcome economic issues of using VDs, Asgarkhani et al. [17] and Kazemi et al. [18] introduced an optimal VDs placement process to reduce the cost of implementing VDs, while this procedure can be applied to those buildings under seismic mainshock effects.

It should be noted that the VDs may be used as retrofitting strategy for damaged buildings, while the modeling of damaged building and implementing the VDs after observing the local damages in the structures are the case of the present study. This research aims to propose a modeling process to implement VDs after damage of buildings under seismic mainshock, and improve their performances for aftershock earthquakes. This procedure considers the effects of pre-damaging in the building, which may increase the failure probability of the building and impose additional financial loss. The following sections try to present an example of using this procedure.

2 Modeling of Structures

The 4-Story and 6-Story frames were designed in accordance with ASCE 7-16 [19] (see also [5, 10, 12] for details of designing process). Figure 1 illustrates the structural details of the 4-, and 6-Story frames.

The structures were modeled in OpenSees [20] software using the procedures that have been employed by Kazemi et al. [21–24]. According to their procedure, to model 3D buildings, due to the symmetric plan of structures, it is possible to use 2D models with the same fundamental period and modal information. They have verified this procedure and used 2D models to facilitate the computational analysis. In addition, the leaning column was employed to represent the gravity columns of building for modeling of the P-delta effects, which plays a crucial role in the lateral behavior of buildings [2–4]. Moreover, the beams of structures were modeled with IMK hinges [10–12] and the columns were assumed to have fiber sections [4, 5].

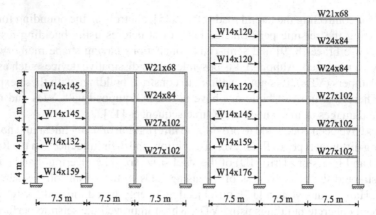

Fig. 1. Structural details of the 4-, and 6-Story frames.

3 Computational Method

Literature review confirmed that VDs can be used to control the RD and ID of the steel structures, in which, this reduction can maintain structure within allowable limitations prescribed by seismic codes. Although implementing VDs for the purpose of retrofitting strategy is a common idea, the correct assessment of ID and RD is a challenging duty in front of structural engineers. Since a pre-damaged building has an initial stage of RD, it should be considered in modeling procedure. It is not easy to model local damages in structural elements, while in each member of building different damage states can be observed due to the strength of elements. Therefore, in this research, a modeling process is proposed to include all damage states of structural members and initial RD due to mainshock earthquake. Figure 2 illustrates the proposed computational method for retrofitting damaged buildings.

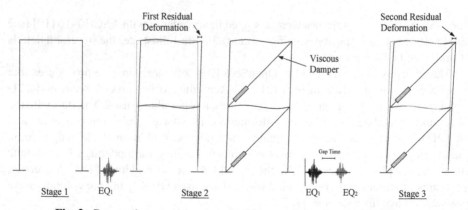

Fig. 2. Proposed computational method for retrofitting damaged buildings.

According to Fig. 2, in first stage, the model of the structure is ready, and in second stage, the first earthquake known as seismic mainshock is applied and first RD is calculated. In second stage, the computational method is applied to the model to implement the VDs as selected retrofitting strategy. In second stage, a Tcl programming code is developed in OpenSees [20] and MATLAB [25] software simultaneously to control the deformed shape model and implement the VDs. It should be noted that the VDs are implemented to a deformed shape model as it looks in Fig. 2. Then, the second earthquake known as aftershock is applied to the structure. To achieve the seismic performance of damaged structure under aftershock records, the Tcl programming code improved with the ability of performing IDAs, in which, three steps of analysis are defined based on the spectral spectrum remarked as $S_a(T_1)$ [26–28]. All of the procedure is automated to reduce the analysis time. The results of the analysis are plotted by MATLAB [25] to have an operator that controls the whole analysis procedure. This software can help the computational method to be repeated for the number of seismic records, and finally, the results of IDA curves are plotted. It is noteworthy that this computational method can be applied to any other methods of retrofitting using dissipative devices, since it is a general method with ability of changing the dissipative device during analysis. The proposed method has the ability of defining different intensity measures (i.e. $S_a(T_1)$) and a wide range of demands (i.e. RD and ID) for any type of structures (i.e. steel and reinforced concrete structures), while increasing the accuracy of the results and reducing computational time.

4 Retrofitting Damaged Building

To present the capability of proposed method, two selected structures were retrofitted with implementing VDs at all floor levels after mainshock earthquake. Figures 3 and 4 compare the results of the deformation of the 4-Story and 6-Story frames in the mainshock-aftershock analysis with and without considering VDs, respectively.

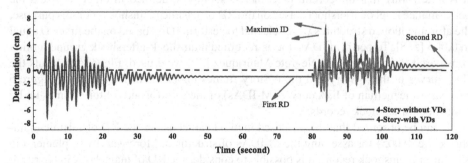

Fig. 3. Comparing the deformation results of the 4-Story frame in the seismic mainshock-aftershock analysis with and without VDs.

The nonlinear dynamic analysis was conducted based on the Northridge record with RSN 949. In the seismic mainshock, there is a permanent deformation known as the first RD that illustrates the remained deformation in the structural elements and floor levels. In the constant part of the performing analysis (i.e. between 60 s to 80 s), the retrofitting procedure was done without stopping the analysis, while taking the first RD into account

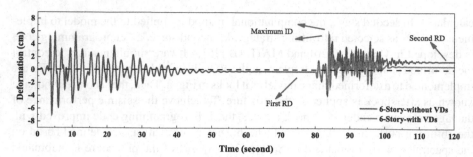

Fig. 4. Comparing the deformation results of the 6-Story frame in the seismic mainshock-aftershock analysis with and without VDs.

(i.e. 0.903 cm and 0.221 cm for the 4-Story and 6-Story frames, respectively). It can be observed that after implementing VDs, there is a significant influence on the values of the second RD. For the 4-Story frame, the second RD significantly decreased by 5.18 times from 0.808 cm to 0.156 cm by implementing VDs. In addition, for the 6-Story frame, the second RD considerably decreased by 2.27 times from 1.347 cm to 0.593 cm by implementing VDs. Therefore, the proposed computational procedure can effectively model the retrofitted structure by taking into account the remained deformation (i.e. RD) as well as the local damages of structure due to mainshock record.

To show the capability of the proposed method, the Tcl programming code was developed to perform Incremental Dynamic Analysis (IDA) that is a well-known method for seismic performance assessment. To perform IDAs, the seismic mainshock is applied to remain at a certain level of RD (i.e. first RD), and then, the aftershock is applied by increasing amplitude of ground motions until the total collapse of structures. All procedure were controlled by MATLAB [25] and results were plotted after analysis. It is noteworthy that the certain level of RD should be defined in order to assess the performance level of aftershock based on the RD of seismic mainshock. For this purpose, the aforementioned structures were selected to perform IDAs based on the first RD equal to 0.005 [5, 8]. To perform IDAs, the as-recorded mainshock-aftershock ground motion considered Ruiz-García and Negrete-Manriquez [29] were used. Figure 5 presents the IDA curves of the 4-Story and the 6-Story frames under mainshock records. Figure 6 compares the median of IDA curves (M-IDAs) of the 4-Story and 6-Story frames under mainshock-aftershock records.

It can be observed that the proposed method can consider the RD of seismic mainshock (i.e. 0.005) for assessing the M-IDAs of structures. Moreover, by implementing VDs after mainshock record, it is possible to consider the RD of mainshock in the result of performance assessment of structures. For instance, in ID of 10%, the aftershock effects reduced the $S_a(T_1)$ values of the 4-Story and 6-Story frames by 5.07% (from 1.468 to 1.398) and 17.24% (from 0.782 to 0.667), respectively. Moreover, implementing VDs improved the seismic performance of the 4-Story and 6-Story damaged frames by 23.965 and 39.13%, respectively.

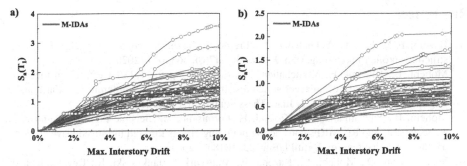

Fig. 5. IDA curves of, a) the 4-Story and, b) the 6-Story frames under mainshock records.

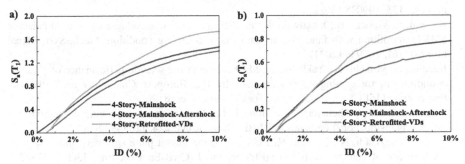

Fig. 6. M-IDAs of, a) the 4-Story and, b) the 6-Story frames under mainshock-aftershock records.

5 Conclusion

This research proposes an effective computational method for retrofitting damaged buildings under seismic mainshock-aftershock sequences. The proposed method can be applied to retrofit steel and reinforced concrete structures assuming different intensity measures and engineering demands. Moreover, a wide range of retrofitting devices can be applied such as VDs and buckling-restrained braces. To show the capability of the developed Tcl code, two structures having four and six-story levels were selected and the numerical nonlinear dynamic analysis and IDAs were conducted. The results of analysis show that the proposed method can provide the seismic performance level of damaged frames based on the seismic mainshock-aftershock sequences. The developed program increases the ability of performing analysis of damaged buildings assuming lateral deformations and local damages of buildings as a result of mainshock record. In addition, the damaged floor level of building is recognized by programming code and can be used for local retrofit instead of retrofitting of the whole structure.

References

1. Kazemi, F., Mohebi, B., Yakhchalian, M.: Predicting the seismic collapse capacity of adjacent structures prone to pounding. Can. J. Civ. Eng. **47**(6), 663–677 (2020)
2. Mohebi, B., Kazemi, F., Yakhchalian, M.: Investigating the P-Delta effects on the seismic collapse capacity of adjacent structures. In: 16th European Conference on Earthquake Engineering (16ECEE), 18–21, June, Thessaloniki, Greece (2018)
3. Mohebi, B., Yazdanpanah, O., Kazemi, F., Formisano, A.: Seismic damage diagnosis in adjacent steel and RC MRFs considering pounding effects through improved wavelet-based damage-sensitive feature. J. Build. Eng. **33**, 101847 (2021)
4. Yazdanpanah, O., Mohebi, B., Kazemi, F., Mansouri, I., Jankowski, R.: Development of fragility curves in adjacent steel moment-resisting frames considering pounding effects through improved wavelet-based refined damage-sensitive feature. Mech. Syst. Signal Process. **173**, 109038 (2022)
5. Kazemi, F., Mohebi, B., Jankowski, R.: Predicting the seismic collapse capacity of adjacent SMRFs retrofitted with fluid viscous dampers in pounding condition. Mech. Syst. Signal Process. **161**, 107939 (2021)
6. Kazemi, F., Mohebi, B., Yakhchalian, M.: Enhancing the seismic performance of adjacent pounding structures using viscous dampers. In: 16th European Conference on Earthquake Engineering (16ECEE), 18–21, June, Thessaloniki, Greece (2018)
7. Asgarkhani, N., Yakhchalian, M., Mohebi, B.: Evaluation of approximate methods for estimating residual drift demands in BRBFs. Eng. Struct. **224**, 110849 (2020)
8. Mohebi, B., Asadi, N., Kazemi, F.: Effects of using gusset plate stiffeners on the seismic performance of concentrically braced frame. Int. J. Civil Environ. Eng. **13**(12), 723–729 (2019)
9. Mohebi, B., Kazemi, F., Yousefi, A.: Seismic response analysis of knee-braced steel frames using Ni-Ti Shape Memory Alloys (SMAs). In: Di Trapani, F., Demartino, C., Marano, G.C., Monti, G. (eds.) Proceedings of the 2022 Eurasian OpenSees Days. EOS 2022. LNCE, vol. 326. Springer, Cham (2023). https://doi.org/10.1007/978-3-031-30125-4_21
10. Kazemi, F., Jankowski, R.: Enhancing seismic performance of rigid and semi-rigid connections equipped with SMA bolts incorporating nonlinear soil-structure interaction. Eng. Struct. **274**, 114896 (2023)
11. Mohebi, B., Kazemi, F., Yousefi, A.: Enhancing seismic performance of semi-rigid connection using shape memory alloy bolts considering nonlinear soil–structure interaction. In: Di Trapani, F., Demartino, C., Marano, G.C., Monti, G. (eds.) Proceedings of the 2022 Eurasian OpenSees Days. EOS 2022. LNCE, vol. 326, pp. 248–256. Springer, Cham (2023). https://doi.org/10.1007/978-3-031-30125-4_22
12. Kazemi, F., Asgarkhani, N., Jankowski, R.: Probabilistic assessment of SMRFs with infill masonry walls incorporating nonlinear soil-structure interaction. Bull. Earthq. Eng. **21**(1), 503–534 (2023)
13. Yahyazadeh, A., Yakhchalian, M.: Probabilistic residual drift assessment of SMRFs with linear and nonlinear viscous dampers. J. Constr. Steel Res. **148**, 409–421 (2018)
14. Deringöl, A.H., Güneyisi, E.M.: Influence of nonlinear fluid viscous dampers in controlling the seismic response of the base-isolated buildings. In: Structures, vol. 34, pp. 1923–1941 (2021)
15. Hareen, C.B., Mohan, S.C.: Energy-based seismic retrofit and design of building frames with passive dampers. Eng. Struct. **250**, 113412 (2022)
16. Pouya, A., Izadinia, M., Memarzadeh, P.: Energy-based collapse assessment of concentrically braced frames under mainshock-aftershock excitations. In: Structures, vol. 47, pp. 925–938 (2023)

17. Asgarkhani, N., Kazemi, F., Jankowski, R.: Optimal retrofit strategy using viscous dampers between adjacent RC and SMRFs prone to earthquake-induced pounding. Arch. Civil Mech. Eng. **23**(1), 7 (2023)
18. Kazemi, F., Asgarkhani, N., Manguri, A., Jankowski, R.: Investigating an optimal computational strategy to retrofit buildings with implementing viscous dampers. In: Groen, D., de Mulatier, C., Paszynski, M., Krzhizhanovskaya, V.V., Dongarra, J.J., Sloot, P.M.A. (eds.) Computational Science – ICCS 2022. ICCS 2022. LNCS, vol. 13351, pp. 184–191. Springer, Cham (2022). https://doi.org/10.1007/978-3-031-08754-7_25
19. Minimum Design Loads for Buildings and Other Structures (ASCE/SEI 7-16), first, second, and third printings. Minimum Design Loads for Buildings and Other Structures (2017)
20. McKenna, F., Fenves, G.L., Filippou, F.C., Scott, M.H.: Open system for earthquake engineering simulation (OpenSees). Pacific Earthquake Engineering Research Center, University of California, Berkeley (2016)
21. Kazemi, F., Jankowski, R.: Machine learning-based prediction of seismic limit-state capacity of steel moment-resisting frames considering soil-structure interaction. Comput. Struct. **274**, 106886 (2023)
22. Kazemi, F., Jankowski, R.: Seismic performance evaluation of steel buckling-restrained braced frames including SMA materials. J. Constr. Steel Res. **201**, 107750 (2023)
23. Kazemi, F., Asgarkhani, N., Jankowski, R.: Predicting seismic response of SMRFs founded on different soil types using machine learning techniques. Eng. Struct. **274**, 114953 (2023)
24. Kazemi, F., Asgarkhani, N., Jankowski, R.: Machine learning-based seismic response and performance assessment of reinforced concrete buildings. Arch. Civil Mech. Eng. **23**(2), 94 (2023)
25. MATLAB/Simulink as a Technical Computing Language.: Engineering Computations and Modeling in MATLAB (2020)
26. Mohebi, B., Kazemi, F., Asgarkhani, N., Ghasemnezhadsani, P., Mohebi, A.: Performance of vector-valued intensity measures for estimating residual drift of steel MRFs with viscous dampers. Int. J. Struct. Civil Eng. Res. **11**(4), 79–83 (2022). https://doi.org/10.18178/ijscer.11.4.79-83
27. Kazemi, F., Mohebi, B., Asgarkhani, N., Yousefi, A.: Advanced scalar-valued intensity measures for residual drift prediction of SMRFs with fluid viscous dampers. Int. J. Struct. Integrity **12**, 20–25 (2023)
28. Kazemi, F., Asgarkhani, N., Jankowski, R.: Machine learning-based seismic fragility and seismic vulnerability assessment of reinforced concrete structures. Soil Dyn. Earthq. Eng. **166**, 107761 (2023)
29. Ruiz-García, J., Negrete-Manriquez, J.C.: Evaluation of drift demands in existing steel frames under as-recorded far-field and near-fault mainshock–aftershock seismic sequences. Eng. Struct. **33**(2), 621–634 (2011)

Hierarchical Classification of Adverse Events Based on Consumer's Comments

Monika Kaczorowska[1,2]($\boxtimes$) ⓘ, Piotr Szymczak[1], and Sergiy Tkachuk[1,3] ⓘ

[1] Reckitt Benckiser Group, Global Data and Analytics, Zajęcza 15, 00-351 Warsaw, Poland
{monika.kaczorowska,piotrwojciech.szymczak,
sergiy.tkachuk}@reckitt.com
[2] Polish-Japanese Academy of Information Technology, Koszykowa 86, 02-008 Warsaw, Poland
[3] Systems Research Institute, Polish Academy of Sciences, Newelska 6, 01-447 Warsaw, Poland

Abstract. This paper focuses on autonomously classifying adverse events based on consumers' comments regarding health and hygiene products. The data, comprising over 152,000 comments, were collected from e-commerce sources and social media. In the present research, we propose a language-independent approach using machine translation, allowing for unified analysis of data from various countries. Furthermore, this study presents a real-life application, making it potentially beneficial for subsequent scientific research and other business applications. A distinguishing feature of our approach is the efficient modeling of colloquial language instead of medical jargon, which is often the focus of adverse event research. Both hierarchical and non-hierarchical classification approaches were tested using Random Forest and XGBoost classifiers. The proposed feature extraction and selection process enabled us to include tokens important to minority classes in the dictionary. The F1 score was utilized to quantitatively assess the quality of classification. Hierarchical classification allowed for faster classification processes than the non-hierarchical approach for the XGBoost classifier. We obtained promising results for XGBoost; however, further research on a wider range of categories is required.

Keywords: classification · adverse event · NLP

1 Introduction

Adverse events are defined as untoward medical occurrences following exposure to a medicine, not necessarily caused by that medicine [1]. Adverse events can potentially be hazardous to humans, causing irreversible changes in the human body and, in extreme cases, death [2]. They pose a significant public issue, affecting human health and life, and causing substantial financial losses [3, 4]. Adverse events can be caused by pharmaceutical products, cosmetics, care products, and cleaning agents [5]. The Center for Food Safety and Applied Nutrition (CFSAN) has provided the Adverse Event Reporting System (AERS), which allows for reporting adverse events and product complaints related to foods, dietary supplements, and cosmetics [6]. Knowledge of adverse event occurrences enables improvements in products, making them safer and more attractive to consumers.

J. Mikyška et al. (Eds.): ICCS 2023, LNCS 14074, pp. 188–195, 2023.
https://doi.org/10.1007/978-3-031-36021-3_17

Nowadays, machine learning techniques are applied in various fields, including adverse event classification [7, 8]. Common approaches are based on using Support Vector Machines [4], Random Forest [7], or Maximum Entropy [9]. Neural networks, such as Convolutional Neural Networks (CNN) [10], attention-based deep neural networks [11], and Bidirectional Encoder Representations from Transformers (BERT) [12], are also employed in adverse event classification. It is worth noting that imbalanced classes are a common problem in this area [13].

Recently, the role of social media and e-commerce portals has become increasingly crucial for expressing opinions among consumers. Social media [14] and e-commerce data sources pose challenges from a data processing perspective, especially for Natural Language Processing. Comments and opinions published on the Internet often contain misspellings and slang expressions. However, activity on social media or e-commerce platforms is an essential part of our lives [15] and can be considered a valuable and underexplored source of information about adverse events [13]. Pattanayak et al. discussed the advantages of applying e-commerce insights in the pharmaceutical industry [16].

Our literature review reveals that approaches to adverse event classification primarily focus on binary classification and are related to adverse events occurring in drugs. Additionally, the conducted analyses concentrate on texts initially written in one language, usually English. Language independence allows for broader analysis of adverse events, taking cultural trends into consideration and enabling quicker detection of problems. In the reviewed scientific papers, authors apply classical classifiers and neural networks; however, to the best of our knowledge, the hierarchical classification approach has not been used for the adverse event classification problem. Hierarchical classification is described as dividing a problem into smaller classification problems [17, 18].

In this paper, we propose the first attempt to conduct hierarchical classification on a large dataset consisting of e-commerce and social media texts. Our research focuses on over-the-counter (OTC) drugs and other health and hygiene products, including sexual well-being products and household chemicals. Our main contributions within the framework of the presented research are:

- performing multiclass classification of adverse events based on consumers' comments using a language-agnostic approach,
- examining both hierarchical and non-hierarchical approaches to classification,
- carrying out classification in conditions as close to production as possible,
- expanding research on adverse event classification to non-drug products.

2 Materials and Methods

2.1 Dataset

The dataset consists of over 152,000 texts gathered from e-commerce sources such as Amazon and Lazada, and social media platforms like Facebook and Twitter. It includes, among others, online product reviews, discussions about usage, and messages directed to brand profiles. This proprietary dataset was collected with the assistance of several third parties for internal processes related to customer relationship management and adverse event reporting. The texts were written in English and other languages, such as Spanish, Russian, Japanese, and Arabic. Approximately 70% of the data were initially written in English, while the remaining 30% were written in non-English languages and machine-translated into English. It is worth noting that the dataset reflects the actual distribution of the data collected, with heavily imbalanced classes – see Fig. 1. The texts were used as collected from the Internet and are unstructured, containing grammar mistakes, misspellings, numbers, and emojis. Non-English observations were machine-translated into English, which can make classification more challenging by introducing translation artifacts and mistakes. However, this approach allows for language-independent classification. Customer relations agents labeled the data in accordance with internal procedures for handling online engagements from customers, ensuring high data quality as trained subject matter experts processed it. A hierarchical labeling system was applied; for example, each observation was assigned a label on each hierarchy level. Figure 1 shows the category tree structure, with the percentage of observations of each class in the adverse-event and non-adverse event sets given in parentheses. The label "Other" means that the observation contains adverse events other than the ones listed.

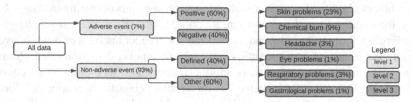

Fig. 1. Structure of labelled data

2.2 Data Processing

The data processing procedure included data cleaning, data translation, feature extraction and selection, and feature normalization. Data cleaning consisted of removing emojis and links. The machine translation process was applied to analyze all gathered data, even if it was not initially written in English. English has been defined as a common language for all comments. Text Translator, using Azure Cognitive Services [19, 20], was applied to translate non-English comments. It is worth noting that machine translation is not a perfect mechanism, and the translation of concise phrases can result in over-translation [21]. To assess the quality of translation, the following pilot experiment was conducted: English language comments were used as a training dataset. The tests were performed separately on English and translated comments. The tests were repeated, and the obtained results showed that classification metrics did not differ, hence the quality of translation was sufficient.

Figure 2 shows the feature extraction flowchart. The extracted features can be divided into two groups. The first group contains a set of general features, calculated once before the classification process. This set includes the number of words in each comment, the number of sentences, the ratio of stop words to all tokens, the number of exclamation signs, the number of questions, the ratio of uppercase to all tokens, mean word length, mean sentence length, and the ratio of numbers to all tokens. The second set of features was calculated before each level of classification. The N most common tokens were extracted for each class, with the union of these extracts being used as a dictionary for vectorization. This approach enables taking into account tokens from all classes. Otherwise, tokens describing small-sized classes could be omitted. In the present research, the first 500 tokens for each class were extracted and vectorized using the Term Frequency-Inverse Document Frequency method (TF-IDF). The Min-Max Scaler, which scales the minimum and maximum values to 0 and 1, was applied. We implemented the software for data analysis and classification in Python, using libraries such as NumPy, scikit-learn, pandas, and Azure-specific libraries.

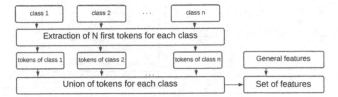

Fig. 2. Feature extraction flowchart.

2.3 Classification

The main aim of the classification process was to predict if the comment mentions an adverse event and, if yes, to assign the type of adverse event. The additional objective was to predict the sentiment of non-adverse events. Two tree structure classification methods were tested: XGBoost and Random Forest. Class weights were provided in the binary classification problem on the first level (i.e., adverse event vs. not adverse event classification) to tackle the class imbalance problem. The rest of the parameters were set empirically. Parameter tuning was conducted using grid search, and the following parameters were tested: learning rate and maximum depth of tree for XGBoost, and tree number for Random Forest. The following parameters were set for XGBoost [22]: learning rate: 0.3, maximum depth of a tree: 6, and scale_pos_weight was used. For Random Forest, the chosen parameters were [23]: tree number: 100, class_weight: 'balanced'. These classification models were chosen for the following reasons: a lower number of parameters (compared to deep learning models), lower computational cost allowing experiments to run in a reasonable amount of time, and clear interpretability. Moreover, literature shows that these classifiers can be applied to text classification problems [24, 25] with imbalanced classes successfully.

Two approaches were tested: the proposed hierarchical approach and the non-hierarchical approach. The hierarchical approach is presented in Fig. 3. In the non-hierarchical approach, all nine classes (tree leaves): seven classes of adverse events and two classes of non-adverse events were considered, and 9-class classification was performed. Hierarchical classification allows performing classification at different levels of the hierarchy tree sequentially. Separate classifiers are trained using 5-fold cross-validation to better estimate the classifier's power. Various classification thresholds are tested to obtain the best threshold, maximizing the F1 score. The dataset was shuffled and divided into train and test datasets in a stratified way, using an 80:20 proportion. The classifiers were evaluated using the following metrics: accuracy, recall, precision, and F1.

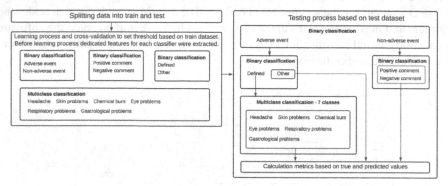

Fig. 3. Hierarchical approach

3   Results

Table 1 presents the results of the classification for hierarchical and non-hierarchical approaches. The table includes the results of nine-class classifications containing seven adverse event classes (six defined classes + seventh – other) and two non-adverse event classes (positive and negative comments). We do not report overall scores because of the class imbalance. The metrics were calculated separately for each class and presented in the table below. We evaluated the model both on training and testing datasets to check if the model is not overfitting. We report only metrics for the test dataset. Precision, recall, and F1 metrics are reported for each class. For adverse events detection, it's crucial to detect as many occurring cases as possible. Because of that, the best metrics for the task are recall and F1 score. The F1 metric was presented for both classifiers. The second column in the table contains information if the class is an adverse event (ADR) or not (NONADR) for top-level binary classification.

The application of hierarchical classification allows adjusting classification thresholds on binary levels in the hierarchy. For XGBoost, the thresholds were set to 0.5 for both binary levels, while for the Random Forest, the threshold was set to 0.3 for the level classifying the occurrence of an adverse event and 0.5 for the level distinguishing between a non-defined adverse event (Other) and the defined types of adverse events. These thresholds were chosen in the course of preliminary tests. As it may be noticed, better results were obtained for the XGBoost classifier, especially for minority classes such as gastrological and respiratory problems. The hierarchical approach has given better results for XGBoost for some classes, such as eye problems, gastrological problems, headaches, and skin problems. Using Random Forest, better results were achieved using the hierarchical approach for all defined adverse event classes.

Table 1. Results of non-hierarchical and hierarchical classification for F1measure

Class	Type	Non-hierarchical		Hierarchical	
		XGBoost	Random Forest	XGBoost	Random Forest
Respiratory problems	ADR	0.54	0.06	0.44	0.35
Chemical burn	ADR	0.62	0.49	0.61	0.52
Eye problems	ADR	0.32	0.22	0.56	0.52
Gastrological problems	ADR	0.14	0.01	0.24	0.08
Headache	ADR	0.55	0.21	0.64	0.62
Skin problems	ADR	0.31	0.47	0.52	0.56
Other	ADR	0.67	0.61	0.63	0.68
Positive comment	NONADR	0.88	0.88	0.87	0.88
Negative comment	NONADR	0.81	0.81	0.79	0.80

Additionally, computation time for both approaches was measured. All calculations were carried out in the same conditions regarding hardware, train, and test datasets. One cycle of the feature extraction and the learning process was considered. Performing hierarchical classification using the XGBoost classifier is over 4.5 times faster than a non-hierarchical approach (around 3 min for hierarchical and 14 min for non-hierarchical). These gains may not seem high, but they will scale with the size of the dataset, the number of classes, and in exhaustive grid searches. In the case of a parameter grid containing a total of 50 combinations, the time gain would rise to 500 min. In the case of Random Forest, the hierarchical approach took around 6 min, while the non-hierarchical approach took 4 min.

4 Discussion and Conclusions

The study aimed to classify adverse events independently of comment language using a hierarchical classification based. The classification was performed based on consumers' comments gathered from e-commerce and social-media sources. Our research presents the classification of adverse events occurring for OTC drugs and other medicinal and hygiene products. The goal of our work was not only to obtain better results of adverse events detection but also to create the framework allowing for easier development of a classification model with other levels and classes. The examination of the language-independent approach has shown that it is possible to analyze comments in various languages by mapping them to English using a machine translation mechanism. To the best of our knowledge, it is the first study performed on such an amount of data independently from comment language, applying hierarchical classification in adverse event problems.

The results showed that the hierarchical approach allows not only to obtain better or similar results but also to carry out calculations in a notably shorter time – over than 4,5 times for XGBoost. Additionally, the hierarchical approach enables threshold adjustment at each binary level, and this approach can be easily extended to more levels. Finding the proper threshold allows us to obtain better results than using the default ones. Hierarchical classification enables the addition of additional levels and classes without starting the learning process from scratch. The application of interpretable machine learning models such as XGBoost and Random Forest allows to create the ranking of features and perform the linguistic analysis of them. Additionally, the hierarchical approach allows for classification with more classes. The application of feature selection technique based on selecting N first tokens for each class allowed us to better tackle class imbalance. Another valuable aspect of our work is the fact that the presented study reflects the real business problem with imbalanced classes. Not all possible categories had been used in the study (class Other) due to insufficient data. This is an area worth exploring in further research. Our proposed method is less time-consuming and gives promising results.

The presented research opens a series of planned studies. We would like to test the approach using more classification algorithms, such as Logistic Regression, Support Vector Machine and Neural Networks. Our present study does not consider the case of multilabel classification (more than one class possible for each of the texts). Further research will also include more detailed class hierarchy levels. We also plan to develop a named entity recognition (NER) model to recognize the drugs (products), indications and individual adverse events as they are reported in social media and e-commerce texts, to support pharmacovigilance practice with detection in search of early product issue signals from the Internet. We can also look to extend the interpretability of the trained models using SHAPley values.

References

1. https://www.ema.europa.eu/en/glossary/adverse-event. Accessed 27 Jan 2023
2. Pirmohamed, M., et al.: Adverse drug reactions as cause of admission to hospital: prospective analysis of 18 820 patients. BMJ **329**(7456), 15–19 (2004)
3. Sarker, A., Gonzalez, G.: Portable automatic text classification for adverse drug reaction detection via multi-corpus training. J. Biomed. Inform. **53**, 196–207 (2015)
4. Zhang, Y., Cui, S., Gao, H.: Adverse drug reaction detection on social media with deep linguistic features. J. Biomed. Inform. **106**, 103437 (2020)
5. Kwa, M., Welty, L.J., Xu, S.: Adverse events reported to the US Food and Drug Administration for cosmetics and personal care products. JAMA Intern. Med. **177**(8), 1202–1204 (2017)
6. CFSAN Adverse Event Reporting System (CAERS) Data Web Posting. https://www.fda.gov/Food/ComplianceEnforcement/ucm494015.htm. Accessed 18 Jan 2023
7. Wang, J., Yu, L.C., Zhang, X.: Explainable detection of adverse drug reaction with imbalanced data distribution. PLoS Comput. Biol. **18**(6), e1010144 (2022)

8. Alhuzali, H., Ananiadou, S.: Improving classification of adverse drug reactions through using sentiment analysis and transfer learning. In: Proceedings of the 18th BioNLP Workshop and Shared Task, pp. 339–347 (2019)
9. Gurulingappa, H., Mateen-Rajpu, A., Toldo, L.: Extraction of potential adverse drug events from medical case reports. J. Biomed. Semant. **3**(1), 1–10 (2012)
10. Miranda, D.S.: Automated detection of adverse drug reactions in the biomedical literature using convolutional neural networks and biomedical word embeddings (2018)
11. Ding, P., Zhou, X., Zhang, X., Wang, J., Lei, Z.: An attentive neural sequence labeling model for adverse drug reactions mentions extraction. IEEE Access **6**, 73305–73315 (2018)
12. Breden, A., Moore, L.: Detecting adverse drug reactions from twitter through domain-specific preprocessing and bert ensembling. arXiv preprint arXiv:2005.06634 (2020)
13. Sloane, R., Osanlou, O., Lewis, D., Bollegala, D., Maskell, S., Pirmohamed, M.: Social media and pharmacovigilance: a review of the opportunities and challenges. Br. J. Clin. Pharmacol. **80**(4), 910–920 (2015)
14. Ginn, R., et al.: Mining Twitter for adverse drug reaction mentions: a corpus and classification benchmark. In: Proceedings of the Fourth Workshop on Building and Evaluating Resources for Health and Biomedical Text Processing, pp. 1–8 (2014)
15. Taher, G.: E-commerce: advantages and limitations. Int. J. Acad. Res. Account. Finan. Manage. Sci. **11**(1), 153–165 (2021)
16. Pattanayak, R.K., Kumar, V.S., Raman, K., Surya, M.M., Pooja, M.R.: E-commerce application with analytics for pharmaceutical industry. In: Ranganathan, G., Fernando, X., Piramuthu, S. (eds.) Soft Computing for Security Applications. Advances in Intelligent Systems and Computing, vol. 1428. Springer, Singapore (2023). https://doi.org/10.1007/978-981-19-3590-9_22
17. Tay, E.: Evaluating Bayesian Hierarchical Models and Decision Criteria for the Detection of Adverse Events in Vaccine Clinical Trials (2022)
18. Freitas, Alex, Carvalho, André: A tutorial on hierarchical classification with applications in bioinformatics. In: Taniar, D. (ed.) Research and Trends in Data Mining Technologies and Applications, pp. 175–208. IGI Global (2007). https://doi.org/10.4018/978-1-59904-271-8.ch007
19. Bisser, S.: Introduction to azure cognitive services. In: Microsoft Conversational AI Platform for Developers, pp. 67–140. Apress, Berkeley, CA (2021). https://doi.org/10.1007/978-1-4842-6837-7_3
20. Satapathi, A., Mishra, A.: Build a multilanguage text translator using azure cognitive services. In: Developing Cloud-Native Solutions with Microsoft Azure and. NET, pp. 231–248. Apress, Berkeley, CA, (2023)
21. Wan, Y., et al.: Challenges of neural machine translation for short texts. Comp. Linguist. **48**(2), 321–342 (2022)
22. Chen, T., Guestrin, C.: Xgboost: a scalable tree boosting system. In: Proceedings of the 22nd ACM SIGKDD International Conference on Knowledge Discovery and Data Mining, pp. 785–794 (2016)
23. Breiman, L.: Random forests. Mach. Learn. **45**(1), 5–32 (2001)
24. Shah, K., Patel, H., Sanghvi, D., Shah, M.: A comparative analysis of logistic regression, random forest and KNN models for the text classification. Augment. Hum. Res. **5**(1), 1–16 (2020)
25. Haumahu, J.P., Permana, S.D.H., Yaddarabullah, Y.: Fake news classification for Indonesian news using Extreme Gradient Boosting (XGBoost). In: IOP Conference Series: Materials Science and Engineering, vol. 1098, no. 5, p. 052081. IOP Publishing (2021)

Estimating Chlorophyll Content from Hyperspectral Data Using Gradient Features

Bogdan Ruszczak[1,4](✉) ⓘ, Agata M. Wijata[2,4](✉) ⓘ, and Jakub Nalepa[3,4](✉) ⓘ

[1] Faculty of Electrical Engineering, Automatic Control and Informatics, Department of Informatics, Opole University of Technology, Opole, Poland
b.ruszczak@po.edu.pl
[2] Faculty of Biomedical Engineering, Silesian University of Technology, Zabrze, Poland
awijata@ieee.org
[3] Faculty of Automatic Control, Electronics and Computer Science, Department of Algorithmics and Software, Silesian University of Technology, Gliwice, Poland
jnalepa@ieee.org
[4] KP Labs, Gliwice, Poland

Abstract. Non-invasive estimation of chlorophyll content in plants plays an important role in precision agriculture. This task may be tackled using hyperspectral imaging that acquires numerous narrow bands of the electromagnetic spectrum, which may reflect subtle features of the plant, and inherently offers spatial scalability. Such imagery is, however, high-dimensional, therefore it is challenging to transfer from the imaging device, store and investigate. We propose a machine learning pipeline for estimating chlorophyll content from hyperspectral data. It benefits from the Savitzky-Golay filtering to smooth the (potentially noisy) spectral curves, and from gradient-based features extracted from such a smoothed signal. The experiments revealed that our approach significantly outperforms the state of the art according to the widely-established estimation quality metrics obtained for four chlorophyll-related parameters.

Keywords: machine learning · chlorophyll content · feature engineering · hyperspectral image · regression

1 Introduction

The agricultural sector has evolved over the years, in response to a growing demand for food, fiber and fuel [13]. The limited availability of land requires targeted management of resource production and leads to the increasing adoption of

---

This work was partially supported by The National Centre for Research and Development of Poland (POIR.04.01.04-00-0009/19). AMW was supported by the Silesian University of Technology, Faculty of Biomedical Engineering grant (07/010/BK_23/1023). JN was supported by the Silesian University of Technology Rector's grant (02/080/RGJ22/0026).

J. Mikyška et al. (Eds.): ICCS 2023, LNCS 14074, pp. 196–203, 2023.
https://doi.org/10.1007/978-3-031-36021-3_18

precision agriculture [17]. In this context, remote sensing can easily become a tool for identifying soil and crop parameters, due to its intrinsic scalability [9,17]. In the case of agriculture, methods using multi- and hyperspectral remote sensing, capturing multispectral and hyperspectral images (MSI and HSI) are used, and non-invasive extraction of the chlorophyll content plays an increasingly important role, as it can ultimately lead to improving agricultural practices [7].

The majority of vegetation indices (VIs) were designed for multispectral sensors [14]. However, the wide bands in such imagery result in limited accuracy in the early detection of negative plant symptoms [1]. The use of HSIs, which are characterized by high spectral resolution, allows for the extraction of more details in the spectral response of an object [10]. Here, estimating chlorophyll content from hyperspectral data is commonly carried out by calculating the value of narrow-band VIs [19]. They include the Normalized Difference Vegetation Index (NDVI), Optimal Soil Adjusted Vegetation Index, Ratio Index and Difference Index [19]. Another parameter is the maximum quantum yield of photochemistry (Fv/Fm) [15]. Also, the Soil and Plant Analyzer Development (SPAD) tool is exploited, which measures the relative level of chlorophyll content in a crop taking into account the level of chlorophyll in the canopy. Finally, the performance index (PI) makes it possible to estimate the level of chlorophyll too [10].

Some approaches operate directly on the HSI data to estimate the chlorophyll content, hence they omit the stage of determining VIs. In such techniques, selected hyperspectral bands are analyzed—the encompass the Red and Near-Infrared (NIR) combination [6], the Blue, Green, Red, and NIR combined range [16], the Blue, Green and Red channels [17], or just the NIR band [4]. The bands undergo feature extraction (often followed by feature selection) in classic machine learning algorithms, whereas deep learning models benefit from automated representation learning over such data. The former group of techniques span across a variety of feature extractors and regression models, including continuous wavelet transforms (CWTs) [17], partial least square regression (PLSR) [18], kernel ridge regression [12] or regression using random forests. On the other hand, convolutional neural networks (CNNs) [14] and generative adversarial nets (GAN) [16] have been utilized for chlorophyll estimation as well.

Unfortunately, the data-driven algorithms are commonly validated over the in-house (private) data, following different validation procedures. This ultimately leads to the reproducibility crisis, and to inability to confront the existing approaches in a fair way [8]. In our recent work, we addressed this research gap and introduced a benchmark dataset, together with the validation procedure and a set of suggested metrics which should be used to quantify the generalization capabilities of machine learning models for chlorophyll estimation [11]. Here, we exploit this dataset and validation procedure to understand the abilities of the proposed processing chain, and to compare its estimation performance with 15 baseline models (for clarity, we focus on the best algorithms from [11]).

We tackle the problem of automated analysis of hyperspectral data using machine learning algorithms in the context of estimating the chlorophyll content (Sect. 2). We show that appropriately designed feature extractors fed into well-established supervised regression models can dramatically enhance their operational capabilities. Our experiments (Sect. 3), performed over the recent dataset and following the validation suggested in [11], indicated that smoothing the spectral curves using the Savitzky-Golay filtering and extracting gradient-based features lead to significant improvements in the estimation quality when compared to the current state of the art. Also, we executed extensive computational experiments to understand the impact of the hyperparameter selection on the regression engine, and to optimize the hyperparameters of the system.

2   Materials and Methods

In this section, we summarize the dataset and the quantitative metrics used to assess the investigated algorithms (Sect. 2.1), together with our machine learning pipeline for estimating chlorophyll content from HSI (Sect. 2.2).

2.1   Dataset

We exploit the CHESS (**CH**lorophyll **ES**timation DataSet) dataset introduced in our recent study [11]—it was collected in the Plant Breeding and Acclimatization Institute -National Research Institute (IHAR-PIB) facility located in Central Poland (Jadwisin, Masovian Voivodeship) during the 2020 campaign (June—July, with three rounds of data acquisition, 4 weeks apart from each other). There were three flights over two sets of 12 plots resulting in 72 HSIs (150 bands, 460–902 nm, with the 2.2 cm ground sampling distance). In the plots, there were two potato varieties planted: *Lady Claire* (12 plots) and *Markies* (12 plots). The image data is accompanied with the in-situ measurements for each plot: (*i*) the SPAD, (*ii*) the maximum quantum yield of the PSII photochemistry (Fv/Fm), (*iii*) the performance of the electron flux to the final PSI electron acceptors, and (*iv*) relative water content (RWC), reflecting the degree of hydration of the leaf's tissue. In [11], we introduced the training-test split in which both subsets are equinumerous, and they are stratified following the distribution of each ground-truth parameter independently, so that both training and test subsets (each containing 36 plots) maintain a similar parameter's distribution.

To quantify the regression performance, we use the metrics, as suggested in [11]: the coefficient of determination R^2 which should be maximized ($\uparrow$; R^2 with one being the perfect score; its negative values indicate a worse fit than the average), mean absolute percentage error (MAPE), mean squared error (MSE) and mean absolute error (MAE)—all those measures should be minimized ($\downarrow$).

2.2   Estimating Chlorophyll Content Using Machine Learning

We exploit a processing chain, in which the input HSI undergoes feature extraction, and the features are fed to a regression model to predict the value of each parameter (we train four independent models). The algorithms at each step of the pipeline can be conveniently replaced by other techniques. We build upon several insights concerning the shape of the median spectral curves extracted for

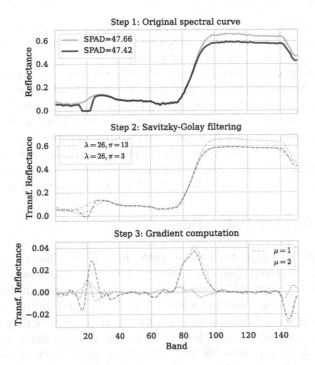

Fig. 1. The flowchart presenting the proposed feature engineering process.

the separate fields of interest (the spectral curves for all pixels are aggregated to generate a single median curve for the field). It is of note that some of the curves extracted for CHESS that should represent similar measurements (e.g., similar ground truth) do not look alike, and there exist noisy curves—this could be related to the difficulties in capturing enough light for selected spectral bands, which could easily lead to a narrower tonal range of the photosensitive camera.

To tackle the issue of the noisiness of spectral curves (and to increase its signal-to-noise ratio through removing high-frequency noise from the signal), the feature extraction stage is preceded by the filtering of spectral data using the Savitzky-Golay filter which may be considered as a generalized moving average filter. We aim at eliminating the influence of random noise and at reducing the drift phenomenon on the spectral reflection coefficient [5]. The Savitzky-Golay filter is given as the discrete convolution (h denotes the signal):

$$y[\pi] = \sum_{m=-\lambda}^{\lambda} h[m] * [\pi - m] = \sum_{m=\pi-\lambda}^{\pi+\lambda} h[\pi - m] * [m], \tag{1}$$

where $2\lambda + 1$ is the length of the approximation interval, and π is the polynomial order, both being the tunable hyperparameters. In Fig. 1, we present the example of the original spectral curves for two fields with similar SPAD values amounting to 47.66 and 47.42 (first row) which underwent Savitzky-Golay filtering for two hyperparameter sets ($\lambda = 26$, $\pi = 13$ and $\lambda = 26$, $\pi = 3$), resulting in slightly different shapes of the smoothed signal (second row). In both cases, the small "noisy" variations of the original curve were removed across the entire spectrum.

Once the original spectral curve is filtered, we extract the gradients which constitute the feature vectors for each field. We build upon the observation that the detection of anomalies in the signal data can be supported by the use of a gradient elaborated over the spectral curve, which allows us to capture the subtle characteristics in such data [2]. The feature extraction may be performed μ times, with $\mu = 0$ denoting the original curve (i.e., the feature vector includes the original reflectance values). Such feature vectors of size $\mathcal{B}$, where $\mathcal{B}$ is the number of hyperspectral bands (here, $\mathcal{B} = 150$), are fed to the regression model.

3   Experimental Validation

We investigate the linear machine learning models: (i) linear regressors with L_1, and (ii) L_2 regularization, (iii) support vector machines with linear kernel, and (iv) the elastic net with regularization, all implemented in Scikit-learn. Savitzky-Golay filtering was implemented in SciPy, and feature extraction in NumPy. We focus on the linear regression models to avoid heavily parameterized techniques—due to this assumption, we were able to extensively evaluate thousands of models in a reasonable time (27 840 model's configurations). At the same time, we maintained the high interpretability of the study.

For each model, we optimized its hyperparameters (this fine-tuning was performed following the 5-fold cross-validation strategy over the training set):

- Linear regression with L1 regularization, for: $\alpha \in \{10^{-15}, 10^{-14}, \ldots, 10^{15}\}$,
- Linear regression with L2 regularization, for: $\alpha \in \{10^{-15}, 10^{-14}, \ldots, 10^{15}\}$,
- Support vector machine with linear kernel, for: $C \in \{2^{-5}, 2^{-4}, \ldots, 2^8\}$, and the maximum number of iterations $\mathcal{I} \in \{500, 1000, 2500\}$,
- Elastic net with L_1 regularization, for $\alpha \in \{10^{-15}, 10^{-14}, \ldots, 10^{15}\}$, and $L_{1ratio} \in \{0.05, 0.1, \ldots, 0.9\}$.

Similarly, the feature extraction is a parameterized step, as it may be performed multiple times. We denoted the number of gradient runs as μ, and by $\mu = 0$ we report the results obtained for the regression models operating over the original curves. Therefore, the full configuration for the performed experimental search was as follows: $\mu \in \{0, 1, 2, 3\}, \lambda \in \{1, 2, \ldots 30\}, \pi \in \{1, 2, \ldots, 29\}$.

In Fig. 2, we depict the impact of π, λ, and μ on the R^2 coefficient for the elastic net models predicting the SPAD parameter. Here, we focused on a single machine learning model (with default parameterization) to verify the importance of signal filtering and feature extraction on the regression capabilities of the algorithm. Albeit the insights learned from this experiment may not be generalizable to other models, we anticipate that a similar trend would be observed, as feature engineering constitutes one of the most important aspects of building machine learning pipelines [3]. We can observe that the exhaustive traversal of the search spaces allows to indicate their most promising regions for the (λ, π)

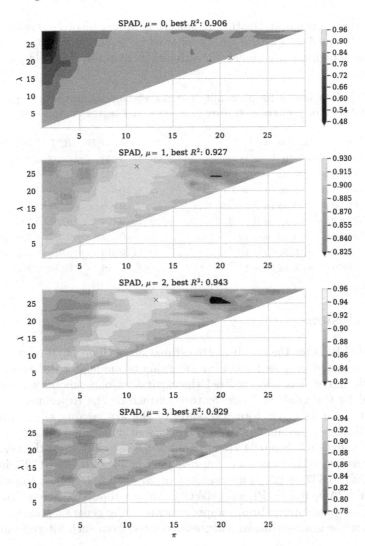

Fig. 2. The R^2 metric obtained for various filtering and feature extraction settings. The light blue × marker indicates the best configuration for each gradient level (μ). (Color figure online)

configurations, which remain consistent for the gradient-extraction levels (μ). However, the exact position of the best parameterization differs across μ's.

The optimized models outperform *Baseline* (the best-known R^2 values from the literature [11])—in Table 1, we report the optimized hyperparameters for the models offering the best regression. The R^2 measure notably increased for all parameters, whereas the regression errors, e.g., MAE, decreased by 53.5%, 23.3%, 30.9%, and 25.0% for SPAD, FvFm, PI, and RWC. The experiments indicated that appropriate feature engineering, which involves Savitzky-Golay filtering of the median spectral curves followed by feature extraction, allows the elaboration of high-quality models for estimating chlorophyll-related parameters.

Table 1. The best results for all quality metrics, elaborated for the best parameterization (Savitzky-Golay filtering, feature extraction and regression models).

Param.	Model configuration	μ	π	λ	MAPE ↓	R^2 ↑	MAE ↓	MSE ↓
SPAD	Elastic net ($\alpha = 10^{-1}$, $L_{1ratio} = 0.5$)	2	26	12	0.035	0.943	0.756	3.012
	Linear regr. with L_2 ($\alpha = 2.5 \times 10^{-5}$)		—		0.072	0.827	1.625	9.095
FvFm	Linear regr. with L_2 ($\alpha = 10^{-3}$)	0	22	17	0.030	0.764	0.016	0.001
	Linear regr. with L_2 ($\alpha = 5 \times 10^{-5}$)		—		0.036	0.727	0.021	0.001
PI	SVM with linear kernel ($C = 2$, $\mathcal{I} = 10^3$)	1	29	20	0.401	0.837	0.194	0.083
	Linear regr. with L_2 ($\alpha = 10^{-11}$)		—		0.532	0.677	0.280	0.169
RWC	Linear regr. with L_1 ($\alpha = 10^{-1}$)	2	22	11	0.010	0.911	0.706	1.089
	Linear regr. with L_2 ($\alpha = 10^{-3}$)		—		0.013	0.859	0.941	1.731

4   Conclusions

We exploited HSIs for the non-invasive estimation of chlorophyll-related parameters in plants and proposed a machine learning technique for this task. To deal with subtle signal noise, we utilized the Savitzky-Golay smoothing filter that is followed by the gradient-based feature extractor. The experiments revealed that our techniques outperform the state of the art, as quantified using four chlorophyll-related parameters. The coefficient of determination (R^2) achieved by our techniques reached 0.943 (compared to 0.827 reported for the best model in [11], therefore we obtained the improvement of 14%), 0.764 (0.727, improvement of 5%), 0.837 (0.667, improvement of 25%), 0.911 (0.859, improvement of 6%) for SPAD, FvFm, PI, and RWC. Also, we showed that employing the Savitzky-Golay smoothing brings improvements in the generalization of a model trained over the gradient-based features extracted from such filtered signal.

References

1. Adão, T., et al.: Hyperspectral imaging: a review on UAV-based sensors, data processing and applications for agriculture and forestry. Remote Sens. 9(11) (2017)
2. Chen, J., Lu, M., Chen, X., Chen, J., Chen, L.: A spectral gradient difference based approach for land cover change detection. ISPRS J. Photogramm. Remote. Sens. **85**, 1–12 (2013)
3. Chicco, D., Oneto, L., Tavazzi, E.: Eleven quick tips for data cleaning and feature engineering. PLoS Comput. Biol. **18**(12), 1–21 (2022)
4. Gorretta, N., Nouri, M., Herrero, A., Gowen, A., Roger, J.M.: Early detection of the fungal disease "apple scab" using SWIR hyperspectral imaging. In: 2019 10th Workshop on Hyperspectral Imaging and Signal Processing: Evolution in Remote Sensing (WHISPERS), pp. 1–4 (2019)

5. Guo, C., et al.: Predicting Fv/Fm and evaluating cotton drought tolerance using hyperspectral and 1D-CNN. Front. Plant Sci. **13**, 3700 (2022)
6. Huynh, N.H., Böer, G., Schramm, H.: Self-attention and generative adversarial networks for algae monitoring. Eur. J. Remote Sens. **55**(1), 10–22 (2022)
7. Jin, X., Li, Z., Feng, H., Ren, Z., Li, S.: Deep neural network algorithm for estimating maize biomass based on simulated sentinel 2A vegetation indices and leaf area index. Crop J. **8**(1), 87–97 (2020)
8. Nalepa, J., Myller, M., Kawulok, M.: Validating hyperspectral image segmentation. IEEE Geosci. Remote Sens. Lett. **16**(8), 1264–1268 (2019)
9. Ponnusamy, V., Natarajan, S.: Precision agriculture using advanced technology of IoT, unmanned aerial vehicle, augmented reality, and machine learning. In: Gupta, D., de Hugo, C., Albuquerque, V., Khanna, A., Mehta, P.L. (eds.) Smart Sensors for Industrial Internet of Things: Challenges, Solutions and Applications, pp. 207–229. Springer International Publishing, Cham (2021). https://doi.org/10.1007/978-3-030-52624-5_14
10. Ruszczak, B., Boguszewska-Mańkowska, D.: Deep potato - the hyperspectral imagery of potato cultivation with reference agronomic measurements dataset: towards potato physiological features modeling. Data Brief **42**, 108087 (2022)
11. Ruszczak, B., Wijata, A.M., Nalepa, J.: Unbiasing the estimation of chlorophyll from hyperspectral images: a benchmark dataset, validation procedure and baseline results. Remote Sens. **14**(21) (2022)
12. Singhal, G., Bansod, B., Mathew, L., Goswami, J., Choudhury, B., Raju, P.: Estimation of leaf chlorophyll concentration in turmeric (curcuma longa) using high-resolution unmanned aerial vehicle imagery based on kernel ridge regression. J. Indian Soc. Remote Sens. **47**, 1–12 (2019)
13. Sishodia, R.P., Ray, R.L., Singh, S.K.: Applications of remote sensing in precision agriculture: a review. Remote Sens. **12**(19) (2020)
14. Wang, J., et al.: Estimating leaf area index and aboveground biomass of grazing pastures using Sentinel-1, Sentinel-2 and Landsat images. ISPRS J. Photogramm. Remote. Sens. **154**, 189–201 (2019)
15. Wen, S., Shi, N., Lu, J., Gao, Q., Yang, H., Gao, Z.: Estimating chlorophyll fluorescence parameters of rice (Oryza sativa L.) based on spectrum transformation and a joint feature extraction algorithm. Agronomy **13**(2) (2023)
16. Yan, T., et al.: Combining multi-dimensional convolutional neural network (CNN) with visualization method for detection of aphis gossypii glover infection in cotton leaves using hyperspectral imaging. Front. Plant Sci. 12 (2021)
17. Yue, J., Zhou, C., Guo, W., Feng, H., Xu, K.: Estimation of winter-wheat aboveground biomass using the wavelet analysis of unmanned aerial vehicle-based digital images and hyperspectral crop canopy images. Int. J. Remote Sens. **42**(5), 1602–1622 (2021)
18. Zhang, J., et al.: Detection of canopy chlorophyll content of corn based on continuous wavelet transform analysis. Remote Sens. **12**(17) (2020)
19. Zhang, Y., et al.: Estimating the maize biomass by crop height and narrowband vegetation indices derived from UAV-based hyperspectral images. Ecol. Ind. **129**, 107985 (2021)

Image Recognition of Plants and Plant Diseases with Transfer Learning and Feature Compression

Marcin Zięba, Konrad Przewłoka, Michał Grela, Kamil Szkoła, and Marcin Kuta(✉) ⓘD

Institute of Computer Science, AGH University of Krakow, Al. Mickiewicza 30, 30-059 Krakow, Poland
mkuta@agh.edu.pl

Abstract. This article introduces an easy to implement kick-starting method for transfer learning of image recognition models, meant specifically for training with limited computational resources. The method has two components: (1) Principal Component Analysis transformations of per-filter representations and (2) explicit storage of compressed features. Apart from these two operations, the latent representation of an image is priorly obtained by transforming it via initial layers of the base (donor) model. Taking these measures saves a lot of computations, hence meaningfully speeding up the development. During further work with models, one can directly use the heavily compressed features instead of the original images each time. Despite having a large portion of the donor model frozen, this method yields satisfactory results in terms of prediction accuracy. Such a procedure can be useful for speeding up the early development stages of new models or lowering the potential cost of deployment.

Keywords: transfer learning · feature extraction · feature compression · plant recognition · plant diseases recognition

1 Introduction

The article presents a method for transfer learning of image recognition models using additional dimensionality reduction of latent features as feature compression. This can be used to create mobile applications for automatic recognition of plants and their diseases based on photos. It required acquiring suitable training datasets and development of efficient and memory-frugal machine learning models.

The proposed approach helps to meet the crucial requirement that the deployed models are self-contained and, consequently, are not dependent on the access to the Internet. This requirement is backed by a reasonable assumption that people taking photos (e.g. farmers in the field, tourists in the mountains) may lack a stable Internet connection. Thus, the main focus was on efficiency of

J. Mikyška et al. (Eds.): ICCS 2023, LNCS 14074, pp. 204–211, 2023.
https://doi.org/10.1007/978-3-031-36021-3_19

the proposed models. Due to limited resources and time constraints, the training process was also optimized.

The visual recognition of plants poses some specific difficulties on its own like similarity of certain species, intra-species variability, different parts of a plant photographed, different growth stages at which the photo of plant is captured and non-obvious image cropping [8].

With transfer learning already being the most time-effective way to start working on computer vision tasks, the method described in this article considers optimizing this process even further, building upon two mechanisms:

- feature extraction using initial layers of a pre-trained model and Principal Component Analysis (PCA) transformation on top of that,
- explicit work on the latent representation of the data (calculating it once and storing in such a form).

Combining these two strategies heavily limits the amount of computation needed to train consecutive models using the compressed features directly as their input. Eventually, one can simply merge feature extractors with the trained model to form a single composite model (or a pipeline).

Six models were prepared, all trained on three datasets: GRASP-125, PlantVillage and PlantDoc. Four proposed models are variations of an approach using two-step feature extraction (exploiting our proposed approach to a full extent) and two models use only single-step feature extraction. We call these steps further (1) primary feature extraction (using initial layers of base model) and (2) secondary feature extraction (using PCA transformation applied in a specific way). Models with only the primary feature extraction achieve the TOP1 test-time accuracy (precise prediction of the true class) of prediction of 81.22% on the GRASP-125 dataset, 97.97% on PlantVillage, 56.24% on PlantDoc (SEResNet-based) and 85.04% on GRASP-125, 98.53% on PlantVillage, 57.55% on PlantDoc (MobileNetV2-based). Meanwhile, models following the approach with two-step extraction result in TOP1 accuracy of 79.58% on GRASP-125, 97.31% on PlantVillage and 51.88% on PlantDoc.

Despite such an aggressive feature extraction (from originally 150528–dimensional problem down to 7840 after primary feature extraction and eventually to 800 after the secondary one), models do not suffer from falling behind too much in terms of the prediction accuracy.

The PCA transformation enabled the creation of a solution that is only slightly worse in terms of performance, but noticeably easier to train compared to more traditional methods (i.e., SEResNet-based and MobileNet-based models). The computational cost of training was further reduced by utilizing the aforementioned latent representation of data. The main advantage of this approach is having a compressed representation of data precomputed ahead of the training itself. This may hurt eventual accuracy of prediction, but enables the use of smaller and simpler models on top of it.

It is important to emphasise that we apply PCA transformation to the data already processed with a deep neural network feature extractor, and not directly to the raw input data itself. Such an approach, to the best of our knowledge, has not been analysed yet. Despite the heavy limitations it may incur in a typical model development (which might be a reason for being overlooked), this method may still be successfully utilized in a transfer learning manner.

2   Related Work

Modern systems for automatic plant species and plant diseases recognition are based on deep convolutional neural networks and transfer learning. Such networks achieve amazing accuracies on a wide range tasks, but their excessive need for memory and computational resources may be prohibitive from deployment on devices with limited resources. One of the main frontiers of advancement in image classification is a strive to minimize the number of parameters a model uses while retaining a reasonable accuracy. Examples of research regarding the high-efficiency image recognition are the mobile networks from MobileNets architectures family: MobileNet [3], MobileNetV2 [9] and MobileNetV3 [2]. The MobileNetV2 model [9] is used in our work as a base model for transfer learning. To reduce computations, MobileNets introduce depthwise separable convolutions, which consist from depthwise convolutional filters and pointwise convolutions.

Speedup in test-time prediction can be achieved by exploiting redundancy between different filters and feature channels. This was obtained with filter factorization, implemented with SVD transformation, and was applied to 15 layer CNN [1]. A related approach was presented in [5]. Our work is related to [1], [5] through the usage of PCA.

3   Datasets

The experiments were conducted on three datasets: the GRASP-125 dataset [6], PlantDoc [10] and PlantVillage [7]. The GRASP-125 and PlantDoc datasets were already split into training and testing sets. To further validate the performance of our models, we created additional validation datasets for these sets. Conversely, the PlantVillage dataset was not originally divided and thus we split it into training, validation, and testing sets for the purpose of evaluating our models.

The collected images were subsequently augmented for the training set to gain a more stable and comprehensive measure of models' predictive power. Image augmentation involved typical operations like random rotation, random cropping, horizontal flip, and random adjustment of brightness, saturation, hue and contrast. Finally, each image was scaled to 224×224 pixels, the format accepted by the utilized models.

GRASP-125 contains 1 6327 images of vascular plants belonging to 125 classes (plant species). The test set contains 1704 images. The validation set contains 12 randomly chosen images per plant from the initial training set, in total 1500 images. The training set contains the remaining 13 119 images.

As a result of the augmentation process, each plant species was represented by 400 training examples and 12 validation examples. The test examples were not changed.

The PlantVillage dataset [7] is dedicated to plant disease detection based on plant leaves. It contains 55 448 images of healthy and unhealthy leaf images, divided into 39 classes by species and disease. As no predefined split was provided, we divided the dataset into test, validation and training sets. The division was stratified, so the original proportions of classes were maintained.

The test set comprised 11 090 samples, constituting 20% of the dataset. The validation set was the same size as the test set (11 090 samples, representing 20% of the dataset). Prior to the data augmentation process, the training set contained 33 268 images, comprising 60% of the original dataset. Following the augmentation process, the number of images in the training set increased to 46 541, from 1000 to 3304 images per class, depending on the class.

PlantDoc [10] is a newer dataset containing 2576 images of healthy or diseased plant leaves belonging to 27 classes (17 diseased, 10 healthy) from 13 plant species. It comes with a predefined train-test split into 2340 images in the training set and 236 images in the test set. From 15% of the original training set, we created the validation set, containing 351 samples. The class distribution of the original dataset was maintained in the validation set. The training set was aggressively augmented to contain between 1037 and 1152 images per class, depending on the class, which gives a total of 28 983 images. This is in contrast to the original dataset, which contained between 44 and 179 images per class before the augmentation process.

4   Proposed Approach

All the further experiments were conducted using the MobileNetV2 model as the base model [9]. This architecture is known to be a good off-the-shelf image classifier, achieving both satisfactory prediction accuracy and computational efficiency. MobileNetV2 has been chosen out of the MobileNet architectures due to the availability of the pre-trained model with weights on a task similar to plant species or plant diseases recognition. The MobileNetV2 model was used mostly as the primary feature extractor – specifically, initial layers making up for 1 364 864 out of all 3 538 984 parameters of the model were utilized.

For the purpose of transfer learning, a model pre-trained on the ImageNet dataset was chosen. This base model most likely did not see an overwhelming majority (if any) of the images from the three datasets used here during its training.

The overall transfer learning procedure for preparing the compressed representation of the dataset, as well as feature extractors (both primary and secondary), consists of the following steps:

1. Specifying of the primary feature extractor (initial layers of some pre-trained model).
2. Encoding the whole dataset using the (primary) feature extractor from the previous step.
3. Saving the compressed examples to a persistent storage.

4. Initialization of separate PCA units, one per each channel (filter output), with an arbitrary number of principal components specified.
5. Fitting the PCA units, passing to each of them only the corresponding channel contents, using only the training dataset (data per channel may require its flattening first).
6. Encoding the whole compressed dataset (training, validation, test sets) from step 3 by applying the learnt PCA transformations (secondary feature extractor).
7. Saving the compressed examples to a persistent storage.

The final representation is a highly compressed version of the original features. Due to the fact that now we have all three datasets in such a form, we can continue working at this abstraction layer as if these were the actual input data. Once some model is finally obtained, it is sufficient to attach it on top of the feature extractors, thus producing a full model capable of working on raw (non-compressed) data. To achieve this, it might be necessary to define a custom model layer for carrying the PCA dimensionality reduction, given the previously learnt parameters of this transformation.

Let n be batch size, c – number of channels, s – size of a channel, and p – number of principal components to be extracted. In a batch, multichannel and multi-PCA-unit case let us assume further the following notation:
$X \in \mathbb{R}^{n \times s \times c}$ – data tensor, $M \in \mathbb{R}^{1 \times s \times c}$ – all empirical means tensor, $W \in \mathbb{R}^{s \times p \times c}$ – all principal components tensor, $Y \in \mathbb{R}^{n \times p \times c}$ – data tensor of reduced dimensionality. Then, for each channel i we have $Y_{:,:,i} = (X_{:,:,i} - M_{0,:,:})W_{:,:,i}$.

For practical use, a custom layer or component should be implemented, so that it is initialized with empirical means and basis vectors of principal components (obtained from PCA units used during training), and during runtime it applies above transformation to its inputs.

Initial layers of convolutional deep learning models for computer vision can be effectively pictured as the extractors of some abstract features. They are successful at learning edges, color gradients and simple shapes, and then using them for further, more high-level, reasoning. Representation of latent features at a specific level of the network gives at our disposal a dozen of abstract feature maps, one per channel. In particular, each channel here is an output of a corresponding filter from the previous layer. Each of these filters is specialized in identifying some specific phenomena. Thus, it may be presumed that intra-channel features are likely to be noticeably correlated with each other. This can result in the suboptimality of the representations. To exploit this fact, a PCA dimensionality reduction can be applied – importantly, a distinct one per each channel. As a result, we obtain a heavily compressed representation of the original input.

Applying the PCA transformation channel-wise is a computationally preferred option compared to the straight-forward use of a single PCA unit. Most notable distinction is the size of the principal axes matrix. Following the earlier notation, we can describe the number of parameters as:

- $c * s * p$ for calculating transformations for each channel separately (in our case $160 * 49 * 5 = 39\,200$),

– $(c*s)*(c*p)$ for calculating transformations for whole input at once (in our case $160*49*160*5 = 6\,272\,000$).

Both approaches reduce the dimensionality to the same extent. The observed discrepancy in the number of parameters originates from disregarding the inter-channel interactions.

It is viable to use Linear Discriminant Analysis (LDA) instead of PCA for dimensionality reduction. The transformation to be executed is analogous to the one depicted for PCA – using empirical means and linear projection matrix in a similar way. The prediction accuracies for both approaches are comparable.

5   Experiments and Results

Three distinctively different model architectures have been eventually determined. One of them is fully built upon the approach presented in Sect. 4 (steps 1–7 of the above procedure – both feature extractors were used) and is presented in four variants. The other two models apply only the primary feature extractor (steps 1–3), providing a good point of reference.

It is essential to keep in mind that the original dimensionality of the problem is 150 528 as the input images are of size 224×224 pixels (3 channels). Primary feature extraction reduces it to 7840 (7×7×160) and the secondary one further decreases the dimensionality to 800 (5×160) due to PCA transformation from 49 features per channel down to 5 principal components.

The implemented architectures are:

– PCA+Dense – The architecture embraces both feature extraction stages applying the full procedure introduced in Sect. 4 – as such, this model is our main object of interest here. It operates on the input dimensionality of size 800. Its input is flattened and fed into a multi-layer perceptron.
– PCA+SepConv – The architecture is structurally similar to PCA+Dense, except for not flattening the extracted features immediately, instead using separable convolutions first.
– LDA+Dense – LDA versions of PCA+Dense architecture.
– LDA+SepConv – LDA versions of PCA+SepConv architecture.
– SEResNet – The architecture is designed as a single block of the Squeeze & Excitation Residual Network (SEResNet) architecture [4]. The choice of the SEResNet was rather arbitrary; nonetheless, it offers a decent predictive power while remaining relatively simple.
– MNv2 – The architecture, consisting of the leftover layers of the original MobileNetV2 [9], is adapted for distinguishing the relevant number of classes (125 for GRASP-125, 39 for PlantVillage, 27 for PlantDoc) instead of 1000 by reducing the number of units at the output layer and adding regularization to mitigate the risk of overfitting.

SEResNet and MNv2 rely only on the primary feature extractor and are kept here for reference. In particular, MNv2 is initialized (where it is possible) with the weights from the original model per-trained on ImageNet – it is likely that this model approximately determines the upper bound of the achievable accuracy in this setting.

The models were trained utilizing a single GPU instance. Each model was trained for 10 epochs. The optimizer used in each case was Nadam (with default Keras parameters: learning_rate=0.001, beta_1=0.9, beta_2=0.999) and the loss function was categorical cross entropy. Primary feature extractor consisted of initial layers of the base model with 1 364 864 parameters in total. Moreover, a learning rate scheduler halving the learning rate when arriving at plateau (with patience of 2 epochs) was used.

The prepared models achieved the accuracies shown in Table 1. Discrepancies between validation set and test set accuracy are due to the reasons mentioned in Sect. 3. Despite utilizing a significant number of data augmentation techniques, the models were unable to fully generalize their classification of images from the PlantDoc dataset due to its diversity.

Table 1. Accuracy of TOP1 and TOP5 prediction for each tested model. TOP1 accuracy means exact prediction of the true class. For TOP5 accuracy, the real class should be among the five most probable outcomes.

	Validation set		Test set	
	TOP1 [%]	TOP5 [%]	TOP1 [%]	TOP5 [%]
GRASP-125				
PCA+Dense	78.45	93.27	78.64	93.34
PCA+SepConv	80.33	92.60	79.99	94.54
LDA+Dense	77.53	92.07	78.64	92.96
LDA+SepConv	78.80	92.47	81.04	93.84
SEResNet	79.33	93.80	81.22	94.84
MNv2	84.27	94.33	85.04	94.89
PlantVillage				
PCA+Dense	97.67	99.92	97.41	99.94
PCA+SepConv	97.65	99.95	97.57	99.94
LDA+Dense	97.24	99.97	97.10	99.90
LDA+SepConv	96.82	99.95	97.13	99.90
SEResNet	98.00	99.96	97.97	99.94
MNv2	98.46	99.95	98.53	99.95
PlantDoc				
PCA+Dense	56.67	88.02	52.43	87.17
PCA+SepConv	52.69	88.31	53.67	88.44
LDA+Dense	52.40	87.44	51.16	85.83
LDA+SepConv	52.97	88.58	50.27	85.00
SEResNet	61.52	90.30	56.24	87.65
MNv2	61.52	88.87	57.55	88.01

6 Conclusions

In the setup of the conducted experiments the dimensionality of the problem was reduced from 150 528 (original photo scaled to 224×224 pixels, 3 channels), to 7840 (7×7×160) after primary feature extraction, and finally to 800 (5×160) after secondary feature extraction (PCA transformation). During these experiments the deterioration of the PCA-reliant model due to extensive feature compression was not as severe as it could be anticipated and satisfactory results could still be provided. Thus, PCA can be successfully used for the extraction of heavily compressed features from the latent representation of the data already processed by a convolutional feature extractor. The proposed procedure is useful for speeding up the early development stages of new models or lowering the potential cost of their deployment.

Acknowledgements. The research presented in this paper was supported by the funds assigned to AGH University of Krakow by the Polish Ministry of Education and Science.

References

1. Denton, E.L., Zaremba, W., Bruna, J., LeCun, Y., Fergus, R.: Exploiting linear structure within convolutional networks for efficient evaluation. In: Annual Conference on Neural Information Processing Systems, NIPS 2014, pp. 1269–1277 (2014)
2. Howard, A., et al.: Searching for mobilenetv3. In: 2019 IEEE/CVF International Conference on Computer Vision, ICCV 2019, pp. 1314–1324 (2019)
3. Howard, A.G., et al.: Mobilenets: efficient convolutional neural networks for mobile vision applications. CoRR abs/1704.04861 (2017)
4. Hu, J., Shen, L., Sun, G.: Squeeze-and-excitation networks. In: 2018 IEEE Conference on Computer Vision and Pattern Recognition, CVPR 2018, pp. 7132–7141 (2018)
5. Jaderberg, M., Vedaldi, A., Zisserman, A.: Speeding up convolutional neural networks with low rank expansions. In: British Machine Vision Conference, BMVC 2014 (2014)
6. Kritsis, K., et al.: Grasp-125: a dataset for Greek vascular plant recognition in natural environment. Sustainability **13**(21) (2021)
7. Mohanty, S.P., Hughes, D.P., Salathé, M.: Using deep learning for image-based plant disease detection. Front. Plant Sci. 7 (2016)
8. Nilsback, M., Zisserman, A.: Automated flower classification over a large number of classes. In: Sixth Indian Conference on Computer Vision, Graphics & Image Processing, ICVGIP 2008, pp. 722–729. IEEE Computer Society (2008)
9. Sandler, M., Howard, A.G., Zhu, M., Zhmoginov, A., Chen, L.: Mobilenetv 2: inverted residuals and linear bottlenecks. In: 2018 IEEE Conference on Computer Vision and Pattern Recognition, CVPR 2018, pp. 4510–4520 (2018)
10. Singh, D., Jain, N., Jain, P., Kayal, P., Kumawat, S., Batra, N.: Plantdoc: a dataset for visual plant disease detection. In: CoDS-COMAD 2020: 7th ACM IKDD CoDS and 25th COMAD, pp. 249–253. ACM (2020)

TwitterEmo: Annotating Emotions and Sentiment in Polish Twitter

Stanisław Bogdanowicz[✉] ⬥, Hanna Cwynar ⬥, Aleksandra Zwierzchowska ⬥,
Cezary Klamra ⬥, Witold Kieraś ⬥, and Łukasz Kobyliński ⬥

Institute of Computer Science, Polish Academy of Sciences, Jana Kazimierza 5, 01-248
Warszawa, Poland
stan.bogdanowicz97@gmail.com, {c.klamra,wkieras,
lkobylinski}@ipipan.waw.pl

Abstract. This article presents TwitterEmo, a new dataset for emotion and sentiment analysis in Polish. TwitterEmo provides a non-domain-specific and colloquial language dataset, which includes Plutchik's eight basic emotions and sentiment annotations for 36,280 tweets collected over a one-year period. Additionally, a sarcasm category is included, making this dataset unique in Polish computational linguistics. Each entry was annotated by at least four annotators. We present the results of the evaluation using several language models, including HerBERT and TrelBERT. The TwitterEmo dataset is a valuable resource for developing and training machine learning models, broadening possible applications of emotion recognition methods in Polish, and contributing to social studies and research on media bias.

Keywords: Natural Language Processing · Emotion Recognition · Sentiment Analysis

1 Introduction

The field of computational linguistics and natural language processing (NLP) increasingly uses social media platforms like Twitter to gather real-world data on user opinions and emotions. Potential restrictions or cancellation of Twitter's unrestricted academic access make the value of such datasets even more significant. This paper presents a new dataset of Polish tweets annotated with emotions, sentiment, and sarcasm, which is a valuable resource for training machine learning models. Unlike existing domain-specific emotion-annotated datasets, TwitterEmo fills a significant gap in the field by providing a non-domain-specific dataset for investigating emotions and sentiment in Polish texts. The dataset covers various topics, making it useful for social studies and research on media bias. Additionally, the dataset includes sarcasm annotations, which is a supplementary characteristic that may contribute to the study of emotions in language. This paper describes the creation and annotation process of the dataset and presents data analysis, including correlations between emotions and the distribution of sentiment over time. Preliminary tests were conducted to evaluate the data in terms of machine learning.

J. Mikyška et al. (Eds.): ICCS 2023, LNCS 14074, pp. 212–220, 2023.
https://doi.org/10.1007/978-3-031-36021-3_20

2   Related Works

A large number of sentiment analysis datasets developed to date have been specifically designed for widely known languages, such as English [4], yet few similar resources exist for lesser-known languages such as Polish.

With regard to sentiment annotation, one notable mention of a Polish resource is the dataset prepared for the sentiment recognition shared task during the PolEval2017 campaign [15]. The dataset consists of 1,550 sentences obtained from consumer reviews from three specific domains. The sentences are annotated at the level of phrases determined by the dependency parser, with each node of the dependency tree receiving one of the three sentiment classes: negative, neutral, and positive. Other instances of sentiment annotated domain-specific corpora for Polish are PolEmo 2.0 [6] and the Wroclaw Corpus of Consumer Reviews [7].

In emotion annotation, it is common to employ Plutchik's [12] Wheel of Emotions, which is exemplified by the collection of texts under the Sentimenti project [8], specifically developed for Polish. Additionally, a set of guidelines [14] has been developed to assist with emotion annotation in Polish consumer reviews.

Furthermore, as preliminary evidence suggests that the occurrence of sarcasm may cause up to a 50% decrease in accuracy in the automatic detection of sentiment in text and recognition of sarcasm based on the occurrence of hashtags may provide inaccurate results [13], manual annotation of sarcasm appears to be a fairly relevant consideration for machine learning applications.

As opposed to aforementioned datasets for Polish, text samples in our dataset come from unspecified and diverse domains. They exhibit a variety of stylistic features and tend to be highly opinionated due to their informal nature, thus rendering them a valuable source for sentiment analysis. To the best of our knowledge, the TwitterEmo dataset is the first for Polish that includes sentiment and emotion annotation along with experimental sarcasm annotation.

3   Data Gathering

3.1   Data Source and Preprocessing

The data used for annotation comes from the period from 01.09.2021 to 31.08.2022. A hundred tweets per day were scraped giving 36,500 entries of up to 280 characters, some of which were excluded in the process of annotation concluding in 36,280 tweets total. A sample of 156 tweets annotated as irrelevant is included in the final dataset. Our dataset comprises tweets collected from an extended version of the set of accounts considered in the Ogrodniczuk and Kopeć [11] study (75%) as well as from the category Twitter Trends (25%). Samples were collected from 36,080 unique accounts. All emoji symbols were removed in order to avoid them from suggesting the emotions. User tags were left for the references to other users' tweets to be detectable. In the published version, user tags were replaced with a unified token '@*anonymized_account*.'

3.2 Annotation Methodology and Guidelines

We followed the methodology used in the creation of similar datasets for Polish to obtain compatible and comparable results. We performed text-level annotations on each tweet separately, with four annotators (five for the first 8,000 tweets) providing their annotations. The annotations were then consolidated, and any discrepancies were discussed during group meetings to achieve a consistent and unambiguous result. In cases where uniformity was impossible, the sentiment was rendered ambiguous. We used Google Spreadsheets to carry out the annotation process, with each annotator allocated an individual sheet. A separate spreadsheet was used to collect annotations from each annotator's sheet, allowing us to identify discrepancies, calculate inter-annotator agreement, and facilitate group annotation.

We devised a set of guidelines for annotating emotions, sentiment, and sarcasm, which serves as an expansion to the previously formulated CLARIN-PL instruction developed for annotating product and service reviews with emotions [14].

Emotions. We adopted the Plutchik's [12] model, which delineates a discrete set of eight basic emotions ('joy', 'sadness', 'trust', 'disgust', 'fear', 'anger', 'surprise', 'anticipation'). For each tweet in the dataset, annotators were asked to assign emotions consistent with the emotions expressed by the author of the text. It was emphasized that contrasting emotions should not be assigned to a single tweet. Moreover, the annotators were instructed to annotate predetermined dyads (e.g., 'love', 'pessimism', 'aggression'). Finally, in the event of a conflict between emotions the annotators were asked to select a predominant emotion or, in the most difficult cases, leave the tweet for group annotation.

Sentiment. The sentiment could take one of the following four values: 'negative', 'positive', 'neutral', or 'ambivalent' (indicated as 'positive' + 'negative'). For each tweet, after annotating emotions, the annotators were instructed to annotate its sentiment, which ideally would follow from the previously annotated emotions ('joy'/'trust' corresponds to 'positive'; 'anticipation'/'surprise' corresponds to 'neutral'; 'fear'/'sadness'/'disgust'/'anger' corresponds to 'negative'). Nevertheless, the annotation of emotions and sentiment remained independent. In ambiguous instances, if a tweet contained both positive and negative sentiment, and neither was predominant, the annotators were allowed to mark both.

Sarcasm. To manually annotate sarcasm, we followed the theoretical approach mentioned in previous studies [5] and identified 'sarcasm' as an inconsistency between the literal content conveyed in the text (positive) and the sentiment intended by the author (negative). In cases where such a contrast could not be detected, annotators could mark the statement as a snide remark/irony (annotated as 'disgust' in the spreadsheet). For every sample in the dataset, annotators marked sarcasm in a binary way, indicating its presence (1) or absence (0).

3.3 Positive Specific Agreement

We used Positive Specific Agreement (PSA) to evaluate the results of manual annotation. The instructions were discussed and adjusted during the annotation process to increase the PSA values. However, due to the linguistic and thematic diversity of the data, it was difficult to provide unambiguous instructions for annotation, which resulted in low agreement between annotators. The level of agreement was higher for sentiment annotations (76.20%) than for emotions (55.29%) or sarcasm (25.27%). The overall PSA for the dataset was 66.31%.

3.4 Rendering the Final Annotation

The rendition of the final dataset was composed of two steps. Firstly, the annotations were automatically summed up given the conditions presented in Table 1. The conflicting pairs of emotions were considered. Hence, a threshold of minimum two consistent annotations of an emotion was set, along with the restriction that the number of annotations of a given emotion needs to be higher than the number of annotations of the opposed emotion. While it was intended for the sentiment to match and follow directly from the emotions it was not always the case. Thus, sentiment was rendered independently. The conditions never rendered a 'None' sentiment.

Table 1. Conditions for automated totaling of annotations of opposed pairs of emotions and sentiment

Result	Conditions	Example
Emotion A	AND(A > B;A > 1)	joy = 3; sadness = 1
Emotion B	AND(A < B;B > 1)	joy = 1; sadness = 2
Conflict	OR(AND(A = 1;B = 1);AND(A = 2;B = 2))	joy = 2; sadness = 2
None	else	joy = 1; sadness = 0
Positive (A)	AND(A > B;A > C;A > 1)	positive = 4; negative = 0; neutral = 1
Negative (B)	AND(A < B;B > C; B > 1)	positive = 0; negative = 3; neutral = 1
Neutral (C)	AND(B < C;A < C;C > 1)	positive = 1; negative = 1; neutral = 2
None	AND(A < 2;B < 2;C < 2)	positive = 1; negative = 1; neutral = 1
Ambivalent	else	positive = 2; negative = 2; neutral = 0

The second step required a group discussion of specific cases. After singular annotations were added up according to the conditions, each case of conflict between emotions was discussed in group as well as each tweet marked with 'sarcasm' and/or 'to be discussed'. Totally a number of 5,207 tweets were dealt with by group annotation, the results of which were written over the results of automatic rendition.

4 Data Analysis

4.1 Overall Results

Table 2 presents the overall results of the annotation process. The most frequent (34,85%) emotion is 'anticipation'. Which may be a result of a very liberal instruction for annotating this emotion – most tweets referring to the future with various attitudes were counted as anticipation. Surprisingly, the least frequent emotion in the dataset is 'fear' with only 0,90% frequency of occurrence.

Table 2. Overall counts and frequencies for each category

Category	Count	Frequency	Category	Count	Frequency
Joy	4168	11,49%	Surprise	2352	6,48%
Sadness	1680	4,63%	Positive	3986	10,99%
Trust	1620	4,47%	Negative	10729	29,57%
Disgust	8361	23,05%	Neutral	**18385**	**50,68%**
Fear	**328**	**0,90%**	Ambivalent	3039	8,38%
Anger	6364	17,54%	Sarcasm	751	2,07%
Anticipation	**12645**	**34,85%**	Irrelevant	156	0,43%

4.2 Co-occurrence of Emotions

We have computed correlations of every pair of emotions (see Fig. 1). The most strongly positively correlated pairs were 'disgust' and 'anger' ($r = 0.42$), 'disgust' and 'sarcasm' ($r = 0.17$), 'joy' and 'trust' ($r = 0.12$), and 'disgust' and 'anticipation' ($r = 0.11$). Combination of 'disgust' and 'anger' corresponds to the predefined 'contempt' dyad, while combination of 'joy' and 'trust' corresponds to the 'love' dyad. The most strongly negatively correlated pairs were 'anticipation' and 'surprise' ($r = -0.18$), 'joy' and 'disgust' ($r = -0.17$), 'joy' and 'anger' ($r = -0.15$) and 'trust' and 'disgust' ($r = -0.11$). These results were fully expected and justified by the instructions given to annotators not to assign conflicting emotions to one text.

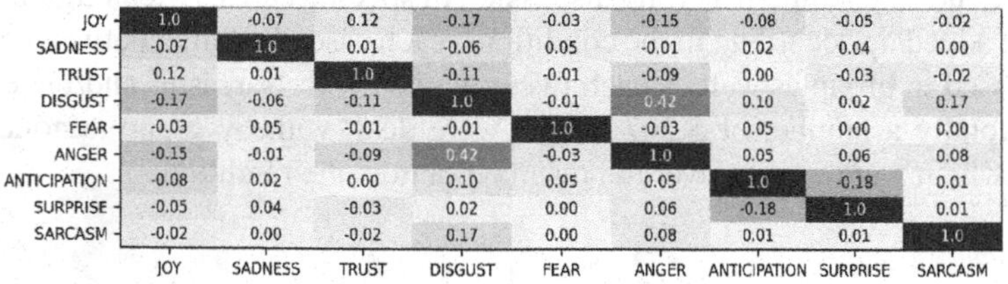

	JOY	SADNESS	TRUST	DISGUST	FEAR	ANGER	ANTICIPATION	SURPRISE	SARCASM
JOY	1.0	-0.07	0.12	-0.17	-0.03	-0.15	-0.08	-0.05	-0.02
SADNESS	-0.07	1.0	0.01	-0.06	0.05	-0.01	0.02	0.04	0.00
TRUST	0.12	0.01	1.0	-0.11	-0.01	-0.09	0.00	-0.03	-0.02
DISGUST	-0.17	-0.06	-0.11	1.0	-0.01	0.42	0.10	0.02	0.17
FEAR	-0.03	0.05	-0.01	-0.01	1.0	-0.03	0.05	0.00	0.00
ANGER	-0.15	-0.01	-0.09	0.42	-0.03	1.0	0.05	0.06	0.08
ANTICIPATION	-0.08	0.02	0.00	0.10	0.05	0.05	1.0	-0.18	0.01
SURPRISE	-0.05	0.04	-0.03	0.02	0.00	0.06	-0.18	1.0	0.01
SARCASM	-0.02	0.00	-0.02	0.17	0.00	0.08	0.01	0.01	1.0

Fig. 1. Correlations of pairs of emotions

4.3 Distribution of Sentiment Over Time

Figure 2 depicts the distribution of annotated sentiment over time. Ambivalent sentiment was assigned to the tweets published after mid-November more frequently due to the fact that the first 8,000 tweets the number of annotators was reduced from 5 to 4. We suspected that the beginning of the Russian invasion of Ukraine might affect the distribution of the sentiment of tweets in subsequent months. However, no such changes can be identified in the graph. However, in tweets posted around New Year's Eve, a significant increase of positive sentiment can be observed.

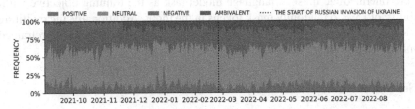

Fig. 2. Distribution of sentiment over time.

4.4 Co-occurrence of Emotions and Sentiment

Because of independent annotation of emotions and sentiment we have examined their co-occurrence (see Fig. 3). Unsurprisingly, positive emotions are positively correlated with 'positive' sentiment ('joy', r = 0.79; 'trust', r = 0.39), and negative emotions, with 'negative' sentiment ('anger', r = 0.62; 'disgust', r = 0.71; 'fear', r = 0.11; 'sadness', r = 0.22). Sarcasm was most strongly correlated with 'negative' sentiment (r = 0.18). Sentiment of tweets containing 'surprise' or 'anticipation' was most frequently annotated as 'negative' or 'neutral' rather than 'positive' or 'ambivalent'.

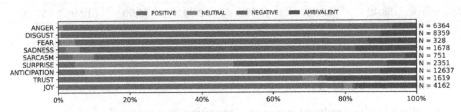

Fig. 3. Distribution of sentiment for every emotion.

5 Experiments

We evaluated three models on our dataset: one for emotion recognition, one for sentiment classification, and a multi-task model that performs both tasks jointly. The F1-score metric was used to measure the predictive performance of emotions and sentiment classifications, with the macro average used to address class imbalance. The models were trained with a fixed learning rate, batch size, and number of epochs. We excluded irrelevant and sarcastic tweets. The dataset was split into train(80%), dev(10%), and test(10%) sets using iterative stratified sampling.

As a main model we adopt HerBERT [9], a Transformer-based model trained specifically for Polish. Because of the specific tweets style, we also utilized HerBERT that was trained using almost 100 million messages extracted from Polish Twitter TrelBERT [1]. This model is available only in the base architecture version, so we decided to train our own large model (HerBERT-large-T), for which we used tweets from the TwitterEmo and PolEval2019 task [10]. We use this set of tweets to fine-tune the HerBERT-large model using the intermediate masked language model task as the training objective with the probability of 15% to randomly mask tokens in the input. Additionally, the multilingual XLM-R model [3] was also taken into consideration, as well as its version pretrained on 198 million multilingual tweets [2].

5.1 Sentiment Analysis

Table 3 reports sentiment classification results. As well as in the following case of emotion recognition, the best model turned out to be HerBERT-large (62.70% average F1-macro).

Table 3. Sentiment classification F1 results on TwitterEmo test set for single-task models

Sentiment	HerBERT base	HerBERT large	TrelBERT	HerBERT large-T	XLM-R base	XLM-R large	XLM-Twitter
Positive	68.89	72.11	**73.87**	72.90	67.19	73.58	69.37
Negative	76.24	**79.27**	78.44	78.54	73.17	78.07	73.29
Neutral	83.86	**84.60**	84.46	83.99	81.33	84.11	82.18
Ambivalent	4.97	**14.83**	9.60	12.56	1.91	10.58	3.56
Macro avg	58.49	**62.70**	61.59	62.00	55.90	61.59	57.10

5.2 Emotion Recognition

Table 4 displays the results of emotion recognition evaluation. The highest average macro F1-score was achieved by the HerBERT-large model, reaching 57.17%. The two emotions with the lowest recognition scores were 'trust' and 'fear,' which is consistent with their low frequency in the dataset, with 4.47% and 0.90% frequencies, respectively (see Table 2). Among the Polish base models, TrelBERT achieved over 3pp higher macro F1-score compared to the model without pretraining on tweets. However, the HerBERT-large-T results were below expectations, possibly due to the small training set. It only marginally improved performance for 'trust,' 'anger,' and 'surprise'. The results of the multilingual XLM-R model were good, but they were still lower than the results of the native models.

Table 4. Emotion recognition F1 results on TwitterEmo test set for single-task models

Emotion	HerBERT base	HerBERT large	TrelBERT	HerBERT large-T	XLM-R base	XLM-R large	XLM-Twitter
Joy	68.26	70.99	72.80	70.85	68.24	72.83	66.33
Sadness	41.88	50.17	43.42	49.03	39.23	47.95	37.35
Trust	11.46	32.52	22.97	33.33	3.59	30.04	8.89
Disgust	67.09	73.21	70.89	72.70	63.43	71.82	64.21
Fear	10.81	40.00	5.71	38.10	0.00	33.33	0.00
Anger	61.51	66.05	63.91	66.38	58.53	66.61	60.42
Anticipation	69.90	71.16	71.57	70.84	69.55	71.60	69.61
Surprise	46.68	53.30	51.96	53.52	41.64	49.28	41.73
Macro avg	47.20	**57.17**	50.40	56.84	43.03	55.43	43.57

6 Conclusions

This paper presents the results of creating a baseline dataset for emotion recognition and sentiment analysis of non-domain-specific and colloquial language tweets in Polish. The dataset covers various topics and spans a full year, allowing for an analysis of emotions and sentiment over time. TwitterEmo fills the gap in the field of emotion and sentiment datasets in Polish by overcoming the limitations of the domain-centered approach and broadening applications of emotion recognition methods in social studies. Annotation of sarcasm introduces novelty to computational linguistics in Polish. The dataset will be made available via an online repository.[1]

Results of preliminary model training on the dataset were presented, showing the HerBERT large model achieving the highest average F1-score. Further research will determine the utility of the dataset in detecting emotions related to specific events and topics in Polish public debate. The dataset can also be used for media bias detection in social media and to detect trends in public opinion based on social media entries. Finally, the absence of emoji symbols in the annotation can be confronted with emotions ascribed to them, which may present interesting results on the use of emoji in Twitter and their potential in emotion recognition research.

Acknowledgements. The project is financed under the 2014–2020 Smart Development Operational Programme, Priority IV: Increasing the scientific and research potential, Measure 4.2: Development of modern research infrastructure of the science sector, No. POIR.04.02.00-00C002/19, "CLARIN - Common Language Resources and Technology Infrastructure" and by the project co-financed by the Minister of Education and Science under the agreement 2022/WK/09.

---

[1] The repository available at https://huggingface.co/datasets/clarin-pl/twitteremo.

References

1. Bartczuk, J., et al.: TrelBERT. https://huggingface.co/deepsense-ai/trelbert. Accessed 1 March 2023
2. Barbieri, F., Espinosa, A., Camacho-Collados, J.: XLM-T: multilingual language models in twitter for sentiment analysis and beyond. In: Proceedings of the Thirteenth Language Resources and Evaluation Conference, pp. 258–266. European Language Resources Association, Marseille, France (2022)
3. Conneau, A., et al.: Unsupervised cross-lingual representation learning at scale. CoRR (2019)
4. Dashtipour, K., et al.: Multilingual sentiment analysis: state of the art and independent comparison of techniques. Cogn. Comput. **8**, 757–771 (2016). https://doi.org/10.1007/s12559-016-9415-7
5. Davidov, D., Tsur, Or., Rappoport, A.: Semi-supervised recognition of sarcastic sentences in Twitter and Amazon. In: Proceedings of the Fourteenth Conference on Computational Natural Language Learning (CoNLL '10). Association for Computational Linguistics, pp. 107–116 (2010)
6. Kocoń, J., Miłkowski, P., Zaśko-Zielińska, M.: Multi-level sentiment analysis of Pol-Emo 2.0: extended corpus of multi-domain consumer reviews. In: Proceedings of the 23rd Conference on Computational Natural Language Learning (CoNLL), pp. 980–991. Association for Computational Linguistics, Hong Kong, China (2019a)
7. Kocoń, J., Zaśko-Zielińska, M., Miłkowski, P., Janz, A., Piasecki, M.: Wrocław Corpus of Consumer Reviews Sentiment. CLARIN-PL (2019c). hdl.handle.net/11321/700
8. Kocoń, J., et al.: Recognition of emotions, polarity and arousal in large-scale multi-domain text reviews. In: Vetulani, Z., Paroubek, P. (eds.) Human Language Technologies as a Challenge for Computer Science and Linguistics, pp. 274–280. Wydawnictwo Nauka i Innowacje, Poznań, Poland (2019d)
9. Mroczkowski, M., Rybak, P., Wróblewska, A., Gawlik, I.: HerBERT: efficiently pretrained transformer-based language model for polish. In: Proceedings of the 8th Workshop on Balto-Slavic Natural Language Processing. pp. 1–10. Association for Computational Linguistics, Kiyv, Ukraine (2021)
10. Ogrodniczuk, M., Kobyliński, Ł.: Proceedings of the PolEval 2019 Workshop. Warszawa, Poland (2019)
11. Ogrodniczuk, M., Kopeć, M.: Lexical correction of polish twitter political data. In: Proceedings of the Joint SIGHUM Workshop on Computational Linguistics for Cultural Heritage, Social Sciences, Humanities and Literature, pp. 115–125. Association for Computational Linguistics, Vancouver, Canada (2017)
12. Plutchik, R.: The nature of emotions: Human emotions have deep evolutionary roots, a fact that may explain their complexity and provide tools for clinical practice. Am. Sci. **89**(4), 344–350 (2001). http://www.jstor.org/stable/27857503
13. Sykora, M., Elayan, S., Jackson, T.W.: A qualitative analysis of sarcasm, irony and related #hashtags on Twitter. Big Data Soc. (2020). https://doi.org/10.1177/2053951720972735
14. Wabnic, K., Zaśko-Zielińska, M., Kaczmarz, E., Matyka, E., Zajączkowska, A., Kocoń, J.: Guidelines for annotating consumer reviews with basic emotions. CLARIN-PL (2022). hdl.handle.net/11321/909
15. Wawer, A., Ogrodniczuk, M.: Results of the poleval 2017 competition: Sentiment analysis shared task. In: 8th Language and Technology Conference: Human Language Technologies as a Challenge for Computer Science and Linguistics (2017)

Timeseries Anomaly Detection Using SAX and Matrix Profiles Based Longest Common Subsequence

Thi Phuong Quyen Nguyen[1]([✉]), Trung Nghia Tran[2], Hoang Ton Nu Huong Giang[3], and Thanh Tung Nguyen[4]

[1] Faculty of Project Management, The University of Danang- University of Science and Technology, 54 Nguyen Luong Bang, Danang, Vietnam
ntpquyen@dut.udn.vn
[2] Faculty of Applied Science, Ho Chi Minh City University of Technology (HCMUT), 268 Ly Thuong Kiet Street, District 10, Ho Chi Minh City, Vietnam
[3] Department of Information Systems and Data Analytics, School of Computing, National University of Singapore, Singapore, Singapore
[4] Faculty of Computer Science and Engineering, Ho Chi Minh City University of Technology (HCMUT), 268 Ly Thuong Kiet Street, District 10, Ho Chi Minh City, Vietnam

Abstract. Similarity search is one of the most popular techniques for time series anomaly detection. This study proposes SAX-MP, a novel similarity search approach that combines Symbolic Aggregate ApproXimation (SAX) and matrix profile (MP). The proposed SAX-MP method consists of two phases. The SAX method is used in the first phase to extract all of the subsequences of a time series, convert them to symbolic strings, and store these strings in an array. In the second phase, the proposed method calculates the MP based on the symbolic strings that are represented for all subsequences extracted in the first phase. Since a subsequence is represented by a symbolic string, the MP is calculated using a distance-based longest common subsequence rather than the z-normalized Euclidean distance. Top-k discords are detected based on the similarity MP. The proposed SAX-MP is implemented on several time series datasets. Experimental results reveal that the SAX-MP method is particularly effective at detecting anomalies when compared to HOT SAX and MP-based methods.

Keywords: Anomaly detection · Matrix profile · SAX · HOT SAX · LCS method

1 Introduction

Anomaly detection is the process of identifying unusual states that occur in the dataset. Anomalies in time series can be caused by a variation in the amplitude of data or an alteration in the shape of the data [1]. Many wasteful costs and damages can be avoided if anomalies are detected early. For instance, real-time transaction data detection aids in the detection of fraudulent Internet transaction activities. Or anomaly detection in industrial manufacturing prevents incidents that may stop the production line.

© The Author(s), under exclusive license to Springer Nature Switzerland AG 2023
J. Mikyška et al. (Eds.): ICCS 2023, LNCS 14074, pp. 221–229, 2023.
https://doi.org/10.1007/978-3-031-36021-3_21

Time series similarity search or discord search is one of the most popular techniques for anomaly detection, which is widely used in a multitude of disciplines, including healthcare systems, energy consumption, industrial process, and so on [2]. Symbolic Aggregate ApproXimation (SAX) [3] is a representative method of similarity search-based approaches. The original SAX method aims to reduce the data dimensionality by converting time series data to a symbolic string. SAX is superior to other representations for detecting anomalies and discovering motifs in data mining [3]. Keogh *et al.* proposed a HOT SAX that employed a heuristic framework that included an outer loop and inner loop to order the SAX sequence for discord discovery [4]. Several extensions of SAX, such as HOT aSAX [5], ISAX [6], HOTiSAX [7], and SAX-ARM [8], were proposed to improve both effectiveness and efficiency concerning SAX and HOT SAX.

Besides, matrix profile (MP) [9, 10] is a measure of similarity between all subsequences of a single time series. MP is also a widely used technique of similarity search-based approach for discord discovery. The MP, which is based on an all-pair-similarity-search on the time series subsequences, is regarded as one of the most efficient methods for comprehensive time series characterization. Assume that there is a time series, T, and a subsequence length, m. The MP presents the z-normalized distance between each subsequence in T and its nearest neighbor. As a result, the MP is used to discover motifs, shapelets, discords, and so on. However, the main disadvantage of the MP is that its complexity rises quadratically with time series length.

Regarding the aforementioned analysis, this study proposes a novel similarity search approach that combines SAX and MP (denoted as SAX-MP) for anomaly detection. The proposed SAX-MP can capitalize on the benefits of both the SAX and the MP while minimizing their drawbacks. First, the MP requires quadratic space concerning the length of time series data, whereas SAX can reduce time series dimensionality. Two conversions are made by SAX: 1) Piecewise Aggregate Approximation (PAA) is employed to reduce the dimensionality of time series from n dimensions to w dimensions, and 2) the subsequences are finally transformed to symbolic representation. Thus, the proposed SAX-MP uses SAX to firstly segment and convert time series data into SAX symbols. Thereafter, the concept of MP is used to calculate the distance between each subsequence and its nearest neighbors. Time series discords are discovered based on the similarity-based MP, with extreme values. In addition, this study employs Longest Common Subsequence (LCS) method [11], which is an effective similarity measure for two strings, to calculate the similarity MP in the proposed SAX-MP.

The paper is structured as follows. Section 2 provides some related works. Section 3 describes the procedure of the proposed SAX-MP method. The numerical results are shown in Sect. 4. Section 5 comes with the conclusion and future research direction.

2 Related Works

This section provides an overview of related studies on time series anomaly detection. The key definitions linked to this work are presented at the beginning of this section.

Definition 1. A *time series* T is a set of real-valued numbers, $T = t_1, t_2, \ldots, t_n$ where n is the length of T.

Definition 2. A *subsequence* $T_{i,m}$ is a subset that contains m continuous variables from T beginning at location i. $T_{i,m} = t_i, t_{i+1}, \ldots, t_{i+m-1}$, where $1 \leq i \leq n - m + 1$.

Definition 3. A *distance profile* D_i is a vector of Euclidean for fix m between a given $T_{i,m}$ and each subsequence in the set of all subsequences. $D_i = [d_{i,1}, d_{i,2}, \ldots, d_{i,n-m+1}]$, where $d_{i,j}$ is the distance between $T_{i,m}$ and $T_{j,m}$, $1 \leq j \leq n - m + 1$.

Definition 4: A *matrix profile MP* is a vector of z-normalized Euclidean distances between each $T_{i,m}$ with its nearest neighbors $T_{j,m}$. MP $= [\min(D_1), \min(D_2), \ldots, \min(D_{n-m+1})$, where D_i is a distance profile that is determined in Definition 3.

Definition 5: A *matrix profile index* I is a vector of integers corresponding to matrix profile MP: $I = I_1, I_2, \ldots, I_{n-m+1}$, where $I_i = j$ if $\min(D_i) = d_{i,j}$.

Definition 6. *Time Series Discord*: Given a time series T and its subsequence $T_{i,m}$, $T_{i,m}$ is defined as a discord of T if it has the largest distance to its nearest neighbors. Regarding the matrix profile's definition, a subsequence $T_{i,m}$, is a kth discord if its matrix profile is the kth the largest distance in the vector of the MP.

2.1 Review of Anomaly Detection-Based SAX Approaches

Anomaly detection, which is the problem of discovering anomalous states in a given dataset, has been a significant issue in the field of data signals research. The detection of anomalies is regarded as the first and most important stage in data analysis to notify the user of unexpected points or trends in a collected dataset. The current approaches for anomaly detection on sensor data can be grouped into several classes such as statistical and probabilistic approaches, pattern-matching approaches, distance-based methods, clustering approaches, predictive methods, and ensemble methods [12]. This study focuses on reviewing some approaches based on the SAX method.

SAX is a symbolic representation of time series that gives an approximation. SAX employs PAA [13], a non-data adaptive representation method, to reduce dimensionality. The PPA technique partitions the data into subsequences of equal length. Each subsequence's mean is calculated and used as a subsequence representation. Then, SAX will convert each subsequence into a symbolic representation. SAX is an efficient and useful method in querying time series subsequences and can also be used in a variety of time series data mining activities, including pattern clustering and classification [2].

Regarding the SAX method, several applications focus on simple and effective signal conversion. Lin et al. [3] employed SAX to compare the differences between data patterns by converting data to strings. Keogh et al. [4] proposed a HOT SAX to search and compare abnormal signals for time series data. Three main advantages of the SAX method are also determined [4]. First, SAX can perform the data dimension reduction which reduces the complexity of subsequent analysis and reduces memory usage. Second, SAX converts data into strings to make data pattern comparison more convenient and easy to interpret. Third, SAX gains an advantage in the definition of the minimum distance formula, which is used to determine the difference between strings. If the distance between two strings is smaller, the shape is closer. For the two closer strings, the greater distance represents the more difference in morphology. SAX simply compares different strings by computing the distance between the strings and using the distance to find the highly repetitive patterns or the most different patterns. The related research employed the SAX method for anomaly detection can be seen in [3–5, 8, 13].

2.2 Review of the Longest Common Subsequence (LCS) Method

Suppose that there are two strings $X = (X_1, X_2, \ldots, X_n), Y = (Y_1, Y_2, \ldots, Y_n)$, and their common subsequence is denoted as $CS(X, Y)$. The longest common subsequence of X and Y, $LCS(X, Y)$ is defined as a $CS(X, Y)$ with maximum length [14]. The concept of LSC is to match two sequences by allowing them to stretch without altering the order of each value in the sequence. Unmatched values discovered throughout the LCS searching and matching process would be excluded (e.g., outliers) without impacting the final LCS outcome. The LCS might reflect the similarity of the two subsequences in this research. The recurrence relations of the LCS method are described as follows:

$$
LCS(X_i, Y_j) = \begin{cases} 0 & \text{if } i = 0 \text{ or } j = 0 \\ LCS(X_{i-1}, Y_{j-1}) + 1 & \text{if } i, j > 0 \text{ and } X_i = Y_j \\ \max\{LCS(X_i, Y_{j-1}), (X_{i-1}, Y_j)\} & \text{if } i, j > 0 \text{ and } X_i \neq Y_j \end{cases} \quad (1)
$$

where $LCS(X_i, Y_j)$ represents the set of LCS of prefixes X_i and Y_j.

3 Proposed Method

This section presents the proposed SAX-MP method that combines SAX and MP for time series anomaly detection. The SAX-MP method consists of two phases. In the first phase, time series data is converted to alphabet strings using the SAX method. The second phase is to calculate the MP based on the LCS technique. Time series discords are found based on the MP's outcome. The SAX-MP procedure is illustrated in Fig. 1.

The proposed method is described detailed as follows. A given time series T is firstly normalized in the proposed SAX-MP method. Then, SAX is employed to convert it into alphabet strings. Regarding the SAX method, two parameters, i.e., alphabet size "a", and word size "w", need to be defined. Referring to the work of HOT-SAX [4], the SAX alphabet size "a" was set as 3 after conducting an empirical experiment on more than 50 datasets. The SAX word size "w" highly depends on time series characteristics. A smaller value of w is preferred for datasets that are typically smooth and slowly altering, whereas a greater value of w is preferred for more complicated time series. Given a length of subsequence (m), a set of all subsequences is extracted by moving the m-length window across the time series. A SAX representation is created based on the selected parameters of "a" and "w". Each subsequence is then converted to alphabet strings and stored in an array A. This whole process is quite similar to the Outer Loop of HOT-SAX. Figure 2 illustrates this converting process.

In the next stage, the MP is calculated based on the set of all subsequences resulting from the first stage. The distance profile D_i is firstly computed based on Definition 3. The distance between two subsequences is calculated by the LCS method. Herein, each subsequence is represented by a symbolic string. Thus, the distance profile D_i based LCS is defined as $D_i = [LCS_{i,1}, LCS_{i,2}, \ldots, LCS_{i,n-m+1}]$, where $LCS_{i,j}$ is the distance between A_i and A_j, $1 \leq j \leq n - m + 1$, A_i and A_j belong to array A. The $LCS_{i,j} = LCS(A_i, A_j)$ is calculated by Eq. (1). Thereafter, the MP-based LCS and index I are obtained using Definitions 4 and 5. Several algorithms were used to compute the MP such as STAMP [9], STOMP, SCRIMP++, and AAMP, with their time complexity being $O(n^2 \log n)$, $O(n^2)$, $O(n^2 \log n/m)$, and $O(n(n - m))$, respectively.

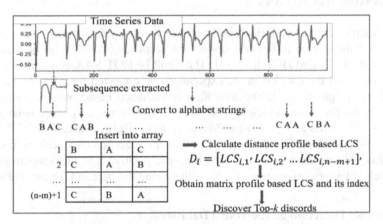

Fig. 1. An illustration of the proposed SAX-MP.

This study inherits the process to perform a pairwise minimum operation in STAMP, STOMP, and SCRIMP algorithms to obtain the matrix profile MP and its index I.

Figure 2 describes the procedure to obtain MP-based LCS. Array A is obtained from the first stage of the proposed SAX-MP. Lines 1 and 2 initialized the matrix profile MP and matrix profile index I. Besides, $idxes$ is initialised as the number of subsequences.. Line 4 calculates the distance profile-based LCS using Eq. (1). Line 5 computes the MP-based LCS and matrix profile index I using the pairwise minimum operation [9]. Finally, the vector MP and corresponding I are obtained.

Time series discords are found based on the obtained matrix profile MP. The maximum value in the MP shows the first discord while its corresponding index I reflects the location of the first discord. According to Definition 6, the top-k discords can be detected where is a user-defined parameter. An anomalous pattern is identified starting at the discord k index and lasting till the end of subsequence length m.

Procedure to find *MP* based LCS
Input: array A (contains n-m+1 subsequences in symbolic representation)
Output: matrix profile *MP* based *LCS* , matrix profile index I

1	Initialization:
2	$MP \leftarrow$ infs, $I \leftarrow$ zeros, *idxes* $\leftarrow n$-m+1
3	for each *idx* in *idxes* do
4	$D \leftarrow$ LCS (A(idx), A) // Calculate distance profile based LCS
5	$MP, I \leftarrow$ Element Wise Min (*MP, I, D, idx*)
6	end for
7	Return *MP, I*

Fig. 2. Algorithm procedure to calculate the MP in the proposed SAX-MP method

4 Experimental Results

This section provides the empirical results to evaluate the performance of the proposed.

SAX-MP method on different datasets. The proposed SAX-MP is compared with the MP method [9], and HOT SAX [4]. The result of HOT SAX is obtained from its original paper. All the datasets in the experiment are collected from a supported page provided by Keogh. E. [15]. The algorithm is evaluated based on two performance metrics: effectiveness, and efficiency. Regarding effectiveness, F_1 score is employed. F_1 score is calculated as $F_1 = ((2 * PR)/(P + R)) * 100\%$, where $P = (TP/(TP + FP)) * 100\%$, and $R = (TP/(TP + FN)) * 100\%$ are precision and recall, respectively. TP, FP, and FN are denoted for true positive, false positive, and false-negative results respectively. F1 scores range from 1 to 0, with 1 being the best and 0 being the worst.

4.1 Anomaly Detection on the Tested Datasets

(I) Space Telemetry dataset

This dataset contains 5000 data points which are divided into five energize cycles. The fifth cycle is remarked as the Poppet pushed considerably out of the solenoid before energizing, indicating the discord. The parameter is set up as $m = 128$, $w = 4$, and $a = 3$. The SAX-MP detects the first discord at the beginning location of 4221, which is quite similar to the result of the HOT SAX method as well as the experts' annotations. The result of the MP method is a little different from HOT SAX and the SAX-MP since the MP can detect the top 3 discords at the beginning locations of 3870, 2872, and 3700, respectively. Thus, the SAX-MP and HOT SAX can effectively detect the anomalies in this dataset while the MP method has some deviation from the real discord.

(II) Electrocardiograms data

The first ECG dataset is stdb_308 which contains two time-series. The length of the sequence is set as $m = 300$ while "w" and "a" are set as 4 and 3, respectively. The proposed SAX-MP finds out the discord at $I_{MP} = 2286$ and $I_{MP} = 2266$ for the datasets stdb308_a and stdb308_b, respectively. This result is consistent with HOT SAX and the ground fact. However, the first discord found by the MP method is quite different from the beginning position of 2680 for the dataset stdb308_a. The second discord of the MP method (at position 2282) is relatively close to the I_{MP}.

The second ECG dataset is mitdb/x_mitdb/x_108, which is more complicated with different types of anomalies. The length of the subsequence is selected as $m = 600$. The proposed SAX-MP method points out the top three discords at 10868, 10022, and 4020, respectively while the top three discords are located at 10871, 10014, and 4017, respectively by the HOT SAX method. The MP method also finds out the top three discords at positions 10060, 11134, and 4368, respectively. The result proves that the proposed SAX-MP method is quite effective in anomaly detection on time series data.

(III) Video surveillance dataset
A video surveillance dataset was compiled from a video of an actor performing various acts with and without a replica gun. The first three discords are founded by the SAX-MP method at locations 2250, 2850, and 290, respectively. The parameter setting of the SAX-MP method is as follows: $m = 250$, $w = 5$, and $a = 3$. The HOT SAX detects 3 potential discords at locations 2250, 2650, and 260, respectively, while the discords discovered by MP are quite similar to the SAX-MP, located at 2192, 2634, and 1942. Note that three anomalous patterns detected by the SAX-MP are covered by the top two discords discovered by the HOT SAX. Besides, compared with the real discords of this dataset in [16], the proposed SAX-MP can detect the majority of anomalies.

Table 1 summarizes the discord discovery result of the SAX-MP, MP, and HOT SAX methods. Figure 3 displays the result on mitdb/x_mitdb/x_108 and video surveillance datasets for illustrations.

Table 1. Result summary of discord discovery

Dataset	Discord length	Top-k discords	Location of discord		
			SAX-MP	MP	HOT SAX
Shuttle	128	1	4221	3780	4228
stdb308_a	300	1	2286	2680	2286
stdb308_b	300	1	2226	2710	2226
mitdb/x_mitdb/x_108	600	1	10868	10060	10871
		2	10022	11134	10014
		3	4020	4368	4017
Video	250	1	2320	2192	2250
		2	2630	2634	2650
		3	2090	1942	260

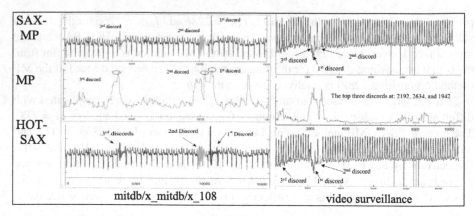

Fig. 3. Experimental result in ECG dataset (mitdb/x_mitdb/x_108)

4.2 Performance Evaluation

To evaluate the performance of the proposed method, the F_1 score and running time are used. Since the abnormal sequences were labeled by domain experts, we considered the results provided in [11] to be correct and these results are used for comparison. The detected results of the proposed SAX-MP are entirely consistent with the experts' discords on Shuttle and the two ECG datasets. Thus, these datasets obtain the F_1 score of 1. For the video dataset, the SAX-MP only achieves an F_1 score of 0.89 while HOT SAX and MP methods adopt the F_1 score as 0.81 and 0.72, respectively. This result proves the effectiveness and efficiency of the proposed method.

5 Conclusion

SAX is a well-known method for reducing time series data dimensions by converting a time series to symbolic strings. The proposed SAX-MP approach, which combines SAX and MP, can take advantage of SAX in reducing the dimensionality of data while lowering the MP's drawback in time complexity. A time series data is firstly segmented using the PPA technique and then converted to symbolic strings. Thereafter, the MP is employed on the converted strings. Instead of using the Euclidean Distance, the proposed SAX-MP utilizes the LCS technique to compute the similarity MP for the strings extracted from the SAX procedure. The similarity MP is used to detect the anomalies in time series data. The experimental results in five time series datasets show that the proposed SAX-MP is extremely effective in discovering the top three discords of the tested datasets.

The length of the subsequence, the SAX word size, and the alphabet size all have an impact on the results of the proposed SAX-MP. Thus, these features can be optimized using some meta-heuristic approaches such as sine-cosine algorithm in future research.

Acknowledgment. This work was funded by Vingroup Joint Stock Company (Vingroup JSC) and supported by Vingroup Innovation Foundation (VINIF) under project code VINIF.2020.DA19. The support is much appreciated.

References

1. Izakian, H., Pedrycz, W.: Anomaly detection in time series data using a fuzzy c-means clustering. In: 2013 Joint IFSA World Congress and NAFIPS Annual Meeting (IFSA/NAFIPS) (2013)
2. Hsiao, K.-J., et al.: Multicriteria similarity-based anomaly detection using Pareto depth analysis. IEEE Trans. Neural Networks Learn. Syst. **27**, 1307–1321 (2015)
3. Lin, J., et al.: A symbolic representation of time series, with implications for streaming algorithms. In: Proceedings of the 8th ACM SIGMOD Workshop on Research issues in data mining and knowledge discovery (2003)
4. Keogh, E., Lin, J., Fu, A.: Hot sax: efficiently finding the most unusual time series subsequence. In: Fifth IEEE International Conference on Data Mining (ICDM 2005) (2005)
5. Pham, N.D., Le, Q.L., Dang. T.K.: HOT aSAX: a novel adaptive symbolic representation for time series discords discovery. In: Nguyen, N.T., Le, M.T., Świątek, J. (eds.) Intelligent Information and Database Systems. ACIIDS 2010. LNCS, vol. 5990. Springer, Berlin, Heidelberg (2010). https://doi.org/10.1007/978-3-642-12145-6_12
6. Sun, Y., et al.: An improvement of symbolic aggregate approximation distance measure for time series. Neurocomputing **138**, 189–198 (2014)
7. Buu, H.T.Q., Anh, D.T.: Time series discord discovery based on iSAX symbolic representation. In: Third International Conference on Knowledge and Systems Engineering (2011)
8. Park, H., Jung, J.-Y.: SAX-ARM: deviant event pattern discovery from multivariate time series using symbolic aggregate approximation and association rule mining. Expert Syst. Appl. **141**, 112950 (2020)
9. Yeh, C.-C.M., et al.: Matrix profile I: all pairs similarity joins for time series: a unifying view that includes motifs, discords and shapelets. In: IEEE 16th International Conference on Data Mining (ICDM) (2016)
10. Yeh, C.-C., et al.: Time series joins, motifs, discords and shapelets: a unifying view that exploits the matrix profile. Data Min. Knowl. Disc. **32**(1), 83–123 (2017). https://doi.org/10.1007/s10618-017-0519-9
11. Hunt, J.W., Szymanski, T.G.: A fast algorithm for computing longest common subsequences. Commun. ACM **20**(5), 350–353 (1977)
12. Cook, A.A., Mısırlı, G., Fan, Z.: Anomaly detection for IoT time-series data: a survey. IEEE Internet Things J. **7**(7), 6481–6494 (2019)
13. Keogh, E., et al.: Dimensionality reduction for fast similarity search in large time series databases. Knowl. Inf. Syst. **3**(3), 263–286 (2001)
14. Iliopoulos, C.S., Rahman, M.S.: Algorithms for computing variants of the longest common subsequence problem. Theoret. Comput. Sci. **395**(2–3), 255–267 (2008)
15. E., K. www.cs.ucr.edu/~eamonn/discords/ (2005)
16. Hu, M., et al.: A novel computational approach for discord search with local recurrence rates in multivariate time series. Inf. Sci. **477**, 220–233 (2019)

Korpusomat.eu: A Multilingual Platform for Building and Analysing Linguistic Corpora

Karol Saputa(✉) ⓘ, Aleksandra Tomaszewska ⓘ,
Natalia Zawadzka-Paluektau ⓘ, Witold Kieraś ⓘ, and Łukasz Kobyliński ⓘ

Institute of Computer Science, Polish Academy of Sciences, Jana Kazimierza 5,
01-248 Warszawa, Poland
{k.saputa,a.tomaszewska,n.zawadzka-paluektau,wkieras,
lkobylinski}@ipipan.waw.pl

Abstract. The paper presents Korpusomat, a new, free web-based platform for effortless building and analysing linguistic data sets (corpora). The aim of Korpusomat is to bridge the gap between corpus linguistics, which requires tools for corpus analysis based on various linguistic annotations, and modern multilingual machine learning-based approaches to text processing. A special focus is placed on multilinguality: the platform currently serves 29 languages, but more can be easily added per user request. We discuss the use of Korpusomat in multidisciplinary research, and present a case study located at the intersection between discourse analysis and migration studies, based on a corpus generated and queried in the application. This provides a general framework for using the platform for research based on automatically annotated corpora, and demonstrates the usefulness of Korpusomat for supporting domain researchers in using computational science in their fields.

Keywords: natural language processing · corpus linguistics · corpus building · discourse studies · migration studies

1 Introduction

Computational technology has revolutionised linguistics by offering powerful tools for language analysis [7, p. 74]. They have progressed through several generations to address user demands and the growing need for larger corpora. First corpus platforms emerged in the late 1970 s with limited usage on mainframe computers in renowned institutions. Second-generation platforms were developed in the 1980 s, introducing concordance and frequency tools on IBM-compatible computers. The third generation facilitated the creation and processing of large corpora on personal computers, increasing accessibility [1]. Today, fourth-generation tools, like Korpusomat, provide advanced features and convenient access to larger data collections. They use multiple export options and features such as Corpus Query Language (CQL), frequency lists, concordance,

J. Mikyška et al. (Eds.): ICCS 2023, LNCS 14074, pp. 230–237, 2023.
https://doi.org/10.1007/978-3-031-36021-3_22

and many more, and can be integrated with other computational methods [3], facilitating the platform selection for a given project. They can be accessed via browsers and store corpora on servers, eliminating the need for local storage and allowing for the processing of larger corpora [1]. The ongoing development of corpus tools contributes to computer science and linguistics, enabling new insights and a deeper understanding of authentic language usage.

Korpusomat[1], a web application designed for building and analysing corpora based on user-provided texts, initially supported only Polish texts [4,5][2]. In this paper, we are presenting a new multilingual version of Korpusomat. Although recent advances in natural language processing have introduced a variety of multilingual high-level text processing frameworks, they still require programming skills and lack the analytical tools typically employed by corpus linguists. The aim of Korpusomat is to bridge this gap between corpus linguistics and modern multilingual machine learning-based approaches to text processing.

2   Korpusomat: Creating Universal Dependencies Corpora

Korpusomat [6][3] aims to address the limitations of corpus tools by offering advanced features, a user-friendly interface, and compatibility with many languages and corpora. It employs two high-level natural language processing libraries, spaCy and Stanza, and the Universal Dependencies (UD) framework [6]. It serves 29 languages and more may be added at users' request as long as text processing frameworks provide models for these languages. With its ability to process files in different formats and convert them to the required encoding, Korpusomat makes building and analysing corpora particularly effortless. Moreover, the tool's integration with the newspaper library allows researchers to add texts from websites, expanding the corpus beyond traditional file import. Korpusomat processes most text data formats[4]. It also allows its users to export data and download the processed texts with annotation layers.

One of the most significant benefits of Korpusomat is the use of the Universal Dependencies (UD) framework. It provides cross-linguistically consistent treebank annotation for over 100 languages. It also enables researchers to easily transfer certain methods from one language to another. A functionality that is unique compared to other corpus tools is the visualization of dependency trees. It is interactive and includes the full utterance containing an example with query results. The results can be modified, for example, punctuation marks may be hidden or the tree may be automatically positioned.

---

[1] The version presented in this paper is available at https://korpusomat.eu.

[2] This older version is still maintained and available at https://korpusomat.pl.

[3] The Description of the Features, Including the Query Language, Is Available in the Documentation at https://korpusomat-eu.readthedocs.io/en/latest/.

[4] A list of possible formats is available in the documentation at http://tika.apache.org/1.17/formats.html.

3   Computational Architecture

The architecture of Korpusomat has a number of requirements regarding the processing of texts and their provision in a searchable, indexed form. In this section we describe the ideas behind the Korpusomat architecture important for the scaling of Korpusomat to the multilingual version. The overall architecture is shown in Fig. 1, with the main components being an API platform, a task queue, a search backend and the processing pipelines described below.

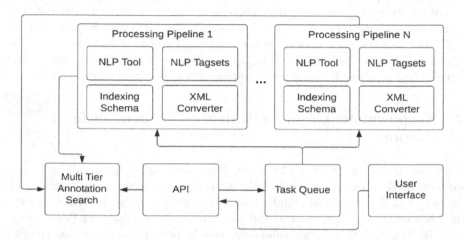

Fig. 1. Overview of the Korpusomat architecture components, showing the elements of the processing pipelines and their interaction with the search backend and the application.

Table 1. The Table shows the time (in minutes) required to build a corpus of three different sizes (in tokens), with the standard deviation in parentheses. The comparison includes two different NLP tools that we currently use.

Corpus size	NLP tool	Build time
147,580	spaCy	5 (0.57)
1,475,800	spaCy	92 (4.2)
14,758,000	spaCy	923 (8.6)
147,563	Stanza	13 (1.5)
1,475,630	Stanza	94 (5.6)

3.1   Processing Pipelines

A processing pipeline (PP) employs a set of components used for adding a text to a corpus (Fig. 1). PPs are interchangeable and versioned thanks to the configurability of their elements: NLP tool used for tagging, extracted tagsets used by the NLP tool, schema of indexing linguistic layers in the search engine, and XML converter of annotated text into indexing schema format.

Due to the multilingual nature of the Korpusomat platform, there is no single tool for all the languages served. We currently use two open source NLP libraries: Stanza and spaCy to provide us with linguistic annotations of texts. Each tool, spaCy and Stanza, can be packaged in a separate, interchangeable PP. It allows for the use of different tools, and the use of one tool for different contexts (e.g. in two different indexing schemes with different linguistic layers available for corpora with these schemes, although using the same NLP tool).

Upgrading the models in the NLP libraries, adding new languages, and other updates require versioning of PPs. Each of the PPs uses semantic versioning, which is linked to the version of the corpora. This ensures that each corpus has a consistent tagging of all its texts.

The specific tagset (e.g. set of named entity tags) extracted from the NLP tool is used for indexing schema generation and then attached to each corpus and used e.g. for the custom query builder of that corpus. An updated version of the PP can use a model with an extended tagset for one of the annotation layers, then there is a mechanism for extracting the tagset from the processing libraries for each language.

3.2   Index and Search

Korpusomat uses MTAS [2] as its corpora search backend. It is a Solr/Lucene search engine extended with the ability to search through annotation layers of texts using the CQL. The indexing scheme specifies the linguistic layers for the corpus, corresponding to the possible queries to the corpus. Each of the corpora has its own indexing scheme based on the chosen PP and its tagset. Each type of linguistic annotation, such as named entities or part-of-speech tags, is treated as a separate annotation layer.

Indexing schemas are automatically generated for the specific PP based on its tagset, and thus versioned to maintain compatibility with older corpora, while automatically allowing for new tags to be indexed, searched, and made visible in the application's query builder.

In addition to linguistic annotations, MTAS indexing includes document metadata, which can provide additional information about individual texts, such as author, publication date, publisher. This metadata can then be used to filter query results or to group results by metadata.

4   Computational Performance Benchmarks

In this section we discuss the performance of the platform based on two of its core features: creating corpora and querying corpora. We show how easily and efficiently Korpusomat can be used as an end-to-end research platform for building (in minutes) and searching (in seconds) corpora. The system is tested on a machine with 16 CPU threads, 47 GB of memory, and SSD storage.

We test the time of adding an ebook in English, containing 147 580 tokens, in three settings: 1, 10, and 100 copies. We measured the time for building the corpora (Table 1) and then querying them for each of these cases.

Query time was measured as the time taken to serve an http query via the API. We analyse four types of queries (respectively in Table 2): (1) all corpus tokens, (2) all corpus sentences, (3) all tokens that are adjectives, and all named entities that contain an adjective.

Queries were run on the three corpora of different sizes. Each query returned 1000 results and was repeated 3 times for 3 consecutive pages of matches. The results are shown in Table 2. We have observed a significantly lower value for the most restrictive query for the smallest corpus, where the second and third pages of results were empty due to the limited number of matches (319).

Table 2. The Table shows a benchmark of search time for different queries and different corpus sizes (in tokens). The results presented are the average time in seconds that the API took to return results, with the standard deviation in parentheses.

Corpus size	Response time for a query			
	[]	<s/>	[upos="ADJ"]	<ne/> containing [upos="ADJ"]
147,580	4.17 (0.056)	7.88 (0.28)	4.14 (0.11)	0.86 (0.74)
1,475,800	4.19 (0.069)	7.58 (0.26)	4.13 (0.11)	4.8 (0.84)
14,758,000	4.92 (0.21)	8.35 (0.33)	4.19 (0.12)	5.01 (0.7)

5 Applying Korpusomat to Multilingual Multidisciplinary Research: A Case Study

This section introduces a brief case study in order to showcase potential uses of Korpusomat for multilingual analysis. The study, which, in a slightly modified version, forms part of a larger project [9], is located at the intersection between migration research and corpus-assisted discourse analysis. Its aim is to contribute to research investigating discursive representations of migrations during the "European refugee crisis" by determining whether there is a mismatch between the amount of media attention received by displaced people of different geographical identities and the actual numbers of refugees who filed for asylum in a European Union country between 2015 and 2018. This is expected to help establish which groups of refugees were backgrounded in European media discourses, despite significant numbers of their actual populations, and which groups were foregrounded and, thus, potentially problematised (as suggested by earlier research [8]).

To this end, Eurostat's statistical data on first time asylum applicants in the EU between 2015 and 2018[5] were contrasted with the results of a corpus

---

[5] https://ec.europa.eu/eurostat/statistics-explained/index.php?title=Asy lum_statistics&oldid=558844.

analysis conducted on three sets of newspaper reports on the "refugee crisis", published in Poland, Spain, and the United Kingdom during the same time period. The corpus contains more than six million tokens in total, with the British and Spanish subsets being more or less equal in size (approx. 2.5 million tokens), whereas the number of tokens in the Polish subcorpus amounts to just over 1.2 million.

As the first step in the analysis, we used Korpusomat's concordancer to determine how British, Polish, and Spanish press referred to the displaced people. Specifically, we verified and compared the frequencies of candidate terms (such as *refugee, immigrant, migrant, asylum-seeker*, etc.). The two most frequent items in each subcorpus (see Table 3) were then selected for the subsequent stage of the analysis, where we searched for adjectives that modify them. Crucially, since UD parsing circumvents differences between languages and tagsets (as discussed above), a single query ([upos="ADJ" & head.lemma="X"], where X stands for the search term) could be used for all three language subcorpora, regardless of syntactic divergences (in English adjectives usually precede the nouns they modify, whereas for Spanish the inverse is true) and inconsistencies (Polish, in turn, has a relatively flexible word order which is, therefore, more difficult to predict).

Table 3. Top terms used with respect to the displaced people.

Corpus	Term	Abs. Freq.	Rel. Freq.
Poland	*uchodźca*	9,604	7,316
	imigrant	3,272	2,568
Spain	*refugiado*	10,959	4,374
	inmigrante	6,075	2,424
United Kingdom	*migrant*	10,467	3,930
	refugee	9,695	3,640

As a result, we obtained lists of adjectives most frequently modifying the search terms (the cut-off point was set at the frequency per million of over 3). We then selected only adjectives denoting geographical identities from each of the six lists, and compared them to Eurostat's data on first time asylum applicants in the EU[6]. These results are presented in Table 4: the nationalities which are foregrounded in the corpus, compared to the "real-life" data, are marked in bold.

The results show that refugees and, to a lesser extent, (im)migrants were often discussed in terms of their geographical identities. At the same time,

---

[6] The top countries of origin of asylum applicants (over 30 thousand applications) were: Syria, Afghanistan, Iraq, Iran, Pakistan, Nigeria, Albania, Eritrea, Russia, Bangladesh, Guinea, Sudan, Turkey, Ukraine, Somalia, Georgia, Ivory Coast, Gambia, Venezuela, Algeria, Mali, Morocco, and Senegal.

Table 4. Top adjectives modifying the search terms.

Poland		Spain		UK	
uchodźca	imigrant	refugiado	inmigrante	refugee	migrant
syryjski	arabski	sirio	subsahariano	Syrian	African
polski	**polski**	afgano	sirio	Afghan	Syrian
czeczeński	syryjski	**palestino**	**africano**	Sudanese	Afghan
afgański	afrykański	rohingyo	**europeo**	Iraqi	Sudanese
palestyński		iraquí	marroquí	Iranian	Eritrean
		eritreo	**magrebí**	Pakistani	Iraqi
		español		**Palestinian**	**Mediterranean**
		somalí		**African**	
				Somali	

there was a significant divergence between the purported nationalities of people labeled refugees versus immigrants/ migrants, especially in the Polish and Spanish corpora, where only one adjective appears on both lists (*syryjski/ sirio* ['Syrian']). In general, refugees tended to be more strongly associated with Syria and Afghanistan (the two top nationalities of first time asylum seekers, according to the Eurostat data), whereas (im)migrants were more likely to be described using more delegitimising vague geographical descriptions, such as *African/ afrykański/ africano, arabski* ['Arab'], or *subsahariano* ['Subsaharan'], despite the fact that among the top ten most numerous groups of asylum-seekers only two were of African origin. At the same time, the newspapers analysed had a tendency to background the specific African nationalities (Nigerians, who filed more than 140 thousand applications for asylum, are a case in point). This may suggest that they were, instead, referred to using the more general modifiers, which demonstrates a conflation of different identities into one. The examined newspapers also tended to foreground nationals of countries which remain in geographical or historical proximity to the countries where they were published (for example, Spanish newspapers paid special attention to Moroccans). On the other hand, migrants crossing the EU borders legally (such as, for instance, Albanians or Russians) were considered to be less newsworthy than those who resorted to irregular means of entry.

While this analysis can only provide a cursory look at the discursive phenomenon in question, it nevertheless points towards some prominent trends in European media representations of migrations during the so-called "refugee crisis". Above all, however, it demonstrates the usefulness of Korpusomat, especially for multilingual studies, where issues of replicability and comparability are of particular concern. Korpusomat is an important step towards reducing the impact of these issues by outflanking (as much as possible) differences between languages and tagsets and, thus, facilitating both the analytical process and the cross-linguistic comparison of results. This might be particularly helpful to users with little to no technical expertise as well as limited experience with linguistic analysis, such as, for instance, scholars from other disciplines wishing to introduce aspects of corpus linguistics into their research.

6 Conclusions and Future Work

Korpusomat is a free platform for creating, processing, and analysing user-created text corpora in a variety of languages. It is scalable and modular, allowing for future user-base growth, and for including new potential processing pipelines. In the future, Korpusomat is also expected to support constituency parsing, parallel corpora, automatic speech recognition systems, and analytical tools such as keywords and terminology extraction, as well as quantitative measures of vocabulary richness.

Acknowledgements. This work was supported by the European Regional Development Fund as a part of 2014–2020 Smart Growth Operational Programme, CLARIN- Common Language Resources and Technology Infrastructure, project no. POIR.04.02.00-00C002/19 and by the project co-financed by the Minister of Education and Science under the agreement 2022/WK/09.

References

1. Anthony, L.: A critical look at software tools in corpus linguistics. Linguist. Res. **30**, 141–161 (2013)
2. Brouwer, M., Brugman, H., Kemps-Snijders, M.: MTAS: a solr/lucene based multi tier annotation search solution (2017)
3. Dunn, J.: Natural language processing for corpus linguistics. Elements in Corpus Linguistics, Cambridge University Press (2022). https://doi.org/10.1017/9781009070447
4. Kieraś, W., Kobyliński, L.: Korpusomat-stan obecny i przyszość projektu. Język Polski **CI**(2), 49–58 (2021)
5. Kiera, W., Kobyliski, L., Ogrodniczuk, M.: Korpusomat – a tool for creating searchable morphosyntactically tagged corpora. Comput. Methods Sci. Technol. **24**(1) (2018). https://doi.org/10.12921/cmst.2018.0000005
6. Kiera, W., Saputa, K., ukasz Kobyliski, Tuora, R.: Korpusomat.eu - user guide (Polish) (2022). https://korpusomat-eu.readthedocs.io/pl/latest/
7. McEnery, T., Brezina, V., Gablasova, D., Banerjee, J.: Corpus linguistics, learner corpora, and SLA: employing technology to analyze language use. Ann. Rev. Appl. Linguist. **39**, 74–92 (2019). https://doi.org/10.1017/S0267190519000096
8. Taylor, C.: Investigating the representation of migrants in the UK and Italian press: a cross-linguistic corpus-assisted discourse analysis. Int. J. Corpus Linguist. **19**(3), 368–400 (2014)
9. Zawadzka-Paluektau, N.: The "European refugee crisis" through a media lens. A cross-national study into press representations of displaced people combining corpus linguistics and argumentation analysis (PhD dissertation). University of Warsaw, University of Seville (forthcoming)

What Will Happen When We Radically Simplify t-SNE and UMAP Visualization Algorithms? Is It Worth Doing So?

Bartosz Minch[✉][iD], Radosław Łazarz[iD], and Witold Dzwinel[iD]

AGH University of Krakow, Kraków, Poland
{minch,lazarz,dzwinel}@agh.edu.pl

Abstract. We investigate how the quality and computational complexity of the golden standards of high-dimensional data (HDD) visualisation - the t-SNE and UMAP algorithms - change with their successive simplifications. We show that by radically reducing the number of the utilised nearest neighbours, introducing binary distances between the samples, and simplifying the loss function, the resulting IVHD algorithm still reconstructs with sufficient precision both local and, particularly, global properties of HDD topology. Although inferior to its competitors for the most moderate data sizes ($M<10^5$ samples), IVHD appears many times faster than state-of-the-art algorithms and reveals its power for multi-million-element datasets for which baseline methods fail in a reasonable computational time.

Keywords: high-dimensional data · data embedding · kNN graph visualization · dimensionality reduction

1 Introduction

In recent years, the explosion of digital datasets resulted in new opportunities and challenges for various fields, such as machine learning, computer vision, bioinformatics, and social network analysis. One of the significant problems related to this data deluge is the high-dimensionality (HD) of the underlying objects, often represented by HD feature vectors with tens, hundreds, or even thousands of dimensions [7]. Their size can be especially burdensome for data analysis, causing increased computational and memory requirements, overfitting, and difficulties in visualisation or interpretation.

To address those issues, researchers developed dimensionality reduction (DR) methods that reduce the number of dimensions in the data while preserving the essential local and global topological properties. DR involves transforming the N-dimensional (N-D) dataset $X = \{x_i\}_{i=1,...M} \in \Re^N$ into its n-dimensional (n-D) representation $Y = \{y_i\}_{i=1,...M} \in \Re^n$, where $N >> n$, and M represents the number of N-D feature vectors x_i (or their corresponding n-D embeddings y_i). This transformation can be perceived as a lossy data compression, achieved by minimizing a loss function $E(|X - Y|)$, where $|.|$ measures the topological dissimilarity between X and Y.

© The Author(s), under exclusive license to Springer Nature Switzerland AG 2023
J. Mikyška et al. (Eds.): ICCS 2023, LNCS 14074, pp. 238–246, 2023.
https://doi.org/10.1007/978-3-031-36021-3_23

In the world of unsupervised HDD embedding, t-SNE [13] and UMAP [8] are among the most widely adopted techniques. The former algorithm resembles the classical Multidimensional Scaling (MDS) [6], but instead of a simple loss function based on the cumulative L2 (or L1) discrepancies between distances in the source and target spaces, t-SNE compares the probability distributions of being a neighbour of each data vector using the Kullback-Leibler (K-L) divergence and summarizes them to calculate the actual embedding error. Those probability distributions reflect the neighbourhood of each data vector in the source and target spaces, with the highest probability assigned to the nearest neighbours (and its value decreasing rapidly in the case of the more distant elements).

On the other hand, UMAP corresponds to the Isomap [6] method, although it has a different conceptual basis than t-SNE and a distinct loss function for error minimization. Instead of calculating the full distance matrix or its Barnes-Hut approximation (as t-SNE does), UMAP focuses on the weights of the nearest neighbours of each data vector and a set of more distant samples. Nevertheless, there is a hidden relationship between t-SNE and UMAP. As demonstrated in [2], a generalisation of negative sampling allows the user to interpolate between embeddings produced by the two methods. Here we show that, additionally, t-SNE and UMAP can be further simplified to a frugal but still efficient approximation.

We have called this approximation the Interactive Visualization of Highdimensional Data (IVHD). It enables visualization of HDD structures in 2D or 3D Euclidean spaces by utilising their k-NN graph representations and the classical MDS loss functions. Moreover, it is assumed that the distances between nodes of the said graph follow the negative sampling principle, i.e. they are set to 0 for each node within the k-NN set — and 1 for m other randomly selected disconnected nodes. In practice, we observed [3,4] that both k and m can be small, usually equal to 1 and 2, respectively. As demonstrated in [4,5], this concept allows visualising both complex networks (e.g., structured, random, scale-free, etc.) and high-dimensional data without changes in the base algorithm. In this paper, we highlight the following contributions:

- The IVHD method can be regarded as a unifying simplification of the t-SNE and UMAP algorithms, achieved through three approximation steps: (1) employing a simpler loss function, (2) utilizing binary distances, and (3) reducing the number of nearest neighbours used in embedding.
- We show that IVHD effectively preserves local and global properties of HDD in 2D embeddings, not only for large datasets but also for small ones, despite its simplicity. Furthermore, IVHD proves to be highly time-efficient when compared to the original methods.
- Additionally, we propose several novel improvements to IVHD (see Sect. 3).

2   Simplifying t-SNE and UMAP

Despite extensive research conducted in the field of HDD visualisation over the past years, new methods continue to emerge regularly [2]. In our research, we specifically investigate publicly available algorithms that have demonstrated superior performance in generating embeddings for datasets of varying sizes, including UMAP [8], IVHD [3], t-SNE, TriMAP, and PaCMAP [13]. Apart from t-SNE, these algorithms involve two key stages: (1) constructing a weighted k nearest neighbour graph and (2) performing an embedding procedure that involves defining a loss function and minimizing it [8]. In this study, we focus primarily on t-SNE and UMAP, which have been the foundation of recent research in this area, and aim to demonstrate that IVHD serves as a more parsimonious approximation for both methods.

2.1   Evaluation Criteria

To verify the properties of the obtained simplifications, we use state-of-the-art quality assessment criteria [10] for unsupervised DR methods, measuring the preservation of the high-dimensional (HD) neighbourhood in the low-dimensional (LD) space. The general consensus is to use the average agreement rate between k-ary neighbourhoods in high and low dimensions. The rank of x_j with respect to x_i in a high-dimensional space is defined as $\rho_{ij} = \left| \{k : \delta_{ik} < \delta_{ij} \vee (\delta_{ik} = \delta_{ij} \wedge 1 \leq k < j \leq N)\} \right|$, where δ_{ij} is the distance between the i-th and j-th data point in HD (d_{ij} denotes an analogous distance in LD, respectively). Similarly, the rank of y_j relative to y_i in the low-dimensional space is equal to $r_{ij} = \left| \{k : d_{ik} < d_{ij} \vee (d_{ik} = d_{ij} \wedge 1 \leq k < j \leq N)\} \right|$. Let $\mathbf{v}_i^k$ and $\mathbf{n}_i^k$ represent the sets of nearest neighbours of x_i and y_i in the high-dimensional and low-dimensional space, with k denoting the number of those neighbours. Now, we define:

$$R_{NX}(k) = \frac{(N-1)\left(\frac{1}{kN}\sum_{i=1}^{N}\left|\mathbf{v}_i^k \cap \mathbf{n}_i^k\right|\right) - k}{N - 1 - k}, \quad G_{NN}(k) = \frac{1}{N}\sum_{i=1}^{N}\frac{\left|j \in \mathbf{n}_i^k\right| - \left|j \in \mathbf{v}_i^k\right|}{k}. \quad (1)$$

$R_{NX}(k)$ quantifies the quality improvement over a random embedding, while $G_{NN}(k)$, measures the average gain (or loss, if negative) considering neighbours of the same class, with a positive value indicating potentially better k-NN classification performance.

2.2   t-SNE with Euclidean and Binary Distances

We investigated whether t-SNE can be viewed as an embedding of an undirected graph, thereby enabling simplification to the IVHD framework. To explore this, we proposed a modified version of t-SNE that uses neighbourhood-limited Euclidean and binary distances instead of the standard probability matrix. This modification allowed us to parametrise t-SNE by the number of nearest neighbours (k) instead of the default perplexity (a critical parameter of t-SNE used to balance the local and global aspects). In this case, k determines the number of nearest data points considered when computing the probability distribution over the pairwise similarities in high-dimensional space. For the binary matrix variant, 1's were inserted to denote neighbours, and 0's otherwise. A similar approach was used for Euclidean distances (refer to [12]), but instead of 1's, the actual Euclidean distance was inserted.

   As illustrated in Fig. 1 (and Fig. 5) in the supplementary materials [12]), for $k \in \{10, 20\}$, our variant of t-SNE achieves a DR quality comparable to the unmodified t-SNE when visualising a small subset (10%) of the MNIST dataset ($M = 7 \cdot 10^3$), as

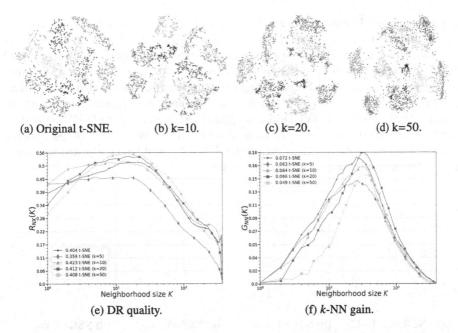

(a) Original t-SNE.   (b) k=10.   (c) k=20.   (d) k=50.

(e) DR quality.   (f) k-NN gain.

Fig. 1. The visualizations and metrics were obtained using a simplified t-SNE method on a 10% subset of the MNIST dataset, with binary distances instead of probabilities. The method was parametrised by the number of nearest neighbours k.

testified by the overlapping metrics curves. Furthermore, the more neighbours were used, the more the quality of visualization improved, as reflected by the higher reaching curves in both discussed graphs, resulting in better AUC values.

Additionally, it should be mentioned that simplifying the t-SNE method into an IVHD-based implementation also consisted of replacing the part of the algorithm that optimizes K-L divergence (a key aspect of DR methods that rely on neighbour embedding) with an optimisation of the MDS-like cost function. As a side effect, it led to obtaining a more streamlined and simplified IVHD method [3,4,11].

2.3 UMAP with a Low Negative Sample Rate and a Small Number of Nearest Neighbours

As mentioned in [2], there is a significant connection between negative sampling (NEG) and noise-contrastive estimation (NCE). UMAP, which uses NEG, can be seen as Neg-t-SNE [2], differing only in the implicit use of a less numerically stable similarity function. A key factor contributing to UMAP's success is its utilization of NEG to refine the Cauchy kernel and its cross-entropy loss function, which distinguishes it from how t-SNE assesses high-dimensional similarities. This refinement allows UMAP to generate more compact clusters and continuous connections between them, as demonstrated in [1]. Another perspective to consider is the similarity between UMAP and IVHD, which, similarly to MDS, aims to preserve pairwise distances or dissimilarities between high and low dimensional data points. To achieve this, UMAP constructs a distance matrix from the original HD data and then finds an LD embedding that minimises the difference between pairwise distances, while also preserving both local and global structures through the construction of a weighted nearest-neighbour graph. Subsequently, it optimizes that LD embedding to retain the aforementioned graph structure. By decreasing the rate of negative sampling and the number of nearest neighbours, we moved UMAP towards an IVHD-like simplification. In contrast to IVHD, UMAP is based on the idea of preserving the local structure of the data in an LD space, rather than just pairwise distances. This means that UMAP is still better adjusted to capture the non-linear relationships between data points.

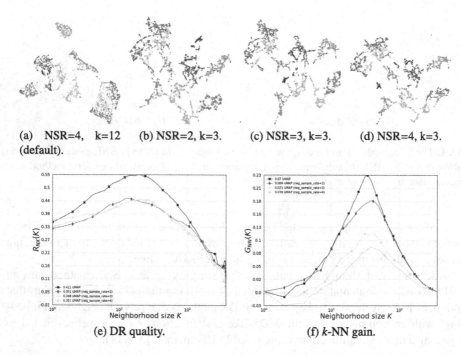

(a) NSR=4, k=12 (default).   (b) NSR=2, k=3.   (c) NSR=3, k=3.   (d) NSR=4, k=3.

(e) DR quality.   (f) k-NN gain.

Fig. 2. UMAP visualizations and metrics were generated using different negative sampling rates and the lowest possible value of k (i.e. $k = 3$) that resulted in meaningful visualizations.

3   Improvements in the IVHD Algorithm

IVHD, as described in [3, 10, 11], exceeds state-of-the-art DR algorithms in computational time by over tenfold in standard DR benchmark datasets [11]. The visualizations obtained are also proficient in reconstructing data separation in large, high-dimensional datasets [12]. Nevertheless, there is potential for improvement in terms of reducing the amount of noise generated between classes. To address this concern, we have developed the following improvements.

Reverse nearest neighbour (RNN) procedure. The query retrieves all points in a HD space that have a given point q as their nearest neighbour. The set of these points is called the influence set of q. It is important to note that a point p being one of the nearest neighbours k to q does not necessarily imply that p is also in q's *RkNN* set.

Manhattan norm employed in the final stages of the embedding procedure, providing a smoothing effect. Using this instead of the conventional Euclidean distance, outliers are drawn closer to the cluster centres. However, it is important to be cautious with the number of steps performed using the L1 metric, as excessive steps may cause the embedding to collapse towards the cluster centres, resulting in distortion of the global structure.

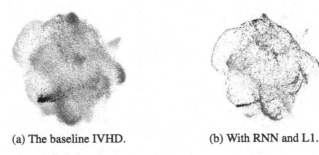

(a) The baseline IVHD.   (b) With RNN and L1.

Fig. 3. A comparison of IVHD on the EMNIST dataset was conducted, employing the L1 norm and RNN mechanisms in the final steps of the embedding procedure.

Figure 3 illustrates the effect of incorporating both mechanisms to provide a generic approach for handling the noise remaining in the visualisation. Quality measurements demonstrate that the discussed upgrades to IVHD enhance the DR quality and the k-NN gain of the obtained embedding. In particular, a clear *suction effect* is observed, where most of the noise is moved from the interstitial space to the clusters. Importantly, the global structure of the visualization is preserved without distortion, as evidenced by the relative positioning of classes remaining unchanged.

The proposed improvements do not introduce significant computational overhead, as the helper graph for RNN is created concurrently (or retrieved from cache) together with the main graph, and interactions between a limited subset of nearest neighbours are calculated. The primary operation that incurs overhead is the search for reverse neighbours based on the two graphs, but the time taken by this procedure is negligible compared to the overall embedding time. It should be noted that all the improvements added to the IVHD method in this study were designed with the consideration of minimizing computational load, given the importance of efficient performance in HDD analysis.

4 Experiments

The primary benefit of IVHD lies in its ability to generate embeddings at a significantly faster rate than baseline methods, once the k-NN graph is stored in the disc cache and the computational time required for its generation can be disregarded. Consequently, users can perform detailed interactive analyses of the multi-scale data structure with a wide range of parameter values and stress function versions without the need to recalculate the said graph. In this regard, we present results obtained for the FMNIST (mid-sized) and the Amazon20M (large-sized) datasets as evidence of the efficiency and effectiveness of the IVHD (with $L1$ and RNN) method. Additional results are provided in the supplementary materials [12].

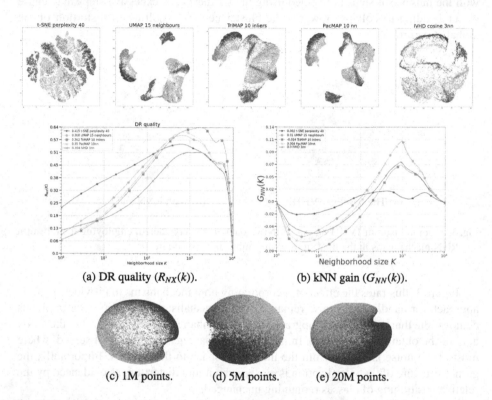

(a) DR quality ($R_{NX}(k)$).   (b) kNN gain ($G_{NN}(k)$).

(c) 1M points.   (d) 5M points.   (e) 20M points.

Fig. 4. Comparison of different DR methods employed on the FMNIST dataset. On the bottom: IVHD visualizations obtained for Amazon20M datasets.

In Fig. 4, we observe that the IVHD applied to Fashion-MNIST forms separate groups of mostly elongated shapes. On the other hand, the MAP-family methods create rounded and clearly separated clusters. Additionally, in t-SNE, some classes are mixed and fragmented. In terms of DR quality, TriMap, UMAP, and PaCMAP are achieving the best results. IVHD surpasses t-SNE only when a large neighborhood is considered ($k > 1000$). Furthermore, IVHD-CUDA [11] was the only method capable of generating visualizations for the Amazon20M dataset in a reasonable time frame of 5 h and 34 min. The generated visualizations clearly depict the separation of the five classes, which comprises book reviews from the Amazon platform. In contrast, other methods, including those implemented in both CPU and GPU environments (e.g. t-SNE CUDA,

Anchor-tSNE [10]), encountered errors or did not generate visualizations even after 12 h of calculations, leading to premature termination. Furthermore, Table 1 corroborates that IVHD stands out as the fastest method among the compared approaches.

Table 1. A selection of timings (measured in seconds) acquired from various datasets with M representing the dataset dimensionality and N representing the number of samples.

	M	N	t-SNE	UMAP	TriMAP	PaCMAP	IVHD
EMNIST	784	103 600	558,67	50,25	123,84	79,21	**34,88**
REUTERS	30	804 409	7457,15	1226,91	1498,96	851,46	**190,47**

5   Conclusions

We demonstrate that IVHD represents a radical simplification of both t-SNE and UMAP, offering a highly efficient platform for fast and interactive HDD visualization. Although it may produce slightly inferior embeddings compared to its competitors for small and moderate data sizes ($M << 10^5$ samples), IVHD remains a remarkably efficient algorithm. It accurately reconstructs both local and global data topology with precision, and its key advantage lies in its computational efficiency, enabling the visualization of large multi-million datasets within a reasonable time frame, where other baseline algorithms fail. We successfully verified the applicability of IVHD to the Amazon20M dataset, highlighting its unique ability to handle such large datasets with minimal resource utilization. Future research will be directed towards further improvements in IVHD, particularly in addressing the challenges of crowding and noise reduction.

Acknowledgments. The research presented in this paper was supported by funds allocated to the AGH University of Krakow by the Polish Ministry of Science and Higher Education. The authors utilized the PL-Grid Infrastructure and computing resources provided by ACK Cyfronet.

Hardware. All CPU implementations were executed in two environments, depending on the scale of the dataset processed. Mid-sized datasets were visualized on Macbook Pro 2.3 GHz 8-Core Intel Core i9, 16 GB 2667 MHz MHz DDR4. Large-sized datasets were processed in GPU/CUDA remote server with Intel Xeon E5-2620 v3 CPU, 8GB GDDR5 NVidia GeForce GTX 1070 GPU, and 252 GB RAM. The source code was compiled using GCC-10.4 and CUDA Toolkit 11.2. Experiments were facilitated by the VisKit C++ library [9] developed by the first author of this work.

References

1. Böhm, J.N., Berens, P., Kobak, D.: Attraction-repulsion spectrum in neighbor embeddings. J. Mach. Learn. Res. **23**(95), 1–32 (2022)
2. Damrich, S., Böhm, J.N., Hamprecht, F.A., Kobak, D.: Contrastive learning unifies t-SNE and UMAP (2022). https://arxiv.org/abs/2206.01816

3. Dzwinel, W., Wcislo, R., Matwin, S.: 2-D embedding of large and high-dimensional data with minimal memory and computational time requirements. arXiv preprint arXiv:1902.01108 (2019)

4. Dzwinel, W., Wcislo, R., Strzoda, M.: ivga: Visualization of the network of historical events. In: Proceedings of the 1st International Conference on Internet of Things and Machine Learning 2017. ACM (2017)

5. Dzwinel, W., Wcisło, R., Czech, W.: ivga: a fast force-directed method for interactive visualization of complex networks. J. Comput. Sci. **21**, 448–459 (2017). ISSN: 1877-7503, https://doi.org/10.1016/j.jocs.2016.09.001, https://www.sciencedirect.com/science/article/abs/pii/S1877750316301430

6. Ghojogh, B., Ghodsi, A., Karray, F., Crowley, M.: Multidimensional scaling, sammon mapping, and isomap: Tutorial and survey (2020). https://arxiv.org/abs/2009.08136

7. Jia, W., Sun, M., Lian, J., Hou, S.: Feature dimensionality reduction: a review. Complex Intell. Syst. **8**, 2663–2693 (2022). https://doi.org/10.1007/s40747-021-00637-x

8. McInnes, L., Healy, J., Melville, J.: UMAP: uniform manifold approximation and projection for dimension reduction. arXiv preprint arXiv:1802.03426 (2018)

9. Minch, B.: Viskit library. https://gitlab.com/bminch/viskit

10. Minch, B.: In search of the most efficient and memory-saving visualization of high dimensional data (2023)

11. Minch, B., Nowak, M.P., Wcislo, R., Dzwinel, W.: GPU-embedding of kNN-graph representing large and high-dimensional data. Computat. Sci. ICCS 2020 **12138**, 322 – 336 (2020)

12. Minch, B., Łazarz, R., Dzwinel, W.: Supplementary materials. https://www.dropbox.com/s/s9wx5bz8wsq1zh3/ICCS_2023_Supplementary_Materials.pdf?dl=0

13. Wang, Y., Huang, H., Rudin, C., Shaposhnik, Y.: Understanding how dimension reduction tools work: An empirical approach to deciphering t-sSNE, UMAP, TriMAP, and PaCMAP for data visualization. J. Mach. Learn. Res. **22**(201), 1–73 (2021)

Fast Electromagnetic Field Pattern Calculation with Fourier Neural Operators

Nattawat Pornthisan[1] and Stefano Markidis[2(✉)]

[1] Chulalongkorn University, Bangkok, Thailand
[2] KTH Royal Institute of Technology, Stockholm, Sweden
markidis@kth.se

Abstract. Calculating the field pattern arising from an array of radiating sources is a central problem in Computational ElectroMagnetics (CEM) and a critical operation for designing and developing antenna systems. Yet, it is a computationally expensive operation when using traditional numerical approaches, including finite-difference in the time and spectral domains. To address this issue, we develop a new data-driven surrogate model for fast and accurate calculation of the field radiation pattern. The method is based on the Fourier Neural Operator (FNO) technique. We show that we achieve a performance improvement of 31x when compared to the performance of the Meep CEM solver when running on a desktop laptop CPU at the cost of a small accuracy loss.

Keywords: Computational Electromagnetics · Fourier Neural Operator · Electromagnetic Field Pattern · Dipole Antenna Array

1 Introduction

Computational ElectroMagnetics (CEM) is a discipline at the intersection of applied mathematics, scientific computing, and electromagnetics theory [4] with critical applications to the design and development of electromagnetics systems [3], such as antenna arrays, waveguides, resonators, radar systems, to mention a few applications. At its heart, CEM comprises several numerical techniques for the solutions of Maxwell's equations for determining the electric and magnetic fields in the vacuum or a medium. These numerical approaches range from simple finite-difference schemes (in the time and frequency domains) to the Finite Element Method (FEM), Method of Moments (MoM), to Montecarlo techniques [11].

In this work, we focus on determining the field pattern generated by the several radiating current sources, a central topic in CEM with applications to the design of antenna arrays. In particular, we aim to solve the field pattern problem of finding the fields produced in response to a source at a single frequency ω. This a central problem in CEM and has a wide range of applications. The solution to this problem can be achieved by several numerical methods [11].

J. Mikyška et al. (Eds.): ICCS 2023, LNCS 14074, pp. 247–255, 2023.
https://doi.org/10.1007/978-3-031-36021-3_24

However, these numerical approaches are computationally expensive as either they require (i) simulating with a constant-frequency source for a long time so that all transient effects, from the source turn-on, vanish or (ii) solving a large linear system with complex-valued terms.

Recently, Machine Learning (ML) data-driven models have been proposed either to substitute or to augment existing traditional techniques [1, 7]. Examples of such emerging ML approaches are neural networks. These methods are data-driven and work in a supervised or semi-supervised fashion – the neural networks take as input a series of input data (for instance, the pixel values of an image) and the associated labels, and an optimization process updates the network weights and biases so that given an unseen input, the neural network produces a result prediction. Minimizing the loss function, which expresses the error between the network prediction and the actual label, drives the optimization step. The final result of the training is a neural network (with its weights and biases) that can be used as a replacement for the simulation tool, employed for training input-output pairs. For this reason, these methods are also called *surrogate* methods. These methods are considerably faster than traditional approaches when considering only the prediction step (as the computationally expensive neural network training is performed offline) at the cost of lower accuracy.

One of the most recent and powerful neural network approaches is the Fourier Neural Operator (FNO) [6]. An FNO is a neural network architecture, combining Deep Learning with Fourier analysis. FNOs are designed to efficiently solve Partial Differential Equations (PDE) using a spectral method based on the Fourier transform. The key idea behind FNOs is to represent the solution to a PDE as a superposition of sinusoidal functions of different frequencies, which can be efficiently computed using the Fast Fourier Transform (FFT). The neural network is then trained to learn the mapping between the input data and the corresponding Fourier coefficients of the solution. The key difference between traditional neural networks and neural operators, like FNO, is that the latter are designed to operate on functions directly [5], rather than on discrete data points. This makes them particularly well-suited for problems that involve continuous functions or signals, such as electromagnetic waves, where the inputs and outputs are functions rather than discrete vectors.

In this work, we design and implement FNOs to calculate the electromagnetic radiation pattern in two-dimensional geometry. In particular, we create the dataset and compare the performance of our surrogate model to the Meep frequency-domain solver [9]. With the use of the FNO, we can able to predict in a fast way the field pattern at a fraction of the cost of the state-of-the-art Meep CEM solver with reasonable accuracy.

2   Methodology

We divide our work into three phases consisting of (i) generating the training of the datasets with state-of-the-art Meep CEM solver (ii) designing the FNO architecture, and (iii) training a neural network. That said, we will present both the technical details and simplifying constraints at each stage in the subsections below.

2.1 Simulation Setup and Dataset Generation with Meep

Our simulation setup and methodology is inspired by Ref. [2]. First, we generate an initial dataset using the state-of-the-art Finite-Difference Time-Domain (FDTD) Meep CEM solver developed at MIT. More specifically, we obtain the field pattern by running the Meep frequency-domain solver that combines an FDTD time step with an iterative solver for complex-value linear system. The specific simulation workflow and setup is shown in Fig. 1 and as follows.

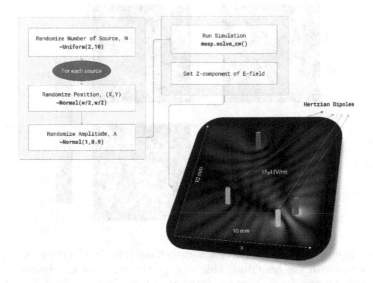

Fig. 1. Setup of Meep simulation for preparing the input data set to train and test the network. The generation process is shown as a stochastic process described in the two blue boxes, while an example of simulation result is shown in the lower-right corner. (Color figure online)

The simulation region is two-dimensional on the x-y plane. The simulation domain is a 10 mm × 10 mm vacuum area centered at the origin, which will be surrounded by a Perfectly Match Layer (PML) of 1mm thick to enable open boundary conditions. The resolution of the simulation is set at 10 pixels per millimeter, totaling a grid size of 140 × 140 grid points. The Meep time step is calculated automatically to satisfy the Courant-Friedrichs-Lewy (CFL) condition. In this work, the simulation result is the field pattern which is the magnitude of the out-of-plane electric field component, E_z in units of V/m.

To create the initial dataset, we run 2,000 Meep simulations with a random number of radiation sources in random locations with random amplitude and obtain the final outcome of our work: the field pattern as the E_z magnitude. More precisely, for each simulation (or item of our dataset), we first select several point sources between two and ten according to a random uniform distribution. Then, for each point source, we uniformly sample their positions within the simulation region and their amplitude from a normal distribution with mean and standard deviation equal to 1.0 mA and 0.9 mA, respectively. All the antennas are z-axis aligned Hertzian dipoles oscillating at the same frequency of 200 GHz. A few examples of field pattern is then obtained by the frequency domain solver provided by Meep. For visualization, we report a few examples of results from the generation process in Fig. 2.

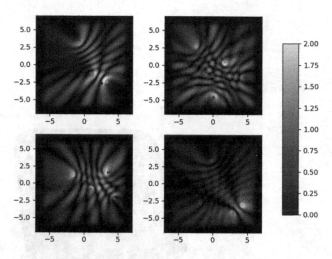

Fig. 2. Example of data generation results from Meep. The dots on the images represent a point source, with the amplitude indicated by the color (in mA units). The Meep-simulated field pattern for each source distribution is shown in the background (in V/m units). Note that the domain also includes the PML cells. (Color figure online)

2.2   Fourier Neural Operator Architecture

To implement our network, we follow the FNO architecture, presented in the seminal paper on FNO [6]. A step-by-step visualization of the architecture can be found in Fig. 3. We implement a neural network consisting of four FNO layers which (i) apply Fast Fourier Transform (F) on the up-projected latent space, (ii) linearly transform the input in the frequency domain (R), and (iii) perform the inverse Fourier (F^{-1}) transform back to a spatial domain. As in Ref. [6], we apply a filter to eliminate the high-frequency: only 16 lower Fourier modes are linearly transformed while the rest are discarded and set to zero. The bias term (W) and activation function (σ) are then added and applied respectively at the end of each Fourier layer.

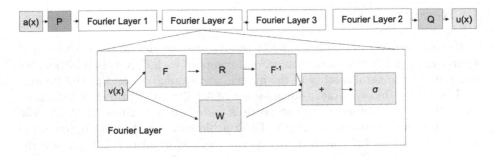

Fig. 3. The FNO architecture [6].

We split the datasets into training and test sets consisting of 1,760 and 240 pairs, respectively. We then train the FNO for 160 epochs (see Fig. 4) using the Adam optimizer with an L_p loss function [6]. After a few trials, we select a suitable learning rate of 10^{-3} and a batch size of 128. The best-performing parameters on the validation set are chosen among all epochs for inference. To compare the performance with FNO, we also train a U-Net [10], a convolutional neural network architecture, commonly used for image segmentation tasks. The Python notebooks for generating the datasets with Meep, and performing the FNO and U-Net training can be found on the GitHub repository[1].

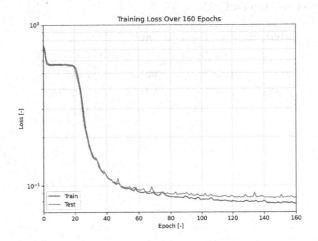

Fig. 4. The loss of the neural network over 160 epochs of optimization on training (blue) and testing (orange) datasets. (Color figure online)

---

[1] https://github.com/winnaries/fno-fastem.

3 Results

In this section, we evaluate the FNO results against ground-truth simulation regarding accuracy and performance, and compare with U-Net. In addition, we also present a generalizability of the model over a range of hand-picked input.

The FNO achieves a mean-square-error of 0.172×10^{-10}, which is 14% more accurate than U-Net model whose score is tested to be 0.207×10^{-3}. For the mean-absolute-percentage-error, the FNO achieves a significantly better score of 0.257, which is 44% more than UNet model's score of 0.456. Qualitatively, it is evident that the FNO can accurately learn the relationship among the sources while also accounting for their amplitudes. The waves' crests are visually indistinguishable to the ground truth (Fig. 5).

We then measure the execution time of the Meep simulation and neural network over 50 inputs on an Ubuntu machine with an AMD Ryzen 7 5800X CPU, and a Nvidia RTX A4000 GPU. The computational performance results are summarized in Table 1. Both methods were given the same input size and evaluated on CPU since Meep does not support acceleration on GPU [8]. The FNO is faster than Meep by 31 times while maintaining high accuracy. When using the Nvidia GPU, the surrogate network is 467x faster than Meep running on the CPU.

Table 1. Average execution time (and its standard deviation) over 50 inputs of Meep field pattern solver on CPU and FNO on CPU and GPU. The speedup ratio is calculated with respect to Meep on CPU.

Method	Running Time, ms		
	CPU/MEEP	CPU/FNO	GPU/FNO
Mean	934.67	34.525	2.0063
±Std	64.671	4.4655	0.22788
Speedup		31x	467x

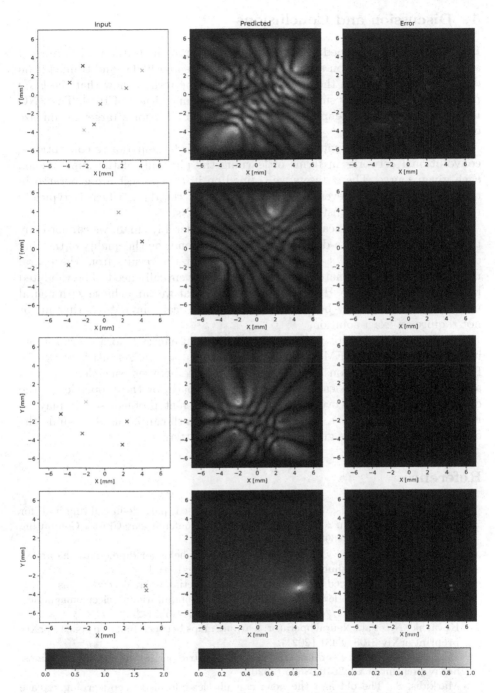

Fig. 5. Example field pattern FNO prediction output (middle) and its corresponding error with respect to the ground-truth (right) given the initial source distribution input (left). Note that the domain also includes the PML cells.

4   Discussion and Conclusions

In this work, we developed a way to efficiently approximate the field pattern of a dipole antenna array on a xy-plane. We implemented FNOs and trained it on datasets generated with the `Meep` CEM solver. The results show that the FNO prediction has relatively small mean-square-error magnitude of 10^{-3}. The FNO is also faster than the `Meep` CEM solver by 31x on CPU for a target region size of 140×140 cells.

By comparing FNO with U-Net results, FNO is far superior to conventional convolutional networks in terms of learning the physical relationship between each given source. The datasets we trained on in this project are generated in the same process. However, we did not augment the data, which is typically required for training transformation-invariant CNNs.

Although the FNO mean-square error is considerably small, we cannot deny that the FNO is still data-driven and partially depends on the quality of training distribution. That said, it is safer to assume that the results from the neural network are an approximate solution that would eventually need to be validated by a conventional solver. However, the speedup that we can achieve with neural networks is undoubtedly beneficial for some large-scale simulations that might not require an exact solution.

The work done in this project needs additional efforts before we can apply it to solve real-world CEM problems, e.g., antenna arrays. As future work, the FNO architecture can consider the phase difference between each element in an antenna array, the physical relationship in 3D space, or the properties of the designated media. Moreover, for an entirely different problem set, it may be worth investigating how a generative neural network can be used to guide the design of antenna arrays given a desirable field pattern.

References

1. Aguilar, X., Markidis, S.: A deep learning-based particle-in-cell method for plasma simulations. In: 2021 IEEE International Conference on Cluster Computing (CLUSTER), pp. 692–697. IEEE (2021)
2. Baas, M.: Using Fastai v2 to approximate electric fields for dipole antenna arrays (2020). https://rf5.github.io/2020/01/02/ml-emags1.html
3. Balanis, C.A.: Advanced engineering electromagnetics. John Wiley & Sons (2012)
4. Bondeson, A., Rylander, T., Ingelström, P.: Computational electromagnetics. Springer (2012). https://doi.org/10.1007/978-1-4614-5351-2
5. Kovachki, N., et al.: Neural operator: learning maps between function spaces. arXiv preprint arXiv:2108.08481 (2021)
6. Li, Z., et al.: Fourier neural operator for parametric partial differential equations. arXiv preprint arXiv:2010.08895 (2020)
7. Markidis, S.: The old and the new: can physics-informed deep-learning replace traditional linear solvers? Front. Big Data **4**, 92 (2021)
8. MEEP - Documentation (2022). https://meep.readthedocs.io/en/latest/FAQ/

9. Oskooi, A.F., al.: Meep: a flexible free-software package for electromagnetic simulations by the FDTD method. Comput. Phys. Commun. **181**(3), 687–702 (2010)
10. Ronneberger, O., Fischer, P., Brox, T.: U-Net: convolutional networks for biomedical image segmentation. In: Navab, N., Hornegger, J., Wells, W.M., Frangi, A.F. (eds.) MICCAI 2015. LNCS, vol. 9351, pp. 234–241. Springer, Cham (2015). https://doi.org/10.1007/978-3-319-24574-4_28
11. Sadiku, M.N.: Numerical techniques in electromagnetics. CRC Press (2000)

Elements of Antirival Accounting
with sNFT

Tommi M Elo[1]($\boxtimes$), Jarno Marttila[2,3], Sergi Cutillas[4], and Esko Hakanen[2]

[1] Department of Management Studies, and Department of Information and
Communications Engineering, Aalto University, Espoo, Finland
tommi.elo@aalto.fi
[2] Department of Industrial Engineering and Management, Aalto University,
Espoo, Finland
esko.hakanen@aalto.fi
[3] Streamr Network, Zug, Switzerland
jarno.marttila@streamr.network
[4] Department of Economic History, Institutions, Politics and World Economy,
University of Barcelona, Barcelona, Spain
sergicutillas@ub.edu

Abstract. Accounting with antirival tokens, i.e., accounting based on
shareable units that gain value with increased use, enables efficient and
effective collective action. However, most currencies are rival tokens which
can naturally represent — and be exchanged to — rival goods, such as a
cup of coffee. Antirival systems of account would be a natural fit for the
economy of antirival goods because the logic of value creation and accounting would be compatible. It is challenging to find an allocatively efficient
price for antirival goods, such as data, measured in rival units of account.
We present an antirival accounting system based on Distributed Ledger
Technology (DLT), where the fundamental operation is *sharing* instead of
exchanging and study it with system dynamics models and simulations.
We illustrate our arguments by presenting a system known as Streamr
Awards that defines three tokens of a fundamentally novel type, shareable
non-fungible token (sNFT). We present the functioning of one of these in
the work allocation of a self-directed online community.

Keywords: sNFT · Antirival goods · System dynamics · Blockchain ·
DLT

1 Introduction

Accounting and accounting systems are a way to increase accountability and
transparency. These map to information security properties for data, integrity,
and availability, respectively [5]. Businesses use accounting methods and financial
accounting systems to maintain checks and balances, reduce risk of fraud, and
be regulatory compliant. These types of accounting systems are commonly seen
as mechanisms to register financial transactions measured in currency values

J. Mikyška et al. (Eds.): ICCS 2023, LNCS 14074, pp. 256–263, 2023.
https://doi.org/10.1007/978-3-031-36021-3_25

where the unit of an account is money [8,14]. Hence, to date, these systems have been predominantly designed for rival resources -most commonly understood as physical resources characterised by scarcity and expendability.

However, we enter into a novel territory with the economic system in which we wish to apply accounting, e.g., in information goods. Such systems and resources are not rival, as they do not deplete in use and can be multiplied practically indefinitely [6]. Consequently, they do not fit the traditional accounting system, and money is proving to be an unsuitable unit of account, as repeated shortcomings of the data markets indicate [10,15]. This paper addresses this shortcoming by describing and exploring the concept of antirival accounting through tokenised efforts of a decentralised open-source community. We ask: how antirival accounting can impact the work of decentralised communities using antirival goods?

Building on previous work [6,10,16], we hypothesise that data markets, in particular, and society, in general, need an efficient socio-technical accounting system that supports the underlying antirival goods. Yet, such antirival accounting systems are far from being standardised arrangements, requiring institutional work [12]. We present a solution implemented by utilising open DLTs to create antirival accounting units that combine the efficiency and security of traditional rival units of account with fundamentally different economic implications.

The proposed solution defines an antirival system with rules and behavioural patterns. We develop a simple theory [4] for the conceptual virtual laboratory [7] of antirival [21] accounting and describe it with system dynamics [19], a complex adaptive systems modelling and simulation methodology. We further explore the arising dynamic feedbacks, and potential value dynamics, of antirival accounting with system dynamics modelling and simulations. Thus, the solution contributes to the development of "antirival institutions" toward a more inclusive digital economy [6].

2   On Open DLTs and Antirival Goods

Accounting can be defined as an information system that measures, processes, and communicates financial and non-financial information about an economic entity [14]. Accounting is a social science representing the economic reality of the enterprise, but it also has the power to create new social realities [8].

There are situations where the use of rival accounting units would be suboptimal. This dilemma exists, e.g., as part of data markets since the goods traded, data and information are of fundamentally different nature than physical items or services [10,16]. For physical items a typical operation is *exchanging*; for data, the typical operation is *sharing*. A blockchain is a growing list of data items, blocks, that are securely linked together using cryptography [13]. An open DLT is public, and anyone can participate in the core activities of the blockchain network, such as reading and writing to the chain. These examples illustrate how data and information are shared, not exchanged.

Economic goods can be classified as rival, nonrival, and antirival based on their subtractability dimension (e.g., [15,16]). *Rival goods* are defined as goods with positive subtractability, which means that their value is transferred and lost

Table 1. Example goods of the extended classical economics taxonomy to information goods [16].

		Subtractability		
		Rival	**Nonrival**	**Antirival**
Excludability	Excludable	private key of a cryptosystem	*uncopyable* electronic book / PDF	access-controlled science journal
	Non-excludable	public key of a cryptosystem	electronic book / PDF	open science journal

if the good is consumed. Meanwhile, nonrival and antirival goods are defined as having neutral and negative subtractability, respectively. Contrary to rivalry, antirivalry is the quality of those goods having negative subtractability. Antirivalry means that the goods having antirival nature *gain value* when given to and shared with others; that is, they have positive network externalities [6]. Furthermore, economic goods can be classified as excludable or non-excludable according to their excludability dimension, i.e., how excludable they are via human decision-making (typically by their owner). Examples of each category of information goods - rival, nonrival, antirival - are presented in Table 1.

An online science journal is antirival since the articles (in) increase value when shared. It is excludable because the journal may require money or some form of membership for access. When open sharing to all interested parties without exclusion is enabled, it is an example of a non-excludable good, an *open science journal*. Should the products be *uncopyable PDFs of scientific papers*, those would be an example of nonrival, excludable goods. Because the data of the book remains nonrival, the uncopyability would make it exclusive to the holder of the (only) copy. Finally, a private key of a public key cryptographic system may be rival because it is essential to use it only by one person or entity and not anyone else. Otherwise, the function of the key is lost; this is an example of a rival yet excludable good since we can keep a private key confidential. The corresponding public part of the same key pair is equally rival but is not excludable since everyone needs to have access to it.

In addition, we follow the argumentation by Olleros [6] that a good itself, in isolation, is not necessarily antirival without the more extensive (social) system, where the good is embedded. The system defines and decides if something is antirival or merely nonrival. For example, an informational component that complements a rival transaction has the potential of being a nonrival or antirival good [16].

3   Methodology

In this paper, we develop a system dynamics model to demonstrate how the work becomes more efficient when directed in a decentralised community utilising the antirival units of account in open DLTs. By applying it in pilot experiments, we have chosen to utilise DLTs to model and implement a representation of an antirival accounting unit as a shareable non-fungible token (sNFT). An sNFT is distinguished from a regular rival NFT via its novel fundamental operation, as they can be shared, not only exchanged, between parties.[1]

Recently, system dynamics has gained more recognition as a prime modelling methodology for a novel branch of economics [3]. System dynamics is commonly utilised to model complex adaptive systems, such as taxation economics [18], and has shown applicability in institutional economics [17]. This paper presents a CLD model of our simulations, illustrating how sNFT tokens can improve the work efficacy in a decentralised community.

As our case, we introduce an active community which underlies and supports the Streamr project. The project is developing a decentralised peer-to-peer network for transmitting real-time data using a topic-based publish-subscribe system. Streamr community is worldwide, and it is built around a strong ethos. The Streamr community produces antirival information goods which form a body of knowledge for the open-source technology. Three types of sNFT tokens were designed to align and reward contributions from the community members to support the goal of a thriving decentralised open-source knowledge creation community. The Streamr Awards tokens, serve as units of account for finding relevant contributions and their producers.[2] The sNFTs form the basis for the CLD model (Fig. 1) and simulations. The model depicts the Streamr Awards system for member contributions in terms of causalities but may be applied to any open knowledge creation community.

4   Model Description

In Fig. 1, starting from the first feedback loop *R1: efficient collaboration*, the more the community has engaged collaborations, the more foundational work rate is increased. *Foundational work* is the work the community was founded for as described by the constitution and inspired by the ethos [11]. The bigger the foundational work rate, the more completed foundational work is achieved and the more significant the amount of identified evaluation work, as all completed foundational work directly defines a work item to be evaluated. *Evaluation work* is the work community does to quality-check the foundational work. Evaluation work, together with endogenous variables *identified evaluation work* and *constitution of the catallaxy* [2], determine how both, the number of contribution sNFT tokens, and body of knowledge develop. The *body of knowledge* is all the vocabulary and processes that make up the knowledge of a particular professional field as defined by the knowledge creation community [9]. Increasing the number of contribution tokens increases the internal blockchain effects, which are assumed to decrease the internal search costs as the open blockchain is a fully openly searchable and reliable joint database describing all the critical internal elements to be searched. Decreasing the internal search costs will lead to a change in the opposite direction of the engaged collaborations, which will

---

[1] ERC-5023: https://eips.ethereum.org/EIPS/eip-5023.

[2] A more detailed description in the Streamr blog: https://blog.streamr.network/streamr-awards-are-here-contribute-and-earn-unique-snfts/.

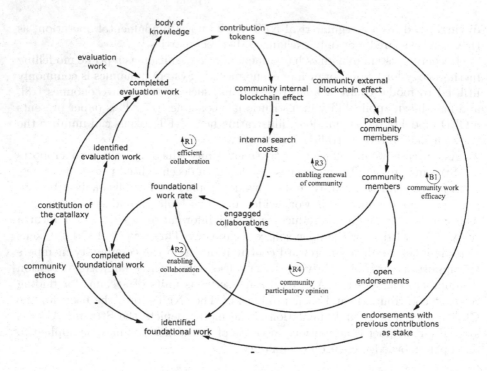

Fig. 1. CLD of the community managed via contribution sNFTs.

thus increase, closing the *R1*, which is named *efficient collaboration* because it is what makes collaboration in the community more efficient each round.

Concentrating on the leftmost part of the diagram, the constitution of the catallaxy also balances foundational work. It is precisely the function of the constitution to frame the work, i.e., the digital catallaxy's constitutional smart contracts, separate from all possible work, the foundational work to be performed. Focusing on the second reinforcing feedback, *R2: enabling collaboration*, the engaged collaborations also increase identified foundational work, as work gets identified synchronously with the collaboration formation. *Engaged collaborations* are collaborations of members working on a piece of foundational work. The more identified foundational work there is, the more manifests as completed foundational work, closing the *R2* feedback by joining it with the *R1*.

The word *effective* needs an outer framing and cannot be described by mere increasing numbers without knowing what such numbers mean. I.e., it needs a context, (ethical) values being attached to those numbers and attaching them to a meaning, an ethos. The constitution is the rules about what the founding fathers found valuable in itself, not quite intrinsically, but instead for some purpose under some value definition, and it describes the actionable purpose of the community. Turning to *R3: enabling renewal of community*, increasing the number of contribution tokens increases the community external blockchain effects. Blockchain effects consist of *transparency effect* and *integrity effect* [5].

Here the word *external* refers to the effects which reach outside the community. The nature of an open Nakamoto blockchain is that it is and remains world readable without the possibility of closure [13]. Thus, the potential community members outside the community will remain in the know about those aspects of the community which are enjoyed by the members. Such an external blockchain effect leads to an increasing number of potential community members. Moreover, the number of community members also increases as new members engage in collaborations which increases *engaged collaborations*.

In the dynamic reinforcing loop *R4: community participatory opinion*, increase in the community members, also increases open endorsements, as people give public encouragement (e.g., thumbs up) to each other. *Open endorsements* are a form of low-commitment participatory commitment work. Open endorsements help to reduce and prioritise the work leading to the effectiveness of the community, instead of mere increasing work rate (efficiency). The loop *R4* is affected by the community constitution, catallaxy. *R4* is a reinforcing loop increasing identified work since it will help speed up the work rate by prioritising what is widely considered most valuable. Finally, there is a goal-seeking loop *B1: community work efficacy*, starting from the contribution tokens. As the number of contribution tokens increases, endorsements with previous contributions as stake increase as community members engage in endorsements based on their and others' contributions. This type of endorsement is a form of high-commitment, high-stakes participatory work to the community. It is based on pre-existing, formally acknowledged deeply community-embodied previous contributions (staking) and others' contributions (basis). The goal-seeking loop *B1* is mediated by the constitution of the catallaxy. The constitution is connected to the identified foundational work with a minus sign because it reduces and selects which work is inside the community and which is outside. Behind it is the hard-to-formalise human interactions related to community ethos, i.e., off-chain governance. The mere compositional structure of the collaboration effort defines the newly identified foundational work in conjunction with the community constitutional catallaxy because these two are the direct reasons causing new work (efficacy).

5   Discussion

Our simulations describe the general dynamics of an sNFT representing contributions in a decentralised community. Via tokenisation, they form a basis for a different type of accounting. Any good can be excluded with the use of force, and even information goods can be excluded with cryptography and secrets. Confidentiality provided by cryptography is based on information asymmetry, with some knowing the key and some not. However, these instruments have not been successful in facilitating efficient markets to share or trade data [10,15]. The *R2* loop of Fig. 1 has the mechanical efficiency to expand very fast. However, it shall remain under the framing of *R3*, *R4* reinforcing, and *B1* balancing loops, which are in turn mediated and controlled by the founding constitution of the catallaxy [2]. The value accumulation of such a system remains under the control of the founding ethos of the community, guiding the embodiment of practical, actionable rules into the constitution of the catallaxy.

A decentralised institution secured with blockchain means sustainable distribution of power. Members of our antirival community contribute to the constitutional catallaxy [2] through exit, voice, or loyalty [1]. The process we define increases efficiency and efficacy within the confines of the pre-existing interpretation of the constitution [2]. So far, such a constitutional process has been known to take place mainly in the context of nation-states. However, later, via the constitutional catallaxy, it can also apply to smaller communities, such as platforms [9]. Streamr Awards allows the members to shape the constitutional catallaxy by contributing directly or by two types of community members endorsing contributions. Previous contributions do not back the first kind of endorsement. The second kind is backed by previous contributions wherein the community member endorsing is included. The second kind is more valuable than the first. In simple terms, the contributions backing the endorsements are part of the post-constitutional game, a process that increases work efficacy (work selection from all possible work) [2]. There are also indirect efficiency effects via *engaged collaborations*. This process is an incremental prioritisation and incorporation of contributions that is fully decentralised without a need for political or centralised negotiation.

Loop *B1* decreases work, i.e., the set of all possible work is larger than the work to be done under this community and in the particular selected collaboration. *Effectiveness* or *efficacy* is a qualitative change in what work is selected to be done, while the efficiency-producing loops merely increase the work rate. Together these components, efficacy, and efficiency, produce increased *allocative efficiency*, which is a better allocation of work and a more efficient execution The direction and selection of work happen in a decentralised manner [20].

We introduced and made progress in developing antirival accounting, a new form of accounting where the fundamental transaction is sharing instead of exchange. We presented an open blockchain-based antirival taxonomy and a further tokenisation example of an antirival token (sNFT). We illustrated our arguments by describing the system dynamics model for Streamr Awards. The sNFT formulation can efficiently and transparently represent values that are considered *externalities* in the rival accounting system. When antirival accounting is utilised in bottom-up digital communities, new egalitarian, decentralised, and heterodox modes of governance become possible. The resulting antirival accounting system seems particularly suitable for the areas of the economy which mainly produce valuable and reliable data, information, knowledge, or wisdom.

Acknowledgements. ATARCA received funding from the EU Horizon 2020 agreement No 964678. The authors thank Prof Pekka Nikander for constitutional entrepreneurship, and Prof Juuso Töyli, Prof Len Malczynski, Dr. Sampsa Ruutu, Prof Heikki Hämmäinen, Prof Raimo Kantola, and Prof Petri Mähönen.

References

1. Berg, A., Berg, C.: Exit, voice, and forking. Cosmos+ Taxis **8**(8), 9 (2017)
2. Berg, A., Berg, C., Novak, M.: Blockchains and constitutional catallaxy. Const. Polit. Econ. **31**(2), 188–204 (2020)
3. Cavana, R.Y., Dangerfield, B.C., Pavlov, O.V., Radzicki, M.J., Wheat, I.D. (eds.): Feedback Economics. CST, Springer, Cham (2021). https://doi.org/10.1007/978-3-030-67190-7
4. Davis, J.P., Eisenhardt, K.M., Bingham, C.B.: Developing theory through simulation methods. Acad. Manag. Rev. **32**(2), 480–499 (2007)
5. Elo, T.M., et al.: Improving IoT federation resiliency with distributed ledger technology. IEEE Access **9**, 161695–161708 (2021)
6. Xavier Olleros, F.: Antirival goods, network effects and the sharing economy — First Monday. First Monday **23**(2) (2018)
7. de Gooyert, V.: Developing dynamic organizational theories; three system dynamics based research strategies. Qual. Quant. **53**(2), 653–666 (2019)
8. Hines, R.D.: Financial accounting: in communicating reality, we construct reality. Acc. Organ. Soc. **13**(3), 251–261 (1988)
9. Kornberger, M., Pflueger, D., Mouritsen, J.: Evaluative infrastructures: accounting for platform organization. Acc. Organ. Soc. **60**, 79–95 (2017)
10. Koutroumpis, P., Leiponen, A., Thomas, L.D.: Markets for data. Ind. Corp. Change **29**(3), 645–660 (2020)
11. Lawrence, T.B., Suddaby, R.: 1.6 institutions and institutional work. The Sage Handbook of Organization Studies, pp. 215–254. Sage Thousand Oaks, CA, USA (2006)
12. Lehenchuk, S., Zhyhlei, I., Syvak, O.: Understanding accounting as a social and institutional practice: possible exit of accounting science from crisis. Account. Financ. Control **3**(1), 11–22 (2020)
13. Nakamoto, S.: Bitcoin: a peer-to-peer electronic cash system. Tech. rep., Internet (Oct 2008). https://bitcoin.org/bitcoin.pdf
14. Needles, B.E., Powers, M., Crosson, S.V.: Accounting principles. South-Western Cengage Learning (2011)
15. Nikander, P., Elo, T.: Will the data markets necessarily fail? A position paper. In: 30th European Regional ITS Conference. International Telecommunications Society (ITS), Helsinki, Finland (2019)
16. Nikander, P., Eloranta, V., Karhu, K., Hiekkanen, K.: Digitalisation, anti-rival compensation and governance: need for experiments. In: Nordic Workshop on Digital Foundations of Business, Operations, and Strategy, p. 7. Aalto University, Espoo, Finland (2020)
17. Radzicki, M.J.: Institutional Economics, Post-Keynesian Economics, and System Dynamics: Three Strands of a Heterodox Economics, p. 156. University of Michigan Press (2008)
18. Saeed, K.: Taxation of fiat money using dynamic control. Systems **10**(3), 84 (2022)
19. Sterman, J.D.: Business Dynamics: Systems Thinking and Modeling for a Complex World. McGraw-Hill Education (2000)
20. Weyl, E.G., Ohlhaver, P., Buterin, V.: Decentralized Society: finding web3's soul. SSRN Electronic Journal (2022)
21. Wood, A.D.: A model to teach non-rival and excludable goods in undergraduate microeconomics. Int. Rev. Econ. Educ. **24**, 28–35 (2017)

Learning Shape-Preserving Autoencoder for the Reconstruction of Functional Data from Noisy Observations

Adam Krzyżak[1,2] , Wojciech Rafajłowicz[3] , and Ewaryst Rafajłowicz[3(✉)]

[1] Department of Computer Science and Software Engineering, Concordia University, 1455 De Maisonneuve Blvd. West, Montreal, QC H3G 1M8, Canada
krzyzak@cs.concordia.ca
[2] Department of Electrical Engineering,
Westpomeranian University of Technology (WUT), Szczecin, Poland
wojciech.rafajlowicz@pwr.edu.pl
[3] Faculty of Information and Communication Technology,
Wroclaw University of Science and Technology, Wrocław, Poland
ewaryst.rafajlowicz@pwr.edu.pl

Abstract. We propose a new autoencoder preserving input functions' general shape (monotonicity, convexity) after their reconstruction without imposing a priori constraints. These properties are inherent to the coefficients of the Bernstein-Durrmeyer polynomials that serve here as theoretical descriptors. Their estimates, computed from noisy observations by the coder, play the role of latent variables. The approach is purely nonparametric, i.e., no prior finite-dimensional model is assumed. The answer to the question of how many latent variables should be used for an acceptable reconstruction accuracy of a family of functional data is inferred from learning based on the proposed approximation of the Akaike Information Criterion. A distinguishing feature of this autoencoder is that the coder and encoder are designed as precomputed and stored matrices with the Bernstein polynomial entries. Thus, after selecting the number of latent variables, the autoencoder usage has a low computational complexity since it is linear with respect to observations. The proposed computational algorithms are tested on real data arising in mechanical engineering when control of damping vibrations is necessary.

Keywords: Functional data · Shape-preservation · Autoencoder design · Bernstein-Durrmeyer polynomials · Nonparametric estimation

1 Introduction

An autoencoder (AE) is a neural network that tries to replicate its input to an output. It consists of a coding algorithm that transforms input vectors into a

---

Part of this research was carried out by the first author during his visit of WUT while on sabbatical leave from Concordia University.

latent vector of a smaller dimension, which is supposed to compress the input information into meaningful form. The latter can be stored and/or transmitted to an encoder that tries to restore the input vector. The idea of reconstructing functions from a smaller number of descriptors that play role of the latent variables arose very early [12], even if the term "autoencoder" was not in common usage at a time.

Autoencoders have been first introduced by Rumelhart, Hinton and Williams in [20]. Their goal was to learn internal informative representation of the data useful in various applications such as clustering and principal component analysis. If encoder and decoder are linear, then latent representation of the resulting linear autoencoder [1] performs Principal Component Analysis (PCA) [16]. This demonstrates that autoencoder is generalization of PCA, which yields latent space in a form of low-dimensional non-linear manifold rather than low-dimensional hyperplane.

Simple autoencoders have a tendency to reconstruct inputs accurately rather than to build latent informative representation. To prevent this undesirable effect additional regularization has been introduced in autoencoders. Sparse autoencoders [14] use L_1 regularization, which induces sparsity of latent representation. Another approach is based on Kullback-Leibler divergence, which is a distance measure between probability distributions. Denoising Autoencoders [5,6,23] carry out regularization by removing additive Gaussian noise added to the input. In Denoising Autoencoders the emphasis is on making them resistant to perturbation of the input, while Contractive Autoencoders [19] put less emphasis on features, which do not play crucial role in decoder reconstruction activity. A real breakthrough in recent years came with the introduction of Variational Autoencoders (VAE) [2,8,11].

Recently, the shape-sensitivity approaches were intensely investigated. Their focus is on estimating descriptors from a signal derivative. An sample of papers representing this direction of research includes [9], [10,15], among others,

The properties of the Bernstein polynomials are well known [13]. Their version exploited here is known as the Bernstein-Durrmeyer polynomials [4,7]. In [17] BDP proved their usefulness for nonparametric regression estimation.

Advantages of the Proposed Autoencoder. The aim of the proposed autoencoder is to reconstruct functions (signals) from observations corrupted by random errors. The empirical version of the Bernstein-Durrmeyer polynomials (BDP) coefficients are used as the latent variables (empirical descriptors) of the autoencoder. Such empirical descriptors proved their usefulness in pattern recognition problems as features of classifiers [18]. Here, we use them for reconstructing noisy signals.

The advantages of the proposed autoencoder can be summarized as follows.

1. Descriptors preserve the monotonicity and convexity of a function to be reconstructed. These properties manifest themselves automatically, i.e., only when they are present in the input function.
2. At the training phase, the number of latent variables of the autoencoder is selected by unsupervised learning.

3. This selection is carried out by the nonlinear algorithm but after its completion the reconstruction process itself has linear computational complexity.
4. There is no need to apply prefiltering or a regularization since BDP are sufficiently "stiff". This simplicity is obtained at a cost of reduced rate of restoration accuracy, but this issue is beyond the scope of this paper.

Shape-preserving properties of the descriptors, when applied to contour parts, can be useful in their understanding [22].

Notice that our autoencoder provides an explicit representation of signals. An appealing alternative is the approach known under the acronym SIREN (see [21] and followers). It offers an implicit representation of signals, images, and even partial differential equations by training a cellular neural network with periodic activation functions. As one can observe from many examples, such networks also have good shape-preserving properties but for simpler tasks with explicitly given noisy observations our approach is sufficiently reliable.

2   Derivation of the Proposed Autoencoder

This section introduces the proposed autoencoder. The proposed descriptors play the role of its latent variables.

Autoencoder Input Vectors. Autoencoder inputs consists of vectors $\mathbf{y} = [y_1, y_2, \ldots, y_n]^{tr}$, where ^{tr} denotes transposition. They can arise as observations y_i's of function f at equidistant points t_i, $i = 1, 2, \ldots, n$ corrupted by random errors ϵ_i i.e.,

$$y_i = f(t_i) + \epsilon_i, \quad \text{or} \quad y_i = f_i + \epsilon_i, \quad i = 1, 2, \ldots, n, \tag{1}$$

or y_i's can arise as noisy observations of a time-series f_i's.

Let $\mathbb{E}$ be the expectation of a random variable, while $\mathbb{V}ar$ denotes its variance. Then, the classic assumptions: $\mathbb{E}(\epsilon_i) = 0$, $\mathbb{V}ar(\epsilon_i) = \sigma^2 < \infty$, $i = 1, 2, \ldots, n$, $\mathbb{E}(\epsilon_i \epsilon_j) = 0$, $i \neq j$, $i, j = 1, 2, \ldots, n$ are imposed, assuming that both σ^2 and the probability distribution of all ϵ_i's are unknown. The only exception from the latter assumption is made when we derive a likelihood function for training the encoder.

For simplicity all functions f considered in this paper are assumed to be defined on $T = [0, 1]$. This interval is covered by the subintervals $T_i = (t_{i-1}, t_i]$, $t_i = i/n$, $i = 1, 2, \ldots, n$ of the lengths $\Delta_n = 1/n$, where $t_0 = 0$.

When a time series is discussed, then its unobserved values f_i's and observations y_i's are also considered to be re-scaled to T interval and associated with points t_i's.

For learning the autoencoder the following sequence $\mathcal{L}_L$ of length $L > 1$ is given: $\mathcal{L}_L \overset{def}{=} \{\mathbf{y}^{(1)}, \mathbf{y}^{(2)}, \ldots, \mathbf{y}^{(L)}\}$, as the only available information, where $\mathbf{y}^{(l)}$'s are n-dimensional column vectors that arise as observations of functions $f_l \in C^0(T)$, $l = 1, 2, \ldots, L$, performed according to (1), where $C^0(T)$ denotes the class of all functions continuous on T.

Autoencoder. Let $(N + 1)$ be the number of latent variables (descriptors) of the autoencoder, where N has to be selected from the range $0 \leq N \leq N_{max} \leq n$ of integers and let $\mathbf{B}_N$ denote $(N + 1) \times n$ matrix with the following rows: $[B_k^{(N)}(t_1), B_k^{(N)}(t_2), \ldots, B_k^{(N)}(t_n)]$, $k = 0, 1, \ldots, N$, where $B_k^{(N)}(t)$ k denotes the Bernstein polynomial of order $N \geq k$. which is given by $B_k^{(N)}(t) = \binom{N}{k} t^k (1 - t)^{N-k}$, $t \in T$, $k = 0, 1, \ldots, N$, where, for simplicity, we assume that $B_k^{(N)}(t) \equiv 0$, if $k < 0$ or $k > N$. In computations, it is more convenient to use its well-known recurrent version.

Then, the general formulation of the autoencoder has the following form:

$$\hat{\mathbf{y}} = (N + 1) \, \Delta_n \, \mathbf{B}_{N_{max}}^{tr} \, \mathbf{I}_{N_{max}}(N) \, \mathbf{B}_{N_{max}} \, \mathbf{y}, \tag{2}$$

where $\mathbf{y}$ is its input that is $n \times 1$ vector of observations (1), while $\hat{\mathbf{y}}$ is its $n \times 1$ output. In (2) $\mathbf{I}_{N_{max}}(N)$ denotes $N_{max} \times N_{max}$ diagonal matrix where $(N+1)$ first elements (counting from the upper left corner) are equal to 1, and $N_{max} - (N+1)$ are equal to zero. Its role is just to filter out proper submatrices from $\mathbf{B}_{N_{max}}$ and its transposition. In actual computations, only these submatrices are involved.

However, matrix $\mathbf{I}_{N_{max}}(N)$ plays an important role in pointing out that the process of learning N based on $\mathcal{L}_L$ is, in fact, nonlinear, but after completing it and substituting the selected $\tilde{N}$, say, into (2), the computations become linear in $\mathbf{y}$, which is important for applying the proposed encoder in real-time. Therefore, we call this autoencoder seemingly linear.

Roughly speaking, $\tilde{N}$ is selected as

$$\tilde{N} = \arg \min_N \left[-\ln(\mathbb{L}(\mathcal{L}_L, N) + penalty(N) \right], \tag{3}$$

where $0 \leq N \leq N_{max}$, $\mathcal{L}_L, N)$ is the likelihood function, while $penalty(N)$ is derived from the Akaike's Information Criterion (AIC) for all $\mathcal{L}_L$.

Observe that in (2) matrix $\mathbf{B}_{N_{max}}$ and its transposition are defined, instead of coming from a learning process, as it is typical for most of the encoders discussed in the literature. Clearly, the number of rows of submatrix $\mathbf{B}_{\tilde{N}}$ depends on learning through $\tilde{N}$ but the formulas for computing their elements are precisely defined.

Properties of the Proposed Autoencoder. To motivate the proposed form of the encoder, it is convenient to split the autoencoder so as to express its latent variables explicitly, namely,

$$\tilde{\mathbf{d}}^{(N)}(\mathbf{y}) = (N + 1) \, \Delta_n \, \mathbf{B}_N \, \mathbf{y}, \quad \hat{\mathbf{y}} = \mathbf{B}_N^{tr} \, \tilde{\mathbf{d}}^{(N)}(\mathbf{y}) \quad l = 1, 2, \ldots, L, \tag{4}$$

where $\tilde{\mathbf{d}}^{(N)}(\mathbf{y})$ is $N \times 1$ vector of the latent variables (subsequently called the vector of empirical descriptors). Explicitly, the elements of $\tilde{\mathbf{d}}^{(N)}(\mathbf{y})$ have the form:

$$\tilde{d}_k^{(N)}(\mathbf{y}) = (N + 1) \, \Delta_n \sum_{i=1}^{n} y_i \, B_k^{(N)}(t_i), \quad k = 0, 1, \ldots, N. \tag{5}$$

From now to the end of this subsection, our explanations concentrate solely on the functional data for which the observations have the form (1), left formula. Taking the expectation of (5) we obtain

$$\mathbb{E}[\tilde{d}_k^{(N)}(\mathbf{y})] = (N+1)\,\Delta_n \sum_{i=1}^{n} f(t_i)\,B_k^{(N)}(t_i) \approx (N+1) \int_T f(t)\,B_k^{(N)}(t)\,dt, \quad (6)$$

$k = 0, 1, \ldots, N$ where $\approx$ denotes the approximation of the Riemann integral by its sum, which is sufficiently accurate for smooth f. The following expressions are further called (theoretical) descriptors of f.

$$d_k^{(N)}(f) \stackrel{def}{=} (N+1) \int_T f(t)\,B_k^{(N)}(t)\,dt = (N+1) < f, , B^{(N)} >, \quad (7)$$

k=0, 1,..., N , where $< .,. >$ denoted the inner product in $L_2(T)$. They serve as reference points for the assessment of $\tilde{d}_k^{(N)}(\mathbf{y})$. Furthermore, $d_k^{(N)}(f)$'s have appealing shape preserving features. Namely,

recovery of constant: if $f = c$ is constant in T, then $d_k(f) = c$, $k = 0, 1, \ldots, , N$
level preservation: if $f(t) \geq c > 0$, $t \in T$, then $d_k(f) \geq c$, $k = 0, 1, \ldots, N$.
monotonicity preservation: if f has continuous derivative in T and $f'(t) > 0$ in T, then also $d_k(f)$, $k = 0, 1, \ldots, N$ is strictly increasing in T,
convexity preservation: if f is twice continuously differentiable in T and $f''(t) > 0$ in T, then sequence $d_k(f)$, $k = 0, 1, \ldots, N$ is strictly convex, i.e., its second order differences are positive.

Proofs of the above properties follow from the well-known formulas for the derivatives of the Bernstein polynomials (see [13]), using integration by parts of their products and the derivatives of f.

It can also be proven that for each $0 \leq k \leq N$ and for every finite and fixed N we have: $\mathbb{V}ar(\tilde{d}_k^{(N)}(\mathbf{y})) \leq \sigma^2 \frac{(N+1)^2}{n}$, while for f continuously differentiable the bias of $\tilde{d}_k^{(N)}(\mathbf{y})$ is of the order $O(N/n)$. Thus, also $\mathbb{E}[\tilde{d}_k^{(N)}(\mathbf{y}) - d_k(f)]^2$ converges to zero as $n \to \infty$.

3   Selecting N Using the Autoencoder

Our aim in this section is to propose the method of selecting the number of descriptors N when n is finite. Contrary to classic problems dedicated to selecting N for samples of a particular f, in our case, we need to select N that is suitable for a family of functions f.

Learning N by Autoencoder. The idea of selecting N is to design an autoencoder such that
a) it obtains $\mathbf{y}^{(l)}$, $l = 1, 2, \ldots, L$ as inputs

b) and subsequently, for each $l = 1, 2, \ldots, L$, converts them into latent variables $\tilde{d}_k^{(N)}(\mathbf{y}^{(l)})$, $k = 0, 1, \ldots, N$, which are the descriptors estimates, computed according to (5),

c) finally, for $l = 1, 2, \ldots, L$ the encoder outputs are computed as follows

$$\tilde{f}_i^{(N)}(\mathbf{y}^{(l)}) = \sum_{k=0}^{N} \tilde{d}_k^{(N)}(\mathbf{y}^{(l)}) B_k^{(N)}(t_i) \quad i = 1, 2, \ldots, n, \tag{8}$$

where $\tilde{f}_i^{(N)}(\mathbf{y}^{(l)})$'s are interpreted as estimates of $f_l(t_i)$'s.

Then, for the whole learning sequence we obtain

$$\tilde{\mathbf{d}}^{(N)}(\mathbf{y}^{(l)}) = (N+1)\,\Delta_n\,\mathbf{B}_N\,\mathbf{y}^{(l)}, \quad \tilde{\mathbf{f}}^{(N)}(\mathbf{y}^{(l)}) = \mathbf{B}_N^{tr}\,\tilde{\mathbf{d}}^{(N)}(\mathbf{y}^{(l)}), \tag{9}$$

$l = 1, 2, \ldots, L,$, where, according to (8), $n \times 1$ vector $\tilde{\mathbf{f}}^{(N)}(\mathbf{y}^{(l)})$ consists of $\tilde{f}_i^{(N)}(\mathbf{y}^{(l)})$, $i = 1, 2, \ldots, n$.

We state a version of the well-known (see [3] for review) Akaike's Information Criterion (AIC) that is suitable for our purposes. The necessity for the AIC generalization comes from selecting N, which is suitable for the whole family of curves.

Assume that the sampling schemes are of the same form as in (1), with i.i.d. noises having $\mathcal{N}(0, \sigma^2)$ distribution. This assumption is made in this section only, since the obtained formulas are interpretable and useful for a large class of probability distributions having zero mean and finite variance.

Corollary 1 (Approximate AIC criterion). *The approximate AIC (AAIC) for selecting N has the following form*

$$AAIC(N) = L\left[2\,\theta\,N + n\ln\left(\sum_{l=1}^{L}||\mathbf{y}^{(l)} - \mathbf{B}_N^{tr}\,\tilde{\mathbf{d}}^{(N)}(\mathbf{y}^{(l)})||^2\right)\right] + C(L,n), \tag{10}$$

where $C(L,n)$ is a constant that may depend on fixed n and L, but not on N, while $0 < \theta \leq 1$ is a correcting factor. The minimizer $\tilde{N}$ of $AAIC(N)$ is considered as an approximately optimal number of descriptors for all family f_l's.

We omit a long proof of this result. The AAIC properly generalizes the AIC to a family of functions in the sense that it is not a simple average of the AIC's for each f_l's and the corresponding sub-models.

Algorithm 1 (for learning the number of descriptors by the AAIC)

Input L, n, $N_{max} < n$, N_{min} (or set $N_{min} = 3$), $\mathbf{y}^{(l)}$, $l = 1, 2, \ldots, L$.

While $N_{max} \leq n$ **do**

Step 1 *Compute $\tilde{N}$ that minimizes the expression in the brackets in (10) over $N_{min} \leq N \leq N_{max}$.*

Step 2 *If $\tilde{N} = N_{max}$, set $N_{min} = N_{max} - 1$, enlarge N_{max} and go to Step 1, otherwize, STOP and output $\tilde{N}$.*

A correcting factor $\theta > 0$ in (10) is in most cases set to 1. However, when observation errors are large, it happens that the minus log-likelihood function decreases rather slowly with N. In such cases, it is reasonable to apply $\theta > 0$ strictly less than 1. Algorithm 1 was tested on synthetic data, providing satisfactory results. We do not display them by the lack of space.

Testing Autoenkoder on Real Data. As a benchmark for testing the proposed descriptors, we selected samples of acceleration signals that are publicly available [24].

Such signals arise when acceleration measurements are made in a bucket wheel excavator operator's cabin or other large machines. The bucket wheel excavator works in a quasi-periodic or a repetitive way since the buckets hit the ground at equidistant time instants. Their positions point out intervals that are considered here as the common domain T of signals to be classified. In our example, each signal is sampled at $n = 1024$ points of T (see Fig. 1) (right panel), where T has the duration of one second.

These samples form vectors $\mathbf{y}^{(l)}$'s. We have only 43 of them, but it is sufficient to run Algorithm 1. The resulting AAIC plot is shown in Fig. 1 (left panel) for $\theta = 0.5$. The minimum is clearly visible at $\tilde{N} = 40$ that is further taken as the number of descriptors used for reconstruction of the acceleration signals. For illustration purposes only, in Fig. 1 (right panel) the signal reconstructed by the autoencoder for $\tilde{N} = 40$ is also shown.

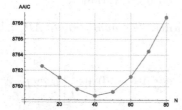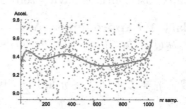

Fig. 1. Left panel – The result of applying Algorithm 1 to the accelerations samples – the AAIC plot for selecting N. Right panel – The result of applying Algorithm 1 to the accelerations samples – reconstruction of one sample by the autoencoder

Conclusions. The proposed autoencoder is based on the Bernstein-Durrmeyer polynomials. It was pointed out that the autoencoder descriptors inherit shape-preserving properties of the Bernstein-Durrmeyer polynomials, such as monotonicity and convexity, provided that no noise is present. These properties can be observed also when observations are corrupted by intensive noise.

Acknowledgements. The authors express their thanks to the anonymous referees for comments clarifying the presentation.

References

1. Baldi, P., Hornik, K.: Neural networks and principal component analysis: learning from examples without local minima. Neural Netw. **2**(1), 53–58 (1989)
2. Bank, D., et al.: Autoencoders. arXiv:2003.05991 (2020)
3. Cavanaugh, J.E., Neath, A.A.: The Akaike information criterion: background, derivation, properties, application, interpretation, and refinements. Wiley Interdisc. Rev. Computat. Stat. **11**(3), e1460 (2019)
4. Chen, W., Ditzian, Z.: Best polynomial and Durrmeyer approximation in $L_p(S)$ Indagationes Mathematicae **2**, 437–452 (1991)
5. Chen, G.Y., Bui, T.D., Krzyżak, A.: Rotation invariant feature extraction using Ridgelet and Fourier transforms. Pattern Anal. Appl. **9**, 83–93 (2006)
6. Chen, G., Krzyżak, A., Qian, S.E.: A new endmember extraction method based on least squares. Can. J. Remote. Sens. **48**(2), 316–326 (2022)
7. Derrienic, M.M.: On multivariate approximation by Bernstein-type polynomials. J. Approx. Theory **45**, 155–166 (1985)
8. Doersch, C.: Tutorial on variational autoencoders. arXiv preprint arXiv:1606.05908 (2016)
9. Gałkowski, T., Krzyżak, A., Filutowicz, Z.: A new approach to detection of changes in multidimensional patterns. J. Artif. Intell. Soft Comput. Res. **10**(2), 125–136 (2020)
10. Harris, T., Tucker, J.D., Li, B., Shand, L.: Elastic depths for detecting shape anomalies in functional data. Technometrics **63**, 1–11 (2020)
11. Kingma, D.P., Welling, M.: An introduction to variational autoencoders. Found. Trends® Mach. Learn. **12**(4), 307–392 (2019)
12. Krzyżak, A., Leung, S.Y., Suen, C.Y.: Reconstruction of two-dimensional patterns from Fourier descriptors. Mach. Vis. Appl. **2**, 123–140 (1989)
13. Lorentz, G.G.: Bernstein Polynomials. American Mathematical Soc. (2013)
14. Luo, W., Li, J., Yang, J., Xu, W., Zhang, J.: Convolutional sparse autoencoders for image classification. IEEE Trans. Neural Netw. Learn. Syst. **297**, 3289–3294 (2017)
15. Marron, J.S., Ramsay, J.O., Sangalli, L.M., Srivastava, A.: Functional data analysis of amplitude and phase variation. Stat. Sci. **30**(4), 468–484 (2015). https://doi.org/10.1214/15-STS524
16. Plaut, E.: From principal subspaces to principal components with linear autoencoders. arXiv preprint arXiv:1804.10253 (2018)
17. Rafajłowicz, E., Skubalska-Rafajłowicz, E.: Nonparametric regression estimation by Bernstein-Durrmeyer polynomials. Tatra Mt. Math. Publ. **17**, 227–239 (1999)
18. Rafajłowicz, W., Rafajłowicz, E., Więckowski J.: Learning functional descriptors based on the bernstein polynomials - preliminary studies, In: Rutkowski, L., Scherer, R., Korytkowski, M., Pedrycz, W., Tadeusiewicz, R., Zurada, J.M. (eds.) ICAISC 2022. LNCS, vol. 13588, pp. 310–321. Springer, Cham (2023). https://doi.org/10.1007/978-3-031-23492-7_27
19. Rifai, S., Vincent, P., Muller, X., Glorot, X., Bengio, Y.: Contractive auto-encoders: explicit invariance during feature extraction. In: Proceedings of the 28th International Conference on Machine Learning, pp. 833–840 (2011)
20. Rumelhart, D.E., Hinton, G.E., Williams, R.J.: Parallel distributed processing: explorations in the microstructure of cognition, **1**, 26 (1986)
21. Sitzmann, V., Martel, J., Bergman, A., Lindell, D., Wetzstein, G.: Implicit neural representations with periodic activation functions. Adv. Neural. Inf. Process. Syst. **33**, 7462–7473 (2020)

22. Tadeusiewicz, R.: Automatic understanding of signals. In: Kłopotek, M.A., Wierz-choń, S.T., Trojanowski, K. (eds.) Intelligent Information Processing and Web Mining. ASC, vol. 25, pp. 577–590. Springer, Heidelberg (2004). https://doi.org/10.1007/978-3-540-39985-8_66

23. Vincent, P.: A connection between score matching and denoising autoencoders. Neural Comput. **23**(7), 1661–1674 (2011)

24. Więckowski, J., Rafajłowicz, W., Moczko, P., Rafajłowicz, E.: Data from vibration measurement in a bucket wheel excavator operator's cabin with the aim of vibrations damping. Data Brief, 106836 (2021). http://dx.doi.org/10.17632/htddgv2p3b.1

Attribute Relevance and Discretisation in Knowledge Discovery: A Study in Stylometric Domain

Urszula Stańczyk[1]([✉])[ID], Beata Zielosko[2][ID], and Grzegorz Baron[1][ID]

[1] Department of Graphics, Computer Vision and Digital Systems,
Silesian University of Technology, Akademicka 2A, 44-100 Gliwice, Poland
{urszula.stanczyk,grzegorz.baron}@polsl.pl
[2] Institute of Computer Science, University of Silesia in Katowice,
Będzińska 39, 41-200 Sosnowiec, Poland
beata.zielosko@us.edu.pl

Abstract. The paper demonstrates the research methodology focused on observations of relations between attribute relevance, displayed by rankings, and discretisation. Instead of transforming all continuous attributes before data exploration, the variables were gradually processed, and the impact of such a change on the performance of a classifier was studied. Considerable experiments carried out on stylometric data illustrate that selective discretisation could be more advantageous to predictive accuracy than some uniform transformation of all features.

Keywords: Discretisation · Attribute ranking · Stylometry

1 Introduction

Feature selection and discretisation are two processes that can have a significant impact on the operations taking place inside the knowledge discovery phase and its outcome. The discretisation of attributes produces changes in their representation and discards some information from available data [5]. As a consequence, it enables the use of inducers that operate only on categorical variables. On the other hand, exploration of transformed data can lead to overlooking some properties or relations existing in continuous features. Thus, discretisation should be treated with caution—it cannot always be assumed to be advantageous [1].

In standard discretisation approaches [5], all variables are processed in some uniform way, and the entire input domain is changed from continuous to discrete. In the methodology presented in this paper and examined through extensive experiments, the discretisation was performed in stages. Starting with the set of all attributes with continuous domains, the variables were next chosen one by one for discretisation, with the selection directed by the observed relevance of features. The importance of attributes was estimated with the help of rankings.

In the proposed methodology, some popular supervised and unsupervised discretisation algorithms were applied to the data, while following the selected

J. Mikyška et al. (Eds.): ICCS 2023, LNCS 14074, pp. 273–281, 2023.
https://doi.org/10.1007/978-3-031-36021-3_27

rankings. The impact of transformation on the performance of the chosen classifiers was studied for a binary authorship attribution task [12], for two datasets with balanced classes, and stylometric features in the lexical category. The experimental results show that the presented non-standard discretisation procedure led to many cases of improved predictions for subsets of features with transformed domains, making selective discretisation worth closer investigation and demonstrating the merits of the procedure for ranking-driven discretisation.

The content of the paper was organised as follows. Section 2 indicates related areas and works. Section 3 gives comments on the discretisation procedure controlled by a ranking of attributes. Section 4 details the experimental setup and presents the results. Section 5 includes concluding remarks.

2   Characteristics of Input Space and Data Mining

In the research presented, important roles were played by characteristics of the input space, algorithms used to estimate attribute importance, possible transformations of their representation, and methods employed for data exploration.

A ranking allows to estimate the importance of attributes and order them by some adopted criterion. In the research, three rankers for features were used: Support Vector Machine (SVM), Wrapper Subset Evaluation with Naive Bayes classifier (WrappB), and Correlation Attribute Evaluation (Corr).

SVM can be an effective tool for the ranking construction process [6]. The attributes are evaluated using an SVM classifier with information on how well each feature contributes to the separation of classes. WrapB uses the induction algorithm along with a statistical re-sampling technique such as k-fold cross-validation to evaluate feature subsets. NB classifier assumes that, within each class, the probability distributions for the attributes are independent of each other, so its performance in domains with redundant features can be improved by removing such variables [9]. Corr takes into account the usefulness of individual features for predicting class label along with the level of inter-correlation among them. It uses Pearson's correlation between a given feature and the class [8].

Discretisation can be considered as a process aiming at a reduction of the number of values of a given continuous variable, by dividing its range into intervals [5], which can be executed in many ways. Unlike supervised discretisation, unsupervised algorithms ignore instance labels during the transformation of attributes. In the proposed methodology, four methods were used: supervised Fayyad and Irani (dsF) [4], and Kononenko (dsK) [10] algorithms, and unsupervised equal width (duw) and equal frequency (duf) binning.

Classification is one of the main tasks in the process of knowledge discovery and pattern recognition. In this work, two state-of-the-art classifiers were used, namely, Bayesian Network (BNet) [11] and Random Forest (RF) [3].

Bayesian Network is a probabilistic model based on Bayes' theory [11]. It is considered as a representation of joint probability distribution over a set of random variables and presented in the form of a directed acyclic graph, where each node corresponds to a random variable and the edges represent probabilistic dependence. Random Forest belongs to the ensemble data mining techniques used for classification [3]. It is a combination of decision trees as predictors such that each tree depends on the values of a random vector, sampled independently from a dataset, and with the same distribution for all trees in the forest. During classification, each tree votes, and the most popular class label is returned.

3   Procedure for Ranking Driven Discretisation

To be unbiased, the methodology for ranking-driven discretisation of attributes required some assumptions and limitations. These elements were as follows.

- Input data and features. Attributes are expected to be continuous, with comparable ranges of values, and should be chosen only on the basis of domain knowledge. A classification task is binary and with balanced classes.
- Ranking mechanisms. The methods to be used must treat all variables as relevant by always assigning a non-zero rank and work in continuous domain.
- Discretisation algorithms. A discretiser needs to work independently on a learner, process variables separately, and ignore any interdependencies among them. Both supervised and unsupervised approaches can be employed.
- Classifiers. The learner needs to be able to operate in both continuous and discrete domains, and capable of discovering knowledge from both forms.
- Starting point. The exploration of data starts in continuous domain. Based on the knowledge discovered in real-valued variables in the train sets, performance is next evaluated by labelling samples in the test sets.
- Steps and direction of processing. Each step involves discretisation of a single attribute, indicated by its ranking position, following either up or down.
- Stopping point. The procedure can be stopped once the entire datasets become discrete. It is also possible to end transformations sooner, when some degradation of performance is detected.

4   Experiments

The experiments started with the construction of the input datasets. Then, the attribute rankings were obtained and used for gradual discretisation of sets. For the selected classifiers, their performance was estimated and investigated.

4.1   Preparation of Input Stylometric Datasets

Two pairs of authors were taken for stylometric analysis, Edith Wharton and Mary Johnston (female writer dataset, F-writers), and Henry James and Thomas Hardy (male writer dataset, M-writers). Long texts of novels were divided into smaller chunks of text of comparable lengths. Over these shorter texts, the frequency of occurrence was calculated for 12 selected function words [15]. They belong to the lexical category of stylometric descriptors widely used in authorship attribution tasks. The words were as follows: as, at, by, if, in, no, of, on, or, so, to, up. When mentioned in the text, the attributes were given in italics.

The preparation resulted in real-valued features, to be employed by some approach to data mining [12]. Due to the specifics of the sample construction process, the input space was stratified. Taking this into account [2], a dataset consisted of a train set and two test sets for performance evaluation. All sets were balanced, including the same number of samples for both classes.

4.2   Rankings of Characteristic Features

Three ranking mechanisms were applied to the data, all implemented in the WEKA [7] workbench, namely WrapB, Corr, and SVM. The obtained orderings of the variables were provided in Table 1, where the highest ranking position was shown on the left and the lowest on the right.

Table 1. Rankings of attributes for the F-writer and M-writer datasets.

F-writer dataset													M-writer dataset											
Ranking position													Ranking position											
1	2	3	4	5	6	7	8	9	10	11	12	Ranker	1	2	3	4	5	6	7	8	9	10	11	12
to	on	of	no	at	if	so	up	in	or	by	as	WrapB	by	if	to	in	so	no	at	of	as	on	up	or
on	to	of	as	by	no	or	so	in	if	at	up	Corr	by	if	in	or	at	of	to	so	no	on	up	as
on	to	of	as	so	by	if	at	up	no	or	in	SVM	by	or	if	at	in	so	no	to	as	on	up	of

The two datasets shared the same features, but their placement in the rankings was different. For a dataset, some similarities could be observed between the rankings. For M-writers, *by* was always the highest ranking, for F-writers, *to* and *on* took the top two positions, and *of* always came third. Lower-ranking positions were more varied. All rankings were followed in ascending order.

4.3   Employed Discretisation Algorithms

In the experiments, all sets were discretised independently [13]. Unsupervised methods (duf and duw) were employed with the number of bins from 2 to 10, so returned 9 variants of data each. Two supervised algorithms (dsF and dsK), gave single variants of the data. These methods rely on the MDL principle and the calculation of entropy, which led to some variables for which one interval was found as representation in a discrete domain. For F-writers for both supervised discretisation algorithms, there were 6 such features. For M-writers and dsK, also 6 variables had single bins, and for dsF this set was expanded to 7 elements.

4.4   Classification Process and Evaluation of Performance

The primary goal of the research was to observe the relations between the importance of features and discretisation, and how the changed representation reflects on classifier performance. As the processing started in the continuous domain and the features were discretised gradually, only at the final step were all variables discrete. With such conditions, the selected classification systems for the most part operated on at least partially continuous data. Two chosen classifiers, BNet and RF, implemented in WEKA, were used with default parameters.

To evaluate inducer performance, classification accuracy was chosen [14], as it is suitable for binary classification with balanced classes, both classes of the same importance and the same cost of misclassification. Due to the stratified input space, cross-validation could not be considered reliable [2]. Instead, test sets were used, and the reported performance is the average obtained from them.

For the Bayesian Network classifier, the performance was shown in Fig. 1, and for Random Forest in Fig. 2. In the included charts, the entire processing path was illustrated, starting with zero discrete attributes and ending with 12 discrete features. For unsupervised methods, the average was calculated over all 9 variants of the data, corresponding to different numbers of constructed bins.

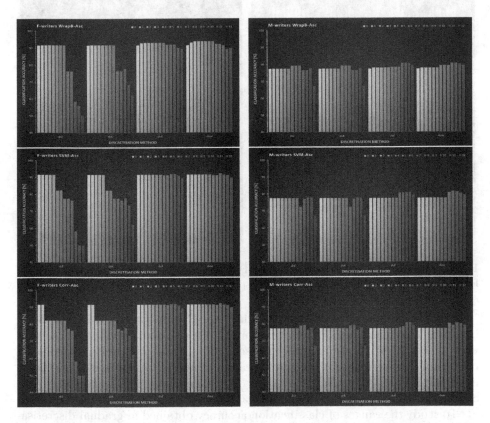

Fig. 1. Performance [%] for the Bayesian Network observed in the discretisation while following the selected rankings. The series specify the number of discretised attributes.

For BNet, both datasets, and all three rankings unsupervised discretisation always brought some improvement for partial transformation of attributes.

Supervised discretisation, in particular for more transformed variables, resulted in cases of degraded performance, especially for F-writers. For the female writer dataset WrapB resulted in the best performance for all discretisation methods, while for M-writers WrapB and Corr came very close, with SVM slightly behind.

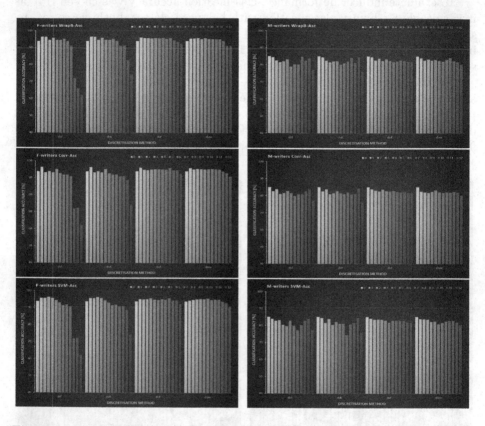

Fig. 2. Performance [%] for the Random Forest observed in the discretisation while following the selected rankings. The series specify the number of discretised attributes.

Because of its mode of operation on data, the RF classifier on the whole fared better in the continuous domain than in the discrete domain when all features were transformed. However, partial discretisation was often advantageous, in particular for F-writers, where, for all rankings and all discretisation methods, improved accuracy was always detected. The maxima observed for rankings were close but again the highest for WrapB, and better for supervised discretisation.

To study the ranges of classification accuracy obtained in gradual discretisation of attributes, the average performance and standard deviation were calculated for the entire transformation process, as shown in Table 2. For reference, the performance of a classifier observed in the continuous domain was also provided. In each row, the highlighted entries correspond to the highest average predictive accuracy for this discrete version for each dataset.

Table 2. Average performance [%] and standard deviation of the Bayesian Network and Random Forest classifiers, for the entire run of the selective discretisation procedure following a ranking (from 1 out of N, to all N discretised attributes).

Discret. method	F-writers			M-writers		
	WrapB	Corr	SVM	WrapB	Corr	SVM
	Bayesian Network (Cont.domain: 91.60)			Bayesian Network (Cont.domain: 77.43)		
dsF	**79.95** ± 16.32	74.83 ± 13.94	76.74 ± 15.71	**76.94** ± 03.20	76.83 ± 03.12	76.19 ± 03.19
dsK	**83.61** ± 10.83	79.44 ± 07.04	81.34 ± 09.12	**76.94** ± 03.20	76.83 ± 03.12	76.19 ± 03.19
duf	**92.20** ± 01.27	91.42 ± 00.67	91.40 ± 00.67	**78.96** ± 01.18	78.27 ± 01.33	78.68 ± 01.59
duw	**92.62** ± 01.66	91.43 ± 00.59	91.40 ± 00.59	**79.06** ± 01.10	78.51 ± 01.40	78.75 ± 01.70
	Random Forest (Cont.domain: 91.96)			Random Forest (Cont.domain: 85.07)		
dsF	**87.70** ± 12.92	87.15 ± 11.40	87.67 ± 11.75	**81.87** ± 02.29	81.57 ± 01.85	81.16 ± 02.38
dsK	**91.68** ± 06.68	91.11 ± 05.68	91.63 ± 05.99	**81.89** ± 02.29	81.63 ± 02.09	81.01 ± 03.03
duf	**95.05** ± 01.08	94.76 ± 00.87	94.82 ± 00.85	82.67 ± 00.85	82.62 ± 00.70	**82.85** ± 00.77
duw	94.43 ± 01.64	**94.54** ± 01.38	94.43 ± 01.33	**82.70** ± 01.02	82.04 ± 00.68	82.35 ± 01.10

Since the averages were calculated over the whole discretisation process and often the transformation of all attributes caused degraded performance, the averages also often fell below the level reported for the continuous domain. However, for the Bayesian Network and unsupervised methods the values were improved, while for Random Forest that was true only for F-writers and duf method. On the other hand, the highest averages were mostly reported for WrapB ranking, for both datasets and both classifiers.

Examination of standard deviation allows to conclude that the highest values, even in two digits, were found when supervised discretisation was applied to the input data, in particular to the female writer dataset. This observation is valid for both studied classifiers and all three rankings. For other discretisation approaches, the values were noticeably smaller, even just fractional.

The statistical characteristics of the process of selective discretisation driven by rankings confirmed earlier observations based on performance trends. In the majority of cases partial discretisation was more advantageous than transformations of entire input domain, showing that attribute relevance incorporated into their processing and discretisation could be beneficial to the knowledge discovery.

5   Conclusions

In the paper a research methodology was demonstrated in which the discretisation process was carried out with gradually expanding the range of transformed variables, selected from the obtained ranking reflecting their relevance. The goal of this processing was to observe the relations between the importance of features and the form of their representation, and how its change can influence the performance of selected classifiers. Extensive experiments were carried out in the stylometric domain, for the binary authorship attribution task. Several rankings and many discretisation methods were investigated and they allowed to discover a remarkable number of cases where partial, instead of complete discretisation of the input data, led to obtaining improved predictive accuracy for the used inducers, making selective discretisation worth deeper study.

Acknowledgements. The research works presented in the paper were carried out within the statutory project of the Department of Graphics, Computer Vision and Digital Systems (RAU-6, 2023), at the Silesian University of Technology, Gliwice, Poland, and at the University of Silesia in Katowice, Sosnowiec, Poland.

References

1. Baron, G., Stańczyk, U.: Performance evaluation for ranking-based discretisation. In: Cristani, M., et al. (eds.) Knowledge-Based and Intelligent Information & Engineering Systems: Proceedings of the 24th International Conference KES-2020, Procedia Computer Science, vol. 176, pp. 3335–3344. Elsevier (2020)
2. Baron, G., Stańczyk, U.: Standard vs. non-standard cross-validation: evaluation of performance in a space with structured distribution of datapoints. In: Wątróbski, J., et al. (eds.) Knowledge-Based and Intelligent Information & Engineering Systems: Proceedings of the 25th International Conference KES-2021, Procedia Computer Science, vol. 192, pp. 1245–1254. Elsevier (2021)
3. Cutler, A., Cutler, D.R., Stevens, J.R.: Random forests. In: Zhang, C., Ma, Y. (eds.) Ensemble Machine Learning: Methods and Applications, pp. 157–175. Springer, NY, US (2012). https://doi.org/10.1007/978-1-4419-9326-7_5
4. Fayyad, U.M., Irani, K.B.: Multi-interval discretization of continuousvalued attributes for classification learning. In: 13th International Joint Conference on Artificial Intelligence, vol. 2, pp. 1022–1027. Morgan Kaufmann Publishers (1993)
5. García, S., Luengo, J., Sáez, J.A., López, V., Herrera, F.: A survey of discretization techniques: taxonomy and empirical analysis in supervised learning. IEEE Trans. Knowl. Data Eng. **25**(4), 734–750 (2013)
6. Guyon, I., Weston, J., Barnhill, S., Vapnik, V.: Gene selection for cancer classification using support vector machines. Mach. Learn. **46**, 389–422 (2002)
7. Hall, M., et al.: The WEKA data mining software: an update. SIGKDD Explor. **11**(1), 10–18 (2009)
8. Hall, M.A.: Correlation-based Feature Subset Selection for Machine Learning. Ph.D. thesis, Department of Computer Science, University of Waikato, Hamilton, New Zealand (1998)
9. Kohavi, R., John, G.H.: Wrappers for feature subset selection. Artif. Intell. **97**(1), 273–324 (1997)
10. Kononenko, I.: On biases in estimating multi-valued attributes. In: 14th International Joint Conference on Artificial Intelligence, pp. 1034–1040 (1995)
11. Sardinha, R., Paes, A., Zaverucha, G.: Revising the structure of Bayesian network classifiers in the presence of missing data. Inf. Sci. **439–440**, 108–124 (2018)
12. Stamatatos, E.: A survey of modern authorship attribution methods. J. Am. Soc. Inform. Sci. Technol. **60**(3), 538–556 (2009)
13. Stańczyk, U., Zielosko, B.: Data irregularities in discretisation of test sets used for evaluation of classification systems: a case study on authorship attribution. Bull. Pol. Acad. Sci. Tech. Sci. **69**(4), 1–12 (2021)

14. Stąpor, K., Ksieniewicz, P., García, S., Woźniak, M.: How to design the fair experimental classifier evaluation. Appl. Soft Comput. **104**, 107219 (2021)
15. Zhao, Y., Zobel, J.: Effective and scalable authorship attribution using function words. In: Lee, G.G., Yamada, A., Meng, H., Myaeng, S.H. (eds.) AIRS 2005. LNCS, vol. 3689, pp. 174–189. Springer, Heidelberg (2005). https://doi.org/10. 1007/11562382_14

Bayesian Networks for Named Entity Prediction in Programming Community Question Answering

Alexey Gorbatovski(✉) and Sergey Kovalchuk

ITMO University, Saint-Petersburg, Russia
{gorbatovski,kovalchuk}@itmo.ru

Abstract. Within this study, we propose a new approach for natural language processing using Bayesian networks to predict and analyze the context and show how this approach can be applied to the Community Question Answering domain, such as Stack Overflow questions. We compared the Bayesian networks with different score metrics, such as the BIC, BDeu, K2, and Chow-Liu trees. Our proposed approach outperforms the baseline model on the precision metric. We also discuss the influence of penalty terms on the structure of Bayesian networks and how they can be used to analyze the relationships between entities. In addition, we examine the visualization of directed acyclic graphs to analyze semantic relationships. The article further identifies issues with detecting certain semantic classes that are separated by the structure of directed acyclic graphs.

Keywords: Bayesian networks · Context prediction · Natural language generation · Natural language processing · Question answering

1 Introduction

Automated solutions in NLP have gained interest in solving problems like text classification, summarization, and generation [1]. However, adding context remains a challenge [2]. This paper proposes a Bayesian approach to predicting context using named entities to recover the meaning of a full text.

The proposed approach utilizes Bayesian networks (BNs) to predict context by using conditional probability distributions (CPDs) of named entities, obtained through named entity recognition (NER). The BN provides a directed acyclic graph (DAG) that shows links between entity classes and identifies significant elements of the programming domain, such as code blocks with error names or class and function entity classes.

This Bayesian approach may prove useful in human code generation quality assessment and community question answering (CQA) domains [3]. While complex neural network architectures such as LSTM or transformers [4] are commonly used to solve such problems, they have limitations such as the need for vast amounts of textual data and time, and the complexity of fine-tuning them [5].

© The Author(s), under exclusive license to Springer Nature Switzerland AG 2023
J. Mikyška et al. (Eds.): ICCS 2023, LNCS 14074, pp. 282–289, 2023.
https://doi.org/10.1007/978-3-031-36021-3_28

In conclusion, the proposed approach presents a solution to the challenges of predicting context by utilizing BNs to predict context using named entities obtained through NER. While there may be errors in the NER model, in an ideal case, BNs specify precise relationships and provide information about semantics and causal relationships [2].

2 Methodology

In this section, we describe different components of the proposed BN approach for context prediction. Figure 1 shows the overall process consists of several parts: 1) Semantic entity recognition by the NER model; 2) Learning the Bayesian network as a causal model; 3) Predicting and evaluating entities in question by title.

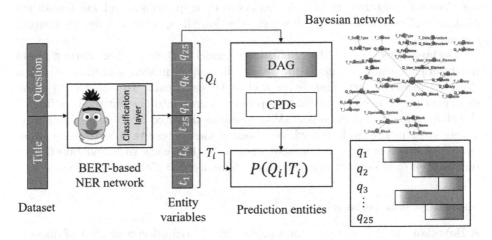

Fig. 1. The overall process of proposed BN approach

2.1 Problem Statement

As shown in Fig. 1, we need to predict the semantically meaningful classes of questions with BN as a multilabel classification problem. For this problem, we have textual data, presented as vectors.

More formally, assume we are given two sets of Questions $Q = \langle Q_1, Q_2, \ldots, Q_N \rangle$ and Titles $T = \langle T_1, T_2, \ldots, T_N \rangle$, where N - is the number of samples in our dataset. For each title $T_i \in T$ we have $k = 25$ dimension vector, $T_i = \langle t_1^i, t_2^i, \ldots, t_k^i \rangle$, where t_k^i represents the k_{th} entity class of the i_{th} title and $t_k^i \in \{0,1\}$, where $t_k^i = 0$ corresponds to the absence of the k_{th} class entity in title, and $t_k^i = 1$ corresponds to the existence of the k_{th} class entity in title. For the questions, it is the same. We solve the multilabel classification problem by predicting for each i_{th} question its entity classes by i_{th} titles entity classes.

2.2 Dataset

The dataset we use is based on 10% of the Stack Overflow[1] Q&A in 2019[2]. For the set of questions we apply the following filtering operations: select questions with the tag "android", select questions with a length of fewer than 200 words and related to the API Usage category proposed by Stefanie et al. [6]. Moreover, we selected questions without links and images, because the information from those types of content is unavailable for Bayesian networks. Thus, we received $N = 707$ pairs of title and question (T_i, Q_i).

2.3 Semantic Entities Recognition

We used CodeBERT [7] to extract domain-specific entities from the text content. This NER model was trained on Stack Overflow data, which is a popular resource for programmers to find answers to their questions, and was fine-tuned to detect 25 entity classes [8]. Each class is domain specific and defines context semantics [9].

Declared precision of the open-source model is 0.60[3], hence markup could not be ideal because of model mistakes. Annotation models sometimes break a word into several parts and define a class for each part. To smooth out these inaccuracies, we decided to combine parts of words into one entity according to the class of the first defined part. While entities detected by the model might be ambiguous, testing the key words of sentences mostly results in correct detection. All pairs are vectorized as one-hot encoding, thus each title and question is represented by a k-dimension vector, as there are $k = 25$ defined classes.

2.4 Bayesian Networks

A Bayesian network models a joint probability distribution over a set of discrete variables. It consists of a directed acyclic graph and a set of conditional probability distributions. The lack of an edge between variables encodes conditional independence. The joint probability distribution can be calculated using the conditional probability distributions. We used the greedy hill climbing algorithm, as the optimal structure is NP-hard problem.

We used three scoring metrics - BIC [10], BDeu, and K2 [11] - to learn Bayesian networks. BIC includes a penalty term for model complexity and produces regularized DAGs. BDeu and K2 are Bayesian Dirichlet scores that use a penalty term based on assumptions of parameter independence, exchangeable data, and Dirichlet prior probabilities. After learning the structure, we pruned the BNs using Chi-Square Test Independence to detect more specific semantic relationships.

---

[1] https://stackoverflow.com.
[2] https://www.kaggle.com/datasets/stackoverflow/stacksample.
[3] https://huggingface.co/mrm8488/codebert-base-finetuned-stackoverflow-ner.

We also used the Chow-Liu Algorithm [12] to find the maximum-likelihood tree-structured graph, where the score is the log-likelihood without a penalty term for graph structure complexity as it is regularized by the tree structure.

For BNs using BIC, BDeu and K2 scores, we predicted question' entities using the Maximum Likelihood Estimation (MLE). A natural estimate for the CPDs is to simply use the relative frequencies for each variable state.

For BNs with tree structures, we tried different probabilistic inference approaches. Algorithms such as Variable Elimination (VE), Gibbs Sampling (GS), Likelihood Weighting (LW) and Rejection Sampling (RS) are detailed in respective articles [13, 14]. Each label in question is predicted by a one-vs-rest strategy by all entities of its title from the pair.

For evaluation, we selected common multilabel classification metrics. We preferred macro and weighted averaging because existing classes are imbalanced and it is important to evaluate each class with its number of instances.

3   Results

In this section, we analyze classification metrics of BNs based on BIC, BDeu, and K2 scores as well as Chow-Liu trees. Each score defines a different structure of DAG, which means different semantic dependencies. We compared DAGs and analyzed the penalty terms of each score and its relationships reflected in graphs, as well as the detected relations.

3.1   Comparison of Evaluation Metrics

We used a common train-test split for evaluation. With the dataset described above, we composed the test dataset as random 30% samples of the whole set.

Table 1. Comparison of evaluation metrics.

	Precision		Recall		F1-score	
Model	Macro	Weighted	Macro	Weighted	Macro	Weighted
CatBoost	0.41	0.58	0.19	**0.35**	0.24	0.41
BIC based	**0.56**	**0.66**	0.20	0.33	0.28	0.42
BDeu based	0.48	0.63	0.20	**0.35**	0.26	**0.43**
K2 based	0.51	**0.66**	**0.24**	0.34	**0.29**	**0.43**
CL trees VE	0.47	0.63	0.21	0.33	0.25	0.41
CL trees LW	0.48	0.63	0.17	0.29	0.22	0.37
CL trees GS	0.41	0.57	0.13	0.25	0.18	0.33
CL trees RS	0.23	0.44	0.07	0.15	0.10	0.22

Table 1 shows the main evaluation results according to the selected classification metrics. We prefer to accentuate precision, because precision of individual classes is most important for information extraction and context prediction, and wrong class predictions cause context misunderstanding.

Our approach shows better precision metrics than the baseline - CatBoost model [15], 0.56 vs 0.41 macro precision and 0.66 vs 0.58 weighted precision, comparing the BIC score-based network and baseline.

We observe the highest precision of the BIC score-based model, but the K2-based model has better recall and comparable precision, making it the best network based on F1-score. BIC regularization is stronger than BDeu and K2-specific penalty terms, leading to fewer detected relationships and lower recall. However, BDeu and K2-based DAGs can classify more instances correctly, resulting in higher recall. Chow-Liu tree-based networks are comparable to other models with Variable Elimination as a sampling algorithm, but this algorithm has the limitation that each node has exactly one parent. Other inference approximations show worse results.

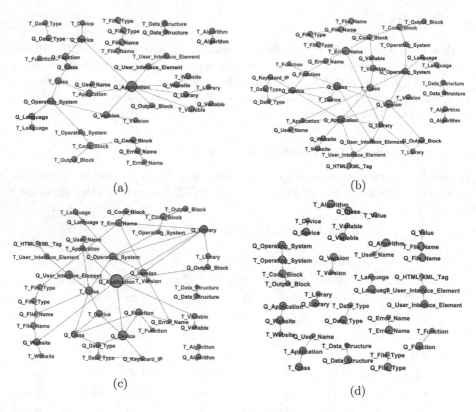

Fig. 2. DAG structures of learned BNs based on a) BIC, b) K2 metric, c) BDeu, d) Chow-Liu trees

3.2   Visual DAG Representation

We visualized DAGs from Bayesian networks to detect relationships between semantic entities in the context. Figure 2 shows the structures learned by K2 (Fig. 2b), BDeu (Fig. 2c), and BIC (Fig. 2a) based networks. K2 and BDeu based graphs detect more relationships and are more complete compared to BIC. Each DAG has semantic links between the same title and question entity classes. The Chow-Liu trees (Fig. 2d) show this well.

Analysis shows different clusters of semantic entities, separating DATA STRUCTURE and ALGORITHM in each graph. FILE NAME and FILE TYPE, as well as CODE BLOCK and OUTPUT BLOCK, are linked, indicating the logic and validity of BN DAG structures.

The tree-structured DAG defines causation from Question ALGORITHM to Title USER NAME and from Title ALGORITHM to Question CLASS without establishing a causal relationship between entities of the same name. These may be outliers due to the imperfect NER model.

3.3   Predictions Analysis

Finally, we compared semantic entities detected by the NER model with those predicted by BN using the K2 metric. Table 2 shows two examples of predictions. In the first example, the predicted entities match the target ones. However, in the second row of Table 2, BN could not detect some semantic instances. Graph

Table 2. Comparison of existing and predicted entities.

Title	Question	Questions entities	Predicted entities
How to send email with attachment using GmailSender in android	I want to know about how to send email with attachment using GmailSender in android.	APPLICATION, OPERATING SYSTEM	APPLICATION, OPERATING SYSTEM
Intel XDK build for previous versions of Android	I have just started developing apps in Intel XDK and was just wondering how to build an app for a specific version of Android OS. The emulator I select "Samsung Galaxy S" is using the version 4.2 of android. My application works fine for Galaxy s3 but not on galaxy Ace 3.2. I could not find a way to add more devices to the emulator list. How can I achieve this. Regards, Shankar.	APPLICATION, OPERATING SYSTEM, VERSION, USER NAME	APPLICATION, OPERATING SYSTEM

(Fig. 2b) reveals that nodes such as VERSION and USERNAME are not directly related to the APPLICATION question, and these entities aren't in the title. Hence, the conditional probability may not have been high enough to consider these entities as part of the question context.

4 Discussion and Conclusion

We found that Bayesian networks are a valuable tool for predicting and analyzing context in the CQA domain. While they can identify entities acceptably, improvements can be made with better recovery tasks [16], optimal search algorithms, and expanding data. Additionally, query expansion techniques based on relevant document feedback are effective in information retrieval systems [17], and Bayesian networks can provide context representation for query reformulation.

Moreover, the proposed approach was tested on a large dataset with the "python" tag. In that case, the quality of precision remains above the baseline of the model, but the DAG has too many relationships, which does not allow a more specific definition of semantic dependencies between title and question entities. It is possible that changing the penalty term may improve the clarity of semantic relationships.

As a result, our new application for Bayesian networks in CQA successfully identified causal semantic relationships for a set of SO questions and related titles. However, we observed that the network struggled with detecting semantic classes that are separated in the DAG structure. Future work includes comparing the performance of Bayesian networks and NER models and exploring the use of BNs and LDA for thematic modeling and information extraction in CQA.

Acknowledgments. This research is financially supported by the Russian Science Foundation, Agreement 17-71-30029 (https://rscf.ru/en/project/17-71-30029/), with co-financing of Bank Saint Petersburg.

References

1. Khurana, D., Koli, A., Khatter, K., Singh, S.: Natural language processing: state of the art, current trends and challenges. Multimedia tools and applications, pp. 1–32 (2022)
2. Santhanam, S.: Context based text-generation using lstm networks. arXiv preprint arXiv:2005.00048 (2020)
3. Kovalchuk, S.V., Lomshakov, V., Aliev, A.: Human perceiving behavior modeling in evaluation of code generation models. In: Proceedings of the 2nd Workshop on Natural Language Generation, Evaluation, and Metrics (GEM), pp. 287–294. Association for Computational Linguistics, Abu Dhabi, United Arab Emirates (Hybrid) (December 2022), https://aclanthology.org/2022.gem-1.24
4. Williams, J., Tadesse, A., Sam, T., Sun, H., Montañez, G.D.: Limits of transfer learning. In: Nicosia, G., Ojha, V., La Malfa, E., Jansen, G., Sciacca, V., Pardalos, P., Giuffrida, G., Umeton, R. (eds.) LOD 2020. LNCS, vol. 12566, pp. 382–393. Springer, Cham (2020). https://doi.org/10.1007/978-3-030-64580-9_32

5. Brown, T.B., et al.: Language models are few-shot learners. ArXiv abs/2005.14165 (2020)
6. Beyer, S., Macho, C., Di Penta, M., Pinzger, M.: What kind of questions do developers ask on stack overflow? a comparison of automated approaches to classify posts into question categories. Empir. Softw. Eng. **25**, 2258–2301 (2020)
7. Feng, Z., et al.: Codebert: a pre-trained model for programming and natural languages. arXiv preprint arXiv:2002.08155 (2020)
8. Tabassum, J., Maddela, M., Xu, W., Ritter, A.: Code and named entity recognition in StackOverflow. In: Proceedings of the 58th Annual Meeting of the Association for Computational Linguistics, pp. 4913–4926. Association for Computational Linguistics, Online (July 2020)
9. Dash, N.S.: Context and contextual word meaning. SKASE Journal of Theoretical Linguistics (2008)
10. Schwarz, G.: Estimating the dimension of a model. The annals of statistics, pp. 461–464 (1978)
11. Heckerman, D., Geiger, D., Chickering, D.M.: Learning bayesian networks: the combination of knowledge and statistical data. Mach. Learn. **20**, 197–243 (1995)
12. Chow, C., Liu, C.: Approximating discrete probability distributions with dependence trees. IEEE Trans. Inf. Theory **14**(3), 462–467 (1968)
13. Koller, D., Friedman, N.: Probabilistic graphical models: principles and techniques. MIT press (2009)
14. Hrycej, T.: Gibbs sampling in bayesian networks. Artif. Intell. **46**(3), 351–363 (1990)
15. Prokhorenkova, L., Gusev, G., Vorobev, A., Dorogush, A.V., Gulin, A.: Catboost: unbiased boosting with categorical features. In: Bengio, S., Wallach, H., Larochelle, H., Grauman, K., Cesa-Bianchi, N., Garnett, R. (eds.) Advances in Neural Information Processing Systems, vol. 31. Curran Associates, Inc. (2018)
16. Kayaalp, M., Cooper, G.F.: A bayesian network scoring metric that is based on globally uniform parameter priors. In: Proceedings of the Eighteenth Conference on Uncertainty in Artificial Intelligence, UAI 2002, pp. 251–258. Morgan Kaufmann Publishers Inc., San Francisco (2002)
17. Kandasamy, S., Cherukuri, A.K.: Query expansion using named entity disambiguation for a question-answering system. Concurrency Comput. Practice Exp. **32**(4), e5119 (2020). https://onlinelibrary.wiley.com/doi/abs/10.1002/cpe.5119, e5119 CPE-18-1119.R1

Similarity-Based Memory Enhanced Joint Entity and Relation Extraction

Witold Kościukiewicz[1,2](✉) [ID], Mateusz Wójcik[1,2] [ID], Tomasz Kajdanowicz[2] [ID], and Adam Gonczarek[1]

[1] Alphamoon Ltd., Wrocław, Poland
witold.kosciukiewicz@alphamoon.ai
[2] Wroclaw University of Science and Technology, Wrocław, Poland

Abstract. Document-level joint entity and relation extraction is a challenging information extraction problem that requires a unified approach where a single neural network performs four sub-tasks: mention detection, coreference resolution, entity classification, and relation extraction. Existing methods often utilize a sequential multi-task learning approach, in which the arbitral decomposition causes the current task to depend only on the previous one, missing the possible existence of the more complex relationships between them. In this paper, we present a multi-task learning framework with bidirectional memory-like dependency between tasks to address those drawbacks and perform the joint problem more accurately. Our empirical studies show that the proposed approach outperforms the existing methods and achieves state-of-the-art results on the BioCreative V CDR corpus.

Keywords: Joint Entity and Relation Extraction · Document-level Relation Extraction · Multi-Task Learning

1 Introduction

In recent years text-based information extraction tasks such as named entity recognition have become more popular, which is closely related to the growing importance of transformer-based Large Language Models (LLMs). Such models are already used as a part of complex document information extraction pipelines. Even though important pieces of information are extracted, these pipelines still lack the ability to detect connections between them. The missing part, which is the relation classification task, was recognized as a significant challenge in recent years. The problem is even harder to solve if tackled with a multi-task method capable of solving both named entity recognition and relation classification task in a single neural network passage.

In this paper, we propose an approach to solve the multi-task problem of the joint document-level entity and relation extraction problem introduced with the DocRED dataset [16]. We follow the already existing line of research of learning a single model to solve all four subtasks: mention detection, coreference resolution, entity classification, and relation extraction. The single model is trained to first detect the spans of text that are the entity mentions and group them into coreference clusters. Those entity clusters are then labeled with the correct entity type and linked to each other by relations. Figure 1 shows an example of a document from the DocRED dataset and a graph of labeled entity clusters that are expected as an output from the model. We introduce the bidirectional memory-like dependency between tasks to address the drawbacks of pipeline-based methods and perform the joint task more accurately.

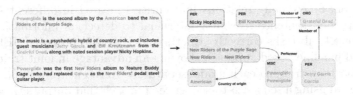

Fig. 1. Visualization of the document-level joint entity and relation extraction task based on the example taken from DocRED. Entity mentions originating from different entity clusters are distinguished by color.

Our contribution can be summarized as follows: (1) we introduce a new approach that solves multi-task learning problems by improving the architecture of the previously proposed pipeline-based method, introducing the memory module to provide bi-directional dependency between tasks (2) we provide evaluation results, which show that our method outperforms the pipeline-based methods and achieves state-of-the-art results on the BioCreative V CDR corpus (3) we propose a novel similarity classifier module solving distance learning problem for document-level joint entity and relation classification serving as a starting point for future work. The code of our solution is available at https://github.com/kosciukiewicz/similarity_based_memory_re.

2   Related Work

Relation Classification. The relation extraction task is commonly approached by using separately trained models for the Named Entity Recognition [10] to detect entities and then detect relations between them. The transformer-based architectures, pre-trained on large text corpora, such as BERT [4], have dominated the field i.e. Baldini Soares et al. [13] uses contextualized input embedding for the relation classification task.

Joint Entity and Relation Extraction. The early end-to-end solutions formulated task joint task as a sequence tagging based on BIO/BILOU scheme. These approaches include solving a table-filling problem proposed by Miwa et al. [11]. Several approaches tried to leverage multi-task learning abilities using attention-based [7] bi-directional LSTM sharing feature encoders between two tasks to improve overall performance. The inability of the BIO/BILOU-based models to assign more than one tag to a token resulted in using the span-based method for joint entity and relation extraction proposed in Lee et al. [8]. Becoming a standard in recent years, this approach was further extended with graph-based methods like DyGIE++ [15] or memory models like TriMF [12] to enhance token span representation to an end-to-end approach for the joint task.

Document-Level Relation Extraction Although the DocRED [16] was originally introduced as relation classification benchmark, the opportunity arose to tackle a more complex joint entity and relation extraction pipelines consisting of mention detection, coreference resolution, entity classification, and relation classification. Since many relations link entities located in different sentences, considering inter-sentence reasoning is crucial to detect all information needed to perform all sub-tasks correctly. Eberts and Ulges [5] proposed JEREX - an end-to-end pipeline-based approach showing an advantage in joint training of all tasks rather than training each model separately. In recent work [2,6], the problem is tackled using a sequence-to-sequence approach that outputs the extracted relation triples consisting of two related entities and relation type as text.

3   Approach

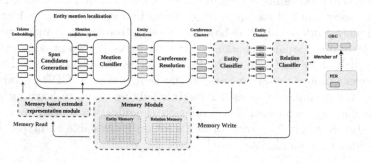

Fig. 2. The proposed architecture based on JEREX [5] enhanced with a feedback loop from entity and relation classifiers to the input of the mentioned classifier step. The novel part of the architecture is highlighted with a gray background and dashed borders.

Our document-level relation extraction framework is inspired by JEREX [5] which consists of four task-specific components: mention extraction ($\mathcal{M}$), coreference resolution ($\mathcal{C}$), entity extraction ($\mathcal{E}$) and relation extraction ($\mathcal{R}$). We change the original one-after-another pipeline architecture, introducing the memory module presented in Fig. 2. The input representations of task-specific models are altered using the memory-based extended representation module that reads the memory using the Memory Read operation. The memory matrices $\mathbf{M}_{\mathcal{E}}$ and $\mathbf{M}_{\mathcal{R}}$, are written by the entity and relation classifier, respectively. That feedback loop allows to share the information with previous steps extending their input by introducing a bi-directional dependency between tasks.

3.1 Memory Reading

Similarly to TriMF [12], our approach memory reading is based on the attention mechanism that extends the input representation with the information read from memory. In our architecture, as shown in Fig. 2 we extend both token embeddings $\mathbf{X}_T$ and mention candidates span representations $\mathbf{X}_S$. For every input representation $\mathbf{X}_i$ where $i \in \{T, S\}$ and memory matrix $\mathbf{M}_j$ where $j \in \{\mathcal{E}, \mathcal{R}\}$, the attention mechanism takes the representation $\mathbf{X}_i \in \mathbb{R}^{n \times h}$ as keys and values, where n denotes the number of representations vectors and h is the embedding size.

As a query, the attention mechanism uses the memory matrix $\mathbf{M}_j \in \mathbb{R}^{m \times s}$ where m denotes the number of memory slots and s is the size of the memory slot. To compute the attention weights vector $\mathbf{a}_{i,j} \in \mathbb{R}^n$ we sum over the memory slots dimension as follows:

$$(\mathbf{a}_{i,j})^\top = \sum_k \text{softmax}(\mathbf{m}_j^{k,:} \mathbf{W}_{i,j}^{read} \mathbf{X}_i^\top) \tag{1}$$

where $\mathbf{W}_{i,j}^{read} \in \mathbb{R}^{s \times h}$ is a learnable parameter matrix for the attention mechanism and $\mathbf{m}_j^{k,:}$ is the k-th row of $\mathbf{M}_j$. The $\mathbf{a}_{i,j}$ vector is then used to weight the $\mathbf{X}_i$ to generate extended input representation $\mathbf{X}'_{i,j}$:

$$\mathbf{X}'_{i,j} = \text{diag}(\mathbf{a}_{i,j})\mathbf{X}_i \tag{2}$$

For each input representation i, the memory reading operation creates two extended representations $\mathbf{X}'_{i,\mathcal{E}}$ and $\mathbf{X}'_{i,\mathcal{R}}$, based on both memory matrices. The final extended representation is then calculated, using the element-wise mean of $\mathbf{X}_i$, $\mathbf{X}'_{i,\mathcal{E}}$ and $\mathbf{X}'_{i,\mathcal{R}}$:

3.2 Memory Writing

Both memory matrices $\mathbf{M}_\mathcal{E}$ and $\mathbf{M}_\mathcal{R}$ store representations for entity and relation categories respectively. Values encoded in those matrices are written using the gradient of the loss function from the associated classifier – the entity classifier for $\mathbf{M}_\mathcal{E}$ and the relation classifier for $\mathbf{M}_\mathcal{R}$. To make the stored representations more precise, the loss depends on the similarity between category embedding and the representation of the instance that belongs to that category according to the instance label. As a result, both entity and relation classifiers rely on similarity function S between input representation and suitable memory matrix. The probability distribution over entity types of entity e_i based on its representation vector $\mathbf{x}_i^e$ is calculated as follows:

$$p(\mathbf{y}_e|e_i) = \text{softmax}(S(\mathbf{x}_i^e, \mathbf{M}_\mathcal{E})) \tag{3}$$

To get the existence probability over relation types for entity pair $p_{i,j}$ represented by entity pair representation $\mathbf{x}_{i,j}^p \in \mathbb{R}^h$ we used the sigmoid function:

$$p(\mathbf{y}_r|p_{i,j}) = \text{sigmoid}(S(\mathbf{x}_{i,j}^p, \mathbf{M}_\mathcal{R})) \tag{4}$$

We define S as bilinear similarity between instance representation $\mathbf{x}$ and memory matrix $\mathbf{M}$ as follows:

$$S(\mathbf{x}, \mathbf{M}) = S_{bilinear}(\mathbf{x}, \mathbf{M}; \mathbf{W}) = \mathbf{M}\mathbf{W}^\top\mathbf{x} \qquad (5)$$

where $\mathbf{W}$ is a learnable parameter matrix. For both entity and relation classifiers, separate learnable bilinear similarity weight matrices are used: $\mathbf{W}_\mathcal{E}^{write} \in \mathbb{R}^{h_e \times s_\mathcal{E}}$ and $\mathbf{W}_\mathcal{R}^{write} \in \mathbb{R}^{h_p \times s_\mathcal{R}}$ where h_e and h_p denote entity and entity pair representation sizes respectively. $s_\mathcal{E}$ and $s_\mathcal{R}$ denote the memory slot size of the entity and relation memory matrices. In our approach number of slots for the memory matrices are equal to the number of types in associated classifiers.

3.3   Training

Finally, our model is trained optimizing the joint loss $\mathcal{L}^{joint}$ which contains the same four, sub-tasks related, loss $\mathcal{L}^j$ weighted with fixed, task-related weight value β_j as in JEREX [5]:

$$\mathcal{L}^{joint} = \beta_\mathcal{M}\mathcal{L}^\mathcal{M} + \beta_\mathcal{C}\mathcal{L}^\mathcal{C} + \beta_\mathcal{E}\mathcal{L}^\mathcal{E} + \beta_\mathcal{R}\mathcal{L}^\mathcal{R}. \qquad (6)$$

We also include the two-stage training approach proposed in TriMF [12], tuning the *memory warm-up proportion* during the hyperparameter search.

4   Experiments

Datasets. We compare the proposed similarity-based memory learning framework to the existing approaches using DocRED [16] dataset which contains over 5000 human-annotated documents from Wikipedia and Wikidata. By design, DocRED dataset was intended to be used as a relation classification benchmark but its hierarchical annotations are perfectly suitable for joint task evaluation. For train, dev, and test split we follow the one provided in JEREX [5]. According to recent work [14], DocRED consists of a significant number of false negative examples. We used dataset splits provided with Re-DocRED [14] which is a re-annotated version of the DocRED dataset. We also provide results on one area-specific corpus annotated in a similar manner as DocRED - BioCreative V CDR [9] that contains 1500 abstracts from PubMed articles. Following the prior work [3,6] we used the original train, dev, and test set split provided with the CDR corpus.

Training. As a pretrained text encoder we used $BERT_{BASE}$ [4]. For the domain-specific BioCreative V CDR dataset we used $SciBERT_{BASE}$ [1] which was trained on scientific papers from Semantic Scholar. All classifiers and memory module parameters were initialized randomly. During training, we used batch size 2, AdamW optimizer with learning rate set to $5e-5$ with linear warm-up for 10% of training steps and linear decay to 0. The stopping criteria for training were set to 20 epochs for all experiments.

Evaluation. During the evaluation we used the strict scenario that assumes the prediction is considered correct only if all subtasks-related predictions are correct. We evaluated our method using micro-averaged F1-score. In Sect. 5 we reported F1-score for a final model evaluated on the test split. As the final model we selected the one that achieved the best F1-score measured on the dev split based on 5 independent runs using different random seeds. Our evaluation technique follows the one proposed in [5,6].

Hyperparameters. All hyperparameters like embedding sizes or multi-task loss weights were adopted from the original work [5] for better direct comparison. Our approach introduces new hyperparameters for which we conducted grid search on the dev split to find the best value. That includes hyperparameters such as *memory warm-up proportion* [12], memory read gradient, number and types of memory modules, and finally the size of memory slots.

5 Results

Table 1. Comparison (F1-score) of our method on the relation extraction task with existing end-to-end systems. * - results from original publications.

Model	CDR	DocRED	Re-DocRED
JEREX [5]	42.88	*40.38	**45.56**
seq2rel [6]	*40.20	*38.20	–
ours	**43.75**	**40.42**	44.37
JEREX$_{pre-training}$	-	41.27	45.81
ours$_{pre-training}$	-	**41.75**	**45.96**

In Table 1 we present a comparison between our approach and existing end-to-end methods on 3 benchmark datasets for joint entity and relation extraction. The provided metric values show that our approach outperforms existing methods on CDR by about 0.9 percent points (pp.), achieving state-of-the-art results. Our method achieves similar results on DocRED and is outperformed by JEREX architecture on Re-DocRED dataset. We argue that the memory warm-up proportion value (0.4) is too small to properly initialize memories with accurate category representation. On the other hand increasing the memory warm-up steps leaves no time to properly train memory read modules. To address this issue we conducted experiments on pre-trained architecture using distantly annotated corpus of DocRED dataset to initialize memory matrices. We did the same pre-training for JEREX and the results show that our approach outperforms the original architecture by up to 0.48 pp. on both DocRED-based datasets.

For the direct comparison with the original architecture we evaluated our memory-enhanced approach with two relation classifiers modules proposed in [5]. Results presented in Table 2 show that our method improves the Global

Table 2. Comparison (F1-score) between our architecture including memory reading module with JEREX using different relation classifier components - Global (*GRC*) and Multi-Instance (*MRC*). ∗ - results from original publications.

Model	CDR	DocRED	Re-DocRED
GRC			
JEREX [5]	42.04	*37.98	43.46
ours	42.04	37.76	43.64
ours + memory	**42.18**	**39.68**	**44.77**
MRC			
JEREX [5]	42.88	*40.38	**45.56**
ours	43.12	**40.68**	44.93
ours + memory	**43.75**	40.42	44.37

Relation Classifier (*GRC*) on every dataset by up to 1.70 pp. We also tested the performance of our method without the memory module - only with distance-based classifiers. Based on the results in Table 2, including a memory module with a feedback loop between tasks, in most cases, improved the final results regardless of the *GRC* or *MRC* module.

6   Conclusions and Future Work

In this paper, we proposed a novel approach for multi-task learning for document-level joint entity and relation extraction tasks. By including memory-like extensions creating a feedback loop between the tasks, we addressed the issues present in the previous architectures. Empirical results show the superiority of our method in performance over other document-level relation extraction methods, achieving state-of-the-art results on BioCreative V CDR corpus. One of the possible directions for future work is further development of the memory module by using different memory read vectors for more meaningful input encoding in enhanced representation module or improving the content written to memory by replacing the bi-linear similarity classifier with different distance-based scoring functions or proposing a different method of writing to memory.

Acknowledgements. The research was conducted under the Implementation Doctorate programme of Polish Ministry of Science and Higher Education and also partially funded by Department of Artificial Intelligence, Wroclaw Tech and by the European Union under the Horizon Europe grant OMINO (grant number 101086321). It was also partially co-funded by the European Regional Development Fund within Measure 1.1. "Enterprise R&D Projects", Sub-measure 1.1.1. "Industrial research and development by companies" as part of The Operational Programme Smart Growth 2014-2020, support contract no. POIR.01.01.01-00-0876/20-00.

References

1. Beltagy, I., Lo, K., Cohan, A.: Scibert: A pretrained language model for scientific text. In: Proceedings of the 2019 Conference on Empirical Methods in Natural Language Processing and the 9th International Joint Conference on Natural Language Processing. ACL (2019)
2. Cabot, P.L.H., Navigli, R.: Rebel: Relation extraction by end-to-end language generation. In: Findings of the Association for Computational Linguistics: EMNLP (2021)
3. Christopoulou, E., Miwa, M., Ananiadou, S.: Connecting the dots: Document-level neural relation extraction with edge-oriented graphs. In: Proceedings of the 2019 Conference on Empirical Methods in Natural Language Processing and the 9th International Joint Conference on Natural Language Processing. ACL (2019)
4. Devlin, J., Chang, M., Lee, K., Toutanova, K.: BERT: pre-training of deep bidirectional transformers for language understanding. In: Proceedings of the 2019 Conference of the North American Chapter of the Association for Computational Linguistics: Human Language Technologies (2019)
5. Eberts, M., Ulges, A.: An end-to-end model for entity-level relation extraction using multi-instance learning. In: Proceedings of the 16th Conference of the European Chapter of the Association for Computational Linguistics: Main Volume (2021)
6. Giorgi, J., Bader, G., Wang, B.: A sequence-to-sequence approach for document-level relation extraction. In: Proceedings of the 21st Workshop on Biomedical Language Processing. pp. 10–25 (2022)
7. Katiyar, A., Cardie, C.: Going out on a limb: Joint extraction of entity mentions and relations without dependency trees. In: Proceedings of the 55th Annual Meeting of the Association for Computational Linguistics (2017)
8. Lee, K., He, L., Lewis, M., Zettlemoyer, L.: End-to-end neural coreference resolution. In: Proceedings of the 2017 Conference on Empirical Methods in Natural Language Processing (2017)
9. Li, J., et al.: Biocreative v cdr task corpus: a resource for chemical disease relation extraction (2016)
10. Li, J., Sun, A., Han, J., Li, C.: A survey on deep learning for named entity recognition. IEEE Trans. on Knowl. and Data Eng. (2022)
11. Miwa, M., Sasaki, Y.: Modeling joint entity and relation extraction with table representation. In: Proceedings of the 2014 Conference on Empirical Methods in Natural Language Processing (2014)
12. Shen, Y., Ma, X., Tang, Y., Lu, W.: A trigger-sense memory flow framework for joint entity and relation extraction. In: Proceedings of the web conference (2021)
13. Soares, L.B., Fitzgerald, N., Ling, J., Kwiatkowski, T.: Matching the blanks: Distributional similarity for relation learning. In: Proceedings of the 57th Annual Meeting of the Association for Computational Linguistics (2019)
14. Tan, Q., Xu, L., Bing, L., Ng, H.T.: Revisiting docred-addressing the overlooked false negative problem in relation extraction. arXiv preprint arXiv:2205.12696 (2022)
15. Wadden, D., Wennberg, U., Luan, Y., Hajishirzi, H.: Entity, relation, and event extraction with contextualized span representations. In: Proceedings of the 2019 Conference on Empirical Methods in Natural Language Processing and the 9th International Joint Conference on Natural Language Processing (2019)
16. Yao, Y., et al.: DocRED: a large-scale document-level relation extraction dataset. In: Proceedings of the 57th Annual Meeting of the Association for Computational Linguistics (2019)

Solving Complex Sequential Decision-Making Problems by Deep Reinforcement Learning with Heuristic Rules

Thanh Thi Nguyen[1](✉), Cuong M. Nguyen[2], Thien Huynh-The[3],
Quoc-Viet Pham[4], Quoc Viet Hung Nguyen[5], Imran Razzak[6],
and Vijay Janapa Reddi[7]

[1] Deakin University, Geelong, VIC, Australia
thanh.nguyen@deakin.edu.au
[2] Université Polytechnique Hauts-de-France, Valenciennes, France
[3] Ho Chi Minh City University of Technology and Education, Ho Chi Minh City,
Vietnam
[4] Trinity College Dublin (TCD), The University of Dublin, Dublin , Ireland
[5] Griffith University, Gold Coast, QLD, Australia
[6] University of New South Wales, Sydney, Australia
[7] Harvard University, Cambridge, MA, USA

Abstract. Deep reinforcement learning (RL) has demonstrated great capabilities in dealing with sequential decision-making problems, but its performance is often bounded by suboptimal solutions in many complex applications. This paper proposes the use of human expertise to increase the performance of deep RL methods. Human domain knowledge is characterized by heuristic rules and they are utilized adaptively to alter either the reward signals or environment states during the learning process of deep RL. This prevents deep RL methods from being trapped in local optimal solutions and computationally expensive training process and thus allowing them to maximize their performance when carrying out designated tasks. The proposed approach is experimented with a video game developed using the Arcade Learning Environment. With the extra information provided at the right time by human experts via heuristic rules, deep RL methods show greater performance compared with circumstances where human knowledge is not used. This implies that our approach of utilizing human expertise for deep RL has helped to increase the performance of deep RL and it has a great potential to be generalized and applied to solve complex real-world decision-making problems efficiently.

Keywords: Complex problems · Reinforcement learning · Sequential decision making · Human expertise · Heuristic rules

1 Introduction

Reinforcement learning (RL) has been applied to solve various real-world sequential decision-making problems such as in robotics, self-driving cars, trading and

J. Mikyška et al. (Eds.): ICCS 2023, LNCS 14074, pp. 298–305, 2023.
https://doi.org/10.1007/978-3-031-36021-3_30

finance, machine translation, healthcare, video games and so on [6]. Prominent challenges for RL methods include long training time and dealing with large state and action spaces. Deep learning has emerged to be a great compliment to RL methods and has enabled them to efficiently handle high-dimensional state and action spaces [3]. The combination of deep learning and RL methods has been termed as deep RL. Deep learning is able to represent high-dimensional data by compact feature sets to facilitate the training process of RL methods when dealing with complex environments. While deep RL methods are able to cope with large-scale problems, their learning process is even more *computationally expensive* and requires a *large number of samples* compared with traditional RL approaches. This directly affects the performance of deep RL methods, especially when applied to problems with complex goals or objectives. One approach to mitigating this issue is incorporating human knowledge into the training process of deep RL methods [10].

On the one hand, humans can communicate a policy to the agent by demonstrating the correct actions in person to complete the tasks. Approaches such as learning from demonstrations or imitation learning can then be employed to learn the policy from the demonstration data [7]. However, the tasks sometimes are too challenging that humans cannot perform properly. Collecting behavioural data from humans is often expensive and erroneous, especially when a large amount of high-quality demonstration data are required. Another approach that is less expensive is for humans to provide feedback to the agent regarding its performance. This kind of guidance may be in the form of evaluative feedback (e.g., policy shaping, reward shaping, intervention) or human preferences [11].

An example of the evaluative feedback approach is presented in [8] where domain knowledge is represented as decision trees to imbue human expertise into a deep RL agent. Humans just need to specify behaviours of agents as high-level instructions via propositional rules without the need for demonstrating the tasks in all states or providing feedback to all actions. That approach helps to improve warm starts in terms of network weights and architecture of deep neural networks. Deep RL agents can start the learning process in a more knowledgeable manner and therefore their learning time is shortened, and their performance is superior to random initialization approaches. Likewise, Dong *et al.* [1] suggested an approach using a Lyapunov function to shape the reward signal in RL. The agent is instigated to reach the region of maximal reward based on the Lyapunov stability analysis. That approach is theoretically proved to have a convergence guarantee without making variance in optimality or biased greedy policy.

Making a full use of a shaped reward function, which is constructed using domain knowledge, may not improve the performance of RL methods because transforming human knowledge into numerical reward values is often subject to human cognitive bias. Hu *et al.* [2] proposed a method to adaptively utilize a given shaping reward function by formulating and solving a bi-level optimization problem. That approach attempts to maximize the use of the beneficial part of the given shaping reward function while ignoring the unbeneficial shaping rewards. This helps to avoid the time-consuming reward tuning process in deep RL applications.

In this study, we focus on complex problems where the objective of the agent must be changed adaptively depending on the status of the agent and the environment. For example, a surveillance unmanned aerial vehicle (UAV), when not under attack, can fly slowly and capture high-quality images of objects in the monitored area. When the UAV senses or recognizes an attack, it needs to automatically fly fast and escape the area quickly. Likewise, if a self-driving car suddenly collides into a crowd of people, it should be able to quickly adapt by changing to a stopping policy (e.g., turning off the engine) rather than continuing to run into the crowd and collide with more people. This paper aims to solve these problems using deep RL methods with heuristic rules. The rules are constructed using human domain knowledge that leads to a change of reward signal or state information adaptively. These changes happening at the right time during the training and execution of the deep RL agent will help it to crack the trap of suboptimal solutions.

2   Modifying A3C to Incorporate Heuristic Rules

We choose the asynchronous advantage actor-critic (A3C) deep RL method [4] to demonstrate the idea of incorporating human expertise. The asynchronous learning architecture of A3C with multiple workers enables it to learn quickly and efficiently by utilizing data from multiple environments. In A3C, multiple workers learn and update global network's parameters asynchronously. Based on the asynchronous updates, the learning process can be parallelized using different threads, which collect experiences independently. Many decorrelated training examples can thus be collected at a time, leading to a reduction of the variance of the learning estimators. Each A3C worker thread interacts with its own environment and updates the global network with its computed gradient. Starting conditions and exploration rates can be chosen differently for the threads to ensure examples collected at a time are adequately varied. A3C has a great advantage compared to the notable deep Q-network (DQN) algorithm [5].

The A3C method has become a benchmark deep RL algorithm due to its efficiency in training and its ability to deal with both discrete and continuous action spaces. The actor produces policies using the feedback from the critic. Both networks improve over time based on their loss function. The value loss function of the critic is:

$$L_1 = \sum (R - V(s))^2 \tag{1}$$

where R is the discounted future reward: $R = r + \gamma V(s')$, with r being an immediate reward of an action a. On the other hand, the policy loss function of the actor is:

$$L_2 = -\log(\pi(a|s)) * A(s) - \beta * H(\pi(s)) \tag{2}$$

where $A(s)$ is the estimated advantage function $A(s) = R - V(s)$ and H is the entropy of the policy π, which is added to improve exploration. The impact of this entropy regularization term is controlled by the hyperparameter β.

Algorithm 1: Modified A3C (Using Heuristic Rules)

Define global shared parameters as θ and θ_v while thread-specific parameters as θ' and θ'_v

Set global shared counter $T = 0$ and thread step counter $t = 1$

repeat

    Reset gradients: $d\theta = 0$ and $d\theta_v = 0$

    Update thread-specific parameters $\theta' = \theta$ and $\theta'_v = \theta_v$

    Set $t_{start} = t$

    Get state s_t

    repeat

        Execute action a_t based on policy $\pi(a_t|s_t;\theta')$

        Receive reward r_t and new state $s_{(t+1)}$

        if state s_t satisfies predetermined conditions

            Change reward $r_t = r_t^*$ or new state $s_{(t+1)} = s_{(t+1)}^*$

        end if

        $t = t + 1; T = T + 1$

    until terminal s_t or $t - t_{start} == t_{max}$

    Set $R = 0$ for terminal s_t or $R = V(s_t, \theta'_v)$ for non-terminal s_t

    for $i \in \{t-1, ..., t_{start}\}$ **do**

        $R = r_i + \gamma R$

        Aggregate gradients wrt θ':

        $d\theta = d\theta + \nabla_{\theta'} \log \pi(a_i|s_i;\theta')(R - V(s_i;\theta'_v)) + \beta\nabla_{\theta'}H(\pi(s_i;\theta'))$

        Aggregate gradients wrt θ'_v: $d\theta_v = d\theta_v + \partial(R - V(s_i;\theta'_v))^2/\partial\theta'_v$

    end for

    Asynchronously update θ using $d\theta$ and θ_v using $d\theta_v$

until $T > T_{max}$

The modified A3C is presented in Algorithm 1 where heuristic rules are injected into the A3C. These rules are constructed based on conditions of the states received by the agent. We perform the experiments to modify the states if some predetermined conditions of the states are met. Specifically, changing the next state $s_{(t+1)} = s_{(t+1)}^*$ (where $s_{(t+1)}^*$ is the modified state) is to recommend areas of the state that the agent needs to focus on in order to complete the designated tasks efficiently. The heuristic rules are devised based on expert knowledge, which is specific for each particular problem. We demonstrate our proposed approach by using the video River Raid game. Details on how human expertise is encoded via heuristic rules are presented in the next section.

3   Heuristic Rules to Encode Human Expertise

3.1   Heuristic Rules for the River Raid Game

The agent is trained to fly a fighter jet over a river and to shoot as many enemies as possible. A screen of the River Raid game is depicted in Fig. 1. This is a vertically scrolling shooter game where the agent scores a point of 30, 60, 80, 100 and 500 when shooting an enemy tanker, a helicopter, a fuel depot, a jet

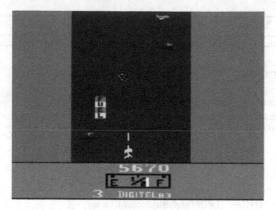

Fig. 1. An illustration of the River Raid game where the fighter jet in yellow colour needs to fly over the river in blue colour. Enemy crafts including helicopters, fuel depots, tankers, jets and bridges are to be shot by the fighter jet to obtain points. The fighter jet (agent) needs to learn to avoid crashing with the riverbank and enemy crafts and also refuel by flying over fuel depots when its fuel level is low. (Color figure online)

and a bridge, respectively. The agent can refuel to full if it decides to fly over instead of shooting a fuel depot. The agent can move left and right, accelerate and decelerate, but cannot manoeuvre up and down the screen. The game is over if the agent (fighter jet) collides with the riverbank or an enemy craft, or if it runs out of fuel.

Because the game objective is to maximise the points, the agent is not instigated to fly over the fuel depots to refuel. It instead attempts to shoot the fuel depots to obtain 80 points. The agent therefore runs out of fuel and the game is over rather quickly. Our observations suggest that when the agent obtains a total score of around 6,000 points, it runs out of fuel and thus the maximum point it can obtain cannot exceed 6,000. Based on this observation, we create a heuristic rule to penalize the agent if it shoots the fuel depots when its fuel level is less than 60% of the full capacity. In that case, the agent will be deducted 80 points instead of getting 80 points if it shoots a fuel depot. Alternatively, if flying over a fuel depot, the agent will be awarded 80 points. It is important to note that this heuristic rule is only applied when the fuel level is less than 60%. This threshold is selected because the fighter jet intuitively does not need to refuel when its fuel level is high and it also does not risk to leave the fuel level too low. In spite of that, other values, e.g., 70%, 50%, 30%, and so on, can be experimented for this fuel level threshold.

3.2   Generalizing to Other Problems

While incorporating heuristic rules into existing deep RL algorithms can be implemented like what we presented in Algorithm 1 (where A3C is chosen as an example), the more challenging part of the overall proposed approach is how to design a heuristic rule for a specific problem. The rule is in the if-then format such as if a predetermined condition is met then some changes have to be made, i.e., change reward signal or next state.

It is important to note that the changes need to be made at the right time to help the agent crack the suboptimal trap. The change is applied to either reward or next state whenever the predetermined condition is met. In general, the expert needs to provide two types of information: what needs to be changed and at what time.

Humans generally do not need to have great expertise in order to realize what causes the suboptimal trap for a specific problem and suggest a heuristic rule for that problem. The rules are normally intuitive to humans. Therefore, in terms of generalization of the proposed approach to other problems, on the one hand, a heuristic rule can not fit all problems, i.e., it is not a one-size-fits-all approach. On the other hand, it is however rather straightforward for a human to create a heuristic rule for a specific problem after observing how that problem plays out.

4   Experimental Results

In this section, we compare our approach that uses human expertise (via the heuristic rule) with two baseline methods: the A3C approach without human expertise (not using the heuristic rule) and the Multi Objective Deep Q-Networks with Decision Values (MODQN-DV) method proposed in [9].

The heuristic rule for the River Raid game is applied to encourage the agent to refuel when its fuel level is low. To implement this heuristic rule, we need to monitor the fuel level of the agent. When the fuel level is less than 60%, the rule is executed to alter the reward signal whenever the agent is facing a fuel depot, i.e., reduce 80 points if shooting it or obtain 80 points if flying over it. When the fuel level is equal to or above 60%, the reward signal is unchanged. For the MODQN-DV approach, two objectives are specified, which includes maximizing the scoring points and maximizing the fuel level. We perform the training process with 100 epochs and one million steps per epoch, leading to a total of 100 million training steps. When the rule is not used, the A3C agent's performance is capped at around 5,000 points even with 100 million training steps. The MODQN-DV method also obtains a maximum score at around 6,000 points. In contrast, the A3C agent obviously achieves higher performance when the rule is applied, i.e., obtaining a score of more than 6,000 points with just 60 million learning steps. The agent can acquire up to 8,000 points if the training process reaches 90 or 100 million steps.

The network parameters are saved at every one million training steps for evaluation purpose, making 100 checkpoints in total. The average rewards obtained using 20,000 evaluation steps at each of 100 checkpoints are presented in Fig. 2 for comparisons. With the human-based heuristic rule, the A3C agent's performance starts to be superior to the experiment without using the rule when the training reaches around 50 million training steps. Therefore, the human domain knowledge has demonstrated its effectiveness in improving the performance of

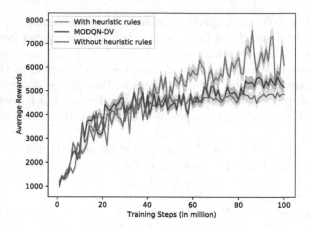

Fig. 2. Evaluation results in terms of average reward of the River Raid game with 20,000 evaluation steps using 100 checkpoints saved during the training process. The standard deep RL A3C algorithm without using human domain expertise (heuristic rules) obtains an average score of 5,000 points maximum even though the training reaches 100 million steps. Likewise, the MODQN-DV method is capped at around 6,000 points. The incorporation of heuristic rules into deep RL has improved its performance by getting the scores of more than 5,000 points after around 50 million training steps. Our approach has thus sped up the learning process of the A3C agent.

deep RL algorithms by avoiding suboptimal solutions, which limit the deep RL score to a maximum of 5,000 points. A video demonstrating the A3C agent playing the River Raid game without human expertise is available at: https://youtu.be/fdvTCC8ffoc. Using the heuristic rule characterizing the human knowledge, the agent is able to learn to pick up fuel depots when its fuel level is low and thus the game is prolonged. The score obtained is much higher as demonstrated in this video: https://youtu.be/LQG7C4NJQRE.

5 Conclusions

Methods used to solve complex sequential decision-making tasks such as RL or imitation learning normally require a large amount of training data and a long training time. Human knowledge on how to solve these tasks is often utilized to prevent them from being trapped in suboptimal solutions and speed up the learning process of these methods. In this paper, we proposed to utilize human expertise via heuristic rules, which are incorporated adaptively into the training of deep RL agents depending on the status of the environment to maximize their performance. The River Raid game has been implemented to demonstrate performance of the proposed approach. Empirical results of this research have highlighted the superiority of deep RL agents when human expertise is integrated. For the games implemented herein, heuristic rules have helped to break the capped performance of the deep RL A3C agents, leading to an increase of the obtained rewards. A future work will evaluate the proposed method with other deep RL algorithms as well as with new games in the Arcade Learning and Gym environments. The experiments in this study are in the game domain; however, extensions of this work can be applied in different industry domains such as controlling self-driving cars, robots, UAVs, and so on. For example, when a self-driving car is going from a sparse pedestrian street to a crowded pedestrian mall, there should be a heuristic rule to trigger the agent to change its policy, e.g., by driving much slower. Likewise, if a deep RL-based autonomous system is under attack, a rule should be applied to transition the system into a safe policy. These real-world examples are far from being solved completely by standard deep RL algorithms, but the incorporation of human expertise to increase capabilities of deep RL as exemplified in this study is an important step towards satisfactory solutions to these problems.

References

1. Dong, Y., Tang, X., Yuan, Y.: Principled reward shaping for reinforcement learning via lyapunov stability theory. Neurocomputing **393**, 83–90 (2020)
2. Hu, Y., et al.: Learning to utilize shaping rewards: a new approach of reward shaping. Adv. Neural. Inf. Process. Syst. **33**, 15931–15941 (2020)
3. Lazaridis, A., Fachantidis, A., Vlahavas, I.: Deep reinforcement learning: a state-of-the-art walkthrough. J. Artif. Intell. Res. **69**, 1421–1471 (2020)
4. Mnih, V., et al.: Asynchronous methods for deep reinforcement learning. In: International Conference on Machine Learning, pp. 1928–1937. PMLR (2016)
5. Mnih, V., et al.: Human-level control through deep reinforcement learning. Nature **518**(7540), 529–533 (2015)
6. Nguyen, T.T., Nguyen, N.D., Nahavandi, S.: Deep reinforcement learning for multiagent systems: a review of challenges, solutions, and applications. IEEE Trans. Cybern. **50**(9), 3826–3839 (2020)
7. Ravichandar, H., Polydoros, A.S., Chernova, S., Billard, A.: Recent advances in robot learning from demonstration. Ann. Rev. Control Robot. Autonomous Syst. **3**, 297–330 (2020)
8. Silva, A., Gombolay, M.: Encoding human domain knowledge to warm start reinforcement learning. In: Proceedings of the AAAI Conference on Artificial Intelligence, vol. 35, pp. 5042–5050 (2021)
9. Tajmajer, T.: Modular multi-objective deep reinforcement learning with decision values. In: 2018 Federated Conference on Computer Science and Information Systems (FedCSIS), pp. 85–93. IEEE (2018)
10. Zhang, P., et al.: KoGuN: accelerating deep reinforcement learning via integrating human suboptimal knowledge. In: The 29th International Conference on International Joint Conferences on Artificial Intelligence, pp. 2291–2297 (2020)
11. Zhang, R., Torabi, F., Guan, L., Ballard, D.H., Stone, P.: Leveraging human guidance for deep reinforcement learning tasks. In: The 28th International Joint Conference on Artificial Intelligence, pp. 6339–6346 (2019)

Weighted Hamming Metric and KNN Classification of Nominal-Continuous Data

Aleksander Denisiuk[✉][iD]

University of Warmia and Mazury in Olsztyn, ul. Słoneczna 54, 10-710 Olsztyn,
Poland
denisiuk@matman.uwm.edu.pl

Abstract. The purpose of the article is to develop a new metric learning
algorithm for combination of continuous and nominal data. We start with
Euclidean metric for continuous and Hamming metric for nominal part
of data. The impact of specific feature is modeled with corresponding
weight in the metric definition. A new algorithm for automatic weights
detection is proposed. The weighted metric is then used in the standard
knn classification algorithm. Series of numerical experiments show that
the algorithm can successfully classify raw, non-normalized data.

Keywords: KNN classification · weighted Hamming metric ·
nominal-continuous data · metric learning

1 Introduction

Similarity is important concept of data science. Most algorithms operate on
groups of similar data, and their results strongly depend on how we define similarity. Mathematical concept that can be used to model similarity is metric.
Assuming that the similarity (metric) is hidden in the data, we get the problem
of determining the metric, i.e. *metric learning*.

The purpose of this article is further development of the metric learning
technics suggested in [4] for combinations of nominal-continuous data. Namely,
instead of unsupervised learning and clustering investigated in [4], here we consider supervised learning and the classification problem.

We do not assume that any additional structure on a dataset is a-priory
known, so we use the Euclidean metric on continuous and the Hamming distance on nominal part of data. These metrices provide the most common and
straightforward way for measuring distances [8].

The main assumption is that each feature has different impact to the structure of classes. We model it with appropriate multipliers in metric (1). So, the
problem is to define these unknown multipliers. We do it by minimizing the total
intra-class squared distance.

As a possible application of our technics we show that the standard knn-
classification algorithm can be improved by using the weighted metric. Let us,
however, note that the proposed approach can be used in any algorithm for
analysis of nominal-continuous data that is based on metric.

© The Author(s), under exclusive license to Springer Nature Switzerland AG 2023
J. Mikyška et al. (Eds.): ICCS 2023, LNCS 14074, pp. 306–313, 2023.
https://doi.org/10.1007/978-3-031-36021-3_31

The paper is organized as follows. In the Sect. 2 we give a brief survey of related works. The algorithm of weights detection is the content of the Sect. 3. The results of numerical experiments are presented in the Sect. 4. Some final conclusions and possible directions for the future work are given in the Sect. 5.

2   Related Works

The metric learning is actively studied in recent years. However, most papers consider either continuous or nominal data. Sometimes nominal data are embedded into continuous space, but this embedding as a rule is arbitrary. In this article we consider nominal and continuous data together in a natural way: the Euclidean metric on continuous part and the Hamming metric on nominal one. Data from each part affects the weights on the other part as well.

One of the first papers in metric learning of continuous data is [17], where a projected gradient descent algorithm for Mahalanobis distance learning is suggested. Two stochastic methods, the neighborhood component analysis and the large margin nearest neighbors were introduced in [6] and [16] respectively. The above methods got further development in later years, see for instance [15].

The authors of [14] used the support vector machine approach to develop an algorithm for the Mahalanobis distance learning. Other approach, the information-theoretic metric learning was introduced in [3].

Let us also mention [7], where the weighted Euclidean distance was considered. In our paper we extend this model to combinations of nominal and continuous data. We refer to works [9] and [2] for more details and references.

The bibliography of metric learning for nominal data is not so extensive. Most of papers assume some additional structure given on nominal data and develop an algorithm for learning of appropriate metric, e.g. tree-editing metric or string-editing metric, see the survey [2].

The Hamming distance itself was considered in [12,18]. The authors defined and optimized projections from continuous features into product of binary sets with the standard Hamming distance. This differs from our context: we assume that the data already have nominal (not only binary) features.

The general space of nominal-continuous data was considered in [4] in context of non-supervised learning. This article is the direct predecessor of our work.

3   Determining the Weights

We consider the data $\mathbb{X} = \{ \mathbf{X}_1, \ldots, \mathbf{X}_M \}$ of M records. Each record consists of two parts $\mathbf{X}_i = (X_i, Y_i)$, where $X_i = (x_i^1, \ldots, x_i^n) \in \mathbb{R}^n$ are the continuous data, and $Y_i = (y_i^1, \ldots, y_i^m)$ are the nominal data, $i = 1, \ldots, M$. We assume that $\mathbb{X}$ is divided into c classes, $\mathbb{X} = C_1 \cup \cdots \cup C_c$, where $c < M$. These classes will be used for learning. The determined weights are used for classification of new records.

The Hamming metric on the set of nominal data is defined as follows:

$$\text{dist}_h(Y_1, Y_2) = \frac{1}{m} \left| \{ \beta = 1, \ldots, m \mid y_1^\beta \neq y_2^\beta \} \right| = \frac{1}{m} \sum_{\beta=1}^m \text{diff}(y_1^\beta, y_2^\beta),$$

where $\operatorname{diff}(t_1, t_2) = \begin{cases} 1 & \text{if } t_1 \neq t_2, \\ 0 & \text{if } t_1 = t_2. \end{cases}$

Introduce the weights vector: $\mathbf{W} = (W, U) = (w_1, \dots, w_n, u_1, \dots u_m)$, where $w_\alpha > 0$, $u_\beta > 0$ for $\alpha = 1, \dots, n$, $\beta = 1, \dots, m$, and assume that classes are formed with respect to the *weighted distance*:

$$\operatorname{dist}^2_{\mathbf{W}}(\mathbf{X_1}, \mathbf{X_2}) = \operatorname{dist}^2_{W,e}(X_1, X_2) + \operatorname{dist}^2_{U,h}(Y_1, Y_2)$$

$$= \sum_{\alpha=1}^{n} w_\alpha^2 (x_1^\alpha - x_2^\alpha)^2 + \left(\sum_{\beta=1}^{m} u_\beta \operatorname{diff}(y_1^\beta, y_2^\beta) \right)^2. \tag{1}$$

To determine the weights vector $\mathbf{W}$ we minimize the total intra-class squared distance:

$$H(\mathbf{W}) = \frac{1}{M^2} \sum_{k=1}^{c} \left(\sum_{\mathbf{X}_i, \mathbf{X}_j \in C_k} \operatorname{dist}^2_{\mathbf{W}}(\mathbf{X}_i, \mathbf{X}_j) \right).$$

The distance (1) and hence the objective function $H(\mathbf{W})$ are homogeneous with respect to $\mathbf{W}$. So, to workout an effective minimizing procedure, we assume that the generalized average of the weights is constant:

$$\left(\frac{1}{n+m} \left(\sum_{\alpha=1}^{n} w_\alpha^r + \sum_{\beta=1}^{m} u_\beta^r \right) \right)^{\frac{1}{r}} = 1, \quad r \in (0, 1).$$

Finally, we get the following constrained minimization problem:

$$\begin{cases} H(W, U) = \dfrac{1}{M^2} \displaystyle\sum_{k=1}^{c} \sum_{i,j \in C_k} \operatorname{dist}^2_{W,U}\big((X_i, Y_i), (X_j, Y_j)\big) \to \min, \\ \displaystyle\sum_{\alpha=1}^{n} w_\alpha^r + \sum_{\beta=1}^{m} u_\beta^r = n + m. \end{cases}$$

We will write $i, j \in C_k$ instead of $\mathbf{X}_i, \mathbf{X}_j \in C_k$ for the sake of simplicity.

To solve this problem we use the method of Lagrange multipliers. The correspondent Lagrange function is as follows:

$$\mathcal{L}(\mathbf{W}, \lambda)$$

$$= \frac{1}{M^2} \sum_{k=1}^{c} \sum_{i,j \in C_k} \operatorname{dist}^2_{W,U}\big((X_i, Y_i), (X_j, Y_j)\big) - \lambda \left(\sum_{\alpha=1}^{n} w_\alpha^r + \sum_{\beta=1}^{m} u_\beta^r - (m+n) \right)$$

$$= \frac{1}{M^2} \sum_{\alpha=1}^{n} w_\alpha^2 \sum_{k=1}^{c} \sum_{i,j \in C_k} (x_\alpha^i - x_\alpha^j)^2 + \frac{1}{M^2} \sum_{k=1}^{c} \sum_{i,j \in C_k} \left(\sum_{\beta=1}^{m} u_\beta \operatorname{diff}(y_\beta^i, y_\beta^j) \right)^2$$

$$- \lambda \left(\sum_{\alpha=1}^{n} w_\alpha^r + \sum_{\beta=1}^{m} u_\beta^r - (m+n) \right).$$

The optimal weight vector is determined from the following conditions:

$$\frac{\partial \mathcal{L}_p(\mathbf{W}, \lambda)}{\partial w^\alpha} = \frac{\partial \mathcal{L}_p(\mathbf{W}, \lambda)}{\partial u^\beta} = \frac{\partial \mathcal{L}_p(\mathbf{W}, \lambda)}{\partial \lambda} = 0, \tag{2}$$

where $\alpha = 1, \ldots, n$, $\beta = 1, \ldots, m$.

The first equation of (2) yields:

$$\frac{2}{M^2} w_\alpha \sum_{k=1}^{c} \sum_{i,j \in C_k} (x_\alpha^i - x_\alpha^j)^2 = \lambda r w_\alpha^{r-1},$$

therefore

$$w_\alpha = \Lambda_r s_\alpha, \tag{3}$$

where

$$\Lambda_r = \left(\frac{\lambda r}{2}\right)^{-\frac{1}{2-r}}, \tag{4}$$

$$s_\alpha = \left(\frac{1}{M^2} \sum_{k=1}^{c} \sum_{i,j \in C_k} (x_\alpha^i - x_\alpha^j)^2\right)^{-\frac{1}{2-r}}. \tag{5}$$

In a similar way, the second equation of (2) implies

$$u_\beta = \Lambda_r z_\beta, \tag{6}$$

where Λ_r is defined in (4), and z_β satisfies the following equation $z_\beta^{r-1} = \sum_{\gamma=1}^{m} A_{\beta\gamma} z_\gamma$, where

$$A_{\beta\gamma} = \frac{1}{M^2} \sum_{k=1}^{c} \sum_{i,j \in C_k} \text{diff}(y_\beta^i, y_\beta^j) \, \text{diff}(y_\gamma^i, y_\gamma^j). \tag{7}$$

The third equation of (2) is $\sum_{\alpha=1}^{n} w_\alpha^r + \sum_{\beta=1}^{m} u_\beta^r = n + m$. Substituting into it (3) and (6) and eliminating Λ_r, one obtains

$$\Lambda_r = \left(\frac{\sum_{\alpha=1}^{n} s_\alpha^r + \sum_{\beta=1}^{m} z_\beta^r}{n + m}\right)^{-\frac{1}{r}}. \tag{8}$$

Following [4] we propose a relaxation iterative method for z_β:

$$z_{\beta,\text{next}} = z_\beta - \tau \left(z_\beta^{r-1} - \sum_{\gamma=1}^{m} A_{\beta\gamma} z_\gamma\right), \tag{9}$$

here τ is a relaxation parameter.

The above considerations are summarized in the Algorithm 3.1.

Remark 1. One can show that for small r every feature has a similar contribution to the total intra-class squared distance. For greater r contribution of each feature is proportional to the value of $s_\alpha(r)$ for continuous features and $z_\beta(r)$ for nominal ones.

Algorithm 3.1. Determining of the metric weights

Require: $\mathbb{X} = \{\mathbf{X}_1, \ldots, \mathbf{X}_M\}$ is the set of records, $C_1, \ldots, C_c$ - the set of classes
Ensure: $\mathbf{W} = (W, U)$ is the optimal weights vector
  Compute s_α, $\alpha = 1, \ldots, n$ with (5)
  Compute matrix A with (7)
  Choose initial z vector as $\mathbf{z} = (1, \ldots, 1)$
  while $\|z_{\text{next}} - z\| > \varepsilon$ **do**
    Compute z_{next} with (9)
  end while
  Compute Λ_r with (8)
  Compute w_α with (3) for $\alpha = 1, \ldots, n$
  Compute u_β with (6) for $\beta = 1, \ldots, m$
  return (W, U)

---

4   Numerical Experiments

To illustrate the concept a few utilities in C, R and Perl have been created. The code is available as a project on Gitlab at https://gitlab.com/adenisiuk/weightedhamming.

We performed tests for values $r = 0.05$, 0.15, 0.35, 0.55, 0.75, and 0.95.

We consider 2 datasets from the UCI Machine Learning Repository [1]: The Australian Credit Approval and The Heart Disease. We tested algorithm on artificial dataset as well.

All the tested datasets were splitted into train (80%) and test (20%) parts.

The purpose of our test is to show that the algorithm improves the standard KNN classification with non-weighted metric. However, we performed also comparison with two powerful classifiers: random forest and SVM. Implementations of these algorithms in R were used in experiments: [10,11].

The continuous data were normalized before testing the KNN algorithm with non-weighted metric and the SVM classifier.

To compare performance of algorithms we used the area under the ROC curve (AUC) as a measure. It is known that AUC is a suitable measure of binary data classification efficiency [5]. To calculate the AUC we used the R implementation [13].

4.1   Australian Credit Approval

The Australian Credit Approval dataset has 6 continuous, 8 nominal attributes, 690 records, and 2 decision categories. The results of experiments are presented in the Table 1. One can see that our algorithm overperforms the standard KNN and its performance is comparable to the random forest and SVM.

Two continuous features in this dataset have values that are bigger than others. The corresponding weights computed by our algorithm are very small and recompense big variance of these two features.

Table 1. Numerical experiments for the *Australian Credit Approval* dataset.

Algorithm	AUC	$H(r)$
Weighted KNN, $r = 0.05$	0.942	166.53
Weighted KNN, $r = 0.15$	0.942	103.93
Weighted KNN, $r = 0.35$	0.938	52.12
Weighted KNN, $r = 0.55$	0.947	31.98
Weighted KNN, $r = 0.75$	0.947	21.15
Weighted KNN, $r = 0.95$	0.942	13.66
Unweighted KNN, normalized data	0.925	
Random forest	0.949	
Support Vector Machine	0.941	

Table 2. The results of numerical experiments for the *Heart Disease* data set.

Algorithm	AUC	$H(r)$
Weighted KNN, $r = 0.05$	0.966	67.56
Weighted KNN, $r = 0.15$	0.969	54.62
Weighted KNN, $r = 0.35$	0.966	36.41
Weighted KNN, $r = 0.55$	0.974	25.68
Weighted KNN, $r = 0.75$	0.979	19.21
Weighted KNN, $r = 0.95$	0.946	14.55
Unweighted KNN, normalized data	0.953	
Random forest	0.941	
Support Vector Machine	0.966	

4.2   Heart Disease

The Heart Disease dataset has 6 continuous, 7 nominal attributes, 370 records, and 2 decision categories. The results are presented in the Table 2. We can see that for most values of r our algorithm overperforms the standard KNN. For $r = 0.66$, $r = 0.55$, and $r = 0.75$ our algorithm overperforms all the tested classifiers. Contrary to the previous dataset, we see that the maximum performance corresponds to middle values of r.

4.3   Artificial Dataset

We have tested our algorithm on artificial dataset as well. The artificial dataset was constructed as follows. Each record has 6 continuous and 6 nominal features. The cardinalities of nominal domains were arbitrary chosen as 2, 12, 3, 13, 4, 14. We use the weighted metric (1) with the following weights: 2500, 1010, 1.5, 5.1, 0.001, 0.0001 for the continuous part and the same for the nominal part. Two classes are two balls with random centers in this metric space. The radius of each ball is equal to $1/\sqrt{1.75}$ of distance between the centers. Analyzed data, 1000 for each class were randomly chosen from corresponding balls.

The Table 3 contains the average rate for 10 generated datasets.

One can see that our algorithm overperforms the standard KNN by 1% and reaches almost 100% performance.

Table 3. The results of numerical experiments for the *Artificial* data set.

Algorithm	AUC	$H(r)$
Weighted KNN, $r = 0.05$	0.997	10213.74
Weighted KNN, $r = 0.15$	0.997	1043.18
Weighted KNN, $r = 0.35$	0.997	127.58
Weighted KNN, $r = 0.55$	0.996	49.96
Weighted KNN, $r = 0.75$	0.995	26.68
Weighted KNN, $r = 0.95$	0.995	13.96
Unweighted KNN, normalized data	0.990	
Random forest	0.999	
Support Vector Machine, normalized data	0.990	

Let us however make a remark concerning the weights of the metric (1). Our algorithm discovers stably the continuous part. That means that discovered weights are proportional to the initial ones. But the nominal weights do not reflect initial weights. There can be two reasons of this phenomen. The first one: our way of modelling impact of nominal weights is not reliable. The second one, that is seemed to be more credible: the Hamming distance is too weak to reflect relations between the nominal data.

5   Conclusion and Future Work

In this article we consider modeling of nominal-continuous data. We used the weighted Euclidean metric on the continuous part and the weighted Hamming metric on the nominal one. A new method for automatic weights detection based on minimizing the total inner-class squared distance is proposed. The detected metric then was used in KNN classification.

Numerical experiments on real and artificial data show that our approach can improve the standard KNN classification algorithm and achieve the AUC performance of such sophisticated classification algorithms as random forest or support vector machine classifier, preserving simplicity of the standard KNN.

Summarizing: the proposed method is an interesting proposition for the classification problem. Moreover, as it was mentioned at the end of Sect. 4, it could get further improvement. Specifically, we plan to consider alternatives to the standard Hamming metric for the nominal part.

Besides, the discovered metric can be used in other algorithms for analysis of nominal-continuous data that are based on similarity.

These issues will be explored and presented in the future.

References

1. Asuncion, A., Newman, D.J.: UCI machine learning repository (2007). http://www.ics.uci.edu/~mlearn/MLRepository.html
2. Bellet, A., Habrard, A., Sebban, M.: Metric learning. Springer Cham (2015). https://doi.org/10.1007/978-3-031-01572-4
3. Davis, J.V., Kulis, B., Jain, P., Sra, S., Dhillon, I.S.: Information-theoretic metric learning. In: Proceedings of the 24th International Conference on Machine Learning, ICML 2007, pp. 209–216. Association for Computing Machinery, New York (2007). https://doi.org/10.1145/1273496.1273523
4. Denisiuk, A., Grabowski, M.: Embedding of the hamming space into a sphere with weighted quadrance metric and c-means clustering of nominal-continuous data. Intell. Data Anal. **22**(6), 1297001314 (2018). https://doi.org/10.3233/IDA-173645
5. Fawcett, T.: An introduction to roc analysis. Pattern Recogn. Lett. **27**(8), 861–874 (2006). rOC Analysis in Pattern Recognition
6. Goldberger, J., Hinton, G.E., Roweis, S., Salakhutdinov, R.R.: Neighbourhood components analysis. In: Saul, L., Weiss, Y., Bottou, L. (eds.) Advances in Neural Information Processing Systems, vol. 17, pp. 513–520. MIT Press (2004)
7. Karayiannis, N.B., Randolph-Gips, M.M.: Non-euclidean c-means clustering algorithms. Intell. Data Anal. **7**(5), 405–425 (2003)
8. Kaufman, L., Rousseeuw, P.J.: Finding Groups in Data: An Introduction to Cluster Analysis, Wiley Series in Probability and Statistics, vol. 344. John Wiley (2008). https://doi.org/10.1002/9780470316801
9. Kulis, B.: Metric learning: a survey. Found. Trends Mach. Learn. **5**(4), 287–364 (2013). https://doi.org/10.1561/2200000019
10. Liaw, A., Wiener, M.: Classification and regression by randomforest. R News **2**(3), 18–22 (2002)
11. Meyer, D., Dimitriadou, E., Hornik, K., Weingessel, A., Leisch, F.: e1071: Misc Functions of the Department of Statistics, Probability Theory Group (Formerly: E1071), TU Wien (2022)
12. Norouzi, M., Fleet, D.J., Salakhutdinov, R.R.: Hamming distance metric learning. In: Pereira, F., Burges, C., Bottou, L., Weinberger, K. (eds.) Advances in Neural Information Processing Systems, vol. 25, pp. 1061–1069. Curran Associates, Inc. (2012)
13. Robin, X., et al.: proc: an open-source package for r and s+ to analyze and compare roc curves. BMC Bioinform. **12**, 77 (2011)
14. Schultz, M., Joachims, T.: Learning a distance metric from relative comparisons. In: Thrun, S., Saul, L., Schölkopf, B. (eds.) Advances in Neural Information Processing Systems, vol. 16, pp. 41–48. MIT Press (2003)
15. Shi, Y., Bellet, A., Sha, F.: Sparse compositional metric learning. In: Proceedings of the AAAI Conference on Artificial Intelligence 28(1) (June 2014). https://doi.org/10.1609/aaai.v28i1.8968
16. Weinberger, K.Q., Saul, L.K.: Distance metric learning for large margin nearest neighbor classification. J. Mach. Learn. Res. **10**(9), 207–244 (2009)
17. Xing, E., Jordan, M., Russell, S.J., Ng, A.: Distance metric learning with application to clustering with side-information. In: Becker, S., Thrun, S., Obermayer, K. (eds.) Advances in Neural Information Processing Systems, vol. 15, pp. 521–528. MIT Press (2002)
18. Zhai, D., et al.: Parametric local multiview hamming distance metric learning. Pattern Recogn. **75**, 250–262 (2018). https://doi.org/10.1016/j.patcog.2017.06.018

Application of Genetic Algorithm to Load Balancing in Networks with a Homogeneous Traffic Flow

Marek Bolanowski[1]([✉])[iD], Alicja Gerka[2][iD], Andrzej Paszkiewicz[1][iD],
Maria Ganzha[3][iD], and Marcin Paprzycki[3][iD]

[1] Rzeszów University of Technology, Rzeszów, Poland
{marekb,andrzejp}@prz.edu.pl
[2] Rzeszów, Poland
alicja.gerka@op.pl
[3] Systems Research Institute Polish Academy of Sciences, Warsaw, Poland
{maria.ganzha,paprzyck}@ibspan.waw.pl

Abstract. The concept of extended cloud requires efficient network infrastructure to support ecosystems reaching form the edge to the cloud(s). Standard network load balancing delivers static solutions that are insufficient for the extended clouds, where network loads change often. To address this issue, a genetic algorithm based load optimizer is proposed and implemented. Next, its performance is experimentally evaluated and it is shown that it outperforms other existing solutions.

Keywords: extended cloud · computer network · load balancing · SDN · routing · IoT · self-adapting networks

1 Introduction

Today, typical "data processing systems" consist of "distributed data sources" and (a) "central cloud(s)". However, the advent of Internet of Things forces changes, since the "decision loop", from the "sensor(s)" to the cloud and back to the "actuator(s)", may take too long for (near-)real time applications. This problem materializes in, so called, *Extended Cloud* (EC). The EC encompasses highly heterogeneous hardware, which is very often managed using Software Defined Networking (SDN) architecture. One of the keys to effective management of SDN network is load balancing achieved, inter alia, by elimination of overloaded links, through dynamic adaptation of the routing policy. This requires an "arbitrator", which collects information about communication requests and dynamically manages routing [14]. This, in turn, allows balancing loads in the aggregation layer, connecting EC elements, including IoT and control devices [17].

Separately, note that classic load balancing algorithms, when applied to the SDN-based networks, are NP-hard [20]. Therefore, a genetic algorithm-based approach to SDN network load balancing (the SDNGALB algorithm), is being

© The Author(s), under exclusive license to Springer Nature Switzerland AG 2023
J. Mikyška et al. (Eds.): ICCS 2023, LNCS 14074, pp. 314–321, 2023.
https://doi.org/10.1007/978-3-031-36021-3_32

proposed [5,10,12,13]. Note that optimization of weights, in MPLS and OSPF networks, is a separate research area. However, it is directly related to the topic discussed here and it is also NP-hard [7,9]. Many works proposed heuristic algorithms to optimize the link weights in the OSFP protocol, to minimize the maximum load of the links [15,16]. Overall, based on a comprehensive analysis of works related to use of standard, and metaheuristics-based, approaches to network load balancing, it can be stated that: (I) found solutions are focused mainly on use of static values of weights for communication links. This results in, temporarily optimal, but static, connection structure, and leaves an open research gap. (II) Special attention should be paid to the possibility of applying the developed solutions in the environment consisting of real network devices. To address found limitations, and to deliver solution applicable to real-world ECs, a genetic algorithm, with high implementation potential, is proposed.

2 Problem Formulation and Proposed Solution

In what follows, computer network will be represented by a directed graph $G(N, E)$, where N is a set of nodes, representing network devices, and E is a set of edges representing network links. Each edge $e_{ij} \in E$ is assigned a weight w_{ij}, the modification of which will affect the current shaping of the routing policy. Moreover, the following assumptions have been made: (1) Communication channels, represented as $e_{ij} \in E$, have the same bandwidth; (2) $G(N, E)$ is a directed graph (capturing asymmetry of flows); (3) Network switches, routers and intermediary nodes are treated as "identical network devices" because, from the point of view of SDN network control, their distinction is irrelevant [18].

Network topology is represented by a graph adjacency matrix $G(N, E)$, denoted as M, with size $N \times N$. For a connection between two nodes i and j, value $e_{ij} = 1$ is assigned, while $e_{ij} = 0$ otherwise. Note that matrix M is not symmetric. To optimize the routing of flows, weights w_{ij} are assigned to the transmission channel. Here, w_{ij} are natural numbers from the range $(1, v)$, where v can take any value. Weight matrices W have size $N \times N$. The weighted adjacency matrix M_W is determined as the Hadamard product [19] of matrices M and W.

$$M_W = M \bullet W = \begin{pmatrix} w_{11} \cdot e_{11} & \cdots & w_{1N} \cdot e_{1N} \\ \vdots & \ddots & \vdots \\ w_{N1} \cdot e_{N1} & \cdots & w_{NN} \cdot e_{NN} \end{pmatrix} \tag{1}$$

For vertex pairs (s, d), where $s, d \in N$, homogeneous traffic flow f specifies requests to transmit information, as represented by a flow matrix F_{sd}.

$$F_{sd} = \begin{pmatrix} p_{11} & \cdots & p_{1N} \\ \vdots & \ddots & \vdots \\ p_{N1} & \cdots & p_{NN} \end{pmatrix} \tag{2}$$

where: $s = d \rightarrow p_{sd} = 0$; $p_{sd} = m \cdot f$, $m \in \mathbb{N}$. Homogeneous flow f corresponds to the granularity of flows in the network, as is the case with queues, e.g. in the

Ethernet network ($f = 64\,\text{kB}$). Total flow p_{sd}, is therefore defined as a multiple of the base flow f. When, a unit value f is assumed, then $p_{sd} = m$. For example, for channels 100 Mb/s and granularity $f = 64\text{kb}$, for any edge $\max m = 1563$. In what follows, such network will be named a *network with homogeneous flow structure*. The flow matrix can change over the life time of the network. The values of the elements of this matrix can also be predicted in advance [8]. The matrix L determines the current link load e_{ij} of the network topology. Here, it is assumed that it will be systematically modified, as a result of the modification of weights in the matrix W and the distribution of flows defined in the matrix F_{sd}. The path between the vertices s, d, for a given flow p_{sd}, will be determined using the Dijkstra algorithm [6]. However, other algorithms can also be used. Thus, the value for l_{ij} is determined as the sum of flows p_{sd} passing through the edge e_{ij}. Taking into account the need of dynamic control of link weights, the problem of network load balancing becomes: seeking a set of link weights W, for which the maximum number of flows passing through the "busiest edge" in the network has been minimized. Therefore, the problem can be formulated as:

$$\min\left(\max\left(l_{ij}\right)\right) \tag{3}$$

Note that, in the algorithm, the load matrix L is represented as a load vector $VL = [l_{11}, l_{12}, ..., l_{1N}, ..., l_{N1}, l_{N2}, ..., l_{NN}]^T$. As noted, the problem of balancing loads of links in the network is NP-hard. Hence, the SDN Genetic Algorithm Load Balancer (SDNGALB) is proposed. It is characterized by low computational complexity; allowing implementation of the balancing algorithm, and actual deployment in production systems.

2.1   SDNGALB Algorithm Description

Initially, values of elements w_{ij} are randomly populated, with natural numbers from the range $(1, v)$. In what follows, $\max v = 9$ was used. However, for very large networks, with high connectivity, a larger range of weights may be needed. However, this must be determined experimentally, or based on the designers' intuition. The following decisions outline the design of SDNGALB (Fig. 1).

1. The chromosome is the weight list VW, of individual network links, obtained from matrix M_W, according to the rule: if for any $x, y \in (1, N)$, $w_{xy} \cdot e_{xy} = 0$ weight is omitted; if $w_{xy} \cdot e_{xy} > 0$ the weight is added to the list (the length of the chromosome is equal to the number of edges in the network graph).
2. The initial chromosomes are randomly generated, from range $(1, v)$. The size of the population is selected experimentally, and is denoted as n.
3. The fitness function (FF) is calculated using formula (3). The FF algorithm is presented, in the form of a pseudocode in Fig. 1.
4. Individuals are ranked on the basis of the values of their fitness function.
5. Pairs of individuals, arranged according to the quality of adaptation, are crossed with each other using the standard one-point method – with the crossing point selected randomly.
6. Mutation occurs with the probability determined by the parameter mp, and consists of drawing a new value (from a specified range) of any gene in the chromosome.

7. Optimization is performed until one of the following stop condition occurs: (a) Stagnation parameter (sz) is reached, i.e. number of solutions, during which the obtained results do not improve; (b) The maximum number of generations (gs) is reached.

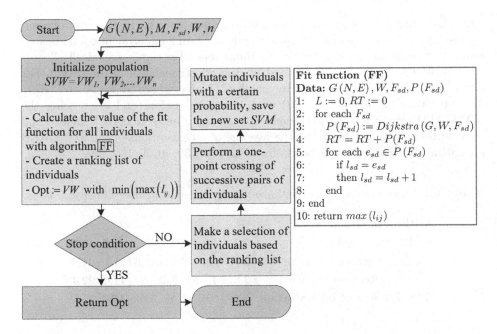

Fig. 1. Block diagram of the o SDNGALB algorithm and Fit Function (FF)

The Fitness Function (FF) algorithm requires a more detailed explanation. For each flow F_{sd}, the shortest path between the source node and the destination node is calculated (using the Dijkstra algorithm, and taking into account matrices M, W, M_W). Optimization of load distribution is achieved by manipulating the link weights in the matrix W (on the basis of which the matrix M_W is built) to eliminate network bottlenecks, by reducing the load on the most frequented edge in the network. For weights W, of chromosome VW, and flows F_{sd}, using the Dijkstra algorithm, the routing table RT is determined. It contains a number of rows equal to the number of non-zero matrix elements F_{sd} and each row contains a list of vertices to be traversed for a given flow p_{sd}, marked as $P(p_{sd})$. For example, a single row in the RT routing table, for a flow between 6 and 0 vertices $F_{6,0}$, might look like $P(F_{6,0}) \rightarrow e_{6,5}, e_{5,3}, e_{3,1}, e_{1,0}$, where $e_{u,d}$ denotes the edge connecting vertices u and d. Then, on the basis of RT, FF algorithm determines the maximum link load in the network $max(l_{ij})$.

The results of operation of SDNGALB is the set of optimal link weights W, in the given network, that allows building a balanced routing table RT, which minimizes the maximum link load in the network. Note that this algorithm can be triggered periodically or when network flow related threshold values are exceeded. The source code of the algorithm is available at the website [3].

3 Experimental Setup and Results

The algorithm was implemented in Python and deployed in the environment consisting of, enterprise class, network devices. During experiments, a dedicated laboratory, configured for Internet of Things related research, was used [2].

The first set of tests was focused on the effectiveness of optimization. For flows in the network represented by graph $G(N, E)$, where $|N| = 10$ and $|E| = 39$, the values of the function (3) were compared before and after using the SDNGALB algorithm. Table 1 shows the mean value of $\max(l_{ij})$, before and after optimization, calculated as the average of 10 executions of the SDNGALB algorithm, for the defined network, for $|F_{sd}| = 20; 30; 40; 50; 100; 200$.

Table 1. Average effectiveness of optimization for different number of network flows.

Arithmetic mean of 10 executions	$|F_{sd}|$					
	20	30	40	50	100	200
$\max(l_{ij})$ before optimisation	5,6	8,6	8.8	11,6	21,8	41,6
$\max(l_{ij})$ after optimisation	2,8	4	4,8	5,8	11,6	22,4
Effectiveness of optimisation	50%	53%	45%	50%	47%	56%

As can be seen, application of SDNGALB resulted in a noticeable (about 50%) reduction of flows in the most heavily loaded links in the tested network.

The next set of experiments was performed to compare the time of reaching the solution using SDNGALB with the exact algorithm (BF), searching the entire solution space. Here, 100 simulations were performed for both algorithms for a network with 4 nodes and 5 edges, and 5 defined flows. The same optimization result, expressed by the value of the function (3), was obtained by both algorithms. For the genetic algorithm the mean the time was 0.0118065 s, while for the BF algorithm the time was 4448.077889 s.

Next, the effectiveness of the proposed solution on networks with different topologies was explored. The speed of obtaining the solution (denoted as Ts, measured in seconds), for $\min(\max(l_{ij}))$ is reported. In the first phase, 1000 simulations were performed, for six different network topologies, in which the initial flows were randomly generated. The characteristics of the networks Net, used in the study, is presented in Table 2, where the connectivity parameter Cn should be understood as the percentage ratio of the number of edges in the tested network to the number of edges in the network with all possible connections.

For experiments reported in Table 2, flows were randomly generated for each individual simulation. The SDNGALB parameters were: mutation probability:

Table 2. Parameters of networks used in the research and average time of reaching the solution by SDNGALB algorithm in seconds

Net	n4e5	n5e11	n6e15	n10e39	n25e219	n50e872
$|N|$	4	5	6	10	25	50
$|E|$	5	11	15	39	219	872
Cn	31%	44%	42%	39%	35%	35%
v	1-5	1-5	1-5	1-9	1-9	1-9
$|F_{sd}|$	5	10	15	20	45	100
Ts	0.012	0.021	0.034	0.093	1.091	10.264

$mp = 10\%$; population size: $n = 50$; generation size: $gs = 500$; stagnation: $sz = 100$. Here, as the network size increases, i.e. from 4 nodes and 5 connections to 50 nodes and 872 connections, the solution time remains within an acceptable range (10s max). Therefore, it can be stipulated that network optimization, based on the SDNGALB algorithm, can be deployed in production network systems.

The final set of experiments compared the performance of the SDNGALB algorithm with two alternative optimizers. The first one – Ant Colony Optimization Load Balancer (ACOLB [11]) uses ant colony optimization to determine optimal routes between nodes. The second one – Dijkstra's Shortest Path Algorithm (DSPA, [4]) is based on the identification of the shortest paths between given nodes, using Djkstra's algorithm, under the assumption that link weights are randomized (using values from a given range) and are not modified later. Here, DSPA did not optimize link weights and served only as a baseline for the execution time length. Experiments were performed on the n10e39 network (with 10 nodes, 39 edges) with the assumption that $|F_{sd}| = 20$. The performance of the ACOLB algorithm was tested with different parameter values (number of ants: 1, 5, 10, 25, 50, 100, 500). For each combination of values for ACOLB, 100 simulations have been run, and the average results of execution time in second (Ex) and $\overline{\max(l_{ij})}$ are reported. The tested algorithms obtained the following results: SDNGLAB: Ex=0,09263 and $\overline{\max(l_{ij})}$=2,792; ACOLB: Ex=0,02229 and $\overline{\max(l_{ij})}$= 31,392, DSPA: Ex=0,00546 and $\overline{\max(l_{ij})}$=5,110. Therefore, it can be concluded that the proposed solution to optimization of link weights, in the network during routing, makes it possible to achieve nearly twice the lower maximum link load in the network, as compared to the performance of the DSPA algorithm. In contrast, the ACOLB does not deliver satisfactory optimization vis-a-vis the proposed solution.

4   Concluding Remarks

In this work the need for efficient SDN network flow optimization has been addressed by means of the dedicated genetic algorithm. The overarching goal was to deliver efficient infrastructure for extended cloud infrastructures, where resources, typically realized as services and microservices, can be highly dispersed. The discussed approach is fast, which should allow to quickly modify the routing table, in response to changing traffic patterns. Results obtained during tests are very encouraging, concerning both the speed and the quality of optimization. Additional details about the approach, including extensive literature review can be found in [1]. As part of further work, (1) scalability of the proposed algorithm will be tested for large, highly distributed, networks; (2) algorithm will be adapted to modify the physical topology of MESH networks; and, (3) possibility of automatic, adaptive tuning of optimizer parameters, using machine learning techniques will be explored.

Acknowledgements. Work of Marek Bolanowski and Andrzej Paszkiewicz is financed by the Minister of Education and Science of the Republic of Poland within the "Regional Initiative of Excellence" program for years 2019-2023. Project number 027/RID/2018/19, amount granted 11 999 900 PLN. Work of Maria Ganzha and Marcin Paprzycki was funded in part by the European Commission, under the Horizon Europe project ASSIST-IoT, grant number 957258.

References

1. https://arxiv.org/abs/2304.09313
2. Research stand IoE. https://zsz.prz.edu.pl/en/research-stand-ioe/about. Accessed: 2023-01-02
3. SDNGALB source code. https://bolanowski.v.prz.edu.pl/download. Accessed: 2023-01-02
4. Babayigit, B., Ulu, B.: Load balancing on software defined networks. In: 2018 2nd International Symposium on Multidisciplinary Studies and Innovative Technologies (ISMSIT), pp. 1–4. IEEE, Ankara (Oct 2018). https://doi.org/10.1109/ISMSIT.2018.8567070
5. Chen, Y.T., Li, C.Y., Wang, K.: A Fast Converging Mechanism for Load Balancing among SDN Multiple Controllers. In: 2018 IEEE Symposium on Computers and Communications (ISCC), pp. 00682–00687. IEEE, Natal (Jun 2018). https://doi.org/10.1109/ISCC.2018.8538552
6. Dijkstra, E.W.: A note on two problems in connexion with graphs. Numer. Math. **1**(1), 269–271 (1959). https://doi.org/10.1007/BF01386390
7. Ericsson, M., Resende, M., Pardalos, P.: A Genetic Algorithm for the Weight Setting Problem in OSPF Routing. J. Comb. Optim. **6**(3), 299–333 (2002). https://doi.org/10.1023/A:1014852026591
8. Gao, K., et al.: Predicting traffic demand matrix by considering inter-flow correlations. In: IEEE INFOCOM 2020 - IEEE Conference on Computer Communications Workshops (INFOCOM WKSHPS), pp. 165–170 (2020). https://doi.org/10.1109/INFOCOMWKSHPS50562.2020.9163001
9. Jain, A., Chaudhari, N.S.: Genetic algorithm for optimizing network load balance in MPLS network. In: 2012 Fourth International Conference on Computational Intelligence and Communication Networks, pp. 122–126. IEEE, Mathura, Uttar Pradesh, India (Nov 2012). https://doi.org/10.1109/CICN.2012.119
10. Jain, P., Sharma, S.K.: A systematic review of nature inspired load balancing algorithm in heterogeneous cloud computing environment. In: 2017 Conference

on Information and Communication Technology (CICT), pp. 1–7. IEEE, Gwalior, India (Nov 2017). https://doi.org/10.1109/INFOCOMTECH.2017.8340645

11. Keskinturk, T., Yildirim, M.B., Barut, M.: An ant colony optimization algorithm for load balancing in parallel machines with sequence-dependent setup times. Comput. Oper. Res. **39**(6), 1225–1235 (2012). https://doi.org/10.1016/j.cor.2010.12.003

12. Li, G., Wang, X., Zhang, Z.: SDN-based load balancing scheme for multi-controller deployment. IEEE Access **7**, 39612–39622 (2019). https://doi.org/10.1109/ACCESS.2019.2906683

13. Mahlab, U., et al.: Entropy-based load-balancing for software-defined elastic optical networks. In: 2017 19th International Conference on Transparent Optical Networks (ICTON), pp. 1–4. IEEE, Girona, Spain (Jul 2017). https://doi.org/10.1109/ICTON.2017.8024847

14. Mazur, D., Paszkiewicz, A., Bolanowski, M., Budzik, G., Oleksy, M.: Analysis of possible SDN use in the rapid prototyping processas part of the Industry 4.0. Bull. Polish Acad. Sci. Tech. Sci. **67**(1), 21–30 (2019). https://doi.org/10.24425/BPAS.2019.127334

15. Mohiuddin, M.A., Khan, S.A., Engelbrecht, A.P.: Fuzzy particle swarm optimization algorithms for the open shortest path first weight setting problem. Appl. Intell. **45**(3), 598–621 (2016). https://doi.org/10.1007/s10489-016-0776-0

16. Mulyana, E., Killat, U.: An Alternative Genetic Algorithm to Optimize OSPF Weights. Internet Traffic Engineering and Traffic Management, pp. 186–192 (Jul 2002)

17. Paszkiewicz, A., Bolanowski, M., Budzik, G., Przeszłowski, L., Oleksy, M.: Process of creating an integrated design and manufacturing environment as part of the structure of industry 4.0. Processes **8**(9), 1019 (2020). https://doi.org/10.3390/pr8091019

18. Smiler. S, K.: OpenFlow cookbook. Quick answers to common problems, Packt Publishing, Birmingham Mumbai, 1. publ edn. (2015)

19. Styan, G.P.: Hadamard products and multivariate statistical analysis. Linear Algebra Appl. **6**, 217–240 (1973). https://doi.org/10.1016/0024-3795(73)90023-2

20. Wang, H., Xu, H., Huang, L., Wang, J., Yang, X.: Load-balancing routing in software defined networks with multiple controllers. Comput. Netw. **141**, 82–91 (2018). https://doi.org/10.1016/j.comnet.2018.05.012

Automatic Structuring of Topics for Natural Language Generation in Community Question Answering in Programming Domain

Lyudmila Rvanova[1,2] and Sergey Kovalchuk[1]

[1] ITMO University, Saint Petersburg, Russia
{alfekka,kovalchuk}@itmo.ru
[2] AIRI, Moscow, Russia

Abstract. The present article describes the methodology for the automatic generation of responses on Stack Overflow using GPT-Neo. Specifically, the formation of a dataset and the selection of appropriate samples for experimentation are expounded upon. Comparisons of the quality of generation for various topics, obtained using thematic modeling of the titles of questions and tags, were carried out. In the absence of consideration of the structures and themes of texts, it can be difficult to train models, so the question is being investigated whether thematic modeling of questions can help in solving the problem. Fine-tuning of GPT-neo for each topic is undertaken as a part of experimental process.

Keywords: Stack Overflow · Question answering · Text generation · Topic modeling

1 Introduction

Generative neural networks are currently widely used and are being actively researched. It is interesting to use generative neural systems in the task of automatically answering questions. Our task is to study the application of generative neural networks for automatic generation of Stack Overflow answers. The complexity of this task lies in the fact that in both answers and questions there are several domains at once: code, natural language, and images.

Currently, generative neural networks, such as GPT-3 [1], are good at general questions, including some factual ones. GPT-3 is an autoregressive transformer model with 175 billion parameters. It based on GPT-2 [2] architecture including pre-normalization, reversible tokenization. This model is different from GPT-2 with their sparse attention patterns in the layers of transformer.
T5 made a breakthrough in multitasking. T5 achieved state-of-the-art result in several NLP tasks, including text generation. This is seq-to-seq transformer pre-trained on a large text corpus.

J. Mikyška et al. (Eds.): ICCS 2023, LNCS 14074, pp. 322–329, 2023.
https://doi.org/10.1007/978-3-031-36021-3_33

Finally, nowadays we have ChatGPT which handles multi-domain responses to questions by being able to generate code along with natural language. Also, this solution has the ability to remember the context and correct errors. At the moment there is no article explaining how the solution works and there is no open source code. Unfortunately, we cannot be sure of the accuracy of the answers of ChatGPT.

In March 2023, OpenAI released the GPT-4 [3] model. According to the OpenAI press release, GPT-4 scores 40% higher than the latest GPT-3 on internal adversarial factuality evaluations by OpenAI. Although this advantage is significant, the developers from Open AI confirm that GPT-4 does not completely solve the problem of generating inappropriate code and inaccurate information. We're investigating how the topic of a question impacts answer quality. Our method uses thematic modeling and fine-tuning of models for each theme, improving accuracy of answers. By identifying relevant topics within a question, we can generate more helpful responses. It is also important to take into account structural differences in different texts within the same domain. For questions of different topics, different response structures are assumed, and if this is not taken into account, it can lead to problems in training the model.

2   Related Works

The paper [4] explores the possibility of identifying low-quality questions on Stack Overflow for their automatic closure. Previous work has explored lexical, voice-based, style-based features. They also proposed a framework that collects semantic information about questions using transformers and explores information from tags and questions using a convolutional graph neural network. This method beat the stat-of-the-art solutions.

In paper [1] few-shot learning for generative neural networks was studied. Most of the SOTA results in text generation are obtained by retraining and fine tuning on thousands and hundreds of thousands of examples. Training on a small number of examples usually cannot beat the results of such a tuning, but when scaled and using a large language model, it can be successful. In this work, authors used GPT-3 with 175 billion parameters.

Researches in this [5] article states that Open-Domain Question Answering achieved good results with combining document-level retrievers with text generation. This approach also called GENQA can affectively answer both factoid and non-factoid questions. They introduce GEN-TYDIQA, an extension of the TyDiQA dataset with well-formed and complete answers for Arabic, Bengali, English, Japanese, and Russian. They translate question to one of appropriate languages, uses retriever, monolingual answer detection and aggreagtion of answer. After that they uses cross-lingual GenQA.

3   Data

We used open sourced Stack Overflow dataset dumps. There are only two files that we used – datadump of Stack Overflow posts dated December 7, 2022 and dump of Stack Overflow comments dated December 6, 2022. The oldest entries date back to 2008. XML files have been converted to a suitable format for us. There are 57721551 records for file with posts and 86754114 for comments.

Our data has been filtered from unnapropriate domains for this experiment, leaving only natural language. For this we simply deleted answers and questions with tags of code and pictures. In our experiments we used only newest questions over the past six months and approved or top-rated answers for them. As the result, we have 25668 questions with answers.

4   Methods

4.1   Thematic Modeling

For our experiments we conducted thematic modeling of questions. There are two types of thematic modeling:

1. **LDA** [6] (Latent Dirichlet Allocation) — generative probabilistic method. We used Tf-Idf to delete stop-words.
2. **CTM** [7] (Correlated topic models) — hierarchical model that allows us to use the correlation of latent topics. It is extension of LDA.

Also we used two types of texts for modeling: question titles and questions tags. We calculated thematic modeling for the number of topics from 1 to 15 in order to find the optimal number of topics for coherence score. To decrease number of words in vocabulary we lemmatized words. Stopwords were removed as common practice in topic modeling. We used TF-IDF for stopwords removing. For LDA we used n-grams with sizes 2 and 3. For CTM we used combination of Bag of Words and BERT base cased embeddings.

For thematic modeling we used short headings of questions.

4.2   Text Generation

In our experiments we used GPT-Neo [8] for text generation. It's GPT-2 like model trained on the Pile dataset. It is transformer-based neural network trained on task of predicting next word in the sequence. GPT-Neo uses local attention for every layer with window size of 256 tokens. We used 1.3 B configuration with 1.3 billion weights. In our experiments we used few-shot learning for inference.

5   Results

For the general dataset, we have the following metrics in the case of generation by GPT-Neo: ROUGE1, ROUGE2, ROUGEL, ROUGELsum, GoogleBLEU, average perplexity, cosine similarity.

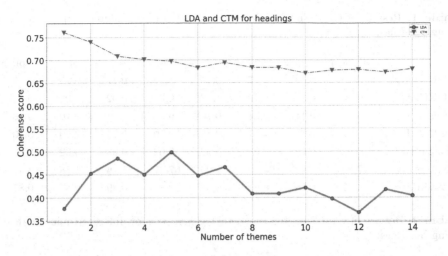

Fig. 1. Thematic modeling with LDA and CTM for headings of questions.

Headings of Questions. Optimal number of topics for CTM is 8, for LDA is 8. In this case coherence score were higher for CTM modeling than for LDA modeling (Fig. 1).

We have next keywords for topics in case of CTM modeling and 8 topics:

1. time, data, database, order, case, unrestand, process – general questions and data
2. version, token, command, project, error, build, package – system administration questions
3. description, norefferimg, alterner image, tr thread, styletextalign, table div, tr tbody – frontend
4. azure, access, token, server, api, client, application – backend and database questions
5. cloudwatch, cloudflare, attachments, collections, sftp, ms teams, kubernetes cluster – DevOps
6. english, inevitably, bcnf, pem string, justify, direct channel, subscrible channel
7. fiscal, deeply, perm strong what, immensely, source directory, looped, alt unloaded
8. classes methods, importing, jsonincludeproperties, java objectives, immensely, taget build, puzzled – java

The last three topics overlap a lot.

For each topic we sampled 2000 questions and generated answers with GPT-Neo. Results are below (Table 1). We used cosine similarity to find out similarity between texts. To measure cosinus similarity we used SentenceTransformers [9] embeddings.

In this case, there are slight fluctuations in perplexity – for questions on Java, its value is better than for system administration, databases and frontend.

Table 1. Results of GPT-Neo few-shot text generation for different topics modeled using headings.

	Topic 1	Topic 2	Topic 3	Topic 4	Topic 5	Topic 6	Topic 7	Topic 8
ROUGE1	0.16	0.16	0.17	0.15	0.15	0.17	0.15	0.16
ROUGE2	0.02	0.02	0.02	0.03	0.01	0.02	0.01	0.02
ROUGEL	0.11	0.11	0.11	0.10	0.10	0.11	0.10	0.10
ROUGELsum	0.11	0.11	0.11	0.10	0.10	0.11	0.10	0.10
Goggle BLEU	0.02	0.02	0.03	0.02	0.02	0.03	0.02	0.02
Avg perplexity	67.04	66.69	65.36	71.09	67.97	54.76	61.78	65.41
Cos similarity	0.87	0.88	0.89	0.88	0.88	0.89	0.87	0.87

Table 2. Results of GPT-Neo fine-tuned text generation for different topics modeled using headings.

	Topic 1	Topic 2	Topic 3	Topic 4	Topic 5	Topic 6	Topic 7	Topic 8
ROUGE1	0.19	0.18	0.18	0.17	0.18	0.19	0.19	0.17
ROUGE2	0.02	0.02	0.02	0.02	0.02	0.03	0.03	0.02
ROUGEL	0.12	0.13	0.12	0.12	0.12	0.13	0.13	0.11
ROUGELsum	0.14	0.14	0.14	0.13	0.14	0.15	0.15	0.13
Goggle BLEU	0.04	0.04	0.04	0.04	0.04	0.04	0.04	0.04
Avg perplexity	42.41	53.18	46.76	50.43	55.26	44.00	56.23	66.47
Cos similarity	0.91	0.91	0.90	0.91	0.91	0.92	0.91	0.91

Other metrics differ less significantly. It makes sense to do another experiment with topic modeling – this time we'll take question tags and run the simulation in the same way.

After experiments with inference of GPT-Neo we fine-tuned models on Stack-Overflow dataset. For each topic we conducted fine-tuning separately with data that was determined as belonging to this topic. We end up with a similar improvement in metrics for all topics, but the responses become more relevant to the topics. When fine-tuning on the basis of all data that is not grouped by topic, the metrics also grow, but the answers do not correspond to the topics of the questions (Table 2).

Practically everywhere except for the last topic, perplexity has decreased. For all topics except the third one, other metrics improved significantly, by 10–40% . The seventh topic showed the best increase in metrics.

Tags of Questions. Optimal number for both of CTM and LDA is 8. For tags coherence score is higher for LDA (Fig. 2)

We have next keywords for topics in case of CTM modeling and 8 topics:

1. android, python selemium, react js, excel formula, html css, my sql, asp net deployment

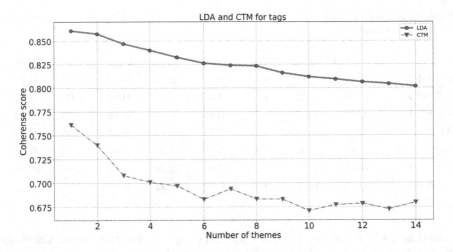

Fig. 2. Thematic modeling with LDA and CTM for tags of questions.

2. visual studio, apache flink streaming, power bidax, python django, google cloud platform, snowflake clout dataplatform, heroku
3. git, github, microsoft teams, sql web service, pycharm, kubernetes, angular, res
4. visual studio code, java android, algorithm, excel, azure, python 3x, excel pivot table
5. flutter dart, mongo db, azure devops, docker, bash, logging, apache kafka consume api
6. unity 3d, c, java androin kotlin, camunda, python algorithm, flutter, java mysql spring
7. python, powerbi, apache kafka, java, td engine, spring, apache spark, apache spark
8. javascript, react, frontend, github, android studio, postgre sql, azure devops

The last three topics overlap a lot.
For each topic we sampled 2000 questions and generated answers with GPT-Neo (Table 3). For tags we also conducted fine-tuning (Table 4).

Table 3. Results of GPT-Neo few-shot text generation for different topics modeled using tags of questions.

	Topic 1	Topic 2	Topic 3	Topic 4	Topic 5	Topic 6	Topic 7	Topic 8
ROUGE1	0.16	0.15	0.15	0.15	0.15	0.17	0.17	0.16
ROUGE2	0.02	0.02	0.01	0.01	0.01	0.02	0.02	0.02
ROUGEL	0.10	0.09	0.09	0.10	0.09	0.11	0.10	0.11
ROUGELsum	0.10	0.09	0.09	0.10	0.09	0.11	0.10	0.11
Goggle BLEU	0.02	0.02	0.02	0.03	0.02	0.03	0.02	0.02
Avg perplexity	63.22	43.40	49.15	48.43	54.76	88.7	84.32	65.52
Cos similarity	0.87	0.86	0.88	0.89	0.89	0.89	0.87	0.87

Table 4. Results of GPT-Neo fine-tuned text generation for different topics modeled using tags of questions.

	Topic 1	Topic 2	Topic 3	Topic 4	Topic 5	Topic 6	Topic 7	Topic 8
ROUGE1	0.19	0.17	0.18	0.19	0.19	0.18	0.18	0.19
ROUGE2	0.03	0.02	0.02	0.03	0.02	0.02	0.03	0.02
ROUGEL	0.13	0.12	0.12	0.13	0.12	0.12	0.13	0.12
ROUGELsum	0.15	0.13	0.14	0.15	0.14	0.14	0.14	0.15
Goggle BLEU	0.04	0.03	0.04	0.04	0.04	0.04	0.04	0.04
Avg perplexity	50.55	66.90	48.35	47.58	60.26	50.59	44.89	48.09
Cos similarity	0.91	0.91	0.92	0.92	0.92	0.92	0.91	0.91

In this cased perplexity decreased for every topic except the second. For every topic metrics improved, especially for the fourth theme.

Despite the large overlap of the last three topics, there is a significant difference in perplexity between them. Topics 2, 3 and 4 showed the best perplexity values, but showed a relatively low ROUGE values.

The perplexity fluctuations are higher than for topic modeling by question headings.

Despite the lower scores, topic modeling for question titles revealed clearer topics than modeling for tags. At the same time, differences in metrics are more significant for modeling by tags, although they are minor. The most visible are the differences in perplexity, for less specific topics the perplexity is lower, for more specialized topics it is higher.

6   Conclusion

GPT-Neo performs well even out of the box, showing good results in semantic similarity. However, for highly specialized topics, perplexity suffers. She also handles questions about software better than questions than questions about programming languages.

Additional training for each topic separately showed an improvement in quality. The answers to the questions are more qualitatively related to the topic of the question in the case of training separately than in the case of additional training on all the data mixed.

It makes sense to continue experimenting with topic modeling, since tags are set by users and may not reflect the essence of the issue, just as headings may be worded incorrectly.

Acknowledgments. This research is supported by Russian Scientific Foundation and Saint Petersburg Scientific Foundation, grant No. 23-28-10069 "Forecasting social well-being in order to optimize the functioning of the urban digital services ecosystem in St. Petersburg" (https://rscf.ru/project/23-28-10069/).

References

1. Brown, T.B., et al.: Language Models are Few-Shot Learners, In: Larochelle, H., Ranzato, M., Hadsell, R., Balcan, M.F., Lin, H. (eds.) Advances Neural Information Processing Systems, vol. 33, pp. 1877–1901. Curran Associates, Inc. (2020)
2. Radford, A., Jeffrey, W., Child, R., Luan, D., Amodei, D., Sutskever, I.: Language models are unsupervised multitask learners. Technical report, OpenAi (2019)
3. OpenAI: GPT-4 Technical Report, eprint arXiv:2303.08774 (2023)
4. Arora, U., Goyal, N., Goel, A., Sachdeva, N., Kumaraguru, P.: Ask It Right! identifying low-quality questions on community question answering services. In: International Joint Conference on Neural Networks (IJCNN), pp. 1–8, Padua, Italy (2022). https://doi.org/10.1109/IJCNN55064.2022.9892454
5. Muller, B., Soldaini, L., Koncel-Kedziorski, R., Lind, E., Moschitti, A.: Cross-lingual open-domain question answering with answer sentence generation. In: Proceedings of the 2nd Conference of the Asia-Pacific Chapter of the Association for Computational Linguistics and the 12th International Joint Conference on Natural Language Processing, vol. 1, pp. 337–353, Association for Computational Linguistics (2022)
6. Blei, D.M., Ng, A.Y., Michael I.: Latent Dirichlet allocation. Jordan J. Mach. Learn. Res. **3**, pp. 993–1022 (2003)
7. Blei, D.M., Lafferty, J.D.: Correlated topic models. In: Weiss, Y., Schölkopf, B., Platt, J., (eds.) Advances in Neural Information Processing Systems (NIPS 2005), vol. 18, pp. 147–154, MIT Press (2005)
8. Black, S., et al.: GPT-NeoX-20B: an open-source autoregressive language model. In: Proceedings of BigScience Episode #5 - Workshop on Challenges & Perspectives in Creating Large Language Models, pp. 95–136, Association for Computational Linguistics (2022). https://doi.org/10.18653/v1/2022.bigscience-1.9
9. Reimers, N., Gurevych, I.: Sentence-BERT: sentence embeddings using Siamese BERT-Networks. In: Proceedings of the 2019 Conference on Empirical Methods in Natural Language Processing and the 9th International Joint Conference on Natural Language Processing (EMNLP-IJCNLP), pp. 3982–3992, Association for Computational Linguistics. https://doi.org/10.18653/v1/D19-1410

Modelling the Interplay Between Chronic Stress and Type 2 Diabetes On-Set

Roland V. Bumbuc[1] , Vehpi Yildirim[2] , and M. Vivek Sheraton[1,3,4]([✉])

[1] Computational Science Lab, Informatics Institute, University of Amsterdam,
UvA - LAB42, Science Park 900, Amsterdam 1090 GH, The Netherlands
v.s.muniraj@uva.nl
[2] Department of Public and Occupational Health, Amsterdam UMC, Meibergdreef 9,
Amsterdam 1105 AZ, The Netherlands
[3] Center for Experimental and Molecular Medicine (CEMM), Amsterdam,
The Netherlands
[4] Amsterdam UMC, Meibergdreef 9, Amsterdam 1105 AZ, The Netherlands

Abstract. Stress has become part of the day-to-day life in the modern
world. A major pathological repercussion of chronic stress (CS) is Type 2
Diabetes (T2D). Modelling T2D as a complex biological system involves
combining under-the-skin and outside-the-skin parameters to properly
define the dynamics involved. In this study, a compartmental model is
built based on the various inter-players that constitute the hallmarks
involved in the progression of this disease. Various compartments that
constitute this model are tested in a glucose-disease progression setting
with the help of an adjacent minimal model. Temporal dynamics of the
glucose-disease progression was simulated to explore the contribution of
different model parameters to T2D onset. The model simulations reveal
CS as a critical modulator of T2D disease progression.

Keywords: Diabetes · Computational modelling · Chronic stress ·
Allostatic Load · Disease progress · *in-silico* tool

1 Introduction

Type 2 Diabetes (T2D) is a slowly progressing metabolic disease characterized
by elevated blood glucose [22]. Hyperglycemia (HG) in individuals can lead to
devastating long-term and even irreversible complications. Metabolic disorders
such as T2D are growing more common across the world, where 6.28% of the
world population have developed T2D symptoms or disease on-set by 2017 [9].
No detailed data was released since, but global forecasts already predict numbers
to reach around 8000±1500 cases of T2D cases per million population by 2040
[9]. With no apparent cure in sight, clinicians can only intervene at early stages
of the disease, through behavioural changes. However, this "chance" at disease
reversal is not fully explored and exploited in the healthcare system due to the
complexity of all the inter-players in T2D pathogenesis [22].

© The Author(s), under exclusive license to Springer Nature Switzerland AG 2023
J. Mikyška et al. (Eds.): ICCS 2023, LNCS 14074, pp. 330–338, 2023.
https://doi.org/10.1007/978-3-031-36021-3_34

CS is linked to the pathogenesis of T2D through a multitude of metabolic ways including the Central Nervous System (CNS) and partially the Peripheral Nervous System (PNS). In the pathogenesis of T2D, insulin production and sensitivity, Glucocorticoids (GC), cortisol and the Hypothalamic-Pituitary -Adrenal(HPA)-Axis play major roles. An in-silico modelling approach to filter the most significant inter-players and the feasible actions that can be taken towards minimizing T2D disease progress (DP) is imperative. Such methodology would be useful in a clinical decision making process after being carefully explored [17]. In this study, a new mechanistic computational model capable of relating each inter-player to its contribution towards T2D on-set and DP is proposed.

The Hallmarks of T2D Related to Chronic Stress

β-cells are responsible for synthesizing, storing and releasing insulin [11]. Glucose [11], Free Fatty Acids (FFAs) [7] and Glucagon-1 [21] play key roles in the process. β-cells can be generated by replication of the existing β-cells depending non-linearly on glucose concentration *in-medium* [20]. Human β-cell proliferative capacity is small and decreases with age, but when metabolic demand is high such as in obesity or during pregnancy, replication may increase [2]. β-cell mass can decrease by undergoing apoptosis (regulated cell death) or necrosis (unregulated cell death), which may be dependent on glucose concentration [20].

In a healthy person, the levels of circulating glucose are well regulated. When plasma glucose increases above 90 mgdL^{-1}, β-cells sense this and produce and secrete insulin, a hormone that triggers glucose absorption by the adipose, liver and muscle tissue, decreasing glucose in circulation. In normal circumstances this is easily kept in balance with a healthy diet [15], if no other disturbances affect the system. During T2D progression, insufficient insulin secretion and insulin resistance give rise to hyperglycemia [1].

In general, when the HPA-axis is activated, it responds by producing and releasing GC, such as cortisol. While this is a healthy and natural reaction to short-term stress, it becomes dysfunctional when the stress signal is prolonged [8], like in the case of CS. Studies show that CS causes HPA-axis dysfunction and increases GC levels [4] and suggest that there is a link between increased GCs and T2D progression, demonstrating how GC excess leads to metabolic dysfunction [3]. Exposure to stressful conditions, an imbalance in effort, psychological traumatic experiences, low socio-economic status or even higher incidence of discrimination can be such examples of stress paradigms that could trigger CS on-set [14]. Continuous recursive activation of the SNS and HPA-axis occurs during CS and can cause physiological long term consequences that may result in accumulated small disturbances signified by "allostatic load" [10]. This concept is now put in use to help operationalize CS into measurable physiological units that allow identification of the relationships between different stressor types and the pathophysiology stages of T2D [6]. There are many hallmarks that can be associated with T2D progression. In this study we suggest 5 hallmarks that empirically were included in many other papers as pro-T2D progression based on Allostatic load [5,6]. These are Insulin Resistance (IR), HG, Low Grade Inflammation (LGI), Hypercortisolism (HC) and Hyperglucagonemia (HGC). These hallmarks, are connected through complex interactions and create feedback loops, which potentiate their impact when CS comes into play.

2   Methods

2.1   Model Definitions

A summary of the model parameters and interconnected compartments is presented in SM-Figure 2 (Supplementary Material[1](SM)-Section 1.2). Computational modelling can help identify the underlying mechanisms of any phenomenon [16], like CS, which can lead to the development of new therapies [18] and interventions [17,18]. As the study CS is complex and extensive, a short review on the computational models that relate it to T2D can be found in SM-Section 1.1. To simulate the healthy and diseased state dynamics we opted to use a simpler and well established minimal model firstly developed by Topp *et al., 2000*, for which the concept of added stress was then implemented by Mohammed *et al., 2019*. This model can emulate some of the connections in our conceptual model, for which we could apply our DP calculations on. The coupled SM-Equations 2, 3 and 4 and the parameters listed in SM-Table 6 (SM-Section 1.3) were used to carry out the simulations in this study.

By simulating the dynamics of Glucose, Insulin and β-cell mass we were able to replicate a healthy a individual dynamics for the first 15 days, only to introduce forced disturbances of the system of ODEs later on with'simulated stress', and to some extent by inclusion of periodic behaviour. We used k_0 from the Compartmental model (Fig. 2) to represent food intake in the form of periodic glucose spikes, that varies within normal range in a non-CS situation and increases above 140 mgdL^{-1} for a chronically stressed individual. This is also the variable that varies between individuals being simulated, as different people have different meals and different peaks of glucose. At each meal time (5 meals per day), the individual receives a glucose peak between 100 mgdL^{-1} and 170 mgdL^{-1} for non-stress situation (larger sample). A mix of the latter with glucose peaks between 225 mgdL^{-1} and 350 mgdL^{-1} for the stress situation (smaller sample). In both cases, glucose pick values a uniformly distributed. There is no simple minimal model that can include all the hallmarks we aim to use for DP calculation, however by using the model developed by Mohammed *et al., 2019* [13] we are able to replicate at least 2 hallmarks.

2.2   Algorithm Definitions

Based on the works of Benthem *et al., 2022* [5], a methodology was built that would make use of threshold values for each of the following hallmarks

---

[1] Supplementary Material (SM) available at Github link.

which allow the calculation of Allostatic load within SM-Algorithm 2 (SM-Section 1.3). These are: Hyperglycemia, where the high Glucose levels at certain time points are monitored; Insulin resistance, by using the Homeostasis model assessment insulin resistance index (HOMA-IR) applying SM-Equation 1 [12]; Low grade inflammation, by using the measure of the output from C_{CRP} (from SM-Figure 2); Hypercortisolism, based on the output of high cortisol level from the Cortisol compartment (C_C) and Hyperglucaconemia, based on outputs from high glucagon level from Glucagon compartment (C_E). The threshold values are the values for which the max peaks for different compartments are "surpassed" at time t, considering a broken elasticity phase where some "wear and tear" is inflicted. By using SM-Algorithm 2 the calculations were ran for Hyperglycemia and Insulin Resistance. This application would correspond to links regarding the $C_G, \beta_{MD}, C_I, C_X$ compartments disregarding constants $k_9, k_{10}, k_{17}, k_{41}, k_{42}, k_{43}, k_{51}, k_{52}, k_{53}, k_{55}, k_{58}$ in the compartmental diagram in SM-Figure 2. A coupled Euler integration method (SM-Algorithm 1) was applied to solve the model SM-Equations 2, 3 and 4. By using the simulated values of Glucose and Insulin, the calculation of HOMA-IR is possible by using SM-Equation 1 (SM-Section 1.3).

Going forward, we hypothesized that each hallmark would be present in this conceptual model. To develop SM-Algorithm 2, we assume that there is a need to quantify damage to the system modelled [5]. At the healthy state, the system remains in stable steady state. Therefore, in order for a change in state to occur, successive damage (W_S) under some weight(w). Moreover, there is always some resistance and resilience to this damage [19], under repair or healing (C_S). When the damage inflicted is higher than the recovery, some threshold T is surpassed and there is some damage to the system in the form of strain (e). To represent that a certain strain value (e) in case some threshold T would be reached or surpassed, a strain event (E_s) is evaluated at time t for each a strain calculation ($S_t(t)$) for each hallmark. Where e_l is low strain, e_i is intermediate strain and e_h represents the high strain of that hallmark towards DP. To calculate the DP for T2D we used SM-Algorithm 2, where $w(e_x)$ is the weight of a certain strain towards DP and x is the equivalent to intensity of the strain (low, intermediate or high) and DP is the T2D progress estimation in % based on a cumulative sum of all counts for all hallmarks.

3 Results

The system was simulated for 100 different cases (Fig. 1, 2 and 3) where only active components were taken into consideration. We resorted to this simple but effective model to extract and test the Event-Driven approach that calculates the DP over 45 days, while the model has a fixed minimal time of one day.

To this experiment were added event-driven meal-like instances of increase in Glucose, firstly in a non-stress setup for 15 days. In Fig. 1 we see the dynamics means corresponding to SM-Equations 2 (A - The dynamics of Glucose over time), 3 (B - The dynamics of Insulin over time) and 4 (C - The dynamics of β-cell mass over time). After 15 days, a CS on-set is induced by adding only

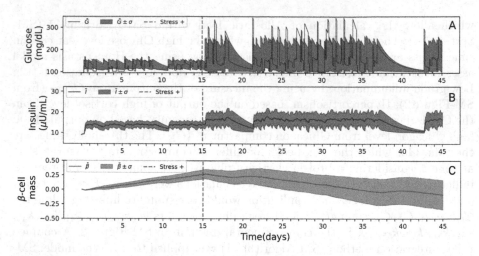

Fig. 1. Glucose(**A**), Insulin(**B**) and β-cell mean mass(**C**) dynamics simulation for 100 different cases during 45 days. The blue, purple and green bold line is the mean of the 100 different cases for Glucose, Insulin and β-cell mean mass, respectively. Each plot containing the bold line, also has an area around, representing the standard deviation between the 100 different cases relative to the mean. (A) Labeled on the y-axis, Glucose concentration in $mgdL^{-1}$ units on an interval between 100 and 310. (B) Labeled on the y-axis, Insulin concentration in $\mu U mL^{-1}$ units on an interval between 5 and 30. (C) Labeled on the y-axis, β-cell mass ratio at an interval between -1 and 1, with 0 being the baseline β-cell mass. All plots are on the same labeled x-axis, the time-span simulated in days. The red intermittent line represents the day of CS induction. The time-step between calculation of one time-point to the other in all the plots is $24/60/60 \approx 0.0066$ days.

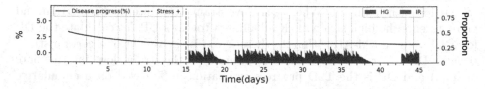

Fig. 2. DP mean simulation for 100 different cases during 45 days.

disease-like values reached after meals (a direct consequence of CS). In Fig. 2 we see the result of direct application of SM-Algorithm 2 to extract DP based on Allostatic load.

On the x-axis (left) labeled as % is the % of DP and on the x-axis (right) labeled as proportion are proportions related to contribution of Hallmark at time t for DP calculation from each hallmark. On the x-axis the time-span simulated in days.

We tested whether SM-Algorithm 2 would be dependent on the time-span of simulation and if more accurate DP would be achieved if 100 different cases were simulated for 10 days(before CS-on set), $\frac{1}{2}$ year, 1 year and 1.5 years (Fig. 3).

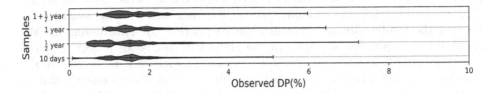

Fig. 3. Simulation of 100 different cases for different time-spans.

Simulation of 100 different cases for different time-spans. Labeled as samples, on the y-axis the different time-span samples are shown, while on the x-axis the time-span simulated in days. Note that the length of each candle shows the variation minimum and maximum peaks and, in red, the density distribution of each sample in terms of DP %. Notice that each candle has a mean DP value for each sample, shown by a vertical line inside the candle plot.

4   Discussions and Conclusions

The main research question was conceptualizing a model that could encompass the inter-players between T2D and CS and calculate the DP. The concept has been proven to work and disease progression calculation has been successfully achieved by using a simpler modelling approach. Biologically relevant results were obtained this way. In Fig. 1A and 1B, we can observe normal behaviour of Glucose and Insulin dynamics, respectively. This is hinted by the peaks, which would show how individuals can sometimes have meals richer in carbohydrates that result in higher glucose and insulin peaks. To accompany this, enclosed by the normal individual dynamics, we see normal increase in mass of β-cells in Fig. 1C. This dynamic ,in reality, has a *plateau* and in this setting in particular, represents that there is no stress to the β-cell mass (it increases). The same can be captured by our SM-Algorithm 2. Notice in Fig. 2, that DP only decreased into a normalized value and no significant strain (under small stress events) can increase this value. Insulin (Fig. 1B) also stays in the normal range of a healthy person and follows Glucose peaks.

After day 15, we emulate CS, where after each meal, the peak values of Glucose and Insulin become aberrant and switch to diseased state values, surpassing disease thresholds and creating strain on the system. This can be clearly seen in Figs. 1A and 1B, where mean Glucose can rise up to 350 mgdL^{-1} or even beyond and Insulin level reach 20 μUmL^{-1}. As per randomization of Glucose values, we see that the mean deviates from the standard deviation in some cases, this is a sign of the difference between individual cases simulated, which is crucial to have when HG states are reached for different reasons. Within this implementation, we see a direct consequence of CS applied to β-cell mass production that fails to keep β-cell mass stable, as shown in Fig. 1C. This would imply that in a case of decrease in β-cell mass, we would also observe a decrease in Insulin secretion. However this is not the case. Induced CS not only damages the β-cell mass, but also indirectly makes the remaining β-cell mass compensate for the loss in mass, keeping the Insulin/Glucose dynamics unchanged (Fig. 3). This finding has the capability to represent one more of the Hallmarks discussed in Sect. 2, under the name of LGI which in most cases is usually ignored as it takes part in the cellular stress. This hallmark was not possible to introduce into our calculation as there are no suggested LGI measures other than C-Reactive Protein (CRP) in a clinical setup. In the future, we intend to fix this by applying separate compartments for Cell Stress(C_{CSG}), LGI(C_{CRP} as an input from data and C_{IG} as a stand-in that holds general body inflammation events) shown in SM-Figure 2 As the CS induced dynamics unfold, in Fig. 2 we can already observe the counts being introduced as proportions for the calculation of DP. Hallmark HG is more evident since Insulin dynamics are able to closely follow HG in order to not count as IR markers. To this effect, we can observe a slowly increasing mean DP depending solely on two hallmarks. Moreover, we questioned whether the mean of DP would eventually find a steady-state if the time-span of simulation would be increased or decreased as well. This is shown in Fig. 3, as we took different time-spans and sampled 100 different cases. Results show that for a sample of 10 days(before CS induction), DP mean is lower than in the case of 1 year and 1.5 years samples. This suggests that the initial 15 days can be used as calibration for the algorithm following introduction of real data. The 0.5 year sample mean is very close to the one for 10 day, but a very big variation between cases is observed which is most probably the main cause for this mean value. Beyond that, the 1 year and 1.5 years samples are the best time-span samples to give closer results of DP to the mean for 45 days in Fig. 2. This suggests that 45 days is not enough as time-span. Variation for 1.5 years in Fig. 3 is also lower, additional testing is needed to verify whether this value is just a minimum value needed for subject following or larger time-span is needed. The experimental setup points towards a successful application and evaluation of SM-Algorithm 2. This indicates that SM-Algorithm 2 has adaptive plasticity to drastic changes in dynamics(favored by nature), producing limiters to DP inherently. This hints at the need of very critical behaviour to change and overcome disease on-set and disease un-set which is suggested in other CS studies. Empirically, the need of real life data is evident for subjects at different CS stages, with or without T2D on-set or even in other disease cases in order to observe differences as well as different hallmarks and fixed clinically observed threshold values to use as indicators for our SM-Algorithm 2. At this stage, *in-silico* simulated data can only provide limited information. However, a methodology is already being developed to account for this need as the conceptual model in SM-Figure 2 now exists. The randomness in the simulations clearly affects the variation of our calculation, nevertheless it is clear that this effect decreases with increased period of simulation and this is essential when searching for most favorable inter-players in disease progression later on. The next steps are being taken towards acquisition of data that can be used to further develop the conceptual model (SM-Figure 2) into a better T2D simulation tool which can be used to further ameliorate Algorithm 2, culminating into an ultimate tool for T2D appraisal and search of key inter-players to which the DP can decrease.

References

1. Anagnostis, A. et al.: The pathogenetic role of cortisol in the metabolic syndrome: a hypothesis. The Journal of Clinical Endocrinology & Metabolism (2009)
2. Bouwens, L., Rooman, I.: Regulation of pancreatic beta-cell mass. Physiological reviews (2005)
3. Chiodini, I., et al.: Association of subclinical hypercortisolism with type 2 diabetes mellitus: a case-control study in hospitalized patients. European journal of endocrinology (2005)
4. Epel, E.S., Crosswell, et al.: More than a feeling: A unified view of stress measurement for population science. Frontiers in neuroendocrinology (2018)
5. Benthem de Grave, R., et al.: From work stress to disease: A computational model. PloS one (2022)
6. Hackett, R.A., Steptoe, A.: Type 2 diabetes mellitus and psychological stress - a modifiable risk factor. Nature Reviews Endocrinology (2017)
7. Itoh, Y., et al.: Free fatty acids regulate insulin secretion from pancreatic β cells through gpr40. Nature (2003)
8. Joseph, J.J., Golden, S.H.: Cortisol dysregulation: the bidirectional link between stress, depression, and type 2 diabetes mellitus. Annals of the New York Academy of Sciences (2017)
9. Khan, M.A., et al.: Epidemiology of type 2 diabetes-global burden of disease and forecasted trends. J. Epidemiology Global Health (2020)
10. Lucassen, P., et al.: Neuropathology of stress. Acta neuropathologica (2014)
11. Marchetti, P., et al.: Pancreatic beta cell identity in humans and the role of type 2 diabetes. Frontiers in cell and developmental biology (2017)
12. Matthews, D.R., et al.: Homeostasis model assessment: insulin resistance and β-cell function from fasting plasma glucose and insulin concentrations in man. Diabetologia (1985)
13. Mohammed, I.I., et al.: Mathematical model for the dynamics of glucose, insulin and β-cell mass under the effect of trauma, excitement and stress. Modeling and Numerical Simulation of Material Science (2019)
14. Nordentoft, M., Rod, et al.: Effort-reward imbalance at work and risk of type 2 diabetes in a national sample of 50,552 workers in denmark: A prospective study linking survey and register data. Journal of psychosomatic research (2020)
15. Roder, P.V., et al.: Pancreatic regulation of glucose homeostasis. Experimental and molecular medicine (2016)
16. Sheraton, M.V., Sloot, P.M.: Parallel performance analysis of bacterial biofilm simulation models. Springer (2018)
17. Sheraton, M., et al.: Emergence of spatio-temporal variations in chemotherapeutic drug efficacy: in-vitro and in-silico 3d tumour spheroid studies. BMC cancer (2020)

R. V. Bumbuc et al.

18. Sheraton, V.M., Ma, S.: Exploring ductal carcinoma in-situ to invasive ductal carcinoma transitions using energy minimization principles. In: Groen, D., de Mulatier, C., Paszynski, M., Krzhizhanovskaya, V.V., Dongarra, J.J., Sloot, P.M.A. (eds) ICCS 2022. LNCS, vol. 13350. Springer, Cham (2022). https://doi.org/10.1007/978-3-031-08751-6_27
19. Sriram, K., et al.: Modeling cortisol dynamics in the neuro-endocrine axis distinguishes normal, depression, and post-traumatic stress disorder (ptsd) in humans. PLoS computational biology (2012)
20. Topp, B., et al.: A model of β-cell mass, insulin, and glucose kinetics: pathways to diabetes. Journal of theoretical biology (2000)
21. Unger, R.H.: Glucagon physiology and pathophysiology in the light of new advances. Diabetologia **28**(8), 574–578 (1985). https://doi.org/10.1007/BF00281991
22. Van Ommen, B., et al.: From diabetes care to diabetes cure-the integration of systems biology, ehealth, and behavioral change. Frontiers in endocrinology (2018)

How to Select Superior Neural Network Simulating Inner City Contaminant Transport? Verification and Validation Techniques

A. Wawrzynczak[1,2]([✉]) [ID] and M. Berendt-Marchel[1] [ID]

[1] Institute of Computer Sciences, Siedlce University, Siedlce, Poland
awawrzynczak@uph.edu.pl
[2] National Centre for Nuclear Research, Swierk-Otwock, Poland

Abstract. Artificial neural networks (ANNs) can learn via experience to solve almost every problem. However, the ANN application in a new task entails a necessity to perform some additional adaptations. First is fitting the ANNs type or structure by applying the various number of hidden layers and neurons in it or using different activation functions or other parameters, allowing ANN to learn the stated task. The second is the validation and verification methods of the ANN quality that should be suited to the stated task. Occasionally the differences between the ANNs output are significant, and it is easy to choose the best network. However, sometimes the differences pronounced by standard performance parameters are minor, and it is difficult to distinguish which ANN has reached the best level of training. This paper presents the results of training the ANN to predict the spatial and temporal evolution of the airborne contaminant over a city domain. Statistical performance measures have validated the trained ANNs performance. Finally, new measures allowing to judge of both time and spatial distribution of the ANN output have been proposed and used to select the prior ANN.

Keywords: Neural network model · Validation methods · Dispersion model

1 Introduction

The study presented in this paper was initiated by the willingness to create an emergency-response system able to localize the airborne toxin source in the urban terrain in real-time. The most probable contamination source location should be indicated based on the concentration data reported by the sensor network. Moreover, the process should be quick to ensure the fast action of the emergency response group. In the literature, the process of the contamination source localization based on the outcome is classified as the backward problem and referred to as source term estimation (STE), e.g., [1]. The goal is to find the best or

© The Author(s), under exclusive license to Springer Nature Switzerland AG 2023
J. Mikyška et al. (Eds.): ICCS 2023, LNCS 14074, pp. 339–347, 2023.
https://doi.org/10.1007/978-3-031-36021-3_35

most likely match between the predicted (by the applied dispersion model) and observed data, i.e., concentration in the sensor location. Consequently, the model parameters space scanning algorithm guided by the likelihood function is used. This requires many thousands of dispersion model runs. In [2] the localization of the contaminant in the highly urbanized terrain using the approximate Bayesian computation algorithm is presented. Although the results are satisfactory, the computational time of the reconstruction is extended. Thus, urban reconstruction in real-time is not possible, even with the distributed system.

The solution might be an application of the trained Artificial Neural Networks (ANNs) in the place of the dispersion model in the STE algorithm. ANNs are learning by example. Thus, they can be skilled with known examples to solve almost any task. Once well-trained ANNs can solve the stated task very quickly. These characteristics make the ANNs an excellent tool in real-time working systems. The ANN must learn to simulate airborne contaminant transport to be used in the emergency response localization system. The contaminant concentration distribution function is multidimensional and depends on spatial coordinates and time. Additionally, its value depends on external parameters like the contaminant source characteristics (location, release rate, release duration), meteorological conditions, and the domain's geometry. The challenge is the urban geometry, which is very complicated as far the wind field structure on which the contaminant is spread is site-dependent. The ANN training is computationally expensive, but once trained, the ANN would be a high-speed tool that estimates the contaminant concentration distribution.

The first results confirming that ANN has the potential to replace the dispersion model in the contaminant source localization systems are presented in [3]. The comparison of various architectures of ANNs in forecasting the contaminant strength correctly is presented in [4]. The results revealed that standard performance measures like correlation R and mean square error are fallible in pointing out the quality of ANN. In none of the mentioned papers the more profound validation of the proposed ANN models was not presented. This work is aimed at fulfilling at least some of these gaps.

Thus, apart from the known statistical measures, we propose the new ones being able to verify the dynamic agreement between the ANN output and the target both in space and time. Moreover, we propose a method to improve the trained ANN quality when a value of zero represents a large proportion of the training data.

2   ANN Model

Feedforward neural networks (FFNNs) are often applied for prediction and function approximation. The first layer of FFNNs consists of the neurons representing the input variables based on which the network should produce the neurons in the output layer. Between the input and output layers, the hidden layers are placed. ANN performance depends on the chosen architecture, i.e., the number of neurons, hidden layers, and the structure of connections. The aim is to teach the ANN to predict the contaminant concentration at a specific time and location for the assumed release scenario. Thus, the structure of the input vector is following $Input_i \equiv \{X_s, Y_s, Q, d, x, y, v, t\}$. Based on the input vector for the contamination source at the coordinates (X_s, Y_s) (in meters within a domain) and release lasting through d seconds with the release rate equal Q under wind blowing from the v direction the trained network should return the output neuron $Output_i \equiv C_i^{S_j(x,y)}(t)$ denoting the concentration C at sensor S_j with coordinates (x, y) in t-seconds after starting the release.

3   Domain, Training, and Testing Dataset. Data Preprocessing

The central part of London was chosen as a domain for the training dataset generation. The reason was a willingness to train the ANN using the real field tracer experiment DAPPLE. Unfortunately, about 600 point concentrations were insufficient to train the ANN properly. Thus, the learning dataset was generated using the QUIC Dispersion Modeling System [5]. The details on the domain and simulations setup are presented in [4] with the difference that the assumed release rate is within interval $Q \in \langle 100mg, 999mg \rangle$.

The obtained set covering about 5×10^7 vectors was divided into training - 66%, validation and testing datasets 17% each. The target function is a multidimensional and time-dependent function. However, the neurons in the input layer do straightforwardly reflect this time dependency. Each input vector corresponds to the concentration for a fixed point in time and space for a unique release scenario. The data included in the training and validation dataset were randomly drawn from the whole dataset. However, the testing dataset was carefully selected to judge how well the trained ANN reflects the time dynamics, i.e., whether the ANNs prediction is correct in subsequent time intervals. Thus, the testing dataset contains the vectors covering the whole 67 simulations. To appropriately validate and compare the ANNs, the same testing dataset was used to estimate the performance measures described in Sect. 4 for all analyzed ANNs architectures.

To give all variables equal weight in the input neuron vector, they have been scaled to the interval $(0, 1)$. In addition, the target concentration was logarithmized [6]. Moreover, the noise was introduced to the target concentration C and release rate Q. These two variables were chosen for noise introduction due to their inseparable connection. The spatial distribution of concentration on the sensors will depend on the strength of the released substance. The noise was introduced after normalizing the ANN input data as $\dot{C} = C \pm \delta \times C$, where δ was drawn uniformly from the interval $\langle 0, 15 \rangle$.

4   ANN Model Validation

The model validation aims to evaluate how useful a model is for a given purpose, thereby increasing confidence in model outputs. The verification and validation

Table 1. The measures calculated for the ANNs models. The ANN colors denote the best values of the measures among other ANNs for training using the noised and original data.

Network	trained using noised data				trained using orginal data			
	24-16-8	24-16-8-4-2	24-20-16-12-8-4-2	48-24-16-8-4	24-16-8	24-16-8-4-2	24-20-16-12-8-4-2	48-24-16-8-4
R training	0.8296	0.8381	0.8591	0.8787	0.8780	0.8814	0.9008	0.9292
R test	0.5881	0.7922	0.7054	0.5549	0.7517	0.7613	0.6765	0.4695
MSE	0.0252	0.0242	0.0213	0.0185	0.0189	0.0184	0.0156	0.0111
RMSE×10^{-7}	28.21	3.25	3.26	3.31	3.25	3.25	4.11	10.79
CE	−73.729	0.0065	−0.0006	−0.0303	0.0071	0.0066	−0.5834	−9.9266
$\rho(d_{ANN}^{1:t}, d_{target}^{1:t})$	0.2027	0.1533	0.1587	0.1795	0.7029	0.6639	0.9616	0.9689
$\overline{MSSDLE}_{\vec{x}}$	0.1508	0.1016	0.1008	0.1251	0.6417	0.6041	0.8924	0.8920
$\overline{MSSDLE}_{\vec{y}}$	0.1460	0.0963	0.0994	0.1211	0.6515	0.5871	0.8886	0.8882
$\overline{MSDLE}$	0.1718	0.1157	0.1181	0.1399	0.6663	0.6611	0.8790	0.8790

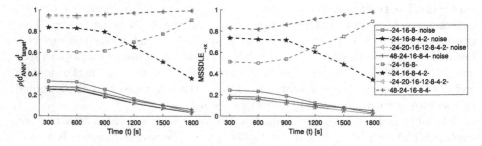

Fig. 1. The measure (a) $\rho(d_{ANN}^t, d_{target}^t)$ (Eq. 1), and (b) $MSSDLE_{\vec{x}}$ (Eq. 3) in subsequent time steps for a few considered ANNs. Profile of the measure $MSSDLE_{\vec{y}}$ is analogous. The solid lines correspond to the networks trained using noised data, while the dashed lines correspond to the ANNs trained using original data.

process is indispensable when the single best model has to be indicated from the subset of models. The selection is more difficult when the primary differences are minor. Here we propose the measures helpful to validate which of the trained ANNs has learned to predict best the spatial and time evolution of the target function.

The most common measure used to judge the level of ANN training is the correlation coefficient R between the actual output and the output predicted by the ANN. Usually, after training different ANNs, the final one is chosen based on the highest value of R. The $R = 1$ denotes the ideal fit. The twin measure is the mean square error $MSE = \frac{1}{n}\sum_{i=1}^{n}(C_i - \hat{C}_i)^2$ denoting difference between the ANN output $\hat{C}_i$ and target C_i. The MSE is typically used as the stopping criterion for the ANN training process. However, using MSE means assuming that the underlying data has been generated from a normal distribution. In reality, a dataset rarely fulfills that requirement. It is better to report $RMSE = \sqrt{\frac{1}{n}\sum_{i=1}^{n}(C_i - \hat{C}_i)^2}$, rather MSE; because RMSE is measured in the same units as the original data, and is thus more representative of the size of a 'typical' error. To distinguish the ANNs quality deeper, more measures should be considered.

Model Selection Measure. The coefficient of efficiency (CE) is one of the model selection measures: $CE = 1 - \frac{\sum_{i=1}^{n}(C_i - \hat{C}_i)^2}{\sum_{i=1}^{n}(C_i - \bar{C})^2}$. CE is intended to range from zero to one, but negative scores are also permitted. The maximum positive score of one represents a perfect model. The negative scores are unbounded, indicating that the model performs worse than a 'no knowledge' model. CE is sensitive to differences in the observed and modeled means and variances.

Measures Estimating the Time Dynamic. The statistical measures described above do not provide information on the quality of the spatial and temporal distribution of the ANN prediction. Therefore, we propose introducing additional measures to verify whether trained ANN can correctly reproduce the concentration gradient (both spatial and time). The focus is put on the agreement in the successive intervals of simulations. The first proposed formula based on a fractional bias is:

$$\rho(d_{ANN}^{1:t}, d_{target}^{1:t}) = \frac{1}{SN} \sum_{j=1}^{SN} \left[\frac{1}{t} \sum_{i=1}^{t} \frac{|C_i^{Sj} - \hat{C}_i^{Sj}|}{C_i^{Sj} + \hat{C}_i^{Sj}} \right], \tag{1}$$

with assumption that if $C_i^{Sj} = 0$ and $\hat{C}_i^{Sj} = 0$ then fraction $\frac{|C_i^{Sj} - \hat{C}_i^{Sj}|}{C_i^{Sj} + \hat{C}_i^{Sj}} = 0$. In Eq. 1 i denotes the subsequent time intervals in which the concentration in Sj point representing the sensor location is estimated. The SN indicates the total number of sensors, $\hat{C}_i^{Sj}$ concentration in time i in point Sj of domain predicted by ANN, while C_i^{Sj} the represents the target concentration. The measure ρ fits into the interval $[0, 1]$. If the ANN model prediction is ideal, then $\rho = 0$, and if the model predictions are completely wrong, it equals 1.

The following measure Mean Squared Derivative Logarithmic Error (MSDLE) is proposed to describe the level of agreement of the target function change in the time between modeled dataset and observed in each point of the 2D space. In each point Sl of the space the $MSDLE$ is calculated as follows:

$$MSDLE(Sl) = \frac{1}{M} \sum_{m=1}^{M} \left[\frac{1}{t-1} \right.$$
$$\left. \sum_{j=2}^{t} \left(\frac{\ln(C_j^{Sl,m} - C_{j-1}^{Sl,m}) - \ln(\hat{C}_j^{Sl,m} - \hat{C}_{j-1}^{Sl,m})}{\ln(C_j^{Sl,m} - C_{j-1}^{Sl,m}) + \ln(\hat{C}_j^{Sl,m} - \hat{C}_{j-1}^{Sl,m})} \right)^2 \right]. \tag{2}$$

The M denotes the number of simulations run over t time steps, $\hat{C}_j^{Sl,m}$ the ANN prediction of the target value in time j at the point Sl for the simulation m. The observed value is $C_j^{Sl,m}$. The result of Eq. 2 is the 2D map of the measure distribution. To characterize the measure by a single value, the averaging over the number of SN space points is performed $\overline{MSDLE} = \frac{1}{SN} \sum_{l=1}^{SN} MSDLE(l)$. The $\overline{MSDLE}$ is scaled to the $[0, 1]$ interval and equals 0 for the ideal model.

The last proposed measure is aimed to represent how well in each time-step t the spatial gradient of the target function is reproduced. The Mean Squared Spatial Derivative Logarithmic Error (MSSDLE) is estimated in two main directions and $\vec{y}$ of the 2D domain. The corresponding formula for $\vec{x}$-direction is following:

$$MSSDLE_{\vec{x}}(t=j) = \frac{1}{M} \sum_{m=1}^{M} \left[\frac{1}{N_L} \sum_{L=1}^{N_L} \left[\frac{1}{N_K - 1} \right. \right.$$

$$\left. \left. \sum_{K=2}^{N_K} \left(\frac{\ln \left(\frac{(C_j^{m(K,L)} - C_j^{m(K-1,L)})}{\Delta K} \right) - \ln \left(\frac{(\hat{C}_j^{m(K,L)} - \hat{C}_j^{m(K-1,L)})}{\Delta K} \right)}{\ln \left(\frac{(C_j^{m(K,L)} - C_j^{m(K-1,L)})}{\Delta K} \right) + \ln \left(\frac{(\hat{C}_j^{m(K,L)} - \hat{C}_j^{m(K-1,L)})}{\Delta K} \right)} \right)^2 \right] \right] \quad (3)$$

The N_K denotes the number of points on a grid in $\vec{x}$ direction. Taking into account the presence of buildings in the domain, the distance between the points of the grid (sensors) $\Delta K ((x_1, y_1), (x_2, y_2)) = |x_2 - x_1|$ is included in the measure. The $MSSDLE_{\vec{y}}(t = j)$ is calculated analogously in the $\vec{y}$ direction. To represent the measure by a single value, the averaging over the time steps of simulations T is performed $\overline{MSSDLE} = \frac{1}{T} \sum_{t=1}^{T} MSSDLE(t)$. The $\overline{MSSDLE}$ is scaled to the $[0, 1]$ interval and equals 0 for the ideal model.

5   Results

We have trained the multiple ANNs using the dataset described in Sect. 3 and Matlab Deep Learning Toolbox. Among tested activation functions in the hidden layers, the *hyperbolic tangent*, and the linear function in the output layer performed the best. The network training was stopped at the lowest possible MSE of the validation test, assuming the upper limit of epochs to 70, with the target MSE set to $1e - 08$ value. We have trained the ANNs with the same architectures using the original and noised datasets. The measures described in Sect. 4 were calculated for each developed ANN model. Table 1 presents the values of the estimated measures for four ANNs with the highest R-value. The differences in R-values are pretty slight. In such cases, selecting the prior ANN is complicated and must be done carefully. The proposed additional measures should help facilitate the selection of the best-trained ANNs. Table 1 is divided into two parts. The left side presents the measures for the ANNs trained using the noised data, and the right side for the same ANNs trained on the original data. Analyzing this table carefully, we can see that we get the highest R ($R = 0.8787$ and $R = 0.9292$) for the training set for the network with the highest number of neurons in the hidden layers, i.e., 48-24-16-8-4. However, the RMSE, representing the level of overall agreement between the target and modeled dataset, are the smallest for the ANN 24-16-8-4-2. The CE also supports the preference of this ANN. Moreover, the CE for the ANN 48-24-16-8-4 is negative, which suggests that this network performs worse than the 'no knowledge' model. It means that the specific task of the ANN model, i.e., correct forecasting of the contaminant concentration's spatial and time gradient, is not achieved. Using the same testing dataset, we verified each ANNs quality by the dynamical measures ρ, $MSSDLE_{x,y}$ and $MSDLE$. The time profile of the ρ and $MSSDLE_x$ for a subset of the analyzed ANNs architectures is presented in Fig. 1. The first look at the figure shows that the networks trained on the noised data (solid lines) perform better than those trained on the original data (dashed lines), regardless of the ANN architecture. Moreover, the ρ value is 2–3 times smaller. The reason is that the ANNs can better learn to forecast small concentrations thanks to introducing the noise. Adding the noise after re-scaling gives a diverse set of small numbers representing the close-to-zero concentrations instead a constant one. The noise contribution to the ANN knowledge is greatly seen in Fig. 2 presenting the 2D distribution of the $MSDLE$ (Eq. 2) for ANNs with the same architectures but trained on original and noised data. The agreement of the ANN trained using noised data is almost perfect, as far as its values are close to zero for nearly the whole domain. The result is much worse for the ANN with the same architecture but trained using original data. The above-described results conclude that the ANN 24-16-8-4-2 trained using the noised data seems to be the best-trained network among the considered ones.

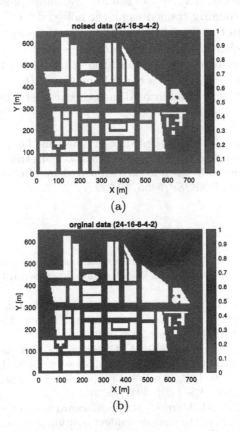

Fig. 2. The measure $MSSDLE$ (Eq. 2) distribution in the 2D city domain for the ANN with hidden layers $24 - 16 - 8 - 4 - 2$ trained on (a) noised and (b) original data.

6 Summary

We have presented the results of training the FFNN to simulate airborne contaminant transport in highly urbanized terrain. The applied ANN structure allowed training the FFNN to simulate the time-dependent nonlinear function in two spatial dimensions. The training dataset consisted of independent vectors representing the point concentration for the assumed release scenario. On the other side, we require the ANN to simulate the contaminant distribution correctly for spatial and time gradients. It occurred that in such a task, classical measures like R and MSE cannot indicate the best ANN reliably. Therefore, we proposed additional measures to verify the quality of the ANN model. Beneficial are the measures estimating the time dynamic of the ANN model like $\rho(d_{ANN}^t, d_{target}^t)$, $MSSDLE_{x,y}$ and $MSDLE$. These measures, as the best one, pointed to the ANN 24-16-8-4-2. This network was not the most extensive and was not characterized by the highest R-value in training. We have presented a significant increase in the ANN quality trained using the noised data. The reason was that the small diversity in the re-scaled target values of concentrations allowed the ANNs to fit the weights in the learning process better.

The presented results lead to the conclusion that the application of ANNs in a new field should be followed by a careful analysis of the verification methods and maybe an adaptation of additional measures as a stopping criterion in the ANN training process.

Acknowledgements. We acknowledge Michael Brown and Los Alamos National Laboratory for the possibility to use the Quick Urban & Industrial Complex Dispersion Modeling System. This work was partially supported by Ministry of Education and Science, project number: DNK/SP/549572/2022.

References

1. Hutchinson, M., Oh, H., Chen, W.: A review of source term estimation methods for atmospheric dispersion events using static or mobile sensors. Inf. Fusion **36**, 130–148 (2017)
2. Kopka, P., Wawrzynczak, A.: Framework for stochastic identification of atmospheric contamination source in an urban area. Atmos. Environ. **195**(1), 63–77 (2018). https://doi.org/10.1016/j.atmosenv.2018.09.035
3. Wawrzynczak, A., Berendt-Marchel, M.: Computation of the airborne contaminant transport in urban area by the artificial neural network. In: Krzhizhanovskaya, V.V., et al. (eds.) ICCS 2020. LNCS, vol. 12138, pp. 401–413. Springer, Cham (2020). https://doi.org/10.1007/978-3-030-50417-5_30
4. Wawrzynczak, A., Berendt-Marchel, M.: Feedforward neural networks in forecasting the spatial distribution of the time-dependent multidimensional functions. In: 2022 International Joint Conference on Neural Networks, IJCNN, Padua, Italy , pp. 1–8 (2022). https://doi.org/10.1109/IJCNN55064.2022.9892001
5. Williams, M.-D., Brown, M.-J., Singh, B., Boswell, D.: QUIC-PLUME theory guide. Los Alamos National Laboratory 43 (2004)

6. Wawrzynczak, A., Berendt-Marchel, M.: Can the artificial neural network be applied to estimate the atmospheric contaminant transport? In: Dimov, I., Fidanova, S. (eds.) Advances in High Performance Computing. HPC 2019. Studies in Computational Intelligence, vol. 902, pp. 132–142. Springer, Cham. (2021) https://doi.org/10.1007/978-3-030-55347-0_12

DeBERTNeXT: A Multimodal Fake News Detection Framework

Kamonashish Saha[✉] and Ziad Kobti

University of Windsor, Ontario N9B 3P4, Canada
{saha91,kobti}@uwindsor.ca

Abstract. With the ease of access and sharing of information on social media platforms, fake news or misinformation has been spreading in different formats, including text, image, audio, and video. Although there have been a lot of approaches to detecting fake news in textual format only, multimodal approaches are less frequent as it is difficult to fully use the information derived from different modalities to achieve high accuracy in a combined format. To tackle these issues, we introduce DeBertNeXT, a multimodal fake news detection model that utilizes textual and visual information from an article for fake news classification. We perform experiments on the immense Fakeddit dataset and two smaller benchmark datasets, Politifact and Gossipcop. Our model outperforms the existing models on the Fakeddit dataset by about 3.80%, Politifact by 2.10% and Gossipcop by 1.00%.

Keywords: Multi-modal · Fake News · DeBERTa · ConvNext

1 Introduction

Fake news gained attention after the 2016 US presidential election when false news spread mainly through social media, which is the primary news source for 14% of Americans [1]. There are fact-checking websites which include Politifact, AltNews, Fact Check, as well as expert-based and crowdsourced methods to verify the news. However, manual methods are time-consuming and inefficient given the vast amount of news generated globally. Automatic methods for detecting false news are becoming increasingly popular. Detecting fake news using only textual content has been heavily researched, but articles with images are retweeted 11 times compared to the ones with only text [7]. Therefore, combining data from various modalities is crucial for classification. Researchers proposed models for multimodal detection, but fake news is usually classified as a binary problem as it is a distortion bias which itself is illustrated as a binary problem [18]. It is challenging to attain good accuracy on large datasets and utilize all the features from different modalities. Hence, we propose DeBERTNeXT which is trained on Fakeddit, Politifact, and Gossipcop datasets where our model outperforms other state-of-the-art models in terms of accuracy and other metrics.

In this paper, we propose DeBERTNeXT which is a transfer learning-based architecture that utilizes textual and visual features for classifying news as real or

J. Mikyška et al. (Eds.): ICCS 2023, LNCS 14074, pp. 348–356, 2023.
https://doi.org/10.1007/978-3-031-36021-3_36

fake. The representations from both modalities are concatenated for classification and it is not dependent on sub-tasks or domain-specific. We trained and tested our model on Fakeddit, Politifact, and Gossipcop datasets where we achieved better classification results in terms of accuracy, precision, recall, and F1 scores compared to other models to the best of our knowledge.

2   Literature Review

Knowledge-graph-based approaches can verify the veracity of the main statements in a news report [18]. Such methods are inefficient for extensive data and depend on an expert to assign the truthfulness of the news [18]. A significant amount of research has been focused on the textual content which uses BERT and RoBERTa [19]. Unimodal approaches are not suitable for multimodal formats and new architectures were subsequently proposed. One baseline multimodal architecture includes SpotFake [22], which uses BERT for learning text features and pre-trained VGG-19 for the image features. SpotFake+ [21] was also introduced later, using pre-trained XLNet and VGG-19 for the combined image and text classification. Another architecture was named Event Adversarial Neural Networks for Multi-Modal Fake News Detection (EANN), which was an end-to-end model for the event discriminator and false news detection [23]. The text representation was attained using CNN, and VGG-19 was used for image representation, where both were later concatenated for classification. Similarly, another model using VGG-19 and bi-directional LSTMs for textual features was introduced, which the authors named Multimodal Variational Autoencoder for Fake News Detection (MVAE) [10]. In addition, FakeNED [15] was introduced, which utilizes finetuned BERT and VGG-19 for binary classification. The baseline multimodal models give good accuracy, but they mainly concentrate on using BERT and VGG architecture, which has some shortcomings of their own. Moreover, the discussed models are trained and tested on small datasets, and some depend on events or require additional preprocessing steps of the dataset. Hence, to tackle these shortcomings, we propose our framework that uses DeBERTa and ConvNeXT which we later fine-tune to train and test on the three different datasets.

3   Methodology

Given a set of n news articles which includes text and image content (multimodal), the data as a collection of a text-image tuple can be represented as:

$$A = (A_i^T, A_i^I)_{i=1}^n \tag{1}$$

where, A_i^T represents textual content, A_i^I represents the image content and n represents the number of news articles. Since we are considering fake news detection as a binary classification problem, we represent labels as $Y = \{0, 1\}$ where 0 represents fake and 1 represents real or true news. For a given set of news

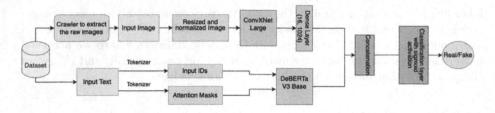

Fig. 1. Model Architecture

records A, a set of features can be extracted from the textual and image information as represented by X_i^T and X_i^I. The objective of multimodal fake news detection is to create a model $F : \{X_i^T, X_i^I\}\epsilon X \rightarrow Y$ in order to deduce the potential labels in news articles A. Hence, the task of the model is to detect whether the news article A is either fake or real such that:

$$f(A) = \begin{cases} 0, & \text{if } A \text{ is fake news} \\ 1, & \text{otherwise} \end{cases} \tag{2}$$

DeBERTNeXT uses Transformer and Convolutional models for text and images of a particular record. For attaining the textual features, the **DeBERTa** (**De**coding-enhanced **BERT** with disentangled **a**ttention) model [5] was used which is based on BERT [3] and RoBERTa [12] models. DeBERTa improves these state-of-the-art models by incorporating disentangled attention mechanism and enhanced mask decoder. For the image part, we have used **ConvNeXT** [13] which is a convolutional model (ConvNet) trained on the large ImageNet dataset (Fig. 1).

Image Inputs: The proposed model consists of the DeBERTa V3 base model for the text inputs and ConvNeXT Large cased [13] for the image input. The ConvNeXT model was configured with the proposed model for transfer learning. All the weights of the ConvNeXT model were used except for the last classification layer from the ImageNet pre-trained version. The last layer of the ConvNeXT model with 1536 dimension output was replaced with a fully connected linear dense layer with 1024 output nodes and was found to give the best results. This was chosen based on several experimentations and was set as a hyperparemeter.

Textual Inputs: The improved DeBERTa V3 [4] consists of 12 layers and a hidden size of 768. It consists of 86M backbone parameters with a vocabulary containing 128K tokens which introduce 98M parameters in the Embedding Layer and was trained using 160 GB data. It takes input IDs and attention masks as inputs and outputs a 768-dimension-long tensor. For both images and text, a batch size of 16 was used.

Concatenation and Classification: The output from the DeBERTa V3 model is concatenated with the ConvNeXT large output along the first axis. The concatenation layer takes 768 dimension output from the DeBERTa V3 model and

1024 dimension output from the ConvNeXT model, producing 1792 dimension output. A connected layer with a sigmoid activation function generates the final classification output. The last layer takes 1792 dimension output from the concatenation layer as input and outputs a value ranging from 0 to 1.

4   Experimentation and Results

Our model was trained on Fakeddit, Politifact, and Gossipcop datasets. Various experiments were conducted manually where we used different hyperparameters to get the best set in order to produce the best performance. The image size was 224×224, and the maximum text length varied by dataset. The Fakeddit dataset was set to a maximum text length of 48 words, while Politifact and Gossipcop to 32 words which were set after analyzing the mean and maximum sequence length in the text inputs. We split the training data into an 80:20 train-validation set and kept the test set separate until training was complete.

Only normalization layers and bias were excluded from weight decay while setting up the AdamW optimizer. For the Fakeddit dataset, a batch size of 16 and the maximum learning rate chosen was $3e^-6$ and was scheduled with the help of the scheduler. The model was trained for 4 epochs and evaluated on the validation set after every epoch. However, for the Politifact and the Gossipcop dataset, the maximum learning rate was chosen to $2.5e^-6$ and was trained for 6 epochs. This was found to be the optimum based on several experiments. Finally, once the training was complete, it was tested on the test set, which was 20% of the respective datasets. The experimentation was carried out in the Google Colaboratory platform using Tesla T4 GPU, and the code is publically available which can be found at https://github.com/Kamonashish/DeBERTNeXT.

Dataset: A crawler was developed to download images from the URLs in the dataset. After filtering GIFs, broken, and distorted images, we extracted 154,644 news records for the Fakeddit dataset [14], 304 for the Politifact dataset, and 8,008 for the Gossipcop dataset. We only used records with both usable text and image data that would refer to the same unique ID of that record.

Data Processing: All the images were extracted using the crawler with the help of the Python libraries: Beautiful Soup and urllib. Once the filtered and refined images were attained, the selected images were reshaped and normalized later during the training phase. Only the URLs were removed for the text since different transformer architectures accept tokens instead of plain English words.

Dataset Pipeline: The data loader pipeline increases loading time significantly as compared to directly loading data to RAM but is necessary as required in case of large datasets when there are RAM constraints. The dataset pipeline accepts the image file path, the text and the label for a particular record in the dataset and processes it to tensors which are accepted by the model. All the images are first loaded from their file path and then reshaped to (224, 224, 3). The reshaping is followed by normalizing the images then converted to tensors. All the text inputs are passed through a model-specific tokenizer which is obtained from the

HuggingFace library [25]. All the textual content in the record is tokenized to the maximum length and the text beyond the maximum length is truncated. Special tokens such as [CLS], [SEP] and [PAD] are used. The tokenizer output is in the form of tensors as input_ids and attention_masks. The input_ids are the tokenizer-processed tokens extracted from text, whereas attention_masks are tensors of 0 and 1 that serve the purpose of allowing the model to use attention on selected tokens. The input_ids and attention_masks are given as output for the text inputs of the DeBerta V3 model. At last, the 2_way_label, which is provided as 0 or 1 in the dataset for fake or real news is also converted to a tensor and given as output to the dataset pipeline.

After training was completed, the model was evaluated on the test set on a variety of classification and evaluation metrics which include accuracy, precision, recall and F1 score. The ROC AUC score, True Positive (TP), True Negative (TN), False Negative (FN) and False Positive (FP) were also noted. Based on our literature review, these were used in calculations for accuracy, recall, precision and F1 score to compare with other models. The best result of three experiments was considered. The model's accuracy and loss plot per epoch for each dataset can be found in .ipynb files in the GitHub repository.

Table 1. Comparison with other models on Fakeddit dataset. (-) indicates that results were not published

Models	Accuracy	Fake News			Real News		
		Precision	Recall	F1	Precision	Recall	F1
VGG 19 + Text-CNN [17]	0.804	0.838	0.749	0.791	0.704	0.728	0.716
VQA* [2]	0.631	0.712	0.512	0.596	0.590	0.693	0.637
NeuralTalk [27]	0.612	0.698	0.610	0.651	0.612	0.712	0.658
att-RNN [6]	0.745	0.798	0.637	0.708	0.627	0.713	0.667
EANN [23]	0.699	0.750	0.628	0.684	0.648	0.720	0.682
MVAE [10]	0.784	0.789	0.699	0.741	0.702	0.717	0.709
FakeNED [15]	0.878	–	–	–	–	–	–
CNN for text and image [16]	0.870	–	–	–	–	–	–
DeepNet [8]	0.864	–	–	–	–	–	–
DistilBERT + VGG 16 [9]	0.604	–	–	–	–	–	–
DeBERTNeXT	**0.912**	**0.910**	**0.950**	**0.930**	**0.917**	**0.854**	**0.884**

From Table 1, the authors fused the outputs from Text-CNN with VGG-19 to attain an overall accuracy of 0.804 [17]. Moreover, they modified and trained the VQA [2], NeuralTalk [27], att-RNN [6], EANN [23] and MVAE [10] on the Fakeddit dataset for binary classification. We include their results for comparison. The authors [17] modified VQA by using a binary class layer as the final layer and renamed it as VQA*. In FakeNED [15], after the textual and visual features were extracted, a step was added where a single one-dimensional tensor was passed to fully connected layers for binary classification, where it attained an accuracy of 0.878 and an F1 score of 0.910. Compared with FakeNED, our model outperforms the accuracy by 3.80% and achieves an F1 score of 0.912 by weighted average. CNN was used for both text and images in [16] where both the feature vectors attained after the convolutional layer were passed through two dense layers with ReLU non-linear activation before being concatenated. In the end, the logsoftmax function is applied to attain a micro-average accuracy of 0.870, precision of 0.880, recall of 0.870 and F1 score of 0.870. Similarly, DeepNet [8], which has ReLU as an activation function and softmax function for the final output layer, attains a precision of 0.894, recall of 0.850, an F1 score of 0.872 and an accuracy of 0.864. As compared to the macro-average, our model also outperforms them as we attain an accuracy of 0.912, precision of 0.913, recall of 0.902 and F1 score of 0.917, respectively.

Table 2. Comparison with other models on Politifact and Gossipcop Dataset. (-) indicates that the results were not published

Models	Dataset: Politifact				Dataset: Gossipcop			
	Accuracy	Precision	Recall	F1	Accuracy	Precision	Recall	F1
SpotFake+ [21]	0.846	–	–	–	0.856	–	–	–
SAFE [29]	0.874	0.889	0.903	0.896	0.838	0.857	0.937	0.859
Cross-Domain Detection [20]	0.840	0.836	0.831	0.835	0.877	0.840	0.832	0.836
att-RNN [6]	0.769	0.735	0.942	0.826	0.743	0.788	0.913	0.846
CMC [24]	0.894	–	–	–	0.893	–	–	–
EANN [23]	0.740	–	–	–	0.860	–	–	–
MVAE [10]	0.673	–	–	–	0.775	–	–	–
SpotFake [22]	0.721	–	–	–	0.807	–	–	–
SceneFND [28]	0.832	–	–	–	0.748	–	–	–
DeBERTNeXT	**0.913**	**0.921**	0.914	**0.915**	**0.902**	**0.914**	0.902	**0.906**

Most of the multimodal models use transfer learning to improve the accuracy of detection. From Table 2, SpotFake+ [21], SpotFake [22], CMC [24] use VGG19 for image content and a transformer-based model such as XLNet [26] and BERT for the textual content. XLNet has a potential for bias [11] and other disadvantages such as more computational cost and longer training time. The authors of SpotFake+ have attained an accuracy of 0.846 for the Politifact dataset and 0.856 on the Gossipcop dataset, which is less as compared to our model. This can be because DeBERTa has some notable advantages over the XLNet model for the textual representations which include higher efficiency, better performance on downstream tasks and improved masking strategy, pre-training techniques and flexibility in model size.

Similar to the textual aspect, our model has notable advantages as it uses ConvXNet over the commonly incorporated VGG 19 used in SpotFake, CMC and Spotfake+ for the image content. Advantages include improved accuracy, scalability, improved regularization, etc. The authors of SpotFake+ have trained the SpotFake, MVAE and EANN models on the Politifact and Gossipcop datasets.

We have used those results for a comparison with our model and it can be seen that our model outperforms all of them in terms of accuracy. SAFE [29] uses Text-CNN architecture for both image and text, but transformer models like DeBERTa perform better due to their bidirectional attention mechanism which allows it to better capture contextual information. SAFE's methodology beats our model's recall result on the Gossipcop dataset, but our model performs better on all other metrics.

5   Conclusion

We presented a new multimodal framework that can detect fake news using images and text in a large and small datasets. Our model utilizes the combined power of the transformer model and the ConvNeXT architecture for textual and image content. As per our literature review, our model achieves a better result than other models trained on the same dataset for binary classification. In future, we plan to incorporate more modalities such as audio, video and extract features to combine them with textual content for a complete all-modal fake news detection framework while achieving satisfying classification results.

References

1. Allcott, H., Gentzkow, M.: Social media and fake news in the 2016 election. J. Econ. Perspect. **31**(2), 211–36 (2017)
2. Antol, S., et al.: VQA: visual question answering. In: Proceedings of the IEEE International Conference on Computer Vision, pp. 2425–2433 (2015)
3. Devlin, J., Chang, M.W., Lee, K., Toutanova, K.: Bert: pre-training of deep bidirectional transformers for language understanding. arXiv preprint:1810.04805 (2018)
4. He, P., Gao, J., Chen, W.: Debertav 3: improving deberta using electra-style pre-training with gradient-disentangled embedding sharing. arXiv preprint arXiv:2111.09543 (2021)
5. He, P., Liu, X., Gao, J., Chen, W.: Deberta: decoding-enhanced Bert with disentangled attention. arXiv preprint:2006.03654 (2020)
6. Jin, Z., Cao, J., Guo, H., Zhang, Y., Luo, J.: Multimodal fusion with recurrent neural networks for rumor detection on microblogs. In: Proceedings of the 25th ACM international conference on Multimedia, pp. 795–816 (2017)
7. Jin, Z., Cao, J., Zhang, Y., Zhou, J., Tian, Q.: Novel visual and stat. Image features for microblogs news verification. IEEE Trans. Multimedia **19**(3), 598–608 (2016)
8. Kaliyar, R.K., Kumar, P., Kumar, M., Narkhede, M., Namboodiri, S., Mishra, S.: Deepnet: an efficient neural network for fake news detection using news-user engagements. In: 2020 5th International Conference on Computing, Communication and Security (ICCCS), pp. 1–6. IEEE (2020)
9. Kalra, S., Kumar, C.H.S., Sharma, Y., Chauhan, G.S.: Multimodal fake news detection on fakeddit dataset using transformer-based architectures. In: Machine Learning, Image Processing, Network Security and Data Sciences: 4th International Conference, MIND 2022, Virtual Event, 19–20 January 2023, Proceedings, Part II. pp. 281–292. Springer, Cham (2023). https://doi.org/10.1007/978-3-031-24367-7_28

10. Khattar, D., Goud, J.S., Gupta, M., Varma, V.: MVAE: multimodal variational autoencoder for fake news detection. In: WWW, pp. 2915–2921 (2019)
11. Kirk, H.R., et al.: Bias out-of-the-box: an empirical analysis of intersectional occupational biases in popular gen. language models. In: Advances in NIPS, vol. 34, pp. 2611–2624 (2021)
12. Liu, Y., et al.: Roberta: a robustly optimized Bert pretraining approach. arXiv preprint:1907.11692 (2019)
13. Liu, Z., Mao, H., Wu, C.Y., Feichtenhofer, C., Darrell, T., Xie, S.: A convnet for the 2020s. In: Proceedings of the IEEE/CVF Conference on Computer Vision and Pattern Recognition, pp. 11976–11986 (2022)
14. Nakamura, K., Levy, S., Wang, W.Y.: r/fakeddit: a new multimodal benchmark dataset for fine-grained fake news detection. arXiv preprint:1911.03854 (2019)
15. Sciucca, L.D., et al.: Fakened: a dl based-system for fake news detection from social media. In: Mazzeo, P.L., Frontoni, E., Sclaroff, S., Distante, C. (eds.) Image Analysis and Processing. ICIAP 2022 Workshops. ICIAP 2022. LNCS, vol. 13373, pp. 303–313. Springer, Cham (2022). https://doi.org/10.1007/978-3-031-13321-3_27
16. Segura-Bedmar, I., Alonso-Bartolome, S.: Multimodal fake news detection. Information **13**(6), 284 (2022)
17. Shao, Y., Sun, J., Zhang, T., Jiang, Y., Ma, J., Li, J.: Fake news detection based on multi-modal classifier ensemble. In: Proceedings of the 1st International Workshop on Multimedia AI against Disinformation, pp. 78–86 (2022)
18. Shu, K., Sliva, A., Wang, S., Tang, J., Liu, H.: Fake news detection on social media: A data mining perspective. ACM SIGKDD exp. newsletter **19**(1), 22–36 (2017)
19. Shushkevich, E., Alexandrov, M., Cardiff, J.: Bert-based classifiers for fake news detection on short and long texts with noisy data: a comp. analysis. In: Sojka, P., Horák, A., Kopeček, I., Pala, K. (eds.) Text, Speech, and Dialogue. TSD 2022. LNCS, vol. 13502, pp. 263–274. Springer, Cham (2022). https://doi.org/10.1007/978-3-031-16270-1_22
20. Silva, A., Luo, L., Karunasekera, S., Leckie, C.: Embracing domain differences in fake news: Cross-domain fake news detection using multi-modal data. In: Proceedings of the AAAI Conference on Artificial Intelligence, vol. 35, pp. 557–565 (2021)
21. Singhal, S., Kabra, A., Sharma, M., Shah, R.R., Chakraborty, T., Kumaraguru, P.: Spotfake+: a multimodal framework for fake news detection via transfer learning (student abstract). In: Proceedings of the AAAI, vol. 34, pp. 13915–13916 (2020)
22. Singhal, S., Shah, R.R., Chakraborty, T., Kumaraguru, P., Satoh, S.: Spotfake: a multi-modal framework for fake news detection. In: 2019 IEEE Fifth International Conference on Multimedia Big Data (BigMM), pp. 39–47. IEEE (2019)
23. Wang, Y., et al.: EANN: event adversarial neural networks for multi-modal fake news detection. In: Proceedings of the 24th ACM SIGKDD International Conference on Knowledge Discovery & Data Mining, pp. 849–857 (2018)
24. Wei, Z., Pan, H., Qiao, L., Niu, X., Dong, P., Li, D.: Cross-modal knowledge distillation in multi-modal fake news detection. In: ICASSP 2022–2022 IEEE International Conference on Acoustics, Speech and Signal Proceedings (ICASSP), pp. 4733–4737. IEEE (2022)
25. Wolf, T., et al.: Transformers: state-of-the-art natural language processing. In: Proceedings of the 2020 Conference on Empirical Methods in Natural Language Processing: System Demonstrations, pp. 38–45 (2020)
26. Yang, Z., Dai, Z., Yang, Y., Carbonell, J., Salakhutdinov, R.R., Le, Q.V.: Xlnet: generalized autoregressive pretraining for language understanding. In: Advances in Neural Information Processing Systems, vol. 32 (2019)

27. Yu, E., Sun, J., Li, J., Chang, X., Han, X.H., Hauptmann, A.G.: Adaptive semi-supervised feature selection for cross-modal retrieval. IEEE Trans. Multimedia **21**(5), 1276–1288 (2018)
28. Zhang, G., Giachanou, A., Rosso, P.: Scenefnd: multimodal fake news detection by modelling scene context information. J. Inf. Sci. 01655515221087683 (2022)
29. Zhou, X., Wu, J., Zafarani, R.: SAFE: similarity-aware multi-modal fake news detection. In: Lauw, H.W., Wong, R.C.-W., Ntoulas, A., Lim, E.-P., Ng, S.-K., Pan, S.J. (eds.) PAKDD 2020. LNCS (LNAI), vol. 12085, pp. 354–367. Springer, Cham (2020). https://doi.org/10.1007/978-3-030-47436-2_27

Optimization and Comparison of Coordinate- and Metric-Based Indexes on GPUs for Distance Similarity Searches

Michael Gowanlock[✉][ID], Benoit Gallet[ID], and Brian Donnelly[ID]

Northern Arizona University, Flagstaff, AZ, USA
{michael.gowanlock,benoit.gallet,brian.donnelly}@nau.edu

Abstract. The distance similarity search (DSS) is a fundamental operation for large-scale data analytics, as it is used to find all points that are within a search distance of a query point. Given that new scientific instruments are generating a tremendous amount of data, it is critical that these searches are highly efficient. Recently, GPU algorithms have been proposed to parallelize the DSS. While most work shows that GPU algorithms largely outperform parallel CPU algorithms, there is no single GPU algorithm that outperforms all other state-of-the-art approaches; therefore, it is not clear which algorithm should be selected based on a dataset/workload. We compare two GPU DSS algorithms: one that indexes directly on the data coordinates, and one that indexes using the distances between data points to a set of reference points. A counterintuitive finding is that the data dimensionality is not a good indicator of which algorithm should be used on a given dataset. We also find that the intrinsic dimensionality (ID) which quantifies structure in the data can be used to parameter tune the algorithms to improve performance over the baselines reported in prior work. Lastly, we find that combining the data dimensionality and ID can be used to select between the best performing GPU algorithm on a dataset.

Keywords: Distance Similarity Search · GPGPU · Metric-based Index

1 Introduction

The distance similarity search (DSS) finds objects within a search distance of points in a dataset. The distance similarity self-join (DSSJ) refers to finding all objects in a dataset within a distance, ϵ, of each other. DSS is a building block of several algorithms, including those used for scientific data analysis [12].

The *search-and-refine* strategy reduces the computational cost of the DSS, where an index prunes the search and generates a set of candidate points, which are then refined using distance calculations to compute the final result set for

This material is based upon work supported by the National Science Foundation under Grant No. 2042155.

each query point. Graphics Processing Units (GPUs) have high computational throughput due to their massive parallelism and are very effective at performing distance calculations. Numerous research on the DSS has demonstrated that the GPU is superior to multi-core CPUs [3,6,7,9,11].

The performance of GPU DSS algorithms is largely a function of data-dependent properties, such as the data distribution, dimensionality, sparsity, and variance. Thus, there is not a single GPU algorithm that is better than all other algorithms, which makes it challenging to select an algorithm to employ on a given workload. The two major classes of indexes are those that employ coordinate- and metric-based indexes [10]. The former constructs an index directly on the coordinates of the data points, whereas the latter uses distances to a set of reference points instead of indexing on the data coordinates.

This paper compares two state-of-the-art GPU algorithms having a coordinate-based index (GDS-JOIN) and a metric-based index (COSS). Because neither algorithm performs best on all datasets, this paper aims to address the following questions: (i) What data properties can be used to determine the number of indexed dimensions (GDS-JOIN) or number of reference points (COSS) to use when processing a given dataset? (ii) Which dataset properties can be used to select whether GDS-JOIN or COSS should be employed on a dataset?

2   Background: Comparison of GDS-JOIN and COSS

In this section, we compare GDS-JOIN [9] to COSS [3]. For more information, we refer the reader to those papers. GDS-JOIN uses a coordinate-based index; in contrast, COSS is a metric-based index which stores points in a grid based on their distance to a set of reference points. GDS-JOIN and COSS are similar as their grid-based indexes are GPU-friendly as they address the drawbacks of the GPU's Single Instruction Multiple Thread execution model. Both GDS-JOIN and COSS use batching schemes to compute batches of query points such that GPU global memory is not exceeded and to hide PCIe data transfer latency.

GDS-JOIN and COSS have a parameter that controls the amount of pruning they perform, such that a trade-off can be reached between index search overhead and the number of distance calculations that are computed. In this paper, this is referred to as k, which is the number of indexed dimensions for GDS-JOIN or the number of reference points for COSS. Thus, the two algorithms have similar GPU kernel designs except that they use coordinate- and metric-based indexing. As we will show in the evaluation, this distinction yields respective strengths and weaknesses which are a function of data-dependent properties.

GDS-JOIN uses two additional optimizations than the preliminary work [9]. We utilize the method by Gowanlock [8] that orders the data points from most work to least work which reduces load imbalance by assigning query points with similar amounts of work to a given warp. This is referred to as REORDER-QUERIES. Instruction-level parallelism (ILP) is also used to hide memory access latency [13]. The DSSJ in high dimensionality performs many distance calculations in the filtering phase of the algorithm. Since the pairwise components of the distance calculation can be computed independently, we exploit ILP by partially computing parts of the distance calculation and storing these partial results in registers. We use ILP, where we define r cached elements that are used to independently compute pair-wise distance calculations. Using these optimizations, GDS-JOIN achieves speedups over the preliminary work [9] in the range 1.82–5.51× across all datasets in Sect. 3.

3   Experimental Evaluation

3.1   Experimental Methodology

All GPU code is written in CUDA. The C/C++ host code is compiled with the GNU compiler and the O3 optimization flag. Our platform has 2×AMD EPYC 7542 2.9 GHz CPUs (64 total cores), 512 GiB of main memory, equipped with an Nvidia A100 GPU with 40 GiB of global memory, using CUDA 11 software.

In all experiments we exclude the time to load the dataset from disk. For all GPU algorithms, we include all other time components, including constructing the index, executing the DSSJ, storing the result set on the host and other host-side operations, and perform all pre-processing optimizations. Thus, we make a fair comparison between approaches. Reported time measurements are averaged over 3 time trials, and data is stored/processed using 64-bit floating point values. Throughout the evaluation, we report the speedup of GDS-JOIN over COSS (or vice versa), defined as the ratio of two response times, $s = T_{\text{COSS}}/T_{\text{GDS-JOIN}}$.

Selectivity of the Experiments: We perform experiments across datasets and ϵ values such that we do not have too few or too many total results. Thus, the values of ϵ should represent values that are pragmatically useful. We define the selectivity of the self-join as $S_D = (|R| - |D|)/|D|$, where $|R|$ is the total result set size. This yields the average number of points within ϵ, excluding a point, p_a, finding itself. We select values of $S_D \sim 0 - 1000$ across all datasets, which is a typical range used in this literature.

Datasets: We use seven real-world datasets that span $n = 18 - 384$ dimensions (Table 1), allowing us to observe how performance varies as a function of dimensionality. We normalize all datasets in the range $[0, 1]$. All datasets except *BigCross* [1] and *Tiny5M*[1] were obtained from the UCI ML repository[2].

Implementation Configurations: All implementations are exact (not approximate) algorithms and are parallelized using the GPU using 64-bit floating point values. **GDS-JOIN** GPU algorithm that uses a coordinate-based index. It is configured using 32 threads per block, as we found it to achieve the best performance on our platform. The default configuration for GDS-JOIN is to use all optimizations (SHORTC, REORDERDIMS, REORDERQUERIES, and ILP), index in $k = 6$ dimensions, and set $r = 8$ as the parameter for the ILP optimization.

---

[1]   https://www.cse.cuhk.edu.hk/systems/hash/gqr/dataset/tiny5m.tar.gz.
[2]   https://archive.ics.uci.edu/ml/index.php.

Table 1. Datasets used in the evaluation, where $|D|$ is the dataset size, and n is the dimensionality. ϵ_{min} and ϵ_{max} refers to the range of search distances used, and S_D^{min} and S_D^{max} refer to their corresponding selectivity values. The algorithms are configured as described in Sect. 3.1. The mean speedup (or slowdown) of GDS-JOIN compared to COSS is shown.

Dataset	$|D|$	n	$[\epsilon_{min}, \epsilon_{max}]$	$[S_D^{min}, S_D^{max}]$	Speedup GDS-JOIN over COSS
SuSy	5,000,000	18	[0.01, 0.021]	[5.17, 1090.45]	3.28
Higgs	11,000,000	28	[0.01, 0.0555]	[0.05, 1009.02]	2.15
WEC	287,999	49	[0.002, 0.007]	[39.46, 1006.39]	0.69
BigCross	11,620,300	57	[0.001, 0.02]	[2.54, 1044.7]	1.69
Census	2,458,285	68	[0.001, 0.01]	[21.64, 1077.6]	1.39
Songs	515,345	90	[0.007, 0.0091]	[126.91, 998.19]	0.70
Tiny5M	5,000,000	384	[0.2, 0.44]	[9.72, 1019.01]	0.34

The source code is publicly available.[3] **COSS** GPU algorithm that employs a metric-based index [3]. In the evaluation, COSS is configured to use 8 threads per point, and $k = 6$ reference points.

3.2   Results

Comparison of Algorithms: Table 1 compares the performance of GDS-JOIN and COSS, where the speedup is computed using the average response times across five different values of ϵ as shown in the table. We observe that GDS-JOIN achieves the greatest speedup on 4 datasets (*SuSy*, *Higgs*, *BigCross*, and *Census*) whereas COSS achieves the greatest speedup on 3 datasets (*WEC*, *Songs*, and *Tiny5M*). The two GPU algorithms have their respective niches, as performance largely depends on data properties. Furthermore, one optimization that must be selected for GDS-JOIN and COSS is the number of indexed dimensions and the number of reference points, respectively. This leads to the following questions: *When should* GDS-JOIN *or* COSS *be employed, how many dimensions (*GDS-JOIN*) or reference points (*COSS*) should be used, and what properties can we use to infer the best algorithm to employ?*

Index Dimensionality Reduction: We examine index dimensionality reduction for GDS-JOIN and COSS. Figure 1 plots the normalized response time of all real-world datasets as a function of k for GDS-JOIN, where a lower time indicates better performance. We normalize to the largest response time (yielding a value of 1 in each plot), such that we can compare the datasets across the same scale. Key observations are as follows: (*i*) Panels (b) and (e) have a parabolic shape and clearly show the trade-off between index search overhead and the number of distance calculations performed. (*ii*) Panels (a), (c), (d), and (f) are

---

[3] https://github.com/mgowanlock/gpu_self_join/.

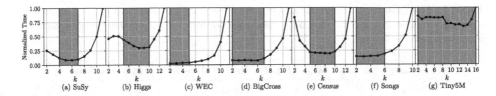

Fig. 1. The normalized response time as a function of k for all real-world datasets, where a lower normalized time is better. GDS-JOIN is executed using the median values of ϵ in Table 1; across (a)–(g) they are as follows: $\{0.0155, 0.03275, 0.0045, 0.0105, 0.0055, 0.00805, 0.32\}$. Shaded regions — the range of values of k that have a response time within 50% of the value of k that yielded the lowest response time.

similar to the above, except that the potential overhead of computing a large number of distance calculations at low values of k is absent. This is because the data is distributed into a sufficient number of grid cells such that a large number of distance calculations can be pruned even at low values of k. This illustrates how the REORDERDIMS optimization combined with indexing $k < n$ dimensions can exploit the *intrinsic dimensionality* of the data; e.g., (f) *Songs* has $n = 90$ dimensions, but yields good performance when indexed in only $k = 3$ dimensions. (*iii*) The shaded regions show values of k that yield a good response time, where k can be selected in a large range and obtain good performance on most datasets. With the exception of *WEC*, $k = 6$ yields respectable performance on all datasets. We carried out the same experiment for COSS where k refers to the number of reference points. The performance behavior between GDS-JOIN and COSS is similar so we omit discussing this.

In summary, indexing in $k < n$ dimensions (GDS-JOIN), or using $k < n$ reference points (COSS) provides a trade-off between distance comparisons and index search overhead. *Finding 1: There is not a direct correlation between data dimensionality and the number indexed dimensions (GDS-JOIN) or reference points (COSS) that should be used. Another metric for understanding data-dependent performance is needed.*

Using Intrinsic Dimensionality (ID) to Select k: The ID is the number of dimensions required to approximately represent a dataset. Intuitively, some data dimensions are correlated, so it is possible to represent the data in fewer than n dimensions. We propose a heuristic for selecting k as described in Eq. 1, where $i \geq 1$ and $n \geq k$.

$$k = 2 + \lceil c \cdot \log_2 i \rceil. \tag{1}$$

In the equation, the base of the log is 2 — conceptually, increasing k to $k + 1$ reduces the number of distance calculations by a factor of two, which yields diminishing returns when increasing k. We take the log of the intrinsic dimensionality (i) because the efficiency of pruning is directly related to the structure of the data, with a low ID being easy to prune with a small k, while a high ID requires a larger value of k to achieve a similar degree of pruning.

Table 2. The rounded intrinsic dimensionality (i) of each dataset, and k_{eqn} which is the selected value of k using Eq. 1. The data dimensionality (n) is shown for comparison. The time ratio is $T_{k=6}/T_{eqn}$.

Dataset	n	i	k_{eqn} (Eq. 1)	Ratio GDS-JOIN	Ratio COSS
SuSy	18	9	5	0.952	1.130
Higgs	28	19	8	1.301	0.882
WEC	49	15	3	1.865	1.932
BigCross	57	3	4	0.926	2.217
Census	68	13	5	1.074	1.481
Songs	90	29	4	1.290	1.426
Tiny5M	384	63	8	0.994	1.011

The notation $\lceil x \rfloor$ is the rounding function, and c is a coefficient where $c = \sqrt{\frac{|D|}{5 \times 10^6}}$. The coefficient is used to scale the number of indexed dimensions as a function of the dataset size $|D|$. This is needed because while the worst case time complexity for a sufficiently large ϵ value is $O(|D|^2)$, in practice, an average query only requires refining a fraction of the total dataset, $|D|$. Therefore, this factor scales with dataset size to limit the number of indexed dimensions when a small dataset is employed and use more dimensions when processing larger datasets. Lastly, because indexing in few dimensions is inexpensive, we index in at least 2 dimensions.

We employ an ID estimator to compute i in Eq. 1 that uses local Principle Component Analysis (PCA) [4,5], where it uses the k-nearest neighbor graph to estimate ID, and we employed the PCA algorithm from the scikit-dimension library[4]. To compute i, we used 100 nearest neighbors on all datasets[5].

Table 2 shows the estimated ID from the PCA method, and the value of k using Eq. 1. We find that this heuristic yields a very good value of k (see Fig. 1). All of the values of k yield an execution time for each dataset in the blue shaded region. Thus, ID is a good tool for determining the number of dimensions that should be indexed.

As described in Sect. 3.1, GDS-JOIN and COSS are configured by indexing in $k = 6$ dimensions and using $k = 6$ reference points, respectively, because those values were found to yield good performance across all datasets in the paper. However, $k = 6$ does not yield the best response time across all datasets. Table 2 shows the ratio of $T_{k=6}$ to T_{eqn}, which refers to the response time ratio when $k = 6$ compared to that given when using k from Eq. 1. For GDS-JOIN, we find that there are three cases where T_{eqn} is slower ($\frac{T_{k=6}}{T_{eqn}} < 1$), but the performance loss is minor. In contrast, there are four cases where T_{eqn} yields a faster response

---

[4] https://scikit-dimension.readthedocs.io/en/latest/.

[5] Due to excessive execution times, we sampled *Higgs* and *Tiny5M* and ensured that 100 neighbors are sufficient across all datasets, and that sampling *Higgs* and *Tiny5M* did not adversely impact ID estimation.

time ($\frac{T_{k=6}}{T_{eqn}} > 1$) where substantial performance gains are achieved on *Higgs*, *WEC*, and *Songs*, yielding a ratio between 1.07–1.87×.

Examining the abovementioned ratio for COSS, we find that there is only one dataset (*Higgs*) where using k_{eqn} yields a slowdown; all other datasets yield a ratio between 1.01–2.22×. Using ID to estimate k reaches a good trade-off between index search overhead and the number of distance comparisons. It is for this reason that other work finds that using $k \approx 6$ reference points yields good performance [2]. We also find that our heuristic yields similar values of k. *Finding 2: The ID is a much better indicator of the number of dimensions that should be indexed than the data dimensionality.*

Table 3. The best GPU algorithm (GDS-JOIN or COSS) from Table 1 is shown, with the values of n and k_{eqn} from Table 2. The ratio of the n/k_{eqn} indicates whether GDS-JOIN or COSS should be employed. We excluded *BigCross* as neither GPU algorithm outperformed the CPU algorithms on that dataset.

Dataset	Best GPU Alg. (Table 1)	n	k_{eqn} (Eq. 1)	n/k_{eqn}
SuSy	GDS-JOIN	18	5	3.60
Higgs	GDS-JOIN	28	8	3.50
WEC	COSS	49	3	16.33
Census	GDS-JOIN	68	5	13.60
Songs	COSS	90	4	22.50
Tiny5M	COSS	384	8	48.00

When Should GDS-JOIN or COSS Be Employed? Both algorithms have distinct niches, but it is not clear from the results thus far under what circumstances GDS-JOIN or COSS should be employed. By definition, metric-based indexes (COSS) are more effective than coordinate-based indexes (GDS-JOIN) when there is less structure in the data indicating lower ID, and so by indexing in the metric space, they are able to better prune the search in instances where coordinate-based indexes cannot.

Table 3 shows the algorithm that achieved the best performance in Table 1. Also reported is n/k_{eqn}, where k_{eqn} is a function of the ID. Intuitively, this indicates how much pruning the index can accomplish relative to all n dimensions. Because COSS outperforms GDS-JOIN when there is less structure in the data (lower ID), we find that when $n/k_{eqn} \geq 16$ COSS should be employed, and likewise when $n/k_{eqn} < 16$ GDS-JOIN should be employed, which indicates that there is more structure in the data, or a higher ID relative to the total number of data dimensions, n. Consequently, one of the GPU algorithms can be selected based on n and k_{eqn}, which is a function of data-dependent properties. *Finding 3: Datasets with greater ID and/or fewer data dimensions are best processed by coordinate-based indexes (GDS-JOIN), whereas datasets with lower ID and/or higher dimensions should be processed by metric-based indexes (COSS).*

4 Discussion and Conclusions

This paper examined fundamental data properties that can be used to: (*i*) improve the performance of GDS-JOIN and COSS; and, (*ii*) determine under which conditions GDS-JOIN or COSS should be employed. Algorithm selection allows for computing DSS more robustly, as we are now able to select an algorithm that performs well depending on the characteristics of a dataset. Using the proposed heuristics, we find that GDS-JOIN or COSS can be selected largely based on the *intrinsic dimensionality* and the number of data dimensions.

Acknowledgements. We thank Ben Karsin for his contributions to the preliminary version of this paper.

References

1. Ackermann, M.R., Märtens, M., Raupach, C., Swierkot, K., Lammersen, C., Sohler, C.: Streamkm++ a clustering algorithm for data streams. J. Exper. Algorithmics (JEA) **17**, 1–2 (2012)
2. Chen, L., Gao, Y., Li, X., Jensen, C.S., Chen, G.: Efficient metric indexing for similarity search. In: 2015 IEEE 31st International Conference on Data Engineering, pp. 591–602. IEEE (2015)
3. Donnelly, B., Gowanlock, M.: A coordinate-oblivious index for high-dimensional distance similarity searches on the GPU. In: Proceedings of the 34th ACM International Conference on Supercomputing, pp. 1–12 (2020)
4. Fan, M., Gu, N., Qiao, H., Zhang, B.: Intrinsic dimension estimation of data by principal component analysis. arXiv preprint arXiv:1002.2050 (2010)
5. Fukunaga, K., Olsen, D.R.: An algorithm for finding intrinsic dimensionality of data. IEEE Trans. Comput. **100**(2), 176–183 (1971)
6. Gallet, B., Gowanlock, M.: Heterogeneous CPU-GPU epsilon grid joins: static and dynamic work partitioning strategies. Data Sci. Eng. **6**(1), 39–62 (2021)
7. Gallet, B., Gowanlock, M.: Leveraging GPU tensor cores for double precision Euclidean distance calculations. In: 2022 IEEE 29th International Conference on High Performance Computing, Data, and Analytics (HiPC), pp. 135–144 (2022)
8. Gowanlock, M.: Hybrid KNN-join: parallel nearest neighbor searches exploiting CPU and GPU architectural features. J. Parallel Distribut. Comput. **149**, 119–137 (2021)
9. Gowanlock, M., Karsin, B.: GPU-accelerated similarity self-join for multi-dimensional data. In: Proceedings of the 15th International Workshop on Data Management on New Hardware, pp. 6:1–6:9. ACM (2019)
10. Hjaltason, G.R., Samet, H.: Index-driven similarity search in metric spaces (survey article). ACM Trans. Database Syst. **28**(4), 517–580 (2003)
11. Lieberman, M.D., Sankaranarayanan, J., Samet, H.: A fast similarity join algorithm using graphics processing units. In: IEEE 24th International Conference on Data Engineering, pp. 1111–1120 (2008)
12. Trilling, D.E., et al.: The solar system notification alert processing system (snaps): design, architecture, and first data release (snapshot1). Astron. J. **165**(3), 111 (2023)
13. Volkov, V.: Better performance at lower occupancy (2010). https://www.nvidia.com/content/GTC-2010/pdfs/2238_GTC2010.pdf. Accessed 1 Jan 2023

Sentiment Analysis Using Machine Learning Approach Based on Feature Extraction for Anxiety Detection

Shoffan Saifullah[1,2]([⊠]) [iD], Rafał Dreżewski[1]([⊠]) [iD], Felix Andika Dwiyanto[1] [iD],
Agus Sasmito Aribowo[2] [iD], and Yuli Fauziah[2] [iD]

[1] Institute of Computer Science, AGH University of Science and Technology,
Kraków, Poland
{saifulla,drezew,dwiyanto}@agh.edu.pl
[2] Department of Informatics, Universitas Pembangunan Nasional Veteran
Yogyakarta, Yogyakarta, Indonesia
{shoffans,sasmito.skom,yuli.fauziah}@upnyk.ac.id

Abstract. In this study, selected machine learning (ML) approaches were used to detect anxiety in Indonesian-language YouTube video comments about COVID-19 and the government's program. The dataset consisted of 9706 comments categorized as positive and negative. The study utilized ML approaches, such as KNN (K-Nearest Neighbors), SVM (Support Vector Machine), DT (Decision Tree), Naïve Bayes (NB), Random Forest (RF), and XG-Boost, to analyze and classify comments as anxious or not anxious. The data was preprocessed by tokenizing, filtering, stemming, tagging, and emoticon conversion. Feature extraction (FE) is performed by CV (count-vectorization), TF-IDF (term frequency-inverse document frequency), Word2Vec (Word Embedding), and HV (Hashing-Vectorizer) algorithms. The 24 of the ML and FE algorithms combinations were used to achieve the best performance in anxiety detection. The combination of RF and CV obtained the best accuracy of 98.4%, which is 14.3% points better than the previous research. In addition, the other ML methods accuracy was above 92% for CV, TF-IDF, and HV, while KNN obtained the lowest accuracy.

Keywords: Anxiety Detection · Machine Learning · Sentiment Analysis · Text Feature Extraction · Text Mining · Model Performance

1 Introduction

Anxiety is a mental disorder [2] related to the nervous system, with characteristics such as considerable and persistent anxiety, excitation of autonomic nervous activity, and excessive alertness. The types of anxiety disorders include, among others, generalized anxiety disorder, panic disorder, and social anxiety disorder [8]. The COVID-19 pandemic caused anxiety and stress for the Indonesian people and government, who implemented programs to reduce the pandemic,

J. Mikyška et al. (Eds.): ICCS 2023, LNCS 14074, pp. 365–372, 2023.
https://doi.org/10.1007/978-3-031-36021-3_38

with a waiver of medical supplies, community assistance, and a waiver of electricity bills. However, the program had many pros and cons comments on social media, which indicated the growth of public anxiety and eventually could cause a global anxiety pandemic if not addressed correctly by future government programs [1].

While psychologists can analyze a person's anxiety, using computer technology we can quickly analyze a large amount of social media data. In this study, the artificial intelligence (AI) and sentiment analysis algorithms are used to detect anxiety [20] based on text processing [17,19] of comments shared on social media concerning COVID-19 [18]. In the proposed approach, data consisting of YouTube comments on Indonesian government COVID-19 programs is processed in sequential steps using machine learning methods such as K-NN, NB, DT, SVM, RF, and XG-boost, and feature extraction methods, such as CV, TF-IDF, HV, and Word2Vec.

This paper consists of five main sections. After the introduction, Sect. 2 presents the related research works. Section 3 explains the proposed method and the research steps. The conducted experiments and obtained results are discussed in Sect. 4. The conclusions based on experimental results are presented in Sect. 5.

2   Related Research

Emotion detection can be performed using the data science approach, text mining algorithms and sentiment analysis methods on text data from social media such as Twitter, YouTube, and Facebook. Negative sentiment on social media around topics such as gender, ethnicity, and religion can be used as input data to detect emotions.

An approach to detecting hate speech in text documents using 2 or 3 labels, and Case-Based Reasoning (CBR) and Naïve Bayes (NB) classification methods was applied to detect emotions [9] and bigotry, achieving an accuracy of over 77% [3]. Several researchers analyzed Arabic language sentiment using Random Forest (RF) and got low accuracy of 72% [15]. However, C4.5, RIPPER, and PART methods increased sentiment classification accuracy to 96% and Support Vector Machine (SVM) and Naïve Bayes (NB) with term frequency-inverse document frequency (TF-IDF) have an accuracy of 82.1% in detecting sentiment analysis [3]. Sentiment analysis has been used to recognize emotions and hate speech in Facebook comments in Italian [9] and English [14] with ML methods, such as the SVM, Recurrent Neural Network (RNN), and Long Short Term Memory (LSTM). A lexicon approach based on a dictionary and sentiment corpus has been used to obtain 73% accuracy in [14]. A binary classifier to distinguish between neutral and hate speech was applied in [10].

Different techniques, like paragraph2vec, Continuous Bag of Words (CBOW), and embedding-binary classifier are also used to perform sentiment analysis [10]. Natural Language Processing (NLP) is utilized [23] to automatically detect emotions about nation, religion, and race [14] in sentiment analysis of Facebook comments [9]. Twitter and YouTube comments can also be used to detect user expectations and identify mass anxiety and fear, such as natural disasters (e.g., earthquakes) and political battles [6,7].

To improve the results of anxiety detection, in this paper we applied several machine learning methods such as K-Nearest Neighbors (KNN), Support Vector Machine (SVM), Decision Tree (DT), Naïve Bayes (NB), Random Forest (RF), and XG-Boost, together with selected feature extraction (FE) methods like count-vectorization (CV), term frequency-inverse document frequency (TF-IDF), Word Embedding (Word2Vec), and Hashing-Vectorizer (HV). Moreover, the ensemble concept was used to ensure the optimal results.

3 Proposed Method

The research presented in this paper builds on previous results using 6 ML methods and 4 FE methods to detect anxiety based on sentiment and emotional analysis of YouTube text comments. The proposed approach involves four main steps (data collection, preprocessing, feature extraction, and classification) to identify anxiety levels based on sentiment analysis of Indonesian language YouTube comments. This study also adopts a prototyping method with system modeling to identify sentiments in emotional data from social media [4,13]. As shown in Fig. 1, the process involves preprocessing, emotion detection based on sentiment analysis, and cross-validation testing.

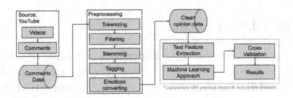

Fig. 1. The sentiment analysis flow using ML algorithms on YouTube comments.

ML methods such as RF and XG-Boost [11,22] are used (along with several other methods like KNN, NB, DT, and SVM to compare the results with [22]) together with selected feature extraction (FE) algorithms. The ML methods results are assessed using the confusion matrix to calculate the performance (accuracy, precision, recall, and F1 Score).

KNN identifies classes based on a distance matrix, and the best K value classification [21] is found using Euclidean Distance (ED) matrix. The second ML method is NB, which uses binary features as vector attributes to identify words based on the probability of their occurrence. The method is robust and can handle noise and missing data. DT is a method, represented as a tree structure, that can be used for data classification and pattern prediction. The relationship between the attribute variable x and the target y is depicted using internal nodes as attribute tests, branches as test results, and outer nodes as labels. SVM are supervised learning models used for classification and regression. SVM requires training and testing phases to find the best hyperplane that acts as a separator of two data classes. It works on high-dimensional datasets and uses a few selected data points to form a model (support vector).

RF uses ensemble learning and can be used to solve regression and classification problems (also based on sentiment analysis [16]). RF can reduce the problems of overfitting and missing data, and it can handle datasets containing categorical variables. Extreme Gradient Boosting (XG-Boost) is a tree-based algorithm that can be applied to classification and regression problems [12]. This algorithm mimics the RF behavior and is combined with gradient descent/boosting. Gradient Boosting is a machine learning concept used to solve regression and classification problems, which produces a prediction model in the form of an ensemble of weak prediction models. XG-Boost is a version of GBM (Gradient Boosting Machine) with some advantages, including accuracy, efficiency and scalability, which works well for applications such as regression, classification, and ranking.

The proposed approach also utilizes selected FE methods (see Fig. 1), including CV, TF-IDF, HV, and Word2Vec, to convert text to its numerical representation that is then used by ML models. The CV method calculates the frequency of occurrence of the detected words, while TF-IDF is a numerical statistic method used for weighting text data. HV transforms a collection of text documents into a matrix of token occurrences. Word2Vec generates a vector representation for each word in a corpus, based on the context in which it appears.

4   Results and Discussion

This section presents and discusses the experimental results of sentiment analysis for anxiety detection using ML approaches. We discuss some key issues like the used dataset and its labeling, the experimental evaluation of the proposed approach, and comparison with the previous research.

The dataset contains a total of 9,706 YouTube comments related to the Indonesian government's COVID-19 program, with 4,862 data from previous research [11,22] and 4,844 newly crawled comments. The comments are labeled as positive ("0") or negative, with negative comments indicating anxiety [5] but not necessarily the hate speech. The dataset was expanded to balance the number of positive and negative comments and avoid overfitting. Currently, the number of positive comments has almost equaled the number of negative comments (the number of positive comments is now about 90% of the number of negative comments, compared to about 50% in previous studies)—see Fig. 2a.

Figure 2a compares Indonesian YouTube comments dataset with those used in previous studies. The dataset is labeled based on the application of sentiment analysis in the process of anxiety detection. The dataset cleaning process involves using "Literature" library and adding stop-words (757) and true-words (13770). Figure 2b shows the sample YouTube comments in Indonesian, categorized as positive or negative.

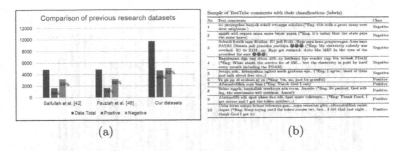

Fig. 2. (a) Comparison with previous research data, and (b) sample datasets.

The experimental results indicated that the dataset requires preprocessing, including tokenizing, filtering (slang words conversion, removing numbers, removing stop-words, removing figures, removing duplicates), stemming, tagging, and emoticon conversion. The dataset is split into the training set (80%) and testing set (20%) using a Python script. Additional rules are added to improve the data cleaning, such as assigning emotional trust, confidence, and anger based on specific criteria. The conversion of emoticons to text indicates the emotional expression. For example, in data in row 3 (presented in Fig. 2b), the emoticon is converted to *"face_with_tears_of_joy."*. The result of preprocessing is clean data (without meaningless or useless text) that can be used as input to the next process.

In this research, we evaluated 24 modeling scenarios using 4 performance metrics of the confusion matrix. Furthermore, the new data was used to compute the confusion matrix, as shown in Fig. 3. The best accuracy of 98.4% was achieved using RF-CV. The other ML methods, such as SVM, DT, and XG-Boost, can identify anxiety with the accuracy of over 92%. Word2Vec has the lowest accuracy when used with most ML methods, except when used with KNN, in which case it performs better than the rest of the FE methods used with KNN. It also obtained better results than the NB-HV method. It is because Word2Vec converts words into vectors and is trained between conditional sentences.

The RF algorithm when used with all FE methods is highly accurate and balanced in terms of precision, recall, and F1 Score. However, RF with Word2Vec has only 81.9% accuracy as compared to the other FE methods (above 96%). In addition, the RF method is superior in detecting anxiety based on sentiment analysis, with consistent performance close to or above 95%. The final results (Fig. 3) showed that SVM, DT, RF, and XG-Boost methods used with CV, TF-IDF, and HV achieved the accuracy close to or above 90%. In addition, KNN and NB obtained lower accuracy compared to other ML methods. The Word2Vec obtained the lowest accuracy in each experiment (less than 82% for all ML models used). However, despite its poor performance, the Word2Vec method is superior to other FE methods when used with KNN algorithm. HV method obtained higher accuracy (94%–96%) when used with the SVM, DT, and RF algorithms.

The research presented in this paper improves the previous results [11,22] on anxiety detection based on sentiment analysis, resulting in enhanced methods

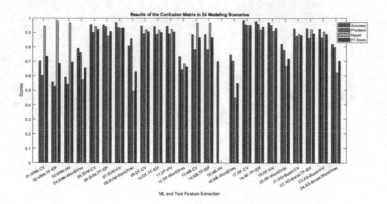

Fig. 3. The performance of ML models used with FE methods (24 scenarios).

and outcomes, as shown in Table 1. In addition, we conducted experiments using previous research data and newly added crawled data, shown in Fig. 2.

Table 1. Comparison of the obtained results (FE: Feature Extraction; Acc: Accuracy; Prec: Precision; Rec: Recall; F1: F1 Score).

No	Method	FE	Saifullah et al. [22]				Initial improvement				Our proposed model			
			Acc	Prec	Rec	F1	Acc	Prec	Rec	F1	Acc	Prec	Rec	F1
1	KNN	CV	0.601	0.468	0.927	–	0.606	0.471	0.924	0.624	0.701	0.602	0.943	0.735
2		TF-IDF	0.391	0.364	0.968	–	0.391	0.364	0.968	0.529	0.557	0.526	0.981	0.685
3		HV	–	–	–	–	0.464	0.395	0.971	0.561	0.589	0.540	0.966	0.693
4		Word2Vec	–	–	–	–	0.687	0.594	0.358	0.446	0.790	0.760	0.572	0.653
5	SVM	CV	0.815	0.797	0.640	–	0.815	0.801	0.634	0.708	0.954	0.899	0.941	0.920
6		TF-IDF	0.799	0.866	0.509	–	0.798	0.866	0.506	0.639	0.953	0.940	0.875	0.906
7		HV	–	–	–	–	0.802	0.854	0.529	0.654	0.968	0.934	0.928	0.931
8		Word2Vec	–	–	–	–	0.687	0.831	0.142	0.243	0.805	0.856	0.493	0.626
9	DT	CV	0.804	0.715	0.738	–	0.8	0.713	0.724	0.719	0.943	0.891	0.918	0.904
10		TF-IDF	0.806	0.720	0.738	–	0.806	0.716	0.747	0.731	0.939	0.885	0.915	0.900
11		HV	–	–	–	–	0.81	0.728	0.738	0.733	0.942	0.888	0.921	0.904
12		Word2Vec	–	–	–	–	0.665	0.526	0.526	0.526	0.732	0.640	0.681	0.660
13	NB	CV	0.823	0.839	0.619	–	0.824	0.839	0.622	0.715	0.885	0.781	0.966	0.864
14		TF-IDF	0.823	0.839	0.619	–	0.824	0.839	0.622	0.715	0.885	0.781	0.966	0.864
15		HV	–	–	–	–	0.646	0	0	0	0.698	0.000	0.000	0.000
16		Word2Vec	–	–	–	–	0.66	0.53	0.331	0.408	0.744	0.701	0.447	0.546
17	RF	CV	0.850	0.786	0.791	–	0.841	0.776	0.773	0.774	0.984	0.954	0.948	0.951
18		TF-IDF	0.826	0.785	0.701	–	0.827	0.778	0.715	0.745	0.976	0.959	0.918	0.938
19		HV	–	–	–	–	0.83	0.81	0.68	0.739	0.969	0.950	0.909	0.929
20		Word2Vec	–	–	–	–	0.73	0.683	0.439	0.535	0.819	0.779	0.666	0.718
21	XG-Boost	CV	0.732	0.895	0.273	–	0.732	0.895	0.273	0.419	0.925	0.874	0.889	0.881
22		TF-IDF	0.747	0.871	0.334	–	0.744	0.868	0.326	0.474	0.928	0.863	0.921	0.891
23		HV	–	–	–	–	0.738	0.83	0.326	0.468	0.924	0.864	0.904	0.884
24		Word2Vec	–	–	–	–	0.739	0.732	0.413	0.528	0.816	0.796	0.621	0.700

The proposed methods achieved better results than previous studies on anxiety detection based on sentiment analysis. The RF-CV method had the highest improvement, with a 13.4% points increase in accuracy (from 85% to 98.4%). The other methods (SVM, DT, NB, and XG-Boost) outperformed previous studies with the accuracy above 85%. The addition of several processes in the preprocessing phase (including stop-words) and the extension of the dataset have improved the proposed method's performance.

5 Conclusions

The research presented in this paper improved the accuracy of methods proposed in [22] by using additional datasets, preprocessing, and text feature extraction to better detect psychological factors. The proposed machine learning based approach is a contribution to the research on detecting anxiety based on sentiment analysis. The best and recommended method is RF-CV, which has obtained the accuracy of 98.4% and consistent precision, recall, and F1 scores with values over 95%. SVM, DT, and XG-Boost methods had also good accuracy, but their performance still needs improvement. The future work will include the application of optimization algorithms and Deep Learning methods.

Acknowledgement. The research presented in this paper was partially supported by the funds of Polish Ministry of Education and Science assigned to AGH University of Science and Technology.

References

1. Ahmad, A.R., Murad, H.R.: The impact of social media on panic during the COVID-19 pandemic in Iraqi Kurdistan: Online questionnaire study. J. Med. Internet Res. **22**(5), e19556 (2020). https://doi.org/10.2196/19556
2. Ahmed, A., et al.: Thematic analysis on user reviews for depression and anxiety chatbot apps: Machine learning approach. JMIR Format. Res. **6**(3), e27654 (2022)
3. Almonayyes, A.: Tweets classification using contextual knowledge and boosting. Int. J. Adv. Electron. Comput. Sci. **4**(4), 87–92 (2017)
4. Bhati, R.: Sentiment analysis a deep survey on methods and approaches. Int. J. Disaster Recovery Bus. Continuity **11**(1), 503–511 (2020)
5. Cahyana, N.H., Saifullah, S., Fauziah, Y., Aribowo, A.S., Drezewski, R.: Semi-supervised text annotation for hate speech detection using k-nearest neighbors and term frequency-inverse document frequency. Int. J. Adv. Comput. Sci. Appl. **13**(10) (2022). https://doi.org/10.14569/ijacsa.2022.0131020
6. Calderón-Monge, E.: Twitter to manage emotions in political marketing. J. Promot. Manag. **23**(3), 359–371 (2017)
7. Chin, D., Zappone, A., Zhao, J.: Analyzing Twitter sentiment of the 2016 presidential candidates. In: Applied Informatics and Technology Innovation Conference (AITIC 2016) (2016)
8. Czornik, M., Malekshahi, A., Mahmoud, W., Wolpert, S., Birbaumer, N.: Psychophysiological treatment of chronic tinnitus: a review. Clin. Psychol. Psychoth. **29**(4), 1236–1253 (2022). https://doi.org/10.1002/cpp.2708

9. Del Vigna, F., Cimino, A., Dell'Orletta, F., Petrocchi, M., Tesconi, M.: Hate me, hate me not: Hate speech detection on Facebook. In: Proceedings of the First Italian Conference on Cybersecurity (ITASEC17), vol. 1816, pp. 86–95 (2017)

10. Djuric, N., Zhou, J., Morris, R., Grbovic, M., Radosavljevic, V., Bhamidipati, N.: Hate speech detection with comment embeddings. In: Proceedings of the 24th International Conference on World Wide Web. ACM (2015)

11. Fauziah, Y., Saifullah, S., Aribowo, A.S.: Design text mining for anxiety detection using machine learning based-on social media data during COVID-19 pandemic. In: Proceeding of LPPM UPN "Veteran" Yogyakarta Conference Series 2020-Engineering and Science Series vol. 1, no. 1, pp. 253–261 (2020)

12. Georganos, S., Grippa, T., Vanhuysse, S., Lennert, M., Shimoni, M., Wolff, E.: Very high resolution object-based land use–land cover urban classification using extreme gradient boosting. IEEE Geosci. Remote Sens. Lett. **15**(4), 607–611 (2018). https://doi.org/10.1109/lgrs.2018.2803259

13. Giannakis, M., Dubey, R., Yan, S., Spanaki, K., Papadopoulos, T.: Social media and sensemaking patterns in new product development: demystifying the customer sentiment. Ann. Oper. Res. 145–175 (2020). https://doi.org/10.1007/s10479-020-03775-6

14. Gitari, N.D., Zhang, Z., Damien, H., Long, J.: A lexicon-based approach for hate speech detection. Int. J. Multimedia Ubiquitous Eng. **10**(4), 215–230 (2015). https://doi.org/10.14257/ijmue.2015.10.4.21

15. Kléma, J., Almonayyes, A.: Automatic categorization of fanatic text using random forests. Kuwait J. Sci. Engrg. **33**(2), 1–18 (2006)

16. Kumar, S., Yadava, M., Roy, P.P.: Fusion of EEG response and sentiment analysis of products review to predict customer satisfaction. Inf. Fusion **52**, 41–52 (2019). https://doi.org/10.1016/j.inffus.2018.11.001

17. Muñoz, S., Iglesias, C.A.: A text classification approach to detect psychological stress combining a lexicon-based feature framework with distributional representations. Inf. Process. Manage. **59**(5), 103011 (2022)

18. Ni, M.Y., et al.: Mental health, risk factors, and social media use during the COVID-19 epidemic and cordon sanitaire among the community and health professionals in wuhan, china: Cross-sectional survey. JMIR Mental Health **7**(5), e19009 (2020). https://doi.org/10.2196/19009

19. Nijhawan, T., Attigeri, G., Ananthakrishna, T.: Stress detection using natural language processing and machine learning over social interactions. J. Big Data **9**(1), 1–24 (2022). https://doi.org/10.1186/s40537-022-00575-6

20. Ragini, J.R., Anand, P.R., Bhaskar, V.: Big data analytics for disaster response and recovery through sentiment analysis. Int. J. Inf. Manage. **42**, 13–24 (2018). https://doi.org/10.1016/j.ijinfomgt.2018.05.004

21. Rezwanul, M., Ali, A., Rahman, A.: Sentiment analysis on twitter data using KNN and SVM. Int. J. Adv. Comput. Sci. Appl. **8**(6) (2017). https://doi.org/10.14569/ijacsa.2017.080603

22. Saifullah, S., Fauziyah, Y., Aribowo, A.S.: Comparison of machine learning for sentiment analysis in detecting anxiety based on social media data. Jurnal Informatika **15**(1), 45 (2021). https://doi.org/10.26555/jifo.v15i1.a20111

23. Schmidt, A., Wiegand, M.: A survey on hate speech detection using natural language processing. In: Proceedings of the Fifth International Workshop on Natural Language Processing for Social Media. Association for Computational Linguistics (2017). https://doi.org/10.18653/v1/w17-1101

Efficiency Analysis for AI Applications in HPC Systems. Case Study: K-Means

Jose Rivas[✉][iD], Alvaro Wong[iD], Remo Suppi[iD], Emilio Luque[iD],
and Dolores Rexachs[iD]

Universitat Autònoma de Barcelona, 08193 Barcelona, Spain
jose.rivas@autonoma.cat,
{alvaro.wong,remo.suppi,emilio.luque,dolores.rexachs}@uab.es

Abstract. Currently, many AI applications require large-scale computing and memory to solve problems. The combination of AI applications and HPC sometimes does not efficiently use the available resources. Furthermore, a lot of idle times is caused by communication, resulting in increased runtime. This paper describes a methodology for AI parallel applications that analyses the performance and efficiency of these applications running on HPC resources to make decisions and select the most appropriate system resources. We validate our proposal by analysing the efficiency of the K-Means application obtaining an efficiency of 99% on a target machine.

Keywords: AI applications · HPC performance and efficiency · PAS2P methodology

1 Introduction

In recent years, the convergence of High Performance Computing (HPC) and Artificial Intelligence (AI) has become increasingly relevant as the advanced HPC technology has enhanced the processing power needed for large-scale AI applications [1,2]. HPC systems have traditionally been used for scientific simulations and modelling. Nevertheless, with the rise of AI, these systems can now be leveraged for complex AI workloads such as deep learning or neural networks.

As these applications become more complex and are used on HPC systems, it is crusial tu ensure that the AI applications are running efficiently. One problem that parallel AI applications face is ensuring that they are utilising the full processing power of HPC systems on which they are executed.

We propose to provide information about these applications' performance and identify segments of the program that could be improved. To achieve this objective, we first evaluate the PAS2P (Parallel Application Signature for Performance Prediction) tool used to predict the performance of a parallel scientific application as well as to analyse if it is possible to generate a model for AI applications.

J. Mikyška et al. (Eds.): ICCS 2023, LNCS 14074, pp. 373–380, 2023.
https://doi.org/10.1007/978-3-031-36021-3_39

The PAS2P methodology [9] is based on characterising the dynamic behaviour of MPI applications on their execution. PAS2P instruments the application to analyse the events it has captured and search for repetitive patterns defined as phases. Each phase is assigned a weight determined by the number of times a pattern repeats. For the performance prediction, PAS2P generates a signature that is constituted by phases. When executed, we obtain the execution times of each phase. By multiplying these times by the weights, we obtain the execution time prediction.

With the PAS2P tool, we can characterise an application in a reduced set of phases, which allows us to focus the efficiency analysis on the phases of the application and later extrapolate this analysis to the entire application. However, to validate our proposal, it is necessary to analyse the application model generated by PAS2P for AI applications in order to analyse the efficiency of AI applications on HPC systems so as to produce a comprehensive report highlighting the areas (phases) in the code that can be improved. In this case, we apply the efficiency analysis to the K-Means application.

This paper is organised as follows: Sect. 2 provides an overview of related works in the realm of AI application performance on HPC systems and previous PAS2P works to characterise scientific applications in HPC environments. Section 3 presents how PAS2P models the AI applications, and the proposed methodology for the efficiency analysis outlines the approach and the three steps taken in the study. Section 4 presents the efficiency analysis results applied to a K-Means application. Finally, in the last section, we offer the conclusions and propose future work.

2 Related Works

There are tools related to the performance of AI parallel applications running on HPC systems. For example, Z. Fink et al. [4], the autors focus on evaluating the performance of two Python parallel programming models: Charm4Py [5], and mpi4py [6]. The authors argue that Python is rapidly becoming a common language in machine learning and scientific computing, and several frameworks scale Python across nodes. However, more needs to be known about their strengths and weaknesses.

N. Alnaasan et al. [7] introduce OMB-Py, a Python micro-benchmark for evaluating MPI library performance on HPC systems. The authors argue that Python has become a dominant programming language in emerging areas such as machine learning, deep learning, and data science. The paper proposes OMB-Py, Python extensions of the open-source OSU Micro-Benchmark (OMB) suite, in order to evaluate the communication performance of MPI-based parallel applications in Python. There are other proposals on the importance of evaluating and improving the performance of AI applications in HPC environments [3]. However, one of the main difficulties with benchmarks is selecting the benchmark most similar to the application you want to evaluate on a specific system. The difference between these works and ours is that with PAS2P, we obtain the application's benchmark (Application Signature) executed in a bounded time representing the application behaviour.

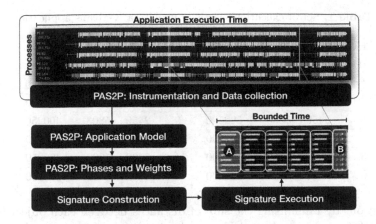

Fig. 1. Overview of PAS2P Methodology.

In previous work on PAS2P [9, 10], methodologies were developed which focused on both applications of SPMD scientific computing and an extension of PAS2P for applications with irregular behaviour, both for HPC environments.

As shown in Fig. 1, PAS2P instruments and runs the application in a target machine, producing trace logs. The data obtained is used to characterise the behaviour of computing and communication events that describe the application behaviour. First, PAS2P assigns a global logical clock to the event to obtain the application model according to the causal relationships between the communication events. Then, once PAS2P has the application model, it identifies and extracts the sequences of the most relevant events defined as phases.

Once we have the phases, PAS2P assigns a weight to each phase, defined as the number of times the phase occurs during the execution. Finally, the signature is represented by a set of phases that can be executed to measure the application performance. For performance prediction, the signature execution in different target systems allows us to measure the execution time of each phase. Therefore it calculates the runtime of the entire application in each of those systems. This is achieved by using equation (1), where the $PhaseET_i$ is the estimated time for each phase i, and W_i is the weight for each phase i.

$$PET = \sum_{i=1}^{n}(PhaseET_i)(W_i) \tag{1}$$

3   Efficiency Analysis Model over Application Phases

With the recent advancements in HPC hardware and the surge in AI, we propose to extend the contributions of the PAS2P methodology to the field of AI. To evaluate the PAS2P application model and propose an efficiency model, we take the AI application K-Means, an unsupervised classification (clustering) algorithm that groups objects into k groups based on their characteristics [8]. Clustering

Table 1. Set of executions carried out with a K-Means application.

Exp ID	Data seed	Centroid seed	Sites per process	Exp ID	Data seed	Centroid seed	Sites per process
1	31359	5803	522500	7	31359	5803	400250
2	31359	18036	522500	8	31359	18036	400250
3	2450	5803	522500	9	2450	5803	400250
4	2450	18036	522500	10	2450	18036	400250
5	19702	5803	522500	11	19702	5803	400250
6	19702	18036	522500	12	19702	18036	400250

is carried out by minimising the sum of distances between each object and the centroid of its group or cluster. The quadratic distance is often used.

There is a need to expand the concept of the performance of a parallel application beyond simply predicting its execution time. The previous PAS2P methodology primarily focuses on predicting execution time as a performance measure. We then propose an extension to the meaning of efficiency by defining the performance of a phase as the ratio between the computational time and its total execution time.

As per our proposal, first, we need to evaluate if PAS2P can characterise the AI application. To validate the characterisation, we instrument and analyse the execution of the K-Means application using PAS2P and construct the application signature. We suppose that this signature accurately predicts the application execution time with the same dataset and conditions as the application execution. In that case, we can validate that the PAS2P methodology generates an application model that represents the application behaviour, in order to prove experimentally our hypothesis: the application has the same structure (same phases) for different datasets and different initial conditions, but with different amounts of repetition (different weights).

We have carried out a set of executions on the K-means application; considering the data and input parameters (K defined as the number of centroids) of the K-means application, we vary the dataset size and the initial number of centroids of each execution. We can control the dataset by changing the random seed to initialise a pseudorandom number generator of both the dataset and the initial centroids. The dimensionality of the data points was set to 16 coordinates, and the algorithm was executed, defining 24 clusters. We conducted our experiments on a 256-process distributed system consisting of 4 compute nodes, each equipped with 64 cores, as shown in Table 1.

The procedure mentioned in [9] is applied using the PAS2P methodology. First, the application is analyzed, and its signature is created for each experiment. Then, we use the signature to predict the execution time for a given machine and a given configuration, according to the values in Table 1. Finally, when we compare the execution time of the application with the time predicted by PAS2P for each of the experiments, an average prediction error of 4.3% is obtained.

Table 2. Weight Variation obtained when K-Means is executed with different datasets.

Exp ID	Phase	Weight	Number of Instructions	Exp ID	Phase	Weight	Number of Instructions
1	1	291	5306507150	2	1	225	5306813225
1	2	290	463	2	2	224	463
1	3	292	463	2	3	226	463
1	4	291	469	2	4	225	469
3	1	285	5306511267	4	1	232	5306812383
3	2	284	463	4	2	231	463
3	3	286	463	4	3	233	463
3	4	285	469	4	4	232	469
5	1	306	5306506593	6	1	230	5306812411
5	2	305	463	6	2	229	463
5	3	307	463	6	3	231	463
5	4	306	469	6	4	230	469
7	1	284	4064938012	8	1	229	4065170526
7	2	283	463	8	2	228	463
7	3	285	463	8	3	230	463
7	4	284	469	8	4	229	469
9	1	294	4064940486	10	1	234	4065172879
9	2	293	463	10	2	233	463
9	3	295	463	10	3	235	463
9	4	294	469	10	4	234	469
11	1	321	4064938896	12	1	232	4065170309
11	2	320	463	12	2	231	463
11	3	322	463	12	3	233	463
11	4	321	469	12	4	232	469

So, experimentally, we can say that PAS2P performs a correct analysis of the phases in such a way as to perform a reduction of an application to a signature of it, with an average reduction of 8.3% in the execution time of the signature in relation to the total execution time.

For the second part of the experiment, we used the PAS2P tool to obtain traces and extract information about the program's structure for each experiment. The analysis of the data, following our hypothesis, is presented in Table 2. The results indicate that the application's structure remains consistent across all cases, with four phases identified. However, the weights of each phase, and the number of times each phase is repeated, vary among the datasets. This observation can be explained by the fact that the application uses the same steps for the experiment but with varying repetitions for each phase.

One proposed objective is to evaluate an application's efficiency on a specific architecture using the application signature. We define computational time as when a process executes computational instructions-the communication time, such as the transmission or reception time of MPI messages, plus the idle time waiting for communication. Therefore we define Phase Execution Time as the sum of computing time plus communication time. Consequently, we define $efficiency_{phase}$ as seen in the following equation.

$$efficiency_{phase} = \frac{computing\_time}{phase\_execution\_time} \tag{2}$$

4   Experimental Results

To evaluate efficiency, we have validated by selecting the Experiment 1, as mentioned in the previous section (Table 2) with the K-Means application. This validation is characterised by a phased analysis in order to identify the efficiency of each phase in relation to the global execution time. The experimentation methodology runs the K-Means application, analyses it, and extracts signatures using PAS2P on a system with a specific architecture. The computing system is composed of 7 compute nodes of 64 cores (AMD Opteron6262 HE processor) with an interconnection network of 40 Gb/s Infiniband. As shown in Table 3, we present the efficiency results for each phase based on the results of Experiment 1 according to Equation (2).

After analysing the K-Means application with PAS2P, the application behaviour is represented by 4 phases, as is shown in Table 3. The first phase (the more relevant phase from the computational point of view) had the highest efficiency, with a value of 99.93%. The second phase had an efficiency of only 11.64%, whilst the third phase had an efficiency of 24.58%. The fourth and final phase had the lowest efficiency, at only 7.28%. In addition to the efficiency results, Table 3 includes the computational time related to the number of instructions and total phase execution time (in ns) and the communication volume for each phase (in MB).

Table 3 shows the efficiency of each phase individually. However, this information is still not enough to know the global impact of each of these phases on the total execution time of the application. Therefore, in order to see the global effects of each phase, we proceed to carry out the procedure described below, resulting in Table 3 from Table 4.

To determine the "Global Computing Time" and the "Total Phase Execution" columns, we multiply the weights of each phase by their respective execution times. Next, we sum up the "Total Phase Execution" values for each phase to obtain the overall "Execution Time" (ET) of the application. To calculate both "the Percentage Global Computational" time and the "Percentage Max. Computation" for each phase, we divide the "Global Computational Time" and the "Total Phase Execution" columns by the total application Execution Time (ET). Lastly, we subtract the two last columns to obtain the values in the "Room for improvement" column.

Global Comp. Time denotes the total duration taken to complete a computing phase, while Total Phase Execution refers to the entire time required for the phase to run. The forth column shows the percentage of total computing time utilised by the phase, indicated as "Global Computational Time [%]". The fifth column (Max. Comp. [%]) represents the maximum possible percentage that the

Table 3. Efficiency for each phase (Results of experiment 1)

Phase	Computing Time [ns]	Phase Execution Time [ns]	Efficiency [%]	Number of Instructions	Communication Volume
1	1.203028e+12	1.203866e+12	99.930416	5306514959	1536
2	5.770710e+06	4.956300e+07	11.643181	463	0
3	1.239423e+07	5.042100e+07	24.581488	464	1536
4	1.236372e+07	1.697620e+08	7.282971	469	4

Table 4. Global efficiency (Results of experiment 1)

Phase	Global Comp Time [ns]	Total Phase Execution [ns]	Global Comp Time[%]	Max. Comp [%]	Room for improvement[%]
1	3.500811e+14	3.503249e+14	99.9080	99.9775	0.0695
2	1.673506e+09	1.437327e+10	0.0004	0.0041	0.0036
3	3.619116e+09	1.472293e+10	0.0010	0.0042	0.0031
4	3.597842e+09	4.940074e+10	0.0010	0.0140	0.0130

phase can achieve if it performs computations for the entire phase duration, and it can be considered the theoretical limit. Finally, the last column (Room for improvement) displays the difference (in percentages) between the actual computational time and the theoretical maximum computational time, reflecting the potential for improvement.

In this scenario, the room for improvement is limited since Phase 1 is already highly efficient and dramatically influences the overall application time. Additionally, it can be noted that the impact of the remaining phases is minimal, as evidenced in the "Max. Computation [%]" column. Therefore, enhancing the efficiency of these less impactful phases may yield few benefits. However, if any of the values in the last column were considerably high, as administrators, we would provide the programmer with a report highlighting the corresponding phase's substantial impact and recommend efforts to improve its efficiency.

5   Conclusions

This work aims to extend the PAS2P methodology's contributions to AI by validating the accuracy of PAS2P in predicting the execution time by taking the K-Means application as a case study. The primary goal is to analyse the K-Means clustering algorithm and demonstrate the hypothesis that, although different datasets or initial conditions may affect the number of repetitions of the same phases, the phases remain the same.

Furthermore, the study aims to calculate the efficiency of each phase, its global impact on performance, as well as the improvement margin. Identifying critical phases and optimising their performance can improve the application's efficiency. This work provides a useful report with the efficiency results of each phase for programmers to take necessary steps toward achieving that goal.

For future work, we will consider establishing mapping policies to improve performance using the methodology presented in this work. As a first approach, we will select three areas of AI for reviewing current tools and their focus. The initial areas of AI with which we will be working are classification algorithms, heuristics, and genetic algorithms.

Acknowledgments. This research has been supported by the Agencia Estatal de Investigacion (AEI), Spain and the Fondo Europeo de Desarrollo Regional (FEDER) UE, under contract PID2020-112496GB-I00 and partially funded by the Fundacion Escuelas Universitarias Gimbernat (EUG).

References

1. Verma, G., et al.: HPCFAIR: enabling FAIR AI for HPC applications. In: 2021 IEEE/ACM Workshop on Machine Learning in High Performance Computing Environments (MLHPC), St. Louis, MO, USA, pp. 58–68 (2021). https://doi.org/10.1109/MLHPC54614.2021.00011
2. Anandkumar, A.: Role of HPC in next-generation AI. In: 2020 IEEE 27th International Conference on High Performance Computing, Data, and Analytics (HiPC), Pune, India, pp. xx–xx (2020). https://doi.org/10.1109/HiPC50609.2020.00010
3. Khan, A., et al.: Hvac: removing I/O bottleneck for large-scale deep learning applications. In: 2022 IEEE International Conference on Cluster Computing (CLUSTER), Heidelberg, Germany, pp. 324–335 (2022). https://doi.org/10.1109/CLUSTER51413.2022.00044
4. Fink, Z., Liu, S., Choi, J., Diener, M., Kale, L.V.: Performance evaluation of Python parallel programming models: Charm4Py and mpi4py. In: 2021 IEEE/ACM 6th International Workshop on Extreme Scale Programming Models and Middleware (ESPM2), pp. 38–44. IEEE, November 2021
5. Charm4py documentation. https://charm4py.readthedocs.io/en/latest/. Accessed 10 Feb 2023
6. MPI for Python documentation. https://mpi4py.readthedocs.io/en/stable/. Accessed 10 Feb 2023
7. Alnaasan, N., Jain, A., Shafi, A., Subramoni, H., Panda, D.K.: OMB-py: python micro-benchmarks for evaluating performance of MPI libraries on HPC systems. In: 2022 IEEE International Parallel and Distributed Processing Symposium Workshops (IPDPSW) (pp. 870–879). IEEE, May 2022
8. Lloyd, S.: Least squares quantization in PCM. IEEE Trans. Inf. Theory **28**(2), 129–137 (1982). https://doi.org/10.1109/TIT.1982.1056489
9. Wong, A., Rexachs, D., Luque, E.: Parallel application signature for performance analysis and prediction. IEEE Trans. Parallel Distrib. Syst. **26**(7), 2009–2019 (2015)
10. Tirado, F., Wong, A., Rexachs, D., Luque, E.: Improving analysis in SPMD applications for performance prediction. In: Arabnia, H.R., et al. (eds.) Advances in Parallel & Distributed Processing, and Applications. TCSCI, pp. 387–404. Springer, Cham (2021). https://doi.org/10.1007/978-3-030-69984-0_29

A New Algorithm for the Closest Pair of Points for Very Large Data Sets Using Exponent Bucketing and Windowing

Vaclav Skala[1]($\boxtimes$)(iD), Alejandro Esteban Martinez[1,2],
David Esteban Martinez[1,2], and Fabio Hernandez Moreno[1,3]

[1] Department of Computer Science and Engineering, University of West Bohemia,
Pilsen, Czech Republic
skala@kiv.zcu.cz
[2] University of A Coruña, A Coruña, Spain
[3] University of Las Palmas de Gran Canaria, Las Palmas, Spain
https://www.VaclavSkala.eu

Abstract. In this contribution, a simple and efficient algorithm for the closest-pair problem in E^1 is described using the preprocessing based on exponent bucketing and exponent windowing respecting accuracy of the floating point representation. The preprocessing is of the $O(N)$ complexity. Experiments made for the uniform distribution proved significant speedup. The proposed approach is applicable for the E^2 case.

Keywords: Closest pair · minimum distance · uniform distribution · data structure · bucketing sort · computational geometry · large data sets · data windowing

1 Introduction

The closest pair problem is a problem of finding two points having minimum mutual distance in the given data set. Brute force algorithms with $O(N^2)$ complexity are used for a small number of points, and for higher number of points algorithms based on sorting have $O(N \lg N)$ complexity. In the case of large data sets[1], the processing time with $O(N \lg N)$ complexity might be prohibitive.

In this contribution, a proposed simple preprocessing of the data set based on the bucketing and exponent windowing is described; similar strategy as in Skala [12–15] and Smolik [16]. The extension to the E^2 case is straightforward using already developed algorithms. The closest pair problem was address in Shamos

---

[1] Data set, where $N \ggg 10^6$. Note that 2 147 483 648 $\doteq$ 2.147 10^9 unsigned distinct values only can be represented in single precision.

Supported by the University of West Bohemia - Institutional research support Martinez, D.E., Martinez, A.E., Moreno, F.H.—students contributed during their Erasmus ACG course at the University of West Bohemia.

J. Mikyška et al. (Eds.): ICCS 2023, LNCS 14074, pp. 381–388, 2023.
https://doi.org/10.1007/978-3-031-36021-3_40

[11], Kuhller [7], Golin [4] and Mavrommatis [8], Pereira [9], Roumelis [10] used space subdivision and Bespamyatnikh [1] used a tree representation, Daescu [2,3] and Katajainen [6] used divide & conquer strategy, Kamousi [5] used a stochastic approach.

In the following, basic strategies for finding a minimum distance of points in E^1 are described.

Brute-force algorithm - The naive approach leads to a brute-force algorithm, which is simple and easy to implement. However, it has $O(N^2)$ computational complexity as it requires $N(N-1)/2$ computational steps. This algorithm cannot be used even for relatively small N due to the algorithm complexity, but can be also used for a higher dimensional case. The algorithm can be speed-up a bit, as d can be computed as $d := \|\mathbf{x}[i] - \mathbf{x}[j]\|^2$ and d_0 set as $d_0 := \sqrt{d_0}$ as the function $f(x) = x^2$ is a monotonically growing function[2].

Algorithm with sorting - In the one dimensional case, i.e. the E^1 case, the given values x_i might be re-ordered into the ascending order. It leads to $O(N \log N)$ computational complexity. Then the ordered data are searched for the minimum distance of two consecutive numbers, i.e. $x_{i+1} - x_i$ as $x_{i+1} \geq x_i \ \forall i$, with $O(N)$ complexity. The above-mentioned algorithms are correct and can be used for computation with "unlimited precision".

However, in real implementations, the used floating point representation has a limited mantissa representation and range of exponents.

Table 1. Bucket length distribution and speed-up of the proposed algorithm

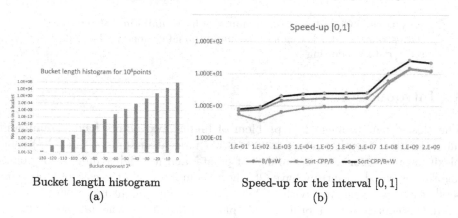

Bucket length histogram
(a)

Speed-up for the interval [0, 1]
(b)

Algorithm with limited mantissa - The computational complexity of the algorithm is $O(N \log N)$ due to the ordering and finding the minimum distance is $O(N)$ complexity as all N values have to be tested, as the smallest difference might be given by the last binary digits in the mantissa.[3]

---

[2] It actually saves $O(N^2)$ computations of the $\sqrt{*}$ function.

[3] Let us consider a sorted sequence $1.01, 1.05, \ldots, 10.0001, 10.0005$, then the minimum distance is 0.0004 not 0.04.

However, in the real data cases, the expected complexity will be smaller, due to values distribution over several binary exponents. Table 1.a presents[4] a histogram of values according to their binary exponent; the uniform distribution $[0, 1]$ is used.

In reality, the IEEE 754 floating point representation or a similar one with a limited mantissa precision is used. It means, that if $d = |x_{i+1} - x_i|$ is the currently found minimum distance, i.e. $d = [m_d, E_d]$, where m_d is the mantissa and E_d is the binary exponent of d, then the stopping criterion for searching the ordered $\mathbf{x}$ values is $x_i > d_0 * 2^{p+1}$, where p is the number of bits of the mantissa, d_0 is already found minimum.

It can be seen, that the limited mantissa precision reduces the computational requirements significantly for the larger range of the binary exponents of values. Table 1.a presents an expected number of points having exponents within exponent buckets for 10^8 of points, if the uniform distribution is used.

2   Proposed Algorithm with $O_{exp}(N)$ Complexity

The bottleneck of the algorithm standard $O(N \log N)$ complexity is the ordering step. However, instead of "standard" sorting algorithms, e.g. heap sort, shall sort, quick sort etc., it is possible to use bucketing by the exponent values[5] instead, which has $O(N)$ complexity, see Fig. 1. The data structure is similar to the standard hashing structure. All values having the same binary exponent are stored in an array-list or in a similar data structure[6], see Fig. 1. The values E_{min} and E_{max} are the *minimum* and *maximum* binary exponents found. It means, that all values are sorted according to their binary exponents, but *unsorted* within the actual bucket, i.e. unordered, if the exponents are equal. The table length is 256, resp. 2048 according to the precision used, i.e. 32-bits, resp. 64-bits.

It should be noted, that even for 10^8 points, the probability for small values of exponents is extremely low. As the mantissa precision is limited the bucket length for very low exponents will be zero or very small.

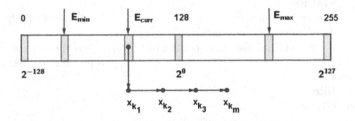

Fig. 1. Bucketing structure - 32-bits

---

[4] Note: 130–120 means exponents interval [-120,...,-111].

[5] the binary exponent is shifted, i.2. a value 2^{-128} has the shifted exponent 0.

[6] The array list, i.e. extensible arrays were used in the actual implementation.

Algorithm 1. Minimum distance with bucketing in E^1

1: **procedure** MINDISTBUCKET($\mathbf{x}, N, d_0$);
2:                             ▷ given set of N points $\mathbf{x}$, $x[i] \geq 0$, distance d_0 found
3:      $E_{min} := maxint$; $E_{max} := minint$;   ▷ initial setting
4:      $p := 24$;   ▷ 32-bits: $p = 24$; 64-bits: $p := 53$
5:      $E_{range} := 256$;   ▷ exponent range $E_{range} := 2048$, if double precision
6:      # preprocessing - buckets construction #
7:      **for** $i := 1$ **to** N **do**
8:          $Ex := \text{Exponent}(x[i])$;   ▷ binary exponent $[0, E_{range}]$
9:          $\text{Add}(x[i], \text{Bucket}(Ex))$;   ▷ add the value to the bucket $Bucket(Ex)$
10:          $E_{min} := \min\{E_{min}, Ex\}$;
11:      **end for**   ▷ all bucket are constructed
12:
13:      $d_0 := \infty$;   ▷ setting a min. distance estimation
14:      $temp := -\infty$;   ▷ initial setting
15:      $i := E_{min}$;   ▷ $Bucket[E_{min}] \neq \emptyset$
16:      $E_{max} := E_{min} + p$;   ▷ E_{max} - upper bound for the windowing
17:      $InWindow := \textbf{true}$;
18:      $PairFound := \textbf{false}$;
19:             ▷ case if the only one point is inside of the exponent window
20:      **while** $(i \leq E_{max}$ **or** **not** $PairFound)$ **and** $i \leq E_{range}$ **do**
21:          **if** $(Bucket[i] \neq \emptyset)$ **and** $InWindow$ **then**
22:               SORT_Bucket(i);   ▷ sorts values in the i-th Bucket
23:               # $[d, temp] := \text{ProcessOneBucket}(i, temp)$; #
24:                    ▷ finds a minimum distance d of $temp$ and values in the $Bucket[i]$
25:                          ▷ $temp$ is last value in the $Bucket[i-1]$
26:               # find a minimum distance in a $\{temp, Bucket[*]\}$, if exists #
27:               **for** $k := 1$ **to** $Bucket.length[i]$ **do**
28:                    $xx := Bucket[i][k]$;   ▷ get the current value
29:                    $d := xx - temp$; $temp := xx$;
30:                    **if** $d_0 > d$ **then**
31:                       $d_0 := d$;
32:                       $PairFound := \textbf{true}$;   ▷ at least one valid pair found
33:                    **end if**
34:               **end for**
35:               $Ex := \text{Exponent}(d_0)$;   ▷ Windowing the exponent
36:               $InWindow := Ex + p < i$;
37:             ▷ STOP, if the exponent Ex of $(d_0 + p) \geq i$; the current exponent i
38:                              ▷ p is the mantissa length+1
39:          **end if**
40:      **end while**
41: **end procedure**

#**SOLVED** - A sequence 10^{23}, $0.1\ 10^0$, $10.001\ 10^{23}$ is handled properly

---

Taking above into consideration, the proposed algorithm based on bucketing and exponent windowing is given by the Algorithm 1, where sorting of buckets is made on a request, i.e. when needed.

The function ProcessOneBucket(i, *temp*) finds a minimum distance within the **sorted** Bucket[i], taking *temp* value as the element before the first element in the Bucket[i]. It should be noted, that there is a "window" in the exponent table long 24 in the case of 32-bits, resp. 53 in the case of 64-bits, in which data are to be processed due to the mantissa limited precision. It leads to significant speed-up, especially for large data interval range.

3 Algorithm Analysis

Let us consider uniform distribution on the interval $[0, 1]$, e.g. using the standard *random*($*$) function. The exponent bucketing is a non-linear space subdivision as the space of values is split non-linearly, i.e. the interval length grows exponentially. In this case, values have the *power distribution* 2^k, $k = -128, \ldots, -1$, or $k = -128, \ldots, 127$, if data generated from the interval $[0, \infty)$. It means, that for small exponent values, fewer elements are stored in a bucket, while for higher exponent values more values are stored in the relevant bucket.

As all values are generated within the interval $[0, 1]$ and the 32-bits precision is used, then each sub-interval is of the length 2^k. It means that if N points are generated uniformly within the interval $[0,1]$, then the interval k contains m_k values, see Eq. 1, where $m_k = 2^k N$, $k = -128, \ldots, -1$ as:

$$\sum_{k=1}^{k=128} 2^{-k} = \frac{1}{2^1} + \frac{1}{2^2} + \frac{1}{2^3} + \ldots + \frac{1}{2^{128}} \doteq 1 \tag{1}$$

As the distance between small values is smaller, there is higher probability, that the minimum distance, i.e. the closest pair will be found faster, see Tab. 1.a.

However, not the whole range of exponents will be used in a real situation and the interval of exponents $[E_{min}, E_{max}]$ can be expected. As the precision of the mantissa is limited to 24 bits in the 32-bits case, a window of 24 exponents is to be processed instead of 128, resp. 256; similarly in the case of 64-bits, the window of 53 exponents is to be processed instead of 1024, resp. 2048.
It should be noted that for $N = 10^{10}$ values, the number of values having the exponent 2^{-128}, i.e. for $\xi = 0$, is $2.9\ 10^{-29}$ only. The lowest non-empty bucket will probably have the non-shifted exponent 33 and the shifted exponent $Ex = 95$[7]. It means that the expected E_{max} will be $E_{max} = 117$, i.e. $k = -33 + 24 = -9$, if the single precision used.

Therefore, in the case of the $[0, 1]$ uniform distribution interval, the last 8 shifted (physical) exponents, i.e. $Ex = [120, 127]$, will not be evaluated. Those buckets contain N_0 values, i.e. approx. 99.609% points, see Eq. 2.

$$N_0 = N \sum_{i=1}^{8} \frac{1}{2^i} \qquad \frac{N_0}{N} = \sum_{i=1}^{8} \frac{1}{2^i} \doteq 99.609375 \tag{2}$$

---

[7] The value $k \doteq 33$ is obtained by solving $10^{10} * 2^k = 1$, which is $k \doteq -33$ and the shifted exponent $Ex = 128 - 33 = 95$.

where i is the non-shifted exponent. Therefore, efficiency of the proposed algorithm grows with the exponent range in the uniform distribution case. The proposed algorithm can be modified for the Gaussian distribution easily.

4   Experimental Results

The implementation of the algorithm described in Algorithm 1 is based on some data profile assumptions. There are two significant factors to be considered:

- the range of data exponents should be higher; if all the data would have the same binary exponent, only one very long bucket would be created,
- the algorithm is intended for larger data sets, i.e. number of points $N > 10^6$.

In the actual implementation an equivalent of the **array-list** was used, which is extended to a double length if needed and data are copied to the new position[8]. It might lead the *copy-paste* extensive use resulting to slow-down. In the case of the uniform distribution, the initial length of a bucket should be set to a recommended length $N\ 2^{-k} * 1.2$ setting used in the experimental evaluation, i.e. $length.Bucket[k] \geq N\ 2^{-k} * 1.2$.

Evaluation. Uniform distribution of values was used with different intervals from $[0, 1]$ to $[0, 10^4]$. Up to $2\ 10^9$ points were generated and efficiency of the proposed algorithm was tested.

Obtained results for the interval $[0, 1]$ are summarized in Tab. 1.b, where ratios of time spent are presented. Notation: $Sort - CPP/B = \frac{time_{Sort-CPP}}{time_{Bucketing}}$, $Sort - CPP/B + W = \frac{time_{Sort-CPP}}{time_{Bucketing+Windowing}}$, and $B/B + W = \frac{time_{Bucketing}}{time_{Bucketing+Windowing}}$

The ratio $Sort - CPP/B$ gives reached speed-up using the exponent bucketing over sort[9] used in C++. It can be seen that there is a speed-up over 1.2 for more than 10^3 points and grows with the range of generated data.

The ratio $Sort - CPP/B + W$ gives reached speed-up using the exponent bucketing over sort used in C++ and the speed-up is over 1.4. It should be noted, that

- speed-up for 10^9 points is over 16 times against if the sort method is used.
- ratio $B/B + W$ clearly shows significant influence of the windowing, which reflects the limited precision of numerical representation.

Notes - As the number of the processed values N is high, there are possible modifications of the algorithm leading to further efficiency improvements, e.g.:

---

[8]   In the case of 10^6 values, over 10^3 bucket extensions were called, but with the 20% additional memory allocation, the extension was called only 7 times.

[9]   Sort-CPP - standard Shell sort in C++.

- finding the E_{min} and E_{max} can be done after the buckets construction; it saves $O(N)$ floating point comparisons, or it can be removed with initial setting $E_{min} := 0$,
- some heuristic strategies can be used, e.g. pick up m values, find the smallest and its exponent Ex, and the second smallest one and determine a first minimum distance estimation. Set $E_{max} := E_x + p$ (E_x is the exponent of already minimum found) as a stopping criterion for building buckets (it eliminated long bucket constructions for higher exponents, which cannot contribute to the minimum distance[10]).

5   Conclusions

In this contribution, an efficient improvement of the minimum distance algorithm E^1 case is presented. It takes a limited precision of the floating point representation into consideration and uses bucketing sort based on exponent's baskets. The presented approach is intended for larger data sets with a higher exponent range. Experiments made proved a significant speed-up of the proposed approach for the uniform distribution. The proposed approach can be extended to the E^2 case using algorithms as proposed in Daescu [2,3], Golin [4], etc. Extensions of the proposed approach for the E^2 and E^3 cases are future work.

Acknowledgments. The author would like to thank colleagues at the University of West Bohemia, Plzen for their comments and suggestions, comments and hints provided, especially to Martin Cervenka and Lukas Rypl for some additional additional counter tests. Thanks also belong to anonymous reviewers for their critical view and recommendations that helped to improve this manuscript.

Responsibilities. Skala, V.: theoretical part, algorithm design, algorithm implementation and verification, manuscript preparation; Esteban Martinez, A., Esteban Martinez, D., Hernandez Moreno, F.: algorithm implementation and experimental verification.

References

1. Bespamyatnikh, S.: An optimal algorithm for closest-pair maintenance. Discrete Comput. Geom. **19**(2), 175–195 (1998). https://doi.org/10.1007/PL00009340
2. Daescu, O., Teo, K.: 2D closest pair problem: a closer look. In: CCCG 2017–29th Canadian Conference on Computational Geometry, Proceedings, pp. 185–190 (2017)
3. Daescu, O., Teo, K.: Two-dimensional closest pair problem: a closer look. Discrete Appl. Math. **287**, 85–96 (2020). https://doi.org/10.1016/j.dam.2020.08.006
4. Golin, M.: Randomized data structures for the dynamic closest-pair problem. SIAM J. Comput. **27**(4), 1036–1072 (1998). https://doi.org/10.1137/S0097539794277718

---

[10] Note, that p depends on the FP precision used.

5. Kamousi, P., Chan, T., Suri, S.: Closest pair and the post office problem for stochastic points. Comput. Geom. Theory Appl. **47**(2 PART B), 214–223 (2014). https://doi.org/10.1016/j.comgeo.2012.10.010
6. Katajainen, J., Koppinen, M., Leipälä, T., Nevalainen, O.: Divide and conquer for the closest-pair problem revisited. Int. J. Comput. Math. **27**(3–4), 121–132 (1989). https://doi.org/10.1080/00207168908803714
7. Khuller, S., Matias, Y.: A simple randomized sieve algorithm for the closest-pair problem. Inf. Comput. **118**(1), 34–37 (1995). https://doi.org/10.1006/inco.1995.1049
8. Mavrommatis, G., Moutafis, P., Corral, A.: Enhancing the slicenbound algorithm for the closest-pairs query with binary space partitioning. In: ACM International Conference Proceeding Series, pp. 107–112 (2021). https://doi.org/10.1145/3503823.3503844
9. Pereira, J., Lobo, F.: An optimized divide-and-conquer algorithm for the closest-pair problem in the planar case. J. Comput. Sci. Technol. **27**(4), 891–896 (2012). https://doi.org/10.1007/s11390-012-1272-6
10. Roumelis, G., Vassilakopoulos, M., Corral, A., Manolopoulos, Y.: A new plane-sweep algorithm for the K-closest-pairs query. In: Geffert, V., Preneel, B., Rovan, B., Štuller, J., Tjoa, A.M. (eds.) SOFSEM 2014. LNCS, vol. 8327, pp. 478–490. Springer, Cham (2014). https://doi.org/10.1007/978-3-319-04298-5_42
11. Shamos, M., Hoey, D.: Closest-point problems. In: Proceedings - Annual IEEE Symposium on Foundations of Computer Science, FOCS 1975-October, pp. 151–162 (1975). https://doi.org/10.1109/SFCS.1975.8
12. Skala, V.: Fast $O_{expected}(N)$ algorithm for finding exact maximum distance in E2 instead of $O(N^2)$ or $O(NlgN)$. AIP Conf. Proc. **1558**, 2496–2499 (2013). https://doi.org/10.1063/1.4826047
13. Skala, V., Cerny, M., Saleh, J.: Simple and efficient acceleration of the smallest enclosing ball for large data sets in e2: Analysis and comparative results. LNCS **13350**, 720–733 (2022). https://doi.org/10.1007/978-3-031-08751-6_52
14. Skala, V., Majdisova, Z.: Fast algorithm for finding maximum distance with space subdivision in E2. LNCS **9218**, 261–274 (2015). https://doi.org/10.1007/978-3-319-21963-9_24
15. Skala, V., Smolik, M.: Simple and fast $oexp(n)$ algorithm for finding an exact maximum distance in E2 instead of $o(n^2)$ or $o(n \lg N)$. LNCS **11619**, 367–380 (2019). https://doi.org/10.1007/978-3-030-24289-3_27
16. Smolik, M., Skala, V.: Efficient speed-up of the smallest enclosing circle algorithm. Informatica **33**(3), 623–633 (2022). https://doi.org/10.15388/22-INFOR477

Heart Rate-Based Identification of Users of IoT Wearables: A Supervised Learning Approach

Sachit A. J. Desa[1](✉), Basem Suleiman[1,2](✉), and Waheeb Yaqub[1]

[1] School of Computer Science, The University of Sydney, Sydney, Australia
{sachit.desa,basem.suleiman,waheeb.faizmohammad}@sydney.edu.au
[2] School of Computer Science and Engineering, The University of New South Wales, Sydney, Australia
b.suleiman@unsw.edu.au

Abstract. Biometric identification from heart rate sequences provides a simple yet effective mechanism, that can neither be reverse engineered nor replicated, to protect user privacy. This study employs a highly efficient time series classification (TSC) algorithm, miniROCKET, to identify users by their heart rate. The approach adopted in this study employs user heart rate data, a simplified form of heart activity, captured during exercise, filtered, and contextualized within exercise routines, for user classification. Results from this study are empirically evaluated on a real-world data set, containing 115,082 workouts across 304 users, by three other state-of-the-art TSC algorithms. Our experiments showed that for 36 users, the variance explained by heart rate feature is 74.0%, when coupled with speed and altitude, the variance increases to 94.0%. For 304 users, the variance explained by heart-rate is 32.8% and increased to 65.9% with contextual features. This exploratory study highlights the potential of heart rate as a biometric identifier. It also underscores how contextual factors, such as speed and altitude change, can improve classification of timeseries data when coupled with smart data preprocessing.

1 Introduction

The popularity of Internet of Things (IoT) devices has been driven by their ability to record biometric and contextual data for use in providing personalised healthcare services. However, this popularity increases the risk of personal data associated with these devices being hacked - a risk magnified by the projected uptake of wearables, forecast to pass six billion by 2025 [1]. Tied to data collected by wearables, and IoT devices in their various forms, is the personal information of individuals that use wearables. Literature surveyed within this area highlighted two critical themes associated with IoT device data: a.) The metadata associated with wearables may be compromised in several instances due to currently employed device protection mechanisms [6], and b.) The limited amount of research, till date, in the privacy field of wearable devices [14].

J. Mikyška et al. (Eds.): ICCS 2023, LNCS 14074, pp. 389–397, 2023.
https://doi.org/10.1007/978-3-031-36021-3_41

The sparse amount of work completed in this domain has established how electrocardiogram (ECG) data, can be used for identification [3]. ECG signals have been showcased to identify individuals with an accuracy upwards of 95% [15]. However, such data is collected within controlled environments and with specialist equipment. The novelty of this study is its use of heart rate data collected by commercial IoT devices, as opposed to specialist equipment (i.e. ECGs).

Currently, IoT devices call for user meta-data to be entered (name, date of birth, etc.) for easy identification. A large number of these devices offer a connection to phone and email services, as well as being a portal to other applications that store user data. Access to this information via passwords, fingerprints, or two-factor authentication can provide hackers with a trove of personal data, as well as information linked to institutions and third parties.

In this paper, we address the challenge of maintaining user privacy by proposing an approach that identifies users via their heart rate, instead of traditional authentication methods that are prone to hacking. The robustness of the proposed approach relies on the difficulty of replicating and reverse-engineering the heart-rate sequence. The goal of this paper is to contribute to the growing adoption of wearables (and personal IoT devices such as smart watches) whilst protecting user privacy. We introduce a new research direction in which we present the findings of an approach that employs key features known to affect heart rate (speed and altitude), and a filtering strategy, to identify users with their wearable devices. The contributions of this study are:

- Exploration of heart rate, collected from IoT wearables, as a unique identifying biometric feature.
- Filtering methods that allow for the most efficient and effective use of biometric data to identify users of wearable devices.
- An empirical study of the performance of time series classification (TSC) algorithms on heart rate data. We consider a real-world data set consisting of 115,082 workouts across 304 users of IoT devices.

2   Related Work

Access to personal information from wearable IoT devices and trusted third parties has been highlighted by several studies [6, 12, 14, 16]. Specifically, activity pattern recognition [12], data driven privacy-setting recommendations [14], and data inference risks from third parties [16], are methods through which privacy breaches occur. Reichherzer et al. (2017) [12] present machine learning techniques that intercept, track, and classify individuals and their fitness activities.

The energy availability for wearable and battery-powered devices is limited, precluding the use of many sophisticated and established authentication mechanisms [2]. Similarly, PINs and biometrics are not suitable due to the limitations of small form factor [12]; biometric authentication such as face, fingerprint, and iris recognition can be counterfeited as they do not check for liveness in a subject [8]. The use of ECG data overcomes this limitation as it is an inherent measure of liveness. Studies have shown that whilst the duration of a heartbeat has been known to vary with stress, anxiety, and time of day, the structure of a heartbeat contains scalar differences to variations in stress [8]. Existing research, including studies conducted by Israel et al. [9], Irvine et al. [7,8], and Biel et al. [3], demonstrate the heartbeat structure to be unique to an individual.

For a set of 20 subjects, Biel et al. [3] use a Soft Independent Modeling of Class Analogy (SIMA) classifier to identify individuals by ECG data. SIMA returned a classification accuracy of 98.0%. In a separate study, Shen et al. [15] found, via feature extraction, that measurements of palm ECG signals on 168 individuals returned an accuracy of 95.3%. A key observation from ECG data was that signals from individuals with a similar age, weight, height, and gender can be significantly different [15]. Whilst ECG data contains multiple dimensions for classification, heart rate is simply a measure of heartbeats per minute. The overarching challenge identified as part of this study, is the use of a simplified measure of heart activity, heart rate, as an encoding mechanism for protecting user privacy. In lieu of data rich signals captured by ECG tests, completed when an individual is at rest, this study uses contextual data, recorded by wearables during exercise, for classification. In using heart rate data, collected by wearables, for user identification, this study aims to bridge the gap between current research that has focused exclusively on ECG data, collected in controlled environments with specialist equipment, and 'real-world' data that is coarse.

3   Datasets for Classification

The data used within this study, originally made available by Endomondo, comprised of 253,020 workouts across 1,104 users. Ni et al. [11] used the Endomondo data to generate a consolidated dataset containing users with a minimum of 10 workouts, restricted features to those relevant to this study, and interpolated time series features to obtain data at regular intervals. As part of this study, features such as latitude, longitude, and timestamp were re-imported, and other redundant features discarded.

Data Sub-setting and Partitioning: An initial exploration of the data found that the number of workouts per user varied significantly (i.e. from 10 to 1301 workouts per user) with the distribution skewed towards users with a lower number of workouts.

In view of achieving a more uniform distribution of workouts per user, reducing the target variable space, and offering sufficient data for each algorithm to be trained, data was subset by the number of records per user. The minimum number of workouts per user was based on a wide ranging study by Ruiz et al. (2021) [13] who completed a multi-variate time series classification study of 18 TSC algorithms over 30 datasets. The largest target variable space consisted of 39 classes, each associated with ~ 170 records. An accuracy of 36.7% was achieved on this dataset. Dataset 1 for this study was filtered to contain 304 users with a minimum of 200 workout records.

Dataset 1 was subset to evaluate and compare the selected algorithms over a successively smaller number of classes and metrics. Subsetting was based on empirical evidence collected from classification tests on the University of California, Riverside (UCR) database. This database contains two heartbeat and three ECG datasets with a maximum of 5 classes. The records per class range from 100 to 1000, with a maximum classification accuracy achieved for a dataset containing 442 records per class. To test the classification sensitivity associated with the number of workout records per user, three datasets were generated, containing a minimum of 350 and a maximum of 560 workout records per user. Each dataset was a subset of its parent, with the test and train partitions of dataset 1 carrying through to dataset 4. The properties of each dataset are presented in Table 1.

Table 1. Dataset used for the current study

Dataset	Records	Min. Records[1]	Users	Size (GB)
1	115,082	200	304	5.9
2	69,111	350	129	3.5
3	59,694	400	104	3.1
4	27,064	560	36	1.4

[1] Denotes the minimum number of exercise records per user

4   Methodology - Heart Rate Based Identification

The network architecture for this study consists of a filtering module, the miniROCKET algorithm, and a feature selection module. The filtering module preprocesses user data to determine the optimal number of activities, shortest date range, and least number of records required to generate a physiological profile. The optimal filtering parameters are applied to the test data to generate a filtered dataset. The miniROCKET algorithm completes classification by contextualising heart-rate with each feature (distance, altitude, etc.) and feature combination. The optimum feature combination is then retained for classification on the test data. For any dataset, the network determines the optimal number of records, shortest date range, and fewest activities needed for classification. The model also ranks and quantifies the contribution of each contextual feature to classification accuracy. Figure 1 presents an overview of the model architecture.

MiniROCKET and Comparative TSC Algorithms: miniROCKET was used as the mainstay algorithm for this study. Ruiz et al. [13], who completed a wide-ranging study on multi-variate timeseries data, found Random Convolutional KErnel Transform (ROCKET) to outperform current time series classification algorithms in accuracy and computational efficiency.

ROCKET employs random convolutional kernels in conjunction with a linear classifier (either ridge regression or logistic regression) for classification. Kernels can be configured based on 5 different parameters: bias, size (length), weights,

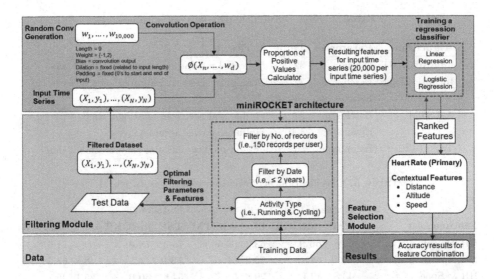

Fig. 1. Model architecture implemented for classification

dilation, and padding. MiniROCKET, an updated version of ROCKET, minimizes the hyperparameter options for each kernel; four of the five hyperparameters are fixed, whilst the weights are selected as either -1 or 2. Fixing hyperameter values ensures that miniROCKET is considerably faster [4].

Time Series Forest (TSF), InceptionTime, and Attention-Long Short-Term Memory (A-LSTM), were employed on a comparative basis. The two deep learning algorithms, Inception Time and A-LSTM, were selected for their accuracy and scalability [10], both of which were key themes underlying this study. TSF was selected for its emphasis on temporal characteristics, which would support the classification of heart-rate sequences, particularly when considering the temporal dependence of heart rate on speed and altitude [5].

Hyperparameter Tuning and Selection: Hyperparameter selection was completed on the test sets of datasets 1, 2, 3, and 4. For deep learning algorithms, InceptionTime and A-LSTM, the batch size and epochs were tuned. The number of epochs had a significant bearing on accuracy and computational time unlike batch size. The dilation parameter was tuned for miniROCKET, and number of estimators were optimized for TSF. For the parameter range over which accuracy remained consistent, hyperparameters were fine-tuned to optimise computational efficiency. Table 2 presents the final parameter range selected for the four algorithms across datasets.

5   Results and Discussion

Table 3 presents the final results, across datasets and feature combinations. The system architecture remained consistent for each of the selected algorithms. It is

Table 2. Final selected hyper-parameter values

Algorithm	Parameter	Value	Criteria
miniROCKET	Dilations	32	Selected for optimized computational time
InceptionTime	Batch size	32	Values selected for improved accuracy
	Epochs	180-220	
A-LSTM	Batch size	32	Influential over computational time
	Epochs	400	Influential over accuracy
TSF	Estimators	50	Fine-tuned to optimize computational time

Table 3. Classiication accuracies across datasets 1 to 4

	Dataset 1				Dataset 2			
Feature List	TSF	mRCKT	IT	ALSTM	TSF	mRCKT	IT	ALSTM
HR, Speed, Alt. & Dist	63.8	**65.9**	63.2	64.6	75.0	**78.8**	69.5	71.6
HR, Speed, & Altitude	**65.0**	64.8	53.8	61.8	**73.0**	72.2	48.2	68.6
HR & Speed	43.8	49.1	29.8	**58.7**	54.6	**60.0**	25.1	45.1
HR	25.1	32.8	7.6	**40.8**	33.9	45.6	12.7	**50.8**
	Dataset 3				Dataset 4			
Feature List	TSF	mRCKT	IT	ALSTM	TSF	mRCKT	IT	ALSTM
HR, Speed, Alt. & Dist	78.4	**82.8**	67.5	78.0	86.7	**94.0**	77.4	89.1
HR, Speed, & Altitude	76.9	**81.2**	59.2	71.3	86.9	**93.4**	59.6	82.6
HR & Speed	57.6	**68.0**	40.0	66.3	72.5	**85.8**	38.2	80.5
HR	36.5	**49.2**	13.7	41.0	54.7	**74.0**	15.5	65.0

to be noted that because speed and time were included, the addition of distance was redundant. However, the distance feature resulted in an improved accuracy for IT, and has therefore been included in the results.

miniROCKET outperformed the baseline models for accuracy across all datasets containing the full set of features. With the exception of IT, other algorithms generated accuracies within 7.3% (±3.6) for datasets with all the features. Based on heart rate alone, which explains upto 40.8% of variance across 306 users, A-LSTM outperformed miniROCKET across datasets 1 and 2. The 'long-short term memory' component of A-LSTM is more effective in sequence detection across a large number of users than the linear classifier employed by miniROCKET. The accuracy of A-LSTM on heart activity is reinforced by Karim et al. [10] who record an accuracy of 95.0% for 5 classes on ECG data.

General trends indicate that accuracy is highly dependent on the number of users. For dataset 1, containing 304 users, the classification accuracy, with the exception of IT, is within 2.1%. The accuracy for dataset 4, containing 36 users, is within 7.3%. These results, presented in Figure 2, are indicative of the inability of algorithms to scale up with an increased number of user classes.

Trends observed in Fig. 2 demonstrate accuracy to be linked with the number of users to be classified. Some linearity is observed in the results. It is noteworthy

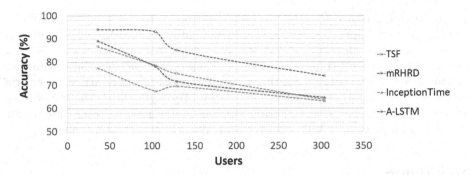

Fig. 2. Accuracy according to the number of users for all features

that there is a slight increase in accuracy recorded by IT between 104 and 129 users. The increase in accuracy, despite the increase in users, is reflective of the documented instability of the Inception Network as a result of the variability from random weight initialization. The current study aimed to overcome this instability by implementing an ensemble of five Inception Networks, with every prediction attributed the same weight, however, this did not completely address the inherent algorithmic instability.

The results of our experimental procedure demonstrate that heart rate explains up to 40.8% of variance across 304 users, the variance explained increases to 66.0% when all features are considered. Similarly, the variance across 36 users is close to 74.0% when heart rate is considered, and increases to 94.0% when all features are accounted for. On average, the addition of speed increased classification accuracy by 15.1% ($\pm$3.3) and altitude increased accuracy by 12.3% ($\pm$3.2). Both speed and altitude are factors known to influence heart rate [5], in the present study their inclusion allowed for an improved classification accuracy.

Classification studies completed using ECG data, for 20 and 168 individuals, recorded accuracies of 98.0% and 95.3% respectively [3,15]. The dimensionality of ECG data, in the form of P, QRS complex, and T waves, is a critical differentiator between users, and device data. The sophistication of currently available wearables, at a consumer level, has not yet evolved to capture underlying heart rate patterns. This is a major challenge that prohibits the use of heart rate data as a bio-metric identifier when a large number of user classes is considered. The addition of speed and altitude to heart rate is more encouraging. With consumer grade wearables unable to collect data that is as rich as ECG signals, the addition of contextual data, and smart preprocessing, can improve user identification.

6 Conclusion and Future Work

This study demonstrates how heart rate, contextualized within within exercise routines and coupled with a smart filtering strategy, can be used as a biometric identifier. Our results establish that heart rate alone can be used to explain up to 40% of variance for 304 users and 74% of variance for 36 users. Whilst the variance explained by heart rate may not sufficiently distinguish between a large number of users, the addition of speed and altitude increases classification accuracy to 66% for 304 users and 94% for 36 users. Future work should consider the inclusion of features representing latent physiological information (i.e. calorie count and hydration) to improve classification. By highlighting the importance of contextual data and smart filtering, this work establishes a strong baseline for researchers to develop a fully operational real-world classification model.

References

1. A survey on wearable technology: history, state-of-the-art and current challenges. Comput. Netw. **193**, 108074 (2021). https://doi.org/10.1016/j.comnet.2021.108074
2. Bellovin, S.M., Merritt, M.: Encrypted key exchange: password-based protocols secure against dictionary attacks. In: Proceedings 1992 IEEE Computer Society Symposium on Research in Security and Privacy (1992)
3. Biel, L., Pettersson, O., Philipson, L., Wide, P.: ECG analysis: a new approach in human identification. Trans. Instrum. Meas., 808–812 (2001)
4. Dempster, A., Schmidt, D.F., Webb, G.I.: minirocket. In: Singapore (ed.) Proceedings of the 27th ACM SIGKDD Conference on Knowledge Discovery and Data Mining (2021)
5. Fleckenstein, D., Ueberschar, O., Wustenfeld, J.C., Rudrich, P., Wolfarth, B.: Effect of uphill running on vo2, heart rate and lactate accumulation on lower body positive pressure treadmills. Sports **9**, 4 (2021)
6. ul Haq, M.E., Malik, M., Azam, M., Naeem, U., Khalid, A., Ghazanfar, M.: Identifying users with wearable sensors based on activity patterns. In: The 11th International Conference on Emerging Ubiquitous Systems and Pervasive Networks. Madeira (2020)
7. Irvine, J., et al.: Heart rate variability: a new biometric for human identification. In: Vegas, L. (ed.) International Conference on Artificial Intelligence (IC-AI'2001) (2001)
8. Irvine, J.M., Israel, S.A., Scruggs, W.T., Worek, W.J.: eigenPulse: robust human identification from cardiovascular function. Pattern Recogn., 3427–35 (2008)
9. Israel, S.A., Irvine, J.M., Cheng, A., Wiederhold, M.D., Wiederhold, B.K.: ECG to identify individuals. Pattern Recogn. **38**(1), 133–142 (2005)
10. Karim, F., Majumdar, S., Darabi, H., Chen, S.: LSTM fully convolutional networks for time series classification. CoRR abs/1709.05206 (2017). http://arxiv.org/abs/1709.05206
11. Ni, J., Muhlstein, L., JulianMcAuley: modeling heart rate and activity data for personalized fitness recommendation. In: WWW 2019: Proceedings of the 2019 World Wide Web Conference, San Francisco (2019)
12. Reichherzer, T., Timm, M., Earley, N., Reyes, N., Kumar, V.: Using machine learning techniques to track individuals & their fitness activities. In: Proceedings of 32nd International Conference on Computers and Their Applications (2017)
13. Ruiz, A., Flynn, M., Large, J., Middlehurst, M., Bagnall, A.: The great multivariate time series classification bake off: a review and experimental evaluation of recent algorithmic advances. Data Min. Knowl. Disc. **35**, 401–49 (2021)

14. Sanchez, O.R., Torre, I., He, Y., Knijnenburg, B.P.: A recommendation approach for user privacy preferences in the fitness domain. User Model. User-Adapted Interact. **30**(3), 513–565 (2019). https://doi.org/10.1007/s11257-019-09246-3
15. Shen, T.W., Tompkins, W.J., Hu, Y.H.: Implementation of a one-lead ECG human identification system on a normal population. J. Eng. Comput. Innov., 12–21 (2011)
16. Torre, I., Koceva, F., Sanchez, O., Adorni, G.: Fitness trackers and wearable devices: how to prevent inference risks? In: 11th International Conference on Body Area Ntwrks (2017)

Transferable Keyword Extraction and Generation with Text-to-Text Language Models

Piotr Pęzik[1,2] , Agnieszka Mikołajczyk[2] , Adam Wawrzyński[2] ,
Filip Żarnecki[2] , Bartłomiej Nitoń[3] , and Maciej Ogrodniczuk[3(✉)]

[1] Faculty of Philology, University of Łódź, Łódź, Poland
[2] VoiceLab, NLP Lab, Gdańsk, Poland
[3] Institute of Computer Science, Polish Academy of Sciences, Warsaw, Poland
`maciej.ogrodniczuk@gmail.com`

Abstract. This paper explores the performance of the T5 text-to-text transfer-transformer language model together with some other generative models on the task of generating keywords from abstracts of scientific papers. Additionally, we evaluate the possibility of transferring keyword extraction and generation models tuned on scientific text collections to labelling news stories. The evaluation is carried out on the English component of the POSMAC corpus, a new corpus whose release is announced in this paper. We compare the intrinsic and extrinsic performance of the models tested, i.e. T5 and mBART, which seem to perform similarly, although the former yields better results when transferred to the domain of news stories. A combination of the POSMAC and InTechOpen corpus seems optimal for the task at hand. We also make a number of observations about the quality and limitations of datasets used for keyword extraction and generation.

Keywords: keyword extraction · T5 language model · POSMAC · Polish

1 Introduction

Author-provided keywords are one of the intrinsic features of scientific articles as a distinct genre of texts. Despite recent advances in information extraction, sets of typically 3 to 5 keywords continue to be widely used to improve automatic retrieval of articles indexed in bibliographic databases. Formally, such keywords are usually noun phrases of varying complexity which may be used verbatim or as variants or derivatives of the wording used in the running text of an article. Some keywords may also denote concepts or descriptors abstracted from the literal content of a text. One implication of this dual nature of scientific keywords is the fact that a purely extractive keyword generation method, which critically depends on the occurrence of keywords in text can rarely produce satisfactorily complete results. Some studies have even proposed a distinction between

Present Keyword Extraction (PKE) and Abstract Keyword Generation (AKG) as separate tasks to reflect those two aspects of labeling texts with keywords [8]. Although this distinction may help evaluate certain solutions optimized for either of these two tasks, it is quite clear that a successful approach to automatically assigning keywords to scientific papers should be both extractive and abstractive as both of these characteristics are used by authors to succinctly describe the topic and domain of such texts. Furthermore, the distinction between PKE and AKG is intrinsically vague as certain keywords are nominalizations or otherwise paraphrased or generalized variants of expressions used in the running text of an article.

This paper focuses on evaluating the performance of the T5 and mBART models on the task of KEG (Keyword Extraction and Generation) from English language scholarly texts. In the initial section of the paper, we discuss the availability of English-language datasets used for KEG and point out some of their peculiarities and limitations. We also introduce the POSMAC corpus, which we believe to be a valuable resource for KEG in English. The subsequent sections of the paper present the evaluation of the aforementioned models on the POSMAC corpus and an extrinsic corpus of news stories.

2   Overview of KEG Datasets

The top section of Table 1 summarizes many openly available datasets proposed for different variants of keyword extraction and generation tasks. We briefly discuss this selection to show how significantly different such datasets can be and justify the choice of corpora used in this paper.

The NUS dataset [12] seems to be an ad hoc collection of 211 scholarly papers from a variety of domains with mostly extractive keywords assigned by 'student volunteers'.

Table 1. A selection of openly available KEG datasets.

Dataset	Type	Documents	Words	Unique words
NUS [12]	Full text	211	1 824 297	42 568
SemEval2010 [4]	Full text	244	2 345 689	53 923
Inspec [5]	Abstracts	2 000	287 908	17 653
Krapivin [6]	Full text	2 305	21 858 324	183 976
KP20k [10]	Abstracts	570 809	104 349 114	701 706
OAGKX [7]	Abstracts	22 674 436	4 237 931 192	18 959 687
InTechOpen	Abstracts	30 418	4 935 962	151 598
POSMAC EN	Abstracts	115 749	13 788 880	165 168
OAG (AMiner) [19]	Abstracts	100 000	19 252 115	699 638
News200	Articles	200	99 081	680

The SemEval2010 corpus "consists of a set of 284 English scientific papers from the ACM Digital Library" which were restricted to three subdomains of computer science and a subset of economic papers [4]. The keywords in this dataset are separated into author- and reader-submitted phrases and they seem to be a mixture of abstractive and extractive descriptors.

The Inspec dataset [5] contains abstracts of 2,000 scientific papers representing two subdomains of computer science. The dataset features two sets of keyphrases, including mostly abstractive keywords from a closed-set vocabulary and uncontrolled, mostly extractive keywords.

Krapivin [6] is a similarly sized collection of 2,000 full articles restricted to the domain of computer science. The keywords were assigned to each paper by its authors and verified by reviewers.

The KP20K corpus is a collection of over 570 000 articles scientific articles also representing the domain of computer science. The keywords assigned to these full-length papers are mostly abstractive and provided by their authors.

The Open Academic Graph corpus [19] and its variant processed for the task of keyword extraction and generation (OAGKX) contain over 9 and 22 million abstracts respectively covering a wide variety of scholarly domains. As such, they may appear to be highly relevant to the task of developing and evaluating KEG solutions which render the remaining datasets described here largely insignificant. However, on closer inspection, the overall quality of the keyword assignments available in OAG/OAGKX calls its usefulness into question. First of all, this dataset seems to contain many automatic, low-quality keywords extracted from the text of abstracts. This can be verified by simply comparing the OAG keyword assignments with the corresponding originally published papers available online. In many cases, a single high-level keyword is assigned to a record and in other cases, dozens of keywords are assigned to an equally sized abstract. Additionally, the OAGKX 'edition' of OAG was tokenized with an NLP pipeline which removed all punctuation and casing. In short, although the inconsistent and occasionally clearly erroneous assignments of keywords may improve the retrieval of documents from this collection, it is not obvious whether OAG or OAGKX can be used to train or even evaluate a KEG classifier. To summarize, openly available KEG datasets differ significantly in terms of the number and size of documents (abstracts/ full texts), quality and type of keyword assignments (abstractive/ extractive), the average number of keywords per text, and the total number of distinct keywords. They are also typically restricted to a handful of domains and in some cases, the keywords they offer were not assigned by the authors or reviewers, but rather by volunteers or, even worse, by algorithms. This variability is further summarized in Table 2.

The bottom part of Table 1 lists the datasets used to evaluate the KEG models described in this paper. The most important of them is the English language subset of the newly released Polish Open Science Metadata Corpus (POSMAC) [13], which was developed in the CURLICAT project[1] [17]. The content of POS-

---

[1] https://curlicat.eu/.

MAC was acquired from the Library of Science (LoS)[2], a platform providing open access to full texts of articles published in over 900 Polish scientific journals and selected scientific books with bibliographic metadata. Over 70% of the metadata records acquired have author-defined keywords. POSMAC combines high-quality keyword assignments with a fairly wide coverage of scholarly domains including non-technical disciplines such as humanities and social sciences.

In addition to this new resource, we compiled a corpus of over 30 000 abstracts of chapters published in open access books crawled from https://www.intechopen.com/. The corpus covers a variety of scholarly disciplines with high-quality keywords assigned by authors of the respective chapters.

The last collection used in this paper is a set of 200 recent news articles published on two websites (http://euronews.com and http://wikinews.org), whose topics range from health, politics to business and sports. We manually assigned a set of keywords to each of these articles to assess the transferability of KEG models trained on scholarly texts.

Table 2. Type of keyword assignment in selected KEG datasets.

Dataset	Average keywords	Keyword types*	Unique KWs	Annotators
NUS	11	Extractive	2 041	Volunteers
SemEval2010	15.5	Abstractive and extractive	3 220	Readers and authors
Inspec	9.5	Extractive	16916	Professional indexers
Krapivin	5	Abstractive	8 728	Authors
KP20k	5	Abstractive	760 652	Authors
OAGKX	4	Unclear	18 959 687	Unclear
POSMAC EN	4.5	Abstractive	198 102	Authors
InTechOpen	4.9	Abstractive	90198	Authors
OAG (AMiner)	4	Unclear	250 899	Unclear

*The predominant type of keywords included.

3   Evaluation of Text-to-text Models for KEG

An increasing number of recent approaches to KGE follow the general trend to use deep neural architectures to address NLP problems, cf. GAN [15], TG-Net [3], Para-Net [20], catSeq [1], corrRNN [2], SetTrans [18], KEA. More specifically, transformer-based architectures have also been used to both extract and abstract keywords from scientific texts, cf. BERT-PKE [8], [11], KeyBART [7].

---

[2] https://bibliotekanauki.pl/.

Table 3. Overall performance of evaluated models on new datasets of scientific and news texts.

Model	Train set	POSMAC			News articles		
		P	R	F_1	P	R	F_1
mT5-base	POSMAC EN	0.265	0.216	0.238	0.260	0.215	0.235
mT5-base	POSMAC EN+InTechOpen	0.276	0.224	0.248	0.249	0.204	0.224
mBART-large	POSMAC EN+InTechOpen	0.270	0.236	**0.252**	0.237	0.213	0.224
mT5-large	POSMAC EN+InTechOpen	0.286	0.223	0.250	0.275	0.222	**0.246**

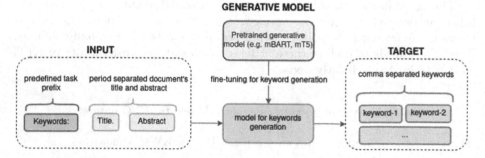

Fig. 1. Training procedure for mBART and Text-To-Text Transfer Transformer model for keywords generation.

In this paper, we focus on applying generative language models, which have recently been successfully applied to a number of NLP tasks. The first of those models is known as T5 [14]. Although its architecture is based on the original encoder-decoder transformer implementation [16], it frames a wide variety of NLP problems as text-to-text operations, where both the input and output are text strings. In the experiments reported in this paper, the input to the mT5[3] variant of T5 is a concatenated title and abstract of a scientific abstract and the text string output is a comma-separated list of lemmatized single- or multiword keywords. For the KGE task at hand, we used an Adam optimizer with 100 warm-up steps, linearly increasing the learning rate from zero to a target of 3e-5. Additionally, we used a multiplicative scheduler that lowered the LR by 0.9 every epoch. The model was trained for ten epochs with a batch size of 32. The maximum input length was set to 512 tokens and the maximum target length was 128. We refer to the resulting KGE model as **mT5kw**. After experimenting with the `no_repeat_ngram_size` and `num_beams` parameters on the development subset of our corpora we found the optimal values of `no_repeat_ngram_size` = 3 and `num_beams` = 4. The general flow of the mT5 and mBART training procedure is shown in Fig. 1.

We compare the results obtained with mT5 with the performance of a KEG model based on mBART, which is a de-noising auto-encoder model pretrained

---

[3] https://huggingface.co/docs/transformers/model_doc/mt5.

on multiple monolingual corpora [9]. As shown in Table 3 (which lists the average micro-precision and recall scores for each model) there is a noticeable advantage in using the larger version of the mT5 compared with the base variant. Additionally, the mT5-based model trained on scholarly texts seems to transfer slightly better to the domain of news articles than mBART-large, for which we observed the highest F_1 score on the source domain of scientific abstracts. Since the two text-to-text models produced 3–5 keywords, there was no need to artificially limit the number of keywords produced by the model. Our qualitative evaluation of the results shows that many of the keywords absent from the gold set seem relevant to the abstract from the test set. One of the most interesting aspects of the mT5 model is its transferability to other domains. The overall results of this paper confirm the conclusions of a separate study (Anonymized et al. 2022), in which compare a selection of approaches to keyword extraction and generation (KEG) for Polish scientific abstracts and concludes that the T5 outperforms purely extractive and abstractive methods and that it is highly transferable to other domains, including transcripts of spoken language. Another clear advantage of T5 is its ability to learn the truecasing and lemmatization of assigned keyphrases, which is of particular value in morphologically complex languages.

Acknowledgements. The work reported here was supported by 1) the European Commission in the CEF Telecom Programme (Action No: 2019-EU-IA-0034, Grant Agreement No: INEA/CEF/ICT/A2019/1926831) and the Polish Ministry of Science and Higher Education: research project 5103/CEF/2020/2, funds for 2020–2022) and 2) "CLARIN - Common Language Resources and Technology Infrastructure", which is part of the 2014–2020 Smart Growth Operational Programme, POIR.04.02.00-00C002/19.

References

1. Chan, H.P., Chen, W., Wang, L., King, I.: Neural keyphrase generation via reinforcement learning with adaptive rewards. In: Proceedings of the 57th Annual Meeting of the Association for Computational Linguistics, pp. 2163–2174. Association for Computational Linguistics, Florence, Italy, July 2019. https://doi.org/10.18653/v1/P19-1208, https://aclanthology.org/P19-1208

2. Chen, J., Zhang, X., Wu, Y., Yan, Z., Li, Z.: Keyphrase generation with correlation constraints. In: Proceedings of the 2018 Conference on Empirical Methods in Natural Language Processing, pp. 4057–4066. Association for Computational Linguistics, Brussels, Belgium, October–November 2018). https://doi.org/10.18653/v1/D18-1439, https://aclanthology.org/D18-1439

3. Chen, W., Gao, Y., Zhang, J., King, I., Lyu, M.R.: Title-guided encoding for keyphrase generation. CoRR abs/1808.08575 (2018). http://arxiv.org/abs/1808.08575

4. Hendrickx, I., et al.: SemEval-2010 task 8: multi-way classification of semantic relations between pairs of nominals. In: Proceedings of the 5th International Workshop on Semantic Evaluation, pp. 33–38. Association for Computational Linguistics, Uppsala, Sweden, July 2010. https://aclanthology.org/S10-1006

5. Hulth, A.: Improved automatic keyword extraction given more linguistic knowledge. In: Proceedings of the 2003 Conference on Empirical Methods in Natural

Language Processing. EMNLP '03, pp. 216–223. Association for Computational Linguistics, USA (2003). https://doi.org/10.3115/1119355.1119383

6. Krapivin, M., Marchese, M.: Large dataset for keyphrase extraction (2009)

7. Kulkarni, M., Mahata, D., Arora, R., Bhowmik, R.: Learning rich representation of keyphrases from text. CoRR abs/2112.08547 (2021). https://arxiv.org/abs/2112.08547

8. Liu, R., Lin, Z., Wang, W.: Keyphrase prediction with pre-trained language model. CoRR abs/2004.10462 (2020). https://arxiv.org/abs/2004.10462

9. Krapivin, M., Autaeu, A., Marchese, M.: Large Dataset for Keyphrase Extraction (2008). http://eprints.biblio.unitn.it/1671/1/disi09055-krapivin-autayeu-marchese.pdf. University of Trento, Dipartimento di Ingegneria e Scienza dell'Informazione, Technical Report # DISI-09-055

10. Meng, R., Zhao, S., Han, S., He, D., Brusilovsky, P., Chi, Y.: Deep keyphrase generation. CoRR abs/1704.06879 (2017). http://arxiv.org/abs/1704.06879

11. Meng, R., Zhao, S., Han, S., He, D., Brusilovsky, P., Chi, Y.: Deep keyphrase generation. In: Proceedings of the 55th Annual Meeting of the Association for Computational Linguistics (Volume 1: Long Papers), pp. 582–592. Association for Computational Linguistics, Vancouver, Canada, July 2017. https://doi.org/10.18653/v1/P17-1054, https://aclanthology.org/P17-1054

12. Nguyen, T.D., Kan, M.-Y.: Keyphrase extraction in scientific publications. In: Goh, D.H.-L., Cao, T.H., Sølvberg, I.T., Rasmussen, E. (eds.) ICADL 2007. LNCS, vol. 4822, pp. 317–326. Springer, Heidelberg (2007). https://doi.org/10.1007/978-3-540-77094-7_41

13. Pęzik, P., Mikołajczyk, A., Wawrzyński, A., Nitoń, B., Ogrodniczuk, M.: Keyword extraction from short texts with a text-to-text Transfer Transformer. In: Szczerbicki, E., Wojtkiewicz, K., Nguyen, S.V., Pietranik, M., Krótkiewicz, M. (eds.) ACIIDS 2022. CCIS, vol. 1716, pp. 530–542. Springer, Singapore (2022). https://doi.org/10.1007/978-981-19-8234-7_41

14. Raffel, C., et al.: Exploring the limits of transfer learning with a unified text-to-text transformer. J. Mach. Learn. Res. 21(140), 1–67 (2020). http://jmlr.org/papers/v21/20-074.html

15. Swaminathan, A., Gupta, R.K., Zhang, H., Mahata, D., Gosangi, R., Shah, R.R.: Keyphrase generation for scientific articles using GANs. CoRR abs/1909.12229 (2019). http://arxiv.org/abs/1909.12229

16. Vaswani, A., et al.: Attention is all you need. In: Guyon, I., et al. (eds.) Advances in Neural Information Processing Systems 30: Proceedings of the Annual Conference on Neural Information Processing Systems (NeurIPS 2017), pp. 5998–6008 (2017). https://proceedings.neurips.cc/paper/2017/hash/3f5ee243547dee91fbd053c1c4a845aa-Abstract.html

17. Váradi, T., et al.: Introducing the CURLICAT corpora: seven-language domain specific annotated corpora from curated sources. In: Proceedings of the Language Resources and Evaluation Conference, pp. 100–108. European Language Resources Association, Marseille, France (2022). https://aclanthology.org/2022.lrec-1.11

18. Ye, J., Gui, T., Luo, Y., Xu, Y., Zhang, Q.: One2Set: generating diverse keyphrases as a set. In: Proceedings of the 59th Annual Meeting of the Association for Computational Linguistics and the 11th International Joint Conference on Natural Language Processing (Volume 1: Long Papers), pp. 4598–4608. Association for Computational Linguistics, Online, August 2021. https://doi.org/10.18653/v1/2021.acl-long.354, https://aclanthology.org/2021.acl-long.354

19. Zhang, F., et al.: Oag: toward linking large-scale heterogeneous entity graphs. In: Proceedings of the 25th ACM SIGKDD International Conference on Knowledge Discovery and Data Mining. KDD '19, pp. 2585–2595. Association for Computing Machinery, New York, NY, USA (2019). https://doi.org/10.1145/3292500.3330785
20. Zhao, J., Zhang, Y.: Incorporating linguistic constraints into keyphrase generation. In: Proceedings of the 57th Annual Meeting of the Association for Computational Linguistics, pp. 5224–5233. Association for Computational Linguistics, Florence, Italy, July 2019. https://doi.org/10.18653/v1/P19-1515, https://aclanthology.org/P19-1515

Adsorption and Thermal Stability of Hydrogen Terminationn on Diamond Surface: A First-Principles Study

Delun Zhou[1], Jinyu Zhang[1], Ruifeng Yue[1,2(✉)], and Yan Wang[1,2,3(✉)]

[1] School of Integrated Circuit, Tsinghua University, Beijing, China
zhoudl18@mails.tsinghua.edu.cn, yuerf@mail.tsinghua.edu.cn
[2] Beijing National Research Center for Information Science and Technology, Beijing, China
[3] Beijing Innovation Center for Future Chips (ICFC), Beijing, China

Abstract. In this paper, we systematically investigated the adsorption characteristics, electronic structure(DOS), band structure and thermal stability of diamond surface with Hydrogen terminals. We found that the most stable adsorption performance may occur on (100) surface. The adsorption stability of hydrogen atom on plane (110) is the second, and the worst on plane (111). A very shallow acceptor level is introduced through Hydrogen termination, explaining the ideal p-type diamond characteristics. The stability of the hydrogen terminal structure decreases as temperature rises. This structure has deteriorated significantly since 400 K, and the instability of the hydrogen-terminated structure on the surface is the root cause of the decrease in the hole concentration of hydrogen-terminated diamond at high temperature.

Keywords: Hydrogen termination · Thermal Stability · Surface Adsorption

1 Introduction

Diamond has been extensively studied due to its outstanding properties since the end of the last century [1–3]. Diamond is not only a superhard material for mechanical cutting, but also a semiconductor material with full potential, and it is even expected to become the ultimate semiconductor material. Diamond performs extremely well in power capability and thermal conductivity, far exceeding the related properties of SiC [4–7].

Doping is a key problem in diamond at present. The performance of bulk doped diamond has not been ideal at present, whether it is single doping (B/P/N/O/S) [8–13] or co-doping (B-S/Li-N/B-O/B-P/B-N) [14–18] is difficult to achieve shallow doping in diamond. The low carrier concentration of diamond at room temperature hinders its wide application [19]. Hydrogen-terminated diamond can form a sufficiently high concentration of holes on the surface, which is an important method to realize p-type diamond [20]. At present, hydrogen surface termination can realize the accumulation of holes on the diamond surface, and the concentration can reach ~10^{13} cm^{-2} at room

J. Mikyška et al. (Eds.): ICCS 2023, LNCS 14074, pp. 406–412, 2023.
https://doi.org/10.1007/978-3-031-36021-3_43

temperature, which is enough to be applied to realize devices such as transistors [21, 22].

The concentration of holes accumulated on the H-terminated diamond surface will decrease drastically due to high temperature [23, 24], which shows that this p-type doping method has the defect of high temperature instability. The reason for this phenomenon is still controversial. There are currently a series of works to increase the thermal stability of H-terminated diamond, mainly through layer deposition at the hydrogen terminal interface. Kueck et al. [25] deposited an AlN thin film by atomic layer deposition (ALD) at 370 °C to maintain 65% of the initial current level. Kasu et al. [26] kept the concentration and mobility of H-terminated p-type channel at 230 °C through Al_2O_3 deposition. According to our calculations in this paper, the hydrogen-terminated structure will be structurally unstable at high temperatures(>400K), which may be the key reason why the surface hole concentration of hydrogen-terminated diamonds decreases at high temperatures. In addition, we further calculated the critical temperature of the hydrogen-terminated stable structure.

2 Calculation Methods

The geometric characteristics of hydrogen-terminated diamond surfaces have been studied through periodic slab using density functional theory (DFT). We adopted the Perdew-Burke-Ernzerhof (PBE) form of the generalized gradient approximation (GGA) through VASP [27, 28], consistent with the setting of our previous work [14, 15]. For calculations, the 64-atom supercell of diamond (2 × 2 × 2) and the Monkhorst-Pack grid of KPOINTS (9 × 9 × 1) are employed. The cut-off energy of the plane wave is set to 500 eV. The convergence criterion of the interatomic force is adjusted to 1×10^{-4} eV after the convergence verification [29].

3 Results and Discussion

3.1 Adsorption Properties of Hydrogen Atoms on Diamond Surface

Surface adsorption characteristics are an important way to describe the stable structure of hydrogen terminal on the diamond surface[30]. We calculated the surface energy and desorption activation energy of hydrogen atoms at (100), (110) and (111) diamond surfaces. Different crystal planes are achieved by setting different slab directions.

The diamond surfaces were calculated through symmetrically terminated slabs, hence, including two interfaces. Since the surfaces with suspension keys are very active, hydrogen passivation is performed on both such interfaces to ensure the accuracy of the calculation results.

The surface energy per area is calculated using the following expression [31]:

$$\gamma = \frac{E_{slab} - E_{bulk}}{2A}$$

where E_{slab} is the total energy of the whole slab containing all Hydrogen atoms and Carbon atoms, E_{bulk} is the bulk energy, which here refers to the energy of pure diamond structure, and A is the surface area. Given that there are two interfaces in a slab, the factor of two in the denominator must be included.

Hydrogen desorption activation energy is defined as [32]

$$E_{desorp} = (E_{bulk} + N_H \cdot \mu_H) - E_{slab}$$

where E_{bulk} is the total energy of the bare diamond surface, E_{slab} is the total energy of the system containing hydrogen atoms, N_H is the number of H atoms on the surfaces, and μ_H is the chemical potential of hydrogen, which is calculated by the total energy of H_2 molecule (Table 1).

Table 1. Surface energy and desorption activation energy of hydrogen terminals on different surfaces

Surface type	Surface energy(eV/A^2)	Desorption activation energy(eV)
(100)	−0.23	9.99
(110)	−0.19	9.59
(111)	−0.11	6.96

In general, the lower the surface energy, the higher the desorption activation energy, indicating that hydrogen atoms adsorbed more firmly on the surface. Combined with the calculation results, hydrogen atom has the lowest surface energy and the highest desorption activation energy on plane (100), which is the most stable adsorption. The adsorption stability of hydrogen atom on plane (110) is the second, and the worst on plane (111). Therefore, from the point of view of surface adsorption stability, h-terminated diamond should be selected (100), which is consistent with the results of current experiments in which (100) h-terminated diamond is used to prepare devices.

3.2 Electronic Structure and Band Structure

We have calculated the band structure and total density of states of H-terminated diamond. Credible p-type characteristic is achieved through hydrogen terminal structures, consistent with experiments results.

It can be seen from the calculation results that there is a very close acceptor energy level near the top of the valence band, which we believe is the introduction of an intermediate energy level due to the introduction of the hydrogen terminal structure. The gap between this intermediate energy level and the valence band maximum (VBM) is so small that a shallow acceptor energy level can be seen, which explains why the hydrogen-terminated structure can realize a ideal p-type diamond (Fig. 1).

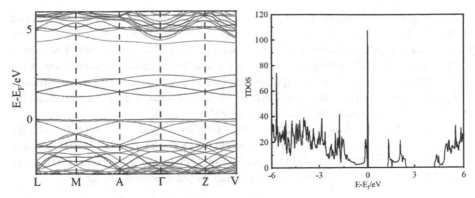

Fig. 1. Band structure of hydrogen-terminated diamond (a); Total density of states(TDOS) of hydrogen-terminated diamond (b)

3.3 Thermal Stability of Hydrogen Atoms on Diamond Surface

Poor thermal stability is a key factor hindering the continued development of H-terminated diamond. It has been found in experiments that the concentration of holes on the surface of H-terminated diamond will decrease at high temperature. The reason for this phenomenon is still controversial. We applied ab initio molecular dynamics methods to calculate the structural changes of hydrogen atoms adsorbed on the surface of (100) diamond at different temperatures (0 K–1000 K) (Fig. 2).

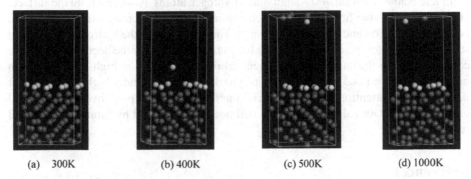

(a)   300K   (b) 400K   (c) 500K   (d) 1000K

Fig. 2. Hydrogen termination structure on diamond surface at different temperatures; (a) at 300 K; (b) at 400 K; (c) at 500 K; (d) at 1000 K

Based on our calculation results, the stability of hydrogen terminal structure deteriorates with the increase of temperature. Below 300 K, the hydrogen terminal can always be stably adsorbed on the diamond surface. At 400 K, the surface structure begins to change, and individual atoms emerge from the diamond surface. Considering the limited duration of molecular dynamics simulation (ps level), this temperature is sufficient to completely change the morphology of hydrogen terminal on diamond surface in practical experiments. At 500 K, the surface structure of hydrogen terminal is no longer stable, and hydrogen atoms have been completely excited out. When the temperature increased further (over 1000 K), the structure of hydrogen terminal on the surface became more chaotic, and the overall flatness and interatomic distance changed greatly. The change of hydrogen terminal surface structure directly affects the cavity concentration of the whole structure surface. In the high temperature environment, the terminal structure of hydrogen is no longer stable, and even completely excited ionization occurs, so the concentration of p-type carrier decreases inevitably.

Our calculations validated the poor thermal stability of hydrogen terminations on the diamond surface. This deficiency is unfavorable for its application in high-temperature and high-power applications, and technical means such as surface passivation are very critical for the application of hydrogen terminals on the diamond surface.

4 Conclusions

In conclusion, H-teminated diamond is currently a p-type diamond implementation method with better performance. The core is to realize the accumulation of holes through the hydrogen terminal on the surface. According to the calculation of surface energy and desorption activation energy, we believe that the hydrogen terminal has the best adsorption characteristics on the (100) surface, followed by the (110) surface, and the worst adsorption characteristics on the (111) surface. The hydrogen-terminated structure introduces an intermediate energy level very close to the top of the valence band, which makes hydrogen-terminated diamonds have shallow acceptor characteristics and can achieve good P-type diamonds. The hydrogen terminal structure remains stable and complete below 300 K; at 400 K, individual hydrogen atoms dissociate from the surface; above 500 K, the stability of the hydrogen terminal structure decreases significantly, and complete excitation and dissociation occurs. The instability of the hydrogen-terminated structure at high temperature may be the fundamental reason for the decrease of the hole concentration on the surface of hydrogen-terminated diamond at high temperature. In order to realize the good performance of p-type doping of diamond at high temperature, it is worth paying attention to the research on improving the stability of hydrogen terminal or bulk doping. Our calculation results still need to be verified by future experimental results.

References

1. Pezzagna, S., Meijer, J.: Quantum computer based on color centers in diamond. Appl. Phys. Rev. **8**(1), 011308 (2021)
2. Arnault, J.C., Saada, S., Ralchenko, V.: Chemical vapor deposition single-crystal diamond: a review. physica status solidi (RRL)–Rapid Res. Let. **16**(1), 2100354 (2022)
3. Bauch, E., Singh, S., Lee, J., et al.: Decoherence of ensembles of nitrogen-vacancy centers in diamond. Phys. Rev. B **102**(13), 134210 (2020)
4. Shen, S., Shen, W., Liu, S., et al.: First-principles calculations of co-doping impurities in diamond. Mater. Today Commun. **23**, 100847 (2019)
5. Li, Y., Liao, X., Guo, X., et al.: Improving thermal conductivity of epoxy-based composites by diamond-graphene binary fillers. Diamond Related Mater. **2022**(126), 126 (2022)

6. Zhang, Z., Lin, C., Yang, X., et al.: Solar-blind imaging based on 2-inch polycrystalline diamond photodetector linear array. Carbon **173**(42), 427–432 (2021)
7. Liu, X., Chen, X., Singh, D.J., et al.: Boron–oxygen complex yields n-type surface layer in semiconducting diamond. In: Proceedings of the National Academy of Sciences (2019)
8. Czelej, K., Piewak, P., Kurzydowski, K.J.: Electronic structure and N-Type doping in diamond from first principles. Mrs Adv. **1**(16), 1093–1098 (2016)
9. Shah, Z.M., Mainwood, A.: A theoretical study of the effect of nitrogen, boron and phosphorus impurities on the growth and morphology of diamond surfaces. Diam. Relat. Mater. **17**(7–10), 1307–1310 (2008)
10. Kato, H., et al.: Diamond bipolar junction transistor device with phosphorus-doped diamond base layer. Diamond Related Mater. 27–28:19–22 (2012)
11. Sque, S.J., Jones, R., Goss, J.P., et al.: Shallow donors in diamond: chalcogens, pnictogens, and their hydrogen complexes. Phys. Rev. Let. **92**(1), 017402 (2004)
12. Prins, J.F.: n-type semiconducting diamond by means of oxygen-ion implantation. Phys. Rev. B **61**(11), 7191–7194 (2000)
13. Kato, H., Makino, T., Yamasaki, S., et al.: n-type diamond growth by phosphorus doping on (001)-oriented surface. MRS Proc. **1039**(40), 6189 (2007)
14. Zhou, D., Tang, L., Geng, Y., et al.: First-principles calculation to N-type Li N Co-doping and Li doping in diamond. Diamond Related Mater. **110**, 108070 (2020)
15. Lin, T., Yue, R., Wang, Y., et al.: N-type B-S co-doping and S doping in diamond from first principles. Carbon Int. J. Sponsor. Am. Carbon Soc. **130**, 458–465 (2018)
16. Shao, Q.Y., Wang, G.W., Zhang, J., et al.: First principles calculation of lithium-phosphorus co-doped diamond.Condensed Matter Phys. **16**(1), 13702: 1–1 (2013)
17. Zhou D , Tang L , Zhang J , et al.: n-type B-N co-doping and N doping in diamond from first principles. In: Groen, D., de Mulatier, C., Paszynski, M., Krzhizhanovskaya, V.V., Dongarra, J.J., Sloot, P.M.A. (eds.) Computational Science – ICCS 2022. ICCS 2022. LNCS, vol. 13350. Springer, Cham (2022). https://doi.org/10.1007/978-3-031-08751-6_38
18. Sun, S., Jia, X., Zhang, Z., et al.: HPHT synthesis of boron and nitrogen co-doped strip-shaped diamond using powder catalyst with additive h-BN. J. Cryst. Growth **377**(aug.15), 22–27 (2013)
19. Yu, C., Zhou, C.J., Guo, J.C., et al.: 650 mW/mm output power density of H-terminated polycrystalline diamond MISFET at 10 GHz. Electron. Lett. **56**(7), 334–335 (2020)
20. Nebel, C.E., Rezek, B., Shin, D., et al.: Surface electronic properties of H-terminated diamond in contact with adsorbates and electrolytes. Physica Status Solidi (a) 203(13), 3273–3298 (2006)
21. Verona, C., Ciccognani, W., Colangeli, S., et al.: V 2 O 5 MISFETs on H-terminated diamond. IEEE Trans. Electron Devices **63**(12), 4647–4653 (2016)
22. Kubovic, M., Janischowsky, K., Kohn, E.: Surface channel MESFETs on nanocrystalline diamond. Diam. Relat. Mater. **14**(3–7), 514–517 (2005)
23. Ye, H., Kasu, M., Ueda, K., et al.: Temperature dependent DC and RF performance of diamond MESFET. Diam. Relat. Mater. **15**(4–8), 787–791 (2006)
24. De Santi, C., Pavanello, L., Nardo, A., et al.: Degradation effects and origin in H-terminated diamond MESFETs.In: Terahertz, R.F. (ed.) Millimeter, and Submillimeter-Wave Technology and Applications XIII. SPIE, vol. 11279, pp. 230–236 (2020)
25. Kueck, D., Leber, P., Schmidt, A., et al.: AlN as passivation for surface channel FETs on H-terminated diamond. Diam. Relat. Mater. **19**(7–9), 932–935 (2010)
26. Kasu, M., Saha, N.C., Oishi, T., et al.: Fabrication of diamond modulation-doped FETs by NO2 delta doping in an Al2O3 gate layer. Appl. Phys. Express **14**(5), 051004 (2021)
27. SanchogarcíA, J.C., Brédas, J.L., Cornil, J.: Assessment of the reliability of the Perdew–Burke–Ernzerhof functionals in the determination of torsional potentials in π-conjugated molecules. Chem. Phys. Lett. **377**(1), 63–68 (2003)

28. Monkhorst, H.J., Pack, J.D.: Special Points for Brillouin-zone Integrations. Phys. Rev. B Condensed matter **13.12**, 5188–5192 (1976)
29. Jones, R., Goss, J.P., Briddon, P.R.: Acceptor level of nitrogen in diamond and the 270-nm absorption band. Phys. Rev. B: Condens. Matter **80**(3), 1132–1136 (2009)
30. Rivero, P., Shelton, W., Meunier, V.: Surface properties of hydrogenated diamond in the presence of adsorbates: a hybrid functional DFT study. Carbon **110**, 469–479 (2016)
31. Ostrovskaya, L., Perevertailo, V., Ralchenko, V., et al.: Wettability and surface energy of oxidized and hydrogen plasma-treated diamond films. Diam. Relat. Mater. **11**(3–6), 845–850 (2002)
32. Carter, G.: Thermal resolution of desorption energy spectra. Vacuum **12**(5), 245–254 (1962)

Exploring Counterfactual Explanations for Predicting Student Success

Farzana Afrin$^{(\boxtimes)}$, Margaret Hamilton, and Charles Thevathyan

School of Computing Technologies, RMIT University,
Melbourne, VIC 3000, Australia
s3862196@student.rmit.edu.au,
{margaret.hamilton,charles.thevathyan}@rmit.edu.au

Abstract. Artificial Intelligence in Education (AIED) offers numerous applications, including student success prediction, which assists educators in identifying the customized support required to improve a student's performance in a course. To make accurate decisions, intelligent algorithms utilized for this task take into account various factors related to student success. Despite their effectiveness, decisions produced by these models can be rendered ineffective by a lack of explainability and trust. Earlier research has endeavored to address these difficulties by employing overarching explainability methods like examining feature significance and dependency analysis. Nevertheless, these approaches fall short of meeting the unique necessities of individual students when it comes to determining the causal effect of distinct features. This paper addresses the aforementioned gap by employing multiple machine learning models on a real-world dataset that includes information on various social media usage purposes and usage times of students, to predict whether they will pass or fail their respective courses. By utilizing Diverse Counterfactual Explanations (DiCE), we conduct a thorough analysis of the model outcomes. Our findings indicate that several social media usage scenarios, if altered, could enable students who would have otherwise received a failing grade to attain a passing grade. Furthermore, we conducted a user study among a group of educators to gather their viewpoints on the use of counterfactuals in explaining the prediction of student success through artificial intelligence.

Keywords: Student success prediction · counterfactual explanation · social media usage behaviour

1 Introduction and Background

Machine learning (ML) is a type of artificial intelligence (AI) that can learn and make predictions based on patterns in data. In the context of predicting student success, ML algorithms can be trained on historical data to identify factors that are associated with student success, such as demographic information, academic

J. Mikyška et al. (Eds.): ICCS 2023, LNCS 14074, pp. 413–420, 2023.
https://doi.org/10.1007/978-3-031-36021-3_44

performance, and behavioral data [9]. These algorithms can then be used to make predictions about which students are likely to succeed or struggle in the future [2, 3].

One of the challenges with using ML in the context of student success prediction is that the models used can be quite complex and difficult to interpret. This means that it can be hard to understand why the model is making certain predictions, which can be problematic for educators and other stakeholders who need to make decisions based on the predictions.

Therefore, it is important to build explainable ML models for student success prediction [1]. Explainability refers to the ability to understand how a model arrived at its predictions. By incorporating explainability into ML models for student success prediction, educators and other stakeholders can better understand the factors that are driving the predictions, which can help them make more informed decisions. This can ultimately lead to more effective interventions and supports for students who may be at risk of struggling, which can improve their chances of success. Additionally, improved explainability can help build trust in the use of ML by ensuring that stakeholders understand the basis for the predictions being made [7]. This can be achieved by using techniques such as feature importance analysis, which can help identify which features in the data were most important in making the prediction. Other techniques that can be used to improve explainability include decision tree analysis, which can provide a visual representation of the decision-making process, and local interpretable model-agnostic explanations (LIME), which can provide explanations for individual predictions. Another approach known as SHapley Additive exPlanations (SHAP) which can be used for both local and global explanations. However, these methods are often inadequate in fulfilling the specific needs of individual students in determining the cause-and-effect relationship of specific features associated with their success.

To address this gap, first we employ a set of state-of-the-art ML models to predict student success by leveraging their social media usage behaviour. Second, we employ a counterfactual approach which allows us to simulate different hypothetical scenarios enhancing the explainability of student success prediction outcomes. In our context of predicting student success, we investigate how changing the value of certain factors may have impacted student success outcomes. In particular, we investigate *"what factors needed to be changed if a failing student had to pass in the course?"*. Additionally, we investigate the practicability of adopting counterfactual explanation in student success prediction. To achieve this, we conduct a real-world user-study comprising educators of a tertiary institution. Specifically, we conduct a short survey to understand their viewpoint on using counterfactual explanations in their decision making process for the given scenario and beyond. The contributions of this paper are as follows.

- Predict student success (in terms of pass or fail) in a course by leveraging students social media usage behaviour. In addition, we provide a counterfactual analysis of the prediction outcomes.
- Conduct a user study to understand the practicability of employing counterfactual explanations in student success prediction.

The organization of the paper includes the modelling and evaluation approaches of student success prediction in Sect. 2, which is followed by a counterfactual generation and analysis in Sect. 3. We report the findings of user-study in Sect. 4. The paper concludes in Sect. 5.

2 Predicting Student Success

2.1 Dataset and Pre-processing

In our study, we made use of an open dataset from [8], which comprised data on the social media activity and final marks of 505 students (221 males and 284 females) who took a compulsory course across multiple disciplines including business, commerce, law, engineering, science, and information technology. The dataset was gathered from a large metropolitan Australian university over the course of three teaching sessions between 2017 and 2018. The dataset we used logs information on the usage times of Facebook, LinkedIn, Snapchat, and Twitter. Notably, the Facebook usage times are further broken down into various purposes of usage including 'Communications with friends and family', 'Enjoyment and entertainment', 'Filling in dead or vacant time', 'Keeping informed about events and news', 'Education and study', 'Work related reasons', 'Arrange a meeting for a group project', 'Class or university work related contact with another student', 'Discuss university work', 'Ask a classmate for help in the class', 'Help manage a group project', 'Collaborate on an assignment in a way my instructor would like', 'Arrange a face-to-face study group'. Note that the breakdown of Facebook usage time is calculated by multiplying the total usage time per day (in minutes) with the extent and likelihood students indicate for different reasons for Facebook usage which is denoted as proxy times in [8]. Additionally, the dataset also provides demographic and background information on the participating students, including their age, gender, and WAM (weighted average mark, which is akin to grade point average). As part of our data pre-processing, we categorized all final exam marks into two groups: pass and fail. Specifically, any marks of 50 or higher were classified as pass, while marks below 50 were labeled as fail. The final dataset is almost balanced where the number of pass and fail labels are 261 and 244 students respectively.

2.2 Modelling Approach and Evaluation

The idea of student success prediction task is to identify whether a student will pass or not in a course by leveraging a set of features representing corresponding information about students' social media usage behaviour, demographic and background. A formal definition of our prediction task can be given as follows:

Let's say we have a set $G_{fe} = \{P, F\}$ that includes two possible final exam grades for n students in their course. In the final dataset, each instance x is a d-dimensional vector of attributes from R^d, which contains information about the students' usage of various social media platforms, demographic details, and

Table 1. Prediction performance by different models

Classification models	Accuracy	F_1-score
GBM classifier	0.74	0.74
LGBM classifier	0.75	0.75
XGBM classifier	0.72	0.72
Logistic regression classifier	0.66	0.66
Random Forest classifier	0.71	0.71
SVC	0.55	0.54

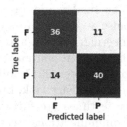

Fig. 1. Confusion matrix showing prediction performance of LGBM.

background information discussed in Sect. 2.1. If we have a prediction function $g(.)$ that can forecast the success (final exam grade) of a student using their d attributes, we can represent $g(.)$ as $g(x_q) : R^d \rightarrow \hat{G}(x_q)$, where $\hat{G}(x_q) \in G_{fe}$ is the predicted final grade for student x_q whose grade was previously unknown.

For our student success prediction experiment, we employ a set of classification models implemented in scikit-learn [5] including Support Vector Classifier (SVC), Random Forest, Logistic regression and three variants of boosting techniques including Gradient Boosting classifier (GBM), Light Gradient Boosting classifier(LGBM), and Extreme Gradient Boosting classifier(XGBM). We randomly split 80% of the data from training these models and the rest are used for testing. The prediction outcomes are evaluated against two metrics: accuracy and F_1 score. As shown in Table 1, the LGBM classifier produces best prediction results in terms of both accuracy and F_1 score. A confusion matrix detailing prediction outcomes of LGBM is given in Fig. 1.

3   Counterfactual Generation and Analysis

For counterfactual generation, we leverage Diverse Counterfactual Explanations (DiCE) for Machine Learning Classifiers [4]. A primary goal of DiCE is to provide explanations for the predictions of ML-based systems that inform decision-making in critically important areas such as finance, healthcare, and education. DiCE approaches the discovery of explanations essentially as an optimization problem, similar to how adversarial examples are identified. As we generate explanations, we require perturbations that change the output of a machine learning model while at the same time being varied and practical to implement. Hence, the DiCE approach facilitates the generation of counterfactual explanations that can be customized for diversity and resemblance to the original input.

Moreover, DiCE allows for the imposition of basic feature constraints to ensure that the counterfactual instances are plausible. In our experiments, the demographic features such as age, gender, and WAM cannot be varied, since that is not practical. Similarly, we set the number of generated counterfactuals to 3, but determining the optimal number is still a research question.

A snapshot of generated diverse counterfactual set (CFs) for an individual failing student is illustrated in Table 2 (see Appendix A). Note that the generated counterfactual explanations are based on our top-performing black-box model LGBM. The first explanation proposes that decreasing the duration of communication with friends and family, as well as minimizing the arrangement time for face-to-face meetings, while increasing the amount of time devoted to project management, may result in a shift from a failing to a passing grade. The second counterfactual explanation recommends allocating more time to studying and education, while reducing the time spent on seeking assistance from classmates through Facebook or scheduling face-to-face meetings. The third counterfactual explanation suggests a substantial decrease in the time spent filling up dead periods and vacant slots, as well as minimizing collaboration time for assignments via Facebook. Although these explanations can be highly beneficial, the feasibility of implementing some of the recommended strategies may be uncertain. As a result, we carried out a user study to assess the effectiveness of the proposed counterfactual explanations, as well as practitioners' perspectives on the utilization of such explanations in their decision-making processes.

4   User Study

Counterfactual explanations can be validated through user studies [6]. In particular, the goal of our user study is to understand the perception of educators towards employing counterfactual explanation in analyzing the student success prediction outcomes. For this purpose, we recruited 18 educators from a tertiary education institution. We provided them with 3 counterfactual explanations corresponding to each individual failing student. Then we asked -

How much do you believe the recommendations provided by DiCE are effective in changing the student success prediction outcome from fail to pass?.

The participants' feedback was collected using a 5-point Likert scale (ranging from 1 for Strongly Disagree to 5 for Strongly Agree) as summarized in Fig. 2 (a). While a majority of the participants agreed with the validity of the generated explanations, some provided negative ratings. Additionally, an optional open text field was provided to gather more detailed responses. It was found that some respondents did not believe certain features (such as communication with friends and family) should be included in the explanation.

We also inquired the participants - *How many counterfactual explanations do you believe should be suggested so that you can implement them effectively and efficiently?.*

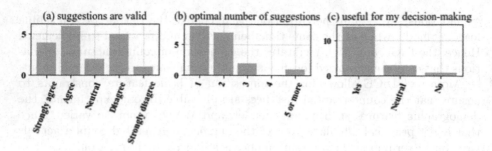

Fig. 2. Summary of user responses. Note: Y-axis denotes frequency of responses

As depicted in Fig. 2 (b), the majority of the participants favored 1–2 suggestions, with only one participant preferring 5 or more. However, this respondent did not indicate any specific reason for the selection.

Furthermore, we asked the participants whether they would be willing to incorporate counterfactual explanations into their decision-making process. As depicted in Fig. 2(c), only one participant from our cohort did not find counterfactual explanations to be very useful.

Three participants from our cohort also provided additional insights on what need to be taken into account for the effective utilization of counterfactual explanations in predicting student success. The responses are given below.

"The utilization of counterfactual explanations can offer multiple benefits. Nonetheless, to determine the most effective interventions or supports for enhancing student success, an expert-in-the-loop approach is necessary, where experienced educators assist in finalizing the list of attributes to be taken into account for a particular student."

"Finalizing interventions for students who are going to receive them is of utmost importance. It is essential to comprehend and address any associated risks, such as the possibility of negative outcomes for a student receiving an intervention."

"It is important to recognize that there may be additional factors beyond the dataset that were not taken into account."

5   Conclusion

This paper uses social media usage behavior of university students to predict whether they will succeed or fail their courses. Using Diverse Counterfactual Explanations (DiCE), this study examines how a student who is failing could pass if certain factors were altered. Additionally, the paper includes an 18-educator user-study to evaluate the practicality of DiCE in predicting student success, and to gather educators' opinions on using counterfactual explanations in their decision-making process.

The educators who participated in the study found counterfactual analysis to be a powerful tool for advancing the prediction of student success. They agreed that by simulating different scenarios and considering the impact of various factors, more effective interventions and support systems could be developed to improve a student's chances of success. The educators also emphasized the importance of including appropriate factors in the collected dataset to achieve more reliable prediction outcomes and explanations. One possibility for future research is to conduct a real-world trial of counterfactual explanations. This could involve creating a scenario in which at-risk students receive a specific intervention, and then comparing their outcomes to similar students who did not receive the intervention. By examining the impact of the intervention on student outcomes, educators could gain insights that could inform future interventions and support. Future research could also explore other datasets, with other explanations, which could help by providing educators with a concrete tool for decision-making.

Acknowledgements. Farzana was supported through an Australian Government's RTP Scholarship.

A   Appendix

Table 2. A snapshot of a set of generated counterfactuals with new outcome 1 (pass) for a query instance with original outcome 0 (fail).

Features	Query Ins.	Diverse CFs		
	Outcome:0	Outcome:1		
	#0	#0	#1	#2
Communications with friends and family	1320	921		
Enjoyment and entertainment	1320			
Filling in dead or vacant time	1320			241
Keeping informed about events and news	1320			
Education and study	1320		1556	
Work related reasons	1320			
Arrange a meeting for a group project	1320			
Class or university work related contact with another student	1320			
Discuss university work	1320			
Ask a classmate for help in the class	990		366	
Help manage a group project	1320	2290		
Collaborate on an assignment in a way my instructor would like	990			497
Arrange a face-to-face study group	1320	613	597	400
Linkedin time	0			
Snapchat time	0			
Twitter time	0			
Age	19			
Gender	1			
WAM	58.67			
FE-Grade	0	1	1	1

References

1. Afrin, F., Hamilton, M., Thevathyan, C.: On the explanation of AI-based student success prediction. In: Groen, D., de Mulatier, C., Paszynski, M., Krzhizhanovskaya, V.V., Dongarra, J.J., Sloot, P.M.A. (eds) ICCS 2022, Part II. LNCS, vol. 13351, pp. 252–258. Springer, Cham (2022). https://doi.org/10.1007/978-3-031-08754-7_34
2. Giunchiglia, F., Zeni, M., Gobbi, E., Bignotti, E., Bison, I.: Mobile social media usage and academic performance. Comput. Hum. Behav. **82**, 177–185 (2018)
3. Liu, Z.: A practical guide to robust multimodal machine learning and its application in education. In: Proceedings of the Fifteenth WSDM, p. 1646. New York, NY, USA (2022)
4. Mothilal, R.K., Sharma, A., Tan, C.: Explaining machine learning classifiers through diverse counterfactual explanations. In: Proceedings of the 2020 Conference on Fairness, Accountability, and Transparency, pp. 607–617 (2020)
5. Pedregosa, F., et al.: Scikit-learn: machine learning in Python. J. Mach. Learn. Res. **12**, 2825–2830 (2011)
6. Spreitzer, N., Haned, H., van der Linden, I.: Evaluating the practicality of counterfactual explanations. In: Workshop on Trustworthy and Socially Responsible Machine Learning, NeurIPS 2022 (2022)
7. Toreini, E., Aitken, M., Coopamootoo, K., Elliott, K., Zelaya, C.G., van Moorsel, A.: The relationship between trust in AI and trustworthy machine learning technologies. In: Proceedings of the 2020 Conference on Fairness, Accountability, and Transparency. FAT* '20, pp. 272–283, New York, NY, USA (2020)
8. Wakefield, J., Frawley, J.K.: How does students' general academic achievement moderate the implications of social networking on specific levels of learning performance? Comput. Educ. **144**, 103694 (2020)
9. Yu, R., Li, Q., Fischer, C., Doroudi, S., Xu, D.: Towards accurate and fair prediction of college success: evaluating different sources of student data. Int. Educ. Data Min. Soc. (2020)

Advances in High-Performance Computational Earth Sciences: Applications and Frameworks

Development of 3D Viscoelastic Crustal Deformation Analysis Solver with Data-Driven Method on GPU

Sota Murakami[1]([✉]), Kohei Fujita[1,2], Tsuyoshi Ichimura[1,2], Takane Hori[3], Muneo Hori[3], Maddegedara Lalith[1], and Naonori Ueda[4]

[1] Earthquake Research Institute and Department of Civil Engineering, The University of Tokyo, Tokyo, Japan
{souta,fujita,ichimura,lalith}@eri.u-tokyo.ac.jp
[2] Center for Computational Science, RIKEN, Kobe, Japan
[3] Japan Agency for Marine-Earth Science and Technology, Yokohama, Japan
{horit,horimune}@jamstec.go.jp
[4] Center for Advanced Intelligence Project, RIKEN, Tokyo, Japan
naonori.ueda@riken.jp

Abstract. In this paper, we developed a 3D viscoelastic analysis solver with a data-driven method on GPUs for fast computation of highly detailed 3D crustal structure models. Here, the initial solution is obtained with high accuracy using a data-driven predictor based on previous time-step results, which reduces the number of multi-grid solver iterations and thus reduces the computation cost. To realize memory saving and high performance on GPUs, the previous time step results are compressed by multiplying a random matrix, and multiple Green's functions are solved simultaneously to improve the memory-bound matrix-vector product kernel. The developed GPU-based solver attained an 8.6-fold speedup from the state-of-art multi-grid solver when measured on compute nodes of AI Bridging Cloud Infrastructure at National Institute of Advanced Industrial Science and Technology. The fast analysis method enabled calculating 372 viscoelastic Green's functions for a large-scale 3D crustal model of the Nankai Trough region with 4.2×10^9 degrees of freedom within 333 s per time step using 160 A100 GPUs, and such results were used to estimate coseismic slip distribution.

Keywords: unstructured finite-element method · data-driven predictor · OpenACC · viscoelastic analysis

1 Introduction

Improvement in the estimation of interplate conditions such as plate sticking and sliding is expected to play an important role in the advancement of source scenarios for large earthquakes. In particular, the estimation of interplate conditions considering viscoelastic deformation is useful for estimating an afterslip and predicting continuous crustal deformation after a large earthquake. In recent years, data required for the advancement of interplate state estimation have been

© The Author(s), under exclusive license to Springer Nature Switzerland AG 2023
J. Mikyška et al. (Eds.): ICCS 2023, LNCS 14074, pp. 423–437, 2023.
https://doi.org/10.1007/978-3-031-36021-3_45

accumulated due to the improvement of the seafloor crustal deformation observation directly above the seismogenic zone (e.g., [16]) and the acquisition of crustal structure data with approximately 1 km resolution by advancement in underground structure exploration. On the other hand, the theoretical solution assuming the crustal structure as a multilayered semi-infinite medium [8] is often used in obtaining the displacement responses at observation points to unit slips (Green's function), which are used in the inverse analysis of interplate conditions. Although the calculation of Green's functions based on a highly detailed three-dimensional (3D) crustal structure model and its use for estimating the interplate state is expected to improve the accuracy of interplate state estimation, this calculation leads to the huge analysis cost comprising 100–1000 cases of large-scale viscoelastic analysis.

Most of the computational cost in viscoelastic crustal deformation analysis is spent on solving the large-scale simultaneous equations obtained by discretizing the crustal structure model. Since a method scalable on a parallel computing environment is essential for conducting large-scale calculations, and since low-frequency components dominate in viscoelastic response, a multi-grid-based solver is considered effective. In fact, multi-grid based conjugate gradient solvers, which use geometric and algebraic multi-grid methods, have been developed and applied to crustal deformation analysis [6,10]. In addition, viscoelastic analysis using these multi-grid solvers has been accelerated using GPUs, enabling forward analysis of viscoelastic response on highly detailed 3D models. On the other hand, further reduction of computation cost is required to realize viscoelastic Green's function calculation, which corresponds to a computation cost of about 100–1000 cases of forward analysis.

In recent years, data-driven methods have been utilized to improve the performance of equation-based methods (e.g., [11]), and their effectiveness in viscoelastic crustal deformation analysis has also been demonstrated [7]. The initial solution to a large simultaneous equation is obtained with high accuracy using a data-driven predictor based on past time-step results, which reduces the number of multi-grid solver iterations and thus reduces the computation cost. Both the data-driven predictor and the multi-grid solver are designed to be scalable, and have been shown to perform well on CPU-based massively parallel computer Fugaku [4], and are expected to be effective on GPU-based systems. In this study, a multi-grid solver with a data-driven predictor for GPU computation environment is developed for fast computation of Green's functions of viscoelastic crustal deformation. Since the data-driven predictor, which learns and predicts solutions based on a large amount of data, hinders the performance of GPUs with relatively small memory capacity, methods enabling a reduction in memory footprint are combined with the data-driven predictor. While the multi-grid solver is also effective on GPUs due to its high scalability, its performance is limited by random access in the sparse matrix-vector computations; we introduce simultaneous computation of multiple Green's functions for a reduction in random access and a further performance improvement. Considering the development cost, we develop the solver using directive-based OpenACC [3].

As an application example, we calculated 372 viscoelastic Green's functions at 333 s per time step for a large-scale 3D crustal model of the Nankai Trough with 4.2×10^9 degrees of freedom using 160 A100 GPUs, and performed inverse estimation of coseismic slip distribution.

The following is the structure of this paper. In Sect. 2, the target viscoelastic crustal deformation analysis is described. Section 3 describes the target multi-grid solver with a data-driven predictor algorithm. Section 4 describes the development of the multi-grid solver with the data-driven predictor on GPUs. Section 5 describes the performance of the solver, and Sect. 6 describes an application example of the proposed analysis method to the Nankai-Trough earthquake. Section 7 summarizes this study.

2 Target Problem

In this study, we model the Earth's crust as a linear viscoelastic body based on the Maxwell model and solve the equations

$$\sigma_{ij,j} + f_i = 0, \tag{1}$$

with

$$\dot{\sigma}_{ij} = \lambda \dot{\epsilon}_{kk} \delta_{ij} + 2\mu \dot{\epsilon}_{ij} - \frac{\mu}{\eta}\left(\sigma_{ij} - \frac{1}{3}\sigma_{kk}\delta_{ij}\right), \tag{2}$$

$$\epsilon_{ij} = \frac{1}{2}(u_{i,j} + u_{j,i}). \tag{3}$$

Here, σ and f are the stress tensor and external force, while $(\dot{\ })$, δ, η, ϵ, and u are the first derivative in time, Kronecker delta, viscosity coefficient, strain tensor, and displacement, respectively. λ and μ are Lame's coefficients. In this study, the governing equations are discretized by the finite-element method, which analytically satisfies the traction-free boundary conditions. Herein, second-order tetrahedral elements are used for the accurate calculation of stress and strain for crust deformation problems with complex geometry and heterogeneous material properties. The time evolution of viscoelastic crustal deformation analysis is computed based on [10] (Algorithm 1). Here, the fault slip is evaluated based on the split-node technique [15]. In general, it is difficult to generate high-quality large-scale 3D finite-element models for complex crustal structure models. In this paper, we construct a 3D finite-element model with unstructured second-order tetrahedral elements by using an automated robust mesh generation method [10]. In Algorithm 1, almost all of the computation time is spent on solving simultaneous equations:

$$\mathbf{K}^v \delta \mathbf{u} = \mathbf{f}, \tag{4}$$

where the degrees of freedom (DOF) of the unknown vector $\delta \mathbf{u}$ becomes large (e.g., the DOF becomes 4.2×10^9 for the application problem shown in this study). Thus, the goal becomes solving Eq. (4) with large DOF in a short time on multi-GPU environments.

Algorithm 1. Algorithm for solving linear viscoelastic response of crust. Here, superscript $()^i$ is the variable in the i-th time step. dt is the time increment. $\mathbf{B}$ is the displacement-strain transformation matrix. $\mathbf{A}$ and $\mathbf{D}$ are matrices indicating material property. $\mathbf{K} = \Sigma_e \int_{\Omega_e} \mathbf{B}^T \mathbf{D} \mathbf{B} d\Omega$. $\mathbf{K}^v = \Sigma_e \int_{\Omega_e} \mathbf{B}^T \mathbf{D}^v \mathbf{B} d\Omega$. $\beta^n = \mathbf{D}^{-1} \mathbf{A} \sigma^n$. $\mathbf{D}^v = (\mathbf{D}^{-1} + \alpha \, dt \, \beta')^{-1}$, where β' is the Jacobian matrix of β and α is the controlling parameter. Ω is the viscoelastic body.

Compute f by split $-$ node technique
Solve $\mathbf{K}\mathbf{u}^1 = \mathbf{f}^1$
$\sigma^1 \Leftarrow \mathbf{D}\mathbf{B}\mathbf{u}^1$
$\delta \mathbf{u}^1 \Leftarrow 0$
$i \Leftarrow 2$
while $i \leq N_t$ do
   $f^i \Leftarrow \Sigma_e \int_{\Omega_e} \mathbf{B}^T (dt\mathbf{D}^v \beta^{i-1} - \sigma^{i-1}) d\Omega + \mathbf{f}^1)$
   Solve $\mathbf{K}^v \delta \mathbf{u}^i = \mathbf{f}^i$ with initial solution $\delta \mathbf{u}^i_{init}$
   $\mathbf{u}^i \Leftarrow \mathbf{u}^{i-1} + \delta \mathbf{u}^i$
   $\sigma^i \Leftarrow \sigma^{i-1} + \mathbf{D}(\mathbf{B}\delta \mathbf{u}^i - dt\beta^i)$
   $i \Leftarrow i + 1$
end while

---

3   Base Multi-grid Solver with Data-Driven Predictor

In this section, we outline the multi-grid solver with the data-driven predictor [7] proposed as a fast solver for Eq. (4), which will be used as a base of the GPU solver developed in this study. A solver algorithm with high single-node peak performance with low computational cost, together with good load-balancing and low communication cost, is required for fast computation of large-scale finite-element models in a massively parallel computing environment. In this solver, a scalable data-driven initial solution predictor is added to a multi-grid solver that fulfills such requirements, leading to a reduction in the number of iterations in the multi-grid solver and thus a reduction in the computation time. Below, we outline the data-driven predictor and the multi-grid based iterative solver.

3.1   Data-Driven Predictor

By using the results of past time steps to accurately predict the initial solution δu^i_{init}, the number of iterations and thus the computation time of the multi-grid solver for solving Eq. (4) can be reduced. The idea of Dynamic Mode Decomposition (DMD) [14] is applied to construct an initial solution predictor suitable for massively parallel computers. Here, computed results up to the $i-1$-th time step are learned to predict the initial solution at the i-th time step. In DMD, an operator that represents time evolution is estimated from time series data, and this operator is used to predict the solution of the next step based on the solution at the current step. Instead of predicting the solution of the entire target domain at once, the target domain is divided into small domains, and the solutions in each domain are predicted within each domain. This enables efficient prediction of the modes including the local time and space components in each domain by only a small number of modes. However, even if a small region is targeted for prediction, it includes trend due to non-stationary time evolution, which is difficult to predict. Therefore, the second-order Adams-Bashforth method is used to predict the trend as

$$\delta \mathbf{u}_{adam}^i \Leftarrow \mathbf{u}^{i-3} - 3\mathbf{u}^{i-1} + 2\mathbf{u}^{i-1}. \tag{5}$$

We apply DMD to $\mathbf{x}^i = \delta \mathbf{u}^i - \delta \mathbf{u}_{adam}^i$ excluding the trend component. This allows $\delta \mathbf{u}^i$ to be predicted with sufficient accuracy from a small number of modes. Specifically, we define a matrix as $\mathbf{X}^{i-1} = [\mathbf{x}^{i-1}, \cdots, \mathbf{x}^{i-s}]$ using the data of previous $s+1$ steps, the time evolution operator $\mathbf{C}$ which satisfies $\mathbf{X}^{i-1} = \mathbf{C}\mathbf{X}^{i-2}$ is estimated from this matrix by the modified Gram-Schmidt method, and the initial solution for the next step is estimated using operator $\mathbf{C}$ as

$$\delta \mathbf{u}_{init}^i \Leftarrow \delta \mathbf{u}_{adam}^i + \mathbf{C}(\delta \mathbf{u}^{i-1} - \delta \mathbf{u}_{adam}^i). \tag{6}$$

The domain in each MPI process is divided into small non-overlapping domains using METIS [2], and the displacement increments for the nodes in each domain are estimated from the time-series data of nodes in the same domain. The algorithm does not require communication between domains, making it scalable in a massively parallel computing environment.

3.2 Multi-grid Solver with Data-Driven Predictor

The prediction results from the data-driven predictor are used for the initial solution of an adaptive conjugate gradient solver with a three-level multi-grid preconditioner. Algorithm 2 shows an overview of the method. In the preconditioner of the adaptive conjugate gradient method, multi-grid models generated by stepwise coarsening of the target finite-element model with second-order tetrahedral elements are used to solve the target model approximately. First, a coarse grid consisting of first-order tetrahedral elements is obtained by removing the edges nodes in second-order tetrahedral elements based on the geometric multi-grid method, and then a further coarsened model is obtained by the algebraic multi-grid method. Although various types of algebraic coarsening are proposed, uniform coarsening is used for maintaining load balance. Using these coarsened models, an approximate solution is obtained for preconditioning of the conjugate gradient method. Hereafter, we refer to the iteration of the original conjugate gradient loop as the outer loop and refer to the iteration of solving the preconditioning equations with another conjugate gradient solver as the inner loop. First, an approximate solution is obtained using the coarsest model (Algorithm 2 line 9; inner loop 2), and using the obtained solution as the initial solution, the approximate solution is updated using the tetrahedral linear element model (Algorithm 2, line 11; inner loop 1). Finally, the solution to the original mesh is obtained (Algorithm 2, line 13; inner loop 0). Inner loops reduce the cost per iteration compared to the original model by reducing the number of unknowns and the nonzero component of the sparse matrix $\mathbf{K}$. In addition, the coarsened

Algorithm 2. The iterative solver to obtain solution $\delta\mathbf{u}$. The input variables are $\mathbf{K}, \bar{\mathbf{K}}_i, \bar{\mathbf{P}}_i, \delta\mathbf{u}, \mathbf{f}, \epsilon, \bar{\epsilon}_i^{in}, N_i$ and N. Here, $\bar{\mathbf{K}}_i$ and $\bar{\mathbf{P}}_i$ represent global stiffness matrices and the mapping matrices between grids. $\bar{N}_i$ and $\bar{\epsilon}_i$ are the threshold values. The other variables are temporal. $(^-)$ represents FP32 variables, while the other variables are in FP64. All computation steps in this solver, except MPI synchronization and scalar coefficient computation, are performed in GPUs.

(a) Outer loop	**(b) Inner loop**
1: predict $\delta\mathbf{u}$ by the data-driven predictor	1: $\bar{\mathbf{e}} \Leftarrow \bar{\mathbf{K}}\bar{\mathbf{u}}$
2: $\mathbf{r} \Leftarrow \mathbf{K}\delta\mathbf{u}$	2: $\bar{\mathbf{e}} \Leftarrow \bar{\mathbf{r}} - \bar{\mathbf{e}}$
3: $\mathbf{r} \Leftarrow \mathbf{f} - \mathbf{r}$	3: $\bar{\beta} \Leftarrow 0$
4: $\beta \Leftarrow 0$	4: $i \Leftarrow 1$
5: $i \Leftarrow 1$	5: **while** $\|\bar{\mathbf{e}}_1\|^2/\|\bar{\mathbf{r}}\|^2 > \bar{\epsilon}$ and
6: **while** $\|\mathbf{r}\|^2/\|\mathbf{f}\|^2 > \epsilon$ **do**	$\quad N > i$ **do**
7: $\quad \bar{\mathbf{u}}_0 \Leftarrow \bar{\mathbf{M}}^{-1}\mathbf{r}$	6: $\quad \bar{\mathbf{z}} \Leftarrow \bar{\mathbf{M}}^{-1}\bar{\mathbf{e}}$
8: $\quad \bar{\mathbf{r}}_2 \Leftarrow \bar{\mathbf{P}}_2^T\bar{\mathbf{P}}_1^T\mathbf{r}$	7: $\quad$ **if** $i \geq 1$ **then**
9: $\quad \bar{\mathbf{u}}_2 \Leftarrow \bar{\mathbf{P}}_2^T\bar{\mathbf{P}}_1^T\bar{\mathbf{u}}_0$	8: $\quad\quad \bar{\beta} \Leftarrow \bar{\rho}_a/\bar{\rho}_b$
10: $\quad$ solve $\bar{\mathbf{u}}_2 = \bar{\mathbf{K}}_2^{-1}\bar{\mathbf{r}}_2$ using (b) with $\bar{\epsilon}_2^{in}$ and	9: $\quad$ **end if**
$\quad\quad N_2$ (* inner loop 2 *)	10: $\quad \bar{\mathbf{p}} \Leftarrow \bar{\mathbf{z}} + \bar{\beta}\bar{\mathbf{p}}$
11: $\quad \bar{\mathbf{u}}_1 \Leftarrow \bar{\mathbf{P}}_2\bar{\mathbf{u}}_2$	11: $\quad \bar{\mathbf{q}} \Leftarrow \bar{\mathbf{K}}\bar{\mathbf{p}}$
12: $\quad$ solve $\bar{\mathbf{u}}_1 = \bar{\mathbf{K}}_1^{-1}\bar{\mathbf{r}}_1$ using (b) with $\bar{\epsilon}_1^{in}$ and	12: $\quad \bar{\gamma} \Leftarrow (\bar{\mathbf{p}}, \bar{\mathbf{q}})$
$\quad\quad N_1$ (* inner loop 1 *)	13: $\quad \alpha \Leftarrow \bar{\rho}_a/\bar{\gamma}$
13: $\quad \bar{\mathbf{u}} \Leftarrow \bar{\mathbf{P}}_1\bar{\mathbf{u}}_1$	14: $\quad \bar{\rho}_b \Leftarrow \bar{\rho}_a$faccou
14: $\quad$ solve $\bar{\mathbf{u}} = \bar{\mathbf{K}}_0^{-1}\bar{\mathbf{r}}_0$ using (b) with $\bar{\epsilon}_0^{in}$ and N_0	15: $\quad \bar{\mathbf{e}} \Leftarrow \bar{\mathbf{e}} - \bar{\alpha}\bar{\mathbf{q}}$
$\quad\quad$ (* inner loop 0 *)	16: $\quad \bar{\mathbf{u}} \Leftarrow \bar{\mathbf{u}} + \bar{\alpha}\bar{\mathbf{p}}$
15: $\quad \mathbf{z} \Leftarrow \bar{\mathbf{u}}_0$	17: $\quad i \Leftarrow i + 1$
16: $\quad$ **if** $i > 1$ **then**	18: **end while**
17: $\quad\quad \gamma \Leftarrow (\mathbf{z}, \mathbf{q})$	
18: $\quad\quad \beta \Leftarrow \gamma/\rho$	
19: $\quad$ **end if**	
20: $\quad \mathbf{p} \Leftarrow \mathbf{z} + \beta\mathbf{p}$	
21: $\quad \mathbf{q} \Leftarrow \mathbf{K}\mathbf{p}_e$	
22: $\quad \rho \Leftarrow (\mathbf{z}, \mathbf{r})$	
23: $\quad \gamma \Leftarrow (\mathbf{p}, \mathbf{q})$	
24: $\quad \alpha \Leftarrow \rho/\gamma$	
25: $\quad \mathbf{r} \Leftarrow \mathbf{r} - \alpha\mathbf{q}$	
26: $\quad \delta\mathbf{u} \Leftarrow \delta\mathbf{u} + \alpha\mathbf{p}$	
27: $\quad i \Leftarrow i + 1$	
28: **end while**	

model allows long-range errors to be solved with fewer iterations. In each inner loop solver, a 3×3 block-Jacobi preconditioned conjugate gradient solver (Algorithm 2,b) with good load-balance and robustness is used. While FP64 is used in the outer loop to guarantee the computational accuracy of the final solution, FP32 is used in inner loops, where only approximate solutions are required. This halves the memory footprint, data transfer size, and communication size in the inner loops, which account for most of the computation time, and is expected to reduce time-to-solution.

4 GPU-Based Multi-grid Solver with Data-Driven Predictor

The multi-grid solver with data-driven predictor, which is designed to be efficient and scalable on massively parallel environments, is also expected to perform well on GPU-based environments. However, GPUs have relatively low memory capacity and memory bandwidth in comparison with its floating point performance, when compared to A64FX CPU-based Fugaku; thus, the data-driven predictor that requires large amounts of memory and matrix-vector products that require large amounts of memory accesses become bottlenecks in GPU performance. Therefore, we developed a multi-grid solver with the data-driven predictor for GPUs based on the previous CPU-based solver while improving the algorithm by reducing the amount of memory usage, memory accesses, and random data accesses.

Considering program development cost and portability, we use OpenACC to port CPU code to the GPU. OpenACC, which enables computation on the GPU by inserting compiler directives into CPU programs, allows porting pre-developed CPU applications to the GPU environment incrementally with relatively little effort. Although native programming models such as CUDA enable detailed tuning of the code to maximize performance on GPUs, it has been shown that by designing algorithms suitable for GPUs, the computation time of an OpenACC implementation is comparable to that of a CUDA implementation (for example, see [20] as an example of crustal deformation analysis using a multi-grid solver).

4.1 Data-Driven Predictor Enhanced by Memory Footprint Reduction Method

In the method of [7], given a data set $\mathbf{X}, \mathbf{Y}$ of sizes $m \times s$ (the number of degrees of freedom in the domain $\times$ time steps), where $\mathbf{X}$ is the input and $\mathbf{Y}$ is the corresponding output, the response $\mathbf{y}$ to another input $\mathbf{x}$ is computed as

$$\mathbf{y} = \mathbf{YUP}^T\mathbf{x}. \tag{7}$$

Here, $\mathbf{P} = \mathbf{XU}$, where $\mathbf{P}$ is a matrix with orthogonal columns and $\mathbf{U}$ is an upper triangular matrix. This orthogonalization $\mathbf{P} = \mathbf{XU}$ is computed by the modified Gram-Schmidt method, but it is not suitable for GPUs with small memory capacity because it requires keeping matrices $\mathbf{X}, \mathbf{Y}$ and another temporary matrix on memory during orthogonalization. In addition, since many times of the inner product is required sequentially for vectors as long as the number of degrees of freedom in the corresponding domain, a large memory access cost is involved. Therefore, in this paper, a random matrix $\mathbf{Q}$ of size $n \times m$ $(m \ll n)$, is used to transform the input data set $\mathbf{X}$ into $\mathbf{X}' \Leftarrow \mathbf{QX}$ and the input value $\mathbf{x}$ into $\mathbf{x}' \Leftarrow \mathbf{Qx}$ (e.g., a $25,745 \times 16$ matrix $\mathbf{X}$ is replaced with a 96×16 matrix $\mathbf{X}'$ in the performance measurement problem), which reduces the computational cost and memory usage for modified Gram-Schmidt orthogonalization. Although predictions based on the transformed data set are an approximation of the original algorithm's predictions, it is known that by taking m sufficiently larger than the number of time steps s used for the prediction, the singular values of $\mathbf{QX}$ and $\mathbf{X}$ coincide with high probability [9]. Therefore, it is possible to estimate $\mathbf{y}$ with almost no reduction in accuracy. In this study, $\mathbf{y}$ is computed as $\mathbf{a} = \mathbf{UP}^T\mathbf{x}'$ at first, and then as $\mathbf{y} = \mathbf{Ya}$. While additional computation for transforming $\mathbf{x}' \Leftarrow \mathbf{Qx}$ is required, its cost is negligible compared to the Gram-Schmidt method on the original problem, and the memory requirement of storing random matrix $\mathbf{Q}$ is also negligible as a common random matrix $\mathbf{Q}$ can be reused for all the small domains in which the data-driven predictor is applied.

4.2   Multi-grid Solver Enhanced by Multi-vector Computation

In the multi-grid solver, the sparse matrix-vector product (SpMV) kernel is the most computationally expensive kernel of each inner loop (Algorithm 2b). In general, the Generalized SpMV (GSpMV) kernel, which computes sparse-matrix dense-matrix products, achieves higher throughput than the SpMV kernels as it corresponds to computing multiple SpMVs by reading the target matrix once, which reduces the amount of memory access. This also leads to a reduction in random memory accesses by allocating the same components of multiple vectors consecutively in the memory address space. This leads to high throughput on GPUs that can access continuous data efficiently. Since the sparse matrix (e.g., $\bar{\mathbf{K}}$ in Algorithm 2b line 11) is constant at any source input in viscoelastic analysis, we calculate four sets of Green's functions simultaneously, thereby replacing the SpMV with the GSpMV. The maximum values for the relative errors in the 4 residual vectors are used for judging the convergence of each loop.

For the outer loop and inner loop 0, the Element-by-Element (EBE) method [17] is used to compute the GSpMV. In the parallel computation of matrix-vector products based on the EBE method, it is necessary to avoid data inconsistency when adding the local matrix-vector product results for each element to the global vector. While coloring of elements can be used to avoid data recurrence on multi-core CPUs, recent NVIDIA GPUs equip hardware-accelerated atomics and have high throughput atomic operations capability. Utilizing this atomic add functionality makes more efficient data access possible compared to the coloring algorithm. In inner loop 1 and inner loop 2, sparse matrices are stored in memory by Block Compressed Row Storage (BCRS) with block size 3 to compute GSpMV.

5   Performance Measurement

5.1   Performance Measurement Settings

Since the performance of the data-driven predictor is highly dependent on the problem characteristics, we evaluate solver performance on the example application problem in Sect. 6. The finite-element model comprises 1.0×10^9 tetrahedral elements with 4.2×10^9 DOF. Setting the time increment as $dt = 86400$

s, we measure the performance of crustal deformation between time step number $21 \leq N_t \leq 30$, where the data-driven predictor can be applicable, as the actual calculation of Green's functions is computed for several to 100 years (100 to 5000 time steps). We solve all problems with relative error tolerance $\epsilon = 10^{-8}$. The tolerances and maximum iterations in the inner loops are set to $(\epsilon_0, \epsilon_1, \epsilon_2) = (0.5, 0.25, 0.15)$ and $(N_0, N_1, N_2) = (30, 80, 300)$, respectively. In the data-driven predictor, the entire domain is divided into 163,840 subdomains, and data of the previous $s = 16$ time steps are used for estimation. The transformation is calculated using a random matrix with $m = 96$.

To demonstrate the effectiveness of the developed method, we compare performance with a 3×3 block-Jacobi preconditioned solver (PCGE) and a multigrid based adaptive conjugate gradient solver (multi-grid solver), both using second-order Adams-Bashforth method for predicting the initial solution[1]. Here, PCGE corresponds to skipping lines 7–13 in Algorithm 2a, and the multi-grid solver corresponds to switching the data-driven predictor in the proposed solver with the Adams-Bashforth method. We also compare the performance of the proposed solver with the multi-grid solver with data-driven predictor on CPU.

Performance was measured on GPU-based supercomputer AI Bridging Cloud Infrastructure (ABCI) [1], which is operated by the National Institute of Advanced Industrial Science and Technology. Each compute node (A) of ABCI has eight NVIDIA Tesla A100 GPUs and two Intel Xeon Platinum 8360Y CPUs (36 cores), and is interconnected with a full bisection bandwidth network (see Table 1). The FP64 peak performance of the GPU is 14.0× (memory bandwidth is 30.4×) of that of the CPU. 16 nodes (128 GPUs) with 1 MPI process per GPU (128 total MPI processes) were used for GPU measurements, and the same number of nodes and processes were used with 9 OpenMP threads per MPI process for CPU measurements.

5.2   GPU Kernel Performance

We measure the performance of the computation kernels which account for most of the execution time of the entire application (Table 2).

As the Gram-Schmidt kernel is memory bandwidth bound, direct porting to GPU led to $4020/248 = 16.2$-fold speedup from the CPU. Attained by directly porting it to (78.9% of memory bandwidth, 1.47% of FP64 peak performance). Furthermore, the reduction in computation by the random matrix transformation led to a further reduction in the time of the Gram-Schmidt kernel. This is due to the reduction of GPU device memory data transfer size from 302 GB to 2.36 GB by use of the proposed method replacing a $25,745 \times 16$ matrix $\mathbf{X}$ with a 96×16 matrix $\mathbf{X}'$. Although this method required computing the random matrix-vector product $\mathbf{Qx}$, it can be performed in 5.38 ms; leading to the overall speedup of the data-driven predictor by 18.9-fold from the direct porting case.

---

[1] Although sophisticated GPU-based methods specialized for viscoelastic crustal deformation analysis and specific GPU architecture is proposed [19], we compare with generally available solvers stated above for readability.

Table 1. Configuration of ABCI Compute Node (A)

		Hardware peak per node
CPU	Intel Xeon Platinum 8360Y	5.529 TFLOPS
	(54 MB Cache, 2.4 GHz, 36 Cores, 72 Threads)×2	
memory	512 GiB DDR4 3200 MHz RDIMM	408 GB/s
GPU	NVIDIA A100 for NVLink	77.6 TFLOPS
	40 GiB HBM2 ×8	12.4 TB/s
Interconnect	InfiniBand HDR (200 Gbps) ×4	100 GB/s

Furthermore, the memory size required for the data-driven predictor was 16.3 GB per GPU for the developed method, which is significantly smaller than the 62.9 GB required for the direct porting method.

Next, we measure the performance of SpMV and GSpMV kernels. While the FP32 peak performance of EBE-based SpMV of inner loop 0 was improved from 10.5% on the CPU to 16.3% on the GPU, due to the large number of registers on GPUs, the use of GSpMV led to 44.3% of FP32 peak on GPU, leading to further improvement in computational performance. While the BCRS-based SpMV in inner loops 1 and 2 are memory-bandwidth bound kernels, conversion of the kernel to GSpMV kernels with 4 vectors reduced the amount of memory access per vector, (1.19 GB to 0.253 GB and 420 MB to 105 MB for inner loop 1 and 2, respectively) resulting in 2.46- and 2.89-fold speedup, respectively, compared to the GPU-based SpMV implementations.

As is seen, the introduction of suitable algorithms for GPUs led to high efficiency on each kernel.

5.3   Solver Performance

We see the effectiveness of the data-driven predictor for a reduction in elapsed time. By use of the data-driven predictor, the initial error ϵ of the second-order Adams-Bashforth method (2.11×10^{-3}) was improved to 2.46×10^{-5}, indicating that prediction can be performed with high accuracy. As a result, the total number of iterations of the multi-grid solver was reduced from 5237 iterations to 1098 iterations. In particular, the number of iterations in inner loop 2 was significantly reduced from 4473 to 936, suggesting that the data-driven predictor has high prediction performance for the low-frequency components. In addition, introducing GSpMV significantly reduces the computation time for the cost dominant matrix-vector products, resulting in 2.01-, 2.12-, and 2.90-fold speedup per iteration for inner loops 0, 1, and 2, respectively. As a result, the developed solver attained an 8.6-fold speedup from a widely used state-of-the-art multi-grid solver (Fig. 1). The multi-grid solver performs well due to its ability to efficiently solve low-frequency errors with the use of fast inner loops (the multi-grid solver's inner loops were 1.59, 9.15, and 15.8-fold faster than the PCGE iterations for

Table 2. Performance of each kernel. Elapsed time is normalized per vector.

computation component	Elapsed time (FLOPS efficiency, Memory Throughput Efficiency)		
	CPU	GPU (direct porting)	GPU (proposed)
BCRS $\bar{K}_2 \bar{u}_2{}^*$	13.2 ms (1.85%, 57.2%)	0.396 ms (2.30%, 65.9%)	0.137 ms (6.60%, 60.4%)
BCRS $\bar{K}_1 \bar{u}_1{}^*$	35.4 ms (1.96%, 57.5%)	0.782 ms (2.52%, 76.2%)	0.318 ms (6.19%, 70.8%)
2nd order EBE $\bar{K}_0 \bar{u}_0{}^*$	145 ms (10.5%, 27.6%)	4.90 ms (16.3%, 64.3%)	1.81 ms (44.3%, 69.0%)
2nd order EBE Ku^{**}	184 ms (8.42%, 39.0%)	8.66 ms (18.5%, 53.2%)	4.63 ms (34.6%, 51.3%)
Gram-Schmidt**	4020 ms (1.04%, 28.5%)	248 ms (1.47%, 78.9%)	2.3 ms (10.9%, 73.2%)
random compression**	–	–	5.38 ms (16.3%, 82.9%)

*ratio to FP32 peak. **ratio to FP64 peak.

the inner loop 0, 1, and 2, respectively), leading to a 191-fold speedup from the standard PCGE solver requiring 10056 iterations and 170 s computation time. Since scalability has been demonstrated for the original CPU-based solver with data-driven predictor, it is expected that the proposed GPU-based solver will also be scalable. The speedup using GPU was 72.5 times when compared with the CPU-based implementation of SCALA22 (64.2 s), which is higher than peak performance and memory bandwidth ratio between CPU and GPU of 14.0- and 30.4-fold, respectively, indicating that the development of algorithms suitable for GPU led to large performance improvements. The introduction of the data-driven predictor enhanced by memory footprint reduction and GSpMV reducing computational cost is expected to be equally effective in CPU implementations.

6 Application Example

To demonstrate the effectiveness of the developed solver, we conducted an inversion analysis on a highly detailed crustal structure model to estimate the coseismic slip for the Nankai Trough earthquake. In this study, only elastic/viscoelastic deformation due to coseismic slip is considered, and crustal deformation due to afterslip and fault locking is not considered. Green's function g_i, which aggregates the displacements at each time and observation point for the unit fault x_i, is calculated by viscoelastic crustal deformation analysis. The observation model using these Green's functions is expressed as

$$d = Ga + e, \qquad (8)$$

where d is the observed data (the observed amount of crustal deformation), $G = [g_1, \cdots, g_n]$, a is a model parameter (the amount of slip in the unit fault x_i), and e is the error following a normal distribution with mean 0 and variance-covariance matrix Σ. Here, the model parameter a is determined by minimizing the objective function,

$$\Phi(a) = (d - Ga)^T \Sigma (d - Ga) + \lambda a^T La + \mu |a|_1, \qquad (9)$$

where $a^T La$ is a term used to constrain the smoothness of the slip distribution [18]. Since the extent of slip cannot be predicted in advance, the basic function

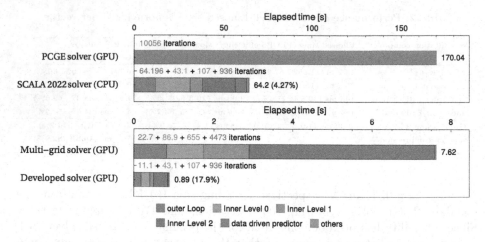

Fig. 1. Elapsed time and required iterations per time step for each solver

of the slip distribution is set wider than the range where slip actually occurs, and the L1 regularization term $|\mathbf{a}|_1$ is used to estimate a sparse slip distribution. The hyperparameters λ and μ are determined by k-fold cross-validation [5].

For the crustal structure data, we use the model based on [12,13]. Based on crustal structure data, the 3D finite-element model of the Japanese island is generated with a target area of 2496 km km × 2496 km km × 1100 km km centered at 135°E, 33.5°N. The viscosity of the continental and oceanic mantle is set to 2.0×10^{18} Pa s. Figure 2 shows the finite-element model generated with the smallest element size $ds = 500$ m. As in the performance measurement problem, $dt = 86400$ s and $N_t = 30$ are used. We introduce unit faults set up in grid form in Hori et al. (only unit faults that are in the Eastern half of the FE model are used). The number of unit faults is 186, and since we consider the slip distribution responses of two components on the fault plane, we calculate $186 \times 2 = 372$ Green's functions.

We set a hypothetical reference coseismic slip distribution shown in Fig. 3a). The direction of the reference seismic slip is assumed to be uniform in the direction of azimuth 125 degrees. Surface displacement is assumed to be observed by the Global Navigation Satellite System (GNSS), GNSS-Acoustic system, and ocean bottom pressure sensors (Fig. 3). The observation noise is not considered, and the displacement obtained from viscoelastic analysis using the reference coseismic slip as input is used as observation data.

In the proposed method, four Green's functions are calculated simultaneously in one set of viscoelastic analyses; thus, 372 Green's functions were calculated in 96 sets of viscoelastic analyses. The overall computation time was 33800 s on 160 GPUs. The computation time for the $21 \le N_t \le 30$ steps measured in the performance measurement was 3330 s s (8.96 s per step/function), which was almost the same time in the performance measurement. Thus, we can see

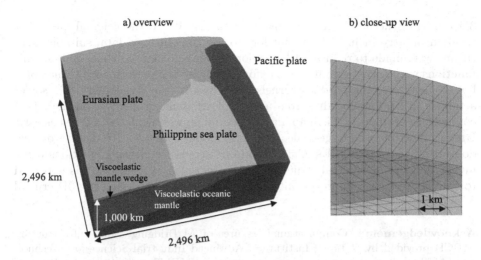

Fig. 2. Generated finite-element model used for the application example. a) Overview and b) close-up view.

that the developed method was robustly effective for the many Green's function inputs.

The estimated coseismic slip distribution is shown in Fig. 3b). The estimated moment magnitude is 8.13, which is almost the same as that of the reference slip (8.11), indicating that the magnitude of the earthquake is almost accurately captured.

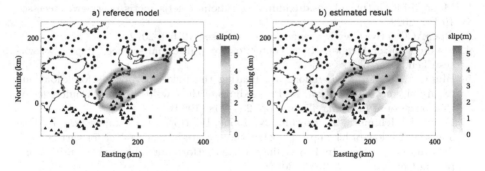

Fig. 3. Coseismic slip distribution in a) reference model and b) estimated results. Black points show observation points.

7 Conclusions

In this study, we developed a multi-grid solver with the data-driven predictor on GPUs for fast computation of the viscoelastic response of a highly detailed

3D crustal structure model for inverse analysis. While the original algorithm resulted in large memory footprint for storing time-history data, suitable algorithms were made to reduce GPU memory usage and elapsed time, and Green's functions were solved simultaneously for improving the performance of memory-bound matrix-vector product kernels. As a result, the developed GPU solver attained an 8.6-fold speedup from a state-of-art multi-grid solver on the ABCI compute environment. As an application example, we calculated 372 viscoelastic Green's functions of a large-scale 3D crustal model with 4.2×10^9 degrees of freedom using 160 A100 GPUs. Calculation of viscoelastic Green's functions using highly detailed 3D crustal structure models enabled by this study is expected to contribute to the improvement of slip estimation considering the 3D crustal structure.

Acknowledgements. Computational resource of AI Bridging Cloud Infrastructure (ABCI) provided by National Institute of Advanced Industrial Science and Technology (AIST) was used. This work was supported by MEXT as "Program for Promoting Researches on the Supercomputer Fugaku" (Large-scale numerical simulation of earthquake generation, wave propagation and soil amplification, JPMXP1020200203). This work was supported by JSPS KAKENHI Grant Numbers 18H05239, 22K12057, 22K18823. This work was supported by MEXT, under its Earthquake and Volcano Hazards Observation and Research Program. This work was supported by JST SPRING, Grant Number JPMJSP2108.

References

1. Computing Resources of AI bridging Cloud Infrastructure. http://abci.ai/en/about_abci/computing_resource.html
2. Metis 5.1.0. http://glaros.dtc.umn.edu/gkhome/metis/metis/overview. Accessed 19 Feb 2022
3. OpenACC. http://www.openacc.org/. Accessed 19 Feb 2022
4. Supercomputer Fugaku, Riken center for computational science. https://www.r-ccs.riken.jp/en/postk/project
5. Blum, A., Kalai, A., Langford, J.: Beating the hold-out: bounds for k-fold and progressive cross-validation. In: Proceedings of the Twelfth Annual Conference on Computational Learning Theory, pp. 203–208 (1999)
6. Fujita, K., Ichimura, T., Koyama, K., Inoue, H., Hori, M., Maddegedara, L.: Fast and scalable low-order implicit unstructured finite-element solver for earth's crust deformation problem. In: Proceedings of the Platform for Advanced Scientific Computing Conference, pp. 1–10 (2017)
7. Fujita, K., et al.: Scalable finite-element viscoelastic crustal deformation analysis accelerated with data-driven method. In: 2022 IEEE/ACM Workshop on Latest Advances in Scalable Algorithms for Large-Scale Heterogeneous Systems (ScalAH), pp. 18–25 (2022). https://doi.org/10.1109/ScalAH56622.2022.00008
8. Fukahata, Y., Matsu'ura, M.: Quasi-static internal deformation due to a dislocation source in a multilayered elastic/viscoelastic half-space and an equivalence theorem. Geophys. J. Int. **166**(1), 418–434 (2006)
9. Halko, N., Martinsson, P.G., Tropp, J.A.: Finding structure with randomness: probabilistic algorithms for constructing approximate matrix decompositions. SIAM Rev. **53**(2), 217–288 (2011)

10. Ichimura, T., et al.: An elastic/viscoelastic finite element analysis method for crustal deformation using a 3-D island-scale high-fidelity model. Geophys. J. Int. **206**(1), 114–129 (2016)
11. Ichimura, T., et al.: 152k-computer-node parallel scalable implicit solver for dynamic nonlinear earthquake simulation. In: International Conference on High Performance Computing in Asia-Pacific Region, pp. 18–29 (2022)
12. Koketsu, K., Miyake, H., Afnimar, Tanaka, Y.: A proposal for a standard procedure of modeling 3-d velocity structures and its application to the Tokyo metropolitan area, japan. Tectonophysics **472**(1), 290–300 (2009). https://doi.org/10.1016/j.tecto.2008.05.037, https://www.sciencedirect.com/science/article/pii/S0040195108002539, deep seismic profiling of the continents and their margins
13. Koketsu, K., Miyake, H., Suzuki, H.: Japan integrated velocity structure model version 1. In: Proceedings of the 15th World Conference on Earthquake Engineering, No. 1773, Lisbon (2012)
14. Kutz, J.N., Brunton, S.L., Brunton, B.W., Proctor, J.L.: Dynamic mode decomposition: data-driven modeling of complex systems. SIAM (2016)
15. Melosh, H.J., Raefsky, A.: A simple and efficient method for introducing faults into finite element computations. Bull. Seismol. Soc. Am. **71**(5), 1391–1400 (1981)
16. Tadokoro, K., Kinugasa, N., Kato, T., Terada, Y.: Experiment of acoustic ranging from GNSS buoy for continuous seafloor crustal deformation measurement. In: AGU Fall Meeting Abstracts, vol. 2018, pp. T41F–0361 (2018)
17. Winget, J., Hughes, T.: Solution algorithms for nonlinear transient heat conduction analysis employing element-by-element iterative strategies. Comput. Methods Appl. Mech. Eng. 711–815 (1985)
18. Yabuki, T., Matsu'Ura, M.: Geodetic data inversion using a Bayesian information criterion for spatial distribution of fault slip. Geophys. J. Int. **109**(2), 363–375 (1992)
19. Yamaguchi, T., et al.: Viscoelastic crustal deformation computation method with reduced random memory accesses for GPU-based computers. In: Shi, Y., et al. (eds.) ICCS 2018. LNCS, vol. 10861, pp. 31–43. Springer, Cham (2018). https://doi.org/10.1007/978-3-319-93701-4_3
20. Yamaguchi, T., Fujita, K., Ichimura, T., Naruse, A., Lalith, M., Hori, M.: GPU implementation of a sophisticated implicit low-order finite element solver with FP21-32-64 computation using OpenACC, pp. 3–24 (2020)

Implementation of Coupled Numerical Analysis of Magnetospheric Dynamics and Spacecraft Charging Phenomena via Code-To-Code Adapter (CoToCoA) Framework

Yohei Miyake[1]([⊠])(iD), Youhei Sunada[1], Yuito Tanaka[1], Kazuya Nakazawa[1], Takeshi Nanri[2], Keiichiro Fukazawa[3], and Yuto Katoh[4]

[1] Kobe University, Kobe 657-8501, Japan
y-miyake@eagle.kobe-u.ac.jp
[2] Kyushu University, Fukuoka 819-0395, Japan
[3] Kyoto University, Kyoto 606-8501, Japan
[4] Tohoku University, Sendai 980-8578, Japan

Abstract. This paper addresses the implementation of a coupled numerical analysis of the Earth's magnetospheric dynamics and spacecraft charging (SC) processes based on our in-house Code-To-Code Adapter (CoToCoA). The basic idea is that the magnetohydrodynamic (MHD) simulation reproduces the global dynamics of the magnetospheric plasma, and its pressure and density data at local spacecraft positions are provided and used for the SC calculations. This allows us to predict spacecraft charging that reflects the dynamic changes of the space environment. CoToCoA defines three types of independent programs: Requester, Worker, and Coupler, which are executed simultaneously in the analysis. Since the MHD side takes the role of invoking the SC analysis, Requester and Worker positions are assigned to the MHD and SC calculations, respectively. Coupler then supervises necessary coordination between them. Physical data exchange between the models is implemented using MPI remote memory access functions. The developed program has been tested to ensure that it works properly as a coupled physical model. The numerical experiments also confirmed that the addition of the SC calculations has a rather small impact on the MHD simulation performance with up to about 500-process executions.

Keywords: Coupled Analysis Framework · Magnetospheric Dynamics · Spacecraft Charging · Space Plasma

1 Introduction

Space plasmas involve complex dynamical processes characterized by a wide range of spatial and temporal scales [1]. One example is the coupling between the global-scale dynamics of the Earth's magnetosphere and micro-scale processes such as plasma wave excitation and energetic particle production. Although these

J. Mikyška et al. (Eds.): ICCS 2023, LNCS 14074, pp. 438–452, 2023.
https://doi.org/10.1007/978-3-031-36021-3_46

phenomena are closely related through magnetic structures, the spatial and temporal scales that characterize them are very different. In numerical simulations of such phenomena, the degree of coarseness in the calculation of each physical process varies greatly [2,3]. Traditionally, most plasma simulation models have evolved through the development and refinement of individual physical models, each dedicated to monoscale elementary processes [4]. Nowadays, the "crossscale coupling" is becoming a key topic of interest. There is a strong demand for simulation techniques that allow us to reproduce the evolution of multi-scale physical systems [5]. This should be achieved by running multiple physical models at different scales simultaneously, and coupling them via intercommunication of key physical quantities.

A standard approach is to embed micro-scale physical models within a macroscale simulation. This approach is best suited for addressing problems where the physical information to be intercommunicated between the models is well defined and verified. Its implementation typically involves bidirectional and highfrequency exchange of physical information, and the coupled models advance their own computations in close synchronization with each other. In general, the approach requires significant development costs to complete the program, but if handled properly, high performance in terms of accuracy and computational efficiency can be expected [6].

There also remain areas of research, where the physical processes involved are themselves fundamental, but the details of how the two different phenomena are coupled are not yet clear or verified. For such targets, one might consider keeping the way of coupling simple (e.g., by assuming only one-way coupling), and wish to investigate the degree of influence of various physical factors by trial and error. The present study aims at such moderate coordination, where multiple governing equation systems are computed in a mostly asynchronous manner, while information is communicated between the models as needed. We have developed the prototype of a Code-To-Code Adapter (CoToCoA) to realize such moderate and flexible coupling, and started to apply it to some physical topics in the field of space plasma [7]. The implementation of a coupled physical system and its performance characteristics strongly depend on how tightly the target physical processes are coupled with each other. Our previous report dealt with a case study where coupled physical models were computed synchronously [7]. In this paper, as a new application case, we report the implementation of spacecraft charging (SC) calculations coupled with the dynamics of the Earth's magnetospheric environment in an asynchronous manner.

2 Physical Models to be Coupled

2.1 Target Physical Phenomena

The coupled physical phenomena addressed in this paper are outlined below. The global-scale model simulates the interaction between the solar wind plasma that is extended out from the Sun and the Earth's dipolar magnetic field. This interaction leads to the formation of the magnetosphere around the Earth. The Sun-facing side of the magnetosphere is compressed by the dynamic pressure of the solar wind, whereas its nightside extends far beyond the Moon's orbit, forming a magnetotail. These physical processes are governed by the magnetohydrodynamic (MHD) equations. The MHD simulation reproduces the time evolution of the plasma macro-parameters over the Earth's magnetosphere.

The surface of a solid body such as a spacecraft collects surrounding plasma particles and gets electrically charged. Since such charging phenomena can lead to spacecraft malfunctions and even failures, this has been an active area of research since about 1980. So far, spacecraft charging has been studied under static plasma conditions, but this study aims to reproduce charging phenomena in a dynamic plasma environment. By extracting the parameter values at a spacecraft position from the above MHD data, the SC solver evaluates the rate of charge accumulation on the spacecraft per unit time (i.e., a current) based on a conventional theory. The ordinary differential equation of the spacecraft potential is then solved with the current values as source terms. This strategy would make it possible to predict SC transitions in conjunction with the dynamical changes in the magnetospheric environment.

2.2   Magnetospheric Simulation

The magneto-hydrodynamic (MHD) equations solved in the magnetospheric global simulation are given as follows:

$$\frac{\partial \rho}{\partial t} = -\nabla \cdot (\boldsymbol{V}\rho) + D\nabla^2 \rho, \tag{1}$$

$$\frac{\partial \boldsymbol{V}}{\partial t} = -(\boldsymbol{V} \cdot \nabla)\boldsymbol{V} - \frac{1}{\rho}\nabla p + \frac{1}{\rho}\boldsymbol{J} \times \boldsymbol{B} + \boldsymbol{g} + \frac{\boldsymbol{\Phi}}{\rho}, \tag{2}$$

$$\frac{\partial p}{\partial t} = -(\boldsymbol{V} \cdot \nabla)p - \gamma p \nabla \cdot \boldsymbol{V} + D_p \nabla^2 p, \tag{3}$$

$$\frac{\partial \boldsymbol{B}}{\partial t} = \nabla \times (\boldsymbol{V} \times \boldsymbol{B}) + \eta \nabla^2 \boldsymbol{B}, \tag{4}$$

$$\boldsymbol{J} = \nabla \times (\boldsymbol{B} - \boldsymbol{B}_{\mathrm{d}}), \tag{5}$$

where ρ, p, $\boldsymbol{V}$, $\boldsymbol{J}$, and $\boldsymbol{B}$ are the plasma density, plasma pressure, velocity vector, current density, and magnetic field, respectively. D and D_p denote the diffusion coefficients of the plasma density and pressure, respectively, $\boldsymbol{g}$ is the gravity term, $\boldsymbol{\Phi}$ is the viscosity term, η is the temperature dependent electrical resistance, and γ is the specific heat constant. The dipole magnetic field $\boldsymbol{B}_{\mathrm{d}}$ is also incorporated to represent the Earth's intrinsic magnetic field.

In developing the coupled analysis, we adopted the MHD program developed by Fukazawa et al. [9]. The code discretizes a 3-dimensional simulation box and defines physical quantities on the Cartesian grid points. The Modified Leap Frog algorithm [8] is used for the time integration of the system equations. The code is fully parallelized and optimized for distributed-memory computer systems via the domain decomposition method. Since a so-called stencil calculation is performed at each position coordinate, the parallel implementation is based on boundary communication using MPI_Sendrecv at the outer edges of the subdomains. The MHD code was evaluated for parallel performance on different computer systems. It achieved a parallel efficiency of 96.5% against 72,000 MPI parallelism on the Fujitsu PRIMERGY CX2550 supercomputer installed at the Kyushu University [10].

2.3 Spacecraft Charging Simulation

Spacecraft charging is generally characterized by the surface potential of the spacecraft with respect to space [11]. A spacecraft usually consists of several components. If the spacecraft is modeled as a multi-conductor electrostatic system, its charge can be described by the following ordinary differential equation.

$$\frac{d\phi_i}{dt} = \frac{1}{C_i} \sum_s I_s(\phi_i, n_0, T_s, \dots) \tag{6}$$

$$(i = 1, \dots, N_{\text{sc\_components}})$$

where ϕ_i and C_i are the potential and capacitance of the i-th spacecraft component, respectively. The I_s, n_0, and T_s are the spacecraft inflow current, plasma density, and plasma temperature of the s-th particle species, respectively. The current term on the right-hand side reflects the effect of charge accumulation due to the motion of charged plasma particles around the spacecraft.

As a rough classification, there are two types of approaches to determine the plasma current on the right-hand side of Eq. 6. One approach is the plasma particle simulation, in which the plasma current is evaluated by numerically tracing the trajectories of charged plasma particles around the spacecraft [12]. While this method allows for sophisticated calculations that take into account the geometric details of the spacecraft, it is better suited for calculations of micro-scale phenomena with short time scales (msec at most). Therefore, it is not easy to work directly with magnetospheric simulations covering long time scales (hours to days).

Another approach is the use of quasi-analytical formulations that relate macro-parameters such as the density and temperature of the surrounding plasma to the values of the plasma current [13]. Although in principle this approach works well only for simple geometries such as spheres and cylinders, in practice it is known to provide a relatively good approximation when the spacecraft size is similar to or smaller than the Debye length. The most important feature is that it is much less computationally expensive than plasma particle simulations. Based on this trait, this quasi-analytical approach is used in the coupled analysis addressed in this study, which facilitates its coupling with MHD simulations for long-term magnetospheric environmental variations. Specifically, the current term on the right-hand side of Eq. 6 is calculated from the plasma density and temperature data at each time. The equation is then numerically integrated by the fourth-order Runge-Kutta method to compute the time evolution of the spacecraft potential.

In the quasi-analytical approach described above, the spacecraft potential is not expressed as a function of spatial coordinates, but only of time. Its parallel

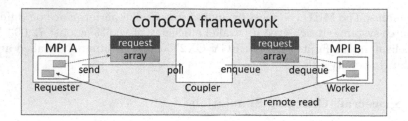

Fig. 1. Overview of code coupling via CoToCoA

implementation is challenging due to the time dependence. Therefore, the spacecraft charging calculations in the present implementation are sequential.

3   Code-To-Code Adapter: CoToCoA

3.1   Overview of the Framework

CoToCoA couples multiple physical models based on the Multiple-Program-Multiple-Data (MPMD) execution model. The concept of inter-code collaboration via CoToCoA is shown in Fig. 1. CoToCoA splits all available MPI processes into three subsets (groups) of processes, and assigns to each group one of the roles of Requester, Coupler, and Worker. Requester and Worker each correspond to different physical models to be coupled. These two roles are performed on different process groups that do not overlap with each other. Thus, CoToCoA assumes that each model is implemented as a standard MPI program and executed within its assigned process group as it was before being coupled. Coupler supervises the entire behavior of the framework, and acts as a mediator between Requester and Worker. The separation of the roles and the exclusive assignment of process groups to the respective roles minimize the effort for program modification.

Requester is responsible for physical computations that affect other physical processes. In the coupled physical systems, it offloads necessary computational tasks (i.e., requests) to Worker. Worker is generally responsible for physical computations that depend on the results of Requester computations. Coupler, to which a single MPI process is always assigned, monitors Requester and controls Worker as described later. CoToCoA has the ability to incorporate multiple Worker programs that are responsible for different computation tasks.

The general behavior of Requester and Workers within the CoToCoA framework is summarized in Fig. 2. Requester starts its own computation immediately after program startup. Workers are initially in an idle state, waiting for computation requests that are issued by Requester and forwarded by Coupler. Upon receiving computation requests, Worker starts its own computations and transitions its state to busy. When Worker has completed the requested computations, it sets its own state back to idle.

Coupler constantly monitors the status of Workers and computation requests issued by Requester. When it detects a computation request issued by Requester,

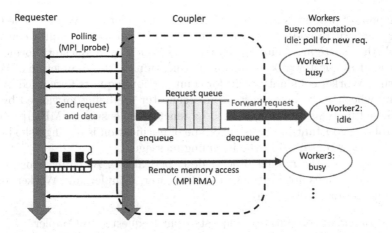

Fig. 2. Typical coordinative behavior of Requester, Coupler, and Workers within the CoToCoA framework.

it delegates the task to one of Workers that are in an idle state. If necessary, Coupler can also perform intermediate processing of the forwarded data on a user-defined basis. If all Workers are in a busy state, Coupler temporarily stores the computation request in a request queue, and waits for one of Workers in a busy state to transition to an idle state. When a worker that has completed its previously assigned work and enters an idle state is detected, it is delegated a new work. By repeating this procedure, CoToCoA sequentially processes the computation requests sent by Requester.

The program termination process is as follows. When Requester has completed all of its computations, it sets itself to an exit state. Coupler sets itself to an exit state when it confirms that Requester has entered an exit state and that its own request queue is empty of unprocessed computation requests. Finally, Workers terminate their own state after confirming that all assigned tasks have been completed and Coupler has entered an exit state. This completes the entire program execution. In the above description, Requester and Coupler do not necessarily know the status of Workers, but they can optionally use a function to check whether all issued requests have been completed on the Worker side or not.

3.2 Asynchronous Control of Coupled Programs

One of the main features of the framework is the handling of asynchronous computation requests, which tend to occur in coupled simulations. CoToCoA implements this through polling, which is hardware independent and can be described by a relatively simple loop statement. The physical models currently being coupled are non-interactive and do not require immediacy or real-time processing of asynchronous requests, so polling processing is sufficient to achieve efficiency.

The polling process is mainly performed by Coupler and Worker. The CoTo-CoA framework provides API functions for this purpose. Coupler constantly polls whether there are computation requests issued by Requester, notifications of the end of Requester processing, and notifications of the completion of Worker processing. Worker constantly polls for computation requests forwarded by Coupler, and whether or not requests from Requester are being forwarded by Coupler. These functions are implemented by function calls such as MPI_Iprobe and MPI_Probe at constant intervals. The type of notification is distinguished by the tag information contained in the incoming message.

In response to the polling process described above, API functions are provided to send various notifications from Requester, Coupler, and Worker, respectively. The major functions are

1. The issuance of computation requests from Requester to Coupler,
2. The forwarding of computation requests from Coupler to Workers,
3. Notifications from Workers to Coupler that computation tasks have been completed,
4. Notification from Requester to Coupler that all requests have been issued,
5. Notification from Coupler to Worker that all requests have been forwarded.

3.3   Inter-code Exchange of Numerical Data

CoToCoA provides several data exchange methods depending on the situation, taking into account that each program executes its own computations asynchronously. The first method is to transfer the necessary physical data at the same time as a computation request is issued. This method can be used in a situation where the data to be sent have already been generated on Requester. In the API functions provided by CoToCoA, this is implemented by standard blocking-type communications such as MPI_Send and MPI_Recv.

In another situation, one program (either Requester or Worker) needs to refer to data generated by the other during its computations. In this case, since Requester and Worker are independent programs performing completely different computations asynchronously, an implementation using MPI functions for one-sided communications would be appropriate. It also allows the user to exploit the advantages of one-sided communications, such as the suppression of synchronous waiting and data copying, as well as compatibility with the RDMA capabilities. CoToCoA provides API functions for such Remote Memory Access (RMA), which is implemented using MPI.

4   Implementation of Coupled Analysis

Using the functionality of the CoToCoA framework described so far, we have implemented a coupled analysis program of magnetospheric dynamics and space-craft charging. The MHD simulation plays the role of Requester and the SC calculation program plays the role of Worker. The skeleton of the developed

Fortran program is shown in Fig. 3. Here, we mainly explain how to use the API functions provided by CoToCoA, and omit the detailed implementation for the physical calculations within each model.

The functions CTCAX_init (X = R, C, or W) are called in Requester, Coupler, and Worker respectively to initiate CoToCoA. The API functions of CoToCoA are named in such a way that the prefixes of the function names easily identify whether the functions are for Requester, Coupler, or Worker. These prefixes are omitted in the following text. The initialization functions handle the construction of various data structures, the grouping and the assignment of processes to each Worker, and the definition of MPI sub-communicators. After the initialization, the CoToCoA function "Regarea_real4" constructs a window object for MPI one-sided communication, for which read/write permissions are set. In this coupled analysis, pressure and density data in spacecraft position coordinates are generated by the magnetospheric MHD simulation. The data are then accessed remotely by Worker.

After the completion of the initialization phase described above, the MHD simulation and the SC calculations are performed in parallel. In fact, the SC calculation on Worker is started based on the calculation request issued by Requester. Coupler polls for the arrival of computation requests from Requester by repeatedly calling the CoToCoA function "Pollreq" in its loop statement. Upon arrival of a computation request from Requester, the request is enqueued by the CoToCoA function "Enqreq". If it finds a Worker in an idle state when the next Pollreq function is called, it forwards the request to that Worker. The Pollreq functions described above are also used to check the completion of Worker and Requester computations.

Before starting the time update loop of the MHD simulation, Requestor calls the CoToCoA function "Sendreq" and issues a computation request to Coupler. At the same time, Requester also sends spacecraft orbit information to Coupler. Coupler then forwards these messages to one of the available Workers. After calling the Sendreq function, Requester immediately starts a time update of its own MHD computation. Although the density and pressure data at the spatial grid points are sequentially updated by the time update, they cannot be cleared until the Worker side has finished reading them. Since the up-to-date version (1.2.3) of the CoToCoA framework does not support this feature, users must implement some mechanism by themselves. There are two possible implementations to guarantee this point. The first is to suspend updating on the Requester side until Worker confirms the completion of reading data on the array. The second method is to copy the time sequence of the required physical data from the time update array and store them in a separate array. In this study, the latter method is adopted because the physical data required by Worker is as small as 4 byte × 2 = 8 byte per time. This method is also consistent with the policy of aiming for an implementation that does not interfere with the progress of the Requester's computations as much as possible. In fact, after the time update loop starts, Requester can perform its computation without interference from Coupler and Worker.

```
┌─ Requester (magnetospheric MHD simulation) ──────────────────────────
│
│  call CTCAR_init( )
│  call CTCAR_regarea_real4(data, size, areaid)
│  ! MHD initialization (omitted herein)
│  call CTCAR_sendreq_withreal4(datint, ndatint, datreal, ndatreal)
│  do step = 1, nsteps
│    ! MHD calculation: temporal integration (omitted herein)
│    data(step, 1:ncomp) = &
│    & field(scx,scy,scz,1:ncomp) ! write data to be offloaded
│  end do
│  ! MHD finalization (omitted herein)
│  call CTCAR_finalize( )
│
└──────────────────────────────────────────────────────────────────────
```

```
┌─ Coupler ─────────────────────────────────────────────────────────────
│
│  call CTCAC_init( )
│  call CTCAC_regarea_real4(areaid)
│  do while ( .true. )
│    call CTCAC_pollreq_withreal4(reqinfo, fromrank, &
│      & datint, ndatint, datreal, ndatreal)
│    if( CTCAC_isfin( ) ) exit
│    progid = SCCHARGE
│    call CTCAC_enqreq_withreal4(reqinfo, progid, &
│      & datint, ndatint, datreal, ndatreal)
│  end do
│  call CTCAC_finalize( )
│
└──────────────────────────────────────────────────────────────────────
```

```
┌─ Worker (spacecraft charging calculations) ───────────────────────────
│
│  call CTCAW_init(progid, procs_per_req)
│  call CTCAW_regarea_real4(areaid)
│  do while ( .true. )
│    call CTCAW_pollreq_withreal4(fromrank, &
│      & datint, ndatint, datreal, ndatreal)
│    if( CTCAW_isfin( ) ) exit
│    ! SC initialization (omitted herein)
│    step = 1
│    do while (step <= nsteps)
│      call CTCAW_readarea_real4(areaid, rank, offset, size, data)
│      if( .not.(find_new_data(data(step, 1))) ) cycle
│      ! SC calculation: temporal integration (omitted herein)
│      step = step + 1
│    end do
│    call CTCAW_complete( )
│  end do
│  call CTCAW_finalize( )
│
└──────────────────────────────────────────────────────────────────────
```

Fig. 3. Skeleton code of the MHD–SC coupled analysis.

Worker performs the SC computation inside a double-loop structure. In the outer loop, it polls for a computation request from Coupler by repeatedly calling the Pollreq function. After detecting the arrival of a computation request, the process enters the inner loop and starts the SC calculation. Since the SC calculation refers to the pressure and density data provided by the MHD at each time, it must always be preceded by the MHD simulation. To accomplish this, Worker periodically accesses and checks the physical data storage array on remote memory managed by Requester. This is implemented by using the CoTo-CoA function "Readarea", which serves as a one-sided communication function. Worker controls its own SC calculations based on the numerical data stored on this remote memory array as follows. At the initial stage of the computation, the memory array is set to a specific value (such as "0.0") to indicate that the computation is incomplete. As the MHD calculation progresses, the "null" values of the array elements reserved for the corresponding time steps are replaced with the obtained plasma pressure and density values. Worker can remotely reference this data to know the progress of the MHD computation. By repeatedly calling Readarea, Worker can always decide whether to proceed with its own SC calculations or wait for Requester to generate plasma macro parameters. This feature provides a mechanism for Worker to control the SC calculations. Finally, when the preset time has been calculated, the Worker calls the CoToCoA function "Complete" to notify Coupler that its task as a Worker is finished.

5   Case Study and Performance Evaluation

This section presents a case study of the coupled analysis performed to verify its validity. We analyzed the charging phenomena of a hypothetical artificial satellite exposed to the Earth's magnetospheric environment on April 5, 2010, when a geomagnetic substorm triggered by a solar flare-driven coronal mass ejection took place [14]. It has been reported that the U.S. satellite Galaxy 15 actually experienced a failure during this substorm. As mentioned in the previous section, the data provided by the MHD computation are the plasma density and pressure. The plasma temperature is derived from these physical parameters and used as an input to the charging calculation. Although the SC calculations require separate electron and ion temperatures, in principle, it is not possible to distinguish between ions and electrons in the MHD simulation. Therefore, the electron and ion temperatures are assumed to be identical.

The experiments were performed on up to 64 nodes of a Fujitsu PRIMERGY CX2550 installed at the Research and Development Center for Information Infrastructure, Kyushu University. Each node of the ITO System A is equipped with two Intel Xeon Gold (Skylake SP) processors and 192 GB of DDR4 memory. To validate the reproduced physical phenomena, a coupled analysis was performed with a simplified configuration considering only currents of the plasma electrons and ions. The computational resources used were 32 nodes of ITO System A. The number of MPI processes assigned to the requester, coupler, and worker were 512, 1, and 1, respectively. The magnetospheric MHD simulation

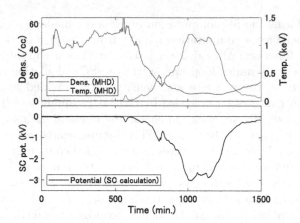

Fig. 4. Numerical results without photo and secondary electron effects. The upper panel shows the plasma density and temperature derived from the MHD simulation. The lower panel shows the spacecraft potential.

uses a 3-dimensional space consisting of $600 \times 400 \times 400 \sim 10^8$ grid points. Figure 4 shows the result of the numerical experiment. The upper panel shows the plasma density and electron temperature evaluated within the MHD simulation, and the lower panel shows the result of the spacecraft potential calculation. A comparison of the time evolution of each quantity shows that the spacecraft potential has an inverse correlation with the electron temperature. In general, when an object is placed in a two-component plasma consisting of electrons and ions, an electron current with a high thermal velocity dominates. As a result, the spacecraft becomes negatively charged at approximately the electron temperature to repel the surrounding electrons. This leads to a current equilibrium between the electrons and the ions. The time evolution of the spacecraft potential obtained in this coupled analysis also reflects this fundamental physical behavior.

In reality, more types of current components are involved in the SC processes, such as photoemission and secondary electron emission, making the processes more complex. The analysis for such a practical situation is shown in Fig. 5, where the individual current values and a secondary electron emission coefficient are displayed. The current magnitudes and the secondary emission coefficient depend on the temperature of the surrounding plasma as well as on the spacecraft potential itself. It follows that each current is closely correlated to the plasma conditions provided by the MHD simulation. A notable feature is the abrupt drop in the spacecraft potential seen at the time period 1024–1060 min.. This period corresponds to the time when the orbiting satellite enters an eclipse by the Earth and the photoelectron emission ceases. Since the photoelectron emission flux (as the current that positively charges the spacecraft) is generally greater than the incoming electron flux, the spacecraft potential fluctuates around a few V positive outside the eclipse. In contrast, the incoming electron flux in turn becomes the dominant component during the eclipse, and the

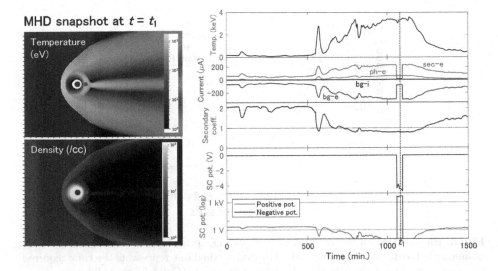

Fig. 5. Numerical results with photo and secondary electron effects. The left panel shows the snapshot of the MHD simulation at $t = t_1$. The right panel shows the time-series result of the SC calculations. The notations: bg-e, bg-i, sec-e, and ph-e in the second plot in the right panel represent the background plasma electron, ion, secondary electron, and photoelectron currents, respectively. The bottom panel shows the spacecraft potential values on a logarithmic scale, with the red line duration being positive and the blue line duration being negative. (Color figure online)

spacecraft potential drops down to a deep negative potential. Such an abrupt change in potential can be detrimental to maintaining the integrity of actual satellite operations. The actual drop in spacecraft potential that occurs during an eclipse varies greatly depending on the temperature of the space plasma as well as the secondary electron emission coefficient. The ability to evaluate the potential based on environmental parameters derived from the physics model-based MHD simulations is a major step forward in building a future SC prediction platform.

Next, a performance evaluation was performed to characterize the computational loads for MHD and SC. We measured the elapsed time required for Requester (magnetospheric MHD simulations), Worker (SC simulations), and the entire coupled analysis. In this evaluation, we changed the degree of parallelism (the number of processes) only for the MHD computations. The number of allocated processes was changed from 32 to 2048 while keeping the problem size constant (i.e., strong scaling mesurement).

Figure 6 shows the results of the performance evaluation. The horizontal axis represents the number of processes assigned to the MHD computation. The total execution time required for the coupled analysis decreases as the number of processes increases in the range from 32 to 512 processes. The execution times for Requester and Worker are nearly identical in the range of 32 to 256 processes.

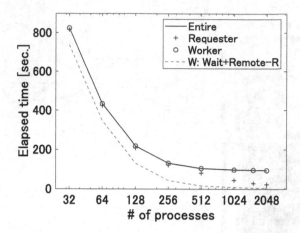

Fig. 6. Elapsed time of the coupled analysis of magnetospheric dynamics and SC phenomena via the developed framework. The pink dashed line represents the time required to "wait" and "read" by RMA in the worker program. (Color figure online)

In the coupled analysis, Worker (SC calculation) continues to attempt remote memory access to Requester without updating the spacecraft potential until the required plasma physics data become available on the MHD side. The processing time of Worker includes this waiting time, and thus the processing speed of Worker is constrained by that of Requester. In fact, in the 32-process execution, the time spent waiting for the Requester accounted for 89.4% of the total execution time. This indicates that by allocating more MPI processes to Requester to speed up it, the waiting time on the worker side can also be reduced accordingly. The time required for Requester and Worker analysis is identical for 32 to 256 process executions. In this regime, the numerical processing time for the SC analysis is less than and hidden in that for the MHD computation.

For 512 to 2048 process executions, there is a discrepancy between the execution times required by Requester and Worker. For the magnetospheric MHD computation, the execution time is reduced up to about 1024 processes, but saturates above this number. This is because the ratio of boundary communication costs increases, whereas the amount of computation within a small region decreases as the number of decompositions increases. Meanwhile, the actual computation processing time of Worker is basically invariant, and the only factor that can be reduced is the overhead part due to waiting and remote memory access. Thus, once the overhead is minimized (3.5% of the total time), the execution time for Worker does not decrease any further.

Through these numerical experiments, it was found that with the experimental setup used in this study, the MHD computation time of 200,000 to 400,000 grid points and the processing time for one time-step update of the spacecraft potential are in balance with each other. It was also confirmed that the coupled implementation has almost no effect on the processing time of the MHD computation. In fact, there is only a difference of less than 10^{-1} seconds for any number of process executions between the MHD processing time during the coupled analysis and the execution time when MHD is run alone. Such overhead is negligible enough when compared to the total processing time of the MHD computation (e.g., ~ 80 seconds for the 512 process execution).

6 Conclusions

A coupled numerical analysis of magnetospheric MHD simulations and spacecraft charging calculations is implemented using an in-house code-to-code coupling framework, CoToCoA. The original simulation codes have been developed independently so far and perform computations with completely different content. Therefore, an asynchronous inter-code coordination is required to realize the coupled analysis. In addition, for high productivity of the coupled analysis, it is beneficial to define a requester-worker relationship between the programs to be coupled. Which model should play the role of Requester or Worker depends on the context of what information and when physical information should be exchanged between models. Based on this idea, we designed the CoToCoA framework, which provides a control program (Coupler) that supervises the inter-code coordination as well as the asynchronous code control, and provides API functions for inter-code data exchange based on remote memory access.

In the case of the coupled analysis of magnetospheric MHD and spacecraft charging simulations, the former is positioned as Requester and the latter as Worker. Worker (SC analysis) uses remote memory access to check the status and progress of Requester (MHD) processing as needed, and to determine whether or not the update of a spacecraft potential can be performed.

The coupled analysis program has been verified on the supercomputer system at the Kyushu University. The developed model successfully reproduces the transient behavior of the spacecraft potential, which is based on the variational plasma macro-parameters generated by the magnetospheric MHD simulation. The execution time required for the coupled analysis was also measured while varying the number of MPI processes assigned to the parallelized magnetospheric MHD calculation. It was confirmed that there is a point between 256- and 512-process executions where the computational cost of the magnetospheric MHD and the SC calculations are balanced.

Since the SC calculation is not yet parallelized in the current implementation, the number of processes used for the MHD computation must be kept below a certain level in order to hide the SC analysis within the processing time of MHD. To make the analysis more flexible by changing the problem size and the amount of computational resources used, it is necessary to introduce time-domain parallelization for the SC analysis, which is left as a future work.

Acknowledgements. This work was supported in part by the Joint Usage and Research Center for Interdisciplinary Large-Scale Information Infrastructure and Innovative High Performance Computing Infrastructure (project numbers: jh210047-NAH, jh220017, jh230042, hp220040, and hp230046), as well as the Japan Society for the Promotion of Science (JSPS) KAKENHI Grant Number JP22K12049. The numerical experiments were carried out using the ITO System at Kyushu University and the Camphor system at Kyoto University.

References

1. Schwartz, S., et al.: Cross-scale: multi-scale coupling in space plasma. Assessment Study Report (2009)
2. Fukazawa, K., Katoh, Y., Nanri, T., Miyake, Y.: Application of cross-reference framework CoToCoA to macro- and micro-scale simulations of planetary magnetospheres. In: Proceedings of 2019 Seventh International Symposium on Computing and Networking Workshops (CANDARW), pp. 121–124 (2019)
3. Katoh, Y., Fukazawa, K., Nanri, T., and Miyake, Y.: Cross-reference simulation by code-to-code adapter (CoToCoA) library for the study of multi-scale physics in planetary magnetospheres. In: Proceedings of 8th International Workshop on Large-scale HPC Application Modernization (LHAM), pp. 223–226 (2021)
4. Matsumoto, H., Omura, Y.: Computer space plasma physics: simulation techniques and software. Terra Scientific Pub, Tokyo (1993)
5. Toth, G., et al.: Space weather modeling framework: a new tool for the space science community. J. Geophys. Res. **110**, A12226 (2005). https://doi.org/10.1029/2005JA011126
6. Sugiyama, T., Kusano, K., Hirose, S., Kageyama, A.: MHD-PIC connection model in a magnetosphere-ionosphere coupling system. J. Plasma Phys. **72**(6), 945–948 (2006)
7. Nanri, T., Katoh, Y., Fukazawa, K., Miyake, Y., Nakazawa, K., Zhow, J., and Sunada, Y.: CoToCoA (Code-To-Code Adapter) version 1.2.2. https://doi.org/10.5281/zenodo.5775280
8. Ogino, T., Walker, R.J., Ashour-Abdalla, M.: A global magnetohydrodynamic simulation of the magnetopause when the interplanetary magnetic field is northward. IEEE Trans. Plasma Sci. **20**, 817–828 (1992)
9. Fukazawa, K., Ogino, T., Walker, R.J.: The configuration and dynamics of the Jovian magnetosphere. J. Geophys. Res. **111**, A10207 (2006)
10. Fukazawa, K., Ueda, M., Inadomi, Y., Aoyagi, M., Umeda, T., and Inoue, K.: Performance analysis of CPU and DRAM power constrained systems with magnetohydrodynamic simulation code. In: Proceedings of 2018 IEEE 20th International Conference High Performance Computing and Communications, pp. 626–631 (2018). https://doi.org/10.1109/HPCC/SmartCity/DSS.2018.00113
11. Whipple, E.C.: Potentials of surfaces in space. Rep. Prog. Phys. **44**(11), 1197–1250 (1981)
12. Miyake, Y., Usui, H.: New electromagnetic particle simulation code for the analysis of spacecraft-plasma interactions. Phys. Plasmas **33**(3), 258–266 (2019)
13. Massaro, M. J., Green, T., and Ling, D.: A charging model for three-axis stabilized spacecraft. In: Proceedings of Spacecraft Charging Techology Conference, pp. 237–269 (1977)
14. Ferguson, D. C., Denig, W. F., Rodriguez, J. V.: Plasma conditions during the Galaxy 15 anomaly and the possibility of ESD from subsurface charging. In: 49th AIAA Aerospace Science Meeting including the New Horizons Forum and Aerospace Exposition, pp. 2011–1061. AIAA (2011)

Artificial Intelligence
and High-Performance Computing
for Advanced Simulations

Improving Group Lasso
for High-Dimensional Categorical Data

Szymon Nowakowski[1,2], Piotr Pokarowski[1], Wojciech Rejchel[3(✉)],
and Agnieszka Sołtys[1]

[1] Institute of Applied Mathematics and Mechanics,
University of Warsaw, Warsaw, Poland
sd.nowakowski2@uw.edu.pl, pokar@mimuw.edu.pl, agnieszkaprochenka@gmail.com
[2] Faculty of Physics, University of Warsaw, Warsaw, Poland
[3] Faculty of Mathematics and Computer Science, Nicolaus Copernicus University,
Toruń, Poland
wrejchel@gmail.com

Abstract. Sparse modeling or model selection with categorical data is challenging even for a moderate number of variables, because roughly one parameter is needed to encode one category or level. The Group Lasso is a well known efficient algorithm for selection of continuous or categorical variables, but all estimates related to a selected factor usually differ. Therefore, a fitted model may not be sparse, which makes the model interpretation difficult. To obtain a sparse solution of the Group Lasso, we propose the following two-step procedure: first, we reduce data dimensionality using the Group Lasso; then, to choose the final model, we use an information criterion on a small family of models prepared by clustering levels of individual factors. In the consequence, our procedure reduces dimensionality of the Group Lasso and strongly improves interpretability of the final model. What is important, this reduction results only in the small increase of the prediction error. In the paper we investigate selection correctness of the algorithm in a sparse high-dimensional scenario. We also test our method on synthetic as well as the real data sets and show that it outperforms the state of the art algorithms with respect to the prediction accuracy, model dimension and execution time. Our procedure is contained in the R package DMRnet and available in the CRAN repository.

Keywords: Information criterion · Model reduction · Penalized likelihood · Regression · Sparse prediction

1 Introduction

Data sets containing categorical variables (factors) are common in statistics and machine learning. Sparse modeling for such data is much more challenging than for those having only numerical variables. There are two main reasons for that:

(i) a factor with k levels is usually encoded as $k - 1$ dummy variables, so $k - 1$ parameters are needed to learn it,

(ii) a reduction of the dimensionality for a factor is much more sophisticated than for a numerical predictor (leave or delete), because we can either delete this factor or merge some of its levels (a number of possibilities grows very quickly with a number of levels). Let us consider as the example the factor corresponding to a continent that a person (client, patient etc.) lives on, or a company (university etc.) is located. This factor has 6 levels (Antarctica is not considered), so there are 203 possibilities to merge its levels (they are usually called *partitions*).

It is really difficult to develop efficient algorithms for categorical data and investigate their statistical properties, when a number of factors and/or a total number of their levels is large. Thus, this topic has not been intensively studied so far and the corresponding literature has been relatively modest. However, categorical data are so common that it had to change. We have found many papers investigating categorical data from the last few years, among others [7, 15, 17–19, 21].

In the paper we consider high-dimensional selection and sparse prediction for categorical data, where a number of active variables is significantly smaller than a learning sample size n and a number of all variables p significantly exceeds n. The goal is to develop a procedure, whose outputs have small prediction errors and easy interpretation. For categorical predictors the latter property means that all non-active factors should be discarded. What is more, if an active factor contains equal levels, then they should be merged.

Neural networks or random forests have good predictive properties, but their outputs are difficult to interpret. They often fail when n is small and/or p is larger than n (cf. the experimental results in Sect. 4). In the paper we focus on such scenarios and consider penalized likelihood methods as a family of Lasso algorithms, which are commonly used for sparse prediction. Some of the methods can discard non-active predictors for high-dimensional data, but they cannot merge the levels of active factors, which strongly limits their intepretability. For instance, the Lasso [22] treats dummy variables as separate, binary predictors, the Group Lasso [25] can only leave or delete a whole factor and the Sparse Group Lasso [20] additionally removes levels of selected factors. The Fused Lasso [23] can merge levels, but only in a simplified case when variables are ordered. These methods significantly reduce a number of parameters, but they do not realize *partition selection*, i.e. they cannot choose models consisting of subsets of numerical variables and partitions of levels of factors.

In the mainstream research on the Lasso-type algorithms, the CAS-ANOVA method [3] fits sparse linear models with the fusion of factor levels using the l_1 penalty imposed on the differences between the parameters corresponding to levels of a factor. The implementation of CAS-ANOVA has been provided in [8, 14]. An alternative approach is a greedy search algorithm, called DMR, from [13]. The growing interest in partition selection has been noticed recently. In [15] a Bayesian method for linear models is introduced based on a prior inducing fusion of levels. Another approach trying to solve the problem from the Bayesian perspective is considered in [7]. The frequentist method using the linear mixed models was presented in [19], where factors were treated as random effects.

A partition selection algorithm called SCOPE, which is based on a regularized likelihood, can be found in [21]. This procedure uses a minimax concave penalty on differences between consecutive, sorted estimators of coefficients for levels of a factor. Finally, tree-based algorithms are applied to categorical data in [18].

Let us note that all the aforementioned partition selection methods are restricted to a classical scenario $p < n$, except SCOPE and DMR. The new implementation of the latter is based on variables screened by the Group Lasso [16]. In this paper, we present an improved as well as simplified version of the DMR algorithm, called PDMR (Plain DMR). Our main contributions are as follows:

1. We propose the following two-step PDMR procedure: first, we reduce data dimensionality using the Group Lasso. Next, a small family of models is constructed by clustering levels of individual factors. The final model is chosen from the family using an information criterion. In DMR the Group Lasso model is refitted and dissimilarity matrices in clustering are computed by likelihood ratio statistics. In PDMR we eliminate this step by employing the Group Lasso coefficients, which simplifies and improves the DMR. The PDMR not only works better in numerical experiments, but we are able to mathematically confirm its properties (Theorem 1).
2. We prove in Theorem 1 that PDMR returns a sparse linear model containing the true model, even if $p >> n$. The proof is based on a new bound of a number of partitions by generalized Poisson moments. It is worth to note that so far there are no theoretical results regarding the correctness of the DMR selection for high-dimensional data, while for SCOPE a weaker property than selection consistency was proved in [21]. Our result is also weaker than selection consistency, but it relates directly to any output of our algorithm, while results from [21, Theorem 6] concern only one of blockwise optima of their objective function. We discuss this issue in Sect. 3.
3. In theoretical considerations, the Lasso-type algorithms are defined for one penalty and return one estimator. However, practical implementations usually use nets of data-driven penalties and return lists of estimators. Our next contribution is an analogous implementation of PDMR. In numerical experiments on simulated and real data, we compare PDMR, DMR and SCOPE. We show that PDMR performs better than SCOPE and DMR with respect to a prediction error and model simplicity/sparsity. Moreover, PDMR is computationally faster than DMR and several dozen times faster than SCOPE. Our procedure is contained in the R package DMRnet [16] and available in the CRAN repository.

In the rest of this paper we describe the considered models and the PDMR algorithm. The main theoretical result, which establishes properties of our method, is given in Sect. 3. Finally, we compare PDMR to the other methods in numerical experiments. The proof of Theorem 1, auxiliary theoretical results and additional descriptions of experiments are relegated to the online supplement[1].

---

[1] https://github.com/SzymonNowakowski/ICCS-2023.

2 Linear Models and the Algorithm

We consider independent data $(y_1, x_{1.}), (y_2, x_{2.}), \ldots, (y_n, x_{n.})$, where $y_i \in \mathbb{R}$ is a response variable and $x_{i.} \in \mathbb{R}^p$ is a vector of predictors. Every vector of predictors $x_{i.}$ can consist of continuous predictors as well as categorical predictors. We arrange them in the following way $x_{i.} = (x_{i1}^T, x_{i2}^T, \ldots, x_{ir}^T).^T$ Suppose that subvector x_{ik} corresponds to a categorical predictor (factor) for some $k \in \{1, \ldots, r\}$. Let a set of levels of this factor be given by $\{0, 1, 2, \ldots, p_k\}$. In that case we usually use $x_{ik} \in \{0, 1\}^{p_k}$, so x_{ik} is a dummy vector corresponding to the k-th predictor of the i-th object in a data set. Notice that we do not include a reference level (say, the zero level) in x_{ik}. The only exception relates to the first factor, whose reference level is contained in x_{i1}. This special level plays a role of an intercept. If necessary, we can rearrange vectors of predictors to have the first factor with $k = 1$. If x_{ik} corresponds to a continuous predictor, then simply $x_{ik} \in \mathbb{R}^{p_k}$ and $p_k = 1$. Therefore, a dimension of $x_{i.}$ is $p = 1 + \sum_{k=1}^{r} p_k$. Finally, let $X = [x_{1.}, \ldots, x_{n.}]^T$ be a $n \times p$ design matrix and by $x_{j,k}$ we denote its column corresponding to the j-th level of the k-th factor.

We consider a linear model

$$y_i = x_{i.}^T \mathring{\beta} + \varepsilon_i \quad \text{for} \quad i = 1, 2, \ldots, n. \tag{1}$$

Coordinates of $\mathring{\beta}$ correspond to coordinates of a vector of predictors, that is $\mathring{\beta} = (\mathring{\beta}_1^T, \mathring{\beta}_2^T, \ldots, \mathring{\beta}_r^T)^T$, where we have $\mathring{\beta}_1 = (\mathring{\beta}_{0,1}, \mathring{\beta}_{1,1}, \ldots, \mathring{\beta}_{p_1,1})^T \in \mathbb{R}^{p_1+1}$ and $\mathring{\beta}_k = (\mathring{\beta}_{1,k}, \mathring{\beta}_{2,k}, \mathring{\beta}_{3,k}, \ldots, \mathring{\beta}_{p_k,k})^T \in \mathbb{R}^{p_k}$ for $k = 2, \ldots, r$. Moreover, we suppose that noise variables ε_i have a *subgaussian distribution* with the same parameter $\sigma > 0$, that is for $i = 1, 2, \ldots, n$ and $u \in \mathbb{R}$ we have

$$\mathbb{E} \exp(u\varepsilon_i) \le \exp(\sigma^2 u^2 / 2). \tag{2}$$

The main examples of subgaussian noise variables are normal variables or those having bounded supports.

2.1 Notations

For $\beta \in \mathbb{R}^p$ and $q \ge 1$ let $|\beta|_q = (\sum_{j=1}^{p} |\beta_j|^q)^{1/q}$ be the ℓ_q norm of β. The only exception is the ℓ_2 norm, for which we use the special notation $||\beta||$.

A feasible model is defined as a sequence $M = (P_1, P_2, \ldots, P_r)$. If the k-th predictor is a factor, then P_k is a particular partition of its levels. If the k-th predictor is continuous, then $P_k \in \{\emptyset, \{k\}\}$. To make the notation coherent and concise we *artificially* augment each $\beta \in \mathbb{R}^p$ by $\beta_{0,k} = 0, k = 2, \ldots, r$. Notice that every β determines a model M_β: if the k-th predictor is a factor, then partition P_k is induced by equalities of coefficients, i.e. $\beta_{j_1,k} = \beta_{j_2,k}, j_1 \ne j_2$ means that levels j_1 and j_2 belong to the same cluster of the k-th factor. If the k-th predictor is continuous, then $P_k = \{k\}$ when $\beta_k \ne 0$ and $P_k = \emptyset$ otherwise.

2.2 The Algorithm

To simplify notations (and without losing the generality), we suppose that all considered predictors are categorical. For estimation of $\overset{\circ}{\beta}$ we consider a quadratic loss function as in the maximum likelihood estimation:

$$\ell(\beta) = \frac{1}{n} \sum_{i=1}^{n} [(x_{i.}^T \beta)^2/2 - y_i x_{i.}^T \beta]. \tag{3}$$

We present the PDMR algorithm, which consists of two steps:
(1) Screening: we compute the weighted Group Lasso

$$\hat{\beta} = \arg\min_{\beta} \ell(\beta) + \lambda \sum_{k=1}^{r} ||W_k \beta_k||,$$

where W_k is a diagonal matrix with $(W_k)_{jj} = ||x_{j,k}||/\sqrt{n}$ playing roles of weights. Such a choice of weights was suggested in the seminal paper [25]. It is also explained in Proposition 1 in the online supplement. Number $\lambda > 0$ is a tuning parameter whose choice is discussed in Sect. 3;
(2) Selection: this step is divided into three parts:
(2a) construction of the nested family of models $\mathcal{M}$: let $\hat{S} = \{1 \leq k \leq r : \hat{\beta}_k \neq 0\}$ and $\hat{\beta}_{0,k} = 0$ for $k \in \hat{S} \setminus \{1\}$. So, $\hat{S}$ is a set of factors which are not discarded by the Group Lasso. For each $k \in \hat{S}$ we separately perform complete linkage hierarchical clustering of levels of those factors. Each clustering starts with a dissimilarity matrix $(D_k)_{j_1,j_2} = |\hat{\beta}_{j_1,k} - \hat{\beta}_{j_2,k}|, 0 \leq j_1, j_2 \leq p_k$. This matrix is consecutively updated as follows: a distance between two clusters A and B of levels of the k-th factor is defined as $\max_{a \in A, b \in B} |\hat{\beta}_{a,k} - \hat{\beta}_{b,k}|$. Each clustering begins with disjoint factor's levels and then two most *similar* clusters are merged. Finally, we obtain the *empty* factor with all levels merged. Cutting heights from this clustering are contained in h_k^T. Then we create a vector h, which consists of elements of a vector $(0, h_1^T, h_2^T, \ldots, h_{\hat{S}}^T)^T$ sorted increasingly. Next, we construct a family $\mathcal{M} = \{M_0 = \hat{S}, M_1, M_2, \ldots, \{\emptyset\}\}$, where M_{j+1} is M_j with one additional merging of appropriate clusters corresponding to the $(j+1)$-th element in h,
(2b) Generalized Information Criterion

$$\hat{M}_{PDMR} = \arg\min_{M \in \mathcal{M}} \ell(\hat{\beta}_M) + \lambda^2/2|M|,$$

where $\hat{\beta}_M$ is a minimum loss estimator over $\mathbb{R}^p$ with constraints determined by a model M and $|M|$ equals to a number of distinct levels in M. Technical details of this constrained minimization is given in Sect. 3 of the online supplementary materials. We also show there that it can be considered as an unconstrained minimization over a smaller space.
(2c) Estimation of parameters in the model $\hat{M}_{PDMR}$:

$$\hat{\beta}_{PDMR} = \arg\min_{\beta_{\hat{M}_{PDMR}}} \ell(\beta_{\hat{M}_{PDMR}})$$

To study data with binary responses we can extend the PDMR algorithm to logistic regression. From the practical point of view this generalization is relatively easy. We apply the Group Lasso for logistic regression with $\ell(\beta) = \sum_{i=1}^{n}[\log(1 + \exp(x_i^T\beta)) - y_i x_i^T\beta]/n$. A similar modification should be done when applying the information criterion step. We have contained it in the R package DMRnet and apply it in Sect. 4. This extension is more difficult from the theoretical perspective, so the result in Sect. 3 is restricted to linear models (1).

3   Statistical Properties of PDMR

In this section we state the main theoretical result concerning our algorithm.

Assume that $M_{\mathring{\beta}}$ is the true model, i.e. the one which is determined by the true parameter $\mathring{\beta}$. In Theorem 1 we establish that our procedure is able to *partially* find $M_{\mathring{\beta}}$. It is possible only if $M_{\mathring{\beta}}$ is sufficiently distinguishable from other models, which is sometimes called *identifiability* of $M_{\mathring{\beta}}$. This issue is quite involved, so we move its precise description to Sect. 2 in the online supplement. Here, the identifiability of $M_{\mathring{\beta}}$ is expressed by a positive value κ such that the larger κ is, the model $M_{\mathring{\beta}}$ is easier to identify. Finally, $|M_{\mathring{\beta}}|$ denotes a number of distinct levels in $M_{\mathring{\beta}}$.

Theorem 1. *Suppose that assumptions* (1) *and* (2) *are satisfied and there exists* $0 < a < 1$ *such that*

$$\frac{2a^{-2}\sigma^2 \log p}{n} \leq \lambda^2 < \frac{\kappa}{16(1 + a)^2}. \tag{4}$$

Then

$$P(\hat{M}_{PDMR} \subsetneq M_{\mathring{\beta}}) \leq 3p \exp\left(-\frac{a^2 n \lambda^2}{2\sigma^2}\right). \tag{5}$$

Theorem 1 states that PDMR computes consistent screening, which means that with high probability it is able to reduce a model returned by the Group Lasso without losing any active variables. The only parameter of PDMR can be chosen as $\lambda^2 = 2a^{-2}\sigma^2 \log p(1 + q)/n$ for some $q > 0$. Then the right-hand side in (5) behaves like $1/p^q$. So, the larger q is, the faster probability in (5) goes to 0. However, increasing q restricts the usefulness of Theorem 1 only to $M_{\mathring{\beta}}$ having large κ. Therefore, the reasonable choice of q would be a small value which still ensures that probability in (5) goes to zero, for instance $q \to 0$ but $q \log p \to \infty$. Notice that it is the same choice of λ as in the *Risk Inflation Criterion* [5].

PDMR can successfully work in the case that p is large. From (4) we see that κ has to be larger than $\log(p)/n$, which goes to zero even if $p = \exp(n^\alpha), \alpha < 1$ for $n \to \infty$. So, as n increases, Theorem 1 can be applied to data having smaller value of κ. Notice that there are no similar theoretical results for other partition selection competitors of PDMR in high-dimensions. Guarantees for DMR from [13] relates only to the $p < n$ scenario. In [21, Theorem 6] it is shown that the output of SCOPE and the true model are some blockwise optima of the SCOPE objective function. This conclusion is very weak, because there might be plenty of such blockwise optima. Consider the function $f(x, y) = |x - y| + |x + y|$, which has one global minimum $(0, 0)$. Notice that the point $(100, 100)$ is one of the blockwise optima of f but it is useless when estimating $(0, 0)$. Therefore, the output of SCOPE might be very poor estimator of $M_{\hat{\beta}}$, while Theorem 1 states that *any* output of PDMR computes consistent screening in high-dimensions.

Calculating a value of κ is difficult. In Sect. 2.1 of the online supplement we show that for the special case that $X^T X$ is an orthogonal matrix, κ can be determined and condition (4) leads to $\Delta^2 \succeq \sigma^2 \log p \max_k p_k / n$, where $\Delta = \min\limits_{1 \le k \le r} \min\limits_{0 \le j_1, j_2 \le p_k : \mathring{\beta}_{j_1,k} \ne \mathring{\beta}_{j_2,k}} |\mathring{\beta}_{j_1,k} - \mathring{\beta}_{j_2,k}|$. The value of Δ states how much distinct levels belonging to the same factor in $M_{\hat{\beta}}$ differ. Therefore, this condition shows a clear relation between characteristics of the data (i.e. $n, p, p_k, \sigma, \Delta$) that gives the sufficient distinguishability of $M_{\hat{\beta}}$.

The proof of Theorem 1 is given in Sect. 2 in online supplementary materials. It can be sketched as follows: first we establish that the Group Lasso is a consistent estimator, which implies that the family $\mathcal{M}$, defined in the step (2a) of PDMR, contains the true set $M_{\hat{\beta}}$. Then we establish that the probability of choosing a submodel of $M_{\hat{\beta}}$ can be expressed as a Touchard polynomial and estimated using the recent combinatorial results from [1]. Proving an upper bound on $P(M_{\hat{\beta}} \subsetneq \hat{M}_{PDMR})$ still remains an open problem.

4 Experiments

We start with the implementation details of PDMR and competing procedures. Then simulated and real data sets are investigated.

In the theoretical analysis of Lasso-type estimators one usually considers only one value of the tuning parameter λ. We have also followed this way in Sect. 3. However, the practical implementations can efficiently return estimators for a data driven net of λ's, as in the R package glmnet [6]. Similarly, using a net of λ's, the Group Lasso and the Group MCP algorithms have been implemented in the R package grpreg [4]. We also propose a net modification of PDMR:

1. For λ belonging to the grid: calculate the Group Lasso estimator $\hat{\beta}(\lambda)$ and then perform complete linkage for each factor and get a nested family of models $M_1(\lambda) \subset M_2(\lambda) \subset \dots$
2. For a fixed model dimension c, select a model M_c from the family $(M_c(\lambda))_\lambda$, which has the minimal prediction loss.
3. Select a final model from the sequence $(M_c)_c$ using the Risk Inflation Criterion (RIC), see [5], i.e. using the tuning parameter $2\sigma^2 \log(p)/n$, which is the same as suggested in Theorem 1. Obviously, σ is unknown, but it can be quite easily estimated: as usual we take an appropriately scaled residual sum of squares on a model returned by the Group Lasso.

The above implementation of PDMR is available in the DMRnet package starting with its 0.3.4 version with algorithm="PDMR" argument.

In experiments we also evaluate the following methods:

(i) DMR, which refits $\hat{\beta}$ after the Group Lasso with the maximum likelihood and computes dissimilarity matrices in the clustering step as the likelihood ratio statistics. It is also contained in the DMRnet package,

(ii) Group Lasso (gL) with a tuning parameter lambda chosen by cross-validation as implemented in cv.grpreg from the R package grpreg with penalty="grLasso",

(iii) Group MCP (gMCP) with a tuning parameter lambda chosen by cross-validation as implemented in cv.grpreg from the R package grpreg with gamma set as the default and penalty="grMCP",

(iv) SCOPE from the R package CatReg [21]. A tuning parameter gamma is chosen as 8 or 32, which is suggested in that paper. We denote them as S-8, S-32, respectively. For real data we consider also the case of binary responses and then gamma is 100 or 250 as in [21]. We denote them S-100 and S-250, respectively,

(v) Random Forest (RF) - we use the randomForest function from the R package randomForest.

All results presented in this section can be reproduced using our codes, which are publicly available at https://github.com/SzymonNowakowski/ICCS-2023.

4.1   Simulation Study

This section contains the experiments with simulated high-dimensional linear models. Design matrices X and parameter vectors $\mathring{\beta}$ are the same as in [21], but additionally we systematically change the signal to noise ratio (SNR).

In the training data we have $n = 500$ observations. Every vector of predictors consists of $r = 100$ factors, each with 24 levels. Thus, after deleting 99 reference levels we obtain $p = 2301$. A design matrix X is generated as in [21], namely: first, we draw matrix Z, whose rows $z_{i\cdot}, i = 1, \ldots, 500$ are independent 100-dimensional vectors having normal distribution $N(0, \Sigma)$. The off-diagonal elements of Σ are chosen such that correlation between $\Phi(z_{ij})$ and $\Phi(z_{ik})$ equals 0.5 for $j \neq k$, where Φ is a cdf of the standard normal distribution. Then we set $x_{ij} = \lceil 24\Phi(z_{ij}) \rceil$. Finally, X is recoded into dummy variables and reference levels of each factors, except the first one, are deleted.

Errors ε_i are independently distributed from $N(0, \sigma^2)$, where σ is chosen in such a way to realize distinct SNR values. The performance of the estimators is measured using the root-mean-square errors (RMSE), which are calculated using the test data consisting of 10^5 observations. To work on the universal scale we divide the RMSE of procedures by the RMSE of the oracle, which knows the true model in advance, i.e. a maximum likelihood estimator computed for the true model. Final results are averages over 200 draws of training and testing data. We consider the following six models, which are the same as in [21, Section 6.1.2]. However, we renumber them with respect to true model dimensions (True MD), i.e. a number of distinct levels among their consecutive factors in $M_{\hat{\beta}}$ (recall that we do not count reference levels of factors, except the first one):

Setting 1: $\mathring{\beta}_k = (0, \ldots, 0, \underbrace{2, \ldots, 2}, \underbrace{4, \ldots, 4})$ for $k = 2, 3$, $\mathring{\beta}_k = (0, \ldots, 0, \underbrace{5, \ldots, 5})$

$\underbrace{}_{7\text{ times}}\;\underbrace{}_{8\text{ times}}\;\underbrace{}_{8\text{ times}}\underbrace{}_{15\text{ times}}\;\underbrace{}_{8\text{ times}}$

for $k = 4, 5, 6$, $\mathring{\beta}_1 = (0, \mathring{\beta}_2)$, and $\mathring{\beta}_k = 0$ otherwise. So, True MD=10,

Setting 2: $\mathring{\beta}_k = (0, \ldots, 0, \underbrace{2, \ldots, 2}, \underbrace{4, \ldots, 4})$ for $k = 2, 3$, and for $k = 4, 5, 6$ we

$\underbrace{}_{7\text{ times}}\;\underbrace{}_{8\text{ times}}\;\underbrace{}_{8\text{ times}}$

have $\mathring{\beta}_k = (0, \ldots, 0, \underbrace{2, \ldots, 2}, \underbrace{4, \ldots, 4})$, $\mathring{\beta}_1 = (0, \mathring{\beta}_2)$, and $\mathring{\beta}_k = 0$ otherwise. So,

$\underbrace{}_{9\text{ times}}\;\underbrace{}_{4\text{ times}}\;\underbrace{}_{10\text{ times}}$

True MD=13,

Setting 3: $\mathring{\beta}_k = (0, \ldots, 0, \underbrace{2, \ldots, 2}, \underbrace{4, \ldots, 4}, \underbrace{6, \ldots, 6})$ for $k = 2, \ldots, 5$, $\mathring{\beta}_1 = (0, \mathring{\beta}_2)$,

$\underbrace{}_{5\text{ times}}\;\underbrace{}_{6\text{ times}}\;\underbrace{}_{6\text{ times}}\;\underbrace{}_{6\text{ times}}$

and $\mathring{\beta}_k = 0$ otherwise. So, True MD=16,

Setting 4: $\mathring{\beta}_k = (0, \ldots, 0, \underbrace{1, \ldots, 1}, \underbrace{2, \ldots, 2}, \underbrace{3, \ldots, 3}, \underbrace{4, \ldots, 4})$ for $k = 2, \ldots, 5$,

$\underbrace{}_{4\text{ times}}\;\underbrace{}_{5\text{ times}}\;\underbrace{}_{4\text{ times}}\;\underbrace{}_{5\text{ times}}\;\underbrace{}_{5\text{ times}}$

$\mathring{\beta}_1 = (0, \mathring{\beta}_2)$, and $\mathring{\beta}_k = 0$ otherwise. So, True MD=21,

Setting 5: $\mathring{\beta}_k = (0, \ldots, 0, \underbrace{2, \ldots, 2}, \underbrace{4, \ldots, 4})$ for $k = 2, \ldots, 10$, $\mathring{\beta}_1 = (0, \mathring{\beta}_2)$, and

$\underbrace{}_{3\text{ times}}\;\underbrace{}_{12\text{ times}}\;\underbrace{}_{8\text{ times}}$

$\mathring{\beta}_k = 0$ otherwise. So, True MD=21,

Setting 6: $\mathring{\beta}_k = (0, \ldots, 0, \underbrace{5, \ldots, 5})$ for $k = 2, \ldots, 25$, $\mathring{\beta}_1 = (0, \mathring{\beta}_2)$, and $\mathring{\beta}_k = 0$

$\underbrace{}_{15\text{ times}}\;\underbrace{}_{8\text{ times}}$

otherwise. So, True MD=26.

In [21, Section 6.1.2] the above Settings 1–6 are numbered as 3, 1, 8, 4, 7 and 5, respectively. In each model we consider distinct SNR values, in particular Settings 2 and 6 from that paper are also studied.

The results of experiments are presented in Figs. 1 and 2. In the former we present the RMSE of procedures divided by the RMSE of the oracle. On the x-axis there are SNR values for the interval $[1, 5]$. In the second figure we observe numbers of distinct levels that are recognized by procedures (model dimension, MD) divided by a number of distinct level in the true model (i.e. True MD). Recall that the goal is to find an easily interpretable model (i.e. MD should be small), which has also good predictive properties. From Figs. 1 and 2 we observe that PDMR is the clear winner. Notice that MD of PDMR is similar to SCOPE-8 and smaller than for other competitors. However, RMSE for SCOPE-8 is larger than for PDMR in Fig. 1. Besides, predictive properties of PDMR are the best (Scenario 1–2) or close to the best (Scenario 3–5). The main competitors of PDMR with respect to prediction are SCOPE-32, Group Lasso and Group MCP. However, Group Lasso and Group MCP do not return interpretable model. Their MD is more than 7 times larger than True MD, so they are not presented in Fig. 2 (except Scenario 4). SCOPE-32 has similar prediction errors to PDMR but it is worse in model interpretation, because its MD is usually 2–3 times larger than for PDMR. Finally, notice that Random Forests fail in our experiments, which confirms the fact that this procedure often works poorly with data having many unknown parameters and small sample sizes.

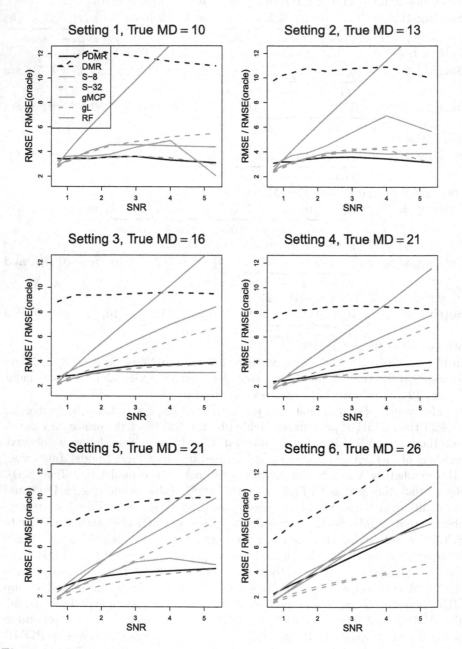

Fig. 1. Relative prediction errors of the considered methods in six settings. SNR - the signal to noise ratio, RMSE - the root-mean-square error. The remaining details are given in the text.

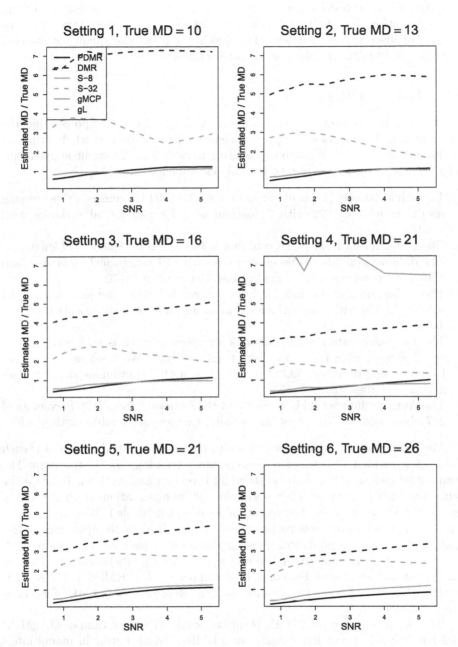

Fig. 2. Relative model dimensions (MD) of the considered methods in six settings. SNR - the signal to noise ratio. The remaining details are given in the text.

In Scenario 6 PDMR works worse than for other settings. It comes from the fact that Scenario 6 does not satisfy the assumptions required for the Group Lasso operation, i.e. the true factors have more parameters in total than the number of observations ($n = 500$). This loss in the prediction quality is observed not only for PDMR, but also for the other methods.

4.2   Real Data Study

We investigate five real data sets: the first two with binary responses and the next three with continuous responses. Here we present only short descriptions of these data sets. A full discussion is given in Sect. 5 in the online supplement. The preprocessing step is also thoroughly explained there.

1. The Adult data set [11] contains data from the 1994 US census. Preprocessing results in $n = 45,222$ with 2 continuous and 8 categorical variables with $p = 93$;
2. The Promoter data set [9,24] contains E. Coli genetic sequences of length 57. The data set consists of 106 observations with 57 categorical variables, each with 4 levels representing 4 nucleotides, thus with $p = 172$;
3. The Airbnb data set is available from *insideairbnb.com*. Preprocessing results in $n = 49,976$ with 765 variables (out of which 3 are categorical) with $p = 40668$;
4. The Insurance data set [10] contains a response, which is an 8-level ordinal variable measuring insurance risk of an applicant. We treat as continuous. Preprocessing results in 59,381 observations with 5 continuous and 108 categorical variables with $p = 823$;
5. The Antigua data set [2] is available at the R package DAAG [12]. It consists of 287 observations with 1 continuous and 4 categorical variables with $p = 58$.

We conduct experiments analogously to the ones described in Sect. 4.1: each data set is divided (200 times repeated) into a training, and testing sets. The training set contains 70% of observations for Promoter and Antigua. To adapt the remaining data to the case $p >> n$ we take only n_i observations to their training sets in the i-th repetition. We present values of n_i's in Table 1. We also show the values of p_i's, which are new parameter space sizes in the i-th repetition. Notice that $p_i < p$, because predictors constant for a given training set get dropped. Estimators are computed on the training sets and their prediction errors (PE) are calculated on testing sets. For continuous responses PE is RMSE as in Sect. 4.1, while for binary responses PE is a misclassification error. PE and MD of methods are given in Fig. 3.

The prediction error of PDMR is similar or slightly worse than of gL, gMCP and RF (for Adult and Promoter), but PDMR is much better in model interpretability. MD of PDMR is at least a few times smaller than for gL and gMCP. PDMR is definitely better than DMR on Insurance and similar on the others. Comparing to SCOPE, PDMR wins on Adult and Promoter, because its MD is smaller and more stable, while their PE are similar. On Antigua and Insurance we have a tie: PDMR has smaller PE, but SCOPE has smaller MD.

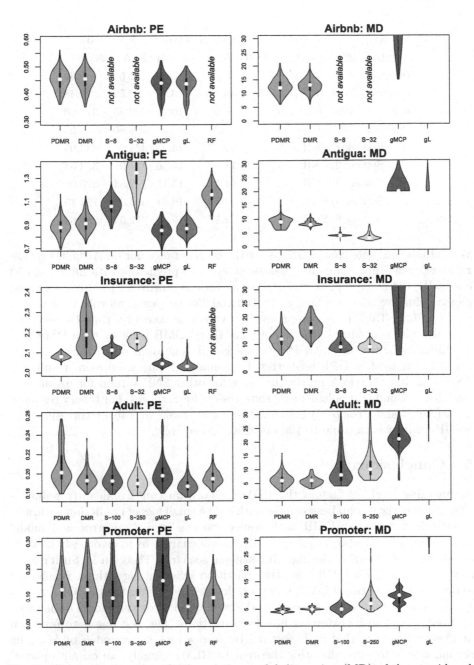

Fig. 3. The prediction error (PE) and the model dimension (MD) of the considered methods for five real data sets. The remaining details are given in the text.

Table 1. Median execution time of procedures (in seconds) and n_i, p_i (real data sets).

Dataset	n_i	p_i range	PDMR	DMR	S-8	S-32
Airbnb	999	$1410, \ldots, 1504$	5.19	5.41	N/A	N/A
Antigua	200	57, 58	0.16	0.19	1.65	0.69
Insurance	1781	$313, \ldots, 365$	46.1	113.82	27.59	25.48
Adult	452	$60, \ldots, 78$	5.66	10.2	300.53	322.96
Promoter	74	172	3.29	4.7	604.86	3616.55
Setting 1, SNR = 3			11.67	15.25	335.6	332.03
Setting 2, SNR = 3			11.57	15.56	371.91	353.68
Setting 3, SNR = 3			11.47	15.11	352.84	370.06
Setting 4, SNR = 3			12.71	16.81	504.25	764.15
Setting 6, SNR = 3			12.17	16.49	712.94	1253.64

We did not complete computations for RF on Insurance nor on Airbnb, because randomForest function cannot handle categorical predictors with more than 53 levels. We did not complete SCOPE computations on Airbnb for any considered gamma value (we tried 4 values: 8, 32, 100 and 250 for gamma, all with no success).

Finally, in Table 1 we present median values of execution time (in seconds) for particular procedures. We can observe that PDMR is faster than DMR and significantly faster than SCOPE. There is only one data set (Insurance) that execution time of SCOPE is shorter. For simulated data we present results for SNR = 3, but they look pretty the same for other SNR values, for instance in Sect. 5 of the online supplement we show results for SNR = 4. For binary response data (Adult and Promoter) columns "S-8" and "S-32" in Table 1 correspond to SCOPE with gamma equal to 100 and 250, respectively.

5   Conclusions

By merging levels of factors the PDMR algorithm can significantly reduce a dimension of the Group Lasso model with only a small loss of prediction accuracy. PDMR is better than DMR with respect to the prediction error and model dimension. PDMR also simplifies DMR, which enables us to rigorously confirm consistency of PDMR in the high-dimensional scenario (Theorem 1). Such results are not available for DMR nor other partition selection algorithms. Numerical experiments show that PDMR is several dozen times faster than SCOPE and is comparable or better with respect to the prediction error and model dimension.

Our paper can be extended in a few directions. The first one is to prove an analogous bound to that in Theorem 1 but concerning supermodels of $M_{\hat{\beta}}$. The second one is to generalize this theorem to GLMs. Finally, we obtain optimal weights for the Group Lasso, which are different from those recommended by the authors of this method (cf. Sect. 2.1 in the supplement). Possibly, the new weights can improve asymptotics of the Group Lasso in the general scenario (not necessarily orthogonal) and its practical performance as well.

References

1. Ahle, T.D.: Sharp and simple bounds for the raw moments of the binomial and poisson distributions. Statist. Probab. Lett. **182**, 109–306 (2022)
2. Andrews, D.F., Herzberg, A.M.: Data. A Collection of Problems from Many Fields for the Student and Research Worker. Springer, New York (1985)
3. Bondell, H.D., Reich, B.J.: Simultaneous factor selection and collapsing levels in anova. Biometrics **65**, 169–177 (2009)
4. Breheny, P., Huang, J.: Group descent algorithms for nonconvex penalized linear and logistic regression models with grouped predictors. Stat. Comput. **25**, 173–187 (2015)
5. Foster, D., George, E.: The risk inflation criterion for multiple regression. Ann. Statist. **22**, 1947–1975 (1994)
6. Friedman, J., Hastie, T., Tibshirani, R.: Regularization paths for generalized linear models via coordinate descent. J. Stat. Softw. **33**, 1–22 (2010)
7. García-Donato, G., Paulo, R.: Variable selection in the presence of factors: a model selection perspective. J. Amer. Statist. Assoc. **117**, 1847–1857 (2022)
8. Gertheiss, J., Tutz, G.: Sparse modeling of categorial explanatory variables. Ann. Appl. Stat. **4**, 2150–2180 (2010)
9. Harley, C., Reynolds, R.: Analysis of e. coli promoter sequences. Nucleic Acids Res. **15**, 2343–2361 (1987)
10. Kaggle: Prudential life insurance assessment (2015). www.kaggle.com/c/ prudential-life-insurance-assessment/data
11. Kohavi, R.: Scaling up the accuracy of naive-bayes classifiers: a decision-tree hybrid. In: Proceedings of KDD, pp. 202–207 (1996)
12. Maindonald, J., Braun, W.: Data Analysis and Graphics Using R. Cambridge University Press (2010)
13. Maj-Kańska, A., Pokarowski, P., Prochenka, A.: Delete or merge regressors for linear model selection. Electron. J. Stat. **9**, 1749–1778 (2015)
14. Oelker, M.R., Gertheiss, J., Tutz, G.: Regularization and model selection with categorical predictors and effect modifiers in generalized linear models. Stat. Modelling **14**, 157–177 (2014)
15. Pauger, D., Wagner, H.: Bayesian effect fusion for categorical predictors. Bayesian Anal. **14**, 341–369 (2019)
16. Prochenka-Soltys, A., Pokarowski, P., Nowakowski, S.: DMRnet: Delete or Merge Regressors Algorithms for Linear and Logistic Model Selection and High-Dimensional Data (2022). https://cran.r-project.org/web/packages/DMRnet
17. Rabinowicz, A., Rosset, S.: Cross-validation for correlated data. J. Amer. Statist. Assoc. **117**, 718–731 (2022)
18. Rabinowicz, A., Rosset, S.: Trees-based models for correlated data. J. Mach. Learn. Res. **23**, 1–31 (2022)
19. Simchoni, G., Rosset, S.: Using random effects to account for high-cardinality categorical features and repeated measures in deep neural networks. In: Proceedings of NIPS, pp. 25111–25122 (2021)
20. Simon, N., Friedman, J., Hastie, T., Tibshirani, R.: A sparse-group lasso. J. Comput. Graph. Stat. **22**, 231–245 (2013)
21. Stokell, B.G., Shah, R.D., Tibshirani, R.J.: Modelling high-dimensional categorical data using nonconvex fusion penalties. J. Roy. Statist. Soc. Ser. B **83**, 579–611 (2021)

22. Tibshirani, R.: Regression shrinkage and selection via the Lasso. J. Roy. Statist. Soc. Ser. B **58**, 267–288 (1996)
23. Tibshirani, R., Saunders, M., Rosset, S., Zhu, J., Knight, K.: Sparsity and smoothness via the fused lasso. J. Roy. Statist. Soc. Ser. B **67**, 91–108 (2005)
24. Towell, G., Shavlik, J., Noordewier, M.: Refinement of approximate domain theories by knowledge-based artificial neural networks. In: Proceedings of AAAI (1990)
25. Yuan, M., Lin, Y.: Model selection and estimation in regression with grouped variables. J. Roy. Statist. Soc. Ser. B **68**, 49–67 (2006)

Actor-Based Scalable Simulation
of N-Body Problem

Kamil Szarek[1,2], Wojciech Turek[1], Łukasz Bratek[2],
Marek Kisiel-Dorohinicki[1], and Aleksander Byrski[1(✉)]

[1] AGH University of Krakow, Al. Mickiewicza 30, 30-059 Krakow, Poland
{wojciech.turek,doroh,olekb}@agh.edu.pl
[2] Cracow University of Technology, ul. Warszawska 24, 31-155 Krakow, Poland
lukasz.bratek@pk.edu.pl

Abstract. Efficient solutions of the N-body problem make it possible
to conduct large-scale physical research on the rules governing our universe. Vast amount of communication needed in order to make each
body acquainted with the information on position of other bodies renders
the accurate solutions very quickly inefficient and unreasonable. Many
approximate approaches have been proposed, and the one introduced in
this paper relies on actor-based concurrency, making the whole design
and implementation significantly easier than using, e.g. MPI. In addition
to presenting three methods, we provide the reader with tangible preliminary results that pave the way for future development of the constructed
simulation system.

Keywords: actor-based concurrency · N-body problem · agent-based
computing and simulation

1 Introduction

The classical N-body problem consists in determining the motion of N objects,
each with a specific mass, position, and initial velocity, and all interacting with
each other by gravity. The very concept of the N-body problem was presented
by Isaac Newton in his work *Philosophiae Naturalis Principia Mathematica*, in
which he formulated the laws underlying classical mechanics. However, an analytical solution of the problem is possible only for a two-body system and for
some small systems with appropriately defined initial conditions. A similar problem to the purely gravitational one can be considered for point electric charges,
provided that magnetic forces and the emission of radiation accompanying the
moving and accelerated charges can be neglected. The concept of the N-body
problem can be extended to other kinds of interaction, including the interaction
of objects with an external background field. There are also more complicated

This research was partially supported by funds from the Polish Ministry of Education
and Science assigned to AGH University of Science and Technology (AB, WT, MKD)
and Cracow University of Science and Technology (ŁB). This research was supported
by PLGrid Infrastructure.

effective interactions, such as those assumed between atoms in molecular dynamics simulations. In any case, irrespective of the particular form of interactions between bodies, for many bodies, a simulation is performed, changing the position of objects with time step, taking into account their interactions. This approach is used in many branches of science: astrophysical analyses (e.g. [15,25]), validation of models in biophysical simulations (e.g. [5]), etc. The critical element for an effective simulation is the knowledge of the entire system required to calculate the force acting on each body.

To date, methods and techniques developed for performing N-body simulations [3] have different advantages and disadvantages, sacrifice accuracy for efficiency, and utilize very particular solutions (like dedicated hardware). For us, the concept of easy-to-implement, natural and appropriately accurate design and implementation of such a system, leveraging the available High Performance Computing (HPC) infrastructure, is paramount. Therefore, in this paper, we show the background and derive the base for utilizing the actor model to create a program that on the example of the classical N-body problem performs a simulation with limited communication and propagation of information about the bodies in the system to achieve scalability and maintain the accuracy of the results. We treat the actor-based approach as very close to the paradigm of agency, which is an interesting and easy-to-apply idea for modeling and implementation of complex, interacting systems(see, e.g., [10,18]).

The actor model of concurrency has already been successfully used to implement high-scalability simulations (see, e.g. [21]), and its advantages include full asynchronicity of message exchange and simplicity of implementation by focusing on modeling communication and agent algorithm. The Akka tool was used to implement the solution, providing a lightweight implementation of the actor with support for efficient handling of a huge number of instances that work simultaneously.

In this paper we start by giving a proper background; then we describe the most important aspects of the proposed system (with several variants of the implementation of the communication) and we provide the reader with sufficient insight into our experimental result.

2   Existing Efficient N-Body Solutions

The classical N-body problem, as originally stated, aims at determining the position changes over time of N objects, knowing their masses, initial positions and velocities. The objects are moving in a three-dimensional space and are not affected by any external force, except for a mutual gravitational interaction. It is assumed that the objects can be described as material points in space governed by Newton's laws of motion and universal gravitation. The vector of force $\vec{F}_i$ exerted on ith body by all other bodies in the system reads

$$\vec{F}_i = G m_i \cdot \sum_{j \neq i} \frac{m_j}{r_{ij}^2} \frac{\vec{r}_{ij}}{r_{ij}}, \qquad \vec{r}_{ij} \equiv \vec{r}_j - \vec{r}_i, \quad r_{ij} \equiv |\vec{r}_{ij}|,, \quad i = 1, 2, \ldots, N$$

where $\vec{r}_k$, $k = 1, 2, \ldots, N$ are position vectors (with respective velocity vectors denoted by $\dot{\vec{r}}_k$), m_k are masses and G is the gravitational constant. To speed up the simulation of largely collisionless systems (like galaxies), the forces can be regularized at short distances. This allows to avoid numerically more demanding computations of very close encounters of concern only in much smaller collisional systems (celestial mechanics, globular clusters). A possible regularisation is to replace r_{ij}^2 with $r_{ij}^2 + \epsilon^2$ in all denominators, where ϵ is a softening length parameter much smaller than the characteristic mean distance between bodies.

In $d = 2, 3$ spatial dimensions, the resulting gravitational dynamics is described by a system of $2dN$ first-order equations

$$\frac{d\vec{v}_i}{dt} = \sum_{j \neq i} Gm_j \frac{\vec{r}_j - \vec{r}_i}{|\vec{r}_j - \vec{r}_i|^3},$$
$$\frac{d\vec{r}_i}{dt} = \vec{v}_i, \qquad i = 1, 2, \ldots, N \qquad (1)$$

for unknown positions and velocities. Although conceptually simple in origin, this is a highly nonlinear and chaotic system of differential equations. It is a textbook problem to find an exact solution for a two-body system ($N = 2$). For more numerous systems ($N \geq 3$) exact finite-form solutions cannot be found with method of first integrals, except for systems with high symmetry (e.g, identical masses constrained at the vertices of a regular polygon, or systems with more complicated periodic orbits offered by N-body choreography strategies [8,22]). There were attempts to rigorously analyze methods using series expansions [1,2,16,19].

When finding numerical solutions, useful are conserved quantities that can play the role of control parameters. There are $(d + 1)(d + 2)/2$ such constants following from space-time symmetries of the considered system (1) which is isolated as a whole: energy E, d components of linear momentum $\vec{P}$, $d(d - 1)/2$ of angular momentum $\vec{L}$ and d components of the center-of-mass motion $\vec{K}$. Upon setting the initial data, all components of $\vec{P}$ and $\vec{K}$ can be set to zero by transforming the initial state to the center of mass frame (this choice of reference frame prevents unbound increase of position components due to motion of the system as a whole during the integration process). Furthermore, for $d = 3$ two components of $\vec{L}$ in this frame can be set to zero by performing a suitable rotation of the system about the center of mass. In this new inertial frame, the new conserved E and the magnitude J of the new $\vec{L}$ will in general be non-zero. The dimensional constants provide a characteristic length scale $J/\sqrt{EM}$ and a characteristic time scale J/E (the scale of the system's rotation period), which together with the total mass M provide a complete set of independent mechanical units characterizing that system. With the help of these or other convenient triad of reference units, it is computationally advantageous that the system of equations be finally rewritten in a rescaled form involving only non-dimensional quantities.

For a small number of particles, the system of Eqs. (1) can be solved with a high degree of precision using the Runge-Kutta methods with adaptive step control [9]. This requires the forces to be computed several times per single time step of the simulation. For a large number of particles, one of the obvious difficulties that arise for a gravitational system characterised by long-ranged forces that cannot be neglected between distant objects is the need to trace and compute all of the $N(N-1)/2$ elementary two-body interactions, and this problem cannot be simply avoided. There are dedicated integration methods for simulations with large numbers of particles, however, they are plagued with at least quadratic complexity, rendering them completely useless when 10^5 or more bodies are considered. Many algorithms have been created to reduce complexity while sacrificing accuracy. They very often use dedicated data structures for the description of the space and bodies located there, where compromises in the data contained affect the nature of algorithms that compute the influences [3].

On the side of the integration method used, the number of times all forces have to be recomputed per a single time step of the simulation can be min-imised at the cost of reducing the accuracy of the integrator. A possible choice for updating/advancing the phase space data at consecutive time steps is the Störmer-Verlet mid point method approach [23].

Essential reduction of the computational time, however, can be achieved only on the side of the algorithm for determining the forces. In the first method applied in the case of large simulations ($N \sim 10^5$), namely Particle-Mesh method, the space is described as a mesh and the particles are placed inside, determining the density of matter over the mesh. Then the discrete Fourier transform is used to solve the Poisson differential equation for the gravitational potential that depicts the influences based on this mesh. Local forces are then determined simply from local gradients of the potential. This method is easy for parallelization, however, it may be inaccurate for a smaller number of bodies because it trades accuracy for efficiency [3].

Efficient reduction of the number of force evaluations is achieved by consid-ering clusterisation of the system to smaller subsystems of nearby bodies. Then the force due to the totality of bodies belonging to one subsystem as exerted onto a body from another subsystem can be approximated by a truncated mul-tipole expansion if the distance between the subsystems meets some criteria (i.e. is large enough). An important and widely used method in this respect is the Barnes-Hut approach [4], which uses a tree structure to describe the space. In the leaves, the data of the single bodies are contained, where the nodes comprise information about the lower-level objects. This setting allows the algorithm not to visit every leave (body), but stop at the nodes that are sufficiently far from the current object, using information about the center of the mass, for com-puting the force. This makes the complexity equal to $O(n \log(n))$, which seems very promising; however, when bodies are distributed among different computing machines, communication can affect the whole complexity to a large extent. Par-ticularly efficient is the Fast Multipole Method, reaching linear complexity for certain accuracy. The accuracy itself is one of the parameters of the algorithm; naturally increasing the accuracy, we lose efficiency. The method was proposed by Greengarg and Rokhlin in 1988 and it divides the space into squares (or cubes in 3D). Instead of building a dynamic tree, a strict structure is imposed, where each level divides the squares (cubes) into four (eight) equal parts [7].

There exist other methods, e.g. hybrid ones (putting together accurate and less accurate methods and balancing the efficiency-accuracy) [3], or the methods focused on leveraging dedicated hardware [13,14] using specially designed modules for computing the gravitational influences. Instead of using dedicated hardware, general purpose one can be adapted, like GPGPU, e.g. one of existing methods put together MPI, OpenMPI and CUDA reaching good speedup and efficiency; however, the limited GPU memory turned out to be a major problem [6]. This project was later developed into [24].

An interesting implementation of the Barnes-Hut algorithm was used in [20], where the authors used the Haskell functional language based on its ability to efficiently manage threads. Another very interesting and accurate implementation of the N-body simulation is [12], which uses the Ruby language and focuses mainly on assessing the accuracy and error of the simulation; however, it avoids delving into matters related to parallelization and HPC.

Each of the methods mentioned here uses selected particular tools and techniques for design and implementation of the N-body simulation. To our knowledge, the existing actor model of concurrency has a great potential in the implementation of scalable HPC grade systems [21], and therefore we would like to focus our proposed approach on this model, providing potential users with a prototype of scalable and reliable implementation. Moreover, thinking about implementations in large scale, we also would like to aggregate certain parts of the bodies (using dedicated, managing agents) and propose first balancing techniques, in order to approach efficient scalability utilizing large, existing HPC infrastructures.

3 The Proposed Actor-Based Models

There are numerous ways to use the actor-based model in the computational tasks considered. At least three of them are worth discussing, as they provide significantly different results, efficiency, and scalability. The three approaches can be briefly characterized as follows:

1. Problem-driven actor model, where each actor represents an entity present in the modeled system. In our case, an actor will be equated with a single body in the simulation. The actors will communicate with each other to exchange information about their state.
2. Computation-driven actor model, where an actor is a computing unit, and the number of actors is more related to the computation efficiency that the model features. Each actor shall be assigned a selected part of the computational problem, which is a set of modeled bodies. The communication will be analogous to the previous case.
3. Limited-communication actor model, where the actors will communicate only with neighbors, other actors responsible for bodies in certain proximity. The information from distant bodies will be forwarded.

The following model descriptions will omit the implementation details regarding initialization, monitoring, and results collection, focusing on the problem model, its mapping to the actor model, and its update algorithm.

3.1 Problem-Driven Actor Model

The most direct way to apply the actor model to the modeling problem is to identify autonomous entities in the modeled system and to assign each entity with a dedicated actor. In the task considered, each body can be represented by a single actor, which knows its *state*: mass, location, and velocity values. The actor is responsible for updating the state, which requires knowledge of other actors' state. Knowledge is exchanged between actors using messages sent directly between all pairs of actors.

The algorithm of a single-body actor is presented in Listing 1.

Algorithm 1. Single-body actor update loop

1: $myState = initialState$
2: $allOtherActors = initialDataWithAddresses$
3: **while** $stopCondition == false$ **do**
4:    **for** $otherActor \in allOtherActors$ **do**
5:       $send(otherActor.address, myState)$
6:    **end for**
7:    **for** $allOtherActors$ **do**
8:       $\{senderAddress, senderState\} = receive()$
9:       $allOtherActors[senderAddress].state = senderState$
10:   **end for**
11:   $force = \vec{0}$
12:   **for** $otherActor \in allOtherActors$ **do**
13:      $force = force + computeForce(otherActor.state, myState)$
14:   **end for**
15:   $myState = computeState(myState, force, \Delta t)$
16: **end while**

Initially, each actor has to be provided with the addresses of all the other actors and with the initial state of the represented body. In each step of the simulation, an actor sends its state to all other actors, collects states from all other actors, and computes the resultant force and its state after Δt. The results of the algorithm are equivalent to the accurate sequential solution.

The computational complexity of the processing performed by a single actor is linearly dependent on the number of all bodies. Additionally, the number of messages sent and received by each actor depends linearly on the size of the simulated system. Therefore, the overall complexity in a single simulation step of a single actor is $\Theta(b)$ and the whole actor system is $\Theta(b^2)$ for b bodies in the simulation. Running the actor system in parallel, it is possible to achieve a computation time proportional to $(b^2)/c$ for c computing cores. It is clearly visible that proper scalability cannot be expected in this case.

3.2   Computation-Driven Actor Model

A more efficient and scalable approach to actor-based modeling of the considered problem is based on the concept of using actors to partition the computation instead of directly representing modeled entities. In the second model, each actor is responsible for processing a set of bodies. Sets are defined at the beginning of the simulation using a clustering algorithm (k-means in our implementation). The communication schema is similar to that before – all actors exchange messages with all other actors in each simulation step. The messages, however, contain only information about the location of the center of mass (CoM) and the mass value of a cluster of bodies. Actors update the state of the assigned bodies using precise data in case of own bodies and the center of mass for bodies in remote clusters. This approach is a common optimization used in the N-body problem, which does not introduce significant errors, provided that the clusters are distant.

The algorithm of a single actor in this case is presented in Listing 2.

Each body-cluster actor is assigned an initial list of bodies that it has to simulate. The optimization based on CoM requires each actor to calculate its CoM at the beginning of each step. The exchange of messages between all pairs of actors remains unchanged. The computation of the change in the state of the assigned bodies requires integrating forces from other assigned bodies and remote clusters $CoMs$.

In a longer simulation, the initial division of bodies between actors may become improper, which can result in significant errors caused by the CoM approximation. If a body approaches another body from a different cluster, its strong mutual impact shall be reduced due to the distance of $CoMs$. Therefore, the bodies migration algorithm has been added to the presented method. It is executed after a fixed number of steps by each of the actors. The actors search for the assigned bodies, which are closer to remote CoM than to the own one. Such bodies are sent to remote actors, which changes their assignment.

In a very long simulation such an approach can lead to imbalance in the number of assigned bodies. This problem can be detected by each actor separately; it is sufficient to compare the current and the initial number of assigned bodies. In such a case, the centralized clustering algorithm has to be executed again.

The total number of actors (a) should be equal to the number of computing cores used $c, c << b$. The number of messages exchanged by each actor is proportional to the total number of actors, a. CoM calculation complexity is proportional to the number of bodies assigned, which is b/a. Computing the state changes requires $b + (b/a)^2$. So, the overall complexity of a single actor is $\Theta(a + b + (b/a)^2)$. The whole actor system can be executed in proportional time on c cores, assuming $a = c$. Therefore, it can be expected that the computation-driven actor model has a better scalability potential than the problem-driven model. Increasing the size of the simulated system proportionally with the number of cores shall not increase the $(b/a)^2$ element. However, the size of each actor's task grows linearly with the number of actors in the system and the number of bodies. It is caused by the communication schema, where all pairs of

Algorithm 2. Body-cluster actor update loop

```
1:  myBodiesState = initialBodiesState
2:  allOtherActors = initialDataWithAddresses
3:  while stopCondition == false do
4:      {myCoM, myMass} = calculateCoM(myBodiesState)
5:      for otherActor ∈ allOtherActors do
6:          send(otherActor.address, {myCoM, myMass})
7:      end for
8:      for allOtherActors do
9:          {senderAddress, {senderCoM, senderMass}} = receive()
10:         allOtherActors[senderAddress].state = {senderCoM, senderMass}
11:     end for
12:     for myBodyState ∈ myBodiesState do
13:         force = 0⃗
14:         for otherActor ∈ allOtherActors do
15:             force = force + computeForce(otherActor.state, myBodyState)
16:         end for
17:         for myOtherBodyState ∈ myBodiesState \ {myBodyState} do
18:             force = force + computeForce(myOtherBodyState, myBodyState)
19:         end for
20:         myBodyState = computeState(myBodyState, force, δt)
21:     end for
22: end while
```

actors communicate directly with each other, and the need for updating each assigned body against each remote cluster.

3.3   Limited-Communication Actor Model

In the aforementioned computation-driven actor model, the need to communicate all pairs of actors is definitely the most significant scalability limiter. To address this issue, another mechanism has been added in the third actor model: actors communicate only with "neighbors", which are the actors managing clusters in predefined proximity. In order to preserve the model correctness, the information about the state of non-neighbor clusters has to be forwarded among all actors, which is done using the same messages. Each message contains the most current state of all known clusters.

The algorithm of a single actor in this case is presented in Listing 3.

The list of neighbor actors is prepared at the beginning of the simulation. The actors can adjust automatically when they detect a change in distance between the clusters. Each actor sends all available data to all neighbors. After receiving the data, an actor checks if it contains information that is newer than already possessed. As a result, the computation of bodies state changes is based on relatively older information for distant bodies, which is another modification to the model with a minor impact on final results.

The number of messages is significantly limited in this case. Each actor sends and receives a constant number of messages, no matter how numerous the actor

---

Algorithm 3. Body-cluster actor update loop

1: $myBodiesState = initialBodiesState$
2: $allOtherActors = initialDataWithAddresses$
3: **while** $stopCondition == false$ **do**
4:     $\{myCoM, myMass\} = calculateCoM(myBodiesState)$
5:     **for** $otherActor \in allOtherActors$ **do**
6:         **if** $otherActor.isNeighbor$ **then**
7:             $send(otherActor.address, \{myCoM, myMass\} \cup allOtherActors)$
8:         **end if**
9:     **end for**
10:     **for** $neighborsCount$ **do**
11:         $receivedActors = receive()$
12:         **for** $receivedActor \in receivedActors$ **do**
13:             **if** $receivedActor.time > allOtherActors[receivedActor.id].time$ **then**
14:                 $allOtherActors[receivedActor.id] = receivedActor$
15:             **end if**
16:         **end for**
17:     **end for**
18:     **for** $myBodyState \in myBodiesState$ **do**
19:         $/* \, body \, state \, update \, same \, as \, in \, Algorithm \, 2 \, */$
20:     **end for**
21: **end while**

---

system is. Nevertheless, the total volume of the messages is even bigger than in the previous approach and the theoretical processing complexity remains the same. However, the messaging layer is less burdened, which typically has a significant impact on final efficiency.

4 Experimental Evaluation

The expected outcome of applying the actor model for the considered simulation task is to achieve a significant performance gain compared to the sequential version. The gain shall result from parallel execution, which is straightforward for actor-based implementation, and from the applied optimizations to the model, which offer better computational complexity. However, optimizations may hinder the accuracy of the model. Therefore, the evaluation will focus on these two aspects: the correctness and the performance.

Implementation has been done using the Akka actor framework for the Scala programming language [17]. All tests were repeated 10 times. In the performance tests 50–100 steps of the simulation were done to measure an average step time.

The experiments were carried out on the Prometheus computing cluster, hosted by the ACK Cyfronet AGH[1]. Prometheus offers 53604 computing cores provided by HP Apollo 8000 servers (24 cores each), which are connected with the 56 Gb/s InfiniBand network. Such resources made it possible to test the system performance and scalability on hundreds of computing cores.

---

[1] https://www.cyfronet.pl/en/computers/15226,artykul,prometheus.html.

4.1 Correctness Evaluation

One of the most common methods for verifying the correctness of the N-body simulation is to monitor the total energy in the simulated system of bodies. With no external forces in the system, the energy should remain constant. In the experiment carried out, 200 bodies were divided into 20 groups, each having 8–20 bodies. The bodies in each group were located at the edges of a regular polygon, which is a theoretically stable figure. The distances between the bodies in groups were significantly smaller than between the groups. The results of the correctness evaluation are presented in Fig. 1.

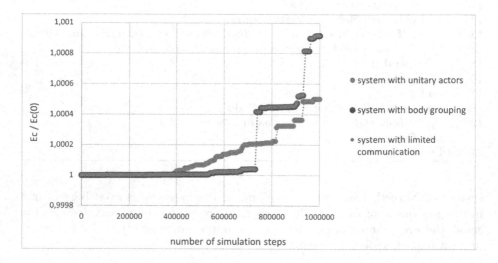

Fig. 1. Comparison of the total energy change of different models

It was expected that the total energy in the system would not be conserved over a large number of steps, due to computer arithmetic inaccuracies. The version with individual actors, which is equivalent to a sequential version, shows an increase in error starting at around 400,000 steps. This is because much more calculations were performed and the error was accumulated. In the actor systems that use clustering, the simulation accuracy is better for a longer time. However, later on the error builds up more rapidly, which may reflect the fact that a small inaccuracy in one cluster affects all others more significantly. However, the error size introduced by clustering optimizations is surprisingly small compared to the classic sequential version, which confirms the usefulness of the proposed models.

4.2 Simulation Performance

To estimate the overall performance of the proposed actor-based simulations, the average time to compute one simulation step has been compared with the classic sequential version. For versions with body grouping, the number of actors increased with the number of bodies. All experiments involved 3 computing nodes of the supercomputer, which provided 48 computing cores. The results are presented in Table 1.

Table 1. Comparison of simulation step execution time (in milliseconds) for different versions of the system

Number of bodies	100	500	1000	5000	10000
Sequential version	1.48	13.6	56.9	1089	3850
Problem-driven actor model	2.40	23.3	55.4	1888	6335
number of clusters	**2**	**5**	**10**	**25**	**50**
Computation-driven actor model	1.37	1.52	4.58	7.47	12.2
Limited-communication model	1.19	1.43	4.51	6.73	10.7

The quadratic increase in the time needed to perform the simulation step in relation to the amount of data for the sequential and problem-driven versions has been expected. It is worth noticing that for 10,000 objects, each step takes 3.85 and 6.34 s, respectively, which makes the approach very inefficient for larger problems. The overhead of using large numbers of actors is also clearly visible – intensive context switching and communication burden the efficiency significantly.

The other two versions, in which the bodies' groups were processed by the actors, behave apparently very well. The time of the simulation step for a small problem reaches about 10ms for both versions and does not change significantly in relation to the change in the size of the problem. This result shows how much the optimization related to the division of bodies into groups and the parallelization of calculations for each of them affects the simulation execution time.

4.3 Strong Scalability

In the strong scalability testing, the problem remains constant while computational resources increase. Only the versions with body grouping were tested with the scenario of 250000 bodies. The bodies were divided into 50 to 1250 groups, each processed by a dedicated actor. Hardware resources allowed each actor to work on a separate computing core. The results are presented in Fig. 2.

The execution time of the simulation step decreases as expected, especially with fewer actors used. This shows that sharing the bodies among more processes effectively reduces the time needed to perform the simulation.

For a given size of the problem, the time decreases to about 750 actors. Further fragmentation of tasks does not improve performance and effectively leads to waste of resources. Here, the time reduction associated with fewer objects per actor is less than the time increase associated with having to process more messages.

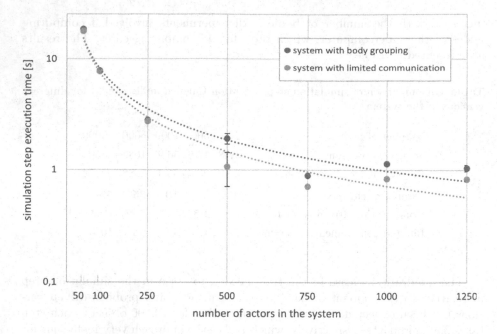

Fig. 2. Strong scalability test results for 250 000 bodies

4.4 Weak Scalability

The weak scalability tests aim to evaluate the system's ability to process growing tasks with growing resources. Typically, the task for each computing core has a constant size in such tests. The task of 100 bodies was defined and tested for 10–2000 actors. The resulting simulation step time is presented in Fig. 3.

The number of 100 bodies per actor has been deliberately selected to demonstrate the significant problem with the scalability of the approach. The computational task performed in each of the actors is constant; however, the number or volume of the messages sent and received grows with the number of actors in the system. This highly undesirable situation, which is forced by the simulation model, is the reason for the growth of the simulation step-time.

The conclusion of the experiments would be to use a proper balance between the number of actors and the number of bodies assigned to each actor. This result is consistent with the theoretical analysis of complexity presented earlier.

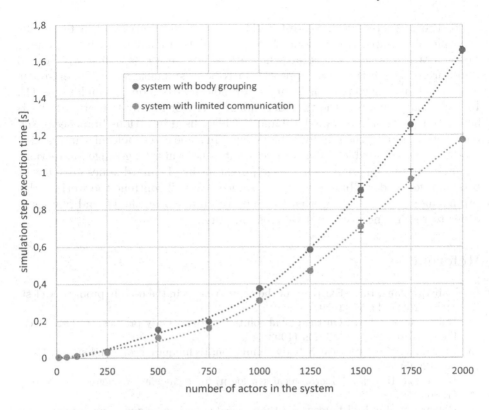

Fig. 3. Strong scalability test results for 100 bodies per actor

5 Conclusion

The proposed method for the simulation of the N-body problem is promising, as we have obtained good quality results, in particular when the bodies were clustered. Moreover, acceptance of a certain delay in delivering the data on the influence of the farther clusters can also be treated as a good direction for further study. Actually using the actor model for concurrent implementation, easy generalization into distributed infrastructure and the "desynchronization" of the communication (cf. [21]) can be treated as the most important aspects of this paper.

As a visible flaw, the results obtained for large number of actors computing small tasks (e.g. 100 bodies) did not turn out to be an efficient case, so these kinds of simulation cannot be sped-up using HPC in such easy way. Further research in this area is needed, leading towards a scale-invariant actor. Its computational task should be constant or depend linearly on the size of the problem fraction assigned to the actor. Thus, improvements to communication methods and perhaps deeper exploration of the desynchronized exchange of messages can lead us further into this interesting research topic.

In the presented work we tested the actor-based approach on a toy-model example of a simple gravitational N-body simulation (that is, with long range-forces). At this stage – as the present work shows – it seems that it will be also appropriate to use the actor approach in N-body simulations with short-range forces such as considered, for example, in molecular dynamics simulations [11]. In molecular dynamics, the time-dependent behavior of a system of particles is simulated at the microscopic level and is widely used in various branches of science and technology, especially in biophysics. In some cases molecular interaction models are used in which inter-atomic or inter-molecular long-range interactions can be neglected and only short-range potentials like Lennard-Jones force law or torsional forces describing bonds, etc. suffice. We will continue the work in this promising area, putting together the expertise in-between the IT and physics in order to reach synergistic and valuable output.

References

1. Babadzanjanz, L.K.: Existence of the continuations in the n-body problem. Celest. Mech. **20**(1), 43–57 (1979)
2. Babadzanjanz, L.K.: On the global solution of the n-body problem. Celest. Mech. Dyn. Astron. **56**(3), 427–449 (1993)
3. Bagla, J.S.: Cosmological N-body simulation: techniques, scope and status. Curr. Sci. (2005)
4. Barnes, J., Hut, P.: A hierarchical o(n log n) force-calculations algorithm. Nature (1986)
5. Bottaro, S., Lindorff-Larsen, K.: Biophysical experiments and biomolecular simulations: a perfect match? Science **361**(6400), 355–360 (2018)
6. Capuzzo-Dolcetta, R., Spera, M., Punzo, D.: A fully parallel, high precision, n-body code running on hybrid computing platforms. Journal of Computational Physics (2012)
7. Carrier, J., Greengard, L., Rokhlin, V.: A fast adaptive multipole algorithm for particle simulations. SIAM J. Sci. Stat. Comput. **9**(4), 669–686 (1988)
8. Chenciner, A., Montgomery, R.: A remarkable periodic solution of the three-body problem in the case of equal masses. Ann. Math. **152**(3), 881–901 (2000)
9. Ellis, R., Perelson, A., Weisse-Bernstein, N.: The computation of the gravitational influences in an n-body system. International Amateur-Professional Photoelectric Photometry Communication, 75 (1999)
10. Faber, L., Pietak, K., Byrski, A., Kisiel-Dorohinicki, M.: Agent-based simulation in age framework. In: Byrski, A., Oplatková, Z., Carvalho, M., Kisiel-Dorohinicki, M. (eds) Advances in Intelligent Modelling and Simulation. SCI, vol. 416, pp. 55–83. Springer, Heidelberg (2012). https://doi.org/10.1007/978-3-642-28888-3_3
11. Hospital, A., Goñi, J.R., Orozco, M., Gelpí, J.L.: Molecular dynamics simulations: advances and applications. Advances and applications in bioinformatics and chemistry, pp, 37–47 (2015)
12. Hut, P., Makino, J.: Moving stars around. The Art of Computational Science (2007)
13. Ito, T., Makino, J., Ebisuzaki, T., Sugimoto, D.: A special-purpose n-body machine grape-1. Computer Physics Communications (1990)
14. Kawai, A., Fukushige, T., Taiji, J.M.M.: Grape-5: a special-purpose computer for n-body simulations. Publications of the Astronomical Society of Japan (2000)

15. Liverts, E., Griv, E., Gedalin, M., Eichler, D.: Dynamical evolution of galaxies: supercomputer N-body simulations. In: Contopoulos, G., Voglis, N. (eds.) Galaxies and Chaos. Lecture Notes in Physics, vol. 626, pp. 340–347. Springer, Heidelberg (2003). https://doi.org/10.1007/978-3-540-45040-5_28

16. Qiu-Dong, W.: The global solution of the n-body problem. Celest. Mech. Dyn. Astron. **50**(1), 73–88 (1990)

17. Roestenburg, R., Williams, R., Bakker, R.: Akka in action. Simon and Schuster (2016)

18. Schaefer, R., Byrski, A., Kolodziej, J., Smolka, M.: An agent-based model of hierarchic genetic search. Comput. Math. Appl. **64**(12), 3763–3776 (2012)

19. Sundman, K.F.: Mémoire sur le problème des trois corps. Acta Mathematica **36**(none), 105–179 (1913)

20. Totoo, P., Loidl, H.-W.: Parallel haskell implementations of the n-body problem. Wiley InterScience (2013)

21. Turek, W.: Erlang-based desynchronized urban traffic simulation for high-performance computing systems. Future Gener. Comput. Syst. **79**, 645–652 (2018)

22. Vanderbei, R.J.: New orbits for the n-body problem. Ann. N. Y. Acad. Sci. **1017**(1), 422–433 (2004)

23. Verlet, L.: Computer "experiments" on classical fluids. i. thermodynamical properties of lennard-jones molecules. Phys. Rev. **159**, 98–103 (1967)

24. Wang, L., et al.: Nbody6++gpu: ready for the gravitational million-body problem. Monthly Notices of the Royal Astronomical Society (2015)

25. Wu, T.: An application of n-body simulation to the rotational motion of solar system bodies. Master's thesis, Miami University (2008)

Intracellular Material Transport Simulation in Neurons Using Isogeometric Analysis and Deep Learning

Angran Li[ID] and Yongjie Jessica Zhang[✉][ID]

Carnegie Mellon University, Pittsburgh, PA 15213, USA
{angranl,jessicaz}@andrew.cmu.edu

Abstract. The intracellular material transport plays a crucial role in supporting a neuron cell's survival and function. The disruption of transport may lead to the onset of various neurodegenerative diseases. Therefore, it is essential to study how neurons regulate the material transport process and have a better understanding of the traffic jam formation. Here, we present to model the neuron material transport process and study the traffic jam phenomena during transport using isogeometric analysis (IGA) and deep learning. We first develop an IGA-based platform for material transport simulation in complex neuron morphologies. A graph neural network (GNN)-based deep learning model is then proposed to learn from the IGA simulation and provide fast material concentration prediction within different neuron morphologies. To study the traffic jam phenomena, we develop a PDE-constrained optimization model to simulate the material transport regulation within neuron and explain the traffic jam caused by reduced number of microtubules (MTs) and MT swirls. A novel IGA-based physics-informed graph neural network (PGNN) is proposed to quickly predict normal and abnormal transport phenomena such as traffic jam in different neuron morphologies. The proposed methods help in discovering several spatial patterns of the transport process and provide key insights into how neurons mediate the material transport process within their complex morphology.

Keywords: Neuron material transport · Isogeometric analysis · Deep learning · Graph neural network · PDE-constrained optimization

1 Introduction

Neurons exhibit striking complexity and diversity in their morphology, which is essential for neuronal functions and biochemical signal transmission. However, it also brings challenges to mediate intracellular material transport since most essential materials for neurons have to experience long-distance transport along axons and dendrites after synthesis in the cell body. In particular, the neuron relies heavily on molecular motors for the fast transport of various materials along the cytoskeletal structure like microtubules (MTs). The disruption of this long-distance transport can induce neurological and neurodegenerative diseases

© The Author(s), under exclusive license to Springer Nature Switzerland AG 2023
J. Mikyška et al. (Eds.): ICCS 2023, LNCS 14074, pp. 486–493, 2023.
https://doi.org/10.1007/978-3-031-36021-3_49

like Huntington's, Parkinson's, and Alzheimer's disease. Therefore, it is essential to study the intracellular transport process in neurons. There have been several mathematical models proposed to simulate and explain certain phenomena during transport but limited to simple 1D or 2D domains without considering the complex neuron morphology [4,7,18]. Here, we present to simulate the intracellular material transport within complex neuron morphologies using isogeometric analysis (IGA) and deep learning (DL). IGA was proposed based on the conventional finite element method (FEM) to directly integrate geometric modeling with numerical simulation [6]. By using the same smooth spline basis functions for both geometrical modeling and numerical solution, IGA can accurately represent a wide range of complex geometry with high-order continuity while offering superior performance over FEM in numerical accuracy and robustness. Therefore, IGA has been successfully applied in shell analysis [2,19], cardiovascular modeling [23,25], neuroscience simulation [15], as well as industrial applications [21,22]. With the advances in IGA, we first develop an IGA-based simulation platform to reconstruct complex 3D neuron geometries and obtain high-fidelity velocity and concentration results during material transport. Though IGA can accurately solve PDEs in complex neuron morphologies, the high computational cost of 3D simulations may limit its application in biomedical field where fast feedback from simulation is necessary. In recent years, DL has been proven successful in solving high-dimensional PDEs [5] and learning the physics behind PDE models [17]. DL also becomes popular in building surrogate model for PDEs since it can provide efficient prediction for complex phenomena [3,9]. To address the limitation of our IGA platform, we develop a DL-based surrogate model based on the simulation platform to improve its computational efficiency. We also propose to solve a PDE-constrained optimization (PDE-CO) problem to study the transport control mechanisms and explain the traffic jam phenomenon within abnormal neurons. We then develop a novel IGA-based physics-informed graph neural network (PGNN) that learns from the PDE-CO transport model and effectively predicts complex normal and abnormal material transport phenomena such as MT-induced traffic jams. Our results provide key insights into how material transport in neurons is mediated by their complex morphology and MT distribution, and help to understand the formation of complex traffic jam.

2   Methodology and Results

We first develop an IGA-based simulation platform for modeling the intracellular material transport within the complex morphology of neurons. The platform consists of two modules: geometric modeling and an IGA solver. In the geometric modeling module, we apply a skeleton-based sweeping method [24–26] to generate all-hexahedral control mesh for the complex neuron morphology. We then construct truncated hierarchical tricubic B-splines (THB-spline3D) [20] on the control mesh to represent the geometry for IGA. Regarding the IGA solver module, we simulate the transport process by generalizing the motor-assisted transport model to 3D and couple the model with Navier-Stokes equations to obtain the accurate velocity field in neurons. Using our IGA solver, we simulate material transport in the complex neuron morphologies from NeuroMorpho.Org [1] and one example is shown in Fig. 1. Our simulation reveals that the geometry of neurons plays an important role in the routing of material transport at junctions of neuron branches and in distributing the transported materials throughout the networks. It provides key insights into how material transport in neurons is mediated by their complex morphology. Our IGA solver can also be extended to solve other PDE models of cellular processes in the neuron morphology. More information about this work can be found in [8].

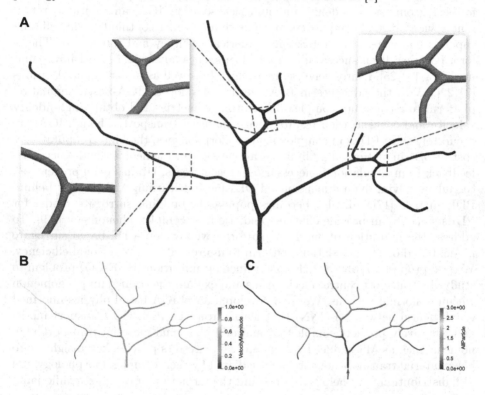

Fig. 1. Simulation of material transport in the neuron morphology of NMO_134036. (A) Hexahedral control mesh of the network with zoom-in details. (B) Velocity field. (C) Concentration distribution of transport materials at 10 s, and the red arrow points to the inlet of material. Unit for color bars: (B) $\mu m/s$ and (C) $mol/\mu m^3$.

Though we can obtain high-fidelity simulation results from the IGA solver, its expensive computational cost limits its application in the biomedical field which needs fast feedback from the simulation. To address this issue, we then develop a DL-based surrogate model to learn from IGA simulation data and provide fast transport prediction in any complex neuron morphology [10]. Instead of using the standard convolutional neural network (CNN) [9], we employ graph neural network (GNN) to tackle the extensive unstructured neuron topologies. Given any neuron geometry, we build a graph representation of the neuron by decomposing the geometry into two basic structures: pipe and bifurcation. We train different GNN simulators for these two basic structures to take simulation parameters and boundary conditions as input and output the spatiotemporal concentration distribution. The residual terms from PDEs are used in training to instruct the model to learn the physics behind simulation data. To reconstruct the original neuron geometry, we train another GNN-based assembly model to connect all the pipes and bifurcations following the graph representation. The well-trained GNN model can predict the dynamical concentration change during the transport process with an average error less than 10% and 120 ∼ 330 times faster compared to IGA simulations. The performance of the proposed method is demonstrated on several 3D neuron trees and one testing example is shown in Fig. 2. The interested reader is referred to [10] for more information of this work.

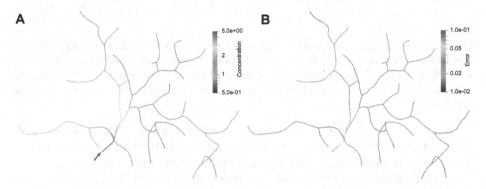

Fig. 2. The concentration prediction of material transport in the neuron morphology of NMO_06846. (A) The predicted concentration result of steady state at $t = 15\ s$. The red arrow points to the inlet of material. Unit for color bars: $mol/\mu m^3$. (B) The nodal error between the ground truth (IGA simulation result) and predicted concentration. Logarithmic scale is used to highlight the distribution pattern. (Color figure online)

In addition, we improve the motor-assisted transport model by considering the active transport control from neuron. We present to solve a novel IGA-based PDE-CO problem that effectively simulates the material transport regulation and investigates the formation of traffic jams and swirls during the transport process in complex neuron structures [11,12]. In particular, we design a new objective function to simulate two transport control mechanisms for (1) mediating the transport velocity field; and (2) avoiding the traffic jam caused by local material accumulation. The control strength can be adjusted through two penalty parameters in the objective function and the impact of these parameters is also studied. We also introduce new simulation parameters to describe the spatial distribution of MTs, which can be used to simulate traffic jams caused by abnormal MTs. In Fig. 3, we present the traffic jam simulation caused by the reduction of MTs in a single pipe geometry. The MT distribution is reduced in the red dashed rectangle region, which leads to a decrease of velocity and the accumulation of material in this local area. The proposed IGA optimization framework is transformative and can be extended to solve other PDE-CO models of cellular processes in complex neurite networks. See more information about this work in [11,12]. Our simulation reveals that the molecular motors and MT structure play fundamental roles in controlling the delivery of material by mediating the transport velocity on MTs.

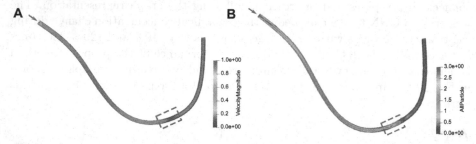

Fig. 3. Traffic jam simulation in a single pipe geometry extracted from NMO_06840. The traffic jam is introduced by reducing MTs in the red dashed rectangle region. (A) Velocity field. (B) Concentration distribution. The red arrows point to the inlet of material. Color bars unit for velocity field: $\mu m/s$ and concentration: $mol/\mu m^2$. (Color figure online)

Based on the PDE-CO model, we develop a novel IGA-based physics-informed graph neural network (PGNN) to quickly predict normal and abnormal transport phenomena such as traffic jam in different neuron geometries. The proposed method learns from the IGA simulation of the intracellular transport process and provides accurate material concentration prediction of normal transport and MT-induced traffic jam. The IGA-based PGNN model contains simulators to handle local prediction of both normal and two MT-induced traffic jams in pipes, as well as another simulator to predict normal transport in bifurcations. Bézier extraction is adopted to incorporate the geometry information into the simulators to accurately compute the physics-informed loss function with PDE residuals. The well-trained model effectively predicts the distribution of transport velocity and material concentration during traffic jam and normal transport with an average error of less than 10% compared to IGA simulations. Using our IGA-based PGNN model, we study abnormal transport processes in different geometries and discover several spatial patterns of the transport process. In Fig. 4, we present the traffic jam predictions caused by the reduction of MTs in a 2D neuron tree. Compared to the IGA simulation results, the sudden decrease of velocity and increase of concentration in the traffic jam region (red dashed circle region) are accurately captured by the PGNN model. The PGNN model is also employed to study traffic jam caused by MT swirls in 3D neuron geometries and successfully captures the unique spatial patterns of transport velocities during traffic jam, such as vortex and reversing streamlines, which explains how the non-uniform MT distributions affect the transport velocity and hinder the smooth delivery of the material. The interested reader is referred to [13] for more information of this work.

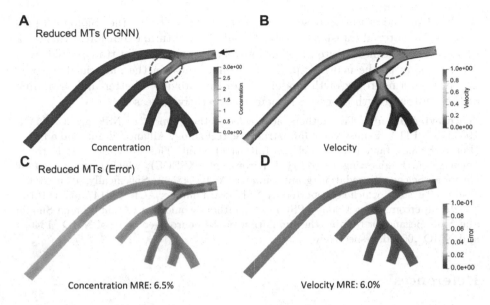

Fig. 4. The comparison between IGA simulation and PGNN prediction of the traffic jam caused by reduced MTs in a 2D neuron tree extracted from NMO_54504. (A, B) The predicted (A) concentration distribution and (B) velocity of the traffic jam. Black arrow points to the material inlet and the red dashed circle labels the traffic jam region. (C, D) The nodal errors of (C) concentration and (D) velocity between IGA and PGNN results. Unit for color bars: Velocity: $\mu m/s$ and Concentration: $mol/\mu m^3$. (Color figure online)

3   Conclusion

In summary, we study the intracellular material transport within complex neuron morphologies using IGA, DL, and PDE-constrained optimization. Our developed computational packages utilize high-performance computing clusters and provide high-fidelity velocity and concentration results within complex neuron morphologies. These results provide references for the comparison with actual transport in real neurons to further answer the question of how neurons deliver the right material to the right destination in a balanced fashion with their complex neurite networks and how the transport may be affected by disease conditions. In the future, there are several interesting directions to extend this work. The current PDE-CO model only considers the influence of traffic jams on the material concentration but neglects its effect on the deformation of neuron geometries. To address this limitation, we can couple the transport model with a structural model and solve a fluid-structure interaction problem to simulate the geometry deformation during traffic jam. It is also necessary to verify our mathematical models by designing comparable biological experiments. For instance, the photoactivation technique can be used to visualize the material transport process and extract the velocity or concentration distribution to compare with simulation results. Employing the proposed transport model to study other biological processes with material transport involved would be a natural extension of this work as well. For instance, neuron elongation relies on the tubulin transported from the cell body to the neurite tip. The coupling between the material transport model with a neuron growth model [14, 16] can provide a better understanding of the neuron growth process and neurodegenerative diseases.

Acknowledgments. The authors acknowledge the support of NSF grant CMMI-1953323, a PITA (Pennsylvania Infrastructure Technology Alliance) grant, and a PMFI (Pennsylvania Manufacturing Fellows Initiative) grant. This work used the Extreme Science and Engineering Discovery Environment (XSEDE), which is supported by National Science Foundation grant number ACI-1548562. Specifically, it used the Bridges-2 system, which is supported by NSF award number ACI-1928147, at the Pittsburgh Supercomputing Center (PSC). The authors would like to thank Faqia Shahid and Sara Begane for running the simulation in the neuron geometry of NMO_134036 and NMO_06846, respectively.

References

1. Ascoli, G.A., Donohue, D.E., Halavi, M.: Neuromorpho.org: a central resource for neuronal morphologies. J. Neurosci. **27**(35), 9247–9251 (2007)
2. Casquero, H., et al.: Seamless integration of design and Kirchhoff-Love shell analysis using analysis-suitable unstructured T-splines. Comput. Methods Appl. Mech. Eng. **360**, 112765 (2020)
3. Farimani, A.B., Gomes, J., Pande, V.S.: Deep learning the physics of transport phenomena. arXiv Preprint:1709.02432 (2017)
4. Friedman, A., Craciun, G.: A model of intracellular transport of particles in an axon. J. Math. Biol. **51**(2), 217–246 (2005)
5. Han, J., Jentzen, A., Weinan, E.: Solving high-dimensional partial differential equations using deep learning. Proc. National Acad. Sci. **115**(34), 8505–8510 (2018)
6. Hughes, T., Cottrell, J., Bazilevs, Y.: Isogeometric analysis: CAD, finite elements, NURBS, exact geometry and mesh refinement. Comput. Methods Appl. Mech. Eng. **194**(39), 4135–4195 (2005)
7. Kuznetsov, A., Avramenko, A.: A macroscopic model of traffic jams in axons. Math. Biosci. **218**(2), 142–152 (2009)
8. Li, A., Chai, X., Yang, G., Zhang, Y.J.: An isogeometric analysis computational platform for material transport simulation in complex neurite networks. Molecular Cellular Biomech. **16**(2), 123–140 (2019)
9. Li, A., Chen, R., Farimani, A.B., Zhang, Y.J.: Reaction diffusion system prediction based on convolutional neural network. Sci. Rep. **10**(1), 1–9 (2020)
10. Li, A., Farimani, A.B., Zhang, Y.J.: Deep learning of material transport in complex neurite networks. Sci. Rep. **11**(1), 1–13 (2021)
11. Li, A., Zhang, Y.J.: Modeling intracellular transport and traffic jam in 3D neurons using PDE-constrained optimization. J. Mech. **38**, 44–59 (2022)
12. Li, A., Zhang, Y.J.: Modeling material transport regulation and traffic jam in neurons using PDE-constrained optimization. Sci. Rep. **12**(1), 1–13 (2022)
13. Li, A., Zhang, Y.J.: Isogeometric analysis-based physics-informed graph neural network for studying traffic jam in neurons. Comput. Methods Appl. Mech. Eng. **403**, 115757 (2023)

14. Liao, A.S., Cui, W., Zhang, Y.J., Webster-Wood, V.A.: Semi-automated quantitative evaluation of neuron developmental morphology in vitro using the change-point test. Neuroinformatics, 1–14 (2022)
15. Pawar, A., Zhang, Y.J.: NeuronSeg_BACH: automated neuron segmentation using B-Spline based active contour and hyperelastic regularization. Commun. Comput. Phys. **28**(3), 1219–1244 (2020)
16. Qian, K., et al.: Modeling neuron growth using isogeometric collocation based phase field method. Sci. Rep. **12**(1), 8120 (2022)
17. Raissi, M., Perdikaris, P., Karniadakis, G.E.: Physics-informed neural networks: a deep learning framework for solving forward and inverse problems involving nonlinear partial differential equations. J. Comput. Phys. **378**, 686–707 (2019)
18. Smith, D., Simmons, R.: Models of motor-assisted transport of intracellular particles. Biophys. J . **80**(1), 45–68 (2001)
19. Wei, X., Li, X., Qian, K., Hughes, T.J., Zhang, Y.J., Casquero, H.: Analysis-suitable unstructured T-splines: multiple extraordinary points per face. Comput. Methods Appl. Mech. Eng. **391**, 114494 (2022)
20. Wei, X., Zhang, Y.J., Hughes, T.J.: Truncated hierarchical tricubic C^0 spline construction on unstructured hexahedral meshes for isogeometric analysis applications. Comput. Math. Appl. **74**(9), 2203–2220 (2017)
21. Yu, Y., Liu, J.G., Zhang, Y.J.: HexDom: polycube-based hexahedral-dominant mesh generation. In: Rebén Sevilla, Simona Perotto, K.M. (ed.) The Edited Volume of Mesh Generation and Adaptation: Cutting-Edge Techniques for the 60th Birthday of Oubay Hassan. SEMA-SIMAI Springer Series. Springer (2021). https://doi.org/10.1007/978-3-030-92540-6_7
22. Yu, Y., Wei, X., Li, A., Liu, J.G., He, J., Zhang, Y.J.: HexGen and Hex2Spline: polycube-based hexahedral mesh generation and spline modeling for isogeometric analysis applications in LS-DYNA. In: Springer INdAM Serie: Proceedings of INdAM Workshop "Geometric Challenges in Isogeometric Analysis". Springer (2020). https://doi.org/10.1007/978-3-030-92313-6_14
23. Yu, Y., Zhang, Y.J., Takizawa, K., Tezduyar, T.E., Sasaki, T.: Anatomically realistic lumen motion representation in patient-specific space-time isogeometric flow analysis of coronary arteries with time-dependent medical-image data. Comput. Mech. **65**(2), 395–404 (2020)
24. Zhang, Y.: Challenges and advances in image-based geometric modeling and mesh generation. In: Zhang, Y. (ed.) Image-Based Geometric Modeling and Mesh Generation, pp. 1–10. Springer (2013). https://doi.org/10.1007/978-94-007-4255-0_1
25. Zhang, Y., Bazilevs, Y., Goswami, S., Bajaj, C.L., Hughes, T.J.: Patient-specific vascular NURBS modeling for isogeometric analysis of blood flow. Comput. Methods Appl. Mech. Eng. **196**(29–30), 2943–2959 (2007)
26. Zhang, Y.J.: Geometric Modeling and Mesh Generation from Scanned Images, vol. 6. CRC Press (2016)

Chemical Mixing Simulations
with Integrated AI Accelerator

Krzysztof Rojek[1]([✉])[ID], Roman Wyrzykowski[1][ID], and Pawel Gepner[2][ID]

[1] Institute of Computer and Information Sciences,
Częstochowa University of Technology, Częstochowa, Poland
krojek@icis.pcz.pl
[2] Faculty of Mechanical and Industrial Engineering,
Warsaw University of Technology, Warsaw, Poland
https://pcz.pl,https://www.pw.edu.pl

Abstract. In this work, we develop a method for integrating an AI model with a CFD solver to predict chemical mixing simulations' output. The proposed AI model is based on a deep neural network with a variational autoencoder that is managed by our AI supervisor. We demonstrate that the developed method allows us to accurately accelerate the steady-state simulations of chemical reactions performed with the MixIT solver from Tridiagonal solutions.

In this paper, we investigate the accuracy and performance of AI-accelerated simulations, considering three different scenarios: i) prediction in cases with the same geometry of mesh as used during training the model, ii) with a modified geometry of tube in which the ingredients are mixed, iii) with a modified geometry of impeller used to mix the ingredients.

Our AI model is trained on a dataset containing 1500 samples of simulated scenarios and can accurately predict the process of chemical mixing under various conditions. We demonstrate that the proposed method achieves accuracy exceeding 90% and reduces the execution time up to 9 times.

Keywords: CFD · chemical mixing · artificial intelligence · machine learning · DNN · HPC

1 Introduction

Artificial intelligence (AI) and machine learning (ML) are rapidly growing fields revolutionizing many areas of science and technology [15], including high-performance computing (HPC) simulations. HPC simulations involve using supercomputers and other advanced computing systems to perform complex calculations and simulations in physics, engineering, biology, etc. AI techniques can be efficiently used by enabling computing platforms to learn from datasets and make accurate extrapolations of simulations to reduce the execution time of intensive solver computations significantly. In particular, machine learning algorithms can be used to analyze and interpret simulation data, identify patterns and trends, and predict outcomes. AI can also be used to optimize simulation parameters and directly predict the results. By leveraging the power of AI/ML, HPC simulations can help researchers and scientists gain new insights and make more informed decisions in a wide range of fields. In this paper, we develop a method for integrating a proposed AI model with the MixIT tool based on the OpenFOAM (Open Field Operation and Manipulation) [9] solver to accelerate chemical mixing simulations.

Chemical mixing simulations are computer-based models used to predict chemical mixtures' behavior under various conditions [15]. These simulations can be used in a variety of industries, including pharmaceuticals, petrochemicals, and food and beverages, to help optimize the production of chemical products and reduce the cost and environmental impact of manufacturing. Overall, chemical mixing simulations are a powerful tool that can help improve chemical manufacturing processes' efficiency while reducing the cost and environmental impact of these processes.

OpenFOAM [9] is a widely used open-source software platform for simulating and analyzing fluid flow and heat transfer. It has been used in a variety of industries, including aerospace, automotive, and chemical processing. One of the main features of OpenFOAM is its highly modular and flexible design, which allows users to easily customize and extend the software to meet their specific needs. OpenFOAM includes a range of solvers and libraries for simulating different types of fluid flow, as well as tools for meshing, visualization, and post-processing of simulation results.

In this paper, we extend the method proposed in our previous papers [14,15] and explore new techniques and models. The previously proposed method has limitations that restrict its applicability. It can not handle modified geometries of the simulated phenomenon with the required accuracy, being able to solve problems within the close family of scenarios used during training. To overcome this limitation, in this work, we develop new methods which enable us to handle a broader range of scenarios. The contributions of our paper are outlined below:

- We develop a method for integrating the AI model with the computational fluid dynamics (CFD) solver that leverages the power of machine learning to improve the accuracy and efficiency of fluid flow predictions.
- We propose an AI model based on the variational autoencoder (VAE) architecture to predict key quantities in CFD simulations, including pressure and velocity, as well as an ML algorithm detecting the steady state and making a decision to stop the simulation.
- The efficiency of our method is demonstrated through a series of experiments, comparing the results of our AI-accelerated simulations to those obtained using traditional CFD methods.
- It is shown that our AI-accelerated simulations can produce accurate results with different tube and impeller geometries in simulated phenomena.
- The performance and accuracy of the proposed solution are investigated for 3D cases with meshes exceeding 1 million cells.

2   Related Work

A significant amount of research has focused on the use of AI techniques in CFD simulations in recent years. These techniques can potentially improve the efficiency and accuracy of CFD simulations and enable the simulation of more complex and realistic problems [2,24].

A common approach is to employ machine learning algorithms to model and predict the behavior of fluids. For example, neural networks were used to predict the aeroelastic response of the coupled system [20], or approximate computational fluid dynamics for modeling turbulent flows [4].

Other researchers have explored using AI techniques to optimize the parameters and settings of CFD simulations [11]. Genetic algorithms and other optimization methods were used to identify the optimal mesh size and solver settings for a given simulation and to adjust these parameters based on the simulation results automatically [1].

The use of AI techniques to analyze and interpret the results of CFD simulations was also a subject of research [27]. For example, clustering algorithms were exploited to group similar flow patterns [16], and classification algorithms were used to identify and classify different types of flow.

There are several ways in which AI techniques are incorporated into HPC simulations. Some common approaches include:

- Machine learning-based models: Machine learning algorithms can be used to model and predict the behavior of systems being simulated by HPC systems [2,14,26]. These models are trained on large amounts of data and used to make accurate and efficient predictions [7], which can be incorporated into HPC simulation.
- Optimization of simulation parameters: AI techniques, such as deep learning and machine learning, are used to optimize the parameters and settings of HPC simulations [13,25]. These techniques can search through a large space of possible parameter values and identify the optimal configurations for a given simulation.
- Data analysis and approximation: AI techniques, such as interpolation with deep learning algorithms, can be used to create generative models able to accurately approximate the training data set of HPC simulations [3]. These techniques identify patterns and trends in the data and provide insights that may not be immediately apparent from the raw data.
- Real-time control: In some cases, AI techniques are used to control HPC simulations in real-time [5]. For example, an AI system is used to adjust the parameters of a simulation based on the current state of the system being simulated.

There has been a growing interest in using AI techniques in weather forecasting simulations in recent years [23]. These techniques can potentially improve the accuracy and reliability of weather forecasts and enable the simulation of more complex and realistic weather scenarios. One common approach is to use machine learning algorithms to model and predict the atmosphere's behavior.

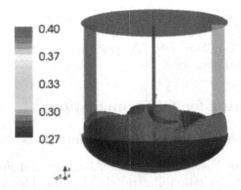

Fig. 1. Tube equipped with a single impeller during the simulation of the magnitude of U [m/s].

For example, neural networks were exploited to model and forecast the evolution of weather systems, such as storms and hurricanes [19]. Other researchers used machine learning techniques to predict the likelihood of extreme weather events, such as floods and droughts. AI techniques have also been used to optimize the parameters and settings of weather forecasting simulations [18]. For example, genetic algorithms and other optimization methods were employed to identify the optimal initial conditions and model configurations for a given simulation [12].

Overall, integrating AI with HPC simulations can greatly improve these simulations' efficiency and accuracy and enable them for more complex and realistic systems.

3   Chemical Mixing Simulations

One of the main benefits of chemical mixing simulations is that they can be used to test different scenarios and variables without the need for expensive and time-consuming physical experiments. This allows researchers and engineers to quickly evaluate the effects of different factors on the resulting mixture, including temperature, pressure, and mixing speed.

Our simulations are performed with a MixIT tool [6], which is based on the OpenFOAM platform. MixIT is the next-generation collaborative mixing analysis tool designed to facilitate comprehensive stirred tank analysis using laboratory and plant data, empirical correlations, and advanced 3D CFD models.

The chemical mixing simulations are based on the standard k-epsilon model [10]. The goal is to compute the converged state of the liquid mixture in a tank equipped with a single impeller and a set of baffles. Based on different settings of the input parameters, we simulate a set of quantities, including the velocity vector field U, pressure scalar field p, turbulent kinetic energy k of the substance, turbulent dynamic viscosity mut, and turbulent kinetic energy dissipation rate ϵ. This paper focuses on predicting the most important quantities, including U

and p. The basic geometry used to train our model is shown in Fig. 1, where we use a cylindrical tube and a single flat impeller. Here we focus on simulations with a mesh of size 1 million cells. The traditional CFD simulation used in our scenario requires 5000 iterations.

4   AI Accelerator for CFD Simulations

4.1   Basic Scheme of AI-Accelerated Simulations

Figure 2b presents the basic scheme of the AI-accelerated simulation versus the conventional non-AI simulation illustrated in Fig. 2a. This scheme includes the initial iterations computed by the CFD solver and the AI-accelerated prediction module. The CFD solver produces results sequentially, iteration by iteration. In the basic scheme, which was used in our previous works [14, 15], the results of initial iterations computed by the solver are sent as input to the AI module, which generates the final results of the simulation.

4.2   New Method of Integrating AI Prediction with a CFD Solver: AI Supervisor

This work proposes a new method for incorporating AI predictions into CFD simulations. Besides the conventional CFD solver, this method involves two other parts. The first one - AI supervisor, is designed to switch between traditional CFD simulation executed by a CFD solver and AI predictions. The second one is the AI accelerator module, responsible for AI predictions. It provides an extrapolation of the simulation to achieve results faster. The overall idea is presented in Fig. 2c.

In this scheme, a traditional CFD simulation is first executed for a specified number of iterations to generate a set of initial data points. These initial data points are then used by a machine learning model, which can accurately predict the fluid flow dynamics in subsequent iterations. Once the machine learning model generates the output, the AI supervisor invokes the traditional CFD solver to resume the simulation on the predicted data. The supervisor continues to switch between CFD and AI parts until a convergence state of the simulation is achieved. The number of iterations required to achieve convergence depends on the complexity of the simulated flow dynamics and the training data quality.

The AI supervisor recognizes the data pattern in simulation and decides if the steady state is achieved. It analyses the output of the CFD simulation and decides whether to invoke the AI accelerator or stop the simulation.

The AI supervisor uses One-Class Support Vector Machine (SVM) [8] model to detect achieving the converged state. This is a type of machine learning algorithm used for anomaly detection. Anomaly detection is the process of identifying observations that deviate significantly from most of the data points, which are considered normal or expected. This algorithm is trained by the data containing the standard deviation of differences between elements of vectors corresponding

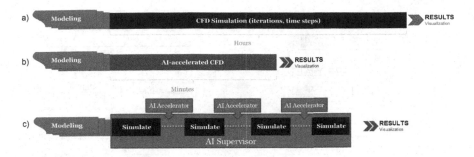

Fig. 2. Comparison of the traditional workflow of CFD simulation (a), the basic scheme of AI-accelerated simulation (b), and the proposed method of AI-accelerated simulation with AI supervisor (c).

to boundary iterations of the last 75 iterations of the CFD simulations. Such data are assumed to contain normal (expected) samples. The algorithm then attempts to build a boundary around these normal examples so that new, unseen data points that fall outside the boundary are classified as anomalies.

The supervisor will stop the simulation and return the results if the predicted output is sufficiently close to the converged state. If the predicted output is not close enough to the converged state, the supervisor will call the AI accelerator to make a prediction and then executes 100 more iterations of a CFD solver. The simulation flow is described by the Algorithm 1.

The proposed method allows us to improve the performance of the simulations compared to the traditional CFD solver while ensuring that the simulation results are reliable and consistent. Using the converged state as a stopping criterion, we can reduce the number of iterations of the simulations.

4.3 AI Model for CFD Acceleration

Our AI model for predicting iterations results is based on a variational autoencoder (VAE) architecture [17]. It is a powerful and flexible neural network architecture well-suited for CFD simulations for two main reasons.

First, the VAE architecture is highly efficient and scalable, allowing us to process large amounts of data quickly and accurately. This is critical in CFD sim-

---

Algorithm 1. Managing a CFD simulation by the AI supervisor

Require: 100 $iterations\ of\ CFD\ solver$
Ensure: $iter_{25}, iter_{50}, iter_{75}, iter_{100}$
   while $OneClassSVM(std(iter_{100} - iter_{25}))$ **do**   ▷ True if anomaly detected
     $iter_{pred} \leftarrow model.predict(iter_{25}, iter_{50}, iter_{75}, iter_{100})$ ▷ Predict steady-state
     $iter_{25}, iter_{50}, iter_{75}, iter_{100} \leftarrow CFD_Solver(iter_{pred})$ ▷ Smooth data with solver
   end while
   Return $iter_{100}$

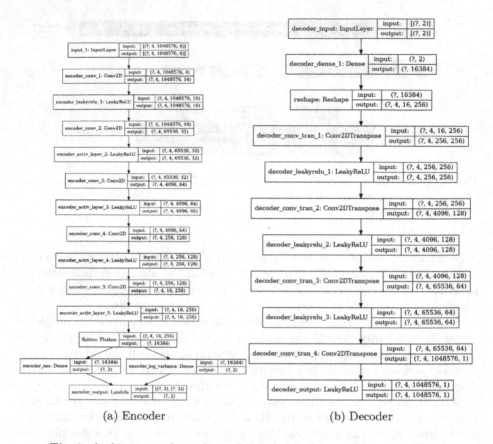

(a) Encoder   (b) Decoder

Fig. 3. Architecture of encoder and decoder models used to create VAE.

ulations, where large datasets and complex computations are commonly encountered. Second, this architecture allows us to learn the underlying structure and patterns in the data without explicit supervision, making it a powerful tool for understanding complex systems.

In our experiments, we use a machine learning pipeline based on the VAE architecture to predict key quantities in CFD simulations, including pressure and velocity. Specifically, we take four iterations of the CFD simulation as input to the model and return a single output. Inputs and outputs are represented by the quantities corresponding to 1 million cells and are used as a separate row of the input/output arrays. As a result, we have four input rows (three for a 3D velocity vector field and one for a scalar pressure field). This approach allows us to accurately predict the system's behavior over time while accounting for the complex and dynamic nature of fluid flows in CFD simulations. Figure 3 shows our encoder and decoder, while the full VAE model is presented in Fig. 4.

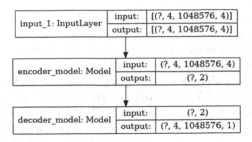

Fig. 4. Full VAE model.

4.4   Dataset

In our simulations, a 3D mesh with 1 million cells is utilized. The mesh was designed to represent the physical characteristics of the system being simulated accurately and to capture the complex fluid flow dynamics in detail.

To feed our network, we use a dataset consisting of results of 50 CFD simulations, each with a different set of parameters and input data, and each requiring 5000 iterations to achieve a steady state. To train our ML model, we create 30 samples for each simulation, resulting in 1500 samples in the dataset. Each sample includes four iterations as input and one iteration as output. Samples generated from a single simulation contain iterations $25 + 100 * i$, $50 + 100 * i$, $75 + 100 * i$, $100 + 100 * i$ (where $i = 0, 1, \ldots, 29$) as input data, and the final 5000-th iteration as output data. All the simulations have the same geometry with a single impeller and cylindrical tube. The differences between the simulations are the results of different liquid levels, rotates per minute (RPM) of the impeller, and viscosity of the mixed substance. The range of RPM varies from 110 to 450, indicating a moderate to the high-speed range. The viscosity range is from 1 to 5000 centipoise (cP) (millipascal x seconds). Finally, the liquid level contains 15 different levels that are linearly distributed across the top half of the tube.

We divided our dataset into two parts for training and validation, with 90% of the data used for training and 10% for validation. This allows us to train our model on a large and diverse set of data while still ensuring that the models can generalize well to new data.

5   Experimental Results

5.1   Testing Platform

The testing platform used to evaluate the performance and accuracy of our AI-accelerated simulations consists of the Intel Xeon Gold 6148 CPU equipped with the NVIDIA V100 GPU dedicated to training the machine learning algorithms. The CPU is used for executing a CFD solver and making AI predictions, while the GPU is dedicated to training the model. During model training and inferencing of the model, the half-precision and single-precision formats are used, respectively. Such a mixed-precision approach [22] is beneficial in our case since half-precision reduces memory usage and increases training speed, while single-precision provides more accurate results during inferencing.

The Intel Xeon Gold 6148 CPU is a server-grade processor that provides 20 cores and 40 threads, with a base clock speed of 2.4 GHz and a turbo frequency of up to 3.7 GHz. The NVIDIA V100 GPU, on the other hand, is a graphics processing unit designed specifically for machine learning and other HPC applications. It includes 5120 CUDA cores and 640 Tensor Cores, with a peak performance of up to 7.8 Tflops for double-precision calculations and 15.7 Tflops for single-precision. It also features 16 GB of HBM2 memory with a memory bandwidth of up to 900 GB/s.

In our experiments, a significant amount of memory is required to handle the large dataset and complex computations required for our simulations. Specifically, 400GB of DRAM memory is used to store and process the data used in our simulations.

5.2 Accuracy Results

Verifying accuracy includes both visualizations and numerical metrics. Visualizations such as contour plots provide qualitative verification of the flow field. We also use statistical metrics [21] such as the root-mean-square error (RMSE) and Pearson and Spearman coefficients to assess the accuracy of our predictions quantitatively.

The experiments in this section are based on three newly created cases. First, we consider the case with the same mesh geometry as used while training the model. Second, we study the case with a modified geometry of the tube in which the ingredients are mixed. Finally, a modified geometry of the impeller used to mix the ingredients is considered.

Figure 5 shows contour plots of the velocity magnitude (U). In relation to cases used during training, this case contains the same geometry of tube and impeller but a different combination of values of RPM, viscosity, and liquid level. For these experiments, we use the parameters from the middle of their respective ranges. This means that the RPM is set to 280, while the viscosity is set to 2500 cP. However, for the liquid level, we use a fully filled tube. By using these parameters, we can expect to see a wide range of fluid properties and accurately measure their behavior in this controlled environment.

Figure 6 presents the simulation results when a different-shaped tube is used. Such a rectangular geometry of the tube was not used during training. The shape of the tube significantly affects the velocity profile of the fluid.

Next, we examine the pressure (p) field in a rectangular tube with flat agitator blades (Fig. 7), which was used during training, and wider agitator blades (Fig. 8), which was not used during training. A wider blade configuration produces a different pressure distribution than a flat blade configuration due to the increased surface area of the blade that comes into contact with the fluid. Comparing the AI-accelerated CFD results for contour plots we conclude that the quality of predictions is acceptable.

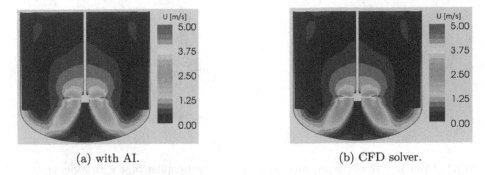

(a) with AI.   (b) CFD solver.

Fig. 5. Contour plot of the velocity magnitude (U) in the cylindrical tube.

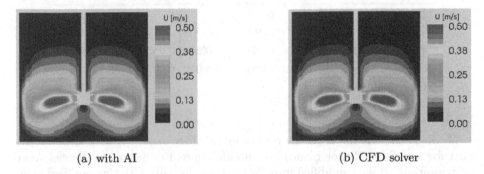

(a) with AI   (b) CFD solver

Fig. 6. Contour plot of the velocity magnitude (U) in the rectangular tube.

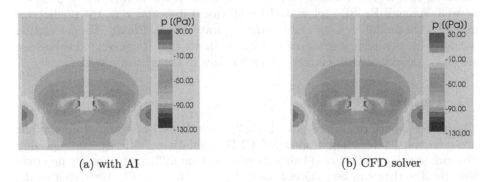

(a) with AI   (b) CFD solver

Fig. 7. Contour plot of the pressure (p) field in the rectangular tube with flat agitator blades.

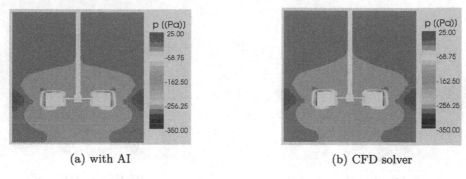

(a) with AI   (b) CFD solver

Fig. 8. Contour plot of the pressure (p) field in the rectangular tube with wider agitator blades.

Table 1. Accuracy results for different cases

Case	No modifications		Impeller modif.		Tube modif.	
Quantity	U	p	U	p	U	p
Pearson coef.	0.918	0.984	0.909	0.975	0.902	0.967
Spearman coef.	0.911	0.966	0.902	0.956	0.895	0.949
RMSE	0.012	0.003	0.012	0.003	0.012	0.003
Histogram equal. [%]	92	95	91	94	91	94

Table 1 contains the accuracy results for all three cases: (i) with the cylindrical tube and flat agitator blades (no modifications in relation to the case used for training), (ii) with modified impeller, and finally, (iii) with the modified tube. The results show a high Pearson correlation, indicating a strong linear relationship between the prediction and accurate results. A high value of the Spearman correlation indicates a strong monotonic relationship between the two variables, meaning that as one variable increases, the other tends to increase or decrease as well. A small RMSE (below 0.02 for all cases) indicates that the predictions made by the model are very close to the accurate values. Finally, the correlation between the two histograms exceeds 90%. It shows that the predicted values are strongly associated with corresponding changes in the accurate values.

5.3   Performance Results

The achieved performance results (Table 2) show that AI-accelerated simulations are generally faster than traditional CFD simulations, although the speedup depends on the specific case being simulated. It is difficult to predict precisely the speedup that can be achieved, as it depends on several factors, such as the complexity of the flow dynamics, the amount and quality of available training data, and the performance of the machine learning algorithms. In some cases, the speedup is up to about 10 times, while in others, it is about 2.4 times. The number of iterations executed by the CFD solver is reduced by 90% for the case with no modifications in the geometry of the mesh, by 76% for the case with the impeller modified, and by 57% for the case with the tube modified. The differences in performance between various cases are illustrated in Fig. 9.

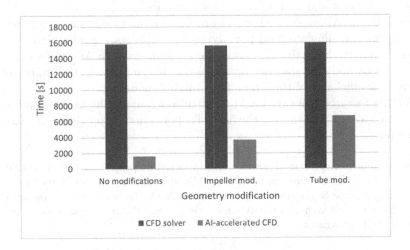

Fig. 9. Execution time comparison between different cases.

The first case with the same geometry as during training is more accelerated than others. Changing the geometry has a significant impact on the speedup of the proposed AI module. This is because the AI model can better learn and make predictions based on patterns that it has encountered before, so simulations that are similar to the ones used during training can be more accurately and faster performed. On the other side, harder-to-predict cases are less accelerated due to the AI supervisor, which executes more iterations to achieve a converged state than in the first case.

Table 2. Performance results for different cases

Case	CFD solver [s]	AI-acc. sim. [s]	Speedup	Sim. reduction [%]
No modifications	15851	1615	9.81	90
Impeller modified	15634	3692	4.23	76
Tube modified	15984	6713	2.38	57

6   Conclusion

The proposed machine learning model for predicting chemical mixing simulations demonstrates the potential for AI techniques to improve the performance and keep the accuracy of these simulations. The proposed model shows accuracy exceeding 90% and allows us to achieve a speedup of up to nine times compared to a traditional CFD solver. The proposed supervised learning techniques used for the model's training can accurately predict the evolution of chemical reactions in the simulations and generalize to new scenarios.

Overall, our AI-accelerated algorithm permits us to significantly reduce the time required to simulate fluid flow dynamics while maintaining high accuracy. By combining traditional CFD simulations' strengths with AI-accelerated simulations' speed and efficiency, we optimize the design of industrial processes and improve the efficiency of the whole fluid mixing modeling.

Further research will explore AI's potential in simulating more complex mixing scenarios and fine-tuning the model's performance on different hardware platforms.

Acknowledgements. This research was supported by project no. 020/RID/2018 /19 financed within the program of the Polish Ministry of Science and Higher Education "Regional Initiative of Excellence" (years 2019 - 2023, the amount of financing 12 000 000 PLN). This research was partly supported by the PLGrid infrastructure at ACK Cyfronet AGH in Krakow.

References

1. Huang, K., Krügener, M., Brown, A., Menhorn, F., Bungartz, H.J., Hartmann, D.: Machine Learning-Based Optimal Mesh Generation in Computational Fluid Dynamics. ArXiv, pp. 1–22 (2021). https://arxiv.org/abs/2102.12923
2. Iserte, S., Macías, A., Martínez-Cuenca, R., Chiva, S., Paredes, R., Quintana-Ortí, E.S.: Accelerating urban scale simulations leveraging local spatial 3D structure. J. Comput. Sci. **62**, 101741 (2022)
3. Kim, B., Azevedo, V.C., Thuerey, N., Kim, T., Gross, M., Solenthaler, B.: Deep fluids: a generative network for parameterized fluid simulations. Comput. Graph. Forum **38**(2), 59–70 (2019)
4. Kochkov, D., Smith, J.A., Alieva, A., Wang, Q., Brenner, M.P., Hoyer, S.: Machine learning-accelerated computational fluid dynamics. Proceed. Nat. Acad. Sci. **118**(21), e2101784118 (2021). https://doi.org/10.1073/pnas.2101784118. https://www.pnas.org/doi/abs/10.1073/pnas.2101784118
5. Kurz, M., Offenhäuser, P., Viola, D., Shcherbakov, O., Resch, M., Beck, A.: Deep reinforcement learning for computational fluid dynamics on HPC systems. J. Comput. Sci. **65**, 101884 (2022). https://doi.org/10.1016/j.jocs.2022.101884. https://www.sciencedirect.com/science/article/pii/S1877750322002435
6. MixIT: the enterprise mixing analysis tool. https://mixing-solution.com/. Accessed 19 Apr 2023
7. Obiols-Sales, O., Vishnu, A., Malaya, N., Chandramowliswharan, A.: CFDNet: a deep learning-based accelerator for fluid simulations. Proceed. Int. Conf. Supercomput. (2020). https://doi.org/10.1145/3392717.3392772

8. One-Class Support Vector Machine (SVM) for Anomaly Detection. https:// grabngoinfo.com/one-class-support-vector-machine-svm-for-anomalydetection/. Accessed 26 Sept 2021
9. OpenFOAM. http://www.openfoam.com/. Accessed 19 Apr 2023
10. OpenFOAM: User Guide: k-epsilon. http://www.openfoam.com/ %20documentation/guides/latest/doc/guide-turbulence-ras-k-epsilon.html. Accessed 28 Feb 2023
11. Panchigar, D., Kar, K., Shukla, S., Mathew, R.M., Chadha, U., Selvaraj, S.K.: Machine learning-based CFD simulations: a review, models, open threats, and future tactics. Neural Comput. Appl. **34**(24), 21677–21700 (2022). https://doi. org/10.1007/s00521-022-07838-6
12. Roh, S., Song, H.J.: Evaluation of Neural Network Emulations for Radiation Parameterization in Cloud Resolving Model. Geophys. Res. Lett. **47**(21), e2020GL089444 (2020)
13. Rojek, K.: Machine learning method for energy reduction by utilizing dynamic mixed precision on GPU-based supercomputers. Concurr. Comput. Practice Exper. **31**(6), e4644 (2019). https://doi.org/10.1002/cpe.4644. https://onlinelibrary.wiley. com/doi/abs/10.1002/cpe.4644
14. Rojek, K., Wyrzykowski, R.: Performance and scalability analysis of AI-accelerated CFD simulations across various computing platforms. In: Singer, J., Elkhatib, Y., Blanco Heras, D., Diehl, P., Brown, N., Ilic, A. (eds.) Euro-Par 2022: Parallel Processing Workshops. Euro-Par 2022. Lecture Notes in Computer Science, vol. 13835. Springer, Cham (2023). https://doi.org/10.1007/978-3-031-31209-0_17
15. Rojek, K., Wyrzykowski, R., Gepner, P.: AI-accelerated CFD simulation based on OpenFOAM and CPU/GPU computing. In: Paszynski, M., Kranzlmüller, D., Krzhizhanovskaya, V.V., Dongarra, J.J., Sloot, P.M.A. (eds.) ICCS 2021. LNCS, vol. 12743, pp. 373–385. Springer, Cham (2021). https://doi.org/10.1007/978-3-030-77964-1_29
16. Savarese, M., Cuoci, A., De Paepe, W., Parente, A.: Machine learning clustering algorithms for the automatic generation of chemical reactor networks from CFD simulations. Fuel **343**, 127945 (2023). https://doi.org/10.1016/j.fuel.2023.127945. https://www.sciencedirect.com/science/article/pii/S0016236123005586
17. Schannen, M., Bachem, O., Lucic, M.: Recent advances in autoencoder-based representation learning. arXiv preprint arXiv:1812.05069 (2018)
18. Song, H.J., Kim, P.S.: Effects of Cloud Microphysics on the Universal Performance of Neural Network Radiation Scheme. Geophys. Res. Lett. **49**(9), e2022GL098601 (2022)
19. Song, H.J., et al.: Benefits of stochastic weight averaging in developing neural network radiation scheme for numerical weather prediction. J. Adv. Model. Earth Syst. **14**(10), e2021MS002921 (2022)
20. Srivastava, S., Damodaran, M., Khoo, B.C.: Machine learning surrogates for predicting response of an aero-structural-sloshing system. arXiv preprint (2019)
21. Statistics How To. https://www.statisticshowto.com/probability-and-statistics/ calculus-based-statistics/. Accessed 28 Feb 2023
22. TensorFlow Core: Mixed precision. http://www.tensorflow.org/guide/mixed_ %20precision. Accessed 19 Apr 2023
23. Tompson, J., Schlachter, K., Sprechmann, P., Perlin, K.: Accelerating Eulerian fluid simulation with convolutional networks. In: Proceedings 34th International Conference Machine Learning, ICML2017 - Vol. 70, pp. 3424–3433 (2017)
24. Vinuesa, R., Brunton, S.L.: The potential of machine learning to enhance computational fluid dynamics. arXiv preprint arXiv:2110.02085 (2021)

25. Wyatt, M.R., Yamamoto, V., Tosi, Z., Karlin, I., Essen, B.V.: Is Disaggrega-
tion possible for HPC Cognitive Simulation? In: 2021 IEEE/ACM Workshop on
Machine Learning in High Performance Computing Environments (MLHPC), pp.
94–105 (2021). https://doi.org/10.1109/MLHPC54614.2021.00014
26. Xiao, D., et al.: A reduced order model for turbulent flows in the urban environment
using machine learning. Build. Environ. **148**, 323–337 (2019)
27. Yu, W., Zhao, F., Yang, W., Xu, H.: Integrated analysis of CFD simulation data
with K-means clustering algorithm for soot formation under varied combustion
conditions. Appl. Thermal Eng. **153**, 299–305 (2019). https://doi.org/10.1016/
j.applthermaleng.2019.03.011. https://www.sciencedirect.com/science/article/pii/
S1359431118349172

Memory-Based Monte Carlo Integration for Solving Partial Differential Equations Using Neural Networks

Carlos Uriarte[1,2]($\boxtimes$)(ID), Jamie M. Taylor[4](ID), David Pardo[1,2,3](ID),
Oscar A. Rodríguez[1,2](ID), and Patrick Vega[5](ID)

[1] Basque Center for Applied Mathematics (BCAM), Bilbao, Spain
{curiarte,orodriguez}@bcamath.org
[2] University of the Basque Country (UPV/EHU), Leioa, Spain
{carlos.uriarte,david.pardo}@ehu.eus
[3] Ikerbasque: Basque Foundation for Science, Bilbao, Spain
jamie.taylor@cunef.edu
[4] CUNEF Universidad, Madrid, Spain
[5] Pontificia Universidad Católica de Valparaíso (PUCV), Valparaíso, Chile
patrick.vega@pucv.cl

Abstract. Monte Carlo integration is a widely used quadrature rule to solve Partial Differential Equations with neural networks due to its ability to guarantee overfitting-free solutions and high-dimensional scalability. However, this stochastic method produces noisy losses and gradients during training, which hinders a proper convergence diagnosis. Typically, this is overcome using an immense (disproportionate) amount of integration points, which deteriorates the training performance. This work proposes a memory-based Monte Carlo integration method that produces accurate integral approximations without requiring the high computational costs of processing large samples during training.

Keywords: Neural Networks · Monte Carlo integration · Optimization

1 Introduction

Over the past decade, neural networks have proven to be a powerful tool in the context of solving Partial Differential Equations (PDEs) [5,8,9,12,18–20]. In such cases, the traditional approach is to reformulate the PDE as a minimization problem, where the loss function is often described as a definite integral and, therefore, approximated or discretized by a quadrature rule. Then, the minimization is performed according to a gradient-based optimization algorithm [6,16].

Using a deterministic quadrature rule with fixed integration points possibly leads to misbehavior of the network away from the integration points (an overfitting problem) [14], leading to large integration errors and, consequently, poor solutions. To overcome this, Monte Carlo integration is a popular and suitable choice of quadrature rule due to the mesh-free and stochastic sampling of

J. Mikyška et al. (Eds.): ICCS 2023, LNCS 14074, pp. 509–516, 2023.
https://doi.org/10.1007/978-3-031-36021-3_51

the integration points during training [2,4,7,10,13]. However, Monte Carlo integration converges as $\mathcal{O}(1/\sqrt{N})$, where N is the number of integration points. Thus, in practice, this may require tens or hundreds of thousands of integration points to obtain an acceptable error per integral approximation —even for one-dimensional problems, which deteriorates the training speed.

In this work, we propose a memory-based approach that approximates definite integrals involving neural networks by taking advantage of the information gained in previous iterations. As long as the expected value of these integrals does not change significantly, this technique reduces the expected integration error and lead to better approximations. Moreover, since gradients are also described in terms of definite integrals, we apply this approach to the gradient computations, which leads to a reinterpretation of the well-known momentum method [11] when we appropriately modify the hyperparameters of the optimizer.

2   Approximation with Neural Networks

Let us consider a well-posed minimization problem,

$$u = \arg \min_{v \in \mathbb{V}} \mathcal{F}(v), \tag{1}$$

where $\mathbb{V}$ denotes the search space of functions with domain Ω, $\mathcal{F} : \mathbb{V} \longrightarrow \mathbb{R}$ is the (exact) loss function governing our minimization problem, and u is the exact solution.

Let $v_\theta : \Omega \longrightarrow \mathbb{R}$ denote a neural network architecture parameterized by the set of trainable parameters θ. We denote the set of all possible realizations of v_θ by

$$\mathbb{V}_\Theta := \{v_\theta : \Omega \longrightarrow \mathbb{R}\}_{\theta \in \Theta}, \tag{2}$$

where Θ is the domain of all admissible parameters θ. Then, a neural network approximation of problem (1) consists in replacing the continuous search space $\mathbb{V}$ with the parameterized space[1] $\mathbb{V}_\Theta$,

$$u \approx \arg \inf_{v_\theta \in \mathbb{V}_\Theta} \mathcal{F}(v_\theta). \tag{3}$$

To carry out the minimization, one typically uses a first-order gradient-descent-based optimization scheme. In the classical case, it is given by the following iterative method:

$$\theta_{t+1} := \theta_t - \eta \frac{\partial \mathcal{F}}{\partial \theta}(v_{\theta_t}), \tag{4}$$

where $\eta > 0$ is the *learning rate*, and θ_t denotes the trainable parameters at the tth iteration.

---

[1] In general, $\mathbb{V}_\Theta$ is non-convex and non-closed, possibly preventing the uniqueness and existence of minimizers (e.g., see Example 2.3 in [1]).

For $\mathcal{F}$ in the form of a definite integral,

$$\mathcal{F}(v_\theta) = \int_\Omega I(v_\theta)(x) \ dx, \tag{5}$$

we approximate it by a quadrature rule, producing a *discrete loss function* $\mathcal{L}$. Taking Monte Carlo integration as the quadrature rule for the loss, we have

$$\mathcal{F}(v_\theta) \approx \mathcal{L}(v_\theta) := \frac{\mathrm{Vol}(\Omega)}{N} \sum_{i=1}^{N} I(v_\theta)(x_i), \tag{6}$$

where $\{x_i\}_{i=1}^{N}$ is a set of N integration points sampled from a random uniform distribution in Ω.

For the gradients, we have

$$\frac{\partial F}{\partial \theta}(v_\theta) \approx g(v_\theta) := \frac{\partial \mathcal{L}}{\partial \theta}(v_\theta) = \frac{\mathrm{Vol}(\Omega)}{N} \sum_{i=1}^{N} \frac{\partial I(v_\theta)}{\partial \theta}(x_i), \tag{7}$$

which yields a discretized version of (4),

$$\theta_{t+1} := \theta_t - \eta g(v_{\theta_t}), \tag{8}$$

known as a *stochastic gradient-descent* (SGD) optimizer [15]. Here, the "stochastic" term means that at different iterations, the set of integration points to evaluate $\partial \mathcal{L}/\partial \theta$ is random[2].

From now on, we write $\mathcal{F}(\theta_t)$, $\frac{\partial \mathcal{F}}{\partial \theta}(\theta_t)$, $\mathcal{L}(\theta_t)$, and $g(\theta_t)$ as simplified versions of $\mathcal{F}(v_{\theta_t})$, $\frac{\partial \mathcal{F}}{\partial \theta}(v_{\theta_t})$, $\mathcal{L}(v_{\theta_t})$, and $g(v_{\theta_t})$, respectively.

3 Memory-Based Integration and Optimization

If we train the network according to (8), we obtain a noisy and oscillatory behavior of the loss and the gradient. This occurs because of the introduced Monte Carlo integration error at each training iteration. Figure 1 (blue curve) illustrates the noisy behavior of Monte Carlo integration in a network with a single trainable parameter that permits exact calculation of $\mathcal{F}$ (black curve).

3.1 Integration

In order to decrease the integration error during training, we replace (6) by the following recurrence process:

$$\mathcal{F}(\theta_t) \approx \mathcal{L}_t := \alpha_t \mathcal{L}(\theta_t) + (1 - \alpha_t)\mathcal{L}_{t-1}, \tag{9a}$$

---

[2] In classical data science, SGD is performed by selecting random (mini-)batches from a finite dataset. In contrast, we have an infinite database Ω, and we select finite random samples from Ω at each iteration, implying that points are never reutilized during training and helping to avoid overfitting.

where $\mathcal{L}(\theta_t)$ is the Monte Carlo estimate at iteration t, and $\{\alpha_t\}_{t\geq 0}$ is a selected sequence of coefficients $0 < \alpha_t \leq 1$ such that $\alpha_0 = 1$. In expanded form,

$$\mathcal{L}_t = \sum_{l=0}^{t} \alpha_l \left(\prod_{s=1}^{t-l} (1 - \alpha_{l+s}) \right) \mathcal{L}(\theta_l), \tag{9b}$$

which shows that the approximation $\mathcal{L}_t$ of $\mathcal{F}(\theta_t)$ is given as a linear combination of the current and all previous Monte Carlo integration estimates. If $\alpha_t = 1$ for all t, we recover the usual Monte Carlo integration case without memory.

Figure 1 (red curve) shows the memory-based loss $\mathcal{L}_t$ evolution along training according to ordinary SGD optimization (8) and selecting $\alpha_t = e^{-0.001t} + 0.001$. In the beginning, we integrate with large errors (α_t is practically one, and therefore, there is hardly any memory in $\mathcal{L}_t$). However, as we progress in training, we increasingly endow memory to $\mathcal{L}_t$ (α_t becomes small), and as a consequence, its integration error decreases. $\mathcal{L}_t$ produces more accurate approximations than $\mathcal{L}(\theta_t)$, allowing better convergence monitoring to, for example, establish proper stopping criteria if desired.

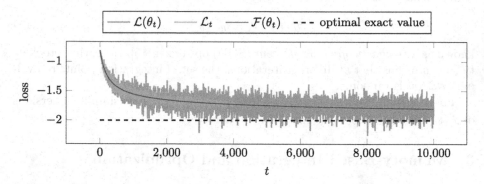

Fig. 1. Training of a single-trainable-parameter network whose architecture permits exact calculation of $\mathcal{F}$. The training is performed according to (8), and thus, $\mathcal{L}(\theta_t)$ is its associated loss. $\mathcal{L}_t$ and $\mathcal{F}(\theta_t)$ are computed for monitoring.

3.2 Optimization

While the proposed scheme may be able to improve the approximation of $\mathcal{F}$, we have that a corresponding SGD scheme using (9) is equivalent to classical SGD with a different learning rate since $\frac{\partial \mathcal{L}_t}{\partial \theta}(\theta_t) = \alpha_t g(\theta_t)$.

However, we can naturally endow the idea of memory-based integration to the gradients, as $g(\theta_t)$ is also obtained via Monte Carlo integration —recall (7),

$$\frac{\partial \mathcal{F}}{\partial \theta}(\theta_t) \approx g_t := \gamma_t g(\theta_t) + (1 - \gamma_t)g_{t-1}, \tag{10a}$$

where $\{\gamma_t\}_{t\geq 0}$ is a selected sequence of coefficients such that $0 < \gamma_t \leq 1$ and $\gamma_0 = 1$. Then, we obtain a memory-based SGD optimizer employing the g_t term instead of $g(\theta_t)$ —recall (8),

$$\theta_{t+1} := \theta_t - \eta g_t. \qquad (10b)$$

In the expanded form, we have

$$g_t = \sum_{l=0}^{t} \gamma_l \left(\prod_{s=1}^{t-l} (1 - \gamma_{l+s}) \right) g(\theta_l). \qquad (10c)$$

Figure 2 shows the gradient evolution during training of the previous single-trainable-parameter model problem. We select $\alpha_t = \gamma_t$ for all t, with the same exponential decay as before. g_t produces more accurate approximations of the exact gradients than $g(\theta_t)$, minimizing the noise.

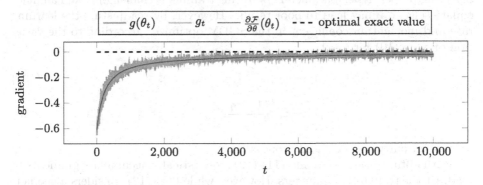

Fig. 2. Gradient evolution of the single-trainable-parameter model problem in Fig. 1. The optimization is performed according to (8) using $g(\theta_t)$, while g_t and $\frac{\partial \mathcal{F}}{\partial \theta}(\theta_t)$ are computed for monitoring.

Proper tuning of the coefficients α_t and γ_t is critical to maximize integration performance. Coefficients should be high (low memory) when the involved integrals vary rapidly (e.g., at the beginning of training). Conversely, when the approximated solution is near equilibrium and the relevant integrals vary slowly, the coefficients should be low (high memory). The optimal tuning of these coefficients to improve training performance will be analyzed in a more extended subsequent work.

4   Relation with the Momentum Method

The SGD optimizer with momentum (SGDM) [11,17] is commonly introduced as the following two-step recursive method:

$$v_{t+1} := \beta v_t - g(\theta_t), \qquad (11a)$$
$$\theta_{t+1} := \theta_t + \eta v_{t+1}, \qquad (11b)$$

where v_t is the *momentum accumulator* initialized by $v_0 = 0$, and $0 \leq \beta < 1$ is the momentum coefficient. If $\beta = 0$, we recover the classical SGD optimizer (8). Rewriting (11) in terms of the scheme $\theta_{t+1} = \theta_t - \eta g_t$, we obtain

$$g_t = g(\theta_t) + \beta g_{t-1} = \sum_{l=0}^{t} \beta^{t-l} g(\theta_l). \tag{11c}$$

A more sophisticated version of SGDM modifies the momentum coefficient during training (see, e.g., [3]), namely, defined as in (11) but replacing β with β_t for some conveniently selected sequence $\{\beta_t\}_{t \geq 1}$. Then, the g_t term results

$$g_t = g(\theta_t) + \beta_t g_{t-1} = \sum_{l=0}^{t} \left(\prod_{s=1}^{t-l} \beta_{l+s} \right) g(\theta_l). \tag{12}$$

Selecting proper hyper-parameters β_t during training is challenging, and an inadequate selection may lead to poor results. However, by readjusting the learning rate and momentum coefficient in the SGDM optimizer according to the variation of γ_t in (10) for $t \geq 1$,

$$\eta_t := \eta_{t-1} \frac{\gamma_t}{\gamma_{t-1}}, \qquad \eta_0 := \eta, \tag{13a}$$

$$\beta_t := \gamma_{t-1} \frac{1 - \gamma_t}{\gamma_t}, \tag{13b}$$

we recover our memory-based proposal (10).

Both optimizations (10) and (11)-(12) stochastically accumulate gradients to readjust the trainable parameters. However, while (11)-(12) considers a geometrically weighted average of past gradients without aiming g_t resemble $\frac{\partial \mathcal{F}}{\partial \theta}(\theta_t)$, our recursive convex combination proposal (10) re-scales the current and prior gradients to g_t deliberately imitate $\frac{\partial \mathcal{F}}{\partial \theta}(\theta_t)$ —recall Fig. 2.

Equation (13) re-interprets the SGDM as an exact-gradient performer by re-scaling the learning rate. In contrast, our memory-based proposal provides the exact-gradient interpretation leaving the learning rate free. In consequence, the learning rate is an independent hyperparameter of the gradient-based optimizer and not an auxiliary element to interpret gradients during training. Moreover, our optimizer is designed to work in parallel with the memory-based loss (9) that approximates the (typically unavailable) exact loss during training.

5   Conclusions and Future Work

In the context of solving PDEs using neural networks, this work addresses a common difficulty when employing Monte Carlo integration: the convergence of the loss and its gradient is noisy due to the large integration errors committed at each training iteration.

To improve integration accuracy (without incurring in prohibitive computational costs), we propose a memory-based iterative method that conveniently accumulates integral estimations in previous iterations when the convergence reaches an equilibrium phase. We show that our resulting method is equivalent to reinterpreting a modified momentum method.

In future work, we shall study the automatic optimal selection of hyper-parameters (α_t, γ_t, and η) of our memory-based algorithm to improve convergence speed in different problems.

Acknowledgments. David Pardo has received funding from: the European Union's Horizon 2020 research and innovation program under the Marie Sklodowska-Curie grant agreement No. 777778 (MATHROCKS); the Spanish Ministry of Science and Innovation projects with references TED2021-132783B-I00, PID2019-108111RB-I00 (FEDER/AEI) and PDC2021-121093-I00 (MCIN/AEI/10.13039/501100011033/Next Generation EU), the "BCAM Severo Ochoa" accreditation of excellence CEX2021-001142-S/MICIN/AEI/10.13039/501100011033; and the Basque Government through the BERC 2022–2025 program, the three Elkartek projects 3KIA (KK-2020/00049), EXPERTIA (KK-2021/00048), and SIGZE (KK-2021/00095), and the Consolidated Research Group MATHMODE (IT1456-22) given by the Department of Education.

Patrick Vega has received funding from: the Chilean National Research and Development Agency (ANID) though the grant ANID FONDECYT No. 3220858.

We would also like to thank the undergraduate students Nicolás Zamorano and Patricio Asenjo, who belong to the Mathematics and Mathematical Civil Engineering bachelor programs of the Pontificia Universidad Católica de Valparaíso and the Universidad de Concepción, respectively, for their concerns and contributions during the elaboration of this work.

References

1. Brevis, I., Muga, I., van der Zee, K.G.: Neural control of discrete weak formulations: Galerkin, least squares & minimal-residual methods with quasi-optimal weights. Comput. Methods Appl. Mech. Eng. **402**, 115716 (2022)
2. Chen, J., Du, R., Li, P., Lyu, L.: Quasi-Monte Carlo sampling for solving partial differential equations by deep neural networks. Numer. Math. Theory Methods Appl. **14**(2), 377–404 (2021)
3. Chen, J., Wolfe, C., Li, Z., Kyrillidis, A.: Demon: improved neural network training with momentum decay. In: ICASSP 2022–2022 IEEE International Conference on Acoustics, Speech and Signal Processing (ICASSP), pp. 3958–3962. IEEE (2022)
4. Dick, J., Kuo, F.Y., Sloan, I.H.: High-dimensional integration: the quasi-Monte Carlo way. Acta Numer **22**, 133–288 (2013)
5. E, W., Yu, B.: The Deep Ritz Method: a deep learning-based numerical algorithm for solving variational problems. Commun. Math. Statist. **6**(1), 1–12 (2018). https://doi.org/10.1007/s40304-018-0127-z
6. Goodfellow, I., Bengio, Y., Courville, A.: Deep learning. MIT press (2016)
7. Grohs, P., Jentzen, A., Salimova, D.: Deep neural network approximations for solutions of PDEs based on Monte Carlo algorithms. Partial Differ. Equ. Appl. **3**(4), Paper No. 45, 41 (2022)
8. Kharazmi, E., Zhang, Z., Karniadakis, G.E.: Variational physics-informed neural networks for solving partial differential equations. arXiv preprint arXiv:1912.00873 (2019)

9. Kharazmi, E., Zhang, Z., Karniadakis, G.E.: hp-VPINNs: variational physics-informed neural networks with domain decomposition. Comput. Methods Applied Mech. Eng. **374**, 113547 (2021)
10. Leobacher, G., Pillichshammer, F.: Introduction to quasi-Monte Carlo integration and applications. Springer, Cham (2014). https://doi.org/10.1007/978-3-319-03425-6
11. Polyak, B.T.: Some methods of speeding up the convergence of iteration methods. USSR Comput. Math. Math. Phys. **4**(5), 1–17 (1964)
12. Raissi, M., Perdikaris, P., Karniadakis, G.E.: Physics-informed neural networks: a deep learning framework for solving forward and inverse problems involving non-linear partial differential equations. J. Comput. Phys. **378**, 686–707 (2019)
13. Reh, M., Gärttner, M.: Variational Monte Carlo approach to partial differential equations with neural networks. Mach. Learn.: Sci. Technol. **3**(4), 04LT02 (2022)
14. Rivera, J.A., Taylor, J.M., Omella, Á.J., Pardo, D.: On quadrature rules for solving partial differential equations using neural networks. Comput. Methods Appl. Mech. Eng. **393**, 114710 (2022)
15. Robbins, H., Monro, S.: A stochastic approximation method. Ann. Math. Stat. **22**, 400–407 (1951)
16. Ruder, S.: An overview of gradient descent optimization algorithms. arXiv preprint arXiv:1609.04747 (2016)
17. Sutskever, I., Martens, J., Dahl, G., Hinton, G.: On the importance of initialization and momentum in deep learning. In: Proceedings of the 30th International Conference on Machine Learning, vol. 28. pp. III-1139-III-1147. ICML'13, JMLR.org (2013)
18. Taylor, J.M., Pardo, D., Muga, I.: A deep fourier residual method for solving PDEs using neural networks. Comput. Methods Appl. Mech. Eng. **405**, 115850 (2023)
19. Uriarte, C., Pardo, D., Muga, I., Muñoz-Matute, J.: A deep double Ritz method (D2RM) for solving partial differential equations using neural networks. Comput. Methods Appl. Mech. Eng. **405**, 115892 (2023)
20. Uriarte, C., Pardo, D., Omella, Á.J.: A finite element based deep learning solver for parametric PDEs. Comput. Methods Appl. Mech. Eng. **391**, 114562 (2022)

Fast Solver for Advection Dominated Diffusion Using Residual Minimization and Neural Networks

Tomasz Służalec[1]([✉])[ID] and Maciej Paszyński[2][ID]

[1] Jagiellonian University, Kraków, Poland
`tomasz.sluzalec@gmail.com`
[2] AGH University of Science and Technology, Kraków, Poland

Abstract. Advection-dominated diffusion is a challenging computational problem that requires special stabilization efforts. Unfortunately, the numerical solution obtained with the commonly used Galerkin method delivers unexpected oscillation resulting in an inaccurate numerical solution. The theoretical background resulting from the famous inf-sup condition tells us that the finite-dimensional test space employed by the Galerkin method does not allow us to reach the supremum necessary for problem stability. We enlarge the test space to overcome this problem. We do it for a fixed trial space. The method that allows us to do so is the residual minimization method. This method, however, requires the solution to a much larger system of linear equations than the standard Galerkin method. We represent the larger test space by its set of optimal test functions, forming a basis of the same dimension as the trial space in the Galerkin method. The resulting Petrov-Galerkin method stabilizes our challenging advection-dominated problem. We train the optimal test functions offline with the neural network to speed up the computations. We also observe that the optimal test functions, usually global, can be approximated with local support functions, resulting in a low computational cost for the solver and a stable numerical solution.

1 Introduction

The neural networks can be introduced as a non-linear function

$$y_{out} = ANN(x_{in}) = \theta_n \sigma(...\sigma(\theta_2 \sigma(\theta_1 x_{in} + \phi_1) + \phi_2) + ...) + \phi_n, \tag{1}$$

where $\{\theta_j\}_{j=1...n}$ are matrices, possibly with different numbers of rows and columns, and $\{\phi_j\}_{j=1...n}$ are bias vectors, possibly with a different number of rows. The selected architecture of the neural network results in a different number and dimensions of matrices and bias vectors. The classical choice for the activation function is the sigmoid function $\sigma(x) = \frac{1}{1+e^{-x}}$, besides several other possibilities (rectified linear unit (ReLU) or leaky ReLU [2,20]).

Recently, in computational science, the classical finite element methods are often augmented with the neural networks [3,8,15,21]. An overview of the application of the Deep Neural Networks into finite element method is presented in [8].

J. Mikyška et al. (Eds.): ICCS 2023, LNCS 14074, pp. 517–531, 2023.
https://doi.org/10.1007/978-3-031-36021-3_52

Paper [15] considers the problem of representing some classes of real-valued uni-
variant approximators with Deep Neural Networks based on the rectified linear
unit (ReLU) activation functions. The space generated by the neural networks
with ReLU activation functions contains the space of the linear finite element
method. The authors claim that the convergence rate of the DNN to the solution is of a similar order to the convergence rate of the classical finite element
method. By taking $(ReLU)^k$ this result can be generated to higher-order finite
element method as well [21]. The Deep Neural Networks can also help guide
refinements in goal-oriented adaptivity [3].

The finite element method utilizes high-order basis functions, e.g., Lagrange
polynomials which are C^0 between finite elements [5,6], or B-spline basis functions, which can be of higher order and continuity [1,4,9].

The simulations of difficult, unstable time-dependent problems, like
advection-dominated diffusion [12,13], high-Reynolds number Navier-Stokes
equations [11], or high-contrast material Maxwell equations, have several important applications in science and engineering. These challenging engineering problems require special stabilization methods, such as Streamline-Petrov Upwind
Galerkin method (SUPG) [10], discontinuous Galerkin method (DG) [14,17], as
well as residual minimization (RM) method [11–13].

2   Methodology

2.1   Galerkin Method

Let us introduce an advection-diffusion problem in the strong form: Find $u \in$
$C^2(0,1)$ such that

$$-\epsilon \frac{d^2u(x)}{dx^2} + 1\frac{du(x)}{dx} = 0, \quad x \in (0,1). \tag{2}$$

The advection part "wind" coefficient is given as 1, and the ϵ represents a diffusion
coefficient. We recall that the problem is numerically unstable for small values
of ϵ; thus solution cannot be correctly approximated by the classical Galerkin
method without a very large computational mesh. The weak formation, suitable
for the Galerkin method, is the following:

$$\int_0^1 -\epsilon \frac{d^2u(x)}{dx^2}v(x)dx + \int_0^1 1\frac{du(x)}{dx}v(x)dx = 0, \quad \forall v \in V. \tag{3}$$

We apply a Cauchy boundary condition

$$-\epsilon\frac{du}{dx}(0) + u(0) = 1.0, \ u(1) = 0, \tag{4}$$

and we rewrite the formula in a simpler form

$$-\epsilon(u'',v)_0 + (u',v)_0 = 0, \quad \forall v \in V. \tag{5}$$

We integrate by parts to get

$$\epsilon(u', v')_0 + (u', v)_0 + u(0)v(0) = v(0) \ \forall v \in V. \tag{6}$$

Now, we select a finite set of test functions, and we formulate a discrete form as:
Find $u_h \in U_h, u_h = \sum_{i=1,\ldots,21} u_i B_i(x)$:

$$b(u_h, v_h) = l(v_h) \quad \forall v_h \in V_h \tag{7}$$
$$b(u_h, v_h) = \epsilon(u'_h, v'_h)_0 + (u'_h, v_h)_0 + u_h(0)v_h(0) \tag{8}$$
$$l(v_h) = v_h(0). \tag{9}$$

In the discrete formulation, our equation is "averaged" by the test functions $v(x)$. This is why the accuracy of the numerical solution u_h depends on the quality of the test space V_h. In the classical Galerkin method, we seek a solution as a linear combination of basis functions. In the Galerkin method, the trial space, where we seek the solution, is equal to the test space. We choose $U_h = V_h$; thus, v_h are the same 21 basis functions that are used to approximate the solution u_h. For example, we choose quadratic B-splines with C^0 separators as they are equivalent to the Lagrange basis functions.

2.2 Petrov-Galerkin Method

More proper results numerical results give Petrov-Galerkin method, where we choose a different basis for a solution and a different basis for testing. Still, our solution is a linear combination of basis functions, for example, linear B-splines. We test with another basis, for example, quadratic B-splines with C^0 separators. The Petrov-Galerkin formulation is similar to the Galerkin method, but the trail space is different than the test space: Find $u_h \in U_h, u_h = \sum_{i=1,\ldots,11} u_i B_i(x)$ such that:

$$b(u_h, v_h) = l(v_h) \quad \forall v_h \in \hat{V}_h \tag{10}$$
$$b(u_h, v_h) = \epsilon(u'_h, v'_h)_0 + (u'_h, v_h)_0 + u_h(0)v_h(0) \tag{11}$$
$$l(v_h) = v_h(0). \tag{12}$$

where v_h contains carefully selected 11 elements of V_h. These test functions are linearly independent, and they are linear combination of the original basis functions from V_h.

$$\hat{V}_h \ni v_i = \{\alpha_1^i w_1 + \cdots + \alpha_{21}^i w_{21} | w_j \subset V_h, \alpha_j^i \in \mathbb{R}\}, i = 1, \ldots, 11 \tag{13}$$

Here V_h is the space of all 21 test functions, and $\hat{V}_h$ are sub-space defined by selected 11 basis functions:

$$11 = \dim U_h = \dim \hat{V}_h \leq \dim V_h = 21. \tag{14}$$

2.3   Advection-Diffusion Problem with Arbitrary Coefficients

When we simulate real-life applications with the advection-diffusion equations, the advection and diffusion coefficients can change from one time step to another. We generalize our problem into arbitrary ϵ coefficient as then it would be more general and could be used in real-life simulation.

For a given $\epsilon \in \mathbb{R}$ find $u_h \in U_h$ such that:

$$\hat{b}(\epsilon, u_h, v_h) = l(v_h) \quad \forall v_h \in \hat{V}_h$$
$$\hat{b}(\epsilon, u_h, v_h) = \epsilon\,(u'_h, v'_h)_0 + (u'_h, v_h)_0 + u_h(0)v_h(0). \tag{15}$$

The problem is difficult to solve since it requires proper space of the optimal test functions $\hat{V}_h$ for stabilization with the Petrov-Galerkin method. The optimal test functions v_h can be different for every ϵ. The problem remains numerically unstable for small ϵ, and we want the optimal test functions to be quickly adapted for a new problem for different ϵ values.

2.4   Theoretical U_h V_h Space Implications and Limitations

To find the optimal test functions we recall that $b(\cdot, \cdot)$ satisfies the following inequality

$$\alpha\|u\|^2 \leq b(u, u) \leq M\|u\|^2, \tag{16}$$

where M is the continuity constant $b(u, v) \leq M\|u\|\|v\|$ and α is coercivity constant $b(u, u) \geq \alpha\|u\|^2$. Since $\alpha \leq M$, the constant is always $\frac{M}{\alpha} \geq 1$.

In the ideal case, we would like to have the approximation error resulting from the Galerkin solution u_h equal to the distance of the trial space from the exact solution.

Cea Lemma *[Céa, Jean (1964). Approximation variationnelle des problèmes aux limites (Ph.D. thesis)]*

$$\|u - u_h\| \leq \frac{M}{\alpha}\mathrm{dist}\{U_h, u\}. \tag{17}$$

Unfortunately, this is only the case if $\frac{M}{\alpha} = 1$. In reality, $\frac{M}{\alpha} \geq 1$, and the best solution in the approximation space can be worse than the distance of the space to the exact solution.

The coercivity constant can be estimated from the following:

Babuška theorem (inf-sup condition) *[Babuška, Ivo (1970). "Error-bounds for finite element method". Numerische Mathematik]*

$$\inf_{u \in U} \sup_{v \in V} \frac{b(u, v)}{\|v\|\|u\|} = \alpha > 0. \tag{18}$$

The problem is that during the simulation, we select finite dimensional spaces $U_h \subset U$ and $V_h \subset V$, and we seek $u_h \in U_h, v_h \in V_h$. Then we have:

$$\inf_{u \in U} \sup_{v \in V} \frac{b(u,v)}{\|v\|\|u\|} = \alpha > \alpha_h = \inf_{u_h \in U_h} \sup_{v_h \in V_h} \frac{b(u_h, v_h)}{\|v_h\|\|u_h\|},$$

$$\frac{M}{\alpha_h} > \frac{M}{\alpha} \geq 1. \tag{19}$$

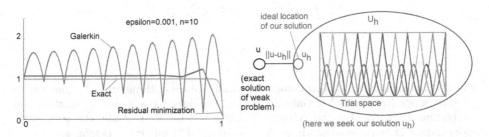

Fig. 1. Comparison of the Galerkin method with trial=test=quadratic B-splines with C^0 separators, and the Petrov-Galerkin method with linear B-splines for trial and quadratic B-splines with C^0 separators for test, and the exact solution. Ideally, the approximation of a solution should be as good as the distance of the space where it lives (trial space) to the exact solution.

The supremum may not be realized in the finite-dimensional subset of the infinite-dimensional test space. This gives the conclusion that we need to test with larger test space V_h, so it realizes the supremum for α, and $\frac{M}{\alpha}$ is closer to 1.

In order to get a better solution, we need to solve in a fixed trial space and test in the larger test space. This approach is used in Petrov-Galerkin $U_h \neq V_h$, but in the case of the "classical" Galerkin method, trial space is equal to test space $U_h = V_h$.

For example, we employ linear B-splines for the trial space $U_h = \{B_{i,1}(x)\}_{1,...,n_x}$, and quadratic B-splines for the test space $V_h = \{B_{i,2}(x)\}_{1,...,N_x}$. We seek the solution in the trial space U_h with 11 basis functions. Our larger test space V_h has 21 basis functions. We need to compute 11 optimal test functions. They are linear combinations of the 21 base functions of V_h.

2.5   Residual Minimization - Optimal Test Functions for Given ϵ

In the residual minimization method we need to prescribe the norm and scalar product, e.g., $g(u,v) = \int_0^1 (uv + u'v')dx$ or $G_{ij} = g(B_{i,2}, B_{j,2}) = \int_0^1 (B_{i,2}B_{j,2} + \frac{dB_{i,2}}{dx}\frac{dB_{j,2}}{dx})dx$ to minimize the residual of the solution (or the numerical error).

522   T. Służalec and M. Paszyński

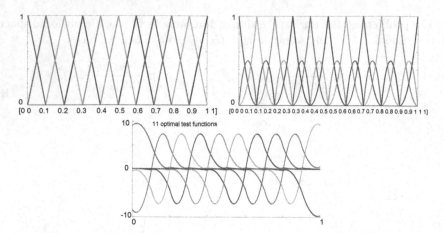

Fig. 2. First plot show 11 linear B-splines, second plot show 21 quadratic B-splines with C^0 separators. In the classical Galerkin method the same basis is used for solution and for testing: $U_h = V_h$. For the Petrov-Galerkin algorithm we carefully choose linearly independent 11 basic functions from the test space of 21 quadratic B-splines.

In the residual minimization problem we minimize the norm under the constrained defined as the solution of our problem

$$\begin{bmatrix} G & B \\ B^T & 0 \end{bmatrix} \begin{bmatrix} r \\ u \end{bmatrix} = \begin{bmatrix} F \\ 0 \end{bmatrix}, \tag{20}$$

where G is the Gram matrix (scalar product matrix) B and F are the problem matrix and right-hand side.

$$G = \begin{bmatrix} g(B_{1,2}, B_{1,2}) & \cdots & g(B_{1,2}, B_{N_x,2}) \\ \vdots & \ddots & \vdots \\ g(B_{N_x,2}, B_{1,2}) & \cdots & g(B_{N_x,2}, B_{N_x,2}) \end{bmatrix}, \; B = \begin{bmatrix} b(B_{1,1}, B_{1,2}) & \cdots & b(B_{n_x,1}, B_{1,2}) \\ \vdots & \ddots & \vdots \\ b(B_{1,1}, B_{N_x,2}) & \cdots & b(B_{n_x,1}, B_{N_x,2}) \end{bmatrix},$$

$$u = \begin{bmatrix} u_1 \\ \vdots \\ u_{n_x} \end{bmatrix}, \quad r = \begin{bmatrix} r_1 \\ \vdots \\ r_{N_x} \end{bmatrix}, \quad F = \begin{bmatrix} l(B_{1,2}) \\ \vdots \\ l(B_{N_x,2}) \end{bmatrix}.$$

where by $B_{i,2}$ we denote basis functions from the test space, and by $B_{i,1}$ we denote basis functions from the trial space. Our solution is $(u_1, ..., u_{n_x})$. Here $(r_1, ..., r_{N_x})$, represents the residual - the local error map.

The residual minimization method is equivalent to the Petrov-Galerkin formulation with the optimal test functions. The coefficients $\{w_i^k\}_{i=1,...,n_x}$ of $k = 1, ..., n_x$ optimal test functions

$\{w^1, \cdots, w^{n_x}\}$ are obtained by solving $Gw = B$.

$$
\begin{bmatrix} g(B_{1,2}, B_{1,2}) & \cdots & g(B_{1,2}, B_{N_x,2}) \\ \vdots & \ddots & \vdots \\ g(B_{N_x,2}, B_{1,2}) & \cdots & g(B_{N_x,2}, B_{N_x,2}) \end{bmatrix} \begin{bmatrix} w_1^1 & w_1^2 & \cdots w_1^{n_x} \\ \vdots & \vdots & \vdots \\ w_{N_x}^1 & w_{N_x}^2 & \cdots w_{N_x}^{n_x} \end{bmatrix} = \begin{bmatrix} b(B_{1,2}, B_{1,1}) & \cdots & b(B_{1,2}, B_{n_x,1}) \\ \vdots & \ddots & \vdots \\ b(B_{n_x,2}, B_{1,1}) & \cdots & b(B_{n_x,2}, B_{n_x,1}) \end{bmatrix}
$$

This gives system of linear equations with multiple right-hand sides. Solving the above system gives the optimal test functions $V_h^{opt} = span\{w^1, \cdots, w^{n_x}\}$ that form a subspace $V_h^{opt} \subset V_h$. The Petrov-Galerkin formulation with the optimal test functions gives the best possible, up to the trial space used, solution. We get this best possible solution by solving:

$$
\begin{bmatrix} b(B_{1,1}, w^1) & \cdots & b(B_{n_x,1}, w^1) \\ \vdots & \ddots & \vdots \\ b(B_{1,1}, w^{n_x}) & \cdots & b(B_{n_x,1}, w^{n_x}) \end{bmatrix} \begin{bmatrix} u_1^{opt} \\ \vdots \\ u_{n_x}^{opt} \end{bmatrix} = \begin{bmatrix} l(w^1) \\ \vdots \\ l(w^{n_x}) \end{bmatrix}, \tag{21}
$$

where $(u_1^{opt}, \cdots, u_{n_x}^{opt})$ is the optimal solution in the base of linear B-splines. This is to fix the LaTeX formatting in front of Sect. 3. This text is hiden using white color.

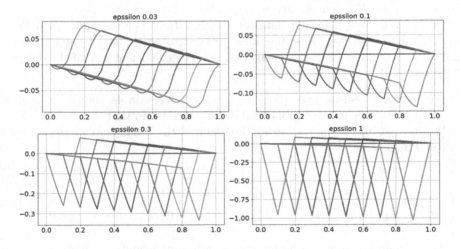

Fig. 3. Test functions for different ϵ calculated using RM method.

3   Numerical Results

3.1   Efficient Numerical Solution Using Artificial Neural Net and Petrov-Galerkin Method

We recall that our problem is formulated with the Petrov-Galerkin method. We seek the solution u, but it requires the knowledge of the optimal V_h for each

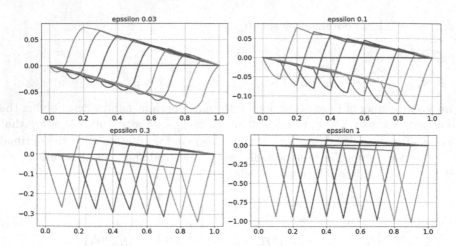

Fig. 4. Test functions for different ϵ calculated using proposed method using simple neural net with 3 hidden layers.

$\epsilon \in \mathbb{R}$. In order to automatically obtain the optimal test functions for a given ϵ, we approximate V_h with the neural networks.

We have to design proper data-set for algorithm. For the given predefined trial $B_{i,1}$ and test $B_{i,2}$ basis functions we need to create set of pairs $(\epsilon, (w_i^k))$ and $(\epsilon, (w^k))$ (mapping from ϵ coefficient into the coefficients w_i^k of the optimal test functions).

- We randomly select $\epsilon \in (0,1)$ - representative sample should be denser towards 0 - we can use exponential distribution to do that.
- We use residual minimization method to find the coefficients of the optimal test functions w^k

$$\begin{bmatrix} g(B_{1,2}, B_{1,2}) \cdots g(B_{1,2}, B_{N_x,2}) \\ \cdots\cdots\cdots \\ g(B_{N_x,2}, B_{1,2}) \cdots g(B_{N_x,2}, B_{N_x,2}) \end{bmatrix} \begin{bmatrix} w_1^1 \cdots w_1^{n_x} \\ \cdots\cdots \\ w_{n_x}^1 \cdots w_{n_x}^{n_x} \end{bmatrix} = \begin{bmatrix} b(B_{1,1}, B_{1,2}) \cdots b(B_{n_x,1}, B_{1,2}) \\ \cdots\cdots\cdots \\ b(B_{1,1}, B_{N_x,2}) \cdots b(B_{n_x,1}, B_{N_x,2}) \end{bmatrix}$$

In order to enhance the online computation we would use artificial neural network to select the proper optimal test functions. Neural network is a function:

$$\text{NN}(x) = A_n \sigma \left(A_{n-1}\sigma(...\sigma(A_1 x + B_1)... + B_{n-1}) + B_n = y, \right. \tag{22}$$

where A_i are the matrices of coefficients A_i^{mn}, B_i are bias vectors with coordinates B_i^m, and σ is the non-linear activation function.

Our aim in first numerical experiment [16] was to check if for given size of trial and test spaces we can construct a function using artificial neural network that gives single w_i^k coefficient for all $w^k \in V_h^{opt}$.

$$\forall \epsilon \in \mathbb{R}_+ \quad \text{NN}_{i,k}(\epsilon) \longrightarrow w_i^k \approx w_i^k, \quad w^k \approx [\text{NN}_{1k}(\epsilon), \cdots, \text{NN}_{n_x k}(\epsilon)] \tag{23}$$

It would be impossible to test and train NN for every ϵ so we have to choose a range of ϵ. We make a mapping that takes logarithm of ϵ onto some predefined closed interval $[-1, 1]$. Such scaling makes the input of NN more sensitise to smaller values of epsilon.

$$scaling(\epsilon) : \forall \epsilon \in E \subset \mathbb{R}_+ \quad \log_{10} \epsilon \longrightarrow [-1, 1] \qquad (24)$$

Let $\epsilon_{min} = \inf E$ and $\epsilon_{max} = \sup E$ then:

$$\text{for } \epsilon < \epsilon_{min}, \ scaling(\epsilon) > 1 \text{ and for } \epsilon > \epsilon_{max}, \ scaling(\epsilon) < -1. \qquad (25)$$

Conclusion from 25 is that we can choose scalled epsilons outside the training set. If our set E have representative values then it would approximate well all the epsilons in $\mathbb{R}_+$. Well designed set E implies proper working outside the training set. Small values of ϵ gives more numerically unstable solutions then bigger values of ϵ. First, we generate the optimal test functions for each ϵ, and we check if NN can approximate well the coefficients of test function. We experiment with the optimal number of layer and neurons. We have found that the optimal approximation is obtained when we construct one neural network for one optimal test function.

$$\forall \epsilon \in E \quad NN_k(scaling(\epsilon)) \longrightarrow v^k = [\omega_1^k, \cdots, \omega_{n_x}^k], \quad v^k \approx w^k \in V_h^{opt} \qquad (26)$$

Our equation for computing the optimal solution with the optimal test functions as provided by the artificial neural networks takes the following form:

$$\begin{bmatrix} b(B_{1,1}, NN_1(\epsilon)) & \cdots & b(B_{n_x,1}, NN_1(\epsilon)) \\ \vdots & \ddots & \vdots \\ b(B_{1,1}, NN_{n_x}(\epsilon)) & \cdots & b(B_{n_x,1}, NN_{n_x}(\epsilon)) \end{bmatrix} \boldsymbol{u} = \begin{bmatrix} l(NN_1(\epsilon)) \\ \vdots \\ l(NN_{n_x}(\epsilon)) \end{bmatrix}. \qquad (27)$$

To keep computations using NN as simple as possible, we propose 3 hidden layer neural network. At input and output, we use the linear activation function and ReLU ($\sigma(x) = \max\{0, x\}$) in hidden layers. We use the Adam optimizer with default settings, and we use the loss function defined as the mean square error. The mean absolute percentage error was also needed in this case as it monitors overall sensitivity for all values. We use Python language with the TensorFlow package compiled to use GPU support. The training procedure takes only 4 seconds to perform 1000 epochs on Nvidia GTX 1650Ti. In the [16], we focused on justification and investigated the aim of using the neural network in coefficient approximation. One or two layers could approximate one single coefficient but performed badly when it came to the whole test function. Too many layers also affect the quality of the coefficients, as neural nets tend to seek patterns. The 4 layer neural network works well in the case of matrix approximation, where there is plenty of repetition of similar values. In the [19], we discussed such cases and the procedure to find the optimal neural net setup (Table 1).

The algorithm for Petrov-Galerkin method enhanced with neural networks providing the optimal test functions:

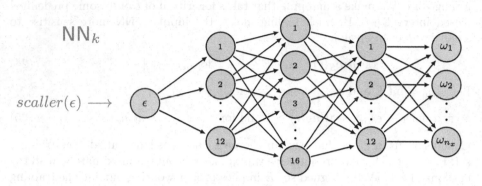

Fig. 5. NN used in computations have 3 hidden dense layers with 12,16 and 12 neurons respectively.

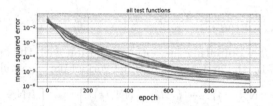

Fig. 6. Learning procedure for all 9 of 11 optimal test functions (for $n = 11$). The first and last test functions are skipped due to the 0 boundary condition.

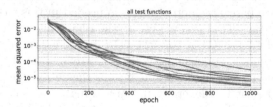

Fig. 7. Interpolation of optimal test functions for a set of epsilons inside the training range that were not used during training. The test set was designed in a way that between every two neighboring epsilons from the train set, one was selected for the test set.

1. Predefine trail U_h and test V_h spaces for solution that satisfies $\dim U_h < \dim V_h$.
2. Construct set E from representative sample of diffusion parameter ϵ
   - for example selection can be done by using exponential distribution from interval $(0, 1)$.
3. Using residual minimization find V_h^{opt}
   - for every $\epsilon \in E$ we calculate optimal test functions by solving $Gw = B$
4. Fit $scaling(\cdot)$ for elements from set E.
5. For every $\epsilon \in E$ construct the data-set consisting set of pairs $(scaller(\epsilon), w^k)$, $w^k \in V_h^{opt}$
6. fit NN_i for every $\epsilon \in E$ with data set $(scaling(\epsilon), w^i)_{i=1,\ldots,n_x}$ to generate coefficients approximating space V_h^{opt}.
7. Simulate equation:
   (a) for a given new $\hat{\epsilon}$ generate approximate w_{opt} by using $v_k = \mathsf{NN}_k(scaller(\hat{\epsilon}))$, $k = 1, \ldots, n_x$
   (b) calculate $u \approx u^{opt}$ by solving the equation:

$$\begin{bmatrix} b(B_{1,1}, v^1) & \cdots & b(B_{n_x,1}, v^1) \\ \vdots & \ddots & \vdots \\ b(B_{1,1}, v^{n_x}) & \cdots & b(B_{n_x,1}, v^{n_x}) \end{bmatrix} \boldsymbol{u} = \begin{bmatrix} l(v^1) \\ \vdots \\ l(v^{n_x}) \end{bmatrix}. \tag{28}$$

Table 1. The table shows relation between number of epochs and the mean squared percentage error (MAPE) for $n = 11$.

MAPE	Number of epochs				
Test function	100	300	1000	3000	10000
2	104.732	26.100	6.229	2.371	0.374
3	110.895	23.452	0.826	0.988	0.224
4	102.360	23.616	4.977	1.811	2.848
5	130.999	15.936	7.905	5.779	0.546
6	11971.054	149.536	14.874	6.198	3.352
7	90.811	37.762	6.205	1.578	2.661
8	98.659	14.200	11.952	1.851	0.034
9	104.788	18.215	10.215	0.561	0.002
10	114.502	25.796	1.870	0.806	0.282

We can check what is the minimum number of samples (epsilons) to train the neural network to solve the equation with satisfactory low error.

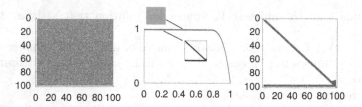

Fig. 8. Trial space linear B-splines, test space quadratic B-splines with C^0 separators, 100 elements, $\epsilon = 0.1$. (Left panel:) Dense matrix from optimal test functions. (Right panel:) Sparse matrix with low rank terms removed. (Middle panel:) Solutions from dense matrix and modified sparse matrix.

3.2   Dealing with Global Optimal Test Functions

The optimal test functions are global, and thus they cannot be used directly for efficient computations, since the Petrov-Galerkin system build with the optimal test functions is global (see left panel in Fig. 8).

Given a trial function $u_h \in U_h$, the corresponding optimal test function v realizes the supremum defining V'-norm of Bu_h, so it satisfies

$$v = argmax_{w \in V} \frac{b(u_h, w)}{\|w\|_V}. \tag{29}$$

Since $b(u_h, v)$ is defined by an integral over the domain Ω of some expression involving products of u_h, v and their gradients, behavior of v outside the support K of u_h is irrelevant to the value of $b(u_h, v)$, but does influence $\|v\|_V$. Let $g = v|_{\partial K}$ be the trace of v on ∂K and consider a function $w \in H^1(\Omega \setminus K)$ such that $w|_{\partial \Omega} = 0$, $w|_{\partial K} = g$. Such w can be extended to $\tilde{w} \in H_0^1(\Omega) = V$ by

$$\tilde{w}(x) = \begin{cases} w(x), & x \notin K \\ v(x), & x \in K \end{cases}. \tag{30}$$

Then $\tilde{w}|_K = v|_K$, and so $b(u_h, \tilde{w}) = b(u_h, v)$. By the maximization property of v, $\|\tilde{w}\|_V \geq \|v\|_V$. As $\tilde{w}$ and v are equal on K, it follows that

$$\int_{\Omega \setminus K} \|\nabla v\|^2 \, dx \leq \int_{\Omega \setminus K} \|\nabla w\|^2 \, dx. \tag{31}$$

and so v restricted to $\Omega \setminus K$ has the minimal L^2-norm of the gradient among all the functions satisfying boundary conditions $\cdot|_{\partial \Omega} = 0$ and $\cdot|_{\partial K} = g$. By Dirichlet's principle, v is the solution of the Laplace equation $\Delta u = 0$ on $\Omega \setminus K$ with these boundary conditions. For example, in 1D case where Ω and K are intervals, solution of such boundary value problem is a linear function, hence optimal test functions outside the support of the corresponding trial function decrease linearly until reaching 0 at the boundary.

Let us illustrate it on the advection-dominated diffusion equations with 100 elements in one dimension. We show on Fig. 9 two exemplary basis functions v_{20} and v_{100} and how they change with ϵ.

The basis of all optimal test functions for 100 elements with trial space of linear B-splines and test space of quadratic B-splines with C^0 separators are presented on left panel in Fig. 10. We can replace this basis by a linear combination, where all except the last two optimal test functions have local supports (they are equal to a very small number on the other parts of the domain, and this number results from the differences of slopes of the original functions and it decreases to zero when we increase the number of elements). Using these basis functions for Petrov-Galerkin formulation results in a sparse matrix that can be factorized in a linear cost (see right panel in Fig. 8). Most of the off-diagonal terms are very small, and their contribution is low rank and can be neglected (neglecting them does not alter the solution, see the middle panel in Fig. 8). This localization of the optimal test functions can be generalized to two and three dimensions.

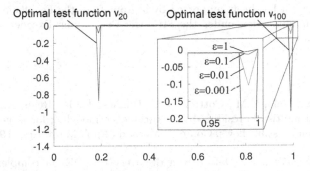

Fig. 9. Trial space with linear B-splines, 100 elements, test space with quadratic B-splines with C^0 separators, 100 elements. Two selected optimal test functions v_{20} and v_{100} for different ϵ.

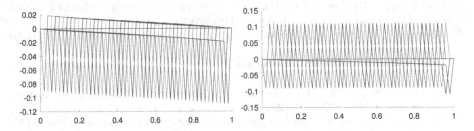

Fig. 10. Trial space linear B-splines, test space quadratic B-splines with C^0 separators, 100 elements. (Left panel:) All optimal test functions. (Right panel:) Optimal test functions replaced by a linear combination (each optimal test function subtracted from the previous one).

4   Conclusions

We showed that the neural network could learn coefficients of the optimal test functions for different parameters of the PDE. They allow for the automatic stabilization of advection-dominated diffusion problems. Moreover, the optimal test functions can be approximated with the function having local support, thus making the matrix of the coefficients of the global test functions sparse. We have verified our methodology using a one-dimensional advection-dominated diffusion problem. Future work will involve the generalization of the method into higher dimensions, as well as replacing the solver with the hierarchical matrices [7,18]. Some preliminary results on the hierarchical matrices solver for two-dimensional problems are discussed in [19].

Acknowledgement. The Authors are thankful for support from the funds assigned to AGH University of Science and Technology by the Polish Ministry of Science and Higher Education. Research project partly supported by program "Excellence initiative – research university" for the AGH University of Science and Technology.

The publication has been supported by a grant from the Faculty of Management and Social Communication under the Strategic Programme Excellence Initiative at Jagiellonian University.

References

1. Bazilevs, Y., Calo, V.M., Cottrell, J.A., Hughes, T.J.R., Reali, A., Scovazzi, G.: Variational multiscale residual-based turbulence modeling for large eddy simulation of incompressible flows. Comput. Methods Appl. Mech. Eng. **197**(1), 173–201 (2007)
2. Berg, J., Nystrom, K.: Data-driven discovery of PDEs in complex datasets. J. Comput. Phys. **384**, 239–252 (2019)
3. Brevis, I., Muga, I., van der Zee, K.: Data-driven finite elements methods: machine learning acceleration of goal-oriented computations (2020). Arxiv arXiv:2003.04485:1–24
4. Cottrell, J.A., Hughes, T.J.R., Bazilevs, Y.: Isogeometric analysis: toward integration of CAD and FEA. John Wiley & Sons (2009)
5. Demkowicz, L.: 2D hp-adaptive finite element package (2Dhp90) version 2.0. Ticam Report, 2:06 (2002)
6. Demkowicz, L., Rachowicz, W., Devloo, P.: A fully automatic hp-adaptivity. J. Sci. Comput. **17**(1–4), 117–142 (2002)
7. Hackbusch, W., Grasedyck, L., Börm, S.: An introduction to hierarchical matrices. Math. Bohem. **127**, 229–241 (2002)
8. Higham, C.F.: Deep learning: an introduction for applied mathematicians. SIAM Rev. **61**(4), 860–891 (2019)
9. Hughes, T.J.R., Cottrell, J.A., Bazilevs, Y.: Isogeometric analysis: CAD, finite elements, NURBS, exact geometry and mesh refinement. Comput. Methods Appl. Mech. Eng. **194**(39), 4135–4195 (2005)
10. Hughes, T.J.R., Franca, L., Mallet, M.: A new finite element formulation for computational fluid dynamics: vi. convergence analysis of the generalized supg formulation for linear time-dependent multidimensional advective-diffusive systems. Comput. Methods Appl. Mech. Eng. **63**(1), 97–112 (1987)

11. Łoś, M., Deng, Q., Muga, I., Paszyński, M., Calo., V.: Isogeometric residual minimization method (iGRM) with direction splitting preconditoner for stationary advection-diffusion problems. Comput. Methods Appl. Mech. Eng. (in press.) (2020). arXiv:1906.06727

12. Łoś, M., Munoz-Matute, J., Muga, I., Paszyński, M.: Isogeometric residual minimization method (iGRM) with direction splitting for non-stationary advection-diffusion problems. Comput. Math. Appl. **79**(2), 213–229 (2019)

13. Łoś, M., Munoz-Matute, J., Muga, I., Paszyński, M.: Isogeometric residual minimization for time-dependent stokes and navier-stokes problems. Comput. Math. Appl. (in press.) (2020). arXiv:2001.00178

14. Łoś, M., Rojas, S., Paszyński, M., Muga, I., Calo, V.: Discontinuous galerkin based isogeometric residual minimization for the stokes problem (DGIRM). In: Invited to the Special Issue of Journal of Computational Science on 20th Anniversary of ICCS Conference (2020)

15. Opschoor, J.A.A., Petersen, P.C., Schwab, C.: Deep ReLU networks and high-order finite element methods. Anal. Appl. **18**(5), 715–770 (2020)

16. Paszyński, M., Służalec, T.: Petrov-galerkin formulation equivallent to the residual minimization method for finding an optimal test functions. ECCOMAS Congress, 5–9 June 2022, Oslo, Norway (2022)

17. Pietro, D., Ern, A.: Mathematical Aspects of Discontinuous Galerkin Methods. Springer (2011). https://doi.org/10.1007/978-3-642-22980-0

18. Schmitz, P., Ying, L.: A fast nested dissection solver for cartesian 3D elliptic problems using hierarchical matrices. J. Comput. Phys., 258:227–245 (2014)

19. Służalec, T., Dobija, M., Paszyńska, A., Muga, I., Paszyński, M.: Automatic stabilization of finite-element simulations using neural networks and hierarchical matrices (2022). arxiv.org/abs/2212.12695

20. Tsihrintzis, G., Sotiropoulos, D.N., Jain, L.C.: Machine learning paradigms: advances in data analytics. Springer (2019). https://doi.org/10.1007/978-3-319-94030-4

21. Jinchao, X.: Finite neuron method and convergence analysis. Commun. Comput. Phys. **28**, 1707–1745 (2020)

Long-Term Prediction of Cloud Resource Usage in High-Performance Computing

Piotr Nawrocki[(✉)] [iD] and Mateusz Smendowski

AGH University of Krakow, Kraków, Poland
`piotr.nawrocki@agh.edu.pl`, `smendowski@student.agh.edu.pl`

Abstract. Cloud computing is gaining popularity in the context of high-performance computing applications. Among other things, the use of cloud resources allows advanced simulations to be carried out in circumstances where local computing resources are limited. At the same time, the use of cloud computing may increase costs. This article presents an original approach which uses anomaly detection and machine learning for predicting cloud resource usage in the long term, making it possible to optimize resource usage (through an appropriate resource reservation plan) and reduce its cost. The solution developed uses the XGBoost model for long-term prediction of cloud resource consumption, which is especially important when these resources are used for advanced long-term simulations. Experiments conducted using real-life data from a production system demonstrate that the use of the XGBoost model developed for prediction allowed the quality of predictions to be improved (by 16%) compared to statistical methods. Moreover, techniques using the XGBoost model were able to predict chaotic changes in resource consumption as opposed to statistical methods.

Keywords: Cloud computing · Resource prediction · High-performance computing · Machine learning

1 Introduction

Cloud computing is increasingly becoming an alternative to supercomputers for some high-performance computing (HPC) applications [6,8]. Among the potential use areas of cloud computing are various advanced simulations such as, for instance, HPC simulations of disease spread or social simulations [21]. Hybrid environments are also being used, where part of the sensitive computation is performed locally and part in a public cloud [14]. Public clouds can be used where the demand for computing resources is greater and cannot be satisfied by local resources. Of course, the cost of using cloud resources in HPC has to be taken into account, which is why the issue of optimizing their use is so important. In this context, it is very important to be able to predict the consumption of cloud resources in order to reserve them optimally. Reserving too many resources may result in increased costs and reserving too few resources may cause problems with simulation execution. Although cloud resource usage prediction (and

J. Mikyška et al. (Eds.): ICCS 2023, LNCS 14074, pp. 532–546, 2023.
https://doi.org/10.1007/978-3-031-36021-3_53

subsequent scheduling) is discussed in the literature, it is almost always in the context of autoscaling and short-term prediction. For HPC applications such as advanced simulations, many tasks may require resources in the long term, and therefore the time horizon of mechanisms such as autoscaling and short-term prediction may prove too short for this type of computing. The long-term prediction solution developed by the authors (using machine learning) allows for a longer prediction horizon, which can be useful for advanced simulations using HPC.

The major contributions of this paper can be summarized as follows:

- designing a solution that provides long-term resource usage predictions for advanced simulations using HPC;
- developing a self-adapting system that can operate in production-grade environments with long-term load changes;
- conducting evaluation using data collected from a real-life production system;
- comparing the results obtained using different prediction techniques based on the XGBoost model with statistical methods.

The rest of this paper is structured as follows: Sect. 2 contains a description of related work, Sect. 3 focuses on the description of a long-term cloud resource usage prediction system, Sect. 4 describes the experiments performed, and Sect. 5 contains the conclusion and further work.

2 Related Work

There are many publications dealing with the use of cloud computing resources for HPC applications. In [7], the author discusses the use of a virtual cluster for this type of computing (using Elastic Computing Cloud – EC2 as an example). Clusters of computer systems are a common HPC architecture. However, small local clusters, although cost-effective, may not be efficient enough. On the other hand, large clusters are more expensive and it is not always possible to provide them with a sufficient sustainable load. Therefore, virtual clusters using cloud resources offer a viable alternative. In this case, public commercial cloud environments provide users with storage and CPU power to build their own dedicated clusters of computers that can be used in scientific computing applications. Based on the experiments performed, the author shows that it is possible to use a virtual cluster to realize various computations including HPC. The author also analyzes the cost of using cloud solutions; however, he does not explore the possibility of optimization of cloud resource consumption.

In [5], the authors discuss scalability, interoperability, and achieving guaranteed Quality of Service (QoS) in a High-Performance Computing Cloud (HPCC) environment. The authors propose a cloud resource management framework to handle a large number of user HPC application requests and manage multiple cloud resources. System tests were performed using a large number of real-world HPC applications. To evaluate the performance of the system proposed, the authors used performance metrics such as response time, the number of requests successfully handled and user satisfaction. What is missing from this work, however, is an analysis of the cost of using cloud resources using the system developed and the possibility of optimizing them.

In [17], the authors present the possibilities of using HPC in the Google Cloud Platform in emergency situations when efficient processing of large amounts of traffic data is required. This allows for effective disaster management using massive data processing and high performance computing. Another application of HPC in cloud computing is presented in [9] where the authors describe a platform for computer vision applications that enables audio/video processing and can use high performance computing cloud resources for this purpose.

An extensive analysis of the possibilities of cloud solutions related to High Performance Computing, among other things, is presented in [2]. The authors analyze the capabilities of the four most popular environments – Amazon Elastic Compute Cloud, Microsoft Azure Cloud, Google Cloud and Oracle Cloud. Today, HPC workloads are increasingly being migrated to the cloud. This is due, among other things, to the flexible nature of the cloud where resources can be expanded and reduced on demand, which optimizes the cost of using cloud computing. At the same time, computationally intensive workloads can be performed efficiently on cloud resources that are connected via a high-speed network. The most suitable cloud model for HPC users is Infrastructure as a Service (IaaS), where it is possible to specify all the details of the infrastructure needed to create clusters. There are many areas where the cloud is being used for HPC, including financial risk simulations, molecular dynamics, weather prediction, and scientific or engineering simulations.

As it can be gathered from the above analysis, the use of cloud resources for HPC is now commonplace. There are also appropriate mechanisms for managing cloud resources so as to adjust them for HPC purposes and, for instance, predict the placement of containers [1]. However, there is a lack of mechanisms to optimally reserve cloud resources for HPC, especially in the long term. One way to conduct such optimization is to use prediction mechanisms to determine what resources will be required in the future for HPC and establish an optimal reservation plan. The most common strategy for using cloud computing resources is to reserve resources at the highest potential demand level, which, however, generates unnecessary costs for unused cloud resources. Predictive resource utilization can prevent this, helping to reserve only the cloud resources needed. However, the majority of research on cloud resource management and optimization concerns autoscaling [4,20], predictive autoscaling (autoscaling with workload forecasting) [18,22] and short-term prediction [3,19]; there are very few studies of long-term prediction.

The authors' earlier works indicate that cloud resource usage prediction makes it possible to create optimal plans for resource reservation, leading to significant savings. Those works included research on using machine learning and adaptive resource planning for cloud-based services [10], anomaly detection in the context of such planning [13], cloud resource usage prediction mechanisms taking into account QoS parameters [11,15], data-driven adaptive cloud resource usage prediction [12] and resource usage cost optimization in cloud computing [16]. However, none of the previous works analyzed HPC needs in the context of cloud computing resources. In the research described in this article, the authors paid particular attention to the need for long-term prediction (relevant to HPC) and used a full data set from more than one year to develop the model. In previous work (such as in [15,16]), only selected data from three months were considered. In addition, the XGBoost model was used, in which prediction was conducted for one-week periods with feature enrichment: using *lagged features* from the previous week, features derived from the Fast Fourier Transform (FFT), moving window statistics and calendar features, taking into account holidays.

3   Long-Term Cloud Resource Usage Prediction System

The prediction system developed was designed to forecast weekly CPU usage by predicting 168 sample resource usage metrics with hourly resolution. The system consists of seven modules and the starting point for its operation is the use of historical data to obtain an initial prediction (Fig. 1).

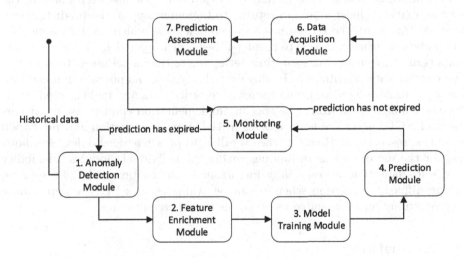

Fig. 1. Prediction system concept

At the start of prediction system operation, historical values of CPU usage metrics are recorded and fed into the *Anomaly Detection Module*. This module is flexible in terms of the ranges of data to which the anomaly detection model is applied. It is noteworthy that anomalies are understood as samples that are outlying in the data, effectively translating into drastic resource consumption peaks. Conceptually, anomaly detection can be applied either to the entire dataset or locally to smaller subsets of data independently. Such a configuration enables conducting multiple simulations and checking whether the anomalous data detected are undesirable or constitute a valuable contribution. Subsequently, the *Feature Enrichment Module* is responsible for enriching data in diverse ways to create an expressive data representation. For time-series, there are several techniques that can be implemented at that level to extract maximum valuable information from the dataset, which will positively impact the results of the prediction model. The training of the prediction model itself is carried out through the *Model Training Module*. After completing the training of the model and its initial evaluation on the validation set, either through walk-forward validation technique or on a separate validation set, long-term prediction for a weekly period is performed. Multi-step prediction can be achieved in various ways, with direct, iterative, and multi-output prediction being among the most popular. After performing the prediction using the Prediction Module, the system transitions from the outer loop to the inner loop. The *Monitoring Module* checks whether the prediction is valid at hourly intervals, which correspond to the granularity intervals of the recorded metrics. If the prediction has not expired and is valid, the resource usage level is recorded and metrics are stored (*Data Acquisition Module*) for the indicated time period, and the prediction previously made is verified against actual usage (*Prediction Assessment Module*). When the *Monitoring Module* detects that the expiration time has been reached, the system exits the inner loop and returns to the outer loop of the prediction system. At this point, the system has acquired new data, which can be scanned for anomalies, enriched and used to re-adapt the prediction model. This is especially important for dynamic, chaotic time series where the model needs to adjust to the current data distribution. In this way, the system can operate in a continuous optimization loop, adapting to newly recorded data and making predictions for the configured horizon. The specific implementation employs Isolation Forest and XGBoost as the main models in the anomaly detection and prediction modules, respectively. However, the overall concept is generic and flexible, allowing for the use of various techniques within the individual modules. Flexibility and self-adaptation are especially important in the context of HPC resources, where different prediction schemes can be evaluated to select the appropriate one regarding future resource reservations for supercomputing.

4   Evaluation

As part of the evaluation of the prediction system, historical records of CPU usage were utilized, with a particular focus on this metric among the various options available from cloud monitoring services. This data is considered crucial in the realm of high-performance computing, alongside Random Access Memory (RAM) usage. The data were collected from a real-life production environment. The recorded CPU usage metrics are closely tied to a timestamp, thus the data are represented in the form of a time series. Ultimately, historical records represent consolidated CPU utilization over a period of exactly 80 weeks, where metric values were registered at 1-h intervals. Moreover, several sequences in the acquired dataset were found to have missing data, which were imputed using a

Simple Moving Average (SMA) approach with a window size equal to the prediction horizon of 168 samples. The starting point for research was analyzing the data, specifically by arbitrarily partitioning them into training, validation, and testing sets. This was done to enable a focused application of methods to the training set, performance evaluation using the validation set, and critical assessment of generalization abilities without unwanted knowledge leakage from the testing set during the training process. Due to high variance in the dataset and significant changes in the characteristic features that amplify the chaotic nature of the time series, effectively distinguishing the sizes of the datasets proved to be a challenge. Ultimately, 10,920 and 2,520 samples were used for the training and testing sets, respectively. Additionally, the last 840 samples from the training set were selected as the validation set. In order to examine the nature of the dataset, a Jarque-Bera test was conducted on the training set (without the validation portion) to assess its normality and skewness. The initial output value (58.96), represented the test statistic of the Jarque-Bera test. This was used to evaluate the deviation of the sample data from a normal distribution, with a higher test statistic indicating a greater deviation from normality. The second value, 1.58×10^{-13}, represents the p-value of the test, which suggests strong evidence against the null hypothesis that the data is normally distributed. Furthermore, the stationary test indicated non-stationarity, thereby implying that the variable under investigation does not exhibit constant statistical properties over time.

The main prediction model utilized in the system developed is XGBoost, which was optimized for evaluation using grid search with selected parameters of 240 estimators, 7 as the maximum depth, and a learning rate of 0.28. To enhance prediction accuracy, various features were incorporated, including lagged features from the preceding week, exogenous features derived from the calendar, such as day of week, month, hour, holidays, and statistical parameters calculated within the scope of a moving window. Two sizes of moving windows (168 and 72) were selected to capture statistics from the last week and the last three days, respectively. The statistical features included classical metrics such as the median and the mean as well as more sophisticated statistics such as interquartile differences, values above the mean, skewness, kurtosis, and features resulting from decomposing the time series into the frequency domain using Fast Fourier Transform. Additionally, cyclical features from all calendar parameters were enriched using the cosine and sine functions. Cumulatively, the feature enrichment phase made it possible to obtain 250 features. The XGBoost algorithm was trained for single-step-ahead prediction. To enable long-term forecasting, an iterative approach was employed where each prediction was used as a training sample in the following step until the end of the prediction horizon. Features were computed for each sample in real time, in a manner preventing information leakage.

In the context of long-term prediction, identifying outlying samples that are understood as drastic and local fluctuations in the resource usage that do not translate into a permanent change in characteristics, and determining whether they contribute to the model as anomalies or provide valuable information are crucial factors. In order to tackle this challenge, we employed an unsupervised Isolation Forest model for evaluation. The parameters of the Isolation Forest model were also determined using grid search, with a selection of 90 estimators and a contamination fraction of 0.05. Other parameters were left at default values to reduce the search space, which, in the case of walk-forward validation, was less computationally complex. The strategy selected for handling outlying samples was their removal. Imputing the mean value of a certain horizon of samples was considered as well, but was not utilized due to the rich feature engineering approach that also took into account statistical parameters calculated within the windows in question.

Prior to evaluating the implemented prediction system based on the XGBoost model, certain reference results were established using statistical methods, which served as a baseline. Typically, naive prediction (repeating the last observed value throughout the entire prediction horizon) or more sophisticated methods such as Simple Exponential Smoothing serve as a solid reference point. However, for comparison purposes, neither of the aforementioned methods nor other methods such as Holt's Exponential Smoothing with a damped trend were used. Although these methods can produce good error metric values, they do not reflect the data characteristics that are crucial for resource demand prediction in HPC. Consequently, Holt-Winter's Seasonal Smoothing methods were used as baselines, namely *Method 1* (with a 1-day seasonality), *Method 2* (with a 3-day seasonality), and finally *Method 3* (with a 1-week seasonality). At the end of each prediction horizon, statistical methods were re-adapted to account for the new data. The listed seasonality periods to be explored were determined by analyzing the autocorrelation function (ACF) and partial autocorrelation function (PACF) as well as by decomposing the data into trend and seasonality, both in terms of additive and multiplicative models. However, at the data exploration stage, it was discovered that no singular principal seasonal/frequency component was present, a finding which was additionally confirmed using the FFT (Fast Fourier Transform) and DWT (Discrete Wavelet Transform) techniques.

Figure 2 presents the forecasting results using three baseline methods. In cases where the time series exhibit stability over a given period, simple methods can adequately reflect utilization values. However, naive methods and the usual statistical models that often assume linear and stationary data, while performing well when the characteristics of the data are stable, may fall short in predicting sudden fluctuations, chaotic increases or decreases in resource utilization, which can be observed in Fig. 3. As far as the analysis of error metrics is concerned, it should be carried out in tandem with the examination of trend characteristics to objectively assess the generalization capabilities of methods as well as their strengths and potential weaknesses where inefficiencies may arise. Table 1 summarizes the error metric values obtained for the entire testing set for baseline methods. *Method 2* has the lowest RMSE (Root Mean Squared Error) and NRMSE (Normalized RMSE) values, which indicate a lower overall error in prediction compared to the other methods. Additionally, *Method 2* has a slightly higher MAE (Mean Absolute Error) compared to *Method 1*. On the other hand, *Method 3* exhibits the poorest performance, except for the MAPE (Mean Absolute Percentage Error), where typically the highest amplitudes and oscillations in forecasts can be observed, and in majority of cases forecasts are below ground truth. The MdAE (Median Absolute Error) indicated a clear advantage for *Method 1*. Finally, based on this analysis, *Method 1* and *Method 2* yield comparably promising results. However, despite their relatively good prediction results in terms of error metrics, these methods completely failed to handle sudden changes in resource consumption. This provides a strong motivation to explore more advanced prediction systems or models, particularly in HPC systems where overprovisioning leads to high costs and underprovisioning results in undesired drops in QoS. The usage of such baselines in production may heighten the risk of user service unavailability.

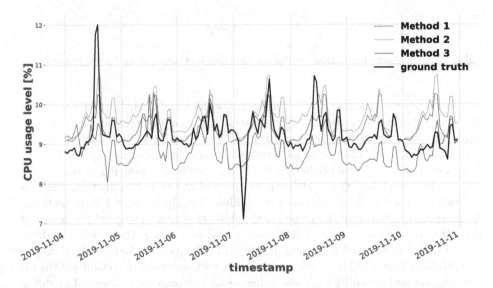

Fig. 2. CPU usage prediction results for baseline methods (first sample period)

Table 1. Error metrics for baseline methods obtained for the entire test set

Reference	RMSE	NRMSE	MAE	MAPE	MdAE
Method 1	2.223	0.141	0.914	0.267	0.327
Method 2	2.168	0.137	0.936	0.276	0.374
Method 3	2.339	0.148	0.999	0.261	0.415

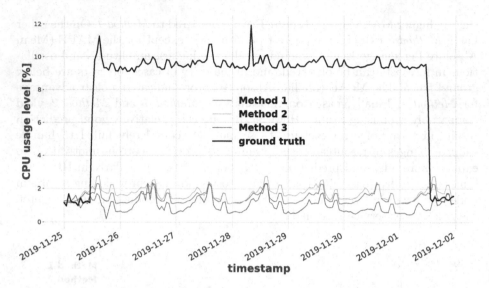

Fig. 3. CPU usage prediction results for baseline methods (second sample period)

Following the conclusions drawn from baselines and their limited applicability, in order to evaluate the prediction system based on the XGBoost model, multiple different simulations and comparisons were conducted, resulting in eleven prediction methods that serve as a robust foundation for drawing conclusions. When evaluating diverse prediction methods, specific components of the prediction system were selectively utilized, as some of the tested prediction methods did not employ the anomaly detection module. Firstly, *Method 4* represents the basic approach, which utilizes a one-week prediction system without the anomaly detection module and without XGBoost model retraining (thus without adapting to newly acquired CPU usage metrics). Subsequently, *Method 5*, *Method 6* and *Method 7* extend the previous approach with anomaly detection within the training set and handling such anomalies using the removal strategy. The differences between these methods concern the scope of anomaly detection: the detection was applied to the entire training set, 4-week portions and weekly portions, respectively. In the context of time series data, it is crucial to determine whether samples identified as anomalies introduce noise and perturbations to the prediction models or rather provide valuable information for generalization purposes. Furthermore, *Method 8* involves refitting the model after the weekly prediction validity period expires, while taking into account newly collected metrics. However, the anomaly detection module was not involved. Subsequently, *Method 9*, *Method 10* and *Method 11* extend the periodical (weekly) re-training approach but include anomaly detection also for the entire training set, 4-week and 1-week portions, respectively. Finally, *Method 12*, *Method 13* and *Method 14* are basically extensions of *Methods 9*, *10* and *11*, but Anomaly Detection is additionally applied to newly collected data before each refit so all data on which the model is trained are scanned against the outliers. Figure 4 indicates that in the case of stable resource usage values, it is evident that the prediction characteristics of the methods applied closely match real-life CPU consumption values from the test set. Particularly noteworthy and inspiring results that lie at the heart of the problem are presented in Fig. 5, where the methods predicted an increase in consumption, approaching the actual recorded consumption, which was entirely missed by the baseline. The predictions dynamically followed the consumption trend, with various methods reacting differently but each responding to the possible change in resource usage trend. This indicates a huge advantage of the implemented prediction model over baselines, which results from its dynamics and reactive attitude.

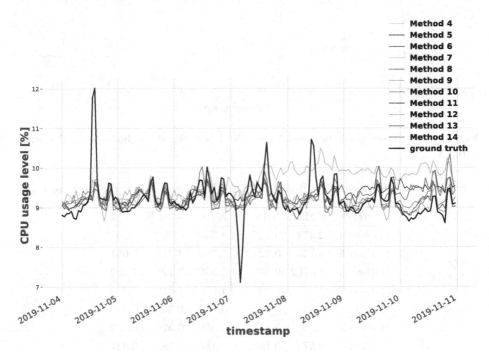

Fig. 4. CPU usage prediction results for XGBoost-based methods (first sample period)

As concerns the error metrics of methods that did not involve periodic adaptation to new data, it was found that anomaly detection within this specific dataset did not contribute to improved accuracy (Table 2). The values identified by the anomaly detection module as outliers were found to have valuable positive impact. Therefore, *Method 4* outperformed *Methods 5, 6* and *7* as removing anomalies led to the removal of valuable information and a dramatic reduction in accuracy. Moreover, the local applicability of anomaly detection also contributed to a significant accumulation of errors because scanning was applied without the broader context that was necessary for these data and for this model. *Method*

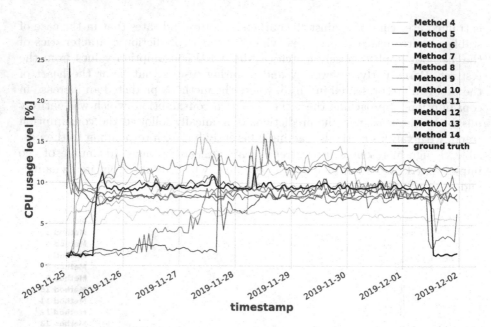

Fig. 5. CPU usage prediction results for XGBoost-based methods (second sample period)

Table 2. Error metrics for XGBoost-based methods obtained for the entire test set

Reference	RMSE	NRMSE	MAE	MAPE	MdAE
Method 4	2.477	0.157	1.355	0.502	0.407
Method 5	6.757	0.428	5.756	3.316	7.051
Method 6	14.762	0.936	9.352	5.121	1.882
Method 7	19.45	1.233	13.565	7.819	5.353
Method 8	1.814	0.115	0.909	0.335	0.335
Method 9	2.701	0.171	1.359	0.526	0.332
Method 10	2.573	0.163	1.185	0.58	0.321
Method 11	2.332	0.148	1.082	0.317	0.325
Method 12	2.627	0.167	1.195	0.546	0.321
Method 13	3.209	0.203	1.851	0.75	0.419
Method 14	3.209	0.192	1.399	0.646	0.334

8, which incorporated re-adaptation to new data and no anomaly detection, proved best in terms of RMSE, NRMSE and MAE metrics. The assessment of this method improved significantly because it uses walk forward validation. In a chaotic time series, validation error may sometimes be smaller than test error due to the variability of data distribution over time and differences between sample distributions among the training, validation and test sets. Chaotic time series are characterized by high variability and unpredictability, and models trained on training data may not be able to generalize correctly to new data. In this case, using a validation set may lead to a situation where the model is strictly optimized for validation data, which may represent only a small subset of test data. To prevent this, cross-validation techniques allow the model to be evaluated on multiple data subsets, reducing the risk of overfitting and allowing a more general model to be obtained. One advantage of this approach is that the model is trained on an increasing amount of data, which allows for continuous improvement over time. However, walk forward cross-validation in connection with grid search turned out be time-consuming, because it requires multiple iterations that involve training and testing the model on successive time windows. Therefore, it was important to carefully consider the appropriate validation method for the specific problem at hand. Refitting the model on an increasing amount of data enabled a better generalization to be to achieved. Additionally, all the subsequent methods incorporated this approach but also used anomaly detection. The results obtained using these methods confirmed that samples identified as anomalies in this particular time series have high information value and should not be discarded from the training set. However, this behavior should be constantly monitored as data characteristics may change over time. Generally, in the case of chaotic time series, in most cases the resource usage metrics recorded from HPC systems are in a form which makes continuous adaptation to new data useful, as data distribution is very dynamic and needs to be taken into account in order to maintain the accuracy of the predictive model.

5 Conclusions

Currently, cloud computing cannot fully replace supercomputers for applications such as advanced simulations using HPC, but it may play a complementary role, allowing calculations to be conducted where local resources are limited. In this work, we present a solution that enables long-term prediction of cloud resource consumption. A longer time horizon of resource consumption prediction is especially important when HPC is used for various simulations, which often run for a longer period of time. The proposed prediction system utilizes the XGBoost model and over a year of historical data to predict CPU resource usage. Experiments are conducted using real-life data from the production system. Their results indicate that proactive resource usage prediction, which models future usage and enables dynamic resource reservation in advance, is superior to reactive approaches. In the traditional approach, resources are reserved at their maximum capacity. In normal usage scenarios without prediction, reservation remains at the highest level, even during periods of low demand. However, with the use of dynamic reservations, predictive capabilities are leveraged to scale resources dynamically in response to fluctuating demand.

It is worth noting that in our system, we have observed the superiority of the prediction system based on XGBoost over traditional statistical methods selected as the baselines. The best prediction method developed was found to outperform the best baseline model by approximately 16% in terms of RMSE. In addition to quantitative evaluation, the model exhibited significantly better qualitative responses to chaotic changes in data characteristics, suggesting that it is a more reliable predictor of system behavior. The output of the prediction system can subsequently be utilized to create a resource reservation plan, which can be used to scale resources.

Standard supervised learning operates under the assumption that each sample is drawn independently from an identical underlying distribution. However, chaotic time series and data from HPC systems are dynamic and often exhibit changes in data distribution, non-linearities, and non-stationarity, which makes accurate predictions difficult. Long-term prediction is a less explored area compared to the more prevalent short-term prediction, especially predicting one step ahead. However, the longer the prediction horizon, the less certain it becomes, but is at the same time more valuable, since it enables more informed decision-making, cost management and optimization of energy consumption. Setting a single universal rule is challenging, but evaluating multiple methods, especially using an incremental approach, allows strong conclusions to be drawn. In-depth exploratory data analysis may result in becoming more familiar with the data and putting forward solid predictions.

In further research, we would like to evaluate several models, such as LSTM (Long short-term memory Neural Network), TCN (Temporal Convolutional Network) and TFT (Temporal Fusion Transformer), to compare their effectiveness and conduct experiments using real HPC monitoring data. Additionally, we would like to explore multivariate prediction to incorporate dependencies between various HPC resource metrics, such as CPU, RAM, I/O, etc. Moreover, our objective is to dynamically evaluate the importance of features and reduce feature dimensionality by selecting a specific number of features for the model and summarizing the remaining features via Principal Component Analysis (PCA) or t-SNE. In addition, long-term prediction is of great significance and critical in HPC as it enables a balance between cost reduction and maintaining high-quality QoS by provisioning an appropriate amount of dynamically scalable resources rather than relying on statically reserved resources.

Acknowledgements. The research presented in this paper was supported by funds from the Polish Ministry of Education and Science allocated to the AGH University of Krakow.

References

1. Abhishek, M.K., Rajeswara Rao, D.: A scalable framework for high-performance computing with cloud. In: Tuba, M., Akashe, S., Joshi, A. (eds.) ICT Systems and Sustainability, pp. 225–236. Springer, Singapore (2022). https://doi.org/10.1007/978-981-16-5987-4_24

2. Aljamal, R., El-Mousa, A., Jubair, F.: A user perspective overview of the top infrastructure as a service and high performance computing cloud service providers. In: 2019 IEEE Jordan International Joint Conference on Electrical Engineering and Information Technology (JEEIT), pp. 244–249 (2019). https://doi.org/10.1109/JEEIT.2019.8717453

3. Chen, X., Wang, H., Ma, Y., Zheng, X., Guo, L.: Self-adaptive resource allocation for cloud-based software services based on iterative QoS prediction model. Fut. Gen. Comput. Syst. **105**, 287–296 (2020). https://doi.org/10.1016/j.future.2019.12.005, https://www.sciencedirect.com/science/article/pii/S0167739X19302894

4. Gari, Y., Monge, D.A., Pacini, E., Mateos, C., Garino, C.G.: Reinforcement learning-based application autoscaling in the cloud: a survey. Eng. Appl. Artif. Intell. **102**, 104288 (2021). https://doi.org/10.1016/j.engappai.2021.104288. ISSN 0952-1976

5. Govindarajan, K., Kumar, V.S., Somasundaram, T.S.: A distributed cloud resource management framework for high-performance computing (HPC) applications. In: 2016 Eighth International Conference on Advanced Computing (ICoAC), pp. 1–6 (2017). https://doi.org/10.1109/ICoAC.2017.7951735

6. Gupta, A., et al.: Evaluating and improving the performance and scheduling of HPC applications in cloud. IEEE Trans. Cloud Comput. **4**(3), 307–321 (2014)

7. Hazelhurst, S.: Scientific computing using virtual high-performance computing: a case study using the amazon elastic computing cloud. SAICSIT 2008, New York, NY, USA, pp. 94–103. Association for Computing Machinery (2008). https://doi.org/10.1145/1456659.1456671, https://doi.org/10.1145/1456659.1456671

8. Kotas, C., Naughton, T., Imam, N.: A comparison of amazon web services and Microsoft azure cloud platforms for high performance computing. In: 2018 IEEE International Conference on Consumer Electronics (ICCE), pp. 1–4 (2018). https://doi.org/10.1109/ICCE.2018.8326349

9. Mahmoudi, S.A., Belarbi, M.A., Mahmoudi, S., Belalem, G., Manneback, P.: Multimedia processing using deep learning technologies, high-performance computing cloud resources, and big data volumes. Concurr. Comput. Pract. Exp. **32**(17), e5699 (2020). https://doi.org/10.1002/cpe.5699, https://onlinelibrary.wiley.com/doi/abs/10.1002/cpe.5699

10. Nawrocki, P., Grzywacz, M., Sniezynski, B.: Adaptive resource planning for cloud-based services using machine learning. J. Parallel Distrib. Comput. **152**, 88–97 (2021). https://doi.org/10.1016/j.jpdc.2021.02.018, https://www.sciencedirect.com/science/article/pii/S0743731521000393

11. Nawrocki, P., Osypanka, P.: Cloud resource demand prediction using machine learning in the context of QoS parameters. J. Grid Comput. **19**(2), 1–20 (2021). https://doi.org/10.1007/s10723-021-09561-3

12. Nawrocki, P., Osypanka, P., Posluszny, B.: Data-driven adaptive prediction of cloud resource usage. J. Grid Comput. **21**(1), 6 (2023)

13. Nawrocki, P., Sus, W.: Anomaly detection in the context of long-term cloud resource usage planning. Knowl. Inf. Syst. **64**(10), 2689–2711 (2022)

14. Netto, M.A., Calheiros, R.N., Rodrigues, E.R., Cunha, R.L., Buyya, R.: HPC cloud for scientific and business applications: taxonomy, vision, and research challenges. ACM Comput. Surv. (CSUR) **51**(1), 1–29 (2018)

15. Osypanka, P., Nawrocki, P.: QoS-aware cloud resource prediction for computing services. IEEE Trans. Serv. Comput. 1–1 (2022). https://doi.org/10.1109/TSC.2022.3164256

16. Osypanka, P., Nawrocki, P.: Resource usage cost optimization in cloud computing using machine learning. IEEE Trans. Cloud Comput. **10**(3), 2079–2089 (2022). https://doi.org/10.1109/TCC.2020.3015769
17. Posey, B., et al.: On-demand urgent high performance computing utilizing the google cloud platform. In: 2019 IEEE/ACM HPC for Urgent Decision Making (UrgentHPC), Los Alamitos, CA, USA, pp. 13–23. IEEE Computer Society, November 2019. https://doi.org/10.1109/UrgentHPC49580.2019.00008, https://doi.ieeecomputersociety.org/10.1109/UrgentHPC49580.2019.00008
18. Radhika, E., Sudha Sadasivam, G.: A review on prediction based autoscaling techniques for heterogeneous applications in cloud environment. Mat. Today Proc. **45**, 2793–2800 (2021). https://doi.org/10.1016/j.matpr.2020.11.789, https://www.sciencedirect.com/science/article/pii/S2214785320394657, international Conference on Advances in Materials Research - 2019
19. Rahmanian, A.A., Ghobaei-Arani, M., Tofighy, S.: A learning automata-based ensemble resource usage prediction algorithm for cloud computing environment. Fut. Gen. Comput. Syst. **79**, 54–71 (2018). https://doi.org/10.1016/j.future.2017.09.049, https://www.sciencedirect.com/science/article/pii/S0167739X17309378
20. Singh, P., Gupta, P., Jyoti, K., Nayyar, A.: Research on auto-scaling of web applications in cloud: survey, trends and future directions. Scalable Comput. Pract. Exp. **20**(2), 399–432 (2019)
21. Wittek, P., Rubio-Campillo, X.: Scalable agent-based modelling with cloud HPC resources for social simulations. In: 4th IEEE International Conference on Cloud Computing Technology and Science Proceedings. pp. 355–362. IEEE (2012)
22. Xue, S., et al.: A meta reinforcement learning approach for predictive autoscaling in the cloud. In: Proceedings of the 28th ACM SIGKDD Conference on Knowledge Discovery and Data Mining. KDD 2022, New York, NY, USA, pp. 4290–4299. Association for Computing Machinery (2022). https://doi.org/10.1145/3534678.3539063, https://doi.org/10.1145/3534678.3539063

Least-Squares Space-Time Formulation for Advection-Diffusion Problem with Efficient Adaptive Solver Based on Matrix Compression

Marcin Łoś[1] , Paulina Sepúlveda[2] , Mateusz Dobija[3,4] ,
and Anna Paszyńska[3(✉)]

[1] Institute of Computer Science, AGH University of Science and Technology,
Al. Mickiewicza 30, Kraków, Poland
los@agh.edu.pl
[2] Instituto de Matemáticas, Pontificia Universidad Católica de Valparaíso, Casilla,
4059 Valparaiso, Chile
paulina.sepulveda@pucv.cl
[3] Faculty of Physics, Astronomy and Applied Computer Science,
Jagiellonian University, ul. prof. Stanisława Łojasiewicza 11, Kraków, Poland
mateusz.dobija@doctoral.uj.edu.pl, anna.paszynska@uj.edu.pl
[4] Doctoral School of Exact and Natural Sciences, Jagiellonian University,
Kraków, Poland

Abstract. We present the hierarchical matrix compression algorithms
to speed up the computations to solve unstable space-time finite element
method. Namely, we focus on the non-stationary time-dependent advection dominated diffusion problem solved by using space-time finite element method. We formulate the problem on the space-time mesh, where
two axes of coordinates system denote the spatial dimension, and the
third axis denotes the temporal dimension. By employing the space-time
mesh, we avoid time iterations, and we solve the problem "at once" by
calling a solver once for the entire mesh. This problem, however, is challenging, and it requires the application of special stabilization methods.
We propose the stabilization method based on least-squares. We derive
the space-time formulation, and solve it using adaptive finite element
method. To speed up the solution process, we compress the matrix of
the space-time formulation using the low-rank compression algorithm.
We show that the compressed matrix allows for quasi-linear computational cost matrix-vector multiplication. Thus, we apply the GMRES
solver with hierarchical matrix-vector multiplications. Summing up, we
propose a quasi-linear computational cost solver for stabilized space-time
formulations of advection dominated diffusion problem.

Keywords: Finite element method · Space-time formulation ·
Isogeometric analysis · H-matrices · Matrix compression · SVD

J. Mikyška et al. (Eds.): ICCS 2023, LNCS 14074, pp. 547–560, 2023.
https://doi.org/10.1007/978-3-031-36021-3_54

1   Introduction

With increasing supercomputer power, the space-time finite element method is becoming more and more popular. The method employs an n-dimensional space-time mesh, with $n-1$ axes corresponding to the spatial dimension, and one axis corresponding to the temporal dimension. One of the advantages of the method is the fact that we can refine the computational mesh in space-time domain. The space-time formulation does not process a sequence of computational meshes from consecutive time moments. We formulate and solve the problem on one big mesh, and we can simultaneously refine the mesh to improve the quality of the solution in space and time.

The problem of developing stabilization methods for space-time finite element is a very important scientific topic nowadays. There are several attempts do develop stabilized FEM solver. Different methods have been employed for this purpose. Paper [1] employs space-time stabilized formulation using an adaptive constrained first-order system with the least squares method. Another space-time discretization for the constrained first-order system least square method (CFOSLS) are discussed in [2]. It is also possible to employ Discontinuous Petrov-Galerkin method for the stabilization of the space-time formulation, as it is illustrated for the Schrödinger equation in [5] and for the acoustic wave propagation in [6,7]. The least-square finite element method has been applied for the stabilization of the parabolic problem in [8].

The most crucial aspect when developing the space time formulations is the computational cost of the solver [10]. Paper [3] summarizes different fast solvers for space-time formulations. In [4] the authors discuss the applications of the algebraic multigrid solvers for an adaptive space-time finite-element discretization in 3D and 4D.

The hierarchical matrices have been introduced by Hackbush [11,12]. They employ the low-rank compression of matrix blocks to speed up the solution process.

In this paper, we present the hierarchical matrix compression algorithms to speed up the computations of difficult, unstable space-time finite element method together with the stabilization method. The described approach is used to solve the stabilized space-time formulations of advection dominated diffusion problem with quasi-linear computational time.

The novelties of our paper are:

- We consider advection-dominated diffusion transient problem formulated in space-time finite elements with the stabilization based on the least squares method.
- We employ adaptive finite element algorithm implemented in FeniCS library refining the elements in the space-time domain.
- We introduce the idea of the hierarchical matrices for the space-time formulation. We generate and compress the matrices resulting from the finite element method discretization using the truncated SVD algorithm applied for blocks of the global matrix.

– We show that the hierarchical matrices can be processed by the GMRES iterative solver one order of magnitude faster than regular matrices.

2   Model Problem

As a model problem, we study the advection-diffusion equation over a domain:

$$\partial_t \phi = \varepsilon \Delta \phi - \beta \cdot \nabla u + f,$$
$$\phi(x, 0) = u_0 \quad for \ x \in \Omega,$$
$$\phi(x, t) = 0 \quad for \ (x, t) \in \partial\Omega \times [0, T].$$

For small $\varepsilon / \|\beta\|$, the problem is advection-dominated and the standard Galerkin method encounters stability issues.

3   Space-Time Formulation

The space-time formulation we employ is a first-order formulation based on the idea of a constrained least squares problem (CFOSLS), and has been first introduced in [2].

3.1   First-Order Formulation

Let $\Omega_T = \Omega \times (0, T)$ denote the space-time domain, and $\Gamma_S = \partial\Omega \times (0, T)$ denote the spatial boundary. We start by writing the equation in the divergence form

$$\partial_t \phi + \mathrm{div}_{\mathsf{x}} \underbrace{(-\varepsilon\nabla\phi + \beta\phi)}_{\mathcal{L}\phi} = f, \tag{1}$$

which allows us to reformulate it as

$$\mathrm{div}_{\mathsf{x},\mathsf{t}} \, \boldsymbol{\sigma} = f, \tag{2}$$

where $\underline{\boldsymbol{\sigma}} = (\mathcal{L}\phi, \phi)$ and $\mathcal{L}\phi = (\mathcal{L}_x\phi, \mathcal{L}_y\phi)$, and $\mathrm{div}_{\mathsf{x},\mathsf{t}}$ denotes the full space-time divergence operator (as opposed to the spatial divergence $\mathrm{div}_{\mathsf{x}}$). Introducing $\underline{\boldsymbol{\sigma}}$ as a new unknown, we can rewrite our equation as a first-order system

$$\begin{cases} \mathrm{div}_{\mathsf{x},\mathsf{t}} \, \underline{\boldsymbol{\sigma}} = f, \\ \underline{\boldsymbol{\sigma}} - \begin{bmatrix} \mathcal{L}\phi \\ \phi \end{bmatrix} = 0, \end{cases} \tag{3}$$

where $\underline{\boldsymbol{\sigma}} \in R = H(\mathrm{div}, \Omega_T)$, $\phi \in V = \{v \in H^1(\Omega_T) : v|_{\Gamma_S} = 0\}$.

For convenience, let us separate components of $\underline{\boldsymbol{\sigma}}$ as $\underline{\boldsymbol{\sigma}} = (\boldsymbol{\sigma}, \sigma_*)$, where σ_* is a scalar function.

3.2   Variational Formulation

Let

$$J(\underline{\sigma}, \phi) = \frac{1}{2} \left\| \underline{\sigma} - \begin{bmatrix} \mathcal{L}\phi \\ \phi \end{bmatrix} \right\|^2 = \frac{1}{2} \| \sigma - \mathcal{L}\phi \|^2. \tag{4}$$

The solution of the system (3) is also a solution of the following minimization problem:

$$\min_{(\underline{\tau}, \omega) \in R \times V} J(\underline{\tau}, \omega) \quad \text{subject to } \mathrm{div}_{\mathrm{x,t}} \, \underline{\tau} = f, \tag{5}$$

since $J(\underline{\sigma}, \phi) = 0$. Applying the Lagrange multipliers method to this constrained minimization problem, we search for the critical points of the functional

$$G(\underline{\sigma}, \phi, \lambda) = \frac{1}{2} \| \sigma - \mathcal{L}\phi \|^2 + \frac{1}{2} \| \sigma_* - \phi \|^2 + (\mathrm{div}_{\mathrm{x,t}} \, \underline{\sigma} - f, \lambda), \tag{6}$$

where $\lambda \in L^2(\Omega_T)$ is the Lagrange multiplier. In this formulation, we need to find $(\underline{\sigma}, \phi, \lambda) \in \mathbf{W}$ such that for all $(\underline{\tau}, \omega, \mu) \in \mathbf{W}$

$$(\sigma - \mathcal{L}\phi, \tau - \mathcal{L}\omega) + (\sigma_* - \phi, \tau_* - \omega) \\ + (\mathrm{div}_{\mathrm{x,t}} \, \underline{\tau}, \lambda) + (\mathrm{div}_{\mathrm{x,t}} \, \underline{\sigma} - f, \mu) = 0. \tag{7}$$

where $\mathbf{W} = R \times V \times L^2(\Omega_T)$. To make the structure of the resulting discrete matrix more apparent, we can rewrite the above equation as

$$\begin{aligned} (\phi, \omega) + (\mathcal{L}\phi, \mathcal{L}\omega) - (\sigma_*, \omega) \quad - (\sigma, \mathcal{L}\omega) \quad &= 0, \\ - (\phi, \tau_*) \quad + (\sigma_*, \tau_*) \quad + (\lambda, \partial_t \tau_*) &= 0, \\ - (\mathcal{L}\phi, \tau) \quad + (\sigma, \tau) \quad + (\lambda, \mathrm{div}_{\mathrm{x}} \, \tau) &= 0, \\ + (\partial_t \sigma_*, \mu) + (\mathrm{div}_{\mathrm{x}} \, \sigma, \mu) \quad &= (f, \mu). \end{aligned}$$

3.3   Discrete Problem

We approximate the independent variables ϕ, σ_x, and σ_y using quadratic B-splines. Let us define $\{u_i\}$ the B-spline basis functions and $\phi^h, \sigma_*^h, \sigma_x^h, \sigma_y^h, \lambda_x^h$ the corresponding vectors of coefficients of the B-spline expansion of ϕ, σ_x, σ_x, respectively. We can write the system in the following matrix structure

$$\begin{bmatrix} M + K & -M & -L_x^T & -L_y^T & 0 \\ -M & M & 0 & 0 & A_t^T \\ -L_x & 0 & M & 0 & A_x^T \\ -L_y & 0 & 0 & M & A_y^T \\ 0 & A_t & A_x & A_y & 0 \end{bmatrix} \begin{bmatrix} \phi^h \\ \sigma_*^h \\ \sigma_x^h \\ \sigma_y^h \\ \lambda^h \end{bmatrix} = \begin{bmatrix} 0 \\ 0 \\ 0 \\ 0 \\ f^h \end{bmatrix}$$

where M represents the mass matrix such that $(M)_{ij} := (u_i, u_j)_{L^2}$, (A_γ), is such that $(A_\gamma)_{ij} = (\partial_\gamma u_i, u_j)$ and L_γ such that $(L_\gamma)_{ij} := (\mathcal{L}_\gamma u_i, u_j)$. Moreover, f^h is also the vector of coefficients related to the expansion of f in the B-spline basis.

4   Matrix Compression

The core of the low-rank matrix compression is the SVD algorithm, illustrated in Fig. 1. The matrix A is decomposed into UDV, namely the matrix of "columns" U, the diagonal matrix of singular values D, and the matrix of "rows" V.

$$A = UDV, \quad [U, D, V] = SVD(B),$$
$$U \in \mathcal{M}^{n \times n}, D - \text{diagonal } m \times n, V \in \mathcal{M}^{m \times m}.$$

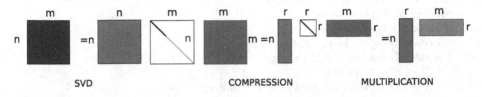

Fig. 1. SVD algorithm for the low-rank matrix compression.

The entries of D (singular values) are sorted in descending order. The diagonal values less than the compression threshold δ are removed together with corresponding columns of U and rows of V. The entries of D (singular values) are sorted in descending order. The diagonal values less than the compression threshold δ are removed together with corresponding columns of U and rows of V. As the result we obtain the low-rank compressed matrix $\mathcal{H}_A$, where $s = rank$ $\mathcal{H}_A = \max\{i : d_{ii} > \delta\}$. The matrix $\mathcal{H}_A$ is the best approximation of A in the Frobenious norm among all the matrices of rank s.

The compression is performed in a recursive way, as expressed by Algorithms 1 and 2. We partition the matrix recursively into blocks, we check how many singular values are larger than the prescribed δ. If the compression of the block with δ results in viewer singular values than the prescribed threshold b, we stop the recursion and store the sub-matrix in a compressed way. Otherwise, we continue with the recursive partitions.

Standard LAPACK subroutine dgesvd for the SVD computations has time complexity $\mathcal{O}(N^3)$ for a square matrix. We, however, employ the truncated SVD that computes only r singular values, having the complexity of $\mathcal{O}(N^2 r)$. Thus, the compression of the space-time matrix has a time complexity of a similar order as the matrix r vectors multiplication.

The compression of the space-time matrix results in a structure presented in Fig. 2.

5   Compressed Matrix-Vector Multiplication

We illustrate in Fig. 3 the process of multiplication of a compressed matrix by s vectors. There are two cases to consider.

Algorithm 1. compress_matrix

Require: $A \in \mathcal{M}^{m \times n}$, δ compression threshold, b maximum rank
1: **if** $A = 0$ **then**
2:    **create new node** v; $v.rank \leftarrow 0$; $v.size \leftarrow size(A)$; **return** v;
3: **end if**
4: $[U, D, V] \leftarrow SVD(A)$; $\sigma \leftarrow diag(D)$;
5: $rank \leftarrow card(\{i : \sigma_i > \delta\})$;
6: **if** $rank < b$ **then**
7:    **create new node** v; $v.rank \leftarrow rank$;
8:    $v.singularvalues \leftarrow \sigma(1 : rank)$;
9:    $v.U \leftarrow U(*, 1 : rank)$;
10:    $v.V \leftarrow D(1 : rank, 1 : rank) * V(1 : rank, *)$;
11:    $v.sons \leftarrow \emptyset$; $v.size \leftarrow size(A)$;
12:    **return** v;
13: **else**
14:    **return** $process\_matrix(A, \delta, b)$;
15: **end if**

Algorithm 2. process_matrix

Require: $A \in \mathcal{M}^{m \times n}$, δ compression threshold, b maximum rank
1: $v \leftarrow create\_node()$
2: $A_{11} \leftarrow A(1 : \frac{m}{2}, 1 : \frac{n}{2})$
3: $A_{12} \leftarrow A(1 : \frac{m}{2}, \frac{n}{2} + 1 : n)$
4: $A_{21} \leftarrow A(\frac{m}{2} + 1 : m, 1 : \frac{n}{2})$
5: $A_{22} \leftarrow A(\frac{m}{2} + 1 : m, \frac{n}{2} : n)$
6: $n_1 \leftarrow compress\_matrix(A_{11}, \delta, b)$
7: $n_2 \leftarrow compress\_matrix(A_{12}, \delta, b)$
8: $n_3 \leftarrow compress\_matrix(A_{21}, \delta, b)$
9: $n_4 \leftarrow compress\_matrix(A_{22}, \delta, b)$
10: $v.sons \leftarrow [n_1, n_2, n_3, n_4]$
11: **return** v

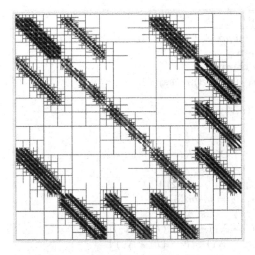

Fig. 2. Exemplary compressed matrix

- The first case is located at the leaves of the compressed matrix, where we multiply the compressed submatrix by a corresponding part of the vector. This is illustrated on the left panel in Fig. 3. In this case, the computational cost of matrix-vector multiplication with compressed matrix and s vectors is $\mathcal{O}(rms + rns)$, when $n = m = N \gg r$ it reduces to $\mathcal{O}(Nrs)$
- The second case is related to the multiplication of a matrix compressed into four SVD blocks by the vector partitioned into two blocks. This is illustrated on the right panel in Fig. 3. We employ the recursive formula $\begin{bmatrix} C_2 * (C_1 * X_1) + D_2 * (D_1 * X_2) \\ E_2 * (E_1 * X_1) + F_2 * (F_1 * X_2) \end{bmatrix}$. The computational cost is $\mathcal{O}(Nrs)$.

The pseudocode is illustrated in Algorithm 3.

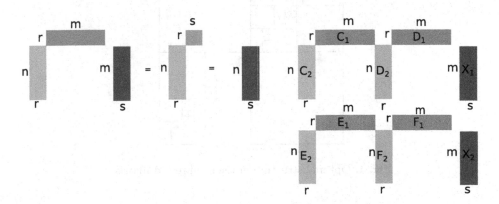

Fig. 3. SVD-compressed matrix multiplication

Algorithm 3. matrix_vectors_multiply

Require: node v, *Compressed matrix* $A(v) \in \mathcal{M}^{m \times n}$, $Y \in \mathcal{M}^{n \times c}$ *vectors to multiply*
1: **if** $v.sons = \emptyset$ **then**
2:     **if** $v.rank = 0$ **then**
3:        **return** $zeros(size(A).rows)$;
4:     **end if**
5:     **return** $v.U * (v.V * X)$;
6: **end if**
7: $rows = size(X).rows$;
8: $X_1 = X(1 : \frac{rows}{2}, *)$; $X_2 = X(\frac{rows}{2} + 1 : size(A).rows, *)$;
9: $C_1 = v.son(1).U$; $C_2 = v.son(1).V$;
10: $D_1 = v.son(2).U$; $D_2 = v.son(2).V$;
11: $E_1 = v.son(3).U$; $E_2 = v.son(3).V$;
12: $F_1 = v.son(4).U$; $F_2 = v.son(4).V$;
13: **return** $\begin{bmatrix} C_2 * (C_1 * X_1) + D_2 * (D_1 * X_2) \\ E_2 * (E_1 * X_1) + F_2 * (F_1 * X_2) \end{bmatrix}$;

The critical from the point of view of the computational cost is the structure of the compressed matrix. If we have the structure of the matrix as presented in Fig. 4, we have the quasi-linear multiplication cost. Namely, at each level, we have 2 leaves and 2 interior nodes

$$C(N) = \underbrace{2\,C(N/2) + 2\mathcal{O}(Nrs/2)}_{\text{multiplication}} + \underbrace{\mathcal{O}(N)}_{\text{addition}}, \quad C(N_0) = \mathcal{O}(N_0 rs)$$

$$\Rightarrow C(N) = \mathcal{O}(N \log N)$$

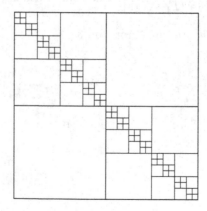

Fig. 4. Optimal structure of the compressed matrix

Fortunatelly, the structure of the space-time problem matrix has this optimal shape in several sub-blocks, see Fig. 2.

6   GMRES Solver

We employ the GMRES iterative solver with hierarchical matrix-vector multiplication, illustrated in Algorithm 4.

Algorithm 4. Pseudo-code of the GMRES algorithm

Require: $\mathcal{H}_A$ compressed matrix, b right-hand-side vector, x_0 starting point

   Compute $r_0 = b - \mathcal{H}_A x_0$

   Compute $v_1 = \frac{r_0}{\|r_0\|}$

   for $j = 1, 2, ..., k$

   Compute $h_{i,j} = (\mathcal{H}_A v_j, v_i)$ for $i = 1, 2, ..., j$

   Compute $\hat{v}_{j+1} = \mathcal{H}_A v_j - \sum_{i=1,...,j} h_{i,j} v_i$

   Compute $h_{j+1,j} = \|\hat{v}_{j+1}\|_2$

   Compute $v_{j+1} = \hat{v}_{j+1}/h_{j+1,j}$

   end for

   Form solution $x_k = x_0 + V_k y_k$ where $V_k = [v_1 ... v_k]$, and y_k minimizes $J(y) =$

$$\|\beta e_1 - \hat{H}_k y\| \text{ where } \hat{H} = \begin{bmatrix} h_{1,1} & h_{1,2} & \cdots & h_{1,k} \\ h_{2,1} & h_{2,2} & \cdots & h_{2,k} \\ 0 & \ddots & \ddots & \vdots \\ \vdots & \ddots & h_{k,k-1} & h_{k,k} \\ 0 & \cdots & 0 & h_{k+1,k} \end{bmatrix}$$

---

7   Numerical Results

For the numerical tests, we use the model problem (2) on a regular domain $\Omega \times (0, T) = (0, 1)^3$ with $\beta = (0, 0.3)$, $\varepsilon = 10^{-5}$, no forcing ($f = 0$), and the initial state u_0 given by $u_0(x) = \psi(10\|x - c\|)$, where $c = (0.5, 0.5)$ and

$$\psi(r) = \begin{cases} (1 - r^2)^2 & \text{for } r \leq 1, \\ 0 & \text{for } r > 1. \end{cases}$$

As a result, the initial state is zero except for a small region in the center of the domain. Tests were performed using basis functions of degree $p = 1$ for all the discrete spaces.

Table 1. Results of the adaptive mesh refinement process

level	DoFs	$J(\sigma U_h, \phi_h)$	solver for A [s]	solver for $\mathcal{H}_A$ [s]
0	3019	0.0137	0.046	0.005
1	11540	0.0039	0.567	0.076
2	25963	0.0052	4.280	0.774
3	63033	0.0040	14.573	2.573
4	163755	0.0031	31.190	5.048
5	359658	0.0023	101.335	20.120
6	730953	0.0017	279.838	45.606

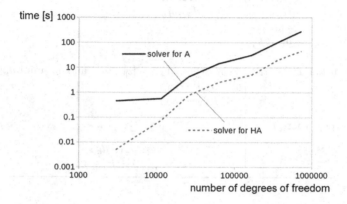

Fig. 5. Execution times for solver without the compression, and after the compression, for a sequence of adaptive grid.

Starting with the coarse initial mesh, we perform adaptive mesh refinements using the value of $J(\boldsymbol{\sigma}_h, \phi_h)$ as the error indicator and the Dörfler marking criterion [9] with $\theta = 0.5$. The improvement of the solutions at particular adaptive iterations is denoted in Fig. 6. Finally, the sequence of generated adaptive space-time meshes is presented in Fig. 7. We also present in Table 1 the convergence of the adaptive space-time finite element method, as well as the execution times for the solver without the compression, and after the compression, for a sequence of adaptive grids. Figure 5 presents execution times for the solver without the compression, and after the compression, for a sequence of adaptive grids.

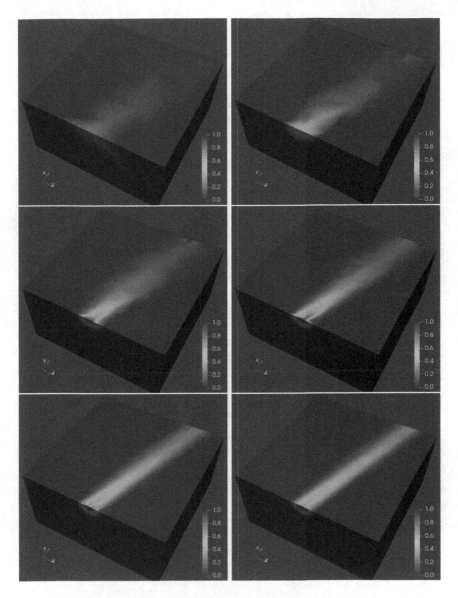

Fig. 6. Solutions in the six consecutive refinement steps

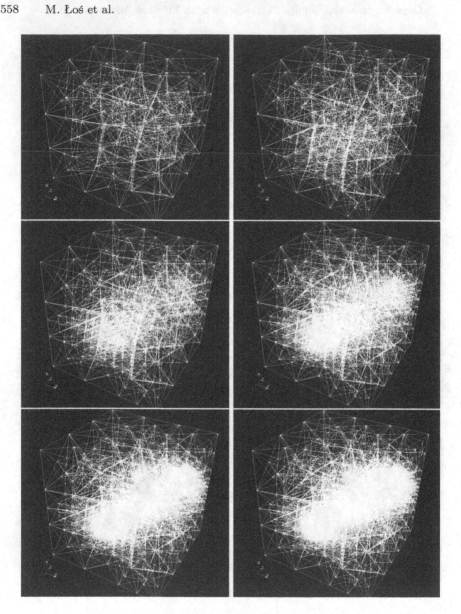

Fig. 7. Mesh in the six consecutive refinement steps

8   Conclusions

In this paper, we considered the space-time formulation of the advection dominated diffusion problem. The stabilization was obtained by introducing the constrained minimization problem with Lagrange multipliers. The obtained system of linear equations was discretized with B-spline basis functions. Next, we employed the low-rank compression of the discrete form of the space-time matrix. The obtained hierarchical form of the space-time matrix has several blocks that has been compressed into a hierarchical "diagonal" form. The hierarchical matrix can be multiplied by the vector in a $\mathcal{O}(NlogN)$ computational cost. This allows to construct a GMRES solver with hierarchical matrices that is one order of magnitude faster than the GMRES solver with the original sparse matrix. Future work may include research on a proper perconditioner for the GMRES solver. For example, we may investigate if the deep compression of the matrix, with a large compression threshold, can be applied as an efficient preconditioner for the iterative solver.

Acknowledgements. Research project supported by the program "Excellence initiative - research university" for the AGH University of Science and Technology.

References

1. Schafelner, A., Vassilevski, P.S.: Numerical results for adaptive (negative norm) constrained first order system least squares formulations. In: Recent Advances in Least-Squares and Discontinuous Petrov-Galerkin Finite Element Methods, Computers & Mathematics with Applications, vol. 95, pp. 256–270 (2021)
2. Voronin, K., Lee, C.S., Neumuller, M., Sepulveda, P., Vassilevski, P.S.: Space-time discretizations using constrained first-order system least squares (CFOSLS). J. Comput. Phys. **373**, 863–876 (2018)
3. Neumuller, M.: Space-time methods: fast solvers and applications, Ph.D. thesis, Graz University of Technology, Institute of Applied Mathematics (2013)
4. Steinbach, O., Yang, H.: Comparison of algebraic multigrid methods for an adaptive space-time finite-element discretization of the heat equation in 3D and 4D. Numer. Linear Algebra Appl. **25**(3), e2143 (2018)
5. Demkowicz, L., Gopalakrishnan, J., Nagaraj, S., Sepulveda, P.: A spacetime DPG method for the Schrodinger equation. SIAM J. Numer. Anal. **55**(4), 1740–1759 (2017)
6. Gopalakrishnan, J., Sepulvedas, P.: A space-time DPG method for the wave equation in multiple dimensions. De Gruyter, Berlin, Boston, Chapter **4**, 117–140 (2019)
7. Ernesti, J., Wieners, C.: A space-time discontinuous Petrov-Galerkin method for acoustic waves. De Gruyter, Berlin, Boston, Chapter **3**, 89–116 (2019)
8. Fuhrer, T., Karkulik, M.: Space-time least-squares finite elements for parabolic equations. Comput. Math. Appl. **9281**, 27–36 (2021)
9. Dorfler, W.: A convergent adaptive algorithm for Poisson's equation. SIAM J. Numer. Anal. **33**(3), 1106–1124 (1996)
10. Skotniczny, M., A. Paszyńska, A., Rojas, S., Paszyński, M.: Complexity of direct and iterative solvers on space-time formulations versus time-marching schemes for h-refined grids towards singularities arXiv:2110.05804, 1–36 (2022)

11. Hackbusch, W.: A sparse matrix arithmetic based on H-matrices. Part I: Introduction to H -matrices. Computing **62**, 89–108 (1999)
12. Hackbusch, W.: Hierarchical Matrices: Algorithms and Analysis. Springer, Heidelberg (2015). https://doi.org/10.1007/978-3-662-47324-5

Towards Understanding of Deep Reinforcement Learning Agents Used in Cloud Resource Management

Andrzej Małota, Paweł Koperek$^{(\boxtimes)}$ (ID), and Włodzimierz Funika (ID)

Faculty of Computer Science, Electronics and Telecommunication,
Institute of Computer Science, AGH University of Krakow,
al. Mickiewicza 30, 30-059 Kraków, Poland
malota.andrzej@gmail.com, pkoperek@gmail.com, funika@agh.edu.pl

Abstract. Cloud computing resource management is a critical component of the modern cloud computing platforms, aimed to manage computing resources for a given application by minimizing the cost of the infrastructure while maintaining a Quality-of-Service (QoS) conditions. This task is usually solved using rule-based policies. Due to their limitations more complex solutions, such as Deep Reinforcement Learning (DRL) agents are being researched. Unfortunately, deploying such agents in a production environment can be seen as risky because of the lack of transparency of DRL decision-making policies. There is no way to know why a certain decision is made. To foster the trust in DRL generated policies it is important to provide means of explaining why certain decisions were made given a specific input. In this paper we present a tool applying the Integrated Gradients (IG) method to Deep Neural Networks used by DRL algorithms. This allowed to obtain feature attributions that show the magnitude and direction of each feature's influence on the agent's decision. We verify the viability of the proposed solution by applying it to a number of sample use cases with different DRL agents.

Keywords: cloud resource management · deep reinforcement learning · explainable artificial intelligence · explainable reinforcement learning · deep neural networks

1 Introduction

In the recent years using cloud computing infrastructure became the dominating approach to provisioning computing resources. Thanks to virtually unlimited resources and usage-based billing, creating the cost-effective applications which automatically adjust the amount of used resources became straightforward. At the same time, the Cloud Service Provider (CSP) can increase the utilization of hardware by using it to serve multiple users in the same time. Managing resources in cloud computing infrastructures is a challenging task. The resource management algorithms can, unfortunately, provide too much or too few resources for

J. Mikyška et al. (Eds.): ICCS 2023, LNCS 14074, pp. 561–575, 2023.
https://doi.org/10.1007/978-3-031-36021-3_55

the services. To mitigate this issue many CSP provide tools which allow for on-demand dynamic resource allocation (auto-scaling). Such tools implement typically a *horizontal scaling* approach in which resources (e.g. Virtual Machine (VM)s) of the same type are added or released [5]. Horizontal scaling can use various methods of triggering scaling actions including rule-based methods [24] or predicting resource requirements using polynomial regression [5]. Nowadays, many methods are based on more advanced techniques such as deep learning and reinforcement learning.

Reinforcement Learning (RL) is a method of learning from interactions with an environment [23]. During a training process, a RL agent learns to map observations into actions through an iterative trial-and-error process. Upon executing an action, the agent receives a positive or negative reward signal and its objective is to maximize the sum of rewards received. Deep Reinforcement Learning (DRL) is one of the subfields of Machine Learning (ML) that combines RL and Deep Learning (DL) by employing Deep Neural Network (DNN), e.g. as various function approximators. In recent years it has attracted much attention and has been successfully applied to many complex domains: playing computer games [15], robotics [18], Natural Language Processing (NLP) in human- machine dialogue [4], traffic signal control [8], or cloud resource management [26].

Applying DNN suffers from the lack of explainability (also called interpretability). It is difficult to *explain* why a certain result has been produced by a DNN. Furthermore, the user of a machine learning algorithm may be legally obliged to provide explanations for certain decisions made with use of those algorithms. That issue can be mitigated by utilizing one of the Explainable Artificial Intelligence (XAI) techniques which aim at providing additional information on how ML algorithms produce their outputs. Whereas explainability can be considered well developed for standard ML models and neural networks [19,21], in the case of RL there are still many issues that need to be resolved in order to enable using it in fields where it is imperative to understand its decisions [11]. Explainable Reinforcement Learning (XRL) is a recently emerged new subfield of XAI which receives a lot of attention. The goal of XRL is to explain how decisions are generated by policies employed by the DRL agents [14]. Such policies can be very complex and difficult to debug. Providing more insight into how they function allows to improve the design of the training environments, reward functions and DNN model architectures.

In this paper we extend the prior work [6,7], where the use of the policy gradient optimization approach for automatic cloud resource provisioning has been studied. We implement a tool that allows to explain the actions of the cloud computing resource management policies trained with the use of DRL. This approach enables post hoc attributing input features to decisions made by the DRL agent which dynamically creates or deletes VMs based on an observed application requirements and resource utilization. The tool is available as an open source project [13]. To the best of authors' knowledge, this paper is the first example of using the Integrated Gradients (IG) XAI technique to interpret DRL agents in the cloud resource management setting.

This paper has the following structure: Sect. 2 presents related work on the use of DRL in cloud resource management and application XAI techniques on DRL agents. Section 3 describes a cloud resource management simulation environment. In Sect. 4 we analyze the experimental results of the interpretation scheme applied in various scenarios. Lastly, Sect. 5 draws conclusions from the experiments and outlines future work.

2   Related Work

2.1   DRL in Cloud Resource Management

In recent years, the usage of DRL in cloud resource provisioning gained significant attention. In [25], authors explore the use of standard RL and DRL for cloud provisioning. They utilize a system where the users specify rewards based on cost and performance to express their goals. In such an environment they compare the use of the tabular-based Q- learning algorithm with the Deep Q-Networks (DQN) approach to achieve the objectives set by the users. The policy trained with the DQN approach achieved the best results.

In their study, authors of [2] utilized the Double Deep Q-Networks (DDQN) algorithm to reduce power consumption in CSP. The proposed DRL-based cloud resource provisioning and task scheduling method consist of two stages. The first one allocates the task to one of the server farms. The second one chooses the exact server to run the task in. The reward function is calculated using the energy cost of the performed action. The proposed *DRL-Cloud* system compared with a round-robin baseline improved the energy cost efficiency while maintaining a low average reject rate.

In [6] three DRL policy gradient optimization methods (Vanilla Policy Gradient (VPG), Proximal Policy Optimization (PPO) and Trust Region Policy Optimization (TRPO)) are used to create a policy used to control the behavior of an autonomous cloud resources management agent. The agent interacts with a simulated cloud computing environment which processes a stream of computing jobs. The environment state is represented by a set of metrics that are calculated in each step of the simulation. The reward function is set up as the negative cost of running the infrastructure with added penalties for breaching the Service-Level-Agreement (SLA) conditions. The policy which achieved the lowest cost was created using the PPO algorithm. In [7], the PPO-based autonomous management agent is compared with the traditional auto-scaling approach available in Amazon Web Services (AWS). As a sample workload an evolutionary experiment, consisting of multiple variable size phases, has been chosen. The policy training has been conducted within a simulated cloud environment. Afterwards the policy has been deployed to a real cloud infrastructure, the AWS Elastic Compute Cloud. The total cost of managed resources was slightly lower (0.7%) when a PPO-trained policy has been used, compared with a threshold-based approach. The trained policy was considered to be able to generalize well enough to be re-used across multiple similar workloads.

2.2   Explainable AI in Deep Reinforcement Learning

DRL has shown great success in solving various sequential decision-making problems, such as playing complicated games or controlling simulated and real-life robots. However, existing DRL agents make decisions in an opaque fashion, taking actions without accompanying explanations. This lack of transparency creates key barriers to establishing trust in an agent's policy and significantly limits the applicability of DRL techniques in critical application fields such as finance, self-driving cars, or cloud resource management [10]. So far, most of the research on the usage of XAI in DRL has been focused on finding the relationship between the agent's action and the input observation at the specific time step - detecting the features that contributed most to the agent's action at that specific time. According to the XAI taxonomies in [1], it is possible to classify all recent studies into two main categories: *post hoc explainability* and *transparent methods*.

When dealing with images as input data, one can provide explanations through saliency maps. A saliency map is a heat map that highlights pixels that hold the most relevant information and the value for each pixel shows the magnitude of its contribution to the Convolutional Neural Network (CNN)'s output. Unfortunately, saliency maps are sensitive to input variations. The authors of [9] aim to fix that with their perturbation-based saliency method applied to agents trained to play Atari games via the Asynchronous-Advantage-Actor-Critic (A3C) algorithm. They conducted a series of investigative explorations aiming to explain how agents made their decisions. First, they identified the key strategies of the three agents that exceed human baselines in their environments. Second, they visualized agents throughout training to see how their policies evolved. Third, they explored the use of saliency for detecting when an agent is earning high rewards for the wrong reasons. This includes a demonstration that the saliency approach allows non-experts to detect such situations. Fourth, they found Atari games where the trained agents performed poorly and used saliency to debug these agents by identifying the basis of their low-quality decisions. The approach presented in [9] is an example of *post hoc explainability*.

In [12] the authors applied a post hoc interpretability to the DRL agent which learned to play the video game CoinRun [3]. They used PPO [20] algorithm, with the agent's policy model being a CNN. *Attribution* shows how the neurons affect each other, usually how the input of the network influences its output. Dimensionality reduction techniques [17] have been applied to the input frames from the game and attributions were calculated using IG. IG computes the gradient of the model's prediction output to its input features. This allows to produce the feature attributions that display the feature's influence on the prediction. Furthermore, IG requires no modification to the original DNN. Authors built an interface for exploring the detected objects that shows the original image and also positive and negative attributions, to explain how the objects influence the agent's value function and policy. Their analysis consists of three main parts: dissecting failure (trying to understand what the agent did wrong and what reasons behind it were), hallucinations (model detected a feature that was not there), model editing (manually editing the model weights so that the agent would ignore certain observations).

In [16], an interpretability analysis of the DRL agent with a recurrent attention model on the games from the ALE environment is performed. The authors observed basic attention patterns using an attention map, which is a scalar matrix representing the relative importance of layer activations at different 2D spatial locations with respect to the target task. It was also discovered that the model attends to task-relevant things in the frame - player, enemies, and score. The ALE environment is predictable, e.g. enemies appear at regular times and in regular configurations. It is important, to ensure that the model truly learns to attend to the objects of interest and act upon the information, rather than to memorize and react only to certain patterns in the game. In order to test it, they injected an enemy object into the observation at an unexpected time and in an unexpected location. They observed that the agent correctly attends to and reacts to the new object. They discovered that the model performs forward planning/scanning - it learns to scan through available paths starting from the player character, making sure there are no obstacles or enemies in the way. When the agent does see the enemy, another path is produced in order to avoid it.

3 Environment Setup

The simulation environment used in our research is a result of the prior work [6]. A fundamental component of the simulation process is implemented with the CloudSim Plus simulation framework [22]. The environment is wrapped with the interface provisioned by the Open AI Gym framework. Figure 1 presents the system architecture.

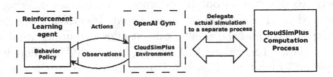

Fig. 1. Components of the system [6].

The workload used in this work is a simple evolutionary experiment which improves the architecture of a network that recognizes handwritten digits. The main objective of the agent is to allocate cloud infrastructure resources in an optimal way to the running workload. The environment in which the agent's training was conducted was simulated to avoid the high costs of provisioning a real computing infrastructure. In order to finish the experiment in a reasonable time, simulation time was sped up and the number of steps per episode was limited.

The reward function R (Eq. 1) equals the negative cost of running the infrastructure added to the SLA penalty. The SLA penalty adds some cost for every second delay in task execution and was calculated by multiplying the number of seconds of the delay of task execution by the penalty value. The cost of running the infrastructure was calculated by multiplying the number of VMs by the hourly cost of using the VM.

$$R = -(\text{NumberOfVMs} * \text{HourlyCostOfVM} + \text{SlaPenalty}) \qquad (1)$$

where:

HourlyCostOfVM = \$0.2,
SlaPenaltyNumberOfSecondsOfDelay * Penalty,
Penalty = 0.00001\$

The environment state is represented by a vector of cloud infrastructure metrics that are being calculated at each time step of the simulation. These metrics include: the number of running virtual machines (vmAllocatedRatio), average RAM utilization (avgMemoryUtilization), 90-th percentile of RAM utilization (p90MemoryUtilization), average CPU utilization (avgCPUUtilization), 90-th percentile of CPU utilization (p90CPUUtilization), total task queue wait time (waitingJobsRatioGlobal), recent task queue wait time (waitingJobsRatioRecent). The set of available actions is limited to the following: do nothing, add or remove a small VM, add or remove a medium VM, add or remove a large VM.

4   Understanding of DRL Agents Used in Cloud Resource Management

In this paper, we develop a tool to interpret the decision-making process of DRL agents used in cloud resource management. Identifying the relationships between the input data and the agent's output allows to understand why certain decisions were made. To achieve this we employ the IG method which shows the magnitude and direction of each feature's influence on the agent's output. It requires minimal modification of the original network and can inform on the cloud infrastructure metrics that influenced the agent's decision-making process. The attribution value for a feature can be positive or negative, indicating its contribution towards or against a certain prediction, respectively. We demonstrate the usefulness of the discussed approach in a few scenarios: attributing input metrics to the action chosen by the policy (for two DRL algorithms: DQN with an MultiLayer Perceptron (MLP) model and PPO with a CNN model), providing a policy summarization, explaining how a policy evolves during training, debugging policy decisions, removing irrelevant features.

4.1 Input Metric Attribution

The DNN model used in the DQN algorithm produces approximations of q-values which denote the value of taking action a in state s. The model produces q-values for each possible action. The final output is chosen using a greedy approach which selects the action with the highest value. Figure 2a presents an overview of the neural network model used in DQN.

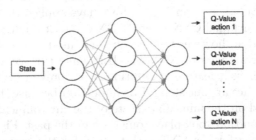

(a) Model used as a policy for the DQN algorithm

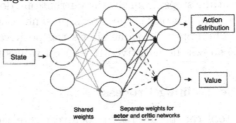

(b) Model used as a policy for the PPO algorithm.

Fig. 2. Simplified diagrams of used DNN models.

The IG can calculate feature attributions for each possible action separately, however, we focus on the action which has the highest q-value, which is then chosen for execution. In the DQN approach we have applied it to a MLP network. The MLP architecture is straight-forward to understand. Due to its simplicity it is also faster to calculate its output's attributions. MLP accepts a one-dimensional (1D) vector of feature observations as an input, which allows to visualize attributions as a simple to read bar-chart. Figure 3 presents the attributions for a sample action chosen by the policy. It consists of two parts. The first one (top) shows the environment state (y-axis presents metric value (0–1)). The second one (bottom) shows the attributions value for the best action (positive in green, negative in red).

The PPO policy model has an actor-critic architecture and produces two outputs. The policy (*actor*) network provides a distribution of the probabilities of the actions possible to execute in a given state. Since an action is selected by

sampling the provided distribution the actor is considered to be directly responsible for choosing the action. The *critic* network estimates the *value* function,
which quantifies the expected reward if the agent follows the policy until the
terminal state. Typically, to optimize the amount of computational resources
required for training, the actor and critic networks are combined into a single
network with two distinct outputs, what is presented in Fig. 2b.

In the context of PPO the IG is used to calculate attributions for the input
features in relation to the action chosen by sampling the distribution. To demonstrate the capabilities of such an approach we have applied the discussed method
to a model which included CNN layers. CNN are best known for their usage in
the image recognition, however they can be also used in the time series prediction. The main advantage of CNN over MLP while performing the attribution
is that CNN can present feature attributions from previous n time steps for the
action that takes place at time step t. It is important because the agent's decision
is not always based on the immediate state of the environment but may depend
upon an observation that took place sometime in the past. Figure 4 presents the
example attributions for the CNN architecture for the best action. It consists
of three plots. The upper left chart shows the frame of the input to the neural
network, x-axis shows time steps, from 0 - the current one down to the - 14 time
step (15-th time step from the past), the y-axis presents a value (0–1) for each
environment metric. The upper right one shows the positive attribution values
while the bottom one shows the negative attribution values for the best action.

4.2   Debugging the Training Process

The interpretability tool can be leveraged to investigate the rationale behind
an agent's decision for a given observation. The aim is to discern whether the
decision was adequate given the observed circumstances and if it was triggered
by the expected inputs. If this is not the case, the tool should help to discover
which feature values might have influenced an incorrect decision.

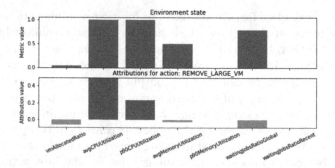

Fig. 3. Attributions for example observation - DQN agent with MLP architecture.
(Color figure online)

In Fig. 3 we observe the attributions underlying the DQN-MLP agent's decision to remove a large VM. Such a result was primarily influenced by the CPU utilization features (avgCPUUtilization, p90CPUUtilization). However, given the context of high CPU utilization, long waiting times for job execution (waitingJobsRatioGlobal), and low resource allocation (vmAllocatedRatio), we can consider removing resources as an incorrect decision. A human operator would have rather avoided taking an action or increased the number of VMs.

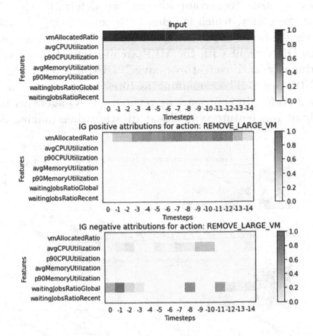

Fig. 4. Attributions for example observation - PPO agent with CNN architecture. Heatmaps are presented in grayscale to improve readability.

In Fig. 4 we present the attributions underlying the PPO-CNN agent's decision to remove a large VM, which at the time was deemed to be the appropriate course of action. The top plot illustrates the state of the environment at the time of the decision. The vmAllocatedRatio feature exhibited consistently high values across all time steps, indicating that most of the available VMs were being utilized, while the values of all other features were close to zero. Examining the middle plot, which illustrates positive attributions (i.e., reasons to make the decision), we can confidently conclude that the decision to remove a large VM was primarily driven by the vmAllocatedRatio feature. The bottom chart shows negative attributions (i.e., reasons to not make the decision), with the waitingJobsRatioGlobal feature - denoting the number of jobs waiting to be executed due to insufficient resources - being assigned a small negative attribution against the decision to remove the large VM. In other words, the decision to remove the large VM was appropriate because at the time, a substantial number of VMs were already in use, and there were very few jobs waiting in the queue to be executed.

4.3   Policy Summarization

To identify general behavioral patterns of the agents, we must examine the policy predictions of the agent across multiple examples, in other words use a *global interpretation approach*. To accomplish this, we determine the global feature importance for the policy, which is achieved by calculating the mean absolute attributions over hundreds of predictions. Figure 5 illustrates the policy summarization for the DQN agent with the MLP architecture. It is apparent that the agent relies primarily on three features - avgCPUUtilization, vmAllocatedRatio, and p90CPUUtilization. The remaining features have little to no influence on the agent's predictions. In contrast, the recurrent PPO agent with MLP policy relies primarily on one feature (vmAllocatedRatio) when making decisions.

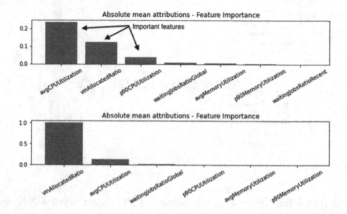

Fig. 5. Absolute mean attributions - Feature Importance - DQN-MLP (top), Recurrent PPO-MLP (bottom).

The feature importance analysis for the CNN based architectures provides more detailed information, allowing to observe the exact mean influence that each feature had at each time step. Figure 6 presents the feature importance for the agents with a CNN architecture. The agent relies primarily on three features: avgCPUUtilization, vmAllocatedRatio, and waitingJobsRatioGlobal. In contrast, the feature importance for the PPO agent with CNN architecture is more balanced, as it makes decisions based on five out of the seven available features across all time steps.

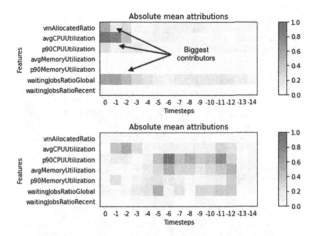

Fig. 6. Absolute mean attributions - Feature Importance - DQN-CNN (top), Recurrent PPO-CNN (bottom). Heatmaps are presented in grayscale to improve readability.

4.4 Evolution of Policies During Training

In the process of training, DRL agents begin with a random policy and adjust their weights to optimize their performance. This unfortunately means that the decisions made during the first training iterations might have a negative effect on the managed system, e.g. the amount of resources might be reduced instead of getting increased. In this study we investigate how the resource management strategy of the DRL agents evolves over time. The insights learnt from this analysis allow to understand whether a policy is making progress during training. Furthermore, in case the results are inadequate, such an analysis can be a starting point for debugging the training process. The presented analysis demonstrates that the agents training in our experiments underwent significant changes in their attributions during the training process.

Figure 7 presents changes in an absolute mean attribution in a sample experiment with training a DQN agent using a policy including CNN layers. The figure includes five plots: attribution in the initial model without any training (top), attribution while the model is being trained (middle three), attribution in the final model after training (bottom). As expected, the initial model's attributions are dispersed across various features and time steps. However, as the training progresses, the policy begins to emerge, and the number of attributed features is being reduced. By the end of the training, the attributions are only present for three features: *vmAllocatedRatio*, *avgCPUUtilization*, and *waitingJobsRatioGlobal*. Furthermore, the policy seemed to have focused solely on the three most recent time steps and began to ignore earlier time steps. These insights suggest that we could reduce the number of past steps included in observations since they are not being utilized. That in turn would result in creating a smaller model and, as a consequence, a faster training.

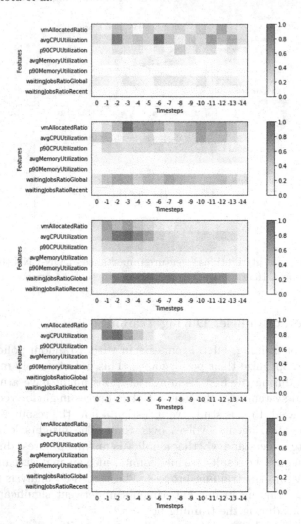

Fig. 7. Mean absolute attributions throughout the training process of the DQN agent with CNN policy. Heatmaps are presented in grayscale to improve readability.

4.5 Removing Irrelevant Features

The global feature importance can be quantified by calculating the absolute mean attributions over a given dataset, and can serve as a useful tool for selecting only the most relevant features for an agent's decision-making. There are several benefits of using a smaller feature set, including a simpler and more compact model, improved interpretability of the agent's decision-making process, and faster training times.

An example of such a situation is presented in Fig. 8. The absolute mean attributions for the Recurrent PPO-MLP agent reveal that the feature *waiting− JobsRatioRecent* has a negligible influence on the agent's decisions.

To validate the feature attributions as a metric for the feature selection, we removed *waitingJobsRatioRecent* and retrained the agent. The bottom chart in Fig. 8 presents the mean attributions of the new agent, which are nearly identical to those of the original agent trained with all the features. The comparable performance of the agent trained with and without the feature *waitingJobsRatioRecent* indicates that the IG and other XRL techniques can effectively aid in the feature selection process for DRL agents.

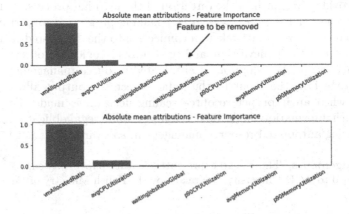

Fig. 8. Absolute mean attributions - Feature Importance - Recurrent PPO-MLP. The top chart presents attributions including the *waitingJobsRatioRecent*. Bottom one presents attributions in the case when that metric is excluded.

5   Conclusions

In this paper an approach allowing to explain the actions of an automated cloud computing resource management policy has been implemented. It was demonstrated how it can help understand why a neural network model returned a particular value (which can be translated, e.g. into a scaling action in the context of the resource management) by applying the IG method. To evaluate its correctness and utility, a number of experiments with training policies using DRL algorithms have been conducted. Two approaches were analyzed: Q-learning (DQN with MLP and CNN models) and policy gradient optimization (PPO with MLP, CNN, recurrent with MLP or CNN feature extractor). As the training environment a simulated cloud computing system processing a sample, dynamic workload has been chosen. The agent embedded in that system was controlled by the policy trained with the use of the mentioned algorithms and could increase or reduce the amount of used resources (virtual machines).

The presented experiments demonstrated how the described approach can be used to identify which input metrics contributed towards making a decision. It is worth noting that such metrics were different depending on the combination of the model architecture and the training algorithm. We have demonstrated that removing a feature with a little to no influence on the predictions did not affect the quality of decisions made by the policy. Additionally we have analyzed how metric changes in time being influenced the policies' output in the case of models that use CNN. Finally, we have demonstrated how Integrated Gradients method can be used to determine whether the decisions made by the policies were taken using the right premises (e.g. if a decision to remove a VM is taken when the usage of available resources is low).

The information obtained with the use of the IG method during the agent's training provided insight into the direction of the learning process. This allowed to confirm whether the policy was in fact making progress during training and that its decisions were in fact taken in connection to the discovered relationships between metrics. We believe that ability to provide evidence of such behavior can help accelerate the adoption of automated resource management systems. Understanding how the input to the policy affects its output allows to debug situations when an improper resource scaling decision is made. Furthermore, tracking such information might be necessary to fulfill legal obligations. Without that deploying automated resource management software might not be allowed.

Acknowledgements. The research presented in this paper was supported by the funds assigned to AGH University of Krakow by the Polish Ministry of Education and Science.

References

1. Barredo Arrieta, A., et al.: Explainable Artificial Intelligence (XAI): Concepts, taxonomies, opportunities and challenges toward responsible AI. Information Fusion **58**, 82–115 (2020). https://doi.org/10.1016/j.inffus.2019.12.012
2. Cheng, M., et al.: DRL-cloud: deep reinforcement learning-based resource provisioning and task scheduling for cloud service providers. In: 2018 23rd Asia and South Pacific Design Automation Conference (ASP-DAC), pp. 129–134 (2018)
3. Cobbe, K., et al.: Quantifying Generalization in Reinforcement Learning (2018). https://doi.org/10.48550/ARXIV.1812.02341
4. Cuayáhuitl, H.: SimpleDS: A Simple Deep Reinforcement Learning Dialogue System (2016). https://doi.org/10.48550/ARXIV.1601.04574
5. Dutta, S., et al.: SmartScale: automatic application scaling in enterprise clouds. In: 2012 IEEE Fifth International Conference on Cloud Computing, pp. 221–228 (2012)
6. Funika, W., Koperek, P.: Evaluating the use of policy gradient optimization approach for automatic cloud resource provisioning. In: Wyrzykowski, R., Deelman, E., Dongarra, J., Karczewski, K. (eds.) PPAM 2019. LNCS, vol. 12043, pp. 467–478. Springer, Cham (2020). https://doi.org/10.1007/978-3-030-43229-4_40
7. Funika, W., Koperek, P., Kitowski, J.: Automatic management of cloud applications with use of proximal policy optimization. In: Krzhizhanovskaya, V.V., et al. (eds.) ICCS 2020. LNCS, vol. 12137, pp. 73–87. Springer, Cham (2020). https://doi.org/10.1007/978-3-030-50371-0_6
8. Gregurić, M., et al.: Application of deep reinforcement learning in traffic signal control: an overview and impact of open traffic data. Appl. Sci. **10**(11) (2020)

9. Greydanus, S., et al.: Visualizing and Understanding Atari Agents (2017). https://doi.org/10.48550/ARXIV.1711.00138
10. Guo, W., et al.: EDGE: Explaining Deep Reinforcement Learning Policies. In: Ranzato, M., Beygelzimer, A., Dauphin, Y., Liang, P., Vaughan, J.W. (eds.) Advances in Neural Information Processing Systems, vol. 34, pp. 12222–12236. Curran Associates, Inc. (2021)
11. Heuillet, A., et al.: Explainability in Deep Reinforcement Learning. CoRR abs/2008.06693 (2020)
12. Hilton, J., et al.: Understanding RL Vision. Distill (2020). https://doi.org/10.23915/distill.00029
13. Małota, A., et al.: Trainloop-driver (2023). https://github.com/andrzejmalota/trainloop-driver/tree/master/examples
14. Milani, S., et al.: A Survey of Explainable Reinforcement Learning (2022). https://doi.org/10.48550/ARXIV.2202.08434
15. Mnih, V., et al.: Playing Atari with deep reinforcement learning. In: NIPS Deep Learning Workshop (2013). http://arxiv.org/abs/1312.5602
16. Mott, A., et al.: Towards Interpretable Reinforcement Learning Using Attention Augmented Agents (2019). https://doi.org/10.48550/ARXIV.1906.02500
17. Olah, C., et al.: The Building Blocks of Interpretability. Distill (2018). https://doi.org/10.23915/distill.00010
18. OpenAI, et al.: Solving Rubik's Cube with a Robot Hand (2019). https://doi.org/10.48550/ARXIV.1910.07113
19. Ribeiro, M.T., et al.: "Why Should I Trust You?": explaining the predictions of any classifier. In: Proceedings of the 22nd ACM SIGKDD International Conference on Knowledge Discovery and Data Mining. KDD 2016, New York, NY, USA, pp. 1135–1144. Association for Computing Machinery (2016)
20. Schulman, J., et al.: Proximal Policy Optimization Algorithms (2017). https://doi.org/10.48550/ARXIV.1707.06347
21. Selvaraju, R.R., Cogswell, M., Das, A., Vedantam, R., Parikh, D., Batra, D.: Grad-CAM: visual explanations from deep networks via gradient-based localization. Int. J. Comput. Vision 128(2), 336–359 (2019). https://doi.org/10.1007/s11263-019-01228-7
22. Campos da Silva Filho, M., et al.: CloudSim plus: a cloud computing simulation framework pursuing software engineering principles for improved modularity, extensibility and correctness. In: 2017 IFIP/IEEE Symposium on Integrated Network and Service Management (IM), pp. 400–406 (2017)
23. Sutton, R.S., Barto, A.G.: Reinforcement Learning: An Introduction, 2nd edn. The MIT Press (2018)
24. Tighe, M., Bauer, M.: Integrating cloud application autoscaling with dynamic VM allocation. In: 2014 IEEE Network Operations and Management Symposium (NOMS), pp. 1–9 (2014)
25. Wang, Z., et al.: Automated Cloud Provisioning on AWS using Deep Reinforcement Learning (2017). https://doi.org/10.48550/ARXIV.1709.04305
26. Zhang, Y., et al.: Intelligent cloud resource management with deep reinforcement learning. IEEE Cloud Comput. 4(6), 60–69 (2017). https://doi.org/10.1109/MCC.2018.1081063

ML-Based Proactive Control of Industrial Processes

Edyta Kuk[1]([⊠]) [iD], Szymon Bobek[2] [iD], and Grzegorz J. Nalepa[2] [iD]

[1] AGH University of Science and Technology, Kraków, Poland
kuk@agh.edu.pl
[2] Faculty of Physics, Astronomy and Applied Computer Science, Institute of Applied Computer Science, and Jagiellonian Human-Centered AI Lab (JAHCAI), and Mark Kac Center for Complex Systems Research, Jagiellonian University, ul. prof. Stanisława Łojasiewicza 11, 30-348 Kraków, Poland

Abstract. This paper discusses the use of optimal control for improving the performance of industrial processes. Industry 4.0 technologies play a crucial role in this approach by providing real-time data from physical devices. Additionally, simulations and virtual sensors allow for proactive control of the process by predicting potential issues and taking measures to prevent them. The paper proposes a new methodology for proactive control based on machine learning techniques that combines physical and virtual sensor data obtained from a simulation model. A deep temporal clustering algorithm is used to identify the process stage, and a control scheme directly dependent on this stage is used to determine the appropriate control actions to be taken. The control scheme is created by an expert human, based on the best industrial practices, making the whole process fully interpretable. The performance of the developed solution is demonstrated using a case study of gas production from an underground reservoir. The results show that the proposed algorithm can provide proactive control, reducing downtime, increasing process reliability, and improving performance.

Keywords: machine learning · artificial intelligence · simulations · optimal control

1 Introduction

Proper process control is crucial to avoid mistakes and downtime that are costly in the case of real-world processes. Optimal control is a key approach that can help industrial processes achieve their maximum potential and improve their overall performance. It involves designing a control algorithm that adjusts the inputs to the process in real-time to achieve the desired output while satisfying certain constraints. Optimal control aims to identify control actions that maximize the desired objective function. The use of Industry 4.0 technologies [1], which offer real-time access to numerous parameters of the process and low-level operational status of the physical devices, can greatly enhance optimal control. Such information can be used to optimize the process in ways that were not possible before. Moreover, simulations play a critical role in the operation of complex industrial processes. They can be used to develop and test control algorithms before implementation helping engineers and operators to make informed decisions, optimize the performance of the process, improve efficiency, and ensure safety [20].

© The Author(s), under exclusive license to Springer Nature Switzerland AG 2023
J. Mikyška et al. (Eds.): ICCS 2023, LNCS 14074, pp. 576–589, 2023.
https://doi.org/10.1007/978-3-031-36021-3_56

In addition to Industry 4.0, simulations can give the possibility to make use of virtual sensors. Virtual sensors provide estimates of process variables that are important for process control but cannot be directly measured by a physical sensor. Virtual sensors based on simulation models enable engineers to estimate process variables that would otherwise be difficult or impossible to measure, and allow for real-time monitoring and control of these variables [4]. The use of virtual sensors based on simulation models provides proactive control of the process, allowing engineers to predict potential issues and take measures to prevent them before they occur, reducing downtime and increasing the reliability of the process [6]. This is different from reactive process control which responds to problems after they have occurred.

Using simulation models vast amounts of synthetic data from industrial processes can be generated, including those that can be measured physically as well as virtual measurements. Machine learning (ML) techniques can be used to analyze this data and develop predictive models that enable proactive control and optimization of the process [17]. One approach is to use simulation models to generate training data for supervised learning algorithms that can predict process behavior and identify optimal control parameters. Unsupervised learning techniques can also be used to identify hidden patterns in the data and provide insights into process behavior [3]. Another approach is to use simulation models to create a digital twin of the process, which can be used to optimize control policies using reinforcement learning algorithms [12]. Overall, the combination of simulations with ML provides a powerful tool for process optimization and control. By leveraging the insights gained from simulation data analysis with ML engineers can make data-driven decisions that improve process performance.

1.1 Paper Contribution

In this paper, we propose a new method for proactive control of industrial processes based on ML techniques. The idea is to combine both physical and virtual sensor data obtained from a simulation model through an ML algorithm to identify the current stage of the industrial process, which is then used to proactively determine appropriate control actions. The proposed approach can be applied to a single asset/process or to multiple assets if the simulation model includes a fleet of them. In this paper, we applied a deep temporal clustering algorithm that employs an autoencoder for temporal dimensionality reduction and a temporal clustering layer for cluster assignment, to combine simulation data and identify the process stage. The appropriate control actions to be taken are then determined according to the current process stage. Although stage identification is fully unsupervised, decision control is defined by a human expert, making the process more transparent and interpretable. This cooperative rather than substitutive role of AI in industrial applications follows the Industry 5.0 paradigm, which places more emphasis on human-AI collaboration rather than black-box automation [18].

1.2 Paper Outline

The rest of the paper is organized as follows: In Sect. 2 we describe papers that cover ML approaches for determining industrial process control and we introduce our motivations. In Sect. 3 we introduce our methodology for establishing proactive process control using ML methods. Next, we apply the developed method to define proactive control of wells on the gas reservoir which is shown in Sect. 4. In Sect. 5 we explore the possible applications of the proposed approach and discuss its potential and limitations. Finally, we summarise our work in Sect. 6.

2   Related Works and Motivation

In industrial practice, numerous sensors are installed at various locations to collect process data. If physical measurements are combined with synthetic data generated by virtual sensors from simulation models of the process the process's historical databases are massive. While there has been an increase in the availability of sensors, the raw data collected from these sensors will eventually become useless without proper data analysis, information extraction, and knowledge exploitation techniques. Data from industrial processes can be analyzed with machine learning techniques to develop predictive models for process control and optimization. In this section, we provide an overview of process control approaches that utilize machine learning methods.

In [7] the Authors discuss the importance of soft sensors, called virtual sensors, in the modern industry as an alternative to physical sensors. The development of soft sensors has been facilitated by achievements in data science, computing, communication technologies, statistical tools, and machine learning techniques. The article also discusses the significance of soft sensing in improving production safety and product quality management from the perspectives of system monitoring, control, and optimization.

In [5] the Authors discuss the challenges of interpreting large volumes of data collected from smart factories, which often contain information from numerous sensors and control equipment. Principal component analysis (PCA) is a common method for summarizing data, but it can be difficult to interpret in the context of fault detection and diagnosis studies. Sparse principal component analysis (SPCA) is a newer technique that can produce PCs with sparse loadings, and the article introduces a method for selecting the number of non-zero loadings in each PC using SPCA. This approach improves the interpretability of PCs while minimizing information loss.

The paper [13] begins by introducing the recent advancements in machine learning and artificial intelligence, which have been driven by advances in statistical learning theory and big data companies' commercial successes. They provide three attributes for process data analytics to make machine learning techniques applicable in process industries. The paper discusses the currently active topics in machine learning that could be opportunities for process data analytics research and development.

The article [15] discusses the most commonly used machine learning techniques in the context of production planning and control. The authors grouped the ML techniques into families to facilitate the analysis of results, and neural networks, Q-Learning, and decision trees were found to be the most frequently used techniques. The high usage of clustering techniques is likely due to the nature of data in manufacturing systems. The study also observed a strong growth in the use of neural networks since 2015, possibly due to the development of specialized frameworks and growing computing power. The results also suggest a growing interest in ensemble learning techniques, possibly at the expense of decision trees since random forests can achieve better performance.

In [21] the Authors perform a diagnosis that uses supervised learning models to detect quality anomalies caused by abnormal process variations. This approach is different from diagnosis based on unsupervised learning models which can only analyze abnormal changes in process variables. The goal of supervised monitoring and diagnosis is to detect quality anomalies before they are measured and confirmed.

The authors of [16] emphasizes that deep learning is highly effective in modelling complex nonlinear processes, but it still faces several challenges in its application to process control. These challenges include difficulties with interpreting the features extracted from data and their relationship to outputs, sensitivity to network hyperparameters, and the influence of the available training dataset's size.

The issue of optimal process control can be generalized to a wide range of statistical process control (SPC) approaches, such as the most commonly used in manufacturing, Six Sigma approach [19]. In the survey paper [14] the authors provide a comprehensive review of machine learning approaches in SPC tasks. While they notice that the applications of ML in that field is growing rapidly, they also point out several challenges that such approaches are facing nowadays. They define several requirements for Industry 4.0/5.0 technologies have to address in order to efficiently implement ML-based system in their process control setups. One of which is the requirement of interpretability of control charts for complex data and data fusion from multiple sources to improve the performance of the quality control mechanism. In following section we describe how our work fits in that requirements.

2.1 Motivation

The papers described above demonstrate the successful application of machine learning techniques in analyzing industrial process data. Prediction and interpretation are the two essential aspects of ML applications to process data analysis, with deep neural networks being better suited for accurate prediction and simpler models preferred for interoperability. However, interpretability is critical for industrial adoption as it helps to establish trust in the analytics algorithms. While deep learning is commonly applied to sensor data, achieving a balance between model complexity and maintainability is necessary to achieve a trade-off between accuracy and complexity. In our work we do not substitute human with black-box machine learning model, but rather include the human expert in the decision process. We approach this challenge by carefully separating technical aspect of the process control from human aspect. The technical aspect is related to data preprocessing, reduction, and unsupervised identification of higher-level concepts (i.e., clusters that represent different stages of the system). The human aspect is related to actual decision making, which is done by the expert in the field. In practice, expert knowledge was encoded in the form of simple rules and automatically executed given the stages of the process discovered in the previous step with a deep learning-based clustering algorithm.

3 Methodology

This paper considers the practical limitations of typical industrial processes that enforce control changes within a discrete-time framework. As a result, we propose dividing the control determination process into smaller subproblems and creating a control scheme that can establish process control within a single time step and then use it in future steps. Furthermore, we assume that this control scheme should be generic enough to be applicable to all components.

We propose an algorithm that analyzes process parameters, such as temperature and pressure, not only at physical sensor locations but also at virtual sensors placed near them in order to proactively respond to process changes. Hence, the proposed procedure assumes the application of a process simulation model that is used to compute features to be analyzed. Then, data from both sources are combined by time series clustering algorithm to determine the current stage of the process, which can be unambiguously interpreted physically. In the proposed procedure these clusters representing the process stages are interpreted by experts based on domain knowledge and physical parameters of the process. The control actions to be taken at each time step are dependent on the determined process stage resulting in proactive process control. In the proposed approach the control scheme's structure is developed by engineers based on industrial practices. It makes it easily understandable for operators and adaptable to any specific case. An algorithm proposed in this work is presented as a block diagram in Fig. 1.

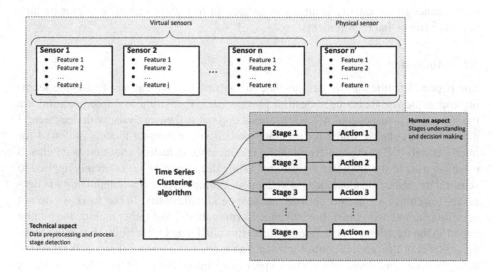

Fig. 1. Block diagram of the proposed ML-based proactive process control.

In this study, we apply deep temporal clustering (DTC) [11] method to define the process stage. The DTC algorithm is a new method for classifying temporal data. It uses advanced deep-learning techniques to reduce the complexity of the data and find patterns that allow the data to be split into different classes. As presented in Fig. 2, the method first uses an autoencoder to reduce the dimensionality of the data, a Convolutional Neural Network (CNN) to identify dominant short-term patterns, and finally a bi-directional Long Short Time Memory neural network (BI-LSTM) to identify longer-term temporal connections between the patterns. The clustering layer then groups the data into classes based on the identified patterns. The neural network architecture employs leaky rectifying linear units (L-ReLUs). The unique aspect of this method is its ability to handle complex temporal data without losing any time-related information, resulting in high classification accuracy.

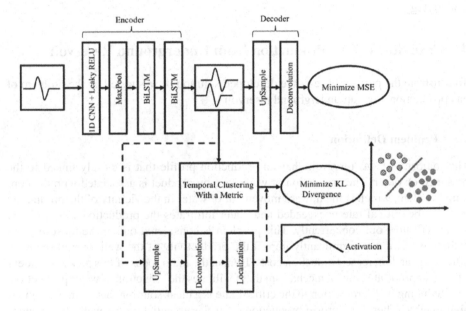

Fig. 2. The overview of the DTC algorithm [11].

The learning process in both the 1D CNN and BI-LSTM models relies on minimizing two cost functions in an interleaved manner [11]. The first cost function is based on the mean square error (MSE) of the reconstructed input sequence obtained from the BI-LSTM latent representation. This cost function ensures that the sequence remains well represented after undergoing dimensionality reduction. The second cost function is derived from the clustering metric, KL divergence. This cost function guarantees that the high-level features defining the subspace spanned by the cluster centroids successfully separate the sequences into distinct clusters exhibiting different spatio-temporal behavior. Optimization of the clustering metric modifies the weights of both the CNN and BI-LSTM models. Consequently, the high-level features encoded by the BI-LSTM model optimally segregate the sequences into clusters, effectively disentangling the spatio-temporal patterns.

To implement this deep learning method, it is necessary to divide the proposed algorithm into two phases: offline learning and online process stage assignment. During the learning phase, the DTC algorithm learns to identify the process stage by analyzing historical values of virtual sensor measurements and physical characteristics. Consequently, the trained algorithm can classify the process stage based on the events that happen around the physical sensors, enabling proactive response.

In the online phase, the trained DTC algorithm analyzes parameter values provided by the simulation model and identifies the current process stage in each time step of the process simulation. Based on the decision scheme proposed, the appropriate control actions to be taken are determined in accordance with the current process stage. These actions are then transferred to the simulation model, resulting in a self-modifying control strategy.

4   Case Study: Gas Production from Underground Reservoir

To illustrate the performance of the developed solution, it was tested on the problem of gas production from an underground reservoir.

4.1   Problem Definition

The majority of gas reservoirs have a production profile that is closely linked to the presence of water. The production of water as a by-product is associated with the conning problem, which is the upward movement of water in the vicinity of the production well. If the critical rate is exceeded and water infiltrates the production well, gas and water will flow out concurrently. This problem is considered one of the most difficult challenges as it can significantly impact the productivity of the well, overall recovery efficiency, and increase production costs. Hence, the main focus of this paper is to identify the appropriate control mechanism that facilitates the reduction of water production.

Reducing well production to the critical rate is a clear solution, but it's not commercially viable. Therefore, proper production control is essential for gas wells that produce water. The production of hydrocarbons is regulated by maintaining a flow rate of wells situated in the reservoir. However, optimizing well control using classical methods is challenging as it is not feasible to establish a functional relationship between the decision variables and the quality indicator being optimized. This is because of the complicated geological structure of hydrocarbon reservoirs, which have complex geometry and spatial heterogeneity of petrophysical parameters. Moreover, the multidimensional physics of fluid flow makes it difficult to describe the system's dynamics. As a result, reservoir simulators are used to model the processes occurring in hydrocarbon reservoirs, which can predict the behavior of the reservoir.

Given the practical constraints of the gas production process that require discrete control changes, it is possible to break down the problem of well control determination into sub-problems. Therefore, modeling a control scheme that determines individual well control in a single time step is sufficient [8]. However, it remains challenging to accurately determine the actions that should be taken at a particular production stage to prevent water infiltration and optimize gas production. In industrial practice, gas wells producing water are typically managed using a reactive control approach that relies on trial and error. However, a proactive control approach that takes preemptive control actions before undesirable changes in flow reach the well can improve the techno-economic efficiency of the production process. To go from reactive to proactive control this paper proposes a solution that utilizes reservoir simulation and virtual active sensors located near the production well. By monitoring reservoir parameters before undesired flow changes reach the well, these virtual sensors enable proactive inflow control and form the foundation for the proposed control system for gas wells conning water.

4.2 Methodology Adjustment for the Well Control

In the case of well control, the idea of the proposed solution is to analyze reservoir parameters surrounding a production well measured by the virtual sensors combined with ground production data. Then, in each time step of the reservoir simulation, use the trained DTC algorithm to determine the current production stage of the well based on the reservoir parameters computed by the reservoir simulator. The control of gas wells that are affected by water influx is then automatically determined based on the production stage of the well.

4.3 Proposed Solution Application

The solution proposed in this paper was evaluated by its application on a real gas reservoir where a water coning problem occurs. The reservoir is modelled with the ECLIPSE reservoir simulator (Schlumberger Limited, Houston, TX, USA) that uses finite difference methods to model fluid flow in complex geological formations and well configurations. It allows for the simulation of a wide range of scenarios and is used for predicting reservoir performance, improving reservoir understanding and optimizing production strategies. The behavior of the reservoir is described by a system of non-linear partial differential equations that describe fluid flow in porous media [2]. The equations are solved numerically using Newton's iterative method.

In the case of the analyzed reservoir, there are 8 production wells, and 3 virtual active sensors are located in close proximity to each well as shown in Fig. 3. The ground production attributes, such as gas and water flow rates, are monitored in each well and the virtual sensors measure water saturation and reservoir pressure in the vicinity of the well to determine its production stage as shown in Fig. 4. These reservoir parameters are analyzed by the DTC model to detect clusters corresponding to the stages of the production process.

The proposed solution was applied to a historical production period to evaluate its effectiveness. In the offline stage, the DTC model was trained using multivariable historical data from the initial 20 years of production. The next 20 years of historical production data were used for validation. Both training and validation data are generated by the numerical reservoir model.

For the DTC model learning, data related to the well with the most complex reservoir situation with respect to water production (the highest water production) were used as it is the most representative. All the analyzed features measured by both physical and virtual sensors are represented as floating point values sampled at regular intervals (in the example provided, every hour). Hence, the input data include 8 attributes measured in 175 200 time steps. The DTC model was trained using the Adam optimizer with a learning rate 0.01. Parameters of the neural network architecture were selected based

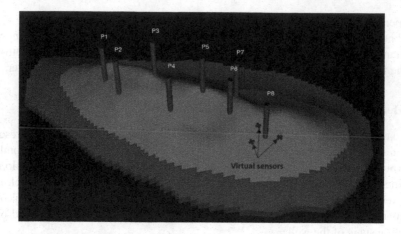

Fig. 3. Visualization of the analyzed reservoir (example of the virtual sensors location for P8 well).

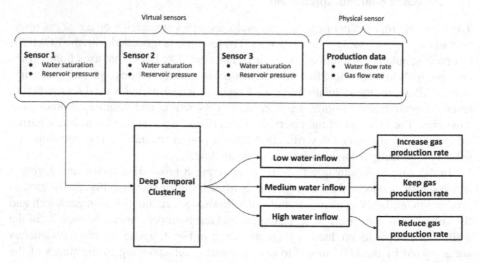

Fig. 4. Block diagram of the proposed control scheme for gas production for the underground reservoir.

on the values proposed in [11]. The output of the DTC model is an array with a class label assigned to each record of input data described above.

In the presented approach, the number of clusters was selected by the human expert based on the characteristics of the analyzed process. In the case of the water conning problem, three distinct stages can be identified in the production process. The first stage involves low water inflow during the initial production period. The second stage is characterized by a gradual increase in water inflow. Finally, when water inflow reaches a high level, it indicates the final stage of production from the well, which is no longer economically viable. Taking into account this phenomenon, the proposed control scheme also includes 3 stages, to keep the solution easily understandable for the opera-tor. When the trained DTC model assigned one of the three clusters to each time step, these clusters were interpreted by experts based on water saturation in the virtual sensor presented in Fig. 5. Cluster 0 represents low water inflow, cluster 1 denotes medium water inflow, and cluster 2 indicates high water inflow. This interpretation facilitates a clear physical understanding of the suggested gas well control.

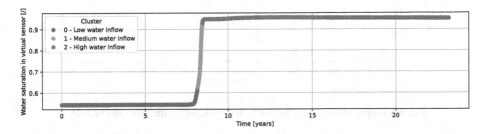

Fig. 5. Block diagram of the proposed ML-based proactive process control.

The online phase was tested using the last 20 years of production data. In this phase, the trained DTC model predicts the production stage of each well at every time step of the reservoir simulation. The well rates are then automatically adjusted according to the decision scheme presented in Fig. 4 that is developed based on engineering knowledge and simulation tests. The decision scheme involves categorizing production wells into three groups that correspond to the interpreted DTC clusters. If a well exhibits high water inflow, its flow rate is decreased. If water inflow is at a medium level, the well performance remains unchanged. However, for wells with low water inflow, the well rate is increased. This approach to determining the appropriate actions to be taken at each production stage results in self-modifying intelligent well control.

4.4 Results

The application of the developed solution resulted in proactive control of the production wells located on the analyzed gas reservoir. This intelligent control allows cumulative water production from the reservoir to be reduced by 15% compared to historical data based on operator experience (Fig. 6). Cumulative gas production remained at the same level, indicating increased production efficiency. The results demonstrate that the devel-oped control strategy accurately manages the water coning phenomenon, reduces oper-ational costs, and increases overall income without incurring extra expenses, as only the control system is changed. Furthermore, the difference between the results obtained using the developed control system and the historical data increases with time, imply-ing that the ML-based solution is more effective in complicated production situations as with time water saturation in the vicinity of the production wells is higher. Thus, the generated control outperforms human control, but the operator can still verify it.

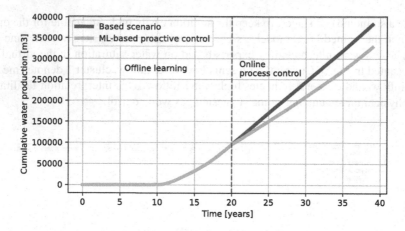

Fig. 6. Comparison of water production depending on the process control).

5   Discussion and Potential Applications

The approach proposed in this paper enables better management of the process under consideration. As data from virtual sensors are incorporated into this procedure, it allows for reacting proactively. However, the limitation of the proposed proactive control is that there is a need to use a process simulation model which may be difficult in the case of real-time applications. However, the trained ML model defining the process stage can be used directly (without simulation) as long as the process simulation model used to train it is valid.

Thanks to the use of advanced machine learning algorithms, the appropriate control can be determined also in situations that are not trivial for a human. As input features from both physical and virtual sensors are combined into a process stage that is physically interpretable it makes the whole decision procedure easily explainable what is the main differentiator of the proposed method.

However, the method presented here is not fully automatic as it requires interaction with a domain expert. In the proposed method, the time series clustering algorithm combines the simulation data into clusters corresponding to process stages, but the meaning of these clusters is given by an expert based on domain knowledge about the analyzed process and values of physical parameters.

In the developed solution, the final structure of the decision scheme is developed by domain experts based on the best industrial practices and domain knowledge. It makes the decision process understandable for the operator, which is especially important considering real applications. In addition, the decision scheme can be adapted to any particular process which makes the proposed method general.

As the proposed solution is generic, it can be applied to process control at different levels of granulation which confirms its scalability. It can be used to define the control of the process as a whole. The proposed approach can be applied also to each component of the process. For example, the temperature of each component of the wind turbine can be measured and the proposed actions can be enough general to be applicable to all components. Another possibility is to apply this approach to a fleet of assets, such as wind turbines for which the same measurements are available. Then, the scheme can be used to define the control of the individual wind turbine and utilized for all of the assets in the fleet as in the presented example of gas wells.

The proposed method can be also adapted as a predictive maintenance tool. In this case, the process stage defined by ML algorithm can be interpreted as the component wear. For an example of a bearing, its temperature can be measured by the physical sensor and virtual sensors can tell about the temperature distribution around this physical measurement to react proactively. Then, the process stage determined based on the ML algorithm, assuming three clusters, can present a component in good condition, a component in wear, or a faulty component. As virtual sensors are utilized the proposed approach makes it possible to proactively predict the moment when maintenance is needed. As the proposed procedure not only defines the process stage but also defines actions to be taken it can also be used as a prescriptive maintenance tool. These actions can regulate operating conditions which allows asset maintenance to be scheduled in order to achieve specific objectives.

Future work of the study reported here includes replacing the simple control actions directly dependent on the process stage with an auto-adaptive decision tree introduced in [9, 10] where control actions are parametrized and can be optimized to reach optimal control. Since the input for this parameterized decision tree comprises simulation data rather than the process stage, the decision-making process is entirely automated while remaining straightforward to interpret. Therefore, it would be reasonable to extend the proposed procedure with an auto-adaptive parameterized decision tree as an automated and directly interpretable method.

6 Conclusion

The paper proposes a new method for proactive control of industrial processes using machine learning techniques. The approach involves combining physical and virtual sensor data obtained from a simulation model to identify the current stage of the process and proactively determine appropriate control actions. The deep temporal clustering algorithm is used to combine simulation data and identify the process stage, and a decision scheme proposed by engineers is used to determine control actions dependent on the identified stage.

The performance of the developed solution was illustrated in the problem of gas production from an underground reservoir. In this case, the proposed solution allowed for managing the considered water coning phenomenon, reducing of unfavourable water inflow, and increasing overall income without incurring extra expenses, as only the control system is changed.

The proposed method is generic and scalable. It can be applied to a single asset or multiple assets if the simulation model includes a fleet of them. It can also be adapted as a predictive and prescriptive maintenance tool. Future work includes optimizing the control actions to reach optimal control.

Acknowledgements. This paper is funded from the XPM (Explainable Predictive Maintenance) project funded by the National Science Center, Poland under CHIST-ERA programme Grant Agreement No. 857925 (NCN UMO-2020/02/Y/ST6/00070).

References

1. Angelopoulos, A., et al.: Tackling faults in the industry 4.0 era-a survey of machine-learning solutions and key aspects. Sensors **20**(1), 109 (2019)
2. Busswell, G., Banergee, R., Thambynayagam, M., Spath, J.: Generalized analytical solution for solution for reservoir problems with multiple wells and boundary conditions, April 2006. https://doi.org/10.2118/99288-MS
3. Casolla, G., Cuomo, S., Schiano Di Cola, V., Piccialli, F.: Exploring unsupervised learning techniques for the internet of things. IEEE Trans. Ind. Inform. **PP**, 1–1 (2019). https://doi.org/10.1109/TII.2019.2941142
4. Dimitrov, N., Göçmen, T.: Virtual sensors for wind turbines with machine learning-based time series models. Wind Energy **25**(9), 1626–1645 (2022)
5. Gajjar, S., Kulahci, M., Palazoglu, A.: Real-time fault detection and diagnosis using sparse principal component analysis. J. Process Control **67**, 112–128 (2018). https://doi.org/10.1016/j.jprocont.2017.03.005, https://www.sciencedirect.com/science/article/pii/S0959152417300677, big Data: Data Science for Process Control and Operations
6. Garcia, E., Montés, N., Llopis, J., Lacasa, A.: Miniterm, a novel virtual sensor for predictive maintenance for the industry 4.0 era. Sensors **22**(16) (2022). https://doi.org/10.3390/s22166222, https://www.mdpi.com/1424-8220/22/16/6222
7. Jiang, Y., Yin, S., Dong, J., Kaynak, O.: A review on soft sensors for monitoring, control, and optimization of industrial processes. IEEE Sensors J. **21**(11), 12868–12881 (2021). https://doi.org/10.1109/JSEN.2020.3033153
8. Kuk, E.: Application of artificial intelligence methods to underground gas storage control. In: SPE Annual Technical Conference and Exhibition, vol. Day 2 Tue, October 01, 2019 (2019). https://doi.org/10.2118/200305-STU, https://doi.org/10.2118/200305-STU, d023S103R025
9. Kuk, E., Stopa, J., Kuk, M., Janiga, D., Wojnarowski, P.: Petroleum reservoir control optimization with the use of the auto-adaptive decision trees. Energies **14**(18) (2021). https://doi.org/10.3390/en14185702, https://www.mdpi.com/1996-1073/14/18/5702
10. Kuk, E., Stopa, J., Kuk, M., Janiga, D., Wojnarowski, P.: Optimal well control based on auto-adaptive decision tree—maximizing energy efficiency in high-nitrogen underground gas storage. Energies **15**(9) (2022). https://doi.org/10.3390/en15093413, https://www.mdpi.com/1996-1073/15/9/3413
11. Madiraju, N.S.: Deep temporal clustering: fully unsupervised learning of time-domain features. Ph.D. thesis, Arizona State University (2018)
12. Malykhina, G.: Digital twin technology as a basis of the industry in future, 416–428, December 2018. https://doi.org/10.15405/epsbs.2018.12.02.45
13. Qin, S.J., Chiang, L.H.: Advances and opportunities in machine learning for process data analytics. Comput. Chem. Eng. **126**, 465–473 (2019). https://doi.org/10.1016/j.compchemeng.2019.04.003, https://www.sciencedirect.com/science/article/pii/S0098135419302248
14. Tran, P.H., Ahmadi Nadi, A., Nguyen, T.H., Tran, K.D., Tran, K.P.: Application of machine learning in statistical process control charts: a survey and perspective. In: Tran, K.P. (ed.) Control Charts and Machine Learning for Anomaly Detection in Manufacturing. SSRE, pp. 7–42. Springer, Cham (2022). https://doi.org/10.1007/978-3-030-83819-5_2

15. Usuga Cadavid, J.P., Lamouri, S., Grabot, B., Pellerin, R., Fortin, A.: Machine learning applied in production planning and control: a state-of-the-art in the era of industry 4.0. J. Intell. Manuf. **31**(6), 1531–1558 (2020). https://doi.org/10.1007/s10845-019-01531-7
16. Vallejo, M., de la Espriella, C., Gómez-Santamaría, J., Ramírez-Barrera, A.F., Delgado-Trejos, E.: Soft metrology based on machine learning: a review. Measur. Sci. Technol. **31**(3), 032001 (2019). https://doi.org/10.1088/1361-6501/ab4b39, https://dx.doi.org/10.1088/1361-6501/ab4b39
17. Vazan, P., Znamenak, J., Juhas, M.: Proactive simulation in production line control. In: 2018 IEEE 13th International Scientific and Technical Conference on Computer Sciences and Information Technologies (CSIT), vol. 1, pp. 52–55 (2018). https://doi.org/10.1109/STC-CSIT.2018.8526675
18. Xu, X., Lu, Y., Vogel-Heuser, B., Wang, L.: Industry 4.0 and industry 5.0-inception, conception and perception. J. Manuf. Syst. **61**, 530–535 (2021). https://doi.org/10.1016/j.jmsy.2021.10.006, https://www.sciencedirect.com/science/article/pii/S0278612521002119
19. Yang, H., et al.: Six-sigma quality management of additive manufacturing. Proc. IEEE (2020). https://doi.org/10.1109/JPROC.2020.3034519
20. Zerilli, J., Knott, A., Maclaurin, J., Gavaghan, C.: Algorithmic decision-making and the control problem. Minds Mach. **29**(4), 555–578 (2019). https://doi.org/10.1007/s11023-019-09513-7
21. Zhu, Q., Qin, S.: Supervised diagnosis of quality and process faults with statistical learning models. In: I&EC Research Revised (2019)

Parallel Algorithm for Concurrent Integration of Three-Dimensional B-Spline Functions

Anna Szyszka[ID] and Maciej Woźniak[(✉)][ID]

AGH University of Science and Technology, Kraków, Poland
`macwozni@agh.edu.pl`

Abstract. In this paper, we discuss the concurrent integration applied to the 3D isogeometric finite element method. It has been proven that integration over individual elements with Gaussian quadrature is independent of each other, and a concurrent algorithm for integrating a single element has been created. The suboptimal integration algorithm over each element is developed as a sequence of basic atomic computational tasks, and the dependency relation between them is identified. We show how to prepare independent sets of tasks that can be automatically executed concurrently on a GPU card. This is done with the help of Diekert's graph, which expresses the dependency between tasks. The execution time of the concurrent GPU integration is compared with the sequential integration executed on CPU.

Keywords: Trace Theory · Concurrency · Isogeometric Finite Element Method · Numerical Integration

1 Introduction

The isogeometric analysis (IGA-FEM) [7] is a modern technique for the integration of geometrical modeling of CAD systems with engineering computations of CAE systems. The IGA-FEM has multiple applications from phase field modeling [9], shear deformable shell theory [3], wind turbine aerodynamics [20], phase separation simulations [15], in compressible hyperelasticity [11], to turbulent flow simulations [6] and biomechanics [5,19]. IGA-FEM computations consist of two phases: (1) generation of the system of linear equations and (2) execution of an external solver algorithm of the global system of linear equations. The generated system of linear equations for elliptic problems is solved with multifrontal direct solvers [12,13] such as MUMPS [2], SuperLU [21] or PaStiX [17], or iterative solvers. It is possible to obtain linear computational cost for two-dimensional h-refined grids with point or edge singularities [1,14,16].

This work is a summary of [23,25] dealing with concurrent integration. We have parallelized the integration routines of the three-dimensional IGA code. This is done by localizing basic undividable tasks and finding sets of tasks that can be executed concurrently [10]. We concentrate on B-spline basis functions employed in a 3D isogeometric finite element method L^2-projection problem. The presented analysis can be applied to any elliptic problem in 3D with analogous results.

J. Mikyška et al. (Eds.): ICCS 2023, LNCS 14074, pp. 590–596, 2023.
https://doi.org/10.1007/978-3-031-36021-3_57

2 Integration Algorithm

2.1 Formulation of the Model Problem

The goal of this study is to evaluate the cost of using different integration methods to build IGA matrices. To illustrate this, we will use the heat equation discretized in time using the forward Euler method and focus specifically on the cost of assembling the Mass matrix. Find $u \in C^1\left((0,T), H^1\left(\Omega\right)\right)$ such that $u = u_0$ at $t = 0$ and, for each $t \in (0,T)$, it holds:

$$\int_\Omega \frac{\partial u}{\partial t} v \, dx = -\int_\Omega \nabla u \cdot \nabla v \, dx, \qquad \forall v \in H^1\left(\Omega\right). \tag{1}$$

For simplicity, we consider a discrete-in-time version of the problem employing the forward Euler method.

$$\int_\Omega u_{n+1} v \, dx = \int_\Omega u_n v \, dx - \Delta_t \int_\Omega \nabla u_n \cdot \nabla v \, dx, \qquad \forall v \in H^1\left(\Omega\right). \tag{2}$$

Then, the matrix element is computed as:

$$A_{\beta,\delta}^\alpha = \sum_{n_1=1}^{P_1} \sum_{n_2=1}^{P_2} \sum_{n_3=1}^{P_3} \omega^{n_1} \omega^{n_2} \omega^{n_3} \, \Pi(x^n) \, J(x^n), \tag{3}$$

This study focuses on using 3D tensor B-spline basis functions with uniform polynomial degree order and regularity on the interior faces of the mesh for ease of demonstration. However, it should be noted that the techniques presented can be easily adapted to other types of basis functions. For the construction of B-spline basis functions, we used Cox-de-Boor recursive formulae [4].

2.2 Algorithms and Computational Cost

We compared two algorithms, the classical integration algorithm and the sum factorization algorithm. In the classical integration algorithm, the local contributions to the matrix on the left side A are represented as the sum of the quadrature points. For the classical integration algorithm, the associated computational cost scales, concerning the polynomial degree p as $\mathcal{O}(p^9)$ [18].

After a relatively simple observation, we can reorganize the integration terms of Eq. (3). In practice, Eq. (3) is written as:

$$A_{\beta,\delta}^\alpha = \sum_{n_3=1}^{P_3} \omega_3^n \, B_j(x_3^n) \, B_{m;p}(x_3^n) \, C(i_2, i_3, j_2, j_3, k_1), \tag{4}$$

where buffer C is given by

$$C(i_2, i_3, j_2, j_3, k_1) = \sum_{n_2=1}^{P_2} \omega_2^n \, B_i(x_2^n) \, B_{l;p}(x_2^n) \underbrace{\sum_{n_1=1}^{P_1} \omega_1^n \, B_h(x_1^n) \, B_{k;p}(x_1^n) \, J(x^n)}_{D(i_3, j_3, k_1, k_2)}.$$

$$\tag{5}$$

With the practical implementation, we end up with three distinct groups of loops and several buffers. As a consequence, the computational cost associated with sum factorization decreases to $\mathcal{O}(p^7)$ [18].

2.3   Concurrency Model for Integration

For both algorithms, we used the identical methodology described in [23]. We introduced four basic types of computational tasks.

- Computational tasks that evaluate the 1D basis function on the element E_α at the coordinate of the quadrature point.
- Computational tasks to evaluate the 3D basis function on the element E_α at the coordinate of the quadrature point.
- Computational tasks that evaluate the value of the product of two basis functions (from the test and trial space) on the element E_α at the quadrature point.
- Computational task that evaluates the value of the integral of the scalar product of two base functions (from the test and trial space) on the element E_α.

Next, we define an alphabet of tasks Σ and a set of dependencies between them D (Fig. 1).

We also applied methodology to the sum factorization algorithm to obtain optimal scheduling and theoretical verification from the trace theory method. Next, we did a series of numerical experiments to measure parallel performance.

All the tasks mentioned in each layer of the Foata Normal Form are meant to be performed on a homogeneous architecture. All tasks within the particular Foata class should take nearly identical amounts of time, and can be effectively scheduled as a common bag. The method should be very useful with practical implementation for large clusters, such as modern supercomputers [8,22,24] with multiple GPUs per computational node (Table 1).

2.4   Results

Table 1. Comparison of integration time with different strategies

algorithm	t_{serial}	t_{OpenMP}	t_{GPU}
classical	11752.54	1125.72 (12 cores)	5.87
sum factorization	394.28	112.65 (4 cores)	10.78

We observed an unexpected performance behavior of the parallel sum factorization. Despite utilizing parallel loops across all elements, it was scaling only up to 4 cores. Beyond 4 cores, there was a plateau in speedup, indicating poor

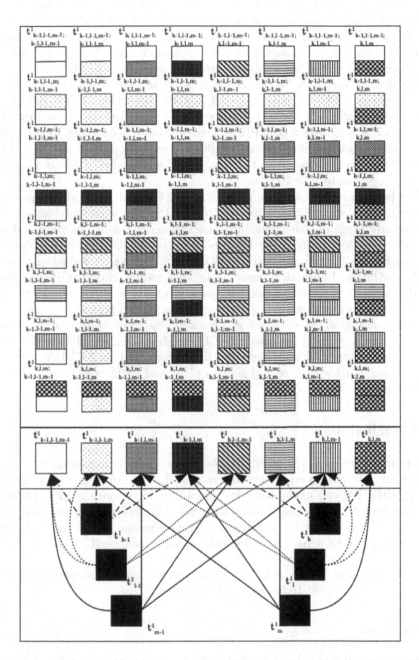

Fig. 1. Part of reduced dependency graph for calculating the frontal matrix using linear basis functions. To maintain the transparency of the chart, the most of relationships are marked with a fill.

performance in a multicore environment. Note that sum factorization requires significantly more memory synchronization compared to the classical method.

We evaluated the performance of classical integration and sum factorization in various scenarios. We focus on $p = 9$ polynomial order basis functions as it was expected to be the most advantageous scenario for sum factorization. We examined three scenarios for a mesh size of: 1) single-core CPU execution, 2) shared-memory CPU computation, and 3) GPU execution. The classical integration on a single core took 11752.54 s, the 12-core OpenMP implementation took 1125.72 s, and estimated GPU implementation was expected to take 5.87 s. For sum factorization integration, single-core execution took 394.28 s, 4-core OpenMP implementation took 112.65 s, and the estimated GPU implementation was expected to take 10.78 s.

The performance of a single core shows that the classical algorithm scales as $\mathcal{O}(p^9)$, while sum factorization as $\mathcal{O}(p^7)$. However, sum factorization does not take advantage of concurrent implementations because of its nature of deep data dependencies. Even in the Figures presented in [23,25] we can observe a single reduction for the classical algorithm, and three such reductions for the sum factorization. In practical implementations, reduction costs $\mathcal{O}(\log n)$, where n is the number of threads/machines/cores from which we reduce. In the case of integration, n would mean the number of quadrature points.

Computations were performed on a Banach Linux workstation equipped with an AMD Ryzen 9 3900X processor, 64GB RAM, a GeForce RTX 2080 SUPER graphic card equipped with 8 gigabytes of memory, and 3072 CUDA cores. The code was compiled with nvcc and gcc, for GPU and CPU, respectively, and -O2 level of optimization.

3   Conclusions

Our approach validates the scheduling for the integration algorithm by utilizing trace theory. We compared the execution of the integration algorithm on a CPU, and a GPU. We can extrapolate its scalability for various elliptic problems. Furthermore, the trace-theory based analysis of concurrency in the integration algorithm can be adapted to different integration methods. The methodology is versatile and can be expanded to include higher-dimensional spaces.

Acknowledgement. The work of Maciej Woźniak was partially financed by the AGH University of Science and Technology Statutory Fund.

References

1. AbouEisha, H., Moshkov, M., Calo, V., Paszyński, M., Goik, D., Jopek, K.: Dynamic programming algorithm for generation of optimal elimination trees for multi-frontal direct solver over h-refined grids. Procedia Comput. Sci. **29**, 947–959 (2014)

2. Amestoy, P.R., Duff, I.S., L'Excellent, J.Y.: Multifrontal parallel distributed symmetric and unsymmetric solvers. Comput. Methods Appl. Mech. Eng. **184**, 501–520 (2000)
3. Benson, D.J., Bazilevs, Y., Hsu, M.C., Hughes, T.J.R.: A large deformation, rotation-free, isogeometric shell. Comput. Methods Appl. Mech. Eng. **200**, 1367–1378 (2011)
4. de Boor, C.: Subroutine package for calculating with b-splines. SIAM J. Numer. Anal. **14**(3), 441–472 (1971)
5. Calo, V.M., Brasher, N.F., Bazilevs, Y., Hughes, T.J.R.: Multiphysics model for blood flow and drug transport with application to patient-specific coronary artery flow. Comput. Mech. **43**, 161–177 (2008)
6. Chang, K., Hughes, T.J.R., Calo, V.M.: Isogeometric variational multiscale large-eddy simulation of fully-developed turbulent flow over a wavy wall. Comput. Fluids **68**, 94–104 (2012)
7. Cottrell, J.A., Hughes, T.J.R., Bazilevs, Y.: Isogeometric Analysis: Toward Integration of CAD and FEA. Wiley (2009)
8. Cyfronet. https://kdm.cyfronet.pl/portal/Main_page: Cyfronet KDM
9. Dedè, L., Borden, M.J., Hughes, T.J.R.: Isogeometric analysis for topology optimization with a phase field model. Arch. Comput. Meth. Eng. **19**, 427–465 (2012)
10. Diekert, V., Rozenberg, G.: The Book of Traces. World Scientific (1995)
11. Duddu, R., Lavier, L.L., Hughes, T.J.R., Calo, V.M.: A finite strain Eulerian formulation for compressible and nearly incompressible hyperelasticity using high-order b-spline finite elements. Int. J. Numer. Meth. Eng. **89**, 762–785 (2012)
12. Duff, I.S., Reid, J.K.: The multifrontal solution of indefinite sparse symmetric linear. ACM Trans. Math. Software **9**, 302–325 (1983)
13. Duff, I.S., Reid, J.K.: The multifrontal solution of unsymmetric sets of linear equations. SIAM J. Sci. Stat. Comput. **5**, 633–641 (1984)
14. Goik, D., Jopek, K., Paszyński, M., Lenharth, A., Nguyen, D., Pingali, K.: Graph grammar based multi-thread multi-frontal direct solver with Galois scheduler. Procedia Comput. Sci. **29**, 960–969 (2014)
15. Gomez, H., Hughes, T.J.R., Nogueira, X., Calo, V.M.: Isogeometric analysis of the isothermal navier-stokes-korteweg equations. Comput. Methods Appl. Mech. Eng. **199**, 1828–1840 (2010)
16. Gurgul, P.: A linear complexity direct solver for h-adaptive grids with point singularities. Procedia Comput. Sci. **29**, 1090–1099 (2014)
17. Hénon, P., Ramet, P., Roman, J.: Pastix: a high-performance parallel direct solver for sparse symmetric definite systems. Parallel Comput. **28**, 301–321 (2002)
18. Hiemstra, R.R., Sangalli, G., Tani, M., Calabrò, F., Hughes, T.J.: Fast formation and assembly of finite element matrices with application to isogeometric linear elasticity. Comput. Methods Appl. Mech. Eng. **355**, 234–260 (2019). https://doi.org/10.1016/j.cma.2019.06.020
19. Hossain, S.S., Hossainy, S.F.A., Bazilevs, Y., Calo, V.M., Hughes, T.J.R.: Mathematical modeling of coupled drug and drug-encapsulated nanoparticle transport in patient-specific coronary artery walls. Comput. Mech. **49**, 213–242 (2012)
20. Hsu, M.C., Akkerman, I., Bazilevs, Y.: High-performance computing of wind turbine aerodynamics using isogeometric analysis. Comput. Fluids **49**, 93–100 (2011)
21. Li, X.S.: An overview of superlu: algorithms, implementation, and user interface. TOMS Trans. Math. Software **31**, 302–325 (2005)
22. ORNL. https://www.olcf.ornl.gov/summit/: Summit, Oak Ridge National Laboratory

23. Szyszka, A., Woźniak, M., Schaefer, R.: Concurrent algorithm for integrating three-dimensional b-spline functions into machines with shared memory such as gpu. Comput. Methods Appl. Mech. Eng. **398**, 115201 (2022). https://doi.org/10.1016/j.cma.2022.115201
24. TACC. https://portal.tacc.utexas.edu/user-guides/stampede2: Stampede2 User Guide
25. Woźniak, M., Szyszka, A., Rojas, S.: A study of efficient concurrent integration methods of b-spline basis functions in IGA-fem. J. Comput. Sci. **64**, 101857 (2022). https://doi.org/10.1016/j.jocs.2022.101857

Biomedical and Bioinformatics Challenges for Computer Science

Resting State Brain Connectivity
Analysis from EEG and FNIRS Signals

Rosmary Blanco[✉], Cemal Koba, and Alessandro Crimi

Sano Centre for Computational Medicine, Computer Vision Data Science Team,
Krakow, Poland
{r.blanco,c.koba,a.crimi}@sanoscience.org

Abstract. Contemporary neuroscience is highly focused on the synergistic use of machine learning and network analysis. Indeed, network neuroscience analysis intensively capitalizes on clustering metrics and statistical tools. In this context, the integrated analysis of functional near-infrared spectroscopy (fNIRS) and electroencephalography (EEG) provides complementary information about the electrical and hemodynamic activity of the brain. Evidence supports the mechanism of the neurovascular coupling mediates brain processing. However, it is not well understood how the specific patterns of neuronal activity are represented by these techniques. Here we have investigated the topological properties of functional networks of the resting-state brain between synchronous EEG and fNIRS connectomes, across frequency bands, using source space analysis, and through graph theoretical approaches. We observed that at global-level analysis small-world topology network features for both modalities. The edge-wise analysis pointed out increased inter-hemispheric connectivity for oxy-hemoglobin compared to EEG, with no differences across the frequency bands. Our results show that graph features extracted from fNIRS can reflect both short- and long-range organization of neural activity, and that is able to characterize the large-scale network in the resting state. Further development of integrated analyses of the two modalities is required to fully benefit from the added value of each modality. However, the present study highlights that multimodal source space analysis approaches can be adopted to study brain functioning in healthy resting states, thus serving as a foundation for future work during tasks and in pathology, with the possibility of obtaining novel comprehensive biomarkers for neurological diseases.

Keywords: EEG · fNIRS · multimodal monitoring · Functional Connectivity · Source-space analysis

1 Introduction

Large-scale functional brain connectivity can be modeled as a network or graph. For example, the synergistic use of cluster-based thresholding within statistical

© The Author(s), under exclusive license to Springer Nature Switzerland AG 2023
J. Mikyška et al. (Eds.): ICCS 2023, LNCS 14074, pp. 599–610, 2023.
https://doi.org/10.1007/978-3-031-36021-3_58

parametric maps could be useful to identify crucial connections in a graph. In the last few years, multimodal monitoring is getting increased interest. Integration of functional near-infrared spectroscopy (fNIRS) and electroencephalography (EEG) can reveal more comprehensive information associated with brain activity, taking advantage of their non-invasiveness, low cost, and portability. Between the two modalities, fNIRS relies on differential measurements of the backscattered light, which is sensitive to oxy- (HbO) and deoxy-hemoglobin (HbR). EEG, on the other hand, captures the electrical brain activity scalp derived from synchronous post-synaptic potentials. The former has high spatial resolution but is highly sensitive to scalp-related (extracerebral) hemoglobin oscillations. The latter allows the tracking of the cerebral dynamics with the temporal detail of the neuronal processes (1 ms) but suffers from volume conduction. Their integration can compensate for their shortcomings and take advantage of their strengths [6,7]. Most concurrent EEG-fNIRS studies focused on the temporal correlation of the time series data between these modalities. However, the two techniques do not have perfect spatiotemporal correspondence, since brain electrical activity and its hemodynamic counterpart are mediated through the neurovascular coupling mechanism. Therefore, comparing the correspondence between the two modalities from a more analytical and standardized perspective can be more informative. In this context, network neuroscience could be an approach to model the actual brain function and study the potential of multimodal approaches in inferring brain functions. To the best of our knowledge, little is known about the fNIRS-based functional connectivity related to EEG functional connectivity of large-scale networks from a graph-theoretical point of view. Therefore, this study aims to explore the topology of brain networks captured by the two modalities across neural oscillatory frequency bands in the resting state (RS) through graph theoretical approaches.

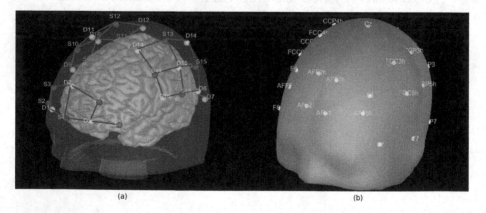

Fig. 1. (a): NIRS optodes location. The red dots are the sources and the green dots are the detectors. (b): EEG electrodes location. (Color figure online)

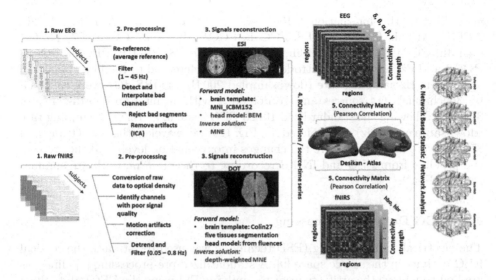

Fig. 2. An overview of the entire pipeline from pre-processing to brain networks for both EEG and fNIRS data. 1. The EEG and fNIRS data recordings for all the subjects. 2. The pre-processing steps for cleaning data. 3. The methods for the reconstruction of the signals in the source space. The Electrical Source Imagine (ESI) technique aims to build a realistic head model (forward model) by segmenting the MRI to estimate the cortical sources (neuronal activity) by solving the inverse problem (estimate of the amplitude of the current dipoles captured by the sensors on the scalp). The Diffuse Optical Tomography (DOT) technique aims to model light propagation within the head and estimate the optical properties of each tissue (forward model) by fluences of light for each optode (source and detector). This model maps the absorption changes along the cortical region (source space) to the scalp sensors (measured as optical density changes). By solving the inverse problem the cortical changes of hemodynamic activity within the brain are estimated. 4. The EEG and fNIRS source-time series were mapped in the same 3D space using an atlas-based approach (Desikan-Killiany). 5. Functional connectivity (Pearson's correlation) estimates the statistical coupling between each ROI of the reconstructed time series. 6. The topology of brain networks captured by the two techniques was compared through graph theoretical approaches. As shown later in Fig. 4.

2   Materials and Methods

Synchronous resting state EEG and fNIRS recordings of healthy adults (n = 29; 28.5 ± 3.7 years) were obtained from an open dataset for hybrid brain-computer interfaces (BCIs) [23]. The recordings feature a 1 min long pre-experiment resting-state data, an experimental paradigm including a motor imagery and a mental arithmetic test. As the focus of this work is RS only, we only processed and examined the RS part of the dataset. The 1-minute RS duration is justified for fNIRS as it was shown that the reliability of the network metrics stabilizes in 1 min of scan [13,17]. EEG data were recorded with 32 electrodes placed according to the international 10–5 system (AFp1, AFp2, AFF1h, AFF2h, AFF5h, AFF6h, F3, F4, F7, F8, FCC3h, FCC4h, FCC5h, FCC6h, T7, T8, Cz, CCP3h, CCP4h, CCP5h, CCP6h, Pz, P3, P4, P7, P8, PPO1h, PPO2h, POO1, POO2 and Fz for ground electrode). Those are referenced to the linked mastoid at 1000 Hz Hz sampling rate (downsampled 200 Hz). fNIRS data were collected by 36 channels (14 sources and 16 detectors with an inter-optode distance of 30 mm), following the standardized 10–20 EEG system, at 12.5 Hz sampling rate (downsampled 10 Hz) as depicted in Fig. 1. Two wavelengths at 760 nm and 850 nm were used to measure the changes in oxygenation levels. All the steps of the pipeline from EEG and fNIRS recording to the brain network are summarized in Fig. 2.

2.1   EEG Data Pre-processing

The electrical source imaging (ESI) technique was used to estimate the cortical EEG activity in the source space [9]. A standardized pre-processing pipeline was applied to remove the artifacts prior to applying ESI using the EEGLab toolbox [4]. The measured EEG data was first re-referenced using a common average reference and filtered (second-order zero-phase Butterworth type IIR filter) with a passband of 1–45 Hz. Bad channels were identified, rejected, and interpolated taking an average of the signal from surrounding electrodes. The signals were then visually inspected to detect and reject segments of data still containing large artifacts. A decomposition analysis by using the fast fixed-point ICA (FastICA) algorithm, was performed to identify and remove artifacts of biological origin from the recordings [15].

2.2   fNIRS Data Pre-processing

We used diffuse optical tomography (DOT) to reconstruct the signals in the source space. More specifically, a combination of the Brainstorm toolbox (i.e., NIRSTORM) and a custom MATLAB script was used. Before the reconstruction, the typical pre-processing pipeline of the fNIRS data was applied. The raw data were converted to optical density (OD-absorption) signals for both wavelengths. The bad channels were removed based on the following criteria: signals had some negative values, flat signals (variance close to 0), and signals had too many flat segments. Then a semi-automatic movement correction was applied: the signals were visually checked for motion artifacts in the form of spikes or baseline shifts and a spline interpolation method was used for the correction. Finally, the OD signals were detrended and band-pass filtered (third-order zero-phase Butterworth IIR filter) with a 0.05–0.8 Hz bandpass.

2.3   Signals Reconstruction

Source EEG and fNIRS data reconstruction were performed using Brainstorm software [24] and custom Matlab scripts [2]. For EEG a multiple-layer head model

(Boundary Element Method-BEM) and an MRI template (MNI-ICBM152) were used to build a realistic head model (forward model) by the OpenMEEG tool, which takes into account the different geometry and conductivity characteristics of the head tissues. The dipoles corresponding to potential brain sources were mapped to the cortical surface parcellated with a high-resolution mesh (15000 vertices). Finally, a leadfield matrix, expressing the scalp potentials corresponding to each single-dipole source configuration, was generated based on the volume conduction model [12].

For the fNIRS, a five tissues segmentation of the Colin27 brain template was used for the sensitivity matrix (forward model) computation. The fluences of light for each optode were estimated using the Monte Carlo simulations with a number of photons equal to 10^8 and projecting the sensitivity values within each vertex of the cortex mesh. The sensitivity values in each voxel of each channel were computed by multiplication of fluence rate distributions using the adjoint method according to the Rytov approximation [22]. Then the NIRS forward model was computed from fluences by projecting the sensitivity values within each vertex of the cortex mesh using the Voronoi-based method, a volume-to-surface interpolation method which preserves sulci-gyral morphology [10].

The inverse problem (i.e., estimation of the source activity given a lead field matrix) was solved by the minimum norm estimate (MNE) method. The MN estimator is of the form:

$$G = BL^T (LL^T + \lambda^2 C)^{-1}, \tag{1}$$

where G is the reconstructed signal along the cortical surface, L is the non-zero element matrix inversely proportional to the norm of the lead field vectors, B is the diagonal of L, C is the noise covariance matrix, and λ is the regularization parameter that sets the balance between reproduction of measured data and suppression of noise. For EEG the standardized low-resolution distributed imaging technique (sLORETA) was applied. While for fNIRS the depth-weighted minimum norm estimate (depth-weighted MNE) method was utilized. This was necessary since standard MNE tends to bias the inverse solution toward the more superficial generators, and the light sensitivity values decrease exponentially with depth.

In order to be comparable, the EEG and fNIRS source-time series were mapped in the same 3D space using an atlas-based approach (Desikan-Killiany) [8]. Since the optodes did not cover the entire scalp, it was necessary to modify the Desikan-Killiany atlas by choosing those ROIs covered by the fNIRS signal, leaving 42 ROIs out of 68 for both modalities. The EEG ROI time series were then decomposed into the typical oscillatory activity by band-pass filtering: $\delta(1 - 4Hz), \theta(4 - 7Hz), \alpha(8 - 15Hz), \beta(15 - 25Hz),$ and $\gamma(25 - 45Hz)$, while fNIRS source reconstructions were converted into oxygenated hemoglobin (HbO) and deoxygenated hemoglobin (HbR) by applying the modified Beer-Lambert transformation [5].

2.4   Connectivity Differences in Brain Networks

The functional connectivity matrices for the 42 ROIS within EEG (for each frequency band) and fNIRS (for HbO and HbR) were computed using Pearson's correlation coefficient, generating 7 42 × 42 connectivity matrices for each subject (5 for each EEG frequency band and 2 for hemodynamic activity of fNRIS). Once the networks have been constructed, their topological features were calculated via *graph/network analysis* [19, 26]. More specifically, small-world index (SWI), global efficiency (GE), clustering coefficient (CC), and characteristic path length (PL) were chosen as the network features of interest.

The small-world topology of a network describes how efficient and cost-effective the network is. A network is said to be a small-world network if SWI > 1. Brain networks with large small-world values are densely locally clustered, and at the same time employ the optimal number of distant connections, in this way the information processing is more efficient with lower information cost. Here the SWI was calculated as $SWI = \frac{CC}{PL}$.

As a measure of functional integration, the GE is defined as the efficiency of information exchange in a parallel system, in which all nodes are able of exchanging information via the shortest paths simultaneously.

The average shortest path length between all pairs of nodes is known as the PL of the network. As a measure of functional segregation, the CC is the fraction of triangles around an individual node reflecting, on average, the prevalence of clustered connectivity around individual nodes. The graph measures were normalized by generating 100 randomized networks (null models of a given network) and preserving the same number of nodes, edges, and degree distribution. Then, for each measure the ratio was calculated as *real metric* over the *matched random metrics*. For the comparison of network topology features, a paired student t-test was used between all the EEG frequency bands and the matched fNIRS metric of $[(\delta, \theta, \alpha, \beta, \gamma$ - HbO), $(\delta, \theta, \alpha, \beta, \gamma$ - HbR)]$, with the significance level set at p < 0.05. Multiple comparison correction was carried out using False Discovery Rate (FDR).

For the *edge-wise analysis*, that compare the connection strength between two regions, the network-based statistic (NBS) method was applied using the python brain connectivity toolbox [26]. For each edge of the 42 × 42 connectivity matrix, two-sample paired t-test was performed independently between each modality (5 frequency bands and 2 hemodynamic responses) and cluster-forming thresholds were applied to form a set of suprathreshold edges. The threshold was chosen based on Hedge's g-statistic effect size (ES) computed between each node of the two matrices, pairing each EEG frequency band FC with each fNIRS FC. The t-statistic corresponding to the Hedge's g score equal to 0.5 (medium ES) was chosen as the critical value (t-stat = 3.0). Finally, an FWER-corrected p-value was ascribed to each component through permutation testing (5,000 permutations). Edges that displayed FWER-corrected p-values below the significance threshold of 0.05 were considered positive results.

The topological characteristics of the graph, at the global and edge level, were calculated using the Matlab Brain Connectivity Toolbox and custom Matlab scripts [19].

3   Results

Topology analyses showed that all the EEG frequency bands $(\delta, \theta, \alpha, \beta, \gamma)$ and fNIRS (HbO and HbR) have SWI > 1, which implies prominent small-world properties in both modalities. A real network would be considered to have the small world characteristics if $CC_{real}/CC_{rand} > 1$ and $PL_{real}/PL_{rand} \approx 1$. It means that compared to random networks, a true human brain network has a larger CC and an approximately identical PL between any pair of nodes in the network. This was demonstrated for all EEG frequency bands, showing significantly higher CC values than HbO, particularly for the lower frequency (δ, θ, α), associated with PL values around 1. That means better EEG networks have clustering ability and small-worldness than HbO. As for HbR, the clustering coefficients were higher with respect to all EEG frequency bands (Fig. 3) Also, significantly higher E values in $\delta, \theta, \alpha, \beta, \gamma$ vs. HbO and HbR, were found (Fig. 4). This means that, for electrical brain activity, the neural information is routed via more globally optimal and shortest paths compared to hemodynamic activity. Thus, they provide faster and more direct information transfer.

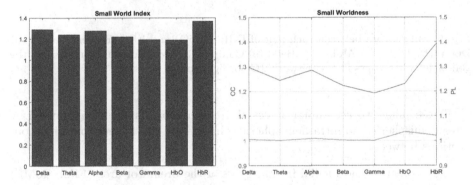

Fig. 3. Left: barplot of Small-World Index (SWI) across EEG (Delta, Theta, Alpha, Beta, Gamma) and fNIRS networks, Right: plot of Global Characteristic Path Length (PL) and Global Clustering Coefficient (CC) for EEG (Delta, Theta, Alpha, Beta, Gamma) and fNIRS (HbO and HbR)

The edge-wise analysis, applied via NBS, identified one subnetwork of increased functional connectivity for HbO compared to EEG for all the frequency bands [δ (p = 0.005), θ (p = 0.005), α (p = 0.004), β (p = 0.004), γ (p = 0.003), Fig. 5, corrected for multiple comparisons] at a pre-defined threshold of 3.5. This subnetwork consisted of 51 edges for δ-HbO connecting 25 nodes, 57 for θ-HbO connecting 23 nodes, 63 for α-HbO connecting 24 nodes, 59 for β-HbO connecting 26 nodes, and 57 for γ-HbO connecting 25 nodes (Table 1).

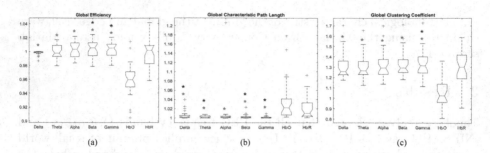

(a)   (b)   (c)

Fig. 4. Boxplot of (a) Global Efficiency (E), (b) Global Characteristic Path Length (PL) and (c) Global Clustering Coefficient (CC) for EEG (Delta, Theta, Alpha, Beta, Gamma) and fNIRS (HbO and HbR). The red asterisk denotes the significant difference between EEG and fNIRS (HbO); The black asterisk denotes the significant difference between EEG and fNIRS (HbR). (Color figure online)

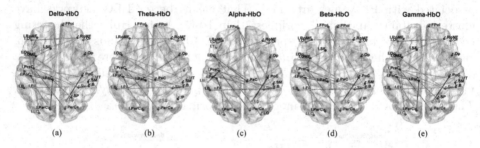

(a)   (b)   (c)   (d)   (e)

Fig. 5. Subnetwork between each pair of fNIRS (HbO) and EEG frequency band: (a) Delta, (b) Theta, (c) Alpha, (d) Beta, (e) Gamma. Red dots denote the nodes and the yellow links the edges of the network. (Color figure online)

These subnetworks, similar across all frequency bands, are characterized by inter-hemispheric and intra-hemispheric connections. The former comprises connections between:

- Right superior and medial frontal gyrus at the premotor area (FPol, RoMF) with both left superior frontal sulcus, at the human frontal eye field (Op, Or, Tr) and left pre- and post-central gyrus at the primary sensory-motor cortex (PreC, PaC, PoC);
- Left sensory-motor cortex (PreC, PaC, PoC) with both the right temporal-occipital regions (MT, B, Fu) and right posterior cingulate/precuneus (PerCa, LO) gyrus;
- Medial frontal gyrus (RoMF) and temporo-occipital cortex (B) with their homotopic regions.

The latter comprises connections between:

- Right frontal eye field (Op, Or, Tr) with right temporal-occipital regions (B)
- Right postcentral gyrus (PoC) with right lingual gyrus (Lg)

- Right temporal-occipital regions with right lingual gyrus (Lg)
- Left superior and medial frontal sulcus (Or, RoMF) with left posterior pre-cuneus (LO)
- Left post-central gyrus (PoC), with left pericalcarine (PerCa) regions.

The full list of the labels from the atlas is given in [8].

Table 1. Summary of the subnetworks with increased functional connectivity compared to EEG frequency bands.

Connection Type	Region 1	Region 2
Inter-hemispheric	Right superior frontal gyrus	Left superior frontal sulcus
		Left pre- and post-central gyrus
	Left sensory-motor cortex	Right temporal-occipital regions
		Right posterior cingulate
Homotopic	Right medial frontal gyrus	Left medial frontal gyrus
	Right temporo-occipital cortex	Left temporo-occipital cortex
Intra-hemispheric	Right frontal eye field	Right temporal-occipital
	Right postcentral gyrus	Right lingual gyrus
	Right temporal-occipital regions	Right lingual gyrus
	Left superior frontal sulcus	left posterior precuneus
	Left post-central gyrus	pericalcarine

4   Discussion

To the best of our knowledge, we are the first to investigate the topological properties of functional networks between synchronous EEG and fNIRS connectomes, across frequency bands, using source space analysis. We have observed a small-world topology network for both modalities, suggesting that the small-worldness is a universal principle for the functional wiring of the human brain regardless of the distinct mechanisms of different imaging techniques. The brain supports both segregated and distributed information processing, a key for cognitive processing, which means that localized activity in specialized regions is spread by coherent oscillations in large-scale distributed systems [19]. Our results have shown a significantly lower CC and large PL for HbO than EEG, indicating a lower specialization ability and a lower ability for parallel information transmission in the HbO network. The network differences between HbO-derived hemodynamic activity and EEG mostly pointed out inter-hemispheric connections, and to a lesser extent intra-hemispheric connections, between prefrontal and temporo-occipital regions. In agreement with the previous study [20], which reported HbO differences during RS in both short-distance ipsilateral connectivity between the prefrontal and occipital regions, and long-distance contralateral between homologous cortical regions. It was suggested that the generation of homologous connectivity is through direct structural connection, while fronto-posterior connectivity may reflect the synchronization of transient neural activation among distant cortical regions [20]. The involvement of different oscillatory frequencies in a cortical network varies as a function of cognitive state [3], and those states often last only a few ms. Since it is accepted that the short- and long-range organization of neural activity by oscillations of different frequencies depends on how those oscillations work in concert and that more than one EEG frequency band has been associated with the same RS network, our data may reflect unknown state changes in RS [14]. Thus, given that the hemodynamic response is delayed by several seconds, fNIRS cannot distinguish between these rapidly changing neural responses, reflecting the sum of several oscillatory network configurations.

The EEG activity was associated with metabolic deactivation in numerous studies. Moosmann et al. (2003) [16] measuring simultaneous EEG-fNIRS RS, found that the alpha activity had a positive correlation with HbR in the occipital cortex. While Koch et al. 2008 [11] reported that a high individual alpha frequency (IAF) peak correlates with a low oxygenation response. They conjectured that the relationship between IAF, neuronal and vascular response depended on the size of the recruited neuronal population. Another possible explanation is that since the oxy-hemoglobin is closely related to local cerebral blood flows, an RS condition correlates with lower metabolic demand [25]. However, it is accepted that functional connectivity maps derived from hemoglobin concentration changes reflect both spontaneous neural activity and systemic physiological contribution. Roughly 94% of the signal measured by a regular fNIRS channel (source-detector distances of 3 cm) reflects these systemic hemodynamic changes, producing low-frequency fluctuations. Thus contributing to a higher proportion of the variance of the HbO signal and to highly correlated brain regions, even after motion artifact correction and pre-whitening were employed, overestimating RS FC [1]. Furthermore, it was observed that oxy-hemoglobin is more heavily contaminated by extracerebral physiology than deoxy-hemoglobin, which has less sensitivity to systemic physiology [21]. To conclude, the results of this study point to the characteristic differences between electrophysiological and hemodynamic networks. Their simultaneous use is suggested since they can give a clearer picture of brain dynamics. However, it is crucial to understand the characteristics of each modality and understand what the differences and similarities mean to interpret them correctly. In this context, combining the brain network topological metrics based on graph theory analysis with multiple machine learning (ML) algorithms could be used to extract significant discriminative features, which could help to reveal the changes in the underlying brain network topological properties. It will be also interesting to study if those graph features will be able to capture transient and localized neuronal activity during task conditions. Future studies will investigate how the technological differences between EEG and fNIRS, when associated with ML, could be useful to extract discriminating features in pathology such as Alzheimer's disease.

Acknowledgements. This research study was conducted retrospectively using human subject data made available in open access by [23]. Ethical approval was not required as confirmed by the license attached with the open-access data.

This publication is supported by the European Union's Horizon 2020 research and innovation program under grant agreement Sano No 857533. This publication is supported by the Sano project carried out within the International Research Agendas program of the Foundation for Polish Science, co-financed by the European Union under the European Regional Development Fund.

References

1. Abdalmalak, A., et al.: Effects of systemic physiology on mapping resting-state networks using functional near-infrared spectroscopy. Front. Neurosci. 16 (2022). https://doi.org/10.3389/fnins.2022.803297
2. Angermann, A., et al.: Matlab-simulink-stateflow. In: MATLAB-Simulink- Stateflow. De Gruyter Oldenbourg (2020)
3. Brookes, M.J., et al.: Investigating the electrophysiological basis of resting state networks using magnetoencephalography. In: Proceedings of the National Academy of Sciences, vol. 108, no. 40, Proceedings of the National Academy of Sciences, 19 Sept. 2011, pp. 16783–16788 (2011). https://doi.org/10.1073/pnas.1112685108
4. Brunner, C., et al.: EEGlab - an open source matlab toolbox for electrophysiological research. Biomed. Eng. (2013). Biomedizinische Technik, Walter de Gruyter GmbH, 7 Jan. 2013. https://doi.org/10.1515/bmt-2013-4182
5. Cai, Z., et al.: Diffuse Optical Reconstructions of fNIRS Data Using Maximum Entropy on the Mean. Cold Spring Harbor Laboratory, 23 February 2021. https://doi.org/10.1101/2021.02.22.432263
6. Chen, Y., et al.: Amplitude of fNIRS resting-state global signal is related to EEG vigilance measures: a simultaneous fNIRS and EEG study. Front. Neurosci. **14** (2020). https://doi.org/10.3389/fnins.2020.560878
7. Chiarelli, A.M., et al.: Simultaneous functional near-infrared spectroscopy and electroencephalography for monitoring of human brain activity and oxygenation: a review. Neurophotonics 4(04), 1 (2017). SPIE-Intl Soc Optical Eng, 22 Aug. 2017, https://doi.org/10.1117/1.nph.4.4.041411
8. Desikan, R.S., et al.: An automated labeling system for subdividing the human cerebral cortex on MRI scans into gyral based regions of interest. NeuroImage **31**(3), 968–980 (2006). https://doi.org/10.1016/j.neuroimage.2006.01.021
9. Hassan, M., Wendling, F.: Electroencephalography source connectivity: aiming for high resolution of brain networks in time and space. In: IEEE Signal Processing Magazine, vol. 35, no. 3, Institute of Electrical and Electronics Engineers (IEEE), May 2018, pp. 81–96 (2018). https://doi.org/10.1109/msp.2017.2777518
10. Hiyoshi, H., Sugihara, K.: Voronoi-based interpolation with higher continuity. In: Proceedings of the Sixteenth Annual Symposium on Computational Geometry, pp. 242–250 (2000)
11. Koch, S.P., et al.: Individual alpha-frequency correlates with amplitude of visual evoked potential and hemodynamic response. NeuroImage, **41**(2), 233–242 (2008). https://doi.org/10.1016/j.neuroimage.2008.02.018
12. Krylova, M.A., Izyurov, I., Gerasimenko, N.Y., Slavytskaya, A., Mikhailova, E.: Human brain networks for visual spatial orientations processing. Fechner Day, p. 85 (2016)

13. Kuntzelman, K., Miskovic, V.: Reliability of graph metrics derived from resting-state human EEG. Psychophysiology, **54**(1), 51–61 (2016). https://doi.org/10.1111/psyp.12600

14. Mantini, D., et al.: Electrophysiological signatures of resting state networks in the human brain. In: Proceedings of the National Academy of Sciences, vol. 104, no. 32, Proceedings of the National Academy of Sciences, 7 August 2007, pp. 13170–13175 (2007). https://doi.org/10.1073/pnas.0700668104

15. Mantini, D., et al.: Improving MEG source localizations: an automated method for complete artifact removal based on independent component analysis. NeuroImage **40**(1), 160–173 (2008). https://doi.org/10.1016/j.neuroimage.2007.11.022

16. Moosmann, M., et al.: Correlates of alpha rhythm in functional magnetic resonance imaging and near infrared spectroscopy. NeuroImage **20**(1), 145–158 (2003). https://doi.org/10.1016/s1053-8119(03)00344-6

17. Niu, H., et al.: Test-retest reliability of graph metrics in functional brain networks: a resting-state fNIRS study. PLoS ONE **8**(9), e72425 (2013). edited by Olaf Sporns, Public Library of Science (PLoS), 9 Sept. 2013. https://doi.org/10.1371/journal.pone.0072425

18. Pollonini, L., et al.: Auditory cortex activation to natural speech and simulated cochlear implant speech measured with functional near-infrared spectroscopy. Hearing Res. **309**, 84–93 (2014). https://doi.org/10.1016/j.heares.2013.11.007

19. Rubinov, M., Sporns, O.: Complex network measures of brain connectivity: uses and interpretations. NeuroImage **52**(3), 1059–1069 (2010). https://doi.org/10.1016/j.neuroimage.2009.10.003

20. Sasai, S., et al.: Frequency-specific functional connectivity in the brain during resting state revealed by NIRS. NeuroImage **56**(1), 252–257 (2011). https://doi.org/10.1016/j.neuroimage.2010.12.075

21. Scholkmann, F., et al.: Systemic physiology augmented functional near-infrared spectroscopy: a powerful approach to study the embodied human brain. Neurophotonics **9**(03) (2022). SPIE-Intl Soc Optical Eng, 11 July 2022. https://doi.org/10.1117/1.nph.9.3.030801

22. Strangman, G., et al.: Non-invasive neuroimaging using near-infrared light. Biol. Psychiatry **52**(7), 679–693 (2002). https://doi.org/10.1016/s0006-3223(02)01550-0

23. Shin, J., et al.: Open access dataset for EEG+NIRS single-trial classification. In: IEEE Transactions on Neural Systems and Rehabilitation Engineering, vol. 25, no. 10, Institute of Electrical and Electronics Engineers (IEEE), October 2017, pp. 1735–1745 (2017). https://doi.org/10.1109/tnsre.2016.2628057

24. Tadel, F., et al.: Brainstorm: a user-friendly application for MEG/EEG analysis. Comput. Intell. Neurosci. **2011**, 1–13 (2011). https://doi.org/10.1155/2011/879716

25. Toronov, V.Y., et al.: A spatial and temporal comparison of hemodynamic signals measured using optical and functional magnetic resonance imaging during activation in the human primary visual cortex. NeuroImage **34**(3), 1136–1148 (2007). https://doi.org/10.1016/j.neuroimage.2006.08.048

26. Zalesky, A., et al.: Network-based statistic: identifying differences in brain networks. NeuroImage **53**(4), 1197–1207 (2010). https://doi.org/10.1016/j.neuroimage.2010.06.041

Anomaly Detection of Motion Capture Data Based on the Autoencoder Approach

Piotr Hasiec[1], Adam Świtoński[1]([✉]) [iD], Henryk Josiński[1] [iD],
and Konrad Wojciechowski[2] [iD]

[1] Department of Computer Graphics, Vision and Digital Systems, Silesian University
of Technology, ul. Akademicka 16, 44-100 Gliwice, Poland
{adam.switonski,henryk.josinski}@polsl.pl
[2] Polish-Japanese Academy of Information Technology, Aleja Legionów 2,
41-902 Bytom, Poland
konrad.wojciechowski@pja.edu.pl

Abstract. Anomalies of gait sequences are detected on the basis of
an autoencoder strategy in which input data are reconstructed from
their embeddings. The denoising dense low-dimensional and sparse high-
dimensional autoencoders are applied for segments of time series representing 3D rotations of the skeletal body parts. The outliers - misre-
constructed time segments - are determined and classified as abnormal
gait fragments. In the validation stage, motion capture data registered
in the virtual reality of the Human Dynamics and Multimodal Interaction Laboratory of the Polish-Japanese Academy of Information Technology equipped with Motek CAREN Extended hardware and software
are used. The scenarios with audio and visual stimuli are prepared to
enforce anomalies during a walk. The acquired data are labeled by a
human, which results in the visible and invisible anomalies extracted. The
neural network representing the autoencoder is trained using anomaly-
free data and validated by the complete ones. AP (Average Precision)
and ROC-AUC (Receiver Operating Characteristic – Area Under Curve)
measures are calculated to assess detection performance. The influences
of the number of neurons of the hidden layer, the length of the analyzed
time segments and the variance of injected Gaussian noise are investigated. The obtained results, with AP = 0.46 and ROC-AUC = 0.71, are
promising.

Keywords: anomaly detection · motion capture · gait analysis ·
autoencoder · CAREN Extended · neural networks

1 Introduction

The motion capture acquisition gives precise measurements of human movements. The attached markers on the anatomically significant body points are

---

P. Hasiec is a student of computer science at the Silesian University of Technology.

J. Mikyška et al. (Eds.): ICCS 2023, LNCS 14074, pp. 611–622, 2023.
https://doi.org/10.1007/978-3-031-36021-3_59

tracked by the calibrated multicamera system. Thus, their 3D positions in the global system for the subsequent time instants are reconstructed. As a result, the relative orientations between adjacent skeletal segments can be established. They are represented by a 3D rotation of joints connecting segments. Motion capture data have a form of multivariate time series of successive poses described by Euler angles or unit quaternions. There are plenty of applications for motion capture registration. Among others, it was successfully used in the diagnosis of movements abnormalities related to selected diseases [16], human gait identification [23], kinematic analysis of body movements performed by sport athlets [6], daily activities classification [2] or assessments of personal nonverbal interactions [5].

Anomalies can be defined as data instances that significantly deviate from the majority of them [18] or patterns that do not conform to expected typical behavior [8]. Their detection is a semi-supervised classification problem. It means that the training set contains only normal data. There are no anomaly instances; their nature is unknown. The problem is similar to the determination of outliers (this term is further used interchangeably with anomaly), the values which are very different from all the others. Anomaly detection is broadly conducted for numerous types of data. It was applied in the detection of host-based and network-based intrusion, banking systems and mobile phone frauds, monitoring of medical and public health, unmasking industrial damages as well as text and sensor data [18].

Although the problem of anomaly detection is extensively studied, there are no publications strictly devoted to gait data and highly precise motion capture measurements, which is the subject of the paper. Two variants of autoencoder strategy – denoising dense low-dimensional and sparse high-dimensional – are selected, adapted and successfully validated in human gait anomaly detection. Moreover, the contribution of the paper is related to the innovative collected dataset, with the registration taking place in the virtual reality of the Motek CAREN Extended laboratory.

2   Related Work

According to [7], multivariate time sequence anomalies can be categorized as point, subsequence or time series ones. It means that a single time instant – point, the consecutive points in time – subsequence, or entire time series constructed by subset of its variables, that behaves unusually when compared to the other values either locally or globally are detected. The model-based approaches are most commonly used to accomplish the task. They rely on the determination of the expected value and its comparison to the actual one. Depending on the difference, the anomaly is identified or not. There are two major variants – estimation and prediction. In the first one, the expected value is computed on the basis of previous and current time instants, while in the second variant, only previous values are taken. The most simple models are based on baseline statistics such as the median [3], local means and standard deviations [9]. Other ones try to estimate the unlikeness of values, assuming, for instance, that data without out-liers have mixed normal distribution [21]. Alternative proposals which analyze the trends of local changes are slope-based. In [22], constraints are established as a maximum and a minimum possible difference between consecutive points and in [24] there is an assumption of insignificant local slope changes.

However, in the estimation approaches of the model-based category, the most broadly applied are the autoencoders [1,10]. They are typically neural networks that learn only the most common and significant features, which can be real-ized by determining low-dimensional embeddings in a hidden layer or sparse connections between neurons. Due to the training being based on anomaly-free data, the anomalies correspond to non-representative features, which means that their precise reconstruction fails. To take into account temporal dependencies, overlapping sliding windows are processed as in [14].

In the prediction approaches, an auto-regressive moving average (ARMA) model [25], convolutional neural networks (CNN) [17] or long-short memory (LSTM) units [11] are used.

Another category of anomaly detection techniques are density-based methods [7]. They assume that in every surrounding of analyzed time instant, there are only a few outliers. Thus, the distances between a given time instant and the ones in its surrounding are calculated. Further, the number of instances with dissimilarities greater than specified threshold value is counted and decides about the anomaly identification.

As regards the methods strictly devoted to subsequence anomalies, discord approaches [13] are quite common. They determine the most unusual subseries by comparing dissimilarities between every pair of subsequences. For time series anomalies, dimensionality reduction techniques can be employed as for instance Principal Component Analysis [12]. In other common variants clustering as k-means algorithm [20] or agglomerative hierarchical one [15] is carried out. In [4] additionally Dynamic Time Warping is utilized to align and compare complete time series.

3   Method

Due to the pioneering application of anomaly detection to gait motion capture data, the baseline and most broadly used approach is selected – the autoencoder strategy. Its structure consists of three main components – encoder, decoder and hidden layer. Input data are reconstructed on the basis of their representation by the hidden layer, typically with lower dimensionality as visualized in Fig. 1. If E and D are encoder and decoder transfer functions to and from the hidden layer, respectively, x and x' denote input and output, the working of the autoencoder is described by the following formula:

$$x' = D(E(x)) \tag{1}$$

It is assumed that normal data are highly correlated and can be represented by low-dimensional embeddings of the hidden layer from which efficient recon-struction is feasible. In the training stage, anomaly-free data are used to establish

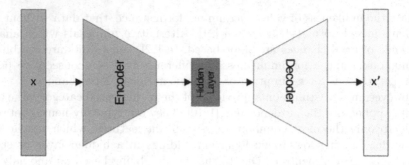

Fig. 1. Autoencoder strategy.

the encoder and decoder workings. Finally, using selected metrics, for instance, Euclidean one, the reconstruction error rE is computed to identify outliers:

$$rE = \|x' - x\| \tag{2}$$

However, the problem of thresholding the rE value on the basis of which anomalies are recognized is not a trivial task. If the normal distribution of the reconstruction error for typical data is naively assumed, the estimated average value increased by two or three standard deviations may be taken. In another possible variant, percentiles of the rE distribution can be used.

Two neural network architectures are chosen for the autoencoder implementation. Primarily, it is a dense low-dimensional variant visualized in Fig. 2a with fewer neurons of the hidden layer than a number of input values. Thus, it works in the standard previously described way – low-dimensional embeddings are determined and reconstruction of the input data is carried out. The second chosen architecture is a sparse high-dimensional network (Fig. 2b) with a greater number of neurons of the hidden layer. To avoid just a simple copy of the input to the output, L1 regularization is applied. It penalizes for high absolute values of neurons' weights. Thus, it tries to determine inactive, sparse connections between neurons with assigned weights close to zero.

To incorporate temporal relationships of motion time series, subsequence anomalies are detected. It is realized by processing by the autoencoders in every detection, fixed-length segments of motion sequences. Moreover, for greater noise resistance, a kind of data augmentation is carried out. During the training stage, input data are modified by a random variable taken from the zero-mean Gaussian distribution with different standard deviations.

4   Dataset

The acquisition took place in the Human Dynamics and Multimodal Interaction Laboratory of the Polish-Japanese Academy of Information Technology. It is equipped with the Computer Assisted Rehabilitation ENvironment Extended

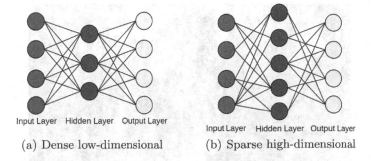

(a) Dense low-dimensional   (b) Sparse high-dimensional

Fig. 2. The concept of the selected autoencoder architectures.

system (CAREN Extended) manufactured by the Motek Company. It integrates a 3D motion capture registration, dual-belt treadmill with 6 degrees of freedom (DoF) motion base and adaptive speed in the range $(0\frac{m}{s}, 15\frac{m}{s})$, ground reaction forces (GRF) and wireless electromyography (EMG) measurements, as well as immersive virtual/augmented reality environment with panoramic video and surround 5.1 audio subsystems. The laboratory is presented in Fig. 3. In the perspective (a) and front (b) views, the treadmill with the registered participant and curved screen are visible. In the bottom view (c), the construction with 6 DoF located below the motion base is shown.

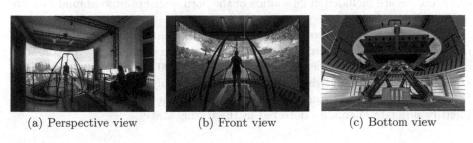

(a) Perspective view   (b) Front view   (c) Bottom view

Fig. 3. Human Dynamics and Multimodal Interaction Laboratory of the Polish-Japanese Academy of Information Technology.

The motion capture camera setup and applied skeleton model are visualized in Fig. 4. However, in the experiments, only segments followed by hip, knee, ankle, and wrist joints are taken into account. They are described by angles triplets expressed in the notation representing flexion/extension, adduction/abduction, and internal/external rotations in sagittal (lateral), frontal (coronal), and transverse planes. It means that body is divided into left/right, anterior/posterior (front/back), and superior/inferior (upper/lower) parts, respectively. There is only one exception – knee joints with single DoF have only a flexion angle. Thus, in total, the pose space is 20-dimensional.

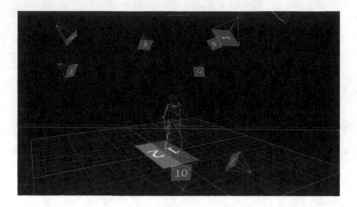

Fig. 4. Mocap camera setup and skeleton model.

To observe gait anomalies, three registration scenarios are proposed. They are located in the virtual reality of a forest, as visualized in Fig. 5. The first one is normal walking without any perturbations and provides anomaly-free data used in the training stage. The second one is related to video stimuli appearing in random time instants. There is a deer crossing the walking path, a simulation of the temporary darkness – the screen is off for one second – and a simulation of dizziness by floating visualization on the screen. In the last scenario, audio disturbances are included. The sounds of the horn, gunshot, and animal roar are generated.

Twelve participants were involved in the registration experiment. Every scenario was repeated three times, the duration of the recordings was approximately one minute, and the frequency of the acquisition 100 Hz. Time instants of the stimuli are exported, and they denote possible anomaly occurrences. Finally, on the basis of video recordings, manual labeling of the data was carried out to identify ultimate anomalies detected in the testing stage.

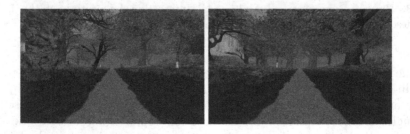

Fig. 5. Prepared virtual forest.

5 Results

For labeled recordings of the scenarios with video and audio stimuli, anomaly detection is carried out by autoencoders trained using the data of the first scenario. On the basis of comparison with ground truth detections, true (TP) and false positives (FP) – numbers of properly and improperly recognized anomalies – as well as true (TN) and false (FN) negatives – the numbers of correctly determined normal instances and unidentified anomalies – are calculated. They are normalized through dividing by the total number of positives and negatives, respectively:

$$TPR = \frac{TP}{TP + FN} \tag{3}$$

$$FPR = \frac{FP}{FP + TN} \tag{4}$$

The balance between TPR and FPR is controlled by applying a threshold for the reconstruction error from Eq. (2). Thus, the curve called Receiver Operating Characteristic (ROC) representing the relationship between TPR and FPR values can be prepared. The area under this curve (ROC-AUC) is a measure of the detection performance.

However, in the faced anomaly identification challenge, the number of true negatives usually is much greater than true positives, which may have an influence on the ROC-AUC value. Therefore another measure is used in our evaluation protocol. It is based on precision and recall, which are true positives normalized by the total number of all positive identifications and all ground truth anomalies, respectively:

$$precision = \frac{TP}{TP + FP} \tag{5}$$

The recall is literally a true positive rate TPR. Once again, the balance between *precision* and *recall* is controlled by the threshold of the reconstruction error. It allows for determining precision for uniformly distributed recall values and calculating the average of them. This is, in fact, the area under the precision-recall curve and it is called average precision (AP). Moreover, to obtain a monotonic dependency, smoothing is carried out in such a way that every precision value is substituted by the maximum calculated for recalls equal to or greater than the current one.

The obtained results for dense low-dimensional and sparse high-dimensional autoencoders are depicted in Table 1 and Table 2, respectively. The lengths of the processed segments SL expressed by a number of time instants are 50 and 100. It means that half- and one-second subsequences are analyzed, which approximately correspond to half of and complete gait cycle with a single and double step performed. The number of inputs and outputs of the neural network is a product of SL and the dimensionality of pose space, which is 20. The angles are stored in degree scale, which means that injected Gaussian noises with standard deviations – 0.05, 0.1, 0.3, 0.5, 0.7 – are insignificant. The achieved AP and

ROC-AUC values for the vast majority of cases are higher than 0.4 and 0.65, respectively. The best performance with AP = 0.462 and ROC-AUC = 0.708 is obtained by sparse autoencoder analyzing the one-second subsequences and having 3000 neurons of the hidden layer. For the considered variants, some general observations can be made: (i) slightly better precision of anomaly detection is achieved by high-dimensional sparse autoencoders and longer subsequences containing 100-time instants, (ii) there is an impact of injected noise on the effectiveness of the training process, but it differs depending on the autoencoder variant, (iii) both taken quality measures – AP and ROC-AUC – are partially correlated.

Table 1. Detection performances – AP and ROC-AUC (denoted as R-AUC) measures – of dense low-dimensional autoencoder with different standard deviations of injected Gaussian noise, number of neurons of the hidden layer D and length of the processed segments SL.

Noise		0.05		0.1		0.3		0.5		0.7	
D	SL	AP	R-AUC	AP	R-AUC	AP	R-AUC	AP	R-AUC	AP	R-AUC
25	50	0.394	0.660	0.405	0.667	0.390	0.654	0.373	0.649	0.343	0.621
50	50	0.408	0.667	0.406	0.667	0.402	0.668	0.382	0.651	0.371	0.646
100	50	0.405	0.677	0.424	0.679	0.405	0.678	0.403	0.666	0.393	0.662
200	50	0.436	0.683	0.428	0.679	0.415	0.672	0.402	0.670	0.393	0.658
400	50	0.423	0.672	0.434	0.682	0.441	0.693	0.430	0.683	0.408	0.675
50	100	0.414	0.677	0.412	0.672	0.416	0.674	0.389	0.678	0.380	0.634
100	100	0.419	0.678	0.436	0.694	0.423	0.679	0.398	0.662	0.393	0.650
200	100	0.430	0.683	0.432	0.693	0.424	0.680	0.430	0.686	0.384	0.646
400	100	0.438	0.688	0.428	0.681	0.442	0.694	0.439	0.686	0.410	0.672

Table 2. Detection performances – AP and ROC-AUC (denoted as R-AUC) measures – of sparse high-dimensional autoencoder with different standard deviations of injected Gaussian noise, number of neurons of the hidden layer D and length of the processed segments SL.

Noise		0.05		0.1		0.3		0.5		0.7	
D	SL	AP	R-AUC	AP	R-AUC	AP	R-AUC	AP	R-AUC	AP	R-AUC
1500	50	0.448	0.703	0.433	0.683	0.443	0.703	0.446	0.697	0.352	0.619
3000	100	0.462	0.708	0.461	0.704	0.417	0.673	0.445	0.700	0.432	0.699
5000	100	0.458	0.706	0.426	0.684	0.412	0.670	0.445	0.692	0.393	0.650
1250	50	0.423	0.670	0.447	0.691	0.413	0.675	0.344	0.617	0.327	0.592
2500	50	0.446	0.690	0.447	0.694	0.439	0.682	0.448	0.703	0.415	0.670
3500	50	0.455	0.691	0.445	0.692	0.434	0.687	0.439	0.700	0.437	0.688

Table 3. Best results – AP and ROC-AUC – obtained by the successive participants for sparse high-dimensional autoencoder ($D = 3000$, $SL = 100$).

Participant	AP	ROC-AUC	Percent of anomalies
1	0.578	0.833	0.082
2	0.541	0.750	0.167
3	0.227	0.695	0.109
4	0.636	0.761	0.204
5	0.396	0.619	0.201
6	0.377	0.646	0.113
7	0.223	0.615	0.113
8	0.454	0.690	0.135
9	0.305	0.709	0.113
10	0.848	0.828	0.383
11	0.739	0.816	0.257
12	0.216	0.531	0.187

The obtained results are at least promising and substantially better than random detection. They differ for successive participants as presented in Table 3 – for some of them, anomalies are recognized pretty efficiently and for others,

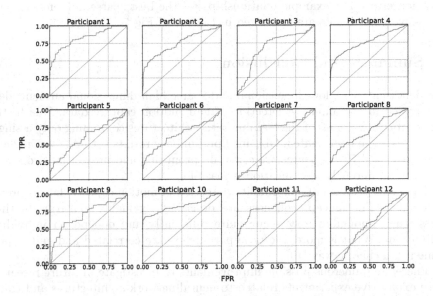

Fig. 6. Receiver Operating Characteristic (ROC) curves (sparse high dimensional autoencoder: $D = 3000$, $SL = 100$, noise std $= 0.05$) for subsequent participants. The blue and red colors correspond to the autoencoder and random guess classifier results, respectively. (Color figure online)

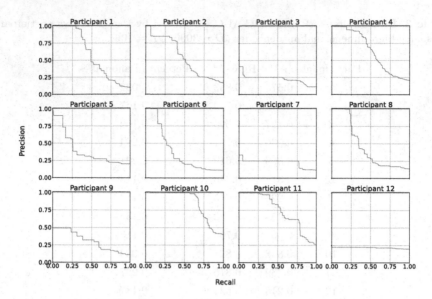

Fig. 7. Precision-recall curves (sparse high dimensional autoencoder: $D = 3000$, $SL = 100$, noise std $= 0.05$) for subsequent participants.

poorly. The balance between TPR and FPR as well as between precision and recall can be controlled by applying different threshold values for the reconstruction error. The example relationships for the best sparse autoencoder and successive participants are visualized in Fig. 6 and Fig. 7.

6   Summary and Conclusions

The denoising dense low-dimensional and sparse high-dimensional autoencoders are applied in the anomaly detection of gait motion capture data. Due to the low number of samples of the training set, architectures containing only a single hidden layer are investigated. Obtained preliminary results with AP $= 0.462$ and ROC-AUC $= 0.708$ measures are certainly promising and prove that process of efficient gait anomaly identification is feasible.

The faced problem of abnormality detection of motion capture data is pretty challenging. The mocap sequences are described by highly multivariate time series, their anomalies may occur quite differently and depend on individual inclinations. What is more, the pose parameters are correlated and gait is performed in a chaotic way [19].

There are numerous possible improvements to investigate in future research. More exhaustive experiments related to neural network architectures and training parameters can be conducted. Moreover, instead of the angles representing segments' orientation in sagittal, frontal and transverse planes, Euler angles with the joint rotation or 3D position of the markers attached to the human body may be taken. The feature selection in the pose space or downsampling in the time domain reduce the number of trainable parameters of the autoencoder, which also may be advantageous. Ultimately, there are plenty of other techniques successfully applied for anomaly detection of similar data. Particularly approaches predicting actual state on the basis of preceding time instants and LSTM networks [11] seem to be reasonable. In addition, it is planned to enlarge the dataset not only with measurements of new participants but also with extra scenarios. The effect of the stimuli will be intensified by their simultaneous occurrence, and new stimuli will be used.

Acknowledgments. This work was supported by the Department of Computer Graphics, Vision, and Digital Systems, under the statutory research project (Rau6, 2023), Silesian University of Technology (Gliwice, Poland).

Ethics statement. During the acquisition process, a non-invasive CAREN Extended system was used. The measurements took place in the Human Dynamics and Multimodal Interaction Laboratory of the Polish-Japanese Academy of Information Technology and were assisted by the certified staff. The registration protocol contains normal gait and it is consistent with the Declaration of Helsinki. All the volunteers participating in the experiments were informed about the rules of acquisition and agreed to use their collected data for research purposes.

References

1. An, J., Cho, S.: Variational autoencoder based anomaly detection using reconstruction probability. Special Lect. IE **2**(1), 1–18 (2015)
2. Barnachon, M., Bouakaz, S., Boufama, B., Guillou, E.: Ongoing human action recognition with motion capture. Pattern Recogn. **47**(1), 238–247 (2014)
3. Basu, S., Meckesheimer, M.: Automatic outlier detection for time series: an application to sensor data. Knowl. Inf. Syst. **11**, 137–154 (2007)
4. Benkabou, S.E., Benabdeslem, K., Canitia, B.: Unsupervised outlier detection for time series by entropy and dynamic time warping. Knowl. Inf. Syst. **54**(2), 463–486 (2018)
5. Bente, G., Novotny, E., Roth, D., Al-Issa, A.: Beyond stereotypes: Analyzing gender and cultural differences in nonverbal rapport. Front. Psychol. **11**, 599703 (2020)
6. Błaszczyszyn, M., Szczęsna, A., Pawlyta, M., Marszałek, M., Karczmit, D.: Kinematic analysis of mae-geri kicks in beginner and advanced kyokushin karate athletes. Int. J. Environ. Res. Public Health **16**(17), 3155 (2019)
7. Blázquez-García, A., Conde, A., Mori, U., Lozano, J.A.: A review on outlier/anomaly detection in time series data. ACM Comput. Surv. (CSUR) **54**(3), 1–33 (2021)
8. Chandola, V., Banerjee, A., Kumar, V.: Anomaly detection: a survey. ACM Comput. Surv. (CSUR) **41**(3), 1–58 (2009)
9. Dani, M.-C., Jollois, F.-X., Nadif, M., Freixo, C.: Adaptive threshold for anomaly detection using time series segmentation. In: Arik, S., Huang, T., Lai, W.K., Liu, Q. (eds.) ICONIP 2015. LNCS, vol. 9491, pp. 82–89. Springer, Cham (2015). https://doi.org/10.1007/978-3-319-26555-1_10

10. Gong, D., et al.: Memorizing normality to detect anomaly: memory-augmented deep autoencoder for unsupervised anomaly detection. In: Proceedings of the IEEE/CVF International Conference on Computer Vision, pp. 1705–1714 (2019)
11. Hundman, K., Constantinou, V., Laporte, C., Colwell, I., Soderstrom, T.: Detecting spacecraft anomalies using LSTMS and nonparametric dynamic thresholding. In: Proceedings of the 24th ACM SIGKDD International Conference on Knowledge Discovery & Data Mining, pp. 387–395 (2018)
12. Hyndman, R.J., Wang, E., Laptev, N.: Large-scale unusual time series detection. In: 2015 IEEE International Conference on Data Mining Workshop (ICDMW), pp. 1616–1619. IEEE (2015)
13. Keogh, E., Lin, J., Fu, A.: Hot sax: efficiently finding the most unusual time series subsequence. In: Fifth IEEE International Conference on Data Mining (ICDM 2005), pp. 8-pp. IEEE (2005)
14. Kieu, T., Yang, B., Jensen, C.S.: Outlier detection for multidimensional time series using deep neural networks. In: 2018 19th IEEE International Conference on Mobile Data Management (MDM), pp. 125–134. IEEE (2018)
15. Lara, J.A., Lizcano, D., Rampérez, V., Soriano, J.: A method for outlier detection based on cluster analysis and visual expert criteria. Expert. Syst. **37**(5), e12473 (2020)
16. Liparoti, M., et al.: Gait abnormalities in minimally disabled people with multiple sclerosis: a 3d-motion analysis study. Multiple Sclerosis Rel. Disord. **29**, 100–107 (2019)
17. Munir, M., Siddiqui, S.A., Dengel, A., Ahmed, S.: Deepant: a deep learning approach for unsupervised anomaly detection in time series. IEEE Access **7**, 1991–2005 (2018)
18. Pang, G., Shen, C., Cao, L., Hengel, A.V.D.: Deep learning for anomaly detection: a review. ACM Comput. Surv. (CSUR) **54**(2), 1–38 (2021)
19. Piórek, M., Josiński, H., Michalczuk, A., Świtoński, A., Szczęsna, A.: Quaternions and joint angles in an analysis of local stability of gait for different variants of walking speed and treadmill slope. Inf. Sci. **384**, 263–280 (2017)
20. Rebbapragada, U., Protopapas, P., Brodley, C.E., Alcock, C.: Finding anomalous periodic time series: an application to catalogs of periodic variable stars. arXiv preprint arXiv:0905.3428 (2009)
21. Reddy, A., et al.: Using gaussian mixture models to detect outliers in seasonal univariate network traffic. In: 2017 IEEE Security and Privacy Workshops (SPW), pp. 229–234. IEEE (2017)
22. Song, S., Zhang, A., Wang, J., Yu, P.S.: Screen: stream data cleaning under speed constraints. In: Proceedings of the 2015 ACM SIGMOD International Conference on Management of Data, pp. 827–841 (2015)
23. Świtoński, A., Josiński, H., Wojciechowski, K.: Dynamic time warping in classification and selection of motion capture data. Multidimension. Syst. Signal Process. **30**, 1437–1468 (2019)
24. Zhang, A., Song, S., Wang, J.: Sequential data cleaning: A statistical approach. In: Proceedings of the 2016 International Conference on Management of Data, pp. 909–924 (2016)
25. Zhang, Y., Hamm, N.A., Meratnia, N., Stein, A., Van de Voort, M., Havinga, P.J.: Statistics-based outlier detection for wireless sensor networks. Int. J. Geogr. Inf. Sci. **26**(8), 1373–1392 (2012)

Influence of the Capillaries Bed in Hyperthermia for Cancer Treatment

Antônio Marchese Bravo Esteves[1], Gustavo Resende Fatigate[2],
Marcelo Lobosco[1,2], and Ruy Freitas Reis[1,2(✉)]

[1] Departamento de Ciência da Computação, Universidade Federal de Juiz de Fora,
Juiz de Fora, Brazil
ruy.reis@ufjf.br
[2] Pós-Graduação em Modelagem Computacional,
Universidade Federal de Juiz de Fora, Juiz de Fora, Brazil

Abstract. This work presents the computational modeling of solid
tumor treatments with hyperthermia using magnetic nanoparticles, considering a bioheat transfer model proposed by Pennes(1948). The simulations consider a tumor seated in a muscle layer. The model was described
with a partial differential equation, and its solution was approximated
using the finite difference method in a heterogeneous porous medium
using a Forward Time Centered Space scheme. Moreover, the Monte
Carlo method was employed to quantify the uncertainties of the quantities of interest (QoI) considered in the simulations. The QoI considers
uncertainties in three different parameters: 1) the angulation of blood
vessels, 2) the magnitude of blood flow, and 3) the number of blood vessels per tissue unit. Since Monte Carlo demands several executions of
the model and solving a partial differential equation in a bi-dimensional
domain demands significant computational time, we use the OpenMP
parallel programming API to speed up the simulations. The results of
the in silico experiments showed that considering the uncertainties presented in the three parameters studied, it is possible to plan hyperthermia treatment to ensure that the entire tumor area reaches the target
temperature that leads to damage.

Keywords: Hyperthermia · Cancer · Bioheat · Uncertainty
Quantification · Monte Carlo

1 Introduction

Cancer is the name given to a large group of diseases that can start in almost any
organ or tissue in the body. According to the World Health Organization [34],
cancer is a leading cause of death worldwide: about 10 million deaths in 2020
were due to cancer. The leading causes of new cases of cancer in 2020 were
breast (2.26 million cases), lung (2.21 million cases), colon and rectum (1.93
million cases), prostate (1.41 million cases), skin (non-melanoma) (1.20 million
cases), and stomach (1.09 million cases). On the other hand, the leading causes
of death due cancer in the same year were lung (1.80 million deaths), colon and
rectum (916,000 deaths), liver (830,000 deaths), stomach (769,000 deaths), and
breast (685,000 deaths) [34].

ⓒ The Author(s), under exclusive license to Springer Nature Switzerland AG 2023
J. Mikyška et al. (Eds.): ICCS 2023, LNCS 14074, pp. 623–637, 2023.
https://doi.org/10.1007/978-3-031-36021-3_60

A correct cancer diagnosis is essential for appropriate and effective treatment because every cancer type requires a specific treatment regimen. In recent years several methods have been developed to fight cancer. The most common treatments are surgery, radiotherapy, and systemic therapy (chemotherapy, hormonal treatments, targeted biological therapies) [20].

One of the treatments that have been studied against cancer is hyperthermia. Hyperthermia was discovered in the 1950s [5], and since then, it has been gaining space as a possible cancer treatment. The main idea behind this treatment is to superheat the tissue to a temperature threshold that induces cell necrosis [17]. Thus, it is possible to destroy tumor cells without a surgical procedure. Currently, hyperthermia is used as a non-invasive therapy against cancer, helping other techniques such as chemotherapy and radiotherapy [6,17]. The hyperthermia is widely used in the treatment of liver [16] and breast tumors [9].

One of the ways to carry out the treatment through hyperthermia is with the application of magnetic nanoparticles [18,19]. Nanoparticles can be applied to irregular and deep tumor tissues, and when exposed to a low-frequency alternating magnetic field, they overheat the site. Most of the heat generated by these nanoparticles is due to Néelian, and Brownian relaxation [18,27]

Despite being discovered in the 50s [5], hyperthermia as a cancer treatment is still in an early stage of development, with open questions that can be answered with the help of mathematical models and computational simulations. Mathematical models were developed and applied to describe the heat in living tissues [12,13,21,24,25]. In this paper, a mathematical model described using partial differential equations (PDEs) is employed to describe the heat dynamics over time due to hyperthermia treatment. A porous media approach [13] is used to describe bioheat according to the properties of the living tissue, such as porosity, density, specific heat, thermal conductivity, metabolism, and velocity field.

The solution of PDEs is still a great challenge for science and engineering once most differential equations do not have an analytical solution. So, these models use numerical methods to solve them. This work employs the explicit Euler's method and the finite difference method (FDM) in a porous heterogeneous medium using a Forward Time Centered Space (FTCS) scheme to solve the mathematical model numerically.

Uncertainties are intrinsic to nature. Uncertainty quantification (UQ) is used to determine the likely results of models with stochastic behavior [7]. This paper uses Monte Carlo (MC) to include stochastic behavior for the UQ [28]. The deterministic model is solved several times, considering the uncertainties in some parameters. In our model, the uncertainties associated with several parameters, such as the number of blood capillaries, the magnitude, and the direction of blood velocity, are described as a density probability function (PDF) based on parameters found in the literature. We consider that both the blood capillaries and their associated values have a stochastic behavior because cancer growth can stimulate the capillaries' angiogenesis in the tumor site [8,37]. Also, they depend on the cancer type, stage, location, and other specificities [11,37].

The numerical resolution of parabolic PDEs demands significant computational effort, and the Monte Carlo method requires solving the model several times. For these reasons, computational execution time becomes an issue. To obtain the simulations results faster, the present study employs a computational parallelization strategy for shared memory architectures using the *OpenMP* Application Programming Interface (API).

We organise this paper as follows. Section 2 describes the bioheat model, numerical approximation and the uncertainty quantification. The results are presented in Sect. 3 and discussed in Sect. 4. Finally, Sect. 5 presents the conclusions and plans for future work.

2 Methods

2.1 Mathematical Model

A porous medium is a material consisting of a solid matrix with an interconnected void [13]. The biological tissues can be seen as a porous medium once they contain dispersed cells separated by voids that allow the flow of nutrients, minerals, and others to reach all cells within the tissue. A fundamental characteristic of this medium is its porosity (ε), defined as the ratio of the void space to the local volume of the medium, and also by its permeability which is a measure of the flow conductivity in this medium.

Heat transfer in human tissues involves complicated processes such as heat conduction in tissues, heat transfer due to blood convection, metabolic heat generation, and others [13]. The dynamics of heat in living tissue is called bioheat. Many scientists have attempted to model bioheat accurately since it is the basis for the human thermoregulation system [30] as well as thermotherapies [31]. The bioheat equation usually expresses the energy transport in a biological system, and the one developed by Pennes [21] is among the earliest models for bioheat.

The application of porous media models for modeling bioheat transfer in human tissues is relatively recent. Xuan and Roetzel [26,35] used the concepts of transport through porous media to model the tissue-blood system composed mainly of tissue cells (solid particles) and interconnected voids that contain arterial or venous blood. They adopted the idea of a local thermal non-equilibrium model as described in the works of Amiri and Vafai [2,3], Alazmi and Vafai [1], and Lee and Vafai [14] to model heat transfer within the tissue and the artery blood. However, Xuan and Roetzel [26,35] indicated that numerous parameters and information are needed to solve the system, which consists of a two-phase energy equation, such as thermal and anatomic properties of the tissue, interstitial convective heat transfer coefficients as well as the velocity field of the blood [13].

Local thermal equilibrium can be used as a good approximation for the temperature field in specific applications with capillaries bed, as shown in the literature[1–3,14]. In this case, blood and tissue temperatures are the same at any location, according to the following equation:

$$[\rho c_p(1 - \varepsilon) + \rho_b c_{p_b}\varepsilon]\,\frac{\partial T}{\partial t} + \varepsilon(\rho c_p)_b \mathbf{u}_b \cdot \nabla T = \nabla\left([\mathbf{k}_s^a + \mathbf{k}_b^a] \cdot \nabla T\right) + Q_m(1-\varepsilon), \quad (1)$$

where $T, \rho, \rho_b, c_p, \varepsilon, c_{p_b}, \mathbf{u}_b, \mathbf{k}_s^a, \mathbf{k}_b^a, Q_m$ are the tissue (and the blood) temperature, the tissue density, the blood density, the specific heat of tissue, the porosity of the tissue, the specific heat of blood, blood velocity field, tissue effective thermal conductivity tensor, blood effective thermal conductivity tensor, and heat generation within the tissue, respectively. Once we are considering an isotropic conduction, $\mathbf{k}_b^a = (1-\varepsilon)k_t$, $\mathbf{k}_b^a = \varepsilon k_b$, k_t and k_b are the tissue and blood thermal conductivities, respectively.

However, modeling the heat generated by the magnetic nanoparticles is necessary. An *in vivo* experimental study done by Salloum [29] and performed on hindlimbs of rats showed that the specific absorption rate (SAR) around an injection site can be approximated by:

$$Q_r(\vec{x}) = \sum_{i=1}^{M} A_i e^{\frac{-r_i^2}{r_{0,i}^2}},\tag{2}$$

where M is the number of nanoparticle injections into the tumor, A is the maximum heat generation rate, r is the distance to the injection point, and r_0 is the hyperthermia coverage radius. Adding this term to Eq. (1), in a well-posed problem, proper boundary conditions on a Lipschitz continuous, and piecewise smooth boundary and initial conditions, the model is given as follows:

$$\begin{cases} \sigma \dfrac{\partial T}{\partial t} + \varepsilon(\rho c_p)_b \mathbf{u}_b \cdot \nabla T = \nabla \cdot (\kappa \cdot \nabla T) + (1-\varepsilon)(Q_m + Q_r) & \text{in } \Omega \times I, \\ T(.,0) = 37 & \text{on } \partial\Omega \times I, \\ \kappa \nabla T \cdot \vec{n} = 0 & \text{in } \Omega, \end{cases}\tag{3}$$

where $\Omega \subset \mathbb{R}^2$ is the spatial domain, $I \subset \mathbb{R}^+$ is the time domain, $T : \Omega \times I \to \mathbb{R}^+$ is the temperature, $\sigma = [\rho c_p(1-\varepsilon) + \rho_b c_{pb}\varepsilon]$, $\kappa = (1-\epsilon)k_t + \epsilon k_b$.

2.2 Numerical Scheme

The numerical method used to solve Eq. (3) is the Finite Difference Method (FDM). We consider a heterogeneous medium and the closed domain Ω discretized into a set of regular points defined by $S_\Omega = \{(x_i, y_j); i = 0, 1, \cdots, N_x; j = 0, 1, \cdots, N_y\}$, where $N_x = N_y = N$ is the number of intervals of length $h_x = h_y = h$. Moreover, the time discretization of the time domain I is partitioned into N_t equal time intervals of length h_t, i.e., $S_I = \{(t_n); n = 0, 1, \cdots, N_t\}$. To obtain the discrete form of the model, we employ an FTCS scheme, resulting in an explicit numerical method. Moreover, we use a first-order upwind scheme to the advection term [15]:

$$T_{i,j}^{n+1} = \frac{h_t}{\sigma}\left(\varphi_{dif}(T^n) - \varphi_{adv}(T^n) + (1-\varepsilon)Q_m + Q_r\right) + T_{i,j}^n,\tag{4}$$

where

$$\varphi_{dif}(T^n) = \frac{\kappa_{i+1/2,j}(T_{i+1,j,k}^n - T_{i,j}^n) - \kappa_{i-1/2,j}(T_{i,j}^n - T_{i-1,j}^n)}{h^2} +$$
$$\frac{\kappa_{i,j+1/2}(T_{i,j+1}^n - T_{i,j}^n) - \kappa_{i,j-1/2}(T_{i,j}^n - T_{i,j-1,k}^n)}{h^2}, \tag{5}$$

where $\kappa_{i+1/2,j,k}$ is the thermal conductivity evaluated at the midpoint. In this paper we consider a piecewise homogeneous media where the thermal conductivity is a discontinuous function, so the thermal conductivity can be estimated using the harmonic mean, $i.e.$:

$$\kappa_{i+1/2,j} \approx \frac{2\kappa_{i,j}\kappa_{i+1,j}}{\kappa_{i,j} + \kappa_{i+1,j}}. \tag{6}$$

Equation (6) assures the continuity of the flux. Furthermore, the thermal conductivity for the other midpoints is evaluated using the same idea.

Furthermore,

$$\varphi_{adv}(T^n) = \varepsilon(\rho c_p)_b \mathbf{u}_{b_{i,j}} \cdot \begin{cases} \dfrac{(u_{i,j}^n - u_{i-1,j}^n)}{h} & \text{for } u_{b_x} > 0 \\ \dfrac{(u_{i+1,j}^n - u_{i,j}^n)}{h} & \text{for } u_{b_x} < 0 \\ \dfrac{(u_{i,j}^n - u_{i,j-1}^n)}{h} & \text{for } u_{b_y} > 0 \\ \dfrac{(u_{i,j+1}^n - u_{i,j}^n)}{h} & \text{for } u_{b_y} < 0 \end{cases} \tag{7}$$

where $\mathbf{u}_b = \begin{pmatrix} u_{b_x} \\ u_{b_y} \end{pmatrix}$.

2.3 Uncertainty Quantification

We employ the Monte Carlo method to quantify two uncertainty scenarios [28]: a) the influence of the capillaries architecture and b) the influence of the blood velocity. The first scenario was divided in two, one considering only the influence of the number of capillaries terminals and the other considering only the influence of the angle of capillaries terminals.

We consider the terminal points of the capillary bed uniformly distributed in the domain for all scenarios. Once cancer growth can stimulate the capillaries' angiogenesis in the tumor site [8, 37], we consider the tumor tissue with 70% of capillaries terminals and the healthy tissue with 50% of capillaries terminals as the base values. Furthermore, the blood velocity also depends on the cancer type, stage, location, and other specificities [11, 37]. So we consider the magnitude of the blood velocity as 3 mm/s in the tumor tissue and 2 mm/s in the healthy tissue as the base values. Finally, we draw the angle for each terminal point using values uniformly distributed between 0 and 2π.

For the first scenario, the influence of the number of capillaries, we drew all terminal points and angles for the entire tissue. For each MC simulation, we use the following expression to determine the number of capillaries:

$$N_c = \begin{cases} 70 \times X \sim U(0.7, 1.3) & \text{for the tumor tissue,} \\ 50 \times X \sim U(0.7, 1.3) & \text{otherwise,} \end{cases}$$

where N_c is the percentage of capillaries terminals in the respective tissue and $X \sim U(a, b)$ is a uniform distribution that generates a random value between a and b.

In the second part of the first scenario, the influence of the capillaries angles, we drew terminal points and base angles (using the proportion of 50% to healthy tissue and 70% to the tumor one). For each MC simulation, we use the following expression to determine the angle of capillaries:

$$\theta_c = \theta \times X \sim U(0.7, 1.3),$$

where θ is the base angle of the capillary and θ_c is the value considering the uncertainties of 30% in the capillary angles.

In the second scenario, the influence of the blood velocity magnitude, we drew terminal points and base angles (using the proportion of 50% to healthy tissue and 70% to the tumor one). We consider 50% of the uncertainty in the magnitude for each MC sample using the following expression:

$$|\mathbf{u}_{b_c}| = |\mathbf{u}_b| \times X \sim U(0.5, 1.5),$$

where $|\mathbf{u}_b|$ is the magnitude of blood velocity (see Table 1) and $|\mathbf{u}_{b_c}|$ is the value considering the uncertainty.

Finally, the Algorithm 1 illustrates the implementation of uncertainty quantification via Monte Carlo using *OpenMP* used in this study. It is worthwhile to notice that `omp parallel` command creates the threads and `omp for` divides the workload among the created threads. Moreover, the time loop has no workload division once data dependence exists between successive time steps.

3   Numerical Results

This section presents the results of the two scenarios evaluated in this work. Both simulation scenarios consider a two-dimensional domain with the tissue consisting only of healthy and tumor cells. The healthy cells represent a piece of muscle tissue. Also, we performed $10,000$ MC simulations of cancer treatment with hyperthermia using magnetic nanoparticles, considering a square domain of length $0.1m$ and a tumor seated in $(x, y) \in [0.04, 0.06] \times [0.04, 0.06]$. The model parameters used for solving Eq. 3 are specified in Table 1. Furthermore, we consider a single nanoparticles injection point seated at $(0.50\,\text{m}, 0.50\,\text{m})$.

The numerical model presented in Sect. 2 was implemented using the C++ programming language and the *OpenMP* API. The code was compiled using *gcc* version 11.3.1 with the optimization flag $-O3$ enabled. The code was executed in a

---

Algorithm 1: Pseudocode of Monte Carlo implementation using *OpenMP*

```
 1 begin
 2      # pragma omp parallel
 3      foreach Monte Carlo Sample do
 4          foreach t_n; n = 0, 1, ⋯ , N_t do
 5              #pragma omp for
 6              foreach (x_i, y_j); i = 0, 1, ⋯ , N_x; j = 0, 1, ⋯ , N_y do
 7                  | evaluate T_{ij}^{n+1} using Equation (4)
 8              end
 9          end
10      end
11 end
```

Table 1. Model parameters values employed in Eq. (3).

Symbol	Unit	Muscle	Tumor
c	$(J/Kg°C)$	4,000	4,000
c_{pb}	$(J/Kg°C)$	4,000	4,000
k	$(W/m°C)$	0.50	0.55
p	(Kg/m^3)	1,000	1,200
ρ_b	(Kg/m^3)	1,000	1,000
ε	-	0.02	0.01
q_m	(W/m^3)	420	4,200
$|\mathbf{u}_b|$	*(mm/s)*	2.0	3.0
A	W/m^3	0.9 × 10^6	
r_0	m	3.1 × 10^{-3}	

2.90 GHz Intel® Core™ *i7*-10700 CPU. Although this CPU has 16 hyperthreading cores, the number of physical cores available is 8, which is the number of threads used during simulations. Finally, the post-processing of the simulations was performed using `matplotlib` library [10] and *ParaView* version 5.10.1 [4].

3.1   Influence of the Capillaries Architecture

The uncertainty quantification simulation results considering the influence of the capillary bed architecture are divided into two scenarios, presented in the following subsections.

The Number of Capillaries. Figure 1A shows the base distribution of the velocity field for the capillary bed considering the uncertainties described in Sect. 2.3, *i.e.*, considering 30% of uncertainty in the number of capillaries terminals.

Figure 1B shows the average temperature obtained after 10,000 MC simulations at t = 50 min. It is worth noting that the white isoline of 43 °C covers the entire tumor (the solid black line), *i.e.*, the tumor tissue reaches the target temperature that causes necrosis.

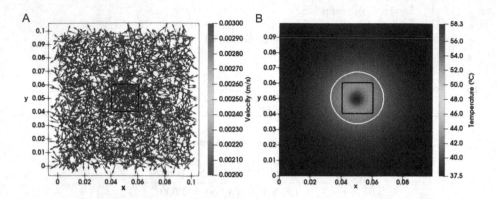

Fig. 1. Simulation of the number of capillaries' influence in the hyperthermia treatment. **A)** Base velocity field considering 100% of capillaries; **B)** Average temperature distribution for 10,000 Monte Carlo simulations at t = 50 min employing Eq. (4) with a tumor located in $(x, y) \in [0.04, 0.06] \times [0.04, 0.06]$ and considering the parameters present in Table 1. The solid white contour represents the temperature of 43 °C, highlighting the damaged area, and the solid black contour is the tumor tissue location.

Figure 2A presents the evolution of the average temperature of the tumor tissue and healthy tissue. It is worthwhile to notice that the temperature inside the tumor is higher than that of the healthy portion. Moreover, even considering the 95% confidence interval (CI), the average temperature of the tumor is higher than 43 °C. On the other hand, Fig. 2B shows that only 5.2% of the healthy tissue reached a temperature of 43 °C or higher while the tumor tissue reached the target temperature with a mean value close to the 95% CI limits.

Angles of Capillaries. This section presents the simulation results obtained after using the uncertainties described in Sect. 2.3 for analyzing the influence of the capillary terminals' angles, *i.e.*, considering 30% of uncertainty in the capillary terminals' angles. Figure 3A shows the base distribution of the velocity field of the capillary bed, considering 70% of blood vessels in cancerous tissue and 50% in healthy tissue.

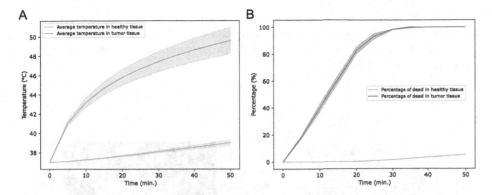

Fig. 2. Uncertainty quantification of the influence of the number of capillaries in the hyperthermia treatment. In both figures, the solid lines represent the mean temperature, and shade regions represent the 95% confidence interval obtained after 10,000 Monte Carlo simulations. **A)** Evolution of temperature obtained by Eq. (4) considering the parameters present in Table 1. The blue line and blue shade represent the average temperature of the tumor, and the yellow line and yellow shade represent the average temperature of healthy tissue; **B)** Percentage of tumor and healthy tissues killed by hyperthermia. The red line and red shaded represent tumor tissue necrosis, and the green line and green shade represent healthy tissue necrosis. (Color figure online)

Figure 3B shows the average temperature obtained after 10,000 MC simulations at t = 50 min. The tumor reaches the target temperature that causes necrosis once the white isoline of 43 °C covers the entire tumor (the solid black line).

Figure 4A presents the evolution of the average tumor and healthy tissues temperature. Again, the temperature inside the tumor is higher than that of the healthy portion. Moreover, even considering the 95% confidence interval, the average temperature of the tumor tissue is higher than 43 °C. On the other hand, Fig. 4B shows that only 5.3% of healthy tissue reached a temperature of 43 °C or higher while the tumor tissue reached the target temperature with 95% CI in [99.5, 100].

3.2   Influence of the Blood Velocity

This section presents the simulation results obtained after using the uncertainties described in Sect. 2.3 for the analysis of the blood velocity magnitude influence, *i.e.*, considering 50% of uncertainty in the magnitude of the blood velocity. Figure 5A shows the base distribution of the velocity field of the capillary bed considering 70% of blood vessels in cancerous tissue and 50% in healthy tissue as well as the blood velocity given by Table 1.

Figure 5B shows the average temperature obtained after 10,000 MC simulations at t = 50 min. The tumor tissue reaches the target temperature that causes necrosis once the white isoline of 43 °C covers the entire tumor (the solid black line).

Figure 6A presents the evolution of the average temperature for the tumor and the healthy tissue. As in the previous scenario, the temperature inside the tumor tissue is higher than that of the healthy portion. Moreover, even considering the

95% confidence interval, the average temperature of the tumor is higher than 43 °C. Figure 6B shows that only 5.1% of healthy tissue reached a temperature of 43 °C or higher while the tumor tissue reached the target temperature with a mean value close to the 95% CI limits.

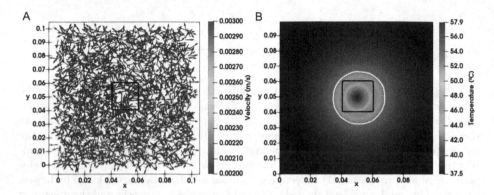

Fig. 3. Simulation of the influence of the capillaries angles in the hyperthermia treatment. **A)** Base velocity field considering 70% of capillaries in the tumor tissue and 50% of capillaries in the healthy tissue; **B)** Average temperature distribution obtained after 10,000 Monte Carlo simulations at t = 50 min employing Equation (4) with a tumor located in $(x, y) \in [0.04, 0.06] \times [0.04, 0.06]$ and considering the parameters present in Table 1. The solid white contour represents the temperature of 43 °C, highlighting the damaged area, and the solid black contour is the tumor location.

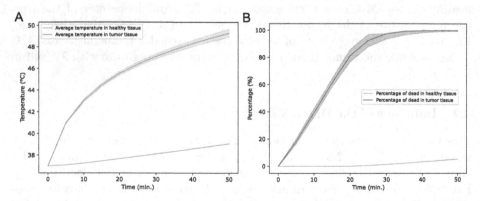

Fig. 4. Uncertainty quantification of the influence of the capillaries angles in the hyperthermia treatment. In both figures, the solid lines represent the mean temperature, and shade regions represent the 95% confidence interval for 10,000 Monte Carlo simulations. **A)** Evolution of temperature determined by Eq. (4) considering the parameters present in Table 1. The blue line and blue shade represent the average temperature of the tumor, and the yellow line and yellow shade represent the average temperature of healthy tissue; **B)** Percentage of tumor and healthy tissues killed by hyperthermia. The red line and red shaded represent tumor tissue necrosis, and the green line and green shade represent healthy tissue necrosis. (Color figure online)

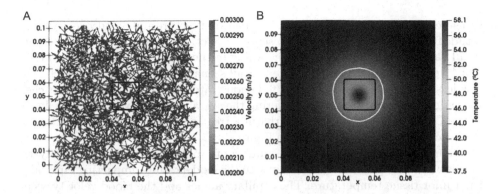

Fig. 5. Simulation of the influence of the blood velocity magnitude in the hyperthermia treatment. **A)** Base velocity field considering 70% of capillaries in the tumor tissue and 50% of capillaries in the healthy tissue; **B)** Average temperature distribution obtained after 10,000 Monte Carlo simulations at t = 50 min employing Eq. (4) with a tumor located in $(x, y) \in [0.04, 0.06] \times [0.04, 0.06]$ and considering the parameters present in Table 1. The solid white contour represents the temperature of 43 °C, highlighting the damaged area, and the solid black contour is the tumor location.

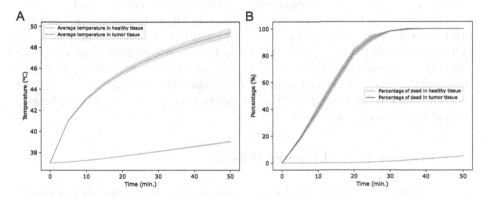

Fig. 6. Uncertainty quantification of the influence of the blood velocity magnitude in the hyperthermia treatment. In both figures, the solid lines represent the mean temperature, and shade regions represent the 95% confidence interval for 10,000 Monte Carlo simulations. **A)** Evolution of temperature determined by the Eq. (4) considering the parameters present in Table 1. The blue line and blue shade represent the average temperature of the tumor, and the yellow line and yellow shade represent the average temperature of healthy tissue; **B)** Percentage of tumor and healthy tissue killed by hyperthermia. The red line and red shaded represent tumor tissue necrosis, and the green line and green shade represent healthy tissue necrosis. (Color figure online)

4   Discussion

Considering the influence of the capillary bed architecture and blood velocity magnitude from the numerical results presented in Sect. 3, it is possible to assume that, in all cases, the average temperature value in the tumor site rises above 43 °C, which leads to its damage. Moreover, in all scenarios, the health tissue damage was less than 6%. Figures 2, 4, and 6 demonstrate that the entire tumor reaches the target temperature when considering the average value. Even when the limits of the CI are considered, the hyperthermia treatment damages the entire tumor site, except in the case illustrated in Figure 4.

Figures 1, 3, and 5 show that the number of capillaries is the variable analyzed in this work that adds more uncertainty to the model results, impacting especially the tumor tissue temperature. The capillary angles and the blood velocity seem to add small amounts of uncertainties to the model resolution, especially to the computation of damage to healthy tissue. These results indicate that a good representation of capillaries may be essential to obtain realistic simulations of the hyperthermia treatment results or, from a medical perspective, should be taken into account in the treatment.

It is essential to notice the limitations of this study. For example, this paper establishes a temperature threshold of $T \geq 43\,°C$ for inducing tissue damage and cell necrosis. But this threshold is considered a threshold temperature to cause cell necrosis within a reasonable duration. During our experiments, we simulated 50 minutes of hyperthermia treatment but did not consider the duration as a parameter for determining hyperthermia success. Also, there is a delay in achieving tissue damage at this temperature, which is not considered in this paper. Additionally, there exist more accurate methods for measuring tumor ablation, such as the Arrhenius models [32, 33]. Lastly, it is important to acknowledge that this study relies on a theoretical model and may not entirely reflect the intricate biological processes that occur in actual tumors. Furthermore, the outcomes of this research may not be immediately applicable to clinical environments, and additional experimental verification is necessary.

5   Conclusions and Future Works

This work presents the results of a two-dimensional simulation of hyperthermia cancer treatment in a heterogeneous porous tissue and UQ analysis via Monte Carlo simulation. The results presented in this study suggest that the capillaries bed architecture significantly influences temperature evolution in the simulated tissue. On the other hand, the target temperature of 43 °C at the final of the treatment presents little influence under the assumption of these uncertainties when considering the parameters used to perform these simulations. This study demonstrates that uncertainty analysis can be a powerful tool for treatment planning once it allows the possibility to perform several *in silico* trials and analyze the best option for a patient-specific scenario, taking into account the experimental uncertainties such as limited accuracy of the measuring apparatus, limitations and simplifications of the experimental procedure and uncontrolled changes to the environment. The results of this paper reinforce that *in silico* medicine might reduce the need for clinical trials with animals and cohort studies with humans.

Although this paper aims to analyze the uncertainty introduced by the values associated with the capillaries bed architecture and the magnitude of the blood velocity parameters in the results of the mathematical model, using a simple approach for this purpose, it is possible to employ different types of probability density functions, for example, fitted from experimental data. Additional studies are needed to determine the uncertain parameters' best probability density function. In addition, the uncertainties introduced by other parameters can be included in the analysis to determine their impact on the results of the mathematical model simulation. Furthermore, it is also essential to consider the study of different tissue layers in the human body, such as skin, muscle, and fat, along with realistic tumor and tissue shapes. Furthermore, we intend to assess the suggested approach on more practical tumors and diverse tissue shapes or even use tumors derived from images specific to a patient as demonstrated in studies available in the literature [22,23,36].

Acknowledgments. The authors would like to express their thanks to CAPES (Finance Code 001 and Projeto CAPES - Processo 88881.708850/2022-01), CNPq (308745/2021-3), FAPEMIG (APQ-02830/17 and APQ-01226-21), FINEP (SOS Equipamentos 2021 AV02 0062/22) and UFJF for funding this work.

References

1. Alazmi, B., Vafai, K.: Constant wall heat flux boundary conditions in porous media under local thermal non-equilibrium conditions. Int. J. Heat Mass Transf. **45**(15), 3071–3087 (2002)
2. Amiri, A., Vafai, K.: Transient analysis of incompressible flow through a packed bed. Int. J. Heat Mass Transf. **41**(24), 4259–4279 (1998)
3. Amiri, A., Vafai, K.: Analysis of dispersion effects and non-thermal equilibrium, non-Darcian, variable porosity incompressible flow through porous media. Int. J. Heat Mass Transf. **37**(6), 939–954 (1994)
4. Ayachit, U.: The ParaView Guide: a parallel visualization application. Kitware, Inc. (2015)
5. Gilchrist, R.K., Medal, R., Shorey, W.D., Hanselman, R.C., Parrott, J.C., Taylor, C.B.: Selective inductive heating of lymph nodes. Ann. Surg. **146**(4), 596–606 (1957)
6. Giustini, A.J., Petryk, A.A., Cassim, S.M., Tate, J.A., Baker, I., Hoopes, P.J.: Magnetic nanoparticle hyperthermia in cancer treatment. Nano Life **1**(01n02), 17–32 (2010). https://doi.org/10.1142/S1793984410000067
7. Guedes, B.R., Lobosco, M., dos Santos, R.W., Reis, R.F.: Uncertainty quantification of tissue damage due to blood velocity in hyperthermia cancer treatments. In: Paszynski, M., Kranzlmüller, D., Krzhizhanovskaya, V.V., Dongarra, J.J., Sloot, P.M.A. (eds.) ICCS 2021. LNCS, vol. 12743, pp. 511–524. Springer, Cham (2021). https://doi.org/10.1007/978-3-030-77964-1_39
8. Hicklin, D.J., Ellis, L.M.: Role of the vascular endothelial growth factor pathway in tumor growth and angiogenesis. J. Clin. Oncol. **23**(5), 1011–1027 (2005)

9. Hilger, I., Hergt, R., Kaiser, W.A.: Towards breast cancer treatment by magnetic heating. J. Magn. Magn. Mater. **293**(1), 314–319 (2005). Proceedings of the Fifth International Conference on Scientific and Clinical Applications of Magnetic Carriers

10. Hunter, J.D.: Matplotlib: a 2D graphics environment. Comput. Sci. Eng. **9**(3), 90–95 (2007). https://doi.org/10.1109/MCSE.2007.55

11. Ishida, H., Hachiga, T., Andoh, T., Akiguchi, S.: In-vivo visualization of melanoma tumor microvessels and blood flow velocity changes accompanying tumor growth. J. Appl. Phys. **112**(10), 104703 (2012)

12. Jiji, L.M.: Heat Conduction. Springer-Verlag, Berlin Heidelberg (2009). https://doi.org/10.1007/978-3-642-01267-9

13. Khaled, A.R., Vafai, K.: The role of porous media in modeling flow and heat transfer in biological tissues. Int. J. Heat Mass Transf. **46**(26), 4989–5003 (2003)

14. Lee, D.Y., Vafai, K.: Analytical characterization and conceptual assessment of solid and fluid temperature differentials in porous media. Int. J. Heat Mass Transf. **42**(3), 423–435 (1999)

15. LeVeque, R.J.: Finite difference methods for ordinary and partial differential equations: steady-state and time-dependent problems. SIAM (2007)

16. Matsuki, H., Yanada, T., Sato, T., Murakami, K., Minakawa, S.: Temperature-sensitive amorphous magnetic flakes for intratissue hyperthermia. Mater. Sci. Eng. A **181**(0), 1366–1368 (1994). Proceedings of the Eighth International Conference on Rapidly Quenched and Metastable Materials: Part 2

17. Minkowycz, W., Sparrow, E.M., Abraham, J.P.: Nanoparticle Heat Transfer And Fluid Flow, vol. 4. CRC Press (2012)

18. Moros, E.: Physics of thermal therapy: fundamentals and clinical applications. CRC Press (2012)

19. Moroz, P., Jones, S., Gray, B.: Magnetically mediated hyperthermia: current status and future directions. Int. J. Hyperth. **18**(4), 267–284 (2002)

20. National cancer institute: types of cancer treatment. https://www.who.int/health-topics/cancer (2023). Accessed 1 Mar 2023

21. Pennes, H.H.: Analysis of tissue and arterial blood temperature in the resting human forearm. J. Appl. Phisiology **1**, 93–122 (1948)

22. Prasad, B., Ha, Y.H., Lee, S.K., Kim, J.K.: Patient-specific simulation for selective liver tumor treatment with noninvasive radiofrequency hyperthermia. J. Mech. Sci. Technol. **30**(12), 5837–5845 (2016). https://doi.org/10.1007/s12206-016-1154-x

23. Rahpeima, R., Lin, C.A.: Numerical study of magnetic hyperthermia ablation of breast tumor on an anatomically realistic breast phantom. PLoS ONE **17**(9), e0274801 (2022)

24. Reis, R.F., dos Santos Loureiro, F., Lobosco, M.: Parameters analysis of a porous medium model for treatment with hyperthermia using OpenMP. J. Phys: Conf. Ser. **633**(1), 012087 (2015)

25. Reis, R.F., dos Santos Loureiro, F., Lobosco, M.: 3D numerical simulations on GPUs of hyperthermia with nanoparticles by a nonlinear bioheat model. J. Comput. Appl. Math. **295**, 35–47 (2016)

26. Roetzel, W., Xuan, Y.: Transient response of the human limb to an external stimulust. Int. J. Heat Mass Transf. **41**(1), 229–239 (1998)

27. Rosensweig, R.: Heating magnetic fluid with alternating magnetic field. J. Magn. Magn. Mater. **252**(0), 370–374 (2002). Proceedings of the 9th International Conference on Magnetic Fluids

28. Rubinstein, R.Y., Kroese, D.P.: Simulation and the Monte Carlo method. Wiley (2016)

29. Salloum, M., Ma, R., Zhu, L.: An In-Vivo experimental study of temperature eleva-
tions in animal tissue during magnetic nanoparticle hyperthermia. Int. J. Hyperth.
24(7), 589–601 (2008)

30. Sanyal, D., Maji, N.: Thermoregulation through skin under variable atmospheric
and physiological conditions. J. Theor. Biol. **208**(4), 451–456 (2001)

31. Sherar, M.D., Gladman, A.S., Davidson, S.R., Trachtenberg, J., Gertner, M.R.: Heli-
cal antenna arrays for interstitial microwave thermal therapy for prostate cancer:
tissue phantom testing and simulations for treatment. Phys. Med. Biol. **46**(7), 1905
(2001)

32. Singh, M.: Incorporating vascular-stasis based blood perfusion to evaluate the thermal signatures of cell-death using modified Arrhenius equation with regeneration
of living tissues during nanoparticle-assisted thermal therapy. Int. Commun. Heat
Mass Transfer **135**, 106046 (2022). https://doi.org/10.1016/j.icheatmasstransfer.
2022.106046

33. Singh, M., Singh, T., Soni, S.: Pre-operative assessment of ablation margins for
variable blood perfusion metrics in a magnetic resonance imaging based complex
breast tumour anatomy: simulation paradigms in thermal therapies. Comput. Methods Programs Biomed. **198**, 105781 (2021). https://doi.org/10.1016/j.cmpb.2020.
105781

34. WHO: World health organization. https://www.who.int/news-room/fact-sheets/
detail/cancer (2023). Accessed 23 Mar 2023

35. Xuan, Y., Roetzel, W.: Bioheat equation of the human thermal system. Chem. Eng.
Technol. Ind. Chem. Plant Equipment Process Eng. Biotechnol. **20**(4), 268–276
(1997)

36. Zastrow, E., Hagness, S.C., Van Veen, B.D.: 3D computational study of non-invasive
patient-specific microwave hyperthermia treatment of breast cancer. Phys. Med.
Biol. **55**(13), 3611 (2010)

37. Zhou, W., Chen, Z., Zhou, Q., Xing, D.: Optical biopsy of melanoma and basal cell
carcinoma progression by noncontact photoacoustic and optical coherence tomography: In vivo multi-parametric characterizing tumor microenvironment. IEEE Trans.
Med. Imaging **39**(6), 1967–1974 (2019)

Phase Correction and Noise-to-Noise Denoising of Diffusion Magnetic Resonance Images Using Neural Networks

Jakub Jurek[1]([✉]) [ID], Andrzej Materka[1] [ID], Kamil Ludwisiak[2], Agata Majos[3] [ID], and Filip Szczepankiewicz[4] [ID]

[1] Institute of Electronics, Lodz University of Technology, Aleja Politechniki 10, 93590 Lodz, Poland
jakub.jurek@p.lodz.pl
[2] Department of Diagnostic Imaging, Independent Public Health Care, Central Clinical Hospital, Medical University of Lodz, Pomorska 251, 92213 Lodz, Poland
[3] Department of Radiological and Isotopic Diagnosis and Therapy, Medical University of Lodz, Pomorska 251, 92213 Lodz, Poland
[4] Medical Radiation Physics, Lund University, Barngatan 4, 22185 Lund, Sweden

Abstract. Diffusion magnetic resonance imaging (dMRI) is an important technique used in neuroimaging. It features a relatively low signal-to-noise ratio (SNR) which poses a challenge, especially at stronger diffusion weighting. A common solution to the resulting poor precision is to average signal from multiple identical measurements. Indeed, averaging the magnitude signal is sufficient if the noise is sampled from a distribution with zero mean value. However, at low SNR, the magnitude signal is increased by the rectified noise floor, such that the accuracy can only be maintained if averaging is performed on the complex signal. Averaging of the complex signal is straightforward in the non-diffusion-weighted images, however, in the presence of diffusion encoding gradients, any motion of the tissue will incur a phase shift in the signal which must be corrected prior to averaging. Instead, they are averaged in the modulus image space, which is associated with the effect of Rician bias. Moreover, repeated acquisitions further increase acquisition times which, in turn, exacerbate the challenges of patient motion. In this paper, we propose a method to correct phase variations using a neural network trained on synthetic MR data. Then, we train another network using the Noise2Noise paradigm to denoise real dMRI of the brain. We show that phase correction made Noise2Noise training possible and that the latter improved the denoising quality over averaging modulus domain images.

Keywords: Phase correction · Diffusion magnetic resonance imaging · Denoising · Convolutional neural networks · Transfer learning

1 Introduction

Diffusion magnetic resonance imaging (dMRI, also diffusion-weighted MRI, DW MRI, DWI) is one of the key tools used both in clinical neurological practice and

J. Mikyška et al. (Eds.): ICCS 2023, LNCS 14074, pp. 638–652, 2023.
https://doi.org/10.1007/978-3-031-36021-3_61

in neuroscientific research [12]. For example, dMRI is used for visualizing the white matter structures of the brain and can characterize features of the tissue microstructure. The main contrast in dMRI comes from the relation between diffusion encoding strength, characterized by the b-value, and the rate at which water in the tissue is diffusing, i.e., in places where the diffusion is fast, the signal is more attenuated and vice versa. High b-values increase the diffusion-based image contrast and can enable more elaborate analysis models, but the trade-off is a lower signal and thus lower SNR. At some b-value, depending on other imaging parameters including resolution, the SNR becomes prohibitively low. Typically, this problem is solved by repeating the same measurement multiple times and averaging the repetitions (the number of which is referred to as NEX).

Although widely used in practice, averaging has significant limitations. First of all, the noise values converge to their mean value, so averaging should be performed at such a step of image reconstruction where noise has zero-mean. Secondly, averaging reduces the standard error of the mean for the signal intensity proportionally to the square root of NEX, but increases the scan time proportionally to the number of repetitions, so it is inefficient. The maximal NEX number may be limited by the hardware, for example to 32 in the Siemens Avanto 1.5T machine. Since scan time is increased by each repetition, the risk of patient motion and patient discomfort grows.

The condition for zero-mean noise is satisfied for complex-valued images before coil signal combination and before computing modulus images. However, in the presence of diffusion encoding gradients, the position of tissue is encoded in the phase. If the position changes due to movement, vibration, pulsation etc., the phase will be incoherent across the object [11]. This is, repeated images ideally have the same intensity modulus, but different values of the complex intensity components. Thus, images acquired with non-zero diffusion weighting cannot be straightforwardly averaged in the complex domain and are instead averaged in the modulus domain. This results in convergence of intensity values to the non-zero mean and formation of the so called noise floor or Rician bias. This effect is not observed for SNR levels greater than 3, but can bias diffusion measures in most noisy image areas where SNR$<$3 [4].

A summary of MR-related phase correction methods was presented in [11]. The problems associated with phase correction in dMRI were analysed in a recent work by Liu et al. [10].

In this paper, we first propose a neural network, called the Phase Shifter, to equalize phase distribution between pairs of image slices taken from two scan repetitions. The Phase Shifter network is trained using transfer learning, based on synthetic echo planar images (EPI) [6]. We showed in a previous work that denoising can be more effectively tackled using neural networks than by averaging [6]. With phase-corrected images of zero-mean noise at hand, we use the Noise2Noise training paradigm [9] to train a N2N Denoiser network without the necessity for noise-free training targets. For both the phase correction and denoising tasks, we use simple and quick-to-train convolutional neural networks [3].

In Sect. 2, we describe the image simulation process, introduce the Phase Shifter and N2N Denoiser networks and how they were trained. In Sect. 3, we show the results of training of both networks as well as their application on brain dMRI. The Phase Shifter is compared to polynomial fitting. Section 4 closes the paper with a discussion of the results.

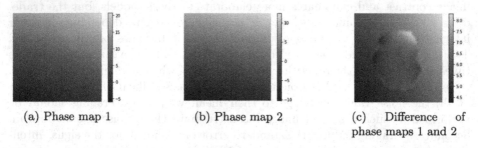

(a) Phase map 1   (b) Phase map 2   (c)   Difference   of
                                                  phase maps 1 and 2

Fig. 1. Example simulated phase maps (a, b) and their difference (c), in radians. Phase values can range between several multiples of π. In this work, the goal is to correct image phase by estimating the phase difference, given images with two non-agreeing phase maps (Eq. 1)

2   Materials and Methods

Experiments were performed using dMRI of the brain acquired with b-values of 0 and 1000 s/mm^2 on a Siemens Avanto 1.5 T machine at the Central Clinical Hospital, Lodz Medical University, Poland. The scanning involved non-accelerated echo-planar images with imaging parameters described in detail in [6]. Synthetic complex MRI images with random phase were simulated based on the BrainWeb phantom [1,2,7,8] as in [6].

2.1   Modelling Synthetic Training Data for the Phase Shifter

We pose phase correction in dMRI as the problem of finding the phase difference map between two repetitions of the same complex-valued image slice. The ideal, non-noisy modulus value of the complex signal is equal for the two repetitions, up to motion, which we assume negligible in an EPI scan involving two successive acquisitions. The individual real and imaginary values across repetitions are different, but contain noise sampled from a zero-mean distribution.

We performed real-MR-data-driven modelling on the BrainWeb volume to obtain a synthetic training dataset with the required properties [6]. The Brain-Web volume was cropped and downsampled to match the resolution of dMRI data. This resulted in a single $160 \times 160 \times 25$ voxels volume. For simplicity in setting the SNR in experiments, the mean value of the tissue in the BrainWeb volume was normalized to 1 using the tissue mask. Phase maps were estimated based on one of the real dMRI scans (Patient 0) by polynomial fitting (PF) [6], cubic within the tissue and linear within background to avoid rapid phase changes in the latter region. In total, 25 phase maps were obtained, one for each slice of the selected dMRI scan. Representative phase maps and their difference are shown in Fig. 1.

The original BrainWeb volume I_m is real valued and thus can be treated as a complex number with zero phase and zero imaginary component. Phase variations within the slices were introduced by:

$$I_c = I_m e^{j\theta} = \text{Re}(I_c) + \text{Im}(I_c)j \tag{1}$$

which yields a complex-valued slices I_c with non-zero real and imaginary parts and phase θ, where θ is a function of pixel coordinates and j is the imaginary unit.

Three different phase maps were imposed on each slice to create triplets and augment the training set (Fig. 2a-2b shows a pair):

$$I_{ci} = I_m e^{j\theta_i}, i = 1, 2, 3 \tag{2}$$

Then, complex noise maps obtained from the scanner were added to the real and imaginary components of all slices. The standard deviation of noise was normalized to 1. EPI regridding effects on noise were taken into account as in [6]. Noisy images with mean SNR $= s$ were calculated as:

$$[I_{ci}^n]^s = s * I_{ci} + N \tag{3}$$

where $[I_{ci}^n]^s$ is the noisy complex image with phase θ_i and SNR s, and N is the zero-mean Gaussian noise. From the triplets of the same slice with different phase, one instance was selected as the one to be corrected and two instances were assumed to be repetitions with reference phase ($i=2$ and $i=1, 3$, respectively, by random choice). The goal of $[I_{c2}^n]^s$ phase correction in this setting is to shift its phase from θ_2 to θ_i:

$$([I_{c2}^n]^s)^{\theta_i'} = [I_{c2}^n]^s e^{-j(\theta_2 - \theta_i)} \tag{4}$$

Note that $([I_{c2}^n]^s)^{\theta_i'}$ has the same phase θ_i but is not equal to $[I_{c2}^n]^s$ due to noise (Fig. 2). By the same argument it satisfies the Noise2Noise criteria for identical image content and zero-mean independent noise. Applying a phase difference computed using the two-argument arctangent, $\text{atan2}([I_{c2}^n]^s) - \text{atan2}([I_{ci}^n]^s)$ (Fig. 3) would yield identical images including their noise values and thus is not suitable. Using two target phase maps for each slice increased the number of training examples to 50. Additionally, SNR values s $= 1, 2, 3, 4, 5, 6, 7, 8, 9$ were considered, resulting in 450 examples in total. Each input training example was setup as a quadruple $\text{Re}([I_{c2}^n]^s), \text{Im}([I_{c2}^n]^s), \text{Re}([I_{ci}^n]^s), \text{Im}([I_{ci}^n]^s)$, with $s = 1, 2, ..., 9$ and $i = 1, 3$. Targets were setup as tuples $\text{Re}([I_{c2}^n]^s)^{\theta_i'}, \text{Im}([I_{c2}^n]^s)^{\theta_i'}$.

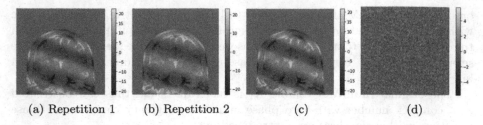

(a) Repetition 1   (b) Repetition 2   (c)   (d)

Fig. 2. Two noisy instances of the real part of same slice with different phase (a, b), Repetition 2 with corrected phase (c), the difference of images (a) and (c) which are equal up to noise (d).

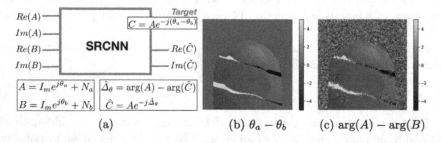

(a)   (b) $\theta_a - \theta_b$ (c) $\arg(A) - \arg(B)$

Fig. 3. The scheme of Phase Shifter operation (a) and an example of clean and noisy phase difference maps computed using ideal phase maps (b) and the two-argument arctangent function (c). Using (c) for phase correction would result in two identical image instances, not suitable for averaging nor for Noise2Noise. The desired phase difference map should be noise-free (b)

2.2 Phase Correction

The Phase Shifter network architecture selected for correcting the phase is based on the SRCNN [3]. SRCNN is three-layered, with the first layer performing 9×9 convolutions 64 times, the second performing 1×1 convolutions 32 times and the final performing 5×5 convolutions 1 time. The output layer in our Phase Shifter adaptation is modified to perform the convolutions twice, so that it outputs the real and imaginary parts of the phase-corrected image as two channels. The input to the network is a quadruple of images as described above in Sect. 2.1. (images A and B in Fig. 3a, with $s = 1$). The expected result of the Phase Shifter is a shift in the phase of the first complex input image so that its (unknown) non-noisy complex value is equal to the non-noisy complex value of the second input image.

The Phase Shifter is setup to first output the phase-corrected complex image estimate $\mathrm{Re}((\widehat{[I^n_{c2}]^s})^{\theta_i{}'}), \mathrm{Im}((\widehat{[I^n_{c2}]^s})^{\theta_i{}'})$ (denoted as $\tilde{C}$ in Fig. 3a). Then, the difference is computed between the phase of the input image to be corrected and of the phase-corrected image at the network output, using the two-argument arctangent function:

$$(\widehat{\theta_2 - \theta_i}) = \text{atan2}([I_{c2}^n]^s) - \text{atan2}(([\widehat{I_{c2}^n}]^s)^{\theta_i}{}') \tag{5}$$

The latter is applied to the input image to be corrected to obtain the final Phase Shifter output (denoted as $\hat{C}$ in Fig. 3a):

$$([\widehat{I_{c2}^n}]^s)^{\theta_i}{}'' = [I_{c2}^n]^s e^{-j(\widehat{\theta_2 - \theta_i})} \tag{6}$$

The computed phase difference maps $(\widehat{\theta_2 - \theta_i})$ (denoted as $\hat{\Delta}_\theta$ in Fig. 3a) may initially contain small errors corresponding to discrepancies in the modulus value, but its application keeps the modulus unchanged in the final output, while the errors are minimized as the network is trained.

2.3 Denoising

The Noise2Noise training paradigm is explained in detail in [9]. The main idea is that training a denoiser with clean targets and L2 loss (mean squared error, MSE loss) has the same minimum as training with zero-mean noisy targets. This approach can be applied with any neural network regression-type architecture. For denoising, we again select the SRCNN architecture. We treat the real and imaginary components of a complex image as separate images to denoise. The N2N Denoiser receives a single noisy image slice as input and outputs a single denoised image slice. Its final layer thus performs one 5×5 convolution. The N2N Denoiser estimates the noise map for each image slice and then subtracts it from the input [13]. It is optimized by minimizing the MSE between the denoised images and the noisy phase-corrected targets. The minimum possible loss value is thus, by definition of the MSE, the variance of noise in the target image. Since the images in our experiments were normalized to unit noise standard deviation [6], the optimum loss value is 1.

2.4 Evaluation Methods

Phase correction accuracy in synthetic BrainWeb images was evaluated using the Phase Shifter loss function (MSE) globally and by comparing MSE maps in selected cases visually. Phase correction accuracy in dMR images was evaluated by visual examination of the difference between target and phase-corrected images. It was also validated by performing Noise2Noise training using phase-corrected images. Denoising accuracy was evaluated using the MSE calculated in reference to clean images for synthetic BrainWeb images at different SNR levels and to noisy images for real dMRI.

3 Results

3.1 Phase Shifter Training and Phase Correction in dMR Images

The Phase Shifter was trained using 450 160×160 synthetic examples using the MSE loss, Adam optimizer, learning rate of 0.001, mini-batch training with batch size of 9 and shuffling, implemented in PyTorch and executed in Google Colaboratory. The validation set was composed of the same image slices, but limited to two phase maps θ_1, θ_2: $\mathrm{Re}([I_{c1}^n]^s)$, $\mathrm{Im}([I_{c1}^n]^s)$, $\mathrm{Re}([I_{c2}^n]^s)$, $\mathrm{Im}([I_{c2}^n]^s)$, and with accordingly modified target: $\mathrm{Re}(([I_{c2}^n]^s)^{\theta_1{}'})$, $\mathrm{Im}(([I_{c2}^n]^s)^{\theta_1{}'})$.

Thus, there were 225 validation examples. Early stopping was implemented as in [5], with patience of 100 epochs, validation loss value checking in each epoch, and minimum delta value of 1e−9. In this setting, the network reached its best performance after 1645 epochs and training took 44 min. Validation mean absolute error (MAE) between the target and estimated phase-corrected images achieved the value of 0.1, which is markedly lower than the standard deviation of the noise (equal to 1). Figures 4 and 5 show chosen training results.

dMR images of the brain were reconstructed from k-space as in [6], which involved the necessary correction steps: regridding and Nyquist ghosting correction, yielding complex-valued volumes from individual receiver coils. Phase correction was performed using the trained Phase Shifter network on these images (Fig. 6). The network was run 100 times on a pair of 25-slice dMRI volumes and processing duration was approximately 0.02 s on average.

Then, polynomial fitting was performed on the same dMRI volumes for comparison. Fitting was run 10 times on each of the volumes. The average run time was approximately 1.55 s. Phase distribution after polynomial correction was more similar to the target than prior to correction, but significant errors were noticed.

These errors were quantified by calculating absolute differences between phase-corrected slices and slices with desired phase. Ideally, these differences should be due to noise only. To decrease the influence of noise on this measure and highlight the errors, phase-correction was performed using polynomial fitting and the Phase Shifter on the set of 32 repeated b=1000 s/mm^2 dMR images (a single coil was selected), i.e. 31 volumes were phase corrected and one was selected as reference. Then, the real and imaginary components of all 25 slices and all 32 repetitions were averaged to decrease random noise variance and preserve deterministic content of the error maps (Fig. 7). Averaging these 1600 slice instances is expected to reduce the noise standard deviation by a factor of 40, from 1 to 0.025. The mean absolute error, calculated over the error maps of Fig. 7, was indeed higher for polynomial fitting than it was for the neural network (1.208 vs. 1.127).

3.2   Averaging Phase-Corrected dMR Images

Phase correction was also validated by averaging over the 32 repeated scans. Figure 8 shows the comparison for the noisy, non-averaged image, averaging without phase correction, averaging after polynomial fitting and after application of the Phase Shifter. For each case, the real and imaginary components are presented as well as the modulus. It can be seen that without phase correction, the intensity of the modulus image greatly deviates from the actual modulus (Fig. 8g and 8c). Both phase correction methods yielded much more similar modulus values. However, it is noticed (Fig. 8l) that polynomial fitting

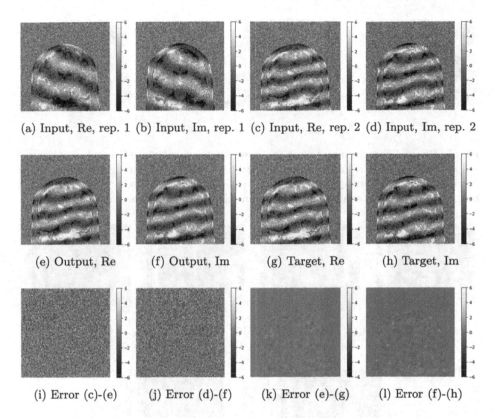

(a) Input, Re, rep. 1 (b) Input, Im, rep. 1 (c) Input, Re, rep. 2 (d) Input, Im, rep. 2

(e) Output, Re   (f) Output, Im   (g) Target, Re   (h) Target, Im

(i) Error (c)-(e)   (j) Error (d)-(f)   (k) Error (e)-(g)   (l) Error (f)-(h)

Fig. 4. Phase shifter training results - a validation example shown at input (a-d) and after Phase Shifter correction compared to targets (e-h). (a-b) is phase-corrected to agree in phase with (c-d). Correction error maps were obtained between the input with target phase and the phase-corrected image (i-j) and between the target and estimated phase-corrected images (k-l). Error maps (i-j) are indistinguishable from noise, which suggests good phase correction. In turn (k-l) show that errors are concentrated within the object and have low value compared to image noise. Re - real part, Im - imaginary part

may be locally inaccurate, leading to artifactual intensity variations. It is also visible for both correction methods that they reduced the noise floor compared to the average of slice moduli (Fig. 8d).

3.3   N2N Denoiser Training and Denoising of Brain dMR Images

The N2N Denoiser was trained on b=1000 s/mm^2 dMRI data of a single patient, processed in four different ways as described below. dMRI were 160×160x25 in size, acquired for three diffusion directions and using four receiver coils. Each time learning was performed on non-cropped slices, which yielded 300 examples

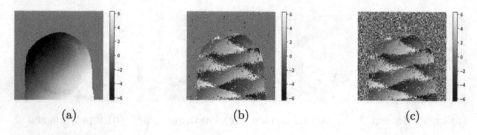

Fig. 5. Phase difference map target (a), neural network estimate (b) and direct calculation from noisy input images (c). The network seems to learn to estimate a denoised version of the input images, so that Eq. 5 yields less noisy phase estimates (b). Still, both phase estimates exhibit phase wrapping due to the use of arctangent

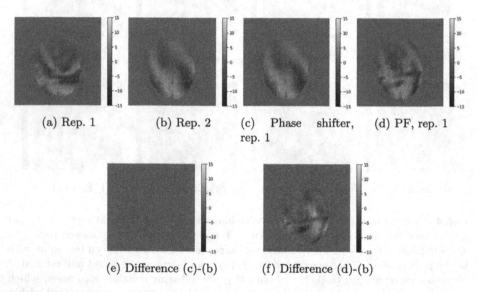

Fig. 6. Example slice from a b=1000 s/mm^2 dMR image, real component. Excellent phase correction is noticed for the Phase Shifter. (e) shows that the correction error is small compared to noise for the Phase Shifter. Errors in phase correction are noticed for polynomial fitting (PF) where it cannot fit the true phase distribution accurately (f).

in total, from which 60 was held out for validation. Hyperparameter settings were MSE loss, Adam optimizer, learning rate of 0.001, mini-batch training with batch size of 9 and shuffling, early stopping with patience of 25 epochs, validation loss value checking in each epoch, and minimum delta value of 1e−6.

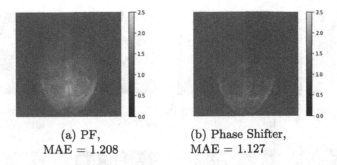

(a) PF,
MAE = 1.208

(b) Phase Shifter,
MAE = 1.127

Fig. 7. Mean absolute error maps obtained for the b = 1000 s/mm^2 dMRI volume by averaging 1600 images (32 repetitions × 25 slices × 2 complex components)

The N2N denoiser was trained using four kinds of training data: a) without phase correction, b) phase correction with polynomial fitting, c) phase correction with the Phase Shifter, d) no phase correction, moduli of the complex slices were taken as examples.

Additional validation sets were created. They did not have any effect on training and were used to study the N2N learning process with the four kinds of training data. One of them comprised 60 slices of real dMR images, with b = 0 s/mm^2. These dMR images were assumed to have identical phase variation across repetitions. Another four validation sets were synthetic BrainWeb images at SNR = 1, 3, 5 and 7. Validation MSE loss was computed two-fold on these: Noise to Noise loss for noisy targets and Noise to Truth loss (N2T) for clean targets.

Training the network on non-phase-corrected data failed due to substantial differences in the signal phase, which violated Noise2Noise conditions.

Figure 9 shows the MSE loss between the network output and the noisy phase-corrected image (N2N loss), for the two correction methods, evaluated on the b = 0 s/mm^2 dMRI. Although these images have higher SNR than b = 1000 s/mm^2 data used for training, the network gradually learns to denoise them. The figure shows that the learning process is simpler after phase correction using the Phase Shifter and that the loss is constantly lower than for polynomial fitting. This difference in the loss values may partly result from distortions introduced to the meaningful content of the image by the network trained on images phase-corrected with polynomial fitting. Similar results were obtained for the synthetic validation data at all tested SNR values, where clean images were used as reference (N2T loss, Fig. 10). The fourth network, trained on modulus images, was evaluated by studying the denoised images.

The four versions of the N2N Denoiser were compared based on visual examination of the denoised b = 1000 s/mm^2 images and in comparison to noisy ones. A single-coil and a multicoil image was considered (Fig. 11).

Comparison shows that both phase correction methods significantly improve the SNR of the image compared to the noisy one and to the results of the N2N denoiser trained on non-phase-corrected data. This is also noticed for the

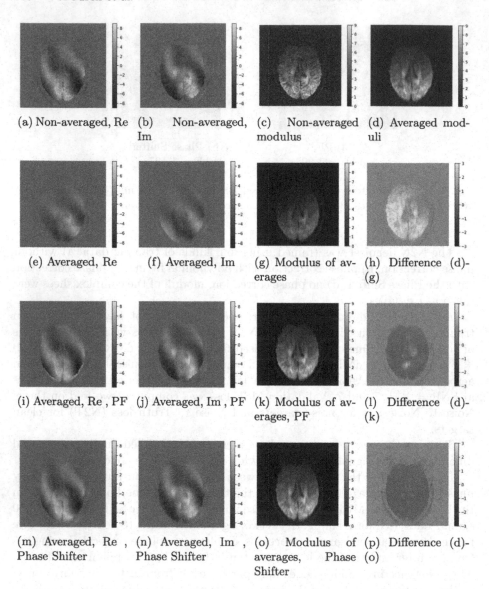

(a) Non-averaged, Re (b) Non-averaged, Im (c) Non-averaged modulus (d) Averaged moduli

(e) Averaged, Re (f) Averaged, Im (g) Modulus of averages (h) Difference (d)-(g)

(i) Averaged, Re , PF (j) Averaged, Im , PF (k) Modulus of averages, PF (l) Difference (d)-(k)

(m) Averaged, Re , Phase Shifter (n) Averaged, Im , Phase Shifter (o) Modulus of averages, Phase Shifter (p) Difference (d)-(o)

Fig. 8. Phase correction results (e-p) after averaging of 32 scan repetitions, compared to a single repetition with desired phase (a-d). Noise floor is reduced due to averaging in the complex domain for both correction methods, but polynomial fitting is not accurate, leading to modulus value variations. Re - real part, Im - imaginary part

network trained on modulus images but Rician bias becomes visible, especially for the multicoil image. Although the N2N Denoiser trained on data that was phase-corrected with polynomial fitting outputs well denoised images, errors in intensity are visible.

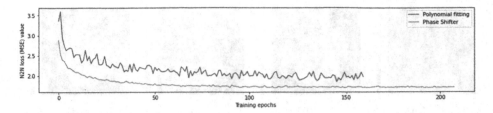

Fig. 9. Validation loss evaluated on the b = 0 s/mm^2 dMRI data while training the N2N denoiser on two kinds of data: corrected with polynomial fitting and the Phase Shifter network

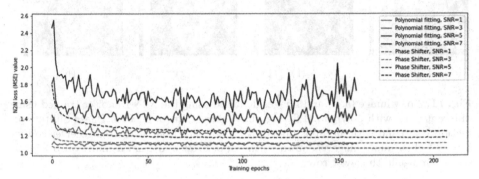

Fig. 10. Validation N2T loss evaluated on the BrainWeb data while training the N2N denoiser on two kinds of data: corrected with polynomial fitting and the Phase Shifter network

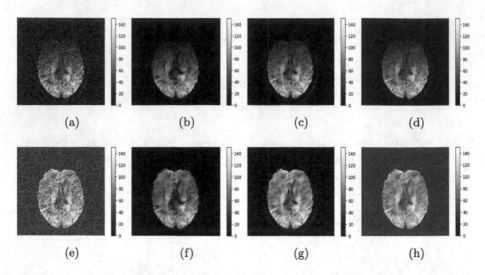

Fig. 11. Noisy image (a, e) and images denoised using the N2N Denoiser trained using different data: with polynomial fitting phase correction (b, f), with Phase Shifter correction (c, g) and with modulus images (d, h). A slice from a selected receiver coil for a b = 1000 s/mm^2 image is shown in the upper row and from a multicoil b = 1000 s/mm^2 image in the lower row

4   Discussion

Phase correction is a necessary step to enable complex-domain averaging of dMR images or, as it was shown, Noise to Noise training. Among the two compared methods, polynomial fitting is slower than neural-network-based correction. It requires brain segmentation, because the true background phase is zero and including the background would yield phase discontinuity that is hard to fit accurately using the cubic polynomial. More advanced fitting can be performed, for example using a higher degree polynomial, Legendre polynomials of Radial Basis Functions, which however is even more time-consuming and may involve additional hyperparameters.

The proposed Phase Shifter network was trained on synthetic data, in which phase maps were approximated using polynomial fitting. Despite this, the neural network was able to capture the hidden dependencies and performed better on real data than polynomial fitting, i.e. it was not limited to phase variations that were easily approximated using the polynomial. Figure 5 shows that the network learns the correct function, i.e. it predicts a denoised phase difference having only noisy complex image pairs as input and a phase-corrected noisy image as target.

Although it may seem more natural to pose the learning problem so that the clean phase maps are the target instead of the phase-corrected noisy images, we encountered significant obstacles using this approach. Namely, the resulting phase maps estimated by the neural network contained non-continuities even if the true phase was continuous, which most probably resulted from phase periodicity. In general, discontinuities of the true phase can also appear naturally due to non-continuous magnetic susceptibility of the tissue, which causes field inhomogeneity. The chosen approach was free of such inaccuracies for continuous true phase and was thus selected. In the future, however, we plan to study in more detail other Phase Shifter training options.

The Phase Shifter uses a two-step approach to estimate the phase-corrected image, in which the first estimate ($\tilde{C}$ in Fig. 3a) is used to compute the phase difference, and then this phase difference is used to compute the final estimate ($\hat{C}$ in Fig. 3a). This solution was applied to ensure that the modulus value of the phase-corrected image was identical to the modulus value of the image to be phase-corrected.

We did not notice phase-correction-related signal loss of the kinde described in [10], in the phase-corrected images obtained using polynomial fitting and the Phase Shifter.

One of the reasons for improving the current network design is other inaccuracies observed from the Phase Shifter phase correction, visible in the results on N2N denoising (Figs. 8, 11). Some background noise was observed to remain in the denoised images, to a greater extent than for polynomial fitting. It is not sure if these distortions are also present within the tissue regions. The probable source is inaccurate estimation of the phase in the background by the Phase Shifter. Although the Phase Shifter network should output zero for the background, it may do so with errors, which may disturb the background noise distribution and lead to noise mean value shifts from zero. In such circumstances, the neural network may learn denoising inaccurately. Alternatively, the network may introduce noise correlation between the N2N inputs and targets. This effect will be more thoroughly studied in our future works. A simple solution would be to use brain masks to exclude the background from phase-shifting, which was the case for polynomial fitting.

Although we tested the phase correction method on b=0 and 1000 s/mm^2 dMRI, it is expected to be applicable, in general, to any b-value or diffusion encoding direction. The limit is probably imposed by decreasing SNR associated with increasing the b-value. Identifying this limit, however, will be the subject of our future studies.

In the future, we also plan to quantify the reduction of bias in the averaged phase-corrected dMRI and the impact of phase correction on image-derived diffusion parameters.

To conclude, neural network phase correction in dMR images is a fast and accurate alternative for polynomial fitting. Training can be performed on simulated images. It enables Noise to Noise denoising in complex dMRI data as well as complex-domain averaging of repeated scans for noise reduction. Future studies are required to study the identified pitfalls and test the phase correction method in more real-life dMRI scenarios.

Acknowledgements. Kamil Gorczewski, Kamil Cepuch and Agata Zawadzka from Siemens Healthcare are warmly thanked for their assistance with MR image acquisition and comments on reconstruction.

References

1. Cocosco, C.A., Kollokian, V., Kwan, R.K.S., Pike, G.B., Evans, A.C.: BrainWeb: online interface to a 3D MRI simulated brain database. Neuroimage **5**, 425 (1997)
2. Collins, D., et al.: Design and construction of a realistic digital brain phantom. IEEE Trans. Med. Imaging **17**(3), 463–468 (1998). https://doi.org/10.1109/42.712135
3. Dong, C., Loy, C.C., He, K., Tang, X.: Image super-resolution using deep convolutional networks. IEEE Trans. Pattern Anal. Mach. Intell. **38**, 295–307 (2016). https://doi.org/10.1109/TPAMI.2015.2439281
4. Gudbjartsson, H., Patz, S.: The Rician distribution of noisy MRI data. Magn. Reson. Med. **34**, 910–914 (1995). https://doi.org/10.1002/mrm.1910340618
5. Jurek, J., Kociński, M., Materka, A., Elgalal, M., Majos, A.: CNN-based superresolution reconstruction of 3d MR images using thick-slice scans. Biocybern. Biomed. Eng. **40**(1), 111–125 (2020). https://doi.org/10.1016/j.bbe.2019.10.003
6. Jurek, J., et al.: Supervised denoising of diffusion-weighted magnetic resonance images using a convolutional neural network and transfer learning. Biocybern. Biomed. Eng. **43**(1), 206–232 (2023). https://doi.org/10.1016/j.bbe.2022.12.006
7. Kwan, R.K.S., Evans, A.C., Pike, G.B.: An extensible MRI simulator for postprocessing evaluation. In: Hohne, K.H., Kikinis, R. (eds.) Visualization in Biomedical Computing. VBC 1996. Lecture Notes in Computer Science, vol. 1131. Springer, Heidelberg (1996). https://doi.org/10.1007/Bfb0046947
8. Kwan, R.S., Evans, A., Pike, G.: MRI simulation-based evaluation of image-processing and classification methods. IEEE Trans. Med. Imaging **18**, 1085–1097 (1999). https://doi.org/10.1109/42.816072
9. Lehtinen, J., et al.: Noise2noise: learning image restoration without clean data. In: Dy, J.G., Krause, A. (eds.) Proceedings of the 35th International Conference on Machine Learning, ICML 2018, Stockholmsmässan, Stockholm, Sweden, 10–15 July 2018. Proceedings of Machine Learning Research, vol. 80, pp. 2971–2980. PMLR (2018). https://proceedings.mlr.press/v80/lehtinen18a.html
10. Liu, F., et al.: Does perfect filtering really guarantee perfect phase correction for diffusion MRI data? Computer. Med. Imag. Graph. **103**, 102160 (2023). https://doi.org/10.1016/j.compmedimag.2022.102160
11. Pizzolato, M., Gilbert, G., Thiran, J.P., Descoteaux, M., Deriche, R.: Adaptive phase correction of diffusion-weighted images. NeuroImage **206**, 116274 (2020). https://doi.org/10.1016/j.neuroimage.2019.116274
12. Tax, C.M., Bastiani, M., Veraart, J., Garyfallidis, E., Irfanoglu, M.O.: What's new and what's next in diffusion MRI preprocessing. NeuroImage **249**, 118830 (2022). https://doi.org/10.1016/j.neuroimage.2021.118830
13. Zhang, K., Zuo, W., Chen, Y., Meng, D., Zhang, L.: Beyond a gaussian denoiser: Residual learning of deep CNN for image denoising. IEEE Trans. Image Process. **26**(7), 3142–3155 (2017). https://doi.org/10.1109/TIP.2017.2662206

CNN-Based Quantification of Blood Vessels Lumen in 3D Images

Andrzej Materka[1]([envelope]) [iD], Jakub Jurek[1] [iD], Marek Kocinski[1,2,3] [iD],
and Artur Klepaczko[1] [iD]

[1] Institute of Electronics, Lodz University of Technology, 90-924 Lodz, Poland
materka@p.lodz.pl
[2] University of Bergen, Bergen, Hordaland, Norway
[3] Mohn Medical Imaging and Visualization Centre, Bergen, Hordaland, Norway
http://www.eletel.p.lodz.pl

Abstract. The aim of this work is to develop a method for automated, fast and accurate geometric modeling of blood vessels from 3D
images, robust to image limited resolution, noise and artefacts. Within
the centerline-radius paradigm, convolutional neural networks (CNNs)
are used to approximate the mapping from the image cross-sections to
vessel lumen parameters. A six-parameter image formation model is utilized to derive conditions for this mapping to exist, and to generate
images for the CNN training, validation and testing. The trained networks are applied to real-life time-of-flight (TOF) magnetic resonance
images (MRI) of a blood-flow phantom. Excellent agreement is observed
between the predictions made by the CNN and those obtained via model
fitting as the reference method. The latter is a few orders of magnitude
slower than the CNN and suffers from local minima problem. The CNN
is also trained and tested on publicly available contrast-enhanced (CE)
computed tomography angiography (CTA) clinical datasets. It accurately predicts the coronary-tree lumen parameters in seconds, compared
to hours needed by human experts. The method can be an aid to vascular
diagnosis and automated annotation of images.

Keywords: Medical image analysis · CNN · Parameter estimation ·
Blood-vessels lumen modeling · Centerline-radius paradigm ·
Uncertainty

1 Introduction

Cardiovascular disease produces immense health and economic burdens globally
[1]. The main types of this disease originate from blocked blood supply to organs
and tissues. The related clinical practices rely on angiography and venography
in multiple imaging modalities to acquire information about human vascularity. Reliable quantification and visualization of vascular structures is important for diagnosis assistance, treatment, surgery planning/execution, pathology

---

Supported by Lodz University of Technology, Institute of Electronics.

J. Mikyška et al. (Eds.): ICCS 2023, LNCS 14074, pp. 653–661, 2023.
https://doi.org/10.1007/978-3-031-36021-3_62

quantification, and evaluation of clinical outcomes in different fields, including laryngology, neurosurgery and ophthalmology [2,3]. Three-dimensional imaging is the main technique used to acquire quantitative information about the vasculature. Example modalities include MR angiography (MRA) which can be flow-dependent (TOF or phase contrast angiography) or flow-independent. An invasive alternative is computed tomography angiography (CTA). They can both be contrast-enhanced (CE) or nonenhanced. Various diseases require the usage of dedicated imaging techniques and image analysis algorithms. The most often considered abnormalities are stenosis, aneurysms and calcifications [2].

The human vasculature exhibits a complex tree-like structure of highly-curved, different-diameter branches. They are closely surrounded by other arteries, veins and tissues. The images feature limited spatial resolution and substantial noise. These factors make the tasks of vessel quantification in 3D images especially challenging for radiologists. Manual vessel labeling is tedious, time-consuming and error-prone [4]. There is a strong need to develop automated techniques for accurate, fast and objective vascularity evaluation [3] – to relieve radiologists of the burden of adding annotations to 3D images, crucial to effectively train data-driven solutions for healthcare. The main difficulties in achieving this task are in limited spatial resolution, intensity variations in regions surrounding the vessels, and imaging artefacts. These nonidealities increase uncertainty of lumen geometry measurements. The aim of this study is to develop a robust algorithm for fast, quantitative, subpixel-accuracy geometric characterization of blood vessels' lumen, given their 3D image.

There are two main approaches to vascular structures segmentation and quantification in 3D images [5,6]:

– direct volumetric lumen segmentation,
– 2D cross-sectional characterization along approximate centerline.

We focus on the second method and use a CNN to estimate of the lumen parameters from image cross-sections. The first approach is usually performed with the use of time-consuming calculations [7] or much faster convolutional neural networks [8]. Still, the segmentation produces images which are a coarse, voxelized representation of the vessel, with a need for postprocessing aimed at smooth approximation of its surface, e.g. for blood-flow simulation.

In our approach, the lumen cross-sections are computed as 2D images on planes perpendicular to the vessel centerline approximated by a smooth curve in 3D space [9] beforehand. This is less troublesome than volumetric segmentation of the lumen; various algorithms are available [10,11]. Normal vectors to the centerline define the cross-section planes. The 2D images on such planes are obtained through the 3D discrete image interpolation and resampling. Alternatively, a lumen cross-section model is fitted to the image data for centerline-based vessel quantification, e.g. with the use of the least-squares (LS) algorithm [9]. This involves long-lasting iterative minimization of a nonlinear error function and is likely to get stuck in its local minima. We will use it as a reference method.

In Sect. 2, we describe the proposed method of lumen cross-section quantification. Section 3 characterizes the 3D image datasets to which the method is applied to investigate its properties. The results are presented in Sect. 4, their discussion and conclusions form the content of Sect. 5.

2   Methods and Materials

To study the problem, it is first assumed that the blood-vessel cross-section forms a circle of radius varying along the centerline. The hypothetical noiseless analog image $F(x, y)$ of the lumen cross-section, at any point (ξ, η, ζ) in the 3D space is a convolution of the imaging system effective impulse response $h(x, y)$ with the function $f(x, y)$ representing the lumen and its background

$$F(x, y) = \int_{-\infty}^{\infty} \int_{-\infty}^{\infty} f(u, v) h(x - u, y - v) dv du \tag{1}$$

where (x, y) are image coordinates on the cross-section plane, $(x, y) = (0, 0)$ at (ξ, η, ζ). The function $f(x, y)$ is constant within the lumen region, surrounded by the background of different, also constant, intensity. The function $h(x, y)$, assumed isotropic Gaussian, combines the effects of the 3D image scanner point-spread function (PSF) and interpolation which precedes 3D image resampling. These assumptions are relevant to many practical situations [12]. Other image formation model options (non-circular shape, non-constant lumen/background intensities) will be accounted for later on in this paper.

The equidistant sampling points on a Cartesian grid are $(x_i^s, y_j^s) = (i\Delta_s, j\Delta_s)$, $i, j \in \{-N, \ldots, 0, \ldots, N\}$, Δ_s denotes the sampling interval, and the cross-section image size is $(2N + 1) \times (2N + 1)$. The image intensity at (i, j) is the sampled convolution (1) multiplied by lumen intensity step b and added to the background intensity a. The circle center is shifted by (d_x, d_y) from the centerline. There are $P = 6$ image model parameters $\boldsymbol{\theta} = (\theta_0, \theta_1, \theta_2, \theta_3, \theta_4, \theta_5) = (a, b, R, w, d_x, d_y)$

$$I(i, j; \boldsymbol{\theta}) = a + bF(i\Delta_s - d_x, j\Delta_s - d_y, R, w). \tag{2}$$

where R and w denote, respectively, the lumen radius and the Gaussian "sigma".

The proposed method uses the CNN as a nonlinear regressor. The input to such network is the vessel cross-section image. Values of the model parameters are predicted at the CNN outputs. The network can be trained to estimate a single parameter or a number of them in $[1, \ldots, P]$. A cascade of three convolutional layers and three dense layers was used in the experiments, Fig. 1, where $N = 7$. The numbers of channels $N1, \ldots, N6$ in the layers depended on the estimated parameter(s) and image noise level.

In the transfer-learning training phase, the assumed parameter values are substituted to (2) to calculate image model intensity on a predefined sampling grid. This results in a "noiseless" image. Gaussian noise ϵ is added to it, to take account of the uncertainty of measurement/reconstruction. Then, the neural network is trained to estimate the model parameters for images at its input. The CNN-predicted parameter values differ from those used to compute the input images. Sum-of-squares of these differences is the goal function minimized in the training process.

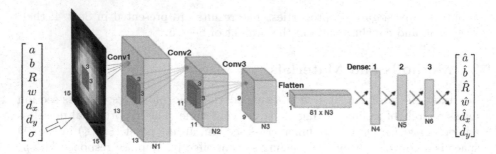

Fig. 1. CNN architecture applied to parameter estimation of blood-vessel lumen visualized in its 15×15-pixel 2D cross-section. The convolutional and dense layers feature ReLU activation functions, the output is linear.

In the recall phase, the neural network is driven by an image of unknown parameters and estimates their values. Ideally, the implicit input-output function $\hat{\theta} = g(\theta)$ of the pipeline in Fig. 1 should be an identity. Including bias and random noise components makes it more realistic:

$$\hat{\theta}_p \cong \theta_p + \beta_p + \nu_p, \qquad p \in \{0, \ldots, P-1\} \tag{3}$$

where θ is the vector of true values of the model parameters. The added terms of bias β and noise ν depend on θ, image noise ϵ, sampling interval Δ_s, image artefacts, model adequacy, and numerical errors. Due to the complexity of the model equations (1) and (2), explicit derivation of the inverse mapping seems to be an impossible task. This justifies the use of a CNN to learn this mapping from example images of known parameters. The CNN ability to reconstruct it is limited by the finite number of layers and count of trainable weights, being another source of the estimator uncertainty. Sensitivity analysis [13] and Monte Carlo simulations were applied to (2) to find parameter subspaces and image sampling patterns where the inverse mapping exists, and to evaluate estimator bias and variance. Importantly, this estimator has much better precision (quantified with the RMS error) than implied by Cramér-Rao lower bound of unbiased predictors. Discussion of these findings extends beyond the scope of this paper.

The performance of the proposed lumen quantification method is assessed through its application to images of real blood vessels and their physical models. The MR-QA123 flow phantom, driven by computer-controlled pump was used to force a steady flow (2.5 ml/s) of blood-mimicking liquid through the lumen of the phantom branches [14]. The U-shaped 8 mm inner diameter pipe of the phantom is used in this work. The phantom was placed in a GE Signa HDxt 1.5T system. Figure 2a shows a maximum-intensity-projection (MIP) of the pipe TOF image. Its voxel dimensions are $(0.82 \times 0.82 \times 1.01)$ mm^3. Image intensity is normalized to $[0, 1]$. To assess the CNN performance on clinical images, 17 annotated datasets available as part of Rotterdam Coronary Artery Algorithms Evaluation Framework [4] are used. The volumes intensity is clipped to $[-300, 800]$ HU and normalized to $[0, 1]$.

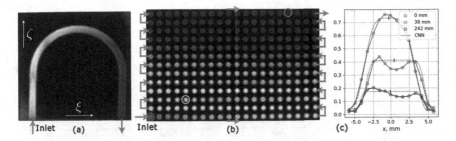

Fig. 2. Scanner-acquired TOF MRI for U-shaped pipe in flow phantom; a) MIP on the $0\xi\zeta$ plane, b) mosaic of lumen cross-sections (not to scale, arrows indicate fluid flow direction), c) 1D intensity profiles at $y = 0$ of sections taken at $d_c = 0, 38$, and 242 mm from inlet (marked respectively by red, green and blue circles in (b) and Fig. 3 upper row), black dotted lines: profiles reconstructed by model (2) supplied with CNN-predicted parameters. (Color figure online)

3   Results

Prior to CNN training, the noise of the phantom images was found to be Gaussian with $PSNR \cong 30$dB. A pseudo-random noise of the corresponding standard deviation was added to each synthesized 2D cross-section in the training, validation, and testing sets. The 15×15-pixel images were computed using (2). For $\Delta_s = 0.82$ mm, the model parameters spanned randomly the ranges: $0 \leq a \leq 0.3$; $0.1 \leq b \leq 1.1$; $1.0 \leq R/\Delta_s \leq 6.0$; $0.3 \leq w/\Delta_s \leq 1.5$; $-1.2 \leq d_x/\Delta_s \leq 1.2$; $-1.2 \leq d_x/\Delta_s \leq 1.2$, with uniform probability distribution. The image counts in the three sets were 60 000, 20 000 and 20 000, respectively.

Five neural networks were trained: four single-output, each for a, b, R, w individual parameters, and one with two outputs for (d_x, d_y). The CNNs were implemented in Keras environment on a desktop computer with 16 GB RAM, Intel^{TM} i5-8300H CPU @2300 MHz under MS Windows 11 OS. Computations were accelerated by NVIDIA GeForce GTX 1050 card with 4 GB GPU memory. Typical learning process (Adam weight optimization up to the time of validation error increase) took less than one hour, for (32,32,32,32,16,8) channels in the CNN layers. Computing the network output required 25 μs on average. The LS method was implemented with *least_squares()* function of the Scipy library, taking a minute to fit the model to a section of the pipe image.

Excellent agreement between the CNN and LS predictions for all six parameters is achieved, Fig. 3. The effect of stripe artefacts is clearly visible for $d_c \in [0, 100]$ mm, Fig. 3 left column (periodically varying lumen intensity b along the centerline). The two estimators differ more in the second half of the pipe, where the TOF SNR is low (due to flow direction reversal w.r.t. the 0ζ axis, Fig. 2a). This concerns especially $\hat{d}_x$ and $\hat{d}_y$ parameters; however, the disparity stays in the subvoxel range.

For each selected point of a coronary artery segment in CE-CTA images, the annotations include three lumen contours marked by respective observers on a

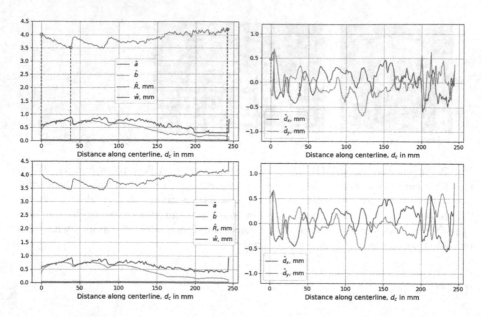

Fig. 3. Plots of estimated lumen parameters for MR TOF image of U-shaped pipe in the flow phantom, flow rate 2.5 ml/s. Left: $\hat{a}, \hat{b}, \hat{R}, \hat{w}$; right: $\hat{d}_x, \hat{d}_y$. Top: CNN, red, green and blue marks correspond to sections encircled in Fig. 2b; bottom: LS model fitting. CNN trained on synthetic images at $PSNR = 30\,\mathrm{dB}$. (Color figure online)

common plane, orthogonal to an agreed centerline, Fig. 4. The contours allow computation of the section equivalent radius $\rho_k^{(o)} = \sqrt{A_k^{(o)}/\pi}$, where $A_k^{(o)}$ denote the area inside the contour, $o = 1, 2, 3$ and k is the section index. Their centers of mass provide coordinates of centerline points according to observers, Fig. 4 red dots. The image intensity is not constant over the background region, Fig. 4. This causes convergence difficulties of LS model fitting, leading to errors and excessive time of computation. The CNN is trained on real, nonideal images to make it insensitive to spurious objects in the background.

Example results of training and testing a CNN for equivalent radius estimation are shown in Fig. 5 for contours marked by Observer #1 on datasets #0, #1, #3, #4, #5, #6 (558 sections). Testing was done on 51 sections of three segments excluded from the training set. The mean absolute error (MAE) over the training and test sets was 0.13 mm and 0.11 mm, respectively. Similarly, MAE was less than 0.09 mm over the training set for each of d_x and d_y parameters, and it did not exceed 0.1 mm for the test set. Thus, subpixel accuracy is achieved for the CE-CTA as well ($\Delta_s = 0.45$ mm in this case).

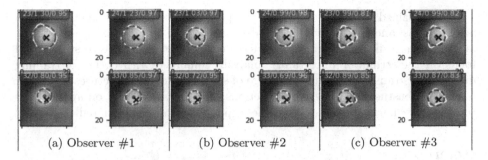

| (a) Observer #1 | (b) Observer #2 | (c) Observer #3 |

Fig. 4. Example lumen contours marked by three observers on sections $k = 23, 24, 32, 33$ of artery segment #8 in CE-CTA dataset #5. Blue cross: agreed reference centerline, white line: manually marked contour, red dot: computed contour centroid, red dashed line: equivalent circle. (Color figure online)

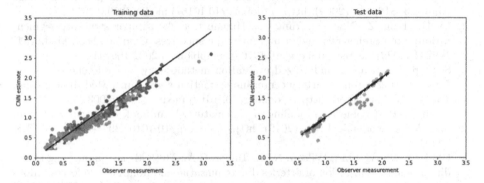

Fig. 5. Scatter plots of training (left) and testing (right) a CNN for estimation of coronary arteries equivalent radius ρ. Horizontal axis: ρ by observer, vertical axis: ρ by the CNN, in mm. Marker color indicates artery segment the input cross-section was sampled from. (Color figure online)

4   Summary and Conclusion

The CNN-based parameter predictions are in excellent agreement with the well-established LS method, for images whose appearance is close to the lumen model of clear object and background (like TOF MR of the flow phantom, Fig. 2bc). Although the time of CNN training can be substantial, the trained CNN is much faster than LS fitting implementation.

Accuracy of the LS algorithm is poor for the CTA dataset, even with the use of cumbersome constrained optimization. Since the CNN estimator can be made robust to deviations from the model assumptions, it shows high, subvoxel accuracy in the case of CTA images as well. In a number of cases, the contours delineated by the observers are apparently placed off the regions of high image intensity. Still, the network can extract such areas in its radius/centerline shift predictions, for further intensity examination, e.g. in search for calcifications.

The obtained results demonstrate the potential usefulness of the CNN as an accurate, fast and robust tool for blood-vessels' lumen quantification – a possible aid to medical diagnosis and automated image annotation. Future studies will focus on the estimator uncertainty analysis, combined with design of training datasets for transfer learning and usage of self-attention architectures to further improve robustness to image imperfections. Collaborative work on applications to other vascularity images, e.g. of the brain, has also been initiated.

References

1. Tsao, C.W., et al.: Heart disease and stroke statistics - 2022 update. Circulation **145**, e153–e639 (2022). https://doi.org/10.1161/CIR.0000000000001052
2. Lesage, D., Angelini, D.E., Bloch, I., Funka-Lea, G.: A review of 3D vessel lumen segmentation techniques: models, features and extraction schemes. Med. Image Anal. **13**, 819–845 (2009). https://doi.org/10.1016/j.media.2009.07.011
3. Li, H., Tang, Z., Nan, Y., Yang, G.: Human tree-like tubular structure segmentation: a comprehensive review and future perspectives. Comput. Biol. Med. **151**, 106241 (2022). https://doi.org/10.1016/j.compbiomed.2022.106241
4. Schaap, M., et al.: Standardized evaluation methodology and reference database for evaluating coronary artery centerline extraction algorithms. Med. Image Anal. **13**(5), 701–714 (2009). https://doi.org/10.1016/j.media.2009.06.003
5. Cheng, C.P.: Geometric Modeling of Vasculature. Handbook of Vascular Motion, pp. 45–66. Academic Press (2019). https://doi.org/10.1016/B978-0-12-815713-8.00004-8
6. Choi, G., Cheng, C.P., Wilson, N.M., Taylor, C.A.: Methods for quantifying three-dimensional deformation of arteries due to pulsatile and nonpulsatile forces: implications for the designs of stenst and stent grafts. Ann. Biomed. Eng. **37**(1), 14–33 (2009). https://doi.org/10.1007/s10439-008-9590-0
7. Sethian, J.A.: Level Set Methods and Fast Marching Methods. Cambridge University Press (2002)
8. Liu, Y., Kwak, H.S., Oh, I.S.: Cerebrovascular segmentation model based on spatial attention-guided 3D inception U-net with multidirectional MIPs. Appl. Sci. **12**, 2288 (2022). https://doi.org/10.3390/app12052288
9. Materka, A., et al.: Automated modeling of tubular blood vessels in 3D MR angiography images. IEEE 9th International Symposium on Image and Signal Processing and Analysis, ISPA 2015, pp. 56–61. IEEE, Zagreb (2015). https://doi.org/10.1109/ISPA.2015.7306032
10. Frangi, A.F., Niessen, W.J., Vincken, K.L., Viergever, M.A.: Multiscale vessel enhancement filtering. In: Wells, W.M., Colchester, A., Delp, S. (eds.) MICCAI 1998. LNCS, vol. 1496, pp. 130–137. Springer, Heidelberg (1998). https://doi.org/10.1007/BFb0056195
11. Wolterink, J.M., van Hamersvelt, R.W., Viergever, M.A., Leiner, T., Išgum, I.: Coronary artery centerline extraction in cardiac CT angiography using a CNN-based orientation classifier. Med. Image Anal. **51**, 46–60 (2019). https://doi.org/10.1016/j.media.2018.10.005
12. Chen, Z., Ning, R.: Three-dimensional point spread function measurement of cone-beam computed tomography system by iterative edge-blurring algorithm. Phys. Med. Biol. **49**, 1865–1880 (2004). https://doi.org/10.1088/0031-9155/49/10/003

13. Materka, A., Mizushina, S.: Parametric signal restoration using artificial neural networks. IEEE Trans. Biomed. Eng. **43**(4), 357–372 (1996). https://doi.org/10.1109/10.486256
14. Klepaczko, A., Szczypinski, P., Dwojakowski, G., Strzelecki, M., Materka, A.: Computer simulation of magnetic resonance angiography imaging: model description and validation. PLoS ONE **9**(4), e93689 (2014). https://doi.org/10.1371/journal.pone.0093689

Tensor Train Subspace Analysis for Classification of Hand Gestures with Surface EMG Signals

Rafał Zdunek[✉][iD]

Faculty of Electronics, Photonics and Microsystems,
Wroclaw University of Science and Technology,
Wybrzeze Wyspianskiego 27, 50-370 Wroclaw, Poland
rafal.zdunek@pwr.edu.pl

Abstract. Processing and classification of surface EMG signals is a challenging computational problem that has received increasing attention for at least two decades. When multichannel EMG signals are transformed into spectrograms, classification can be performed using multilinear features that can be extracted from a set of spectrograms by various tensor decomposition methods. In this study, we propose to use one of the most efficient tensor network models, i.e. the tensor train decomposition method and to combine it with the tensor subspace analysis to extract the more discriminant 2D features from multi-way data. Numerical experiments, carried out on surface EMG signals registered during hand gesture actions, demonstrated that the proposed feature extraction method outperforms well-known tensor decomposition methods in terms of classification accuracy.

Keywords: Tensor train decomposition · Tensor subspace analysis · sEMG signal classification · hand gesture recognition

1 Introduction

Hand gestures are a form of non-verbal communication that is typically used by deaf-mute people but can also be used for human-computer interaction (HCI) through multiple mediums, such as vision, audition, tactition, etc. Surface electromyography (sEMG) is a technique for measuring the electrical activity produced by the skeletal muscles. The sEMG signals are associated with the movements of the healthy hand and fingers, and they can clearly represent the will of the user to control various devices, including computer desktops, hand prosthesis, exoskeletons, etc. Such signals can also contain meaningful information on static and dynamic hand gestures. Regarding this possibility, many sEMG-based HCI systems have been designed; see, e.g. [4,14,15].

Classification of sEMG signals is a challenging computational problem, mainly due to many sources of disturbance, such as motion artifacts, electrocardiographic artifacts, crosstalk, ambient noise, and inherent measurement device noise, which results in highly noisy observations, where the desired signal may be deeply hidden in the observations. Hence, raw sEMG must be adequately processed to extract the most relevant features for classification.

J. Mikyška et al. (Eds.): ICCS 2023, LNCS 14074, pp. 662–669, 2023.
https://doi.org/10.1007/978-3-031-36021-3_63

In this study, we propose a new computational method to extract relevant features from sEMG signals that contain information on hand gestures. We used sEMG signals coming from the forearm and acquired in the dynamic gaming environment presented in [13]. Sliding a window in the time domain, and applying the short-time Fourier transform (STFT) to each signal in the slide, we get a set of spectrograms that can be collected in the form of a multiway array, referred to as an input tensor. The modes of the tensor represent the frequency, time, channel, and ordering of samples. Due to this representation, multi-modal features can be next extracted using various tensor decomposition methods. This concept was developed in [17], where several well-known matrix and tensor decomposition methods were compared for sEMG data collected from subjects performing various grasping movements. This study is based on the similar methodology, but we present here a new approach to multilinear feature extraction that is based on the tensor train (TT) decomposition method, additionally enhanced by the information on the geometry of observed samples. Moreover, the experiments are performed on different sEMG signals.

The remainder of this paper is organized as follows. In Sect. 2, we briefly discuss the related works. Section 3 presents the proposed method, preceded by the fundamentals of the TT model. The experimental results are described in Sect. 4. Finally, the conclusions are drawn in Sect. 5.

2   Related Works

There are three basic approaches to extract features from sEMG signals that are outlined in the following domains: time, frequency, or time-frequency. Time-domain features, such as mean absolute value (MAV), slope sign changes (SSC), waveform lengths (WL), and zero crossings (ZC) [8], are useful for real-time processing and offer increased classification accuracy with respect to raw data. However, the analysis of strongly perturbed and nonstationary weak signals in the time domain is certainly not so efficient as in the time-frequency domain. There are many computational approaches for extracting time-varying frequency features from sEMG signals, in particular the continuous wavelet transform (CWT) that leads to scalograms and STFT that leads to spectrograms. In this study, we used SIFT because it is more suitable for processing nonstationary and highly noisy signals. Reviews of feature extraction methods for the classification of sEMG can be found in [1,9,16].

The TT model, known as the matrix product state (MPS) in quantum physics [12] and quantum chemistry [10], has gained a great popularity in the area of computational sciences, machine learning, and signal processing [2,3]. However, to our best knowledge, it has never been used for feature extraction from sEMG signals. The TT undoubtedly combines multiple advantages that are important for our purpose. Therefore, the motivations behind the use of this model in our study can be stated as follows. First, it provides 2D features that capture multilinear information from all modes, which is more useful for classification than single 1D representations. Second, it is a multi-rank procedure that is more efficient than the standard CP decomposition but its computational and storage complexities are lower than for the standard Tucker decomposition. Third, its multimodal structure fits very well for augmented data. The last and probably the most important for this study, the TT model can be easily combined with the tensor subspace analysis (TSA) method [6], which is very profitable for classification tasks.

3   Methods

Notation: Multi-way arrays, matrices, vectors, and scalars are denoted by calligraphic uppercase letters (e.g., $\mathcal{Y}$), boldface uppercase letters (e.g., $\boldsymbol{Y}$), lowercase boldface letters (e.g., $\boldsymbol{y}$) and unbolded letters (e.g., y), respectively. Multi-way arrays will also be equivalently referred to as tensors [11].

3.1   Tensor Train Model

Let $\mathcal{Y} \in \mathbb{R}^{I_1 \times I_2 \times \cdots \times I_N}$ be the N-way array that is referred to as the N-th order tensor. The TT decomposition of $\mathcal{Y}$ can be expressed by the following model:

$$\mathcal{Y} = \sum_{r_1=1}^{R_1} \sum_{r_2=1}^{R_2} \cdots \sum_{r_{N-1}=1}^{R_{N-1}} \mathcal{Y}_1(1,:,r_1) \circ \mathcal{Y}_2(r_1,:,r_2) \circ \ldots \circ \mathcal{Y}_N(r_{N-1},:,1) \quad (1)$$

where $\forall n : \mathcal{Y}_n \in \mathbb{R}^{R_{n-1} \times I_n \times R_n}$ is the n-th core tensor, and symbol $\circ$ stands for the outer product. Set $\{R_0, R_1, \ldots, R_{N-1}, R_N\}$ determine the TT-ranks, where $R_0 = R_N = 1$. Cores $\mathcal{Y}_1 \in \mathbb{R}^{I_1 \times R_1}$ and $\mathcal{Y}_N \in \mathbb{R}^{R_{N-1} \times I_N}$ are the boundary matrices. Core tensors capture mode-related features, where $\mathcal{Y}_n$ contains the features related to the n-th mode of $\mathcal{Y}$.

Assuming $I_1 = \ldots = I_N = I$ and $R_1 = \ldots = R_N = R$, the storage complexities of the CANDECOM/PARAFAC (CP), Tucker, and TT decomposition models can be approximated by $\mathcal{O}(NIR)$, $\mathcal{O}(NIR + R^N)$, and $\mathcal{O}(NIR^2)$.

3.2   Proposed Method

Let $\mathcal{Y} = [y_{i_1,\ldots,i_N}] \in \mathbb{R}^{I_1 \times \cdots \times I_N}$ be the N-th order tensor of observed samples ordered along the n-th mode, where $1 < n < N$. If the samples are ordered along the first or last mode, then $\mathcal{Y}$ should be permuted accordingly. We assume that $n = [N/2]$, where $[\cdot]$ rounds to the nearest integer number. Furthermore, due to the computational complexity of the TT decomposition, the first and last modes should contain the largest numbers of entries with respect to the other modes; expect the mode of ordering the samples.

Note that the TT model in (1) can be equivalently expressed in the element-wise notation:

$$y_{i_1,\ldots,i_N} = \boldsymbol{y}_1(i_1)^T \boldsymbol{Y}_2(i_2) \cdots \boldsymbol{Y}_{N-1}(i_{N-1}) \boldsymbol{y}_N(i_N), \quad (2)$$

where $\boldsymbol{y}_1(i_1) = \mathcal{Y}_1(1, i_1, :) \in \mathbb{R}^{R_1}$ and $\boldsymbol{y}_N(i_N) = \mathcal{Y}_N(:, i_N, 1) \in \mathbb{R}^{R_N}$ are the i_1-th row of $\mathcal{Y}_1$ and i_N-th column of $\mathcal{Y}_N$, respectively, and $\boldsymbol{Y}_n(i_n) \in \mathbb{R}^{R_{n-1} \times R_n}$ are lateral slices of core tensor $\mathcal{Y}_n$ for $i_n = 1, \ldots, I_n$ and $n = 2, \ldots, N - 1$.

A tensor train subspace is determined by cores $\{\mathcal{Y}_1, \ldots, \mathcal{Y}_N\}$ and is invariant with respect to the orthogonal transformations between modes. However, we slightly alleviate this condition by taking the following approximations:

$$
\begin{aligned}
y_{i_1, \ldots, i_N} &= \boldsymbol{y}_{(\leq n-1)}^T \boldsymbol{Y}_n(i_n) \boldsymbol{y}_{(\geq n+1)} = \boldsymbol{y}_{(\leq n-1)}^T \boldsymbol{U} \boldsymbol{U}^T \boldsymbol{Y}_n(i_n) \boldsymbol{V} \boldsymbol{V}^T \boldsymbol{y}_{(\geq n+1)} \\
&\cong \boldsymbol{y}_{(\leq n-1)}^T \tilde{\boldsymbol{U}} \left(\tilde{\boldsymbol{U}}^T \boldsymbol{Y}_n(i_n) \tilde{\boldsymbol{V}} \right) \tilde{\boldsymbol{V}}^T \boldsymbol{y}_{(\geq n+1)} \\
&= \tilde{\boldsymbol{y}}_{(\leq n-1)}^T \tilde{\boldsymbol{Y}}_n(i_n) \tilde{\boldsymbol{y}}_{(\geq n+1)},
\end{aligned}
\tag{3}
$$

where $\boldsymbol{y}_{(\leq n-1)} = \boldsymbol{y}_1(i_1)^T \boldsymbol{Y}_2(i_2) \cdots \boldsymbol{Y}_{n-1}(i_{n-1}) \in \mathbb{R}^{R_{n-1}}$ aggregates the features from the first to the $(n-1)$-th mode, and the features from the $(n+1)$-th mode to the last are represented by $\boldsymbol{y}_{(\geq n+1)} = \boldsymbol{Y}_{n+1}(i_{n+1}) \cdots \boldsymbol{Y}_{N-1}(i_{N-1}) \boldsymbol{y}_N(i_N) \in \mathbb{R}^{R_n}$. Matrices $\boldsymbol{U} \in \mathbb{R}^{R_{n-1} \times R_{n-1}}$ and $\boldsymbol{V} \in \mathbb{R}^{R_n \times R_n}$ are orthogonal matrices, i.e. $\boldsymbol{U} \boldsymbol{U}^T = \boldsymbol{I}_{R_{n-1}}$ and $\boldsymbol{V} \boldsymbol{V}^T = \boldsymbol{I}_{R_n}$. Projection matrices $\tilde{\boldsymbol{U}} \in \mathbb{R}^{R_{n-1} \times J_1}$ and $\tilde{\boldsymbol{V}} \in \mathbb{R}^{R_n \times J_2}$ are created from $\boldsymbol{U}$ and $\boldsymbol{V}$ by selecting the columns that correspond to the most significant eigenvalues, where $J_1 < R_{n-1}$ and $J_2 < R_n$. Thus, $\tilde{\boldsymbol{Y}}_n(i_n) \in \mathbb{R}^{J_1 \times J_2}$ contains the features of $\boldsymbol{Y}_n(i_n)$ projected onto the bases spanned by $\tilde{\boldsymbol{U}}$ and $\tilde{\boldsymbol{V}}$.

To calculate the projection matrices, the concept of the TSA method [6] can be applied, which leads to the following optimization problem:

$$
\min_{\tilde{\boldsymbol{U}}, \tilde{\boldsymbol{V}}} \frac{1}{2} \sum_{i_n^{(1)}=1}^{I_n} \sum_{i_n^{(2)}=1}^{I_n} s_{i_n^{(1)}, i_n^{(2)}} \| \tilde{\boldsymbol{Y}}_n(i_n^{(1)}) - \tilde{\boldsymbol{Y}}_n(i_n^{(2)}) \|_F^2,
\tag{4}
$$

where $\boldsymbol{S} = [s_{i_n^{(1)}, i_n^{(2)}}] \in \mathbb{R}^{I_n \times I_n}$ is the weight matrix that can be expressed by the heat kernel:

$$
s_{i_n^{(1)}, i_n^{(2)}} = \begin{cases} \exp \left\{ -\dfrac{\| \mathcal{Y}(i_n^{(1)}) - \mathcal{Y}(i_n^{(2)}) \|_F^2}{\tau} \right\}, & \text{if } \mathcal{Y}(i_n^{(1)}) \text{ and } \mathcal{Y}(i_n^{(2)}) \\ & \text{belong to the same class} \\ 0 & \text{otherwise} \end{cases}.
\tag{5}
$$

for $\tau > 0$. Samples $\left\{ \mathcal{Y}(i_n^{(1)}), \mathcal{Y}(i_n^{(2)}) \right\} \in \mathbb{R}^{I_1 \times \ldots \times I_{n-1} \times I_{n+1} \times \ldots \times I_N}$ form $\mathcal{Y} = \text{cat}\{\mathcal{Y}(i_n = 1), \ldots, \mathcal{Y}(i_n = I_n), 1\}$ for $i_n^{(1)}, i_n^{(2)} \in \{1, \ldots, I_n\}$.

Following straightforward computations, the objective function in (4) takes the form:

$$
\Psi = \text{tr}\{\tilde{\boldsymbol{V}}^T \boldsymbol{D}_V \tilde{\boldsymbol{V}}\} - \text{tr}\{\tilde{\boldsymbol{V}}^T \boldsymbol{S}_V \tilde{\boldsymbol{V}}\},
\tag{6}
$$

where

$$
\boldsymbol{D}_V = \sum_{i_n^{(1)}=1}^{I_n} d_{i_n^{(1)}} \boldsymbol{Y}_n(i_n^{(1)})^T \tilde{\boldsymbol{U}} \tilde{\boldsymbol{U}}^T \boldsymbol{Y}_n(i_n^{(1)}) \in \mathbb{R}^{R_n \times R_n}
\tag{7}
$$

with $d_{i_n^{(1)}} = \sum_{i_n^{(2)}=1}^{I_n} s_{i_n^{(1)},i_n^{(2)}}$,

$$S_V = \sum_{i_n^{(1)}=1}^{I_n} \sum_{i_n^{(2)}=1}^{I_n} s_{i_n^{(1)},i_n^{(2)}} Y_n(i_n^{(1)})^T \tilde{U}\tilde{U}^T Y_n(i_n^{(2)}) \in \mathbb{R}^{R_n \times R_n}. \tag{8}$$

According to the assumptions of the spectral graph theory and locality preserving projections (LPP) [7], problem (4) with respect to $\tilde{V}$ can be reformulated in terms of the generalized Rayleigh quotient:

$$\max_{\tilde{V}} \frac{\mathrm{tr}\{\tilde{V}^T (S_V - D_V)\tilde{V}\}}{\mathrm{tr}\{\tilde{V}^T D_V \tilde{V}\}}. \tag{9}$$

The solution to (9) can be approximated by computing generalized eigenvectors that are associated with the J_2 most significant eigenvalues of the pair of matrices $\{(D_V - S_V), D_V\}$.

Similarly, the solution to (4) with respect to $\tilde{U}$ can be found following a similar computational strategy. The computations of both factors require the use of an alternating optimization procedure.

To obtain core tensors $\{\mathcal{Y}_1, \ldots, \mathcal{Y}_N\}$ from the training samples in $\mathcal{Y}$, we used the both-side recursive update algorithm (Algorithm 1) from [5]. A similar approach was used to compute the projection of the testing samples onto the tensor space spanned by the training factors (Algorithm 2 in [5]). The proposed procedure for extracting 2D features is summarized in Algorithm 1.

Algorithm 1. TSA-TT feature extraction

Input : $\mathcal{Y} \in \mathbb{R}^{I_1 \times \cdots \times I_N}$ – N-way array of multilinear training samples collected along the n-th mode $(1 < n < N)$, $\mathcal{Y}^{(test)} \in \mathbb{R}^{I_1 \times \cdots \times I_{n-1} \times I_K \times I_{n+1} \times \cdots \times I_N}$ – N-way array of I_K multilinear testing samples, $J = [J_1, \ldots, J_{N-1}]$ – TT-ranks, (J_1, J_2) – TSA ranks

Output: $\mathcal{Z} \in \mathbb{R}^{I_n \times J_1 \times J_2}$ – estimated 2D training features,
$\mathcal{Z}^{(test)} \in \mathbb{R}^{I_K \times J_1 \times J_2}$ – estimated 2D testing features

Compute weight matrix S according to (5),
Compute $[\mathcal{Y}_1, \ldots, \mathcal{Y}_N] = \mathtt{TT-SVD}(\mathcal{Y}, J)$;   // Algorithm 1 from [5]
Compute $\tilde{U}$ and $\tilde{V}$ by solving problem (4),
Run projection: $\mathcal{Z}(i_n, :, :) = \tilde{U}^T \mathcal{Y}_n(i_n)\tilde{V} \in \mathbb{R}^{J_1 \times J_2}$ for $i_n = 1, \ldots, I_n$,
Compute $[\mathcal{Y}_n^{(test)}] = \mathtt{Tensorproj}(\mathcal{Y}^{(test)}, \mathcal{Y}_1, \ldots, \mathcal{Y}_N)$;   // Algorithm 2 from [5]
Run projection: $\mathcal{Z}^{(test)}(i_k, :, :) = \tilde{U}^T \mathcal{Y}_n(i_k)^{(test)}\tilde{V} \in \mathbb{R}^{J_1 \times J_2}$ for $i_k = 1, \ldots, I_K$,

---

Remark 1. Assuming $J = J_1 = J_2$, the computational complexity of the TSA methods can be roughly estimated as $\mathcal{O}\left(4KIR_{n-1}R_n I_n^2\right)$, where K is the number of alternating iterations. The TT core tensors in the $\mathtt{TT-SVD}$ algorithm can be computed with $\mathcal{O}(\sum_{n=1}^N \prod_{r=1}^n R_r \prod_{p=n}^N I_p)$. Therefore, if the number of training samples is large, TSA may predominate in the overall computational complexity. However, I_n is rather a small number in the application discussed.

4   Results and Discussion

Experiments were conducted on publicly available sEMG signals [13] from the UCI Machine Learning Repository[1]. The 8 channel sEMG signals were collected from 36 healthy subjects, aged 18 to 47 years, during PC gaming activity. The MYO Thalmic bracelet was equipped with 8 sensors equally spaced around the forearm. The subjects performed the following hand gestures: 1 – hand clenched in a fist, 2 – wrist flexion, 3 – wrist extension, 4 – radial deviations, 5 – ulnar deviations, 6 – extended palm. Two series of experiments are carried out, and each gesture lasts for 3 s.

The temporal signals from each channel were converted into spectrograms using the Matlab `spectrogram` function with the parameters: window length – 64 samples, overlapping – 54 samples, frequency sampling points – 256. The spectrograms were then partitioned into non-overlapping blocks of 64 samples with respect to the time axis. The blocks in which the signal energy was below a given threshold were neglected. Other blocks were processed with local averaging 2D filtration in the log scale, and then ordered in the form of 4-way array $\mathcal{Y} \in \mathbb{R}^{129 \times 64 \times 8 \times 36}$, where the successive modes correspond to the frequency bins, the temporal resolution, the channels, and the number of labeled samples. The data were divided into the training and testing sets according to the 4-fold cross-validation (CV) rule.

The following algorithms were compared: HO-SVD (Higher-order singular value decomposition) [17], HALS-NTF (Hierarchical alternating least squares nonnegative tensor factorization) [3], SVD-TT (Tensor train based on singular value decomposition) [5], TT-SVD-TSA (proposed method). Negative input entries for HALS-NTF were replaced with the zero value. Due to the nonconvexity issue of the HALS-NTF, and the randomness of the CV splitting, the Monte Carlo (MC) analysis was carried out with 20 runs for each algorithm.

The Hinton diagrams of the confusion matrices obtained with the algorithms tested in the first series with the first subject are illustrated in Fig. 1. The average accuracy is given in parentheses in the caption of the figure. The elapsed time of the algorithms tested for both series with the first subject is depicted in Fig. 2(a). The range of the standard deviation is marked with the whiskers. The

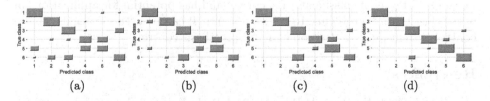

Fig. 1. Hinton diagrams of the confusion matrices obtained with the algorithms: (a) HO-SVD ($Acc = 66.11\%$), (b) HALS-NTF ($Acc = 75\%$), (c) SVD-TT ($Acc = 80.97\%$), (d) TSA-TT ($Acc = 92.5\%$). The accuracy is given in parentheses.

---

[1] https://archive.ics.uci.edu/ml/datasets/EMG+data+for+gestures.

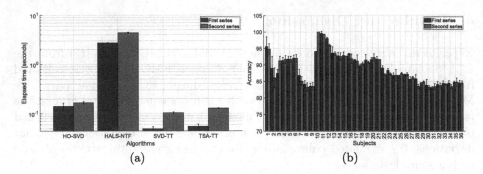

Fig. 2. (a) Elapsed time (in seconds) for all the methods in both series with the first subject, (b) accuracy of TSA-TT for all subjects and both series.

average accuracy obtained with TSA-TT in both series and with all the subjects is presented in Fig. 2(b).

5   Conclusions

The results presented in Fig. 1 demonstrate that TT-based methods lead to higher classification accuracy of sEMG signals than standard tensor factorization models, such as HO-SVD and HALS-NTF. Furthermore, the proposed method (TSA-TT) outperforms all the tested methods. The most challenging task is to classify an extended palm, especially with HO-SVD, but this is not a problem for TSA-TT. The efficiency of TSA-TT results from two aspects: first, due to the use of the TT model which better captures multilinear relationships, and second, it results from the incorporation of discriminant information on the geometry of training samples. Interestingly, the use of TSA did not affect the computational complexity considerably, as shown in Fig. 2(a). Among the algorithms tested, only HALS-NTF is substantially slower. The proposed method was also used in both series of experiments with all the subjects (see Fig. 2(b)). The results show that the accuracy of classification depends quite strongly on the subject. More research is needed to explain this phenomenon.

In summary, the proposed method significantly outperforms the other tested methods in terms of classification accuracy and has potential for applications in other classification problems. In future research, it can be used in various remote control systems, e.g. in desktop or hand prosthesis control.

References

1. Chowdhury, R.H., Reaz, M.B.I., Ali, M.A.B.M., Bakar, A.A.A., Chellappan, K., Chang, T.G.: Surface electromyography signal processing and classification techniques. Sensors **13**(9), 12431–12466 (2013)
2. Cichocki, A., Lee, N., Oseledets, I.V., Phan, A.H., Zhao, Q., Mandic, D.P.: Tensor networks for dimensionality reduction and large-scale optimization: part 1 low-rank tensor decompositions. Found. Trends Mach. Learn. **9**(4–5), 249–429 (2016)

3. Cichocki, A., et al.: Tensor networks for dimensionality reduction and large-scale optimization: part 2 applications and future perspectives. Found. Trends Mach. Learn. **9**(6), 431–673 (2017)
4. Côté-Allard, U., et al.: Deep learning for electromyographic hand gesture signal classification using transfer learning. IEEE Trans. Neural Syst. Rehabil. Eng. **27**(4), 760–771 (2019)
5. Fonał, K., Zdunek, R.: Distributed and randomized tensor train decomposition for feature extraction. In: IJCNN, pp. 1–8 (2019)
6. He, X., Cai, D., Niyogi, P.: Tensor subspace analysis. In: Weiss, Y., Schölkopf, B., Platt, J. (eds.) Advances in Neural Information Processing Systems, vol. 18. MIT Press (2005)
7. He, X., Niyogi, P.: Locality preserving projections. In: In Advances in Neural Information Processing Systems, vol. 16. MIT Press (2003)
8. Hudgins, B., Parker, P.A., Scott, R.N.: A new strategy for multifunction myoelectric control. IEEE Trans. Biomed. Eng. **40**, 82–94 (1993)
9. Inam, S., et al.: A brief review of strategies used for EMG signal classification. In: 2021 International Conference on Artificial Intelligence (ICAI), pp. 140–145 (2021)
10. Khoromskaia, V., Khoromskij, B.N.: Tensor numerical methods in quantum chemistry: from hartreefock to excitation energies. Phys. Chem. Chem. Phys. **17**, 31491–31509 (2015)
11. Kolda, T.G., Bader, B.W.: Tensor decompositions and applications. SIAM Rev. **51**(3), 455–500 (2009)
12. Lanyon, B.P., et al.: Efficient tomography of a quantum many-body system. Nat. Phys. **13**, 1158–1162 (2017)
13. Lobov, S., Krilova, N., Kastalskiy, I., Kazantsev, V., Makarov, V.A.: Latent factors limiting the performance of semg-interfaces. Sensors **18**(4) (2018)
14. Ozdemir, M.A., Kisa, D.H., Guren, O., Akan, A.: Hand gesture classification using time-frequency images and transfer learning based on CNN. Biomed. Sig. Process. Control **77**, 103787 (2022)
15. Qi, J., Jiang, G., Li, G., Sun, Y., Tao, B.: Intelligent human-computer interaction based on surface EMG gesture recognition. IEEE Access **7**, 61378–61387 (2019)
16. Sultana, A., Ahmed, F., Alam, M.S.: A systematic review on surface electromyography-based classification system for identifying hand and finger movements. Healthc. Anal. **3**, 100126 (2023)
17. Wołczowski, A., Zdunek, R.: Electromyography and mechanomyography signal recognition: experimental analysis using multi-way array decomposition methods. Biocybern. Biomed. Eng. **37**(1), 103–113 (2017)

A New Statistical Approach to Image Reconstruction with Rebinning for the X-Ray CT Scanners with Flying Focal Spot Tube

Piotr Pluta$^{(\boxtimes)}$ and Robert Cierniak

Department of Intelligent Computer Systems, Czestochowa University of Technology,
Czestochowa, Poland
piotr.pluta@pcz.pl
https://www.kisi.pcz.pl

Abstract. This paper presents an original approach to the image reconstruction problem for spiral CT scanners where the FFS (Flying Focal Spot) technology in an x-ray tube is implemented. The geometry of those scanners causes problems for CT systems based on traditional (FDK) reconstruction methods. Therefore, we propose rebinning strategy, i.e. a scheme with abstract parallel geometry of x-rays, where these problems do not occur. As a consequence, we can reconstruct an image from projections with a non-equiangular distribution, present in the Flying Focal Spot technology. Our method is based on statistical model-based iterative reconstruction (MBIR), where the reconstruction problem is formulated as a shift-invariant system. The statistical fundamentals of the proposed method allow for a reduction of the x-ray dose absorbed by patients during examinations. Performed simulations showed that our method overcomes the traditional approach regarding the quality of the obtained images and an x-ray dose needed to complete an examination procedure.

Keywords: X-ray computed tomography · flying focal spot · statistical method · image reconstruction from projections · rebinning

1 Introduction

At the beginning of the previous century, the spiral scanner with an X-ray tube with a flying focal spot (FFS) has been introduced [3]. This new technique aims at increasing the density of simultaneously acquired slices in the longitudinal direction and the sampling density of the integral lines in the reconstruction planes. This technique allows for view-by-view deflections of the focal spot in the rotational α-direction (αFFS) and in the longitudinal z-direction (zFFS). Thanks to this, the quality of the reconstructed images can be improved, mainly by reducing the windmill artifacts in z-direction (zFFS) and by decreasing the influence of the aliasing effect in the reconstruction plane(αFFS). The FFS implementation excludes the usage of traditional methods to reconstruct an image

© The Author(s), under exclusive license to Springer Nature Switzerland AG 2023
J. Mikyška et al. (Eds.): ICCS 2023, LNCS 14074, pp. 670–677, 2023.
https://doi.org/10.1007/978-3-031-36021-3_64

from projections. So, in practice, manufacturers decided primarily to modify the adaptive multiple plane reconstruction (AMPR) method in this case (see, e.g. [4]). The AMPR technique is a nutating reconstruction method and, specifically, is an evolution of the advanced single slice rebinning (ASSR) algorithm [1,5]. Generally, nutating methods have several serious drawbacks, particularly equispaced resolution of the slices in z-direction is very difficult to obtain due to the constant change in the position of successive reconstruction planes.

Meanwhile, CT scanner manufacturers began a challenge to develop methods of reducing the X-ray dose absorbed by patients during their examinations. Generally recognized strategy in this direction is to keep the image quality high, when radiation intensity is kept at a defined low level. It is possible by using appropriately formulated methods that are able to suppress noise which appears at the decreased intensity of x-rays. The most advanced conceptually are approaches based on a probabilistic model of the individual measurements in which the reconstruction problem is reformulated into an optimization problem (MBIR - Model-Based Iterative Reconstruction methods).

Unfortunately, the MBIR methods used commercially [8], have some problems with computational complexity because it is approximately proportional to N^4, where N is the image resolution. That approach is based on discrete-to-discrete (D-D) data model, i.e. both the reconstructed image and the measurements during reconstruction problem formulation in discrete forms are considered. Our approach has some significant advantages over the methods based on the D-D model. Firstly, the forward model is formulated here as a shift-invariant system. That allows for implementing an FFT algorithm in the most computationally demanding parts of the reconstruction algorithm. Secondly, our approach reduces the scheme of transforming the X-rays from the cone-beam geometry of the scanner to a parallel geometry, where the problem of not equiangular x-ray naturally does not occur. Finally, the reconstruction process can be carried out in only one plane in 2D space, thus greatly simplifying the reconstruction problem. Additionally, every scan of the body can be obtained separately.

The analysis of the method described in this article is strongly based on the construction principles of a CT scanner with a multifocal x-ray tube.

2   Flying Focus Spot Technique

The FFS technique utilizes the specific design of an X-ray tube, where it is possible to deflect the electron beam (using an electric field) before it hits the anode of the X-ray tube. This mechanism allows for view-by-view deflections of the focal spot for X-rays emitted from that anode. As a result of the FFS, it is possible to obtain the greater density of lines of rays used in the reconstruction process, both in the plane of the reconstructed image and along the z axis around which the projection system rotates. These geometrical conditions are shown in Figs. 1 and 2.

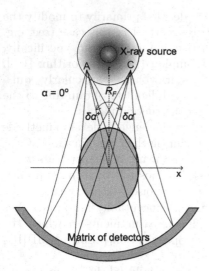

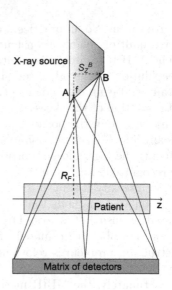

Fig. 1. Scheme of densification of rays in the reconstruction plane (αFFS).

Fig. 2. Geometry of X-rays along the z axis (zFFS).

3   Reconstruction Algorithm

The reconstruction method dedicated for the flying focal spot technique proposed by us in this paper, is based on the maximum-likelihood (ML) estimation (see e.g. [7]). Usually, the objective of this kind of approach to the reconstruction problem is formulated according to a discrete-to-discrete (D-D) data model. In opposition to that approach, we have designed an optimization method. According to the originally formulated by us statistical approach to the reconstruction problem, it is possible to present a practical model-based iterative reconstruction procedure, as follows:

$$\mu_{\min} = \arg\min_{\mu} \left(\sum_{i=1}^{I} \sum_{j=1}^{J} \left(\sum_{\bar{i}} \sum_{\bar{j}} \mu^* \left(x_{\bar{i}}, y_{\bar{j}} \right) \cdot h_{\Delta i, \Delta j} - \tilde{\mu} \left(x_i, y_j \right) \right)^2 \right), \quad (1)$$

and $\tilde{\mu}(i, j)$ is an image obtained by way of a back-projection operation, in the following way:

$$\tilde{\mu} \left(x_i, y_j \right) = \Delta_\alpha \sum_\theta \dot{p} \left(s_{ij}, \alpha_\psi \right), \quad (2)$$

where $\dot{p}(s_{ij}, \alpha_\psi)$ are measurements performed using parallel beams, and:

$$h_{\Delta i, \Delta j} = \Delta_\alpha \sum_{\psi=0}^{\Psi-1} int \left(\Delta i \cos \psi \Delta_\alpha + \Delta j \sin \psi \Delta_\alpha \right), \quad (3)$$

and $int\,(\Delta s)$ is an interpolation function (we used the linear interpolation function).

In our reconstruction procedure one can distinguish two parts. In the first one, we prepare and perform rebinning of all cone beam X-ray raw data from CT scanners with FFS to X-rays of parallel geometry. This operation is described in detail in Subsect. 3.1. After rebinning we determine a set of parallel projections which are ready to use for back-projection operation (see relation 2). Secondly, we use a statistical iterative procedure for 2D images, which computational complexity is $8log_24N^2$, where N is the dimension of the reconstruction image. Finally set of 2D views can be successfully used to obtain a 3D image.

3.1 Rebinning Operation

During the rebinning process we will describe every X-ray from cone-beam by elementary parameters, namely following specific points: Focus, SemiIsocenter, Detector [6]. Each of these parameters have coordinates in 3D space. This way, it is possible to change the description of equiangular X-rays using a set of three points in 3D. Next, we can compute all necessary calculations based on only those elementary components. This redefining can be described as follows:

$$xray(F_x^A, F_y^A, F_z^A, F_x^B, F_y^B, F_z^B, Q_x, Q_y, Q_z, D_x, D_y, D_z, p, s_m, \alpha_\psi), \qquad (4)$$

During the rebinning process, we will describe every X-ray from the cone-beam by elementary parameters, namely, the following specific points: Focus, SemiIsocenter, Detector [6]. Each of these parameters has coordinates in 3D space. This way, it is possible to change the description of equiangular X-rays using a set of three points in 3D. Next, we can compute all necessary calculations based on only those elementary components. This redefining can be described as follows:

$$xray(F_x^A, F_y^A, F_z^A, F_x^B, F_y^B, F_z^B, Q_x, Q_y, Q_z, D_x, D_y, D_z, p, s_m, \alpha_\psi), \qquad (5)$$

where points F determining Focus, points Q determining SemiIsocenter, points D determining Detector, all of those points are represented in 3D (x,y,z), p is value of projection, s_m and α_ψ described the same X-ray but in parallel projection. The parameters are depicted in Fig. 3.

After the above calculations, the virtual X-rays have to be chosen as the most suitable real X-ray. This ray is chosen in a comparing procedure. For this purpose, we chose the best Focus in the following way:

$$f_x = -(R_F + \Delta R_F^T) \cdot \sin(\alpha + \Delta\alpha^T); \qquad (6)$$

$$f_y = (R_F + \Delta R_F^T) \cdot \cos(\alpha + \Delta\alpha^T); \qquad (7)$$

$$f_z = z_0 + \Delta z^T; \qquad (8)$$

where the T is one of two focuses, R_F is the distance between the focus and isocenter. The comparing procedure is based on calculating the ζ angle between

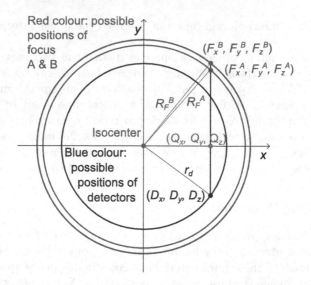

Fig. 3. Scheme of determined X-ray points.

real focus and virtual focus with connection to SemiIsocenter:

$$\zeta = \arccos\left(\frac{\hat{w}_x \cdot \hat{v}_x + \hat{w}_y \cdot \hat{v}_y}{\sqrt{\hat{w}_x^2 + \hat{w}_y^2} \cdot \sqrt{\hat{v}_x^2 + \hat{v}_y^2}}\right), \tag{9}$$

where:

$$\hat{v}_x = f_x^T - Q_x; \hat{v}_y = f_y^T - Q_y; \hat{w}_x = F_x^T - Q_x; \hat{w}_y = F_y^T - Q_y. \tag{10}$$

After choosing the two nearest focuses we try to determine the best detectors. There are some problems regarding the fact that we can not use the square function because its definition is unsuitable for the line parallel to y-axis. That is why we use a more complicated formula for the collinearity of three points in space. Thanks to this formula we can calculate the real detector position that is being searched for (D_x, D_y):

$$\begin{cases} (D_x - f_{x_0})^2 + (D_y - f_{y_0})^2 = R_{FD}^2 \\ (Q_x - f_x)(D_y - f_y) - (Q_y - f_y)(D_x - f_x) = 0 \end{cases}, \tag{11}$$

After finding this detector, we can calculate the value for a virtual detector using the three-linear interpolation function. Finally, the determined virtual parallel X-ray could be used for back-projection operation to create an image for an iterative reconstruction procedure (described in the next subsection). These virtual measurements can be used for filtered back-projection and the creation of a starting image for the iterative reconstruction procedure (see e.g. [2]). As a result of the transfer of the most demanding calculations into the frequency domain, it is possible to reduce the computational complexity from N^4 to $8log_24N^2$ where N is the dimension of the reconstruction image.

4   Results

This section is devoted to presentation of the results of our research based in the first part on mathematical simulation data and second part for the physical tomographic data that the authors of the work obtained by participating in the Low-Dose Grand Challenge. The physical data has been prepared artificially to simulate a quarter-dose of the X-ray radiation.

Our primary experiments are based on the *Shepp-Logan* mathematical data model. That model consists of ellipses or ellipsoids, depending on the two-dimensional or three-dimensional geometric system under consideration. They are the most simple geometric figures used to build mathematical phantoms. The ellipse/ellipsoid phantom allows for obtaining projection values for all rays at any projection angle. If simulating projection systems use a conical beam of

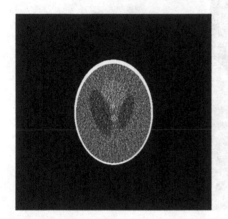

Fig. 4. Traditional reconstructed image; Noise: large; MSE: 766.691; NRMSE: 0.108585.

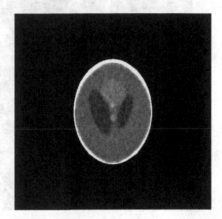

Fig. 5. Iterative reconstructed image; Noise: large; Iteration: 5000; MSE: 246.7224; NRMSE: 0.06159765.

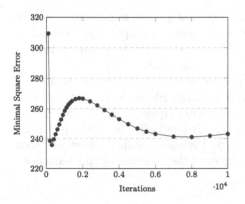

Fig. 6. MSE plot for large noise

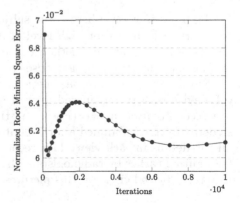

Fig. 7. NRMSE plot for large noise

radiation spiralling around the patient, it is necessary to define a mathematical phantom in 3D. It means that the equations for ellipses (flat figures) should be replaced with the equation for an ellipsoid (three-dimensional solid) using the following relations. These 3D mathematical model of head in our experiments was used.

Figures 4 and 5 show a visual comparison of the reconstructed mathematical sections of the mathematical model for $z = 0$. Figure 4 was made using the traditional method, while Fig. 5 was reconstructed using an iterative-statistical method, both with a high level of noise. We also presented the Minimal Square Error (MSE) plot - Fig. 6 and Normalised Root Minimal Square Error (NRMSE) plot - Fig. 7 booth are for iterative method.

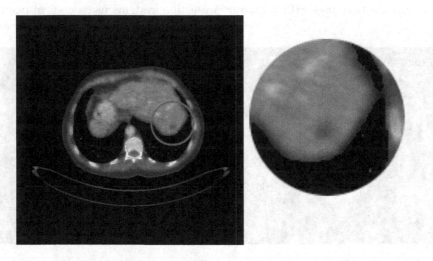

Fig. 8. Reconstruction view of the first patient. Z 109; Iterations 5000; Dose $1/4$. (Color figure online)

An important issue here is the subjective visual quality improvement of the reconstructed images for high level of noise presented in input data. Obtained results indicate that this method can be dedicated to increasing the usefulness of diagnostic images using low-dose computed tomography.

Next, we conducted reconstruction experiments using data from a commercial CT device with the simulated quarter-dose intensity of the X-rays. We present the reconstruction image (Fig. 8) at the simulated quarter dose. This reconstructed image is obtained after 5000 iteration. It is usually enough to recognize all the details in each view. This reconstructed image contains a form of lesion (recognized earlier by radiologists). The places of the presence of this pathology is marked by red circle on this pictures.

5 Conclusion

An original complete statistical iterative reconstruction method with a rebinning method has been presented. The proposed by us algorithm can be applied in scanners with a flying focal spot here. Experiments conducted by us have proved that our reconstruction approach is relatively fast (about 5 s for all operations in middle efficient GPU), mainly thanks to applying of an FFT algorithm during the most demanding calculations in the iterative reconstruction procedure.

Acknowledgements. The project financed under the program of the Polish Minister of Science and Higher Education under the name "Regional Initiative of Excellence" in the years 2019 - 2023 project number 020/RID/2018/19 the amount of financing PLN 12,000,000.

References

1. Cierniak, R., Pluta, P., Kaźmierczak, A.: A practical statistical approach to the reconstruction problem using a single slice rebinning method. J. Artif. Intell. Soft Comput. Res. **10**(2), 137–149 (2020). https://doi.org/10.2478/jaiscr-2020-0010
2. Cierniak, R., et al.: A new statistical reconstruction method for the computed tomography using an x-ray tube with flying focal spot. J. Artif. Intell. Soft Comput. Res. **11**(4), 271–286 (2021). https://doi.org/10.2478/jaiscr-2021-0016
3. Flohr, T.G., Stierstorfer, K., Ulzheimer, S., Bruder, H., Primak, A.N., McCollough, C.H.: Image reconstruction and image quality evaluation for a 64-slice CT scanner with z-flying focal spot. Med. Phys. **32**(8), 2536–2547 (2005). https://doi.org/10.1118/1.1949787
4. Flohr, T., Stierstorfer, K., Bruder, H., Simon, J., Polacin, A., Schaller, S.: Image reconstruction and image quality evaluation for a 16-slice CT scanner. Med. Phys. **30**(5), 832–845 (2003). https://doi.org/10.1118/1.1562168
5. Kachelrieß, M., Schaller, S., Kalender, W.A.: Advanced single-slice rebinning in cone-beam spiral CT. Med. Phys. **27**(4), 754–772 (2000). https://doi.org/10.1118/1.598938
6. Pluta, P.: A new approach to statistical iterative reconstruction algorithm for a CT scanner with flying focal spot using a rebinning method, pp. 286–299 (2023). https://doi.org/10.1007/978-3-031-23480-4_24
7. Sauer, K., Bouman, C.: A local update strategy for iterative reconstruction from projections. IEEE Trans. Sig. Process. **41**(2), 534–548 (1993). https://doi.org/10.1109/78.193196
8. Yu, Z., Thibault, J.B., Bouman, C.A., Sauer, K.D., Hsieh, J.: Fast model-based x-ray CT reconstruction using spatially nonhomogeneous ICD optimization. IEEE Trans. Image Process. **20**(1), 161–175 (2011). https://doi.org/10.1109/TIP.2010.2058811

Investigating the Sentiment in Italian Long-COVID Narrations

Maria Chiara Martinis[✉] , Ileana Scarpino[✉] , Chiara Zucco ,
and Mario Cannataro

Data Analytics Research Center, Department of Medical and Surgical Sciences,
University Magna Graecia of Catanzaro, Catanzaro, Italy
{martinis,ileana.scarpino,chiara.zucco,cannataro}@unicz.it

Abstract. Through an overview of the history of the disease, Narrative
Medicine (NM) aims to define and implement an effective, appropriate
and shared treatment path. In the context of COVID-19, several blogs
were produced, among those the "Sindrome Post COVID-19" contains
narratives related to the COVID-19 pandemic. In the present study, different analysis techniques were applied to a dataset extracted from such
"Sindrome Post COVID-19" blog. The first step of the analysis was to
test the VADER polarity extraction tool. Then the analysis was extended
through the application of Topic Modeling, using Latent Dirichlet Allocation (LDA).

The results were compared to verify the correlations between the
polarity score obtained through VADER and the extracted topics
through LDA. The results showed a predominantly negative polarity consistent with the mostly negative topics represented by words on post virus
symptoms. The results obtained derive from three different approaches
applied to the COVID narrative dataset. The first part of the analysis
corresponds to polarity extraction using the VADER software, where,
from the score, polarity was inferred by dichotomizing the overall score.
In the second part, topic modeling through LDA was applied, extracting
a number of topics equal to three. The third phase is based on the objective of finding a qualitative relationship between the polarity extracted
with VADER and the latent topics with LDA, considering it a semi-
supervised problem. In the end, the presence of polarized topics was
explored and thus a correspondence between sentiment and topic was
found.

Keywords: Sentiment Analysis · Polarity Detection · Narrative
Medicine · VADER · Topic Modeling · Latent Dirichlet Allocation

1   Introduction

The COVID-19 pandemic had a strong impact not only on people's physical
health, but also on their mental health [6,11,21]. Social networks were widely

---

M. C. Martinis and I. Scarpino—Contributed equally to this work.

J. Mikyška et al. (Eds.): ICCS 2023, LNCS 14074, pp. 678–690, 2023.
https://doi.org/10.1007/978-3-031-36021-3_65

used during the isolation period as they allowed people to tell about their experience in dealing with the pandemic, sharing their physical and emotional state.

The large amount of data available for collection and analysis has therefore allowed the scientific community to apply Natural Language Processing (NLP) techniques, specifically sentiment analysis, to synthesize and investigate various aspects related to the COVID-19 pandemic narrative.

There are several studies in the literature on topic extraction in tweets related to COVID-19 pandemic [2, 4, 30]. In particular, the study of Medford et al. [14] collected tweets related to COVID-19 in January 2020, during the initial phase of the outbreak, measuring the themes and frequency of keywords related to infection prevention practices, identifying the polarity of sentiment and predominant emotions in the tweeting and conducting a topic modeling of the topics of discussion over time through Latent Dirichlet Allocation.

Sentiment analysis techniques [32] were applied to identify topics and trends in Twitter users' sentiment about the pandemic. The authors' conclusions assert that the use of the social network has allowed activating sometimes coping mechanism to combat feelings of isolation related to long-term social distancing [29] and sometimes exacerbating concern and disinformation [20]. The evaluation of the most discussed topics by people who have dealt with COVID-19 disease can lead to field applications of text mining techniques useful in various fields, e.g. polarity analysis, which involves a classification into three labels: positive, negative and neutral, is useful to identify the depth of emotions expressed by users.

In Natural Language Processing, the polarity detection task falls under the umbrella of automatic text classification. Typically, the approaches for polarity classification are divided into two macro-categories: the lexicon-based approaches and the machine learning-based approaches [16]. Lexicon-based approaches start from the assumption that the overall polarity of the text unit can be inferred from the polarity of the words found in the text unit, using a set of predefined rules. Machine learning-based approaches aim to classify the polarity of a text by training a machine learning algorithm with a set of already tagged examples [32]. An example of such a tool is VADER, a lexicon-based sentiment analysis tool specifically adapted to sentiments expressed in social media, which also works well on texts from other domains [7].

Mathematical and computer models are applied in research to analyze communications about health problems. Detecting emotions from users' reflections is useful for the purpose of diagnosis of diseases and other medical emergencies or epidemics [9, 25, 27]. Particular interest is given to the narratives of patients with Post-Acute Sequelae of COVID-19 (PASC). Post-Acute Sequelae of COVID-19, also known as "post-COVID-19 syndrome" or "Long-COVID," refers to those convalescent individuals in whom prolonged and often debilitating sequelae persist [15]. The clinical symptomatology includes several main manifestations, often affecting different organ systems. In addition to fatigue, malaise and dyspnoea, PASC patients may also be affected by a number of psychiatric disorders, including depression, anxiety, and post-traumatic stress disorder [26].

Some studies found evidence of mood and cognitive impairment that urgently requires the development of targeted therapies and telemedicine support [1]. In others, prevalence estimation was conducted to identify PASC and predict fluctuations in the number of people with persistent symptoms over time [18].

With the huge popularity of social media platforms, humans express their thoughts and feelings more openly than ever before, therefore sentiment analysis has quickly become a hot research field that may help monitor and understand opinions and emotions in different application domains, as for instance marketing and financial analysis, customer services, recommendation systems, but also clinical applications [10].

The present work focuses on the application of Natural Language Processing (NLP) and Sentiment Analysis (SA) methods, in particular polarity detection VADER tool to extract sentiment from Narrative Medicine textual sources containing narratives of subjects with post-covid syndrome (PASC). In a previous work [24], the main focus was investigation of Topic Modeling Techniques to extract meaningful insights in Italian Long COVID narrations. The automatic collection was performed through web scraping of NM Italian blogs, i.e. "Sindrome Post COVID-19"[1] that produced a dataset composed of 73 narratives related to PASC.

The rest of the paper is organized as follows. In Sect. 2 the background information about NM and NLP is introduced, and an overview of polarity detection and topic modeling application on COVID-19 related text is discussed. Section 3 describes the data collection procedure, the dataset used and the analysis methodology. Section 4 discusses the result of the analysis. Finally, Sect. 5 concludes the paper and outlines future works.

2   Background on Sentiment Analysis of COVID-19 Texts

Narrative Medicine is a field of medicine that helps all professionals in the health care system to carefully accommodate the experiences of people living with a disease and their caregivers through research and clinical practice. To analyze written and oral narratives, e.g. extracting polarity from texts and other relevant information, polarity detection and topic modeling techniques are adopted. Sentiment Analysis is an emerging methodology that contributes to the understanding of human emotions from social media reviews. In [12] the authors conducted the global sentiment analysis of tweets related to coronavirus and how the sentiment of people in different countries varied over time. Twitter is one of the most commonly used social media platforms, and it showed a massive increase in tweets related to coronavirus, including positive, negative and neutral tweets. Early detection of COVID-19 sentiment from collected tweets allowed for better understanding and management of the pandemic.

In another study [8], researchers analyzed emotions extracted from tweets for sentiment classification using various feature sets and classifiers. Tweets were

---

[1] https://www.sindromepostcovid19.it/.

classified into positive, negative and neutral sentiment classes. The performance of machine learning (ML) and deep learning (DL) classifiers is evaluated using different evaluation metrics (accuracy, precision, recall and F1 score).

In [28] the tweets were analyzed to understand the mood of citizens and students during the COVID-19 pandemic. Sentiment analysis was conducted in relation to the age of users, and the results showed that the entity of tweets was higher in young people during the pandemic.

In [31] authors reviewed the Twitter opinion of nine states of the United States (U.S.) on COVID-19 from April 1 to April 5, 2020 by developing the most popular methods of machine learning and deep learning to predict users' sentiments toward COVID-19 based on tweets. In the work, coding techniques were compared with two separate word embedding systems Word2vec and Glo2vec. Opinions on the introduction of the COVID-19 vaccine varied among people; some showed concern about safety, others supported vaccination as an effective mitigation strategy.

In [19], researchers examined the level of hesitation of the COVID-19 vaccine as part of the effort to combat the infectious virus. The pros and cons of vaccination have become a key topic on social media platforms. The results show that hesitation toward the COVID-19 vaccine is gradually decreasing over time, suggesting that society's positive views of vaccination have gradually increased.

In the context of COVID-19, storytelling is an additional resource for dealing with past and present events during the pandemic, as it provides concrete tools to health care professionals to support health prevention and treatment.

Long COVID is not only a new disease, but also a disease that, perhaps unique in recent history, emerged largely without the patient's clinician acting as a witness or sounding board. Since COVID-19 turns out to be an occupational disease in that it has disproportionately affected health care workers, a socio-narratological approach is necessary in the Long COVID field since both patient trust and obligation on the part of physicians come into play. The story of Long COVID as it unfolded in 2020, includes an overarching metanarrative of absent listeners: the collective failure - arguably for good reasons - of clinicians to acknowledge, interpret or act on their patients' stories and plights [22].

In [23, 24] standard Text Mining (TM) techniques were applied for topic modeling to characterize narrative medicine texts written on COVID-19. To extract polarity in texts written in English, the original VADER software was used, a version that was extended to the Italian language, thus creating VADER-IT [13].

3   Materials and Methods

The aim of the present study is to sentiment analysis and text mining approaches for the analysis of a blog about the COVID-19 disease, named "Sindrome post COVID-19"[2], containing narrative of patients affected by Long-COVID (in short Long-COVID testimonies). The dataset of "Sindrome post COVID-19" can be

---

[2] https://www.sindromepostcovid19.it/.

considered an important attempt to create a database to collect the testimonies of those who have had, or are having, PASC syndrome. People who shared their experiences were asked to provide specific information about their symptoms after the test was negative.

3.1   Long-COVID PASC Dataset

In a recent work [24], we analyzed a dataset extracted by "Sindrome Post COVID-19" containing 73 narrative texts to Post-acute Sequelae of COVID-19 (PASC), also known as Long-COVID syndrome.

For the present study the texts were automatically collected and translated by API Google into English to allow the use of original VADER.

Both data collection and data analysis were performed in Python, a high-level programming language. Python's "Beautiful Soap" and "requests" libraries have been used to automatically collect the testimonies and to store the data in a single CSV file.

For instance, to collect data from the "Sindrome post COVID-19" blog, the Python code:

- collects all the URL of the testimonies per page,
- connects to the page and, by parsing the HTML document:
  - extracts only the textual narration,
  - performs a basic text cleaning,
  - tags it as "PASC",
- the extracted information is stored in a CSV file format.

The described process is then iterated through all the narrations posted on the blog. The information is stored in a CSV text file for the subsequent analysis, storing the data sorted by chronological order from the most recent to the oldest post.

Figure 1 shows an example of text extracted from the dataset.

I started having fever and general malaise on 20 oct.
I discover I am positive on the 24th.
I start immediately with antibiotic, cortisone and eparine.
The constant fever at 38.7 for about 10 days,
so the doctor advises me to go to the PS where they discover
the bilateral interstitial pneumonia.
After 4 days in the emergency room
(assisted superficially due to overcrowding and unpreparedness by the hospital).
I sign the release and I ask to return home.
The symptoms remain about 10 days about 10 days
(fatigue, chest pain, impossibility of breathing deeply).
Finally on November 12, but the respiratory problems remain.
After a visit from the pneumologist and cardiologist,
after more than 3 months I still have respiratory
and moderate insufficiency pericarditis that to date (February 2).
I am still treating"

Ho iniziato ad avere febbre e malessere generale il 20 ottobre.
Scopro di essere positivo il 24.
Inizio subito con antibiotico, cortisone ed eparina.
La febbre costante a 38,7 per circa 10 giorni,
così il medico mi consiglia di andare al PS dove scoprono
la polmonite interstiziale bilaterale.
Dopo 4 giorni in pronto soccorso
(assistito superficialmente per sovraffollamento e impreparazione da parte dell'ospedale)
firmo la liberatoria e chiedo di tornare a casa.
I sintomi rimangono circa 10 giorni circa
(stanchezza, dolore al petto, impossibilità di respirare profondamente).
Finalmente il 12 novembre, ma i problemi respiratori rimangono.
Dopo una visita dallo pneumologo e dal cardiologo,
dopo più di 3 mesi ho ancora una pericardite respiratoria
e una moderata insufficienza che ad oggi (2 febbraio).
Mi sto ancora curando".

Fig. 1. Example of extracted text fragment.

3.2   Sentiment Analysis Using VADER

Sentiment analysis is useful for detecting positive, negative or neutral sentiment in text. It is often used by companies to detect sentiment in social data, assess brand reputation, and understand customers. Several pre-trained language models have also been proposed in the biomedical field, such as BioBert [4], ClinicalBert [5], Medicines [5], calBert [5], MedBert [6], and PubMedBert [7].

A standard approach to build pre-trained language models in the biomedical field consists of training a general BERT model on biomedical texts [8].

We performed a polarity-based sentiment analysis of the previous Long-COVID PASC dataset by using the VADER TOOL. First, each text was translated in English using the API google (Python library), then the VADER tool[3] was applied to each of the 73 texts and a polarity value (ranking between -1,0,1) was associated to each of them. A lexicon-based approach for extracting polarity from written texts in the English language is the VADER tool, namely the Valence Aware Dictionary and sEntiment Reasoner [7]. VADER combines highly curated lexicon resources with rule-based modeling consisting of 5 humans validated rules to predict a polarity score from textual input.

Several are the advantages of a VADER-like approach for clinical applications. One of the most important is that VADER does not require a training phase. Consequently, it can be easily applied in both low-resource domains and in fields where a limited number of data can be collected at a time, as usually happens in clinical trials.

The original VADER has been proven to work well on short text, making it usable to analyze open answers to questionnaires, which still are the most widespread method for carrying out investigations in the clinical field.

Moreover, a lexicon-based approach is generally faster than other approaches and, therefore, it may be suited for near real time applications. In addition, by exploiting general "parsimonious" rules, a VADER-like approach is domain-agnostic and easily customizable to different domains by extending the existing dictionary with domain-specific words. Finally, it can be considered a white box model, which makes it highly interpretable and highly adaptable to different languages.

The starting point of the VADER system is a generalizable, valence-based, human-curated gold standard sentiment lexicon, built on top of three well-established lexicons, i.e. LIWC [17], General Inquirer and ANEW [3], expanded with a set of lexical features commonly used in social media, which included emoji, for a total of 9,000 English terms subsequently annotated in a $[-4, 4]$ polarity range through the Amazon Mechanical Turk's crowd-sourcing service. The second core step of the VADER engine is the identification of some general grammatical and syntactical heuristics to identify semantic shifters.

---

[3] https://github.com/cjhutto/vaderSentiment.

3.3 Topic Modeling Using LDA

We performed a topic model analysis of the previous Long-COVID PASC dataset by iterative appling LDA[4]. A topic can be seen as a collection of representative words in a text, that helps to identify what is the subject or the subjects of a document. Latent Dirichlet Allocation (LDA) [5] is a generative probabilistic model commonly used for the identification of latent topic in textual corpora. The starting point of an LDA model is that a document is represented as a bag-of-words. Each single word is represented as a pair of values: the first represents its position, the second is the number of occurrences of the word itself within the document. The assumption under an LDA model is that each document in a corpus can be modeled as a mixture of a finite number of topics with a certain probability, while a topic can be characterized by a distribution over words. More in details, assume that there are k topics across all documents. Le $\mathbf{w}$ be a document in a corpus $\mathcal{D}$ and let consider:

- $\theta \sim Dir(\alpha)$, be a mixture of k topics, having a Dirichlet probability distribution where α is the per-document topic distribution;
- a topic $z_n \sim Multinomial(\theta)$, where n represents the number of words that define a topic;

then, given the topic z_n, a word w_n is sampled from $p(w_n|z_n, \beta)$, and β represents the per-topic word distribution. Then, the probability of $\mathbf{w}$ containing n words can be described as (Eq. 1):

$$p(\mathbf{w}) = \int \left(\prod_{n=1}^{N} \sum_{z_n=1}^{k} p(w_n|z_n, \beta) \right) p(\theta, \alpha) d\theta \tag{1}$$

4   Results and Discussion

To gain useful insights about the collected data, a preliminary exploratory data analysis was performed by summarizing data through suitable visualization.

In this section, we present the results of three different approaches applied on the COVID-19 narrative dataset. First part of the analysis corresponds to the extraction of the polarity using the VADER software, in which, starting from the score, the polarity was inferred by dichotomizing the overall score. In the second part, the topic modeling was applied through LDA, extracting a number of topics equal to three. The third phase is based on the objective of finding a qualitative relationship between the polarity extracted with VADER and the latent topics with LDA, considering this as a semi-supervised problem. Once polarity was considered as a real label of the various texts, the presence of polarized topics was explored and therefore a correspondence between the sentiment and the topic was found.

---

[4] https://github.com/topics/lda-topic-modeling.

4.1 Exploratory Data Analysis

The longest document among PASC narrative counts 660 words and an interquartile range of 366.5 words [24]. The most used words in the documents are shown through a word cloud, as shown in Fig. 2.

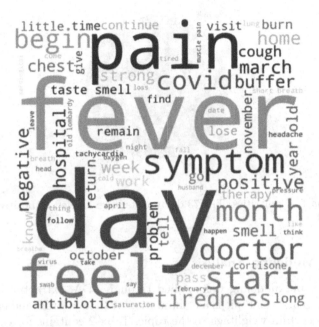

Fig. 2. Word cloud showing the most frequent tokens in the document after preprocessing. Tokens with the largest font size are the most frequent.

The corpus was pre-processed with standard NLP techniques, i.e., tokenization, stop word removal, and lemmatization and then the word cloud was built by a Python word cloud generator. Terms as "fever", "doctor", "pain", "covid", "day" are the most frequent words in the corpus, followed by "tiredness", "symptom", "doctor", "antibiotic", "positive", "negative".

4.2 Polarity Analysis

Initially the dataset did not have a polarity score. VADER has been used on the dataset containing 73 narrations to extract the polarity by associating a score equal to 1 for positive polarity and -1 for negative polarity. The results show that, by calculating the occurrence of positive and negative polarities, VADER assigned positive polarity (POS) to 12 documents and negative polarity (NEG) to 61 documents, as demonstrated in Table 1. The application of VADER revealed more negative polarities than positive ones, as shown in Fig. 3.

Table 1. The total number of documents is 73, 12 tagged as positive and 61 as negative.

No. of Documents	POS	NEG
73	12	61

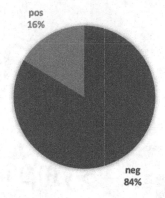

Fig. 3. Pie chart of polarities. The 73 documents were labeled as positive or negative using VADER.

4.3   Topic Modeling Analysis

In this task, topic modeling was applied through LDA, extracting a number of topics equals to 3. Each topic is a combination of 10 keywords, and each keyword contributes a certain weightage to the topic. Table 2 contains for each topic the keywords with the associated weight.

To continue the analysis, the number of documents associated with the main topic to which they belong was obtained using the python library *gensim*, as shown in Table 3.

To verify the presence of polarized topics, the 12 documents with positive polarities were selected (see Fig. 3), obtaining the assignment of the polarities to the specific topic, as shown in Table 4.

Then, the analysis of polarized topics was carried out again with reference to the 61 PASC testimonials with negative polarities, as shown in Table 5.

The results show that the 7 positive documents are associated with topic 0, 4 are associated with topic 1 and the remaining document with topic 2.

The results show that the split is 1/3 per topic. In fact 20 negative documents are associated with topic 0, other 20 with topic 1 and the remaining 21 documents with topic 2.

Finally, in the last stage, the keywords of each topic were entered into a list to be analyzed with VADER. Each was associated not only with positive negative polarity scores, but also with neutral ones. Note that although there is a distinction in positive and negative labels regarding the texts, the topics have in descending order predominantly a neutral, negative polarity, while positive equal to 0 for each topic, as shown in Table 6.

Table 2. Keywords per topic with weights.

Topic ID	Keywords and weights
Topic 0	0.019 · fever + 0.014 · doctor + 0.030 · pain + 0.012 · year + 0.011 · home + 0.011 · hospital + 0.011 · start + 0.010 · symptom + 0.009 · begin + 0.009 · march
Topic 1	0.029 · pain + 0.013 · start + 0.013 · year + 0.012 * fever + 0.010 · negative + 0.010 · chest + 0.009 · covid + 0.009 · feel + 0.009 · smell + 0.009 · month
Topic 2	0.022 · pain + 0.017 · smell + 0.016 · feel + 0.015 · year + 0.015 · fever + 0.015 · covid + 0.013 · taste + 0.011 · negative + 0.011 · month + 0.010 · symptom

Table 3. The total number of texts for the associated topic.

Topic	N. Texts	%
0	27	37%
1	24	33%
2	22	30%

Table 4. Documents with positive polarity, partitioned with respect to the 3 topics.

Topic	Positive N. Texts	%
0	7	59%
1	4	33%
2	1	8%

Table 5. Documents with negative polarity, partitioned with respect to the 3 topics.

Topic	Negative N. Texts	%
0	20	33%
1	21	34%
2	21	34%

Table 6. Polarity score and class polarity for each topic. Polarity scores were extracted with VADER, while polarity classes are the result of a discretization process which associates the positive polarity (POS) for scores greater than 0 and negative polarity (NEG) for scores less than 1.

Topic	POLARITY SCORE	CLASS POLARITY
0	−0.510	NEG
1	−0.79	NEG
2	−0.79	NEG

5   Conclusions and Future Work

In the present work, some sentiment analysis and topic modeling methods have been applied to 73 Italian narrative medicine texts of patients with post-acute sequelae of COVID-19, i.e. PASC. After carrying out the automatic collection and translation in English of the texts, VADER software was applied in order to extract the polarity of the 73 narratives, resulting in 61 negative and 12 positive documents. The results of this approach seem more consistent with the theme of Long-COVID, as it seems intuitively more correct that the testimonies have a predominantly negative polarity.

With the aim of finding a connection between the extracted polarities and possible topics contained in texts, topic modeling was applied to extract the latent topics through the LDA model. In order to conduct a qualitative analysis and find a relationship between the polarity and the related topics, it was checked whether the extracted topics mainly mention negative or positive words.

However, the results did not show significant differences between what emerges from the polarity analysis and from the topic extraction, since in both approaches the prevailing polarity is the negative one. It can be concluded that the analysis made at the document level is not exactly representative of the true feeling of the texts.

Better results may be obtained by analyzing the texts with a sentiment analysis approach not of the "document level" type, but of the "sentence/aspect level" type. In future work we plan to extend the dataset and consider the sentence level approach to perform sentiment analysis.

An other interesting future development would be to apply Named Entity Recognition analysis on the proposed dataset by finding a possible correlation with the polarity analysis.

References

1. Agrusta, M., Cenci, C.: Telemedicine and digital narrative medicine for the customization of the diagnostic-therapeutic path at the time of COVID 19 (2021)
2. Boon-Itt, S., Skunkan, Y.: Public perception of the COVID-19 pandemic on Twitter: sentiment analysis and topic modeling study. JMIR Publ. Health Surveill. **6**(4), e21978 (2020)

3. Bradley, M.M., Lang, P.J.: Affective norms for English words (anew): instruction manual and affective ratings. Technical report C-1, the center for research in psychophysiology ... (1999)
4. Chandrasekaran, R., Mehta, V., Valkunde, T., Moustakas, E.: Topics, trends, and sentiments of tweets about the COVID-19 pandemic: temporal infoveillance study. J. Med. Internet Res. **22**(10), e22624 (2020)
5. Hidayatullah, A.F., Aditya, S.K., Gardini, S.T., et al.: Topic modeling of weather and climate condition on twitter using latent Dirichlet allocation (LDA). In: IOP Conference Series: Materials Science and Engineering, vol. 482, p. 012033. IOP Publishing (2019)
6. Hossain, M.M., et al.: Epidemiology of mental health problems in COVID-19: a review. F1000Research **9** (2020)
7. Hutto, C., Gilbert, E.: Vader: a parsimonious rule-based model for sentiment analysis of social media text. In: Proceedings of the International AAAI Conference on Web and Social Media, vol. 8, pp. 216–225 (2014)
8. Jalil, Z., et al.: COVID-19 related sentiment analysis using state-of-the-art machine learning and deep learning techniques. Front. Publ. Health **9**, 2276 (2022)
9. Kim, L., Fast, S.M., Markuzon, N.: Incorporating media data into a model of infectious disease transmission. PLoS ONE **14**(2), e0197646 (2019)
10. Kiritchenko, S., Zhu, X., Mohammad, S.M.: Sentiment analysis of short informal texts. J. Artif. Intell. Res. **50**, 723–762 (2014)
11. Maison, D., Jaworska, D., Adamczyk, D., Affeltowicz, D.: The challenges arising from the COVID-19 pandemic and the way people deal with them. A qualitative longitudinal study. PloS ONE **16**(10), e0258133 (2021)
12. Mansoor, M., Gurumurthy, K., Prasad, V., et al.: Global sentiment analysis of COVID-19 tweets over time. arXiv preprint arXiv:2010.14234 (2020)
13. Martinis, M.C., Zucco, C., Cannataro, M.: An Italian lexicon-based sentiment analysis approach for medical applications. In: Proceedings of the 13th ACM International Conference on Bioinformatics, Computational Biology and Health Informatics, pp. 1–4 (2022)
14. Medford, R.J., Saleh, S.N., Sumarsono, A., Perl, T.M., Lehmann, C.U.: An "infodemic": leveraging high-volume twitter data to understand early public sentiment for the coronavirus disease 2019 outbreak. In: Open Forum Infectious Diseases, vol. 7, p. ofaa258. Oxford University Press US (2020)
15. Mehandru, S., Merad, M.: Pathological sequelae of long-haul COVID. Nat. Immunol. **23**(2), 194–202 (2022)
16. Minaee, S., Kalchbrenner, N., Cambria, E., Nikzad, N., Chenaghlu, M., Gao, J.: Deep learning-based text classification: a comprehensive review. ACM Comput. Surv. (CSUR) **54**(3), 1–40 (2021)
17. Pennebaker, J.W., Boyd, R.L., Jordan, K., Blackburn, K.: The development and psychometric properties of liwc2015. Technical report (2015)
18. Pye, A., Roberts, S.R., Blennerhassett, A., Iqbal, H., Beenstock, J., Iqbal, Z.: A public health approach to estimating the need for long COVID services. J. Publ. Health (2021)
19. Qorib, M., Oladunni, T., Denis, M., Ososanya, E., Cotae, P.: Covid-19 vaccine hesitancy: text mining, sentiment analysis and machine learning on COVID-19 vaccination Twitter dataset. Expert Syst. Appl. **212**, 118715 (2023)
20. Rosenberg, H., Syed, S., Rezaie, S.: The Twitter pandemic: the critical role of twitter in the dissemination of medical information and misinformation during the COVID-19 pandemic. Can. J. Emergency Med. **22**(4), 418–421 (2020)

21. Rossi, R., et al.: COVID-19 pandemic and lockdown measures impact on mental health among the general population in Italy. Front. Psychiatry, 790 (2020)
22. Rushforth, A., Ladds, E., Wieringa, S., Taylor, S., Husain, L., Greenhalgh, T.: Long COVID-the illness narratives. Soc. Sci. Med. **286**, 114326 (2021)
23. Scarpino, I., Zucco, C., Cannataro, M.: Characterization of long COVID using text mining on narrative medicine texts. In: 2021 IEEE International Conference on Bioinformatics and Biomedicine (BIBM), pp. 2022–2027. IEEE (2021)
24. Scarpino, I., Zucco, C., Vallelunga, R., Luzza, F., Cannataro, M.: Investigating topic modeling techniques to extract meaningful insights in Italian long COVID narration. Biotech **11**(3), 41 (2022)
25. Sun, C., Yang, W., Arino, J., Khan, K.: Effect of media-induced social distancing on disease transmission in a two patch setting. Math. Biosci. **230**(2), 87–95 (2011)
26. Taquet, M., Luciano, S., Geddes, J.R., Harrison, P.J.: Bidirectional associations between COVID-19 and psychiatric disorder: retrospective cohort studies of 62 354 covid-19 cases in the USA. Lancet Psychiatry **8**(2), 130–140 (2021)
27. Tchuenche, J.M., Bauch, C.T.: Dynamics of an infectious disease where media coverage influences transmission. Int. Scholar. Res. Not. **2012** (2012)
28. Umair, A., Masciari, E., Habib Ullah, M.H.: Sentimental analysis applications and approaches during COVID-19: a survey. In: Proceedings of the 25th International Database Engineering and Applications Symposium, pp. 304–308 (2021)
29. Valdez, D., Ten Thij, M., Bathina, K., Rutter, L.A., Bollen, J.: Social media insights into us mental health during the COVID-19 pandemic: longitudinal analysis of Twitter data. J. Med. Internet Res. **22**(12), e21418 (2020)
30. Wicke, P., Bolognesi, M.M.: COVID-19 discourse on Twitter: how the topics, sentiments, subjectivity, and figurative frames changed over time. Front. Commun. **6**, 45 (2021)
31. Yeasmin, N., et al.: Analysis and prediction of user sentiment on COVID-19 pandemic using tweets. Big Data Cogn. Comput. **6**(2), 65 (2022)
32. Zucco, C., Calabrese, B., Agapito, G., Guzzi, P.H., Cannataro, M.: Sentiment analysis for mining texts and social networks data: methods and tools. Wiley Interdisc. Rev. Data Min. Knowl. Discov. **10**(1), e1333 (2020)

Toward the Human Nanoscale Connectome: Neuronal Morphology Format, Modeling, and Storage Requirement Estimation

Wieslaw L. Nowinski[(✉)]

Sano Centre for Computational Personalised Medicine, Krakow, Poland
w.nowinski@sanoscience.org
http://www.wieslawnowinski.com

Abstract. The human brain is an enormous scientific challenge. Knowledge of the complete map of neural connections (connectome) is essential for understanding how neural circuits encode information and the brain works in health and disease. Nanoscale connectomes are created for a few small animals but not for human. Moreover, existing models and data formats for neuron morphology description are "merely" at the microscale.

This work (1) formulates a complete set of morphologic parameters of the entire neuron at the nanoscale and introduces a new neuronal nanoscale data format; (2) proposes four geometric neuronal models: straight wireframe, enhanced wireframe, straight polygonal, and enhanced polygonal, based on the introduced neuronal format; and (3) estimates storage required for these neuronal models.

The straight wireframe model requires 18 petabytes (PB). The parabolic wireframe model needs 36 PB and the cubic model 54 PB. The straight polygonal model requires 24 PB. The parabolic polygonal model needs 48 PB and the cubic model 72 PB.

To my best knowledge, this is the first work providing for the human brain (1) the complete set of neuronal morphology parameters, (2) neuronal nanoscale data format, and (3) storage requirement estimation for volumetric and geometric neuronal morphology models at the micro and nanoscales. This work opens an avenue in human brain nanoscale modeling enabling the estimation of computing resources required for the calculation of the nanoscale connectome.

Keywords: Human Brain · Modeling · Storage · Big Data · Neuron · Dendrite · Axon · Nanoscale · Connectome

1 Introduction

The human brain empowers each of us with enormous functionality enabling perception, locomotion, behavior, emotion, and higher functions, such as cognition, attention, motivation, learning, language, and memory. It performs automatic functions for monitoring, controlling, and maintaining our whole body as well as ensuring survival. The brain determines our personality and is the source of our creativity in problem solving, construction of tools, doing research, development of technology, and artistic creation.

© The Author(s), under exclusive license to Springer Nature Switzerland AG 2023
J. Mikyška et al. (Eds.): ICCS 2023, LNCS 14074, pp. 691–698, 2023.
https://doi.org/10.1007/978-3-031-36021-3_66

Understanding the brain structure, function, and dysfunction is an enormous scientific challenge, a critical social need, and a great market opportunity. Though the average human brain has a volume of only 1,400 cm^3 it is the most complex living organ in the known universe with approximately a hundred billion neurons and a thousand trillion connections. Moreover, societies are aging. One-third of the world's adult population suffers from neurological diseases. Brain diseases are the most common and account for 13% of all diseases. The cost of neurological diseases is huge and increasing. Hence, brain research is the next huge technological wave after the space conquest and numerous large-scale projects are underway to uncover the brain's mysteries [1].

One of the key challenges is to generate a connectome at various spatial scales, a complete map of neural connections (circuits) in the brain. This knowledge is essential for understanding how neural circuits encode information and, consequently, how the brain works in health and disease [2]. This will also enable the development of applications in non-medical areas such as neuromorphic computing, artificial intelligence, intelligent machines, and energy-efficient computation. Nanoscale connectomes have been completed only for very small brains including nematode *Caenorhabditis elegans* [3], larva *Ciona intestinalis* [4], and *Drosophila* [5]. At the macroscale, the human connectome has been developed by providing anatomical and functional connectivity [6]. Anatomical connectivity employs diffusion magnetic resonance imaging techniques including diffusion tensor and diffusion spectrum imaging whereas functional connectivity exploits functional magnetic resonance imaging. However, the human nanoscale connectome has not yet been created and here I address the feasibility of producing one.

One of the most critical obstacles in obtaining complete human brain maps at micro and nanoscales is a prohibitively long overall acquisition time. For instance, using the same brain imaging protocol as was employed for *Drosophila* [5] would take an estimated 17 million years to image the whole human brain [7]. Another coarse estimate assesses that the complete reconstruction of a single 1 mm^3 of the cortex would take 10,000 man-years [8] and cost above $100 million [2].

Owing to the progress in imaging, the acquisition time substantially decreases. In particular, by employing synchrotron X-ray tomography the whole human brain acquisition time at the sub-cellular level is estimated to be reduced to a few years [9]. This promising modality, similar to standard computed tomography (CT), provides tissue imaging but with a short wavelength and much higher spatial resolution up to a single synapse [10]. For instance, by employing synchrotron tomography, whole brain imaging was performed for about 16 h at a 25 μm spatial resolution with 16-bit voxels [11]. This amount of spatial resolution enables imaging of neuronal cell bodies, however, it is insufficient to demonstrate complete neurons with their dendrites and axons. For example, for the human globus pallidus, the average cell body diameter is 33 μm, the proximal dendrite diameter is 4 μm, and the distal dendrite diameter is between 0.3–0.5 μm. Generally, the average diameter of the axon is 1 μm and the synapses, which functionally connect neurons, are at the level of 20–40 nm.

There are several formats to store and share neuron morphologies, including SWC, Eutectic, and Neurolucida. SWC is the most widely used format and it is also embedded in the state-of-the-art neuronal modeling employed in probably the most popular *NeuroMorpho.Org* repository of more than 140,000 neuronal reconstructions. It automatically calculates 21 standard morphometric parameters including the soma surface, number of branches, length, volume, angles, topological asymmetry, fractal dimension, and taper rate [12]. Though this kind of neural modeling is effective for neural tree comparison and cell characterization at the mesoscale, it is not sufficient to fully describe neurons at the nanoscale enabling the determination of synapses and neural circuits. Therefore, a new neuron morphology format at the nanoscale is required.

Our goal here is three-fold, namely, to (1) formulate a complete set of morphologic parameters of the entire neuron at the nanoscale and introduce a new neuronal nanoscale data format; (2) propose four neuronal geometric models based on the introduced neuronal format, and (3) estimate the storage required for the proposed neuronal models.

2 Neuron Morphology and Types

The neuron is the smallest functional processing unit in the nervous system. Neurons are not homogenous and diverse in terms of morphology, connectivity, neuroelectrophysiology, molecular, and genetic properties. Morphologically the neuron comprises a soma (cell body) with two processes (neurites), the dendrites and the axon. The dendrites, which play the role of receptors, receive inputs from other neurons and conduct the impulses to the soma. The axon (or rather the axonal region) acts as a neuron's projector and relays impulses to other cells.

The dendrites form a set of dendritic trunks (stems) each with a dendritic tree. Each tree comprises branches along which dendritic spines with postsynaptic terminals are located. The axonal region contains the hillock which is the soma-axon connector and continues as the axon proper (axonal trunk or stem) terminating as an axonal tree with branches that contain presynaptic terminals.

Neurons are highly variable and no two neurons are the same. Some examples of neuronal types are multipolar (projection and inter) neurons, pyramidal cells, Purkinje cells, or bipolar neurons. It has been evident that there is a plethora of neuronal types and though their total number is still unknown it could be as high as 1000 cell types [13]. Identifying the different brain cell types to determine their roles in health and disease is of great importance [2]. Toward achieving this goal a whole-brain cell atlas is under development [14] that integrates, morphological, physiological, and molecular annotations of neuronal cell types for a comprehensive characterization of cell types, their distributions, and patterns of connectivity.

3 Neuronal Morphologic Parameters and Nanoscale Data Format

To characterize the morphology at the nanoscale, the four groups of parameters are required for the neuron, soma, dendrites, and axon. The proposed neuronal morphologic data format at the nanoscale (nN format) is the following:

NEURON
> Neuron ID
> SOMA
>> Center coordinates
>> Surface descriptor
> DENDRITES
>> Number of dendritic trunks
>> For each trunk
>>> Trunk ID
>>> Proximal coordinates
>>> Proximal diameter
>>> Dendritic tree root coordinates
>>> Dendritic tree root diameter
>>> Number of bi(multi)furcations in the dendritic tree
>>> Number of terminals in the dendritic tree
>>> For each dendritic tree bi(multi)furcation
>>>> Dendritic tree bi(multi)furcation ID
>>>> Dendritic tree bi(multi)furcation coordinates
>>>> Dendritic tree bi(multi)furcation diameter
>>> For each dendritic tree terminal
>>>> Dendritic tree terminal ID
>>>> Dendritic tree terminal coordinates
>>>> Dendritic tree terminal diameter
> AXON
>> Axon ID
>> Hillock proximal (at soma) coordinates
>> Hillock proximal diameter
>> Axonal trunk proximal coordinates
>> Axonal trunk proximal diameter
>> Axonal tree root coordinates
>> Axonal tree root diameter
>> Number of bi(multi)furcations in the axonal tree
>> Number of terminals in the axonal tree
>> For each axonal tree bi(multi)furcation
>>> Axonal tree bi(multi)furcation ID
>>> Axonal tree bi(multi)furcation coordinates
>>> Axonal tree bi(multi)furcation diameter
>> For each axonal tree terminal
>>> Axonal tree terminal ID
>>> Axonal tree terminal coordinates
>>> Axonal tree terminal diameter

4 Neuronal Morphology Geometric Models

Based on the introduced nano neuron (nN) data format, we consider four neuronal morphology geometric models: straight wireframe, enhanced wireframe, straight polygonal, and enhanced polygonal.

The straight wireframe neuronal model is the simplest. The soma is represented as a center point, the neuronal branches as straight line segments with the start and end points being bifurcations, and the bifurcations and presynaptic and postsynaptic terminals as points.

In the enhanced wireframe neuronal model, in comparison to the straight wireframe one, each branch, besides its start and end points, is determined by additional intermediate points. Hence, a branch forms a polyline segment. Alternatively, the branch points can be connected by the cardinal splines forming a curved branch. With a single intermediate point, the branch is parabolic, and with two points cubic.

The straight polygonal neuronal model, in comparison to the straight wireframe one, requires the knowledge of diameters at the bifurcations and terminals. In this model, the soma, dendrites, and axons are modeled as polygonal surfaces. The soma can have a predefined shape, such as a sphere or pyramid, or a free shape. In the latter case, the soma can be created via iso-surfacing by employing, e.g., the Marching Cubes algorithm [15]. The dendritic and axonal branches are modeled as cylinders or truncated cones.

The enhanced polygonal neuronal model, in comparison to the straight polygonal one, requires the determination of intermediate points considered the centers of cross-sections along with their corresponding diameters. To get a more accurate and better quality of branch surfaces they can be modeled as tubular segments created by subdivision with centerline smoothing and diameter outlier removal [16].

5 Storage Requirements for Volumetric and Geometric Neuronal Morphology Models

Here we consider storage requirements for two classes of neuronal models (1) volumetric with a sampling-dependent size containing the raw (unprocessed) synchrotron tomographic volumetric data, meaning the volumetric data after their reconstruction from projections; and (2) geometric with all complete neurons that have already been extracted from the volumetric data along with their calculated neuronal parameters as specified in the nN format.

5.1 Volumetric Neuronal Models

Let us first consider the case when the average brain of $1,400 \text{ cm}^3$ is acquired with $1 \text{ }\mu\text{m}^3$ spatial resolution with each voxel (sample) of 16-bit intensity resolution. This spatial sampling rate is sufficient to demonstrate the cell bodies. Then, the complete average brain of $1 \text{ }\mu\text{m}^3$ voxels requires $1.4 \times 10^3 \times 10^{12} \times 2 = 2.8 \times 10^{15}$ meaning 2.8 petabytes (PB) of storage.

However, this $1 \text{ }\mu\text{m}$ resolution is not sufficient to fully demonstrate the axons of $1 \text{ }\mu\text{m}$ diameter and according to the Nyquist sampling theorem, this sampling resolution shall be no lower than 0.5 μm. Hence, for 0.5 μm spatial sampling resolution, the required storage is increased 8 times to 22.4 PB.

To demonstrate the synapses, which are approximately 20–40 nm wide, the sampling rate shall be at least 10 nm to get the smallest synapses. So, the complete average brain of 10 nm^3 voxels requires $1.4 \times 10^3 \times 10^{18} \times 2 = 2.8 \times 10^{21}$ meaning 2,800 exabytes (EB) of storage.

5.2   Geometric Neuronal Models

For storage estimation, we take the number of neurons of one hundred billion and that of connections of one thousand trillion [17], meaning that each neuron has on average 10,000 connections. We also assume that the neurites form perfect binary trees.

Let us first consider the straight wireframe neuronal model. To store a neuron's unique identifier out of $10^{11} = 2^{36.5}$ neurons, 5 bytes (B) are needed. To store the soma, 3 center point coordinates are required. Three bytes (approximately) are needed to store a single value in the range of spatial resolutions from 1 μm ($2^{19.9}$) to 20 nm ($2^{25.6}$). Hence, the soma requires 9 B of storage. Each terminal point has 3 coordinates and requires 9 B of storage. As on average a single neuron has 10,000 terminal points, then all the neuronal terminal points need 90 KB. The number of bifurcations equals the number of terminal points minus one, as in the perfect binary tree $n = 2l - 1$, where n is the number of nodes and l is the number of leaves (i.e., including the terminal points). Each bifurcation has 3 coordinates and needs 9 B. Hence, all bifurcations require approximately 90 KB. Neglecting a small storage requirement for the neuron's identifier in comparison to that for the terminal points and bifurcations, the total storage needed for a single neuron equals approximately 180 KB. Hence, all 10^{11} neurons each of ~180×10^3 KB require approximately 18 PB of storage.

The enhanced wireframe neuronal model additionally requires intermediate points for neurite branches. Axons may need more intermediate points as they are generally much longer than dendrites. The number of these points is variable and, preferably, shall be curvature-dependent. In the perfect binary tree the number of branches $b = 2n - 2l$. Since $n = 2l - 1$, then $b = 2l - 2$ meaning that the number of branches approximately doubles that of terminals. Consider the case where each neurite branch has one intermediate point meaning that this branch requires the additional 9 B and is represented by a parabolic segment. Then, the additional storage required for a single neuron is ~2×9 B $\times 10^4 \approx 180$ KB and 18 PB for the entire brain. Hence, the parabolic wireframe model requires approximately $18 + 18 = 36$ PB of storage, the cubic wireframe model 54 PB, and every additional intermediate branch point increases the required storage by 18 PB.

Let us consider the straight polygonal neuronal model. Then, the branches are modeled as cylinders or truncated cones, and let us assume that the soma has a spherical shape. As these regular surfaces with a huge number of polygons can be calculated on demand, there is no need to store the polygonal vertices and the normals. However, the bifurcation and terminal diameters shall be taken into account. Then, to store the soma, 3 center point coordinates and 1 diameter are required meaning 12 B. Each terminal is characterized by 3 coordinates and 1 diameter, so all the terminals need 120 KB. Each bifurcation is characterized by 3 coordinates and 1 diameter, thus all bifurcations require approximately 120 KB. Neglecting a small storage requirement for the neuron's identifier and soma, the total storage needed for all 10^{11} neurons each of ~240×10^3 KB amounts to approximately 24 PB.

In the enhanced polygonal neuronal model a single intermediate cross-section with its center point and diameter requires 12 B, so the additional storage needed for a single neuron is ~2×12 B $\times 10^4 \approx 240$ KB and 24 PB for the entire brain. Hence, the parabolic polygonal model requires approximately $24 + 24 = 48$ PB of storage, the cubic wireframe model 72 PB, and every additional intermediate branch cross-section increases the required storage by 24 PB.

6 Discussion

One of the key challenges in neuroscience is to create the nanoscale human connectome. Towards this objective, we here present a complete nanoscale morphometric model of the neuron from which the synapses can subsequently be extracted, the neuronal circuits formed, and the nanoscale connectome produced. To my best knowledge, this is the first work providing for the human brain (1) the complete set of neuronal morphology parameters, (2) neuronal nanoscale data format, and (3) storage requirement estimation for both volumetric and geometric neuronal morphology models at the micro and nanoscales.

Solving neuroscience problems requires high-performance computing and big data [18]. Here we quantitatively address this data bigness in the context of the human nanoscale connectome. The size of the raw volumetric data is spatial resolution-dependent and ranges from 2.8 PB for 1 μm spatial resolution to 2,800 EB for 10 nm resolution.

The size of the straight wireframe neuronal model is approximately 18 PB. The parabolic wireframe model needs approximately 36 PB, the cubic wireframe model 54 PB, and every additional intermediate branch point increases the required storage by 18 PB. The straight polygonal model requires approximately 24 PB. The parabolic polygonal model needs 48 PB, the cubic polygonal model 72 PB, and every additional intermediate branch cross-section increases the required storage by 24 PB. Note that the branches in the parabolic models are planar with a constant curvature sign which limitations are overcome with the cubic models. The sizes of the geometric models, in contrast to those of the volumetric ones, remain constant in the range of the considered here spatial resolutions from 1 μm to 20 nm.

The sizes of the volumetric and geometric models are comparable at the spatial level enabling axon identification whereas at the level of synapses the geometric model hugely outperforms the volumetric model in terms of size. Moreover, the geometric model is easier to automatically process, enhance, extend, mine, visualize, and interact with.

The future work includes storage estimation for the neuronal circuits and the assessment of high-performance computing resources necessary to process neurons and neural circuits. This in turn may lead to the development of a human brain atlas at macro, meso, micro, and nano scales as proposed in [19].

References

1. Nowinski, W.L.: Advances in neuroanatomy through brain atlasing. Anatomia **2**(1), 28–42 (2023)
2. BRAIN Initiative BRAIN Working Group. BRAIN 2025. A Scientific Vision. NIH (2014). https://www.braininitiative.nih.gov/pdf/BRAIN2025_508C.pdf. Accessed 17 Feb 2023
3. White, J.G., Southgate, E., Thomson, J.N., Brenner, S.: The structure of the nervous system of the nematode Caenorhabditis elegans. Philosop. Trans. Royal Soc. London. Ser. B, Biol. Sci. **314**(1165), 1–340 (1986)
4. Ryan, K., Lu, Z., Meinertzhagen, I.A.: The CNS connectome of a tadpole larva of Ciona intestinalis (L.) highlights sidedness in the brain of a chordate sibling. Elife **5**, e16962 (2016). https://doi.org/10.7554/eLife.16962
5. Chiang, A.S., Lin, C.Y., et al.: Three-dimensional reconstruction of brain-wide wiring networks in Drosophila at single-cell resolution. Curr Biol. **21**(1), 1–11 (2011)
6. Van Essen, D.C.: Cartography and connectomes. Neuron **80**, 775–790 (2013)
7. Landhuis, E.: Neuroscience: big brain, big data. Nature **541**, 559–561 (2017)
8. Plaza, S.M., Scheffer, L.K., Chklovskii, D.B.: Toward large-scale connectome reconstructions. Curr. Opin. Neurobiol. **25C**, 201–210 (2014)
9. Chin, A.-L., Yang, S.-M., Chen, H.-H., et al.: A synchrotron X-ray imaging strategy to map large animal brains. Chin. J. Phys. **65**, 24–32 (2020)
10. Hwu, Y., Margaritondo, G., Chiang, A.S.: Q&A: Why use synchrotron x-ray tomography for multi-scale connectome mapping? BMC Biol. **15**(1), 122 (2017). https://doi.org/10.1186/s12915-017-8
11. Walsh, C.L., Tafforeau, P., Wagner, W.L., et al.: Imaging intact human organs with local resolution of cellular structures using hierarchical phase-contrast tomography. Nat. Meth. **18**(12), 1532–1541 (2021)
12. Akram, M.A., Ljungquist, B., Ascoli, G.A.: Efficient metadata mining of web-accessible neural morphologies. Prog. Biophys. Mol. Biol. **168**, 94–102 (2022)
13. Allen Cell Types Database. Technical white paper: overview. https://help.brain-map.org/display/celltypes/Documentation. Accessed 17 Feb 2023
14. Ecker, J.R., Geschwind, D.H., Kriegstein, A.R., et al.: The brain initiative cell census consortium: lessons learned toward generating a comprehensive brain cell Atlas. Neuron **96**(3), 542–557 (2017)
15. Lorensen, W., Cline, H.: Marching cubes: a high resolution 3-D surface construction algorithm. Comput. Graph. **21**(4), 163–169 (1987)
16. Volkau, I., Zheng, W., Aziz, A., Baimouratov, R., Nowinski, W.L.: Geometric modeling of the human normal cerebral arterial system. IEEE Trans. Med. Imag. **24**(4), 529–539 (2005)
17. DeWeerdt, S.: How to map the brain. Nature **571**, S6–S8 (2019)
18. Chen, S., He, Z., Han, X., et al.: How big data and high-performance computing drive brain science. Genom. Proteom. Bioinform. **17**(4), 381–392 (2019)
19. Nowinski, W.L.: Towards an architecture of a multi-purpose, user-extendable reference human brain atlas. Neuroinformatics **20**, 405–426 (2022)

Replacing the FitzHugh-Nagumo Electrophysiology Model by Physics-Informed Neural Networks

Yan Barbosa Werneck[1], Rodrigo Weber dos Santos[1(✉)] [iD],
Bernardo Martins Rocha[1] [iD], and Rafael Sachetto Oliveira[2] [iD]

[1] Graduate Program in Computational Modeling, Federal University of Juiz de Fora,
Juiz de Fora, Brazil
rodrigo.weber@ufjf.br
[2] Graduate Program in Computer Science, Federal University of São Jõao del Rei,
São Jão del Rei, Brazil

Abstract. This paper presents a novel approach to replace the
FitzHugh-Nagumo (FHN) model with physics-informed neural networks
(PINNs). The FHN model, a system of two ordinary differential equations, is widely used in electrophysiology and neurophysiology to simulate
cell action potentials. However, in tasks such as whole-organ electrophysiology modeling and drug testing, the numerical solution of millions of cell
models is computationally expensive. To address this, we propose using
PINNs to accurately approximate the two variables of the FHN model at
any time instant, any initial condition, and a wide range of parameters.
In particular, this eliminates the need for causality after training. We
employed time window marching and increased point cloud density on
transition regions to improve the training of the neural network due to
nonlinearity, sharp transitions, unstable equilibrium, and bifurcations of
parameters. The PINNs were generated using NVIDIA's Modulus framework, allowing efficient deployment on modern GPUs. Our results show
that the generated PINNs could reproduce FHN solutions with average
numerical errors below 0.5%, making them a promising lightweight computational model for electrophysiology and neurophysiology research.

Keywords: Neurophysiology · Computational Electrophysiology ·
FitzHugh-Nagumo · Physics-Informed Neural Networks

1 Introduction

Computational electrophysiology and neurophysiology are fields that study the
electrical activity of cells and tissues, such as those in the heart or the brain. The
Hodgkin-Huxley (HH) model, proposed in 1952 [4], was a breakthrough in these
fields as it was the first model to explain the electrical activity of neurons. The
FitzHugh-Nagumo (FHN) model, proposed in 1961 [2], is a simplified version of
the HH model that retains its essential features while being much more computationally efficient. The FHN model is a system of two ordinary differential

J. Mikyška et al. (Eds.): ICCS 2023, LNCS 14074, pp. 699–713, 2023.
https://doi.org/10.1007/978-3-031-36021-3_67

equations that describe the dynamics of a cell action potential (AP). Because of its simplicity, the FHN model has become one of the most important building blocks in computational electrophysiology and neurophysiology, and it is widely used in simulations of neuronal and cardiac activity.

Despite its advantages, the numerical solution of millions of cell action potential models based on the FHN model is still computationally expensive, especially for large-scale simulations, such as whole-organ electrophysiology modeling [7], and drug testing and development [8]. Therefore, there is a need for lightweight computational models, also called emulators, proxies, or surrogates, that can replace the original cell model.

In this work, we present a new approach to replace the FHN model by physics-informed neural networks. This machine-learning technique uses mathematical equations that describe the phenomena during the model development or training phase. The resulting neural network model takes as input the time, initial conditions, and parameters of the FHN model and provides accurate approximations for the two variables of the original equations at the specified time.

Different from classical numerical methods for transient equations, causality is not needed after the network is trained. The computations of $u(t_1)$ and $u(t_2)$ by the network do not have to respect any relation between the time instants t_1 and t_2. For instance, t_1 can be greater than t_2. This further speeds up the calculation of the solutions since fine numerical discretizations are no longer needed.

Due to the nonlinearity of the model, sharp transitions, unstable equilibria of the solutions, and bifurcations of parameters, two techniques were used to improve the training phase of the neural network: Time window marching and increasing point cloud density on transition regions. The new physics-informed neural networks were generated with the NVIDIA Modulus framework [6] and are optimized for deployment in modern GPUs. The results show that the neural networks were able to reproduce the FHN solutions with average numerical errors below 0.5%. This approach has the potential to significantly reduce the computational cost of large-scale simulations in electrophysiology and neurophysiology.

2   Background

The use of neural networks for electrophysiology modeling has become increasingly important and interesting in recent years, particularly for complex and expensive cardiac electrophysiology models. One notable study in this area is [3], which demonstrated the ability of neural networks to emulate the intricate spatial and temporal dynamics of tissue action potential (AP) propagation.

Recently, physics-informed neural networks have emerged as a promising approach to improving data-driven models [1]. Physics-informed neural networks (PINNs) combine deep learning with physical laws to improve data-driven models. In a PINN, a neural network is trained to predict the solution to a physical problem while enforcing the governing differential equations to capture the underlying physical behavior of the system being modeled. This approach has been applied to various fields such as fluid mechanics, heat transfer, and electrophysiology modeling.

In the context of cardiac electrophysiology, several recent works have demonstrated the potential of PINNs informed by electrophysiological dynamics, using the relationships described by the Eikonal equation, to generate accurate predictions for cell activation time. For example, Bin et al. [12] and Sahli et al. [11] both presented PINN-based methods for predicting cell activation time with reasonable accuracy and performance. Additionally, Ruiz-Baier et al. [10] proposed a method to estimate the cardiac fiber architecture using physics-informed neural networks.

This concept has also been extended to more complex cell activation models. In [3], the authors used PINNs to reconstruct the entire spatiotemporal evolution of the monodomain model, accurately predicting action potential propagation in 1D and 2D meshes with only a sparse amount of data in the training. The resulting PINNs were also used to perform the inverse estimation of parameters using both *in silico* and *in vitro* data, showcasing their potential for clinical applications.

Overall, the use of neural networks and PINNs in electrophysiology modeling has shown great promise in recent years and has the potential to significantly improve our understanding and prediction of complex electrophysiological phenomena. The networks can prove to be a more efficient replacement for the numerical solvers in order to facilitate studies requiring large-scale simulations, such as whole organ simulations of the heart and brain in fine detail.

3   Methods

3.1   FitzHugh-Nagumo Model

The FitzHugh-Nagumo model is one the simplest model capable of describing the dynamics of excitable cells such as neurons and cardiac cells. It consists of two coupled ordinary differential equations (ODEs) that capture the dynamics of the cell through the activation and recovery variables. The FHN model is given by:

$$\frac{dU}{dt} = KU(U - \alpha)(1 - U) - W, \tag{1}$$

$$\frac{dW}{dt} = \epsilon(\beta U - 0.8W). \tag{2}$$

The model is particularly useful in studying the activation and all-or-nothing behaviors of the action potential generation dynamic in cardiac cells with few equations and parameters. The above equations were numerically solved with the classical Euler method in order to generate data for the training and validation phases of the neural networks.

In this study, the following model parameters are used: $\epsilon = 0.5$, $\alpha = 0.4$, and $\beta = 0.2$. The remaining parameter K as well as the initial conditions for U and W are allowed to vary.

3.2   Neural Network Architecture

In this work, the neural networks were developed with Modulus, NVIDIA's framework for neural networks, capable of incorporating data-driven and physics-informed constraints. Modulus provides a high-level interface for training and deploying PINNS and traditional neural networks. It can handle complex model geometries and parameterization to produce surrogate models for various high-dimensional problems. It is designed to be used with NVIDIA's high-end GPUs to accelerate training and evaluation.

The PINNs in this study were used to replace the FHN model, replicating the temporal evolution of the solutions in the whole parameter space, and for any initial conditions. The proposed NN model has the following form:

$$u_{net}(U_0, W_0, K, t) = U, W. \tag{3}$$

To handle this problem a network topology was designed using a total of 10 hidden layers, each one with 300 nodes, all fully connected, i.e., a Multi-Layer Perceptron (MLP) architecture. The hidden layers connect the input layer, which consists of time, the initial conditions, and the model parameter, to an output layer with nodes for the variables U and V.

The neural network proposed in this study was trained using a loss function that incorporates both physics-informed and data-driven constraints, providing the network with actual data and prior knowledge of the system dynamics. As a result, the network produces more accurate solutions and converges faster.

The physics constraints are incorporated by accounting for the consistency of the solutions predicted by the network in relation to the system's known governing equations, described by the FHN model:

$$f_u(U, W) = KU(U - 0.4)(1 - U) - W \tag{4}$$

$$f_w(U, W) = 0.5(0.2U - 0.8W) \tag{5}$$

The loss function comprises the sum of losses for each variable, evaluated in a domain using a specified batch size and aggregated using the L_2 norm. This interior loss function, L_i is expressed as:

$$L_I(U_0, W_0, K, t) = \frac{1}{N_{batch}} \left(\sum^{batch} \frac{\partial u_{net}(U_0, W_0, K, t)}{\partial t} - f(U, W) \right)^2, \tag{6}$$

where $batch$ is a meta-parameter used to define the number of points used to evaluate the above expression during the training phase. In particular, these points will be randomly chosen.

The resulting constraint requires the derivative of the output of the model with respect to the input parameters, such as time. In order to calculate it, the network u_{net} is assumed to be differentiable, which is a reasonable assumption since the activation functions are infinitely differentiable. The derivatives are then calculated by Modulus using Automatic Differentiation [9].

Additionally, to ensure accurate solutions, another constraint is imposed on the neural network to enforce proper initial conditions. This is achieved by measuring the error of the network's prediction at the initial time in relation to a constant or parameterized initial state. The initial condition constraint, L_B has the following form:

$$L_B(U_0, W_0, K) = \frac{1}{N_{batch}} \sum_{}^{batch} (u_{net}(U_0, W_0, K, t = 0) - (U0, W0))^2 \quad (7)$$

Finally, a data-driven constraint is also introduced. It evaluates the similarity of the network predictions in relation to data generated using numerical results of the FHN equations. Similarly, the data constraint, L_D consists of the L_2 aggregation of the error function in a number of points defined by the batch size:

$$L_D = \frac{1}{N_{batch}} \sum_{}^{N_{batch}} \frac{1}{2}(u_{net}(U_0, W_0, K, t) - u_{solver}(U_0, W_0, K, t))^2. \quad (8)$$

For this, a training set was assembled containing the time evolution of different solutions in the parameter space, explored on a regular grid manner, with 10 homogeneously spaced values for each parameter. The solutions are obtained using the simple forward Euler numerical solver with a time step of 0.01.

The total loss function was formulated simply by the weighted sum of the previous constraints, and is given by:

$$L_T = \lambda_D L_D + \lambda_I L_I + \lambda_B W L_B. \quad (9)$$

The λ coefficients control the relative weighting of each constraint in the total loss. In this implementation, the initial condition constraint is heavily imposed with a much larger weighting ($\lambda_B > \lambda_I, \lambda_D$) to ensure its effectiveness in eliminating non-viable solutions. However, this does not mean the solution is overly constrained since the initial condition constraint only evaluates at $t = 0$. By giving greater weight to the initial condition constraint, the network can learn the correct initial conditions for the system, which is crucial for generating reliable solutions. At the same time, the other constraints are still crucial for ensuring that the solutions are consistent with the governing equations and the available data.

A training protocol was formulated for PINNs with a fixed amount of training steps (3000). The loss function was minimized using the classical Adam optimizer. An exponential decaying function was used for the learning rate:

$$l_r = l_{r0}\gamma^{step}. \quad (10)$$

This allows the network to learn more aggressively in the initial training steps and only make minor adjustments to fine-tune the solutions in the final steps.

3.3   Advanced Techniques

In this work, we use PINNs to replace the FHN model with increasing dimensions of parameterization. As a result, the complexity of the problem increases at each stage. To tackle the more challenging problems in the later stages, we implemented two techniques to enhance model training efficiency and improve the overall accuracy, which are described next.

Time Window Marching. Large time domains, with heterogeneous scales of solution behavior, pose additional challenges during the training phases. In our case, we have a rapid activation, a slow recovery, followed by a long steady state rest, resulting in vastly different behaviors across time.

To mitigate this issue, we implemented the time window marching technique, which is illustrated in Fig. 1. The idea of this technique is to divide the time domain into smaller windows and solve them separately. Different networks are trained separately for each time window.

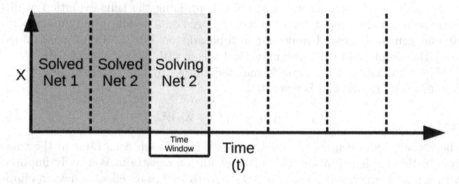

Fig. 1. Time window scheme, each training window consists of a region of the time domain to be trained separately (adapted from [6]).

An additional constraint is considered to connect the solution of each window to the next, which imposes the continuity of the solution on the interface of the time windows. The constraint is given by:

$$L_W(Y) = \frac{1}{N_{batch}} \sum_{}^{batch} (u_{net}(U_0, W_0, K, t)^{prevW} - u_{net}(U_0, W_0, K, t)^{nextW})^2.$$

(11)

The points are sampled at the window interface, i.e., the final time of the previous window, and the initial condition of the next. This constraint forces the initial conditions of the next window and thus a weight similar to the one of the initial condition constraint was adopted.

Overall, time window marching is a helpful technique to handle large and complex time domains, especially when the solution behavior is highly heterogeneous across time. It also helps reduce the computational cost and memory usage required for solving the problem.

Increasing Point Cloud Density on Transition Regions. Another common source of complexity, and consequently also a source of error, are the regions in the parameter space where the solution behavior changes rapidly. This happens, for instance, when the solution is near unstable equilibria or when the parameters are near bifurcations. An example of this complexity can be seen in the activation threshold, generated by the term $(u - \alpha)$ of the FHN equations. Solutions with U_0 above the threshold, known as supra-threshold, trigger an action potential, whereas solutions below the threshold, known as sub-threshold, only decay to the steady state. Due to the scarcity of data in these critical regions, the resulting predictions can often contain significant errors.

This problem is addressed by constraints implemented to evaluate additional points in such transition regions, forcing the network to learn more about the solution in those regions with more physics-generated data. This approach accelerates convergence at a small cost, proportional to the batch training size used in the transition regions.

The constraint used has the same form as the regular psychs constraints but is enforced only on points within the critical region. It also has a separate λ coefficient for the weight.

3.4 Validation

In order to assess the quality of the trained networks, validation sets were also generated. This generation was done by sampling the numerical solutions for different parameter sets than the ones used in training and using the same numerical solver (forward Euler). The validation loss was also calculated with the L_2 error for the whole domain, which compares numerical solutions and the predictions of the network.

4 Results

4.1 Basic Scenario with Time as the Single Parameter

The first proposed problem is a simple time-only parameterization of the FHN model. Requiring the network to predict the solution of the model for the whole time domain for specific initial conditions and set of parameters. The proposed model has the form:

$$u_{net}(t) \approx FHN(t). \tag{12}$$

The training was done by minimizing the total loss based only on physics-informed constraints of the interior of the time domain, $t \in (0, 10)$ with a batch size of 3000 points, and the boundary of the time domain, $t = 0$, with a batch size of 500.

For such a simple problem, data was unnecessary, and a very good match was obtained using only physics and the time window technique. The time domain was divided into 20-time windows, each with the size of a time unit, and training separately for 3000 steps. This might be a little costly, computationally wise, but allows the network to train for this particular problem only with the prior knowledge of the mathematical equations that describe the dynamics of the system.

Figure 2 shows how the network is capable of solving the problem for the whole time domain, closely matching the numerical solver.

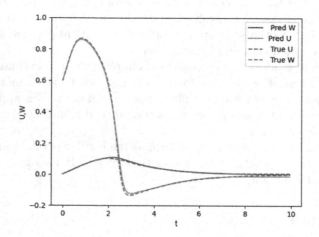

Fig. 2. Scenario with one input parameter: time. Predictions of the neural network for the $FHN(t)$ model (pred) compared to the numerical solution using the forward Euler method (true).

The resulting network prediction achieved a mean error of 0.005, with a relative error of around 0.5%. The maximum error was 0.002. For such a simple problem, high precision is expected; what makes this interesting is that the network managed such precision without the need for data, with only prior knowledge of the system dynamics and the initial state.

4.2   PINN Parameterized by Time and the Initial Condition for U

The next step was to increase the problem's complexity by parameterizing part of the initial condition. The initial condition U_0 of the solution was parameterized, requiring the network to predict solutions for any initial condition U_0 and for any time instant, t. The proposed network to solve the problem is described by:

$$u_{net}(t, U_0), \approx FHN(t, U_0).\tag{13}$$

In this case, the increase in complexity required adding a data-driven constraint, as the network did not produce a good fit with only physics-driven constraints and with the limitation, for the sake of comparison, of 3000 iterations. The total loss function contains information on particular solutions, data-driven, and on the governing equations of the FHN model, physics-informed. Additionally, a constraint for initial conditions is also imposed. The time domain was divided into 10 time windows for this problem.

The network was trained with a batch size of 1000 for the data-driven constraint and 2000 for the physics-driven constraint, with the same 500 batch size for the initial condition. It was also set to train for 3000 steps per window.

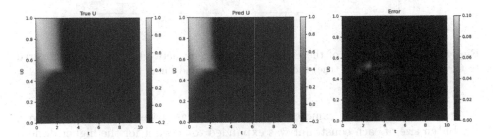

Fig. 3. Neural network's solutions to the $FHN(t, U_0)$ problem. The first column shows the true solution, while in the second the network prediction is presented, and finally, in the third column the error between the two is displayed. In this scenario, t and U_0 parameterize the neural network.

In this case, the network solution had a mean error of 0.0012 for the U variable, representing a good approximation with less than 0.2% relative error. Furthermore, by analyzing individual solutions, as shown in Fig. 4, one can see how the network manages to replicate the action potential dynamics of the model.

The maximum error in the solution space is 0.04, as shown in Fig. 3. It occurs in a very localized region near the threshold $\alpha = 0.4$. This region contains solutions very distinct since this is an unstable equilibrium.

4.3   PINNs for Any Initial Conditions of the FHN Equations (U_0, W_0) and Time

Now the network has to solve the whole temporal evolution of the FHN model for a fixed set of parameters and any initial condition. Effectively, replacing the numerical solver for a given set of model parameters. The proposed model has the form:

$$u_{net}(t, U_0, W_0) = U, W(t, U_0, W_0). \tag{14}$$

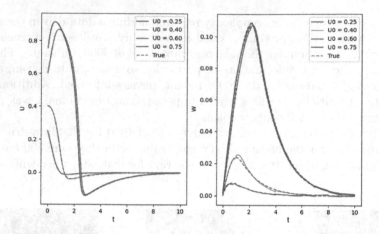

Fig. 4. Specific solutions of the $FHN(t, U_0)$ problem: network predictions compared to the numerical scheme (true) indicated by the dashed line.

In this problem, the same constraints of the last case were used. However, the batch size of each constraint was expanded to account for the higher dimensionality of the problem, which, of course, results in a higher training time when compared to previous cases.

Both the physics and the data constraints had a λ weighting of one and the IC constraint was heavily enforced with a weighting of 1000. The network was trained with batch sizes of 7000 for the data constraints, 5000 for the physics constraints, and 500 for the boundary constraints.

Figure 5 shows how the network is capable of replicating the model in the whole solution space. And that most of the error is still localized in the region immediately above the threshold.

Figure 6 presents specific network predictions, which had a mean error of only 0.001 and a maximum error of 0.01, meaning a relative mean and maximum error of 0.1% and 1%, respectively. From the results, we note that the neural network yields accurate enough results.

4.4   PINNs Replace the FHN Model Parameterized by Any Initial Conditions, the Parameter K, and Time

Our final experiments further increase the complexity of the problem by adding an extra dimension for the parameterization of the FHN model (the parameter K), which represents the velocity of the dynamics of the action potential. Now the network not only has to solve for any initial condition, but it also has to produce different solutions for the K parameter. The proposed model has the following form:

$$u_{net}(t, U_0, W_0, K) = U, W(t, U_0, W_0, K). \tag{15}$$

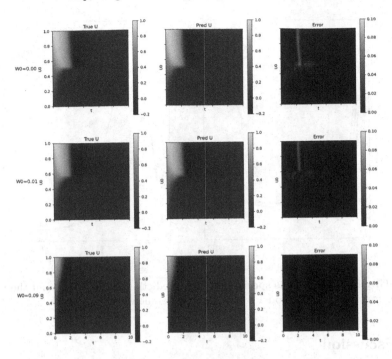

Fig. 5. Colormaps showing the evaluations of the Neural Network, FHN numerical solution and the respective error for three different values of W_0. U_0 varies along the vertical axis and time along the horizontal axis. (Color figure online)

For this problem, initially, the error in the critical region near the unsteady equilibrium was too high. Therefore, an additional constraint was explicitly added for evaluating the consistency of the solution in this region, i.e., for $U_0 \in (0.45, 0.65)$. This constraint double-enforces the solution of the FHN equations in this region.

The physics constraints were evaluated with a batch size of 2000 for the constraint enforced in the whole domain and 1000 for the one enforced on the critical region. Furthermore, the initial condition constraint had a batch size of 500, and the data-driven had a batch size of 2000. The time domain was divided in 10 windows, like in the previous examples. In this case, data and physics constraints had a λ weighting of one and the IC constraint was heavily enforced with a weight of 1000.

The network's performance for this problem is worse than the previous examples, with a maximum error of 0.18 (18% of maximum relative error). However, we can see in Fig. 7 that this high error is highly localized. The relative mean error measured was 0.1%, showing that the network has good overall accuracy.

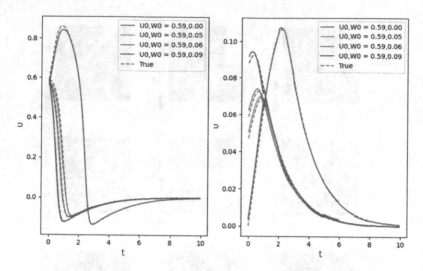

Fig. 6. Graphics showing the specific network predictions (solid lines) and the numerical solutions of the FHN model (dashed) for different values of W_0.

5   Discussion

Table 1 summarizes all the presented results. Note that in each different scenario presented the generation of the PINNs took less than one hour of execution time on a single NVIDIA Tesla T4 with 16GB of memory. Also, note that the number of advanced techniques used to improve the results steadily increased from only physics-informed constraints and the time window technique (PINNs parameterized only by time) to physics-informed and data-driven constraints, time window technique and an increased point cloud density on a critical region near an unsteady equilibrium (PINNs parameterized by time, the initial conditions, and the parameter K of the FHN model).

It's important to note that although several techniques were employed to improve model accuracy and training rate there is much more room to improve. One conceivable way is to explore the use of different neural network architectures. Recurrent architectures can be particularly suited for this problem due to their time-aware nature that propagates the solution through time, at the cost of being traditionally more expensive than MLP [13]. Another architecture worth mentioning is the Fourier-based one [5]. By exploring different neural network architectures, it may be possible to find more efficient and effective models for solving this problem. However, it is important to carefully consider the trade-offs between model accuracy and computational cost, which is a topic of future research.

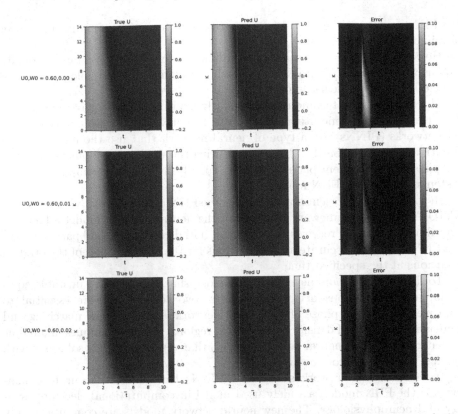

Fig. 7. Figures showing the evaluations of the Neural Network in the whole solution space, and its respective error for 3 parameter sets, U_0, W_0, and K. The results show a reduced localized error after the special constraint was imposed in the region near the unsteady equilibrium.

Table 1. Table showing each studied scenario, if physics or data constraints were used, the performance techniques employed, the PINNs training time, and the resulting accuracy.

Input parameters	Mean Error	Max error	Training time	Employed techniques
Time	0.0058	0.041	23 min	Physics + Window
Time, U_0	0.0012	0.042	12 min	Physics + Data + Window
Time, U_0 and W_0	0.0014	0.113	43 min	Physics + Data + Window
Time, U_0, W_0 and K	0.0007	0.164	25 min	All previous + region

6   Conclusions

Together with the Hodgkin-Huxley model, the FitzHugh-Nagumo model is an essential building block in computational electrophysiology and neurophysiology. The FHN is a simple model based on two ordinary differential equations that can reproduce the dynamics of a cell action potential. Despite its simplicity, the FHN model is widely used in many research fields, including neuroscience, physiology, and cardiology.

However, the numerical solution of millions of cell models is computationally expensive, particularly in tasks such as whole-organ electrophysiology modeling for the brain or the heart and drug testing and development.

Machine learning techniques, particularly neural networks, have emerged as promising tools for generating emulators in recent years. Physics-informed neural networks (PINNs) are a type of neural network that use the mathematical equations that describe the phenomena during the model development or training phase, making them particularly well-suited for modeling complex physical systems, such as the FHN model.

In this work, we present how to replace the FHN model with PINNs, using the NVIDIA Modulus framework to generate the neural network models. The new neural network model receives input time, t, initial conditions, and parameters of the FHN and offers accurate approximations for the two variables of the original equations at the specified time.

To deal with the nonlinearity of the model, sharp transitions, unstable equilibria, and bifurcations of the parameters, two techniques were essential to improve the training phase of the neural network: Time window marching and increasing point cloud density on transition regions. These techniques allowed us to generate accurate neural network models that reproduce FHN solutions with average numerical errors below 0.5%.

In conclusion, our work shows how PINNs can generate accurate emulators for the FHN model, a widely used model in computational electrophysiology and neurophysiology. The new neural network models are computationally lightweight, making them well-suited for whole-organ electrophysiology modeling, drug testing, and development. Moreover, our work demonstrates the potential of PINNs as a tool for emulating other complex physical models and highlights the importance of combining data-driven and physics-based approaches in machine learning for scientific applications.

Acknowledgements. This work was supported by NVIDIA (project "Patient-specific models of the heart for precision medicine", NVIDIA Academic Hardware Grant Program), the Federal University of Juiz de Fora, Brazil, through the scholarship "Coordenação de Aperfeiçoamento de Pessoal de Nível Superior" (CAPES) -Brazil-Finance Code 001; by Conselho Nacional de Desenvolvimento Científico e Tecnológico (CNPq)-Brazil Grant numbers 423278/2021-5, 308745/2021-3 and 310722/2021-7; by Fundação de Amparo à Pesquisa do Estado de Minas Gerais (FAPEMIG) TEC APQ 01340/18.

References

1. Cai, S., Mao, Z., Wang, Z., Yin, M., Karniadakis, G.E.: Physics-informed neural networks (PINNs) for fluid mechanics: a review. Acta. Mech. Sin. **37**(12), 1727–1738 (2021)
2. FitzHugh, R.: Impulses and physiological states in theoretical models of nerve membrane. Biophys. J . **1**(6), 445–466 (1961)

3. Herrero Martin, C., et al.: Ep-PINNS: cardiac electrophysiology characterisation using physics-informed neural networks. Front. Cardiovasc. Med. **8**, 2179 (2022)

4. Hodgkin, A., Huxley, A.: A quantitative description of membrane current and its application to conduction and excitation in nerve. J. Physiol. **117**(4), 500–544 (1952)

5. Ngom, M., Marin, O.: Fourier neural networks as function approximators and differential equation solvers. Stat. Anal. Data Min. ASA Data Sci. J. **14**(6), 647–661 (2021)

6. NVIDIA: Nvidia modulus (2022). https://www.nvidia.com/en-us/docs/developer/modulus/

7. Oliveira, R.S., et al.: Ectopic beats arise from micro-reentries near infarct regions in simulations of a patient-specific heart model. Sci. Rep. **8**(1), 1–14 (2018)

8. Passini, E., et al.: Human in silico drug trials demonstrate higher accuracy than animal models in predicting clinical pro-arrhythmic cardiotoxicity. Front. Physiol. 668 (2017)

9. Paszke, A., et al.: Automatic differentiation in PyTorch (2017)

10. Ruiz Herrera, C., Grandits, T., Plank, G., Perdikaris, P., Sahli Costabal, F., Pezzuto, S.: Physics-informed neural networks to learn cardiac fiber orientation from multiple electroanatomical maps. Eng. Comput. **38**(5), 3957–3973 (2022)

11. Sahli Costabal, F., Yang, Y., Perdikaris, P., Hurtado, D.E., Kuhl, E.: Physics-informed neural networks for cardiac activation mapping. Front. Phys. **8**, 42 (2020)

12. bin Waheed, U., Haghighat, E., Alkhalifah, T., Song, C., Hao, Q.: Pinneik: Eikonal solution using physics-informed neural networks. Comput. Geosci. **155**, 104833 (2021)

13. Wu, B., Hennigh, O., Kautz, J., Choudhry, S., Byeon, W.: Physics informed RNN-DCT networks for time-dependent partial differential equations. In: Groen, D., de Mulatier, C., Paszynski, M., Krzhizhanovskaya, V.V., Dongarra, J.J., Sloot, P.M.A. (eds.) ICCS 2022, Part II. LNCS, vol. 13351, pp. 372–379. Springer, Cham (2022). https://doi.org/10.1007/978-3-031-08754-7_45

Sensitivity Analysis of a Two-Compartmental Differential Equation Mathematical Model of MS Using Parallel Programming

Matheus A. M. de Paula[1,2], Gustavo G. Silva[2], Marcelo Lobosco[1,2],
and Bárbara M. Quintela[1,2(✉)]

[1] Graduate Program in Computational Modelling (PGMC), Federal University
of Juiz de Fora (UFJF), Juiz de Fora-MG, Brazil
barbara.quintela@ufjf.br

[2] FISIOCOMP, Computational Physiology and High-Performance Computing
Laboratory, Federal University of Juiz de Fora (UFJF), Juiz de Fora-MG, Brazil

Abstract. Multiple Sclerosis (MS) is a neurodegenerative disease that
involves a complex sequence of events in distinct spatiotemporal scales
for which the cause is not completely understood. The representation of
such biological phenomena using mathematical models can be useful to
gain insights and test hypotheses to improve the understanding of the
disease and find new courses of action to either prevent it or treat it
with fewer collateral effects. To represent all stages of the disease, such
mathematical models are frequently computationally demanding. This
work presents a comparison of parallel programming strategies to optimize the execution time of a spatiotemporal two-compartmental mathematical model to represent plaque formation in MS and apply the best
strategy found to perform the sensitivity analysis of the model.

Keywords: Computational Biology · Parallel Programming · Multiple
Sclerosis

1 Introduction

Multiple Sclerosis (MS) is a neurodegenerative disease with onset in early adult-
hood, and the most common form is Relapse-Remitting MS (RRMS), characterized by symptom attacks and remission phases [4]. Mathematical models have
been applied to other brain diseases [1,9,13], and several models have been proposed for MS, including models of self-tolerance [7], drug effects [16], disease progression [15], and initial damage by the immune system. Our previous paper [3]
proposed a new model to represent the damage caused by MS, influenced by the
dynamics of immune system activation in the nearest lymph node. The main
objective of this work is to identify critical model parameters in each phase of
MS using sensitivity analysis. However, sensitivity analysis is a computationally
expensive study, and since MS is a long-term disease, the model needs to simulate disease progression over long periods, requiring a parallel implementation
to reduce the computation time.

© The Author(s), under exclusive license to Springer Nature Switzerland AG 2023
J. Mikyška et al. (Eds.): ICCS 2023, LNCS 14074, pp. 714–721, 2023.
https://doi.org/10.1007/978-3-031-36021-3_68

This paper is organized as follows. Section 2 briefly reviews the main characteristics of our mathematical model. The same section also presents the techniques used to develop a parallel version of the code and the sensitivity analysis used in this work. Section 3 presents and discusses the results. Finally, Sect. 4 presents our conclusions and plans for future works.

2 Methods

2.1 Mathematical Model

MS is characterized by an infiltration of lymphocytes across the blood-brain barrier into the brain. The microglia is stimulated to attack the oligodendrocytes. After being destroyed, brain parenchymal cells are captured by dendritic cells. Thus, the dendritic cells become activated and migrate to the lymph node, acting as presenting antigen cells (APC) and stimulating adaptive immune system [10, 14]. Then, $CD8^+$ T cells and antibodies migrate to the brain parenchyma [12]. $CD8^+$ T cell stimulates microglia and attacks brain cells, while antibodies do the opsonization of brain cells [11,17].

The mathematical model represented in Fig. 1 comprises two distinct compartments: the brain parenchyma (i.e., tissue) and the peripheral lymph node [3].

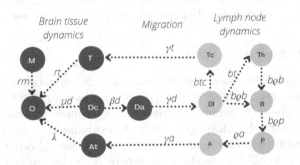

Fig. 1. Two-compartment mathematical model for MS. Dendritic cells migrate to the peripheral lymph node and activate the T and B cells recruiting antibodies. Adapted from [3].

The model employs a set of 6 PDEs to depict the dynamics of RRMS in the spatial domain Ω and temporal domain I. These PDEs correspond to the following entities: Microglia (M), Oligodendrocytes (O), Conventional Dendritic Cells (D_C), Activated Dendritic Cells (D_A), Antibodies (A_t), and $CD8^+$ T Cells (T). Additionally, a set of 6 ODEs, defined in the temporal domain I, represents the dynamics in the lymph node, namely: Activated Dendritic cells (D^L), $CD8^+$ T cells (T_C^L), $CD4^+$ T cells (T_H^L), B Cells (B), Plasma cells (P), and Antibodies (A_t^L). Details of the mathematical model can be consulted in our previous work [3].

2.2 Numerical Implementation

The code was implemented in C, utilizing the Finite Difference Method to solve the system of Partial Differential Equations. First-order central difference was used to numerically solve the diffusion, while chemotaxis was resolved using both first-order central difference and up-wind-down-wind. The system of ordinary differential equations was solved using the Explicit Euler method.

For our simulations, we assumed a two-dimensional domain of $20mm \times 20mm$, and the simulation time was set to represent 4 weeks of autoimmune response after the onset of microglia activation. All simulation parameters were obtained from our previous work [3].

2.3 Sensitivity Analysis

Sensitivity analysis (SA) [19] evaluates how changes in input parameters affect model outputs, helping researchers identify the most influential factors and improve model accuracy. SA is essential in assessing the robustness and reliability of mathematical models, especially those involving complex systems with many uncertain input parameters. Various methods are available to perform SA, including the one-factor-at-a-time (OFAT) approach, Sobol method, Latin hypercube sampling, and Fourier Amplitude Sensitivity Test (FAST) [2,6,18,20]. In this work, the second-order Sobol method was used, with bounds set for each parameter, and samples generated to compute the model for each sample.

2.4 Parallel Implementation

While the second-order Sobol method can provide valuable insights into the behavior of complex models, it also comes with a significant computational demand due to the large number of model runs required to estimate the sensitivity indices accurately. More specifically, to evaluate the importance of each input variable and the interactions between them, the second-order Sobol method requires at least $N(2D + 2)$ model runs, where N is the number of samples to generate, and D is the number of parameters of the model. This means that the number of model runs increases rapidly as the number of input variables and the maximum order of interaction increase. With $D = 34$ parameters and $N = 2048$, the second-order Sobol method requires the execution of $143,360$ instances of the model, which hurts performance. To overcome this issue, we implemented two parallel versions of the code, one using MPI and the other OpenMP.

The parallel solution for a coupled system of ODEs and PDEs can lead to additional communication and synchronization overheads, data dependency, and memory management issues that can impact the efficiency and accuracy of the parallelized version of the code.

In our mathematical coupled model, the exchange of data between the two compartments requires the use of synchronization primitives, *i.e.*, data must be exchanged, at each time step, between the system of PDEs, which solves the dynamics that occur in the brain tissue, and the system of ODE, which solves the dynamics on the lymph node. In other words, before the next time-step starts, it is necessary to calculate the average concentration of those populations of cells in the tissue (PDEs) that migrate to the lymph node, so that the computation of the ODEs system in the next time-step starts with the populations updated. The same occurs with the antigen population, which must be updated before the PDEs system starts its next time-step computation. Finally, the advection and diffusion terms also require synchronization at each time step.

Solving the PDEs with OpenMP (OMP). To solve the model using OMP, the algorithm first creates a team of threads. Next, the ODEs system, which represents the lymph node compartment, is solved sequentially. Then, OMP is utilized to divide the tissue domain into slices and assign each slice to a previously created thread, which will solve the PDEs in parallel in the assigned part of the domain. Since all points of the mesh require the same amount of work, we kept the default static schedule. Finally, we calculate the average concentrations of cell populations in the tissue, which will be used to solve the ODEs system in the next time step.

Solving the PDEs with Message Passing Interface (MPI). To implement a parallel version of the code using MPI, the domain must be divided into smaller subdomains, and each subdomain must be assigned to a separate MPI process. Each process performs local computations for the ODEs system and solves the PDEs within its designated domain using locally available information. Afterwards, the resulting solutions must be communicated to neighboring domains to update the boundary conditions, as depicted in Fig. 2. This communication is accomplished using `MPI_Send` and `MPI_Receive` operations. The process of solving the PDEs system and updating the boundary conditions is repeated iteratively until a converged solution is obtained.

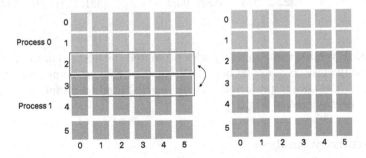

Fig. 2. To properly synchronize the mesh division between two processes, it is essential for each process to exchange borders with its neighboring processes at every time step.

3   Results and Discussion

3.1   Computational Environment

The numerical model presented in our previous work [3] was implemented using the C programming language. The code was compiled using *gcc* version 9.4.0 with the optimization flag $-O3$ enabled. In this work, the second-order Sobol method was implemented with the help of the Python library SALib [5,8] executed with Python 3.9.13. The code was executed in a 3.30 GHz Intel® Core™ $i5$-12600 CPU with 16 GB of main memory. The number of physical cores available is 6, which is the maximum number of threads/processes used during simulations.

3.2   Results for the Parallel Execution

Table 1 presents the average wall clock time obtained for ten executions of each parallel version of the code. Each parallel version was executed using two, four, and six threads/processes. The Table also presents the standard deviation and the 95% confidence interval for each execution. As a reference, the sequential version of the code executes in 734±4.73 s. The spatial discretization used in this comparison was $h_x = 0.5$ mm, and the time-step used was $h_t = 2 \times 10^{-6}$ days.

Table 1. Average execution time, standard deviation and 95% confidence interval (in seconds).

	Average		Standard Deviation		Confidence Interval	
# of Threads	MPI	OMP	MPI	OMP	MPI	OMP
2	420	386	6.13	5.50	$(418, 423)$	$(382, 390)$
4	308	246	7.17	1.92	$(303, 309)$	$(245, 248)$
6	322.2	213	0.87	2.02	$(321, 323)$	$(211, 214)$

The OMP version running on 6 nodes presented the best performance, achieving a speedup close to 3.5. The best MPI version achieved a speedup of 2.4 in 4 nodes. This difference in performance can be attributed to communication costs, such as those depicted in Fig. 2.

3.3   Sensitivity Analysis

We have run the SA considering all 34 parameters of the model. The OpenMP version with 6 threads was used to solve each sample generated by the Sobol method. The bounds used for each parameter by the Sobol method are ±10% the baseline values presented in our previous work [3]. The quantity of interest (QoI) is the total concentration of destroyed oligodendrocytes 28 days after the start of the microglia attack. To take advantage of the embarrassingly parallel nature of the SA analysis, where each run with a set of parameters is independent of other runs, we divided the execution into multiple instances that could be run in parallel on multiple machines.

Figure 3 presents the results for the first-order SA. The parameters that have more impact on the destruction of the brain cells are related to the increase in the microglia population: the rate of proliferation (μ_m), diffusion (d_M), and chemotaxis (χ). The decay rate of microglia (c_M) has a lower impact when compared to the other microglia terms, but it is also relevant to prevent/increase the destruction of oligodendrocytes.

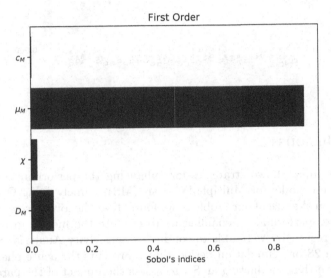

Fig. 3. Results from the first-order sensitivity analysis show that the parameters most relevant to oligodendrocyte destruction are those related to microglia proliferation and diffusion, while the other parameters appear to have no significant relevance.

Figure 4 shows the covariance of parameters obtained using the second-order Sobol method. As we can observe, there is a high covariance between μ_m and d_M. Additionally, the decay rate of microglia (c_M) is also relevant when combined with the parameters of the equations for lymph nodes and the decay rate of dendritic cells in the tissue.

The first and second-order Sobol's indices suggest a significant dependence of oligodendrocyte destruction on the concentration of microglia. Our current model configuration indicates a limited impact of other immune system cells on the disease dynamic. Therefore, it can be concluded that microglia' concentration plays a crucial role in determining the destruction of oligodendrocytes.

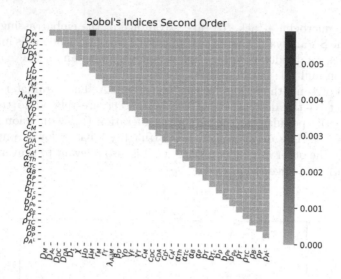

Fig. 4. Parameters covariance.

4   Conclusions

Our study compared two strategies for enhancing the performance of a differential equation model for Multiple Sclerosis (MS), namely using OpenMP and MPI. Based on the size of our problem, we found that the former strategy demonstrated better performance, enabling us to execute the model up to 3.5 times faster. As a result, we were able to conduct a Sensitivity Analysis (SA) for the model over a 28-day simulation period that represents the acute phase of MS.

We believe that conducting an SA to assess the impact of the parameters on chronic MS, which can last for years, would be a valuable next step. Our study contributes to the current understanding of MS dynamics by providing insight into the potential impact of different strategies for enhancing computational performance.

Acknowledgements. This work has been supported by UFJF, by CAPES - Finance Code 001; by CNPq - Grant number 308745/2021-3; and by FAPEMIG Grant number APQ 02830/17 and APQ-02513-22; by FINEP SOS Equipamentos 2021 AV02 0062/22.

References

1. Broome, T.M., Coleman, R.A.: A mathematical model of cell death in multiple sclerosis. J. Neurosci. Methods **201**(2), 420–425 (2011)
2. Cukier, R.I., Fortuin, C.M., Shuler, K.E., et al.: Study of the sensitivity of coupled reaction systems to uncertainties in rate coefficients. I theory. J. Chem. Phys. **59**(8), 3873–3878 (1973). https://doi.org/10.1063/1.1680571

3. de Paula, M.A.M., de Melo Quintela, B., Lobosco, M.: On the use of a coupled mathematical model for understanding the dynamics of multiple sclerosis. J. Comput. Appl. Math. **428**, 115163 (2023). https://doi.org/10.1016/j.cam.2023.115163
4. Giovannoni, G., Butzkueven, H., Dhib-Jalbut, S., et al.: Brain health: time matters in multiple sclerosis. Multiple Sclerosis Rel. Disord. **9**, S5–S48 (2016)
5. Herman, J., Usher, W.: SALib: An open-source python library for sensitivity analysis. J. Open Source Software **2**(9) (2017). https://doi.org/10.21105/joss.00097
6. Iman, R.L.: Uncertainty and sensitivity analysis for computer modeling applications. ASME Aerosp Div Publ AD., ASME, New York, NY(USA), **28**, 153–168 (1992)
7. Iwami, S., Takeuchi, Y., Miura, Y., et al.: Dynamical properties of autoimmune disease models: tolerance, flare-up, dormancy. J. Theor. Biol. **246**(4), 646–659 (2007)
8. Iwanaga, T., Usher, W., Herman, J.: Toward SALib 2.0: advancing the accessibility and interpretability of global sensitivity analyses. Socio-Environ. Syst. Modell. **4**, 18155 (2022). https://doi.org/10.18174/sesmo.18155
9. Khonsari, R.H., Calvez, V.: The origins of concentric demyelination: self-organization in the human brain. PLoS ONE **2**(1), e150 (2007)
10. Kinzel, S., Weber, M.S.: B cell-directed therapeutics in multiple sclerosis: rationale and clinical evidence. CNS Drugs **30**(12), 1137–1148 (2016)
11. Lassmann, H.: Multiple sclerosis pathology. Cold Spring Harb. Perspect. Med. **8**(3), a028936 (2018)
12. Lazibat, I.: Multiple sclerosis: new aspects of immunopathogenesis. Acta Clin. Croat. **57**(2) (2018)
13. Luca, M.: Chemotactic signaling, microglia, and Alzheimer's disease senile plaques: is there a connection. Bull. Math. Biol. **65**(4), 693–730 (2003)
14. Ludewig, P., Gallizioli, M., Urra, X., et al.: Dendritic cells in brain diseases. Biochim. Biophys. Acta Mol. Basis Dis. **1862**(3), 352–367 (2016)
15. Pappalardo, F., Russo, G., Pennisi, M., et al.: The potential of computational modeling to predict disease course and treatment response in patients with relapsing multiple sclerosis. Cells **9**(3), 586 (2020)
16. Pernice, S., Follia, L., Maglione, A., et al.: Computational modeling of the immune response in multiple sclerosis using epimod framework. BMC Bioinform. **21**(S17), 550 (2020)
17. Salou, M., Nicol, B., Garcia, A., et al.: Involvement of CD8+ T cells in multiple sclerosis. Front. Immunol. 6 (2015)
18. Saltelli, A., Annoni, P., Azzini, I., et al.: Variance based sensitivity analysis of model output. design and estimator for the total sensitivity index. Comput. Phys. Commun. **181**(2), 259–270 (2010). https://doi.org/10.1016/j.cpc.2009.09.018
19. Saltelli, A., Tarantola, S., Campolongo, F., et al.: Sensitivity analysis in practice: a guide to assessing scientific models, vol. 1. Wiley Online Library (2004)
20. Sobol', I.: Global sensitivity indices for nonlinear mathematical models and their monte carlo estimates. Math. Comput. Simul. **55**(1), 271–280 (2001). https://doi.org/10.1016/S0378-4754(00)00270-6, the Second IMACS Seminar on Monte Carlo Methods

Hierarchical Relative Expression Analysis in Multi-omics Data Classification

Marcin Czajkowski$^{(\boxtimes)}$, Krzysztof Jurczuk, and Marek Kretowski

Faculty of Computer Science, Bialystok University of Technology, Wiejska 45a,
15-351 Bialystok, Poland
{m.czajkowski,k.jurczuk,m.kretowski}@pb.edu.pl

Abstract. This study aims to develop new classifiers that can effectively integrate and analyze biomedical data obtained from various sources through high-throughput technologies. The use of explainable models is particularly important as they offer insights into the relationships and patterns within the data, which leads to a better understanding of the underlying processes.

The objective of this research is to examine the effectiveness of decision trees combined with Relative eXpression Analysis (RXA) for classifying multi-omics data. Several concepts for integrating separated data are verified, based on different pair relationships between the features. Within the study, we propose a multi-test approach that combines linked top-scoring pairs from different omics in each internal node of the hierarchical classification model. To address the significant computational challenges raised by RXA, the most time-consuming aspects are parallelized using a GPU. The proposed solution was experimentally validated using single and multi-omics datasets. The results show that the proposed concept generates more accurate and interpretable predictions than commonly used tree-based solutions.

Keywords: Relative expression analysis · Decision trees · Multi-omics data · Classification

1 Introduction

Comprehensive multi-omics analysis refers to the simultaneous study of multiple types of omics data, such as genomics, proteomics, and metabolomics, in order to gain a more complete understanding of biological systems [10]. This type of analysis can provide a holistic view of the advanced interactions within a biological system by combining data from different omics platforms and modalities. However, multi-omics data is typically high-dimensional, making it difficult to analyze and interpret. Traditional machine learning algorithms for biomedical data tend to prioritize prediction accuracy and use complex predictive models, which can impede the discovery of new biological understanding and hinder practical applications [1]. Currently, there is a strong need for simple, interpretable models that can aid in understanding and identifying relationships between specific features, and enhance biomarker discovery.

© The Author(s), under exclusive license to Springer Nature Switzerland AG 2023
J. Mikyška et al. (Eds.): ICCS 2023, LNCS 14074, pp. 722–729, 2023.
https://doi.org/10.1007/978-3-031-36021-3_69

This research focuses on the development of computational methods for biomedical analysis within the field of interpretable and explainable machine learning. The main idea is not only to perform predictions efficiently, but also provide insight, which is the ultimate goal of data-driven biology. To address this challenge, we enhanced Decision Tree (DT) [11], well-known white-box approach [1], to unlock its potential in contemporary biological data analysis. We propose DTs with splits composed of several simple tests which are based on Relative eXpression Analysis (RXA) concept [7]. In the rest of this article, we will refer to such group of tests as multi-test.

In contrast to DT, Relative eXpression Analysis (RXA) was originally developed for a specific task of identifying connections among a small group of genes [9]. RXA methods focus on analyzing the relative order of gene expressions, as opposed to their raw values, which makes them robust to various factors such as methodological and technical issues, biases, and normalization procedures. The most commonly used method within RXA is top scoring pair (TSP) analysis which examines the pairwise ordering relationships between two genes within the same sample. These pairs of genes can be viewed as "biological switches" that are linked to regulatory patterns or other aspects of gene expression networks. There are several extensions of TSP within RXA, including increasing the number of pairs in the prediction model, extending the relationships to more than two genes, and hierarchical interactions between the genes. TSP and its extensions have been successful in real-world applications due to their straightforward biological interpretation. Their effectiveness has also been recognized in other fields such as proteomics and metabolomics, but they have not yet been applied for multi-omics analysis.

A simple straightforward application of the TSP solution is not possible due to the following reasons:

– TSP is designed for binary classification, however, with the use of DT it can be successfully applied for multi-class problems [3];
– high computational complexity of the RXA solution which significally limits the size of the analyzed data;
– lack of more advanced inter-gene relations especially in the context of multi-omics data.

Some of the aforementioned issues have been already addressed in various TSP extensions. In one of them called Relative eXpression Classification Tree (RXCT) [5], the top-scoring pairs are applied as a splitting rules in a top-down induced DT. The search for pairwise relationships is paralellized using the GPU which significally improves the speed of the creating the prediction model.

In this study, we use the RXCT solution [5] as a baseline and extend the RXA concept to the multi-omics analysis. In the proposed solution called Relative Multi-test Classification Tree (RMCT), we introduce the multi-test splitting rule [4] which can be viewed as a collection composed of multiple pairwise comparisons. The feature space of each pair of attributes is limited to a single omic and preserve the clarity in interpretation by not mixing "apples and oranges". The general idea is to use multi-tests in which top-scoring-pairs are tightly linked

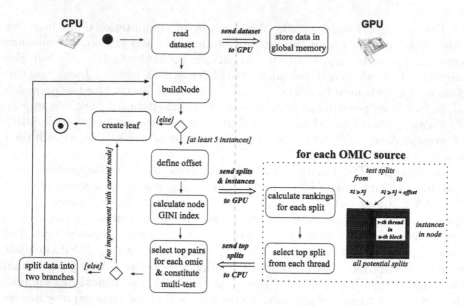

Fig. 1. General flowchart of a RMCT algorithm.

and could participate in a common pathway. The measure of similarity between the pairwise unit tests pairs that compose multi-test is the number of observations routed in the same way. It can be viewed as a mix of horizontal and vertical integrations of the multi-omics sources in each internal node of the tree.

Experimental validation of RMCT was performed on six single and two multi-omics datasets from three omics sources: gene expression, DNA methylation and miRNA. In order to evaluate the performance of the proposed concept accuracy and F1 weighted score were used.

2   Relative Multi-test Classification Tree

The new RMCT solution utilizes the RXCT algorithm as a foundation. In this section some basic steps such as DT induction and TSP creation are briefly mentioned. Next, we focus on our contribution and highlight the differences between RMCT and the RXCT system.

2.1   Overview

The general flowchart of our GPU-accelerated RMCT is illustrated in Fig. 1. It can be seen that the DT induction is run in a sequential manner on a CPU, and the most time-consuming operation is performed in parallel on a GPU. This ensures that the parallelization does not alter the behavior of the original algorithm.

The overall structure of the proposed solution is based on a typical top-down induced binary classification tree. The greedy search starts with the root node, where the locally optimal split (multi-test) is searched. Then the training instances are redirected to the newly created nodes and this process is repeated for each node till there is a noticable improvement in the Gini index (default 5%). We propose using a two-stage scoring method to enhance performance. Initially, a screening and scoring process is done using the GPU, and then the top results are further evaluated by the CPU.

As it is illustrated in Fig. 1, the data is initially transferred from the CPU's main memory to the GPU's device memory, allowing each thread block to access it. This process is done only once before beginning the tree induction, as later on only the indexes of instances that are present in a calculated node are sent. The CPU launches GPU functions, known as kernels, separately for each omic source. Each thread on the device is given an equal amount of relations, referred to as offset, to compute. The algorithm scores all possible pairwise relations from a single source and uses the Gini index to calculate the scores. This is done to make the algorithm suitable for multi-class analysis. After all the thread blocks finish their calculations, the results are transferred from the GPU's memory to the CPU's memory.

2.2 Constituting the Multi-test

The final tree node split is built on the CPU on the basis of the results from the GPU computation. We have studied three ways of constituting multi-test by taking into account two most common strategies [15] of integrating multi-omics data (see Fig. 2) and our own hybrid approach.

 (i) The horizontal integration treats each type of omics measurements equally. It can be viewed as a direct extension from single-omics data analysis to integrative analysis, where important associations between multi-level omics measurements have been identified simultaneously in a joint model. We realize horizontal integration by building the multi-test split with top-scoring-pair from each single-omic source.
 (ii) The hierarchical integration incorporates the prior knowledge of the regulatory relationship among different platforms of omics data. The integration methods are developed to more closely reflect the nature of multidimensional data. Alike in horizontal integration, we use multi-test with each pair from single-omic source. However, instead of looking at the highest scoring pairs, we focus on ones that are associated with the same class (surrogate test). This way, the multi-test stores hierarchical interactions between each source and becomes a collection of rules with similar patterns.
 (iii) The proposed hybrid method simplifies the guidelines in (ii) by eliminating the requirement to utilize steam from every source. Even though the number of pairs in the multi-test stays the same, we permit the possibility that in certain parts of the tree the data may be split using pairs from only 2 or 1 data sources. This way the hybrid approach can also work with

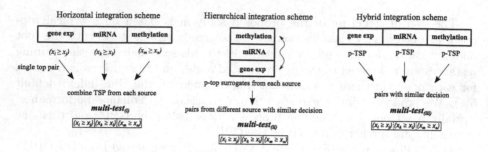

Fig. 2. Integration schemes of multi-omics data to constitute a split in internal node.

the single-omic data. The attributes that make up the pairs cannot be repeated, and similarity between the pairs remains the primary criterion for including them in the multi-test.

Finally, the splitting criterion is guided by a majority voting mechanism in which all pair components of the multi-test have the same weight.

3   Experiments

In this section, we experimentally validate the proposed RMCT approach and confront its results with popular counterparts.

3.1   Datasets and Setup

In our experiments we used datasets from Multi-Omics Cancer Benchmark TCGA Preprocessed Data repository [12]. From the list of datasets we have selected two: Glioblastoma with 4 classes (Classical: 71 instances, Mesenchymal: 84, Neural: 47, Proneural: 72) and Sarcoma with 5 classes (DDLPS: 71, LMS: 105, MFH: 29, MFS: 21, UPS: 21) as they have the largest number of patients and clinical data with defined class labels. Each dataset consists of three types of omics data: gene expression, DNA methylation, and miRNA expression. We also perform single-omics analysis in which algorithms are tested on each data source seperately. Due to the performance reasons, the Relief-F [13] feature selection was applied and the number of selected genes was arbitrarily limited to the top 500 for each omic, thus 1500 attributes for multi-omic datasets.

We evaluate the performance of the proposed RMCT solution with three multi-test variants denoted as $RMCT_i$, $RMCT_{ii}$, $RMCT_{iii}$ (see Sect. 2.2) against:

- RXCT [5], the predecessor of the RMCT approach;
- C4.5 [14]: popular state-of-the-art tree learner (Weka implementation [8] under name J48);
- JRip: rule learner - repeated incremental pruning to produce error reduction (RIPPER) [2].

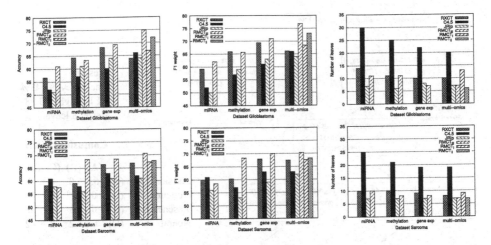

Fig. 3. The performance of the algorithms in terms of accuracy, F1 weight and size on the multi-omics Glioblastoma and Sarcoma databases and their single-omic sets.

A standard 10-fold cross-validation technique is used to evaluate the performance of the proposed solutions. The evaluation is based on accuracy, F1 weight score (a modified version of the F1 macro algorithm that takes into account the imbalance in the samples) and the number of tree leaves/rules. Due to the multi-class nature of the datasets, it was not possible to compare the RMCT with other TSP-family solutions. However, in previous research [5], we found that the RXCT algorithm outperformed other popular TSP-family algorithms on 8 real-life cancer-related datasets that concern binary classification. It should be noted that in case of single-omic data, results are only provided for $RMCT_{iii}$.

3.2 Results

Figure 3 presents a summary of the classification performance for the proposed solution, the RMCT, and its competitors for both multi-omics datasets and their individual components. The results show that RMCT outperforms the predecessor RXCT algorithm and popular white box classifiers such as the decision tree C4.5 and JRip learner. Analysis of the results using the Friedman test revealed statistically significant differences between the algorithms (p-value < 0.05) in terms of accuracy. According to Dunn's multiple comparison test [6], the RMCT with a hybrid multi-test variant ($RMCT_{iii}$) was able to significantly outperform the other algorithms on both multi-omics datasets and most of the single-omic sets. On average, most of the tested classifiers showed improved results when using multi-omics data. However, in some cases, using single-omics data resulted in more distinct patterns that distinguished between classes. In all cases, the classification models built on gene expression datasets were more accurate than those built on miRNA data. The highest prediction performance on all datasets was achieved by the RMCT algorithm with a hybrid multi-test variant ($RMCT_{iii}$).

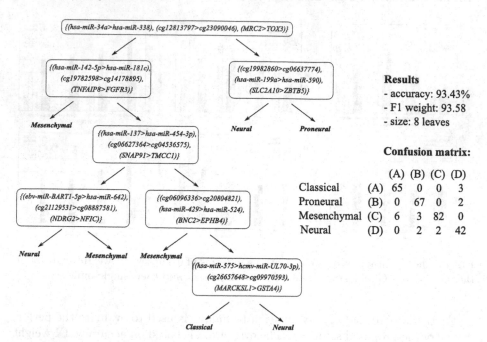

Results
- accuracy: 93.43%
- F1 weight: 93.58
- size: 8 leaves

Confusion matrix:

		(A)	(B)	(C)	(D)
Classical	(A)	65	0	0	3
Proneural	(B)	0	67	0	2
Mesenchymal	(C)	6	3	82	0
Neural	(D)	0	2	2	42

Fig. 4. An example multi-test DT induced by $RMCT$ for the Glioblastoma multi-omics dataset together with prediction results and a confusion matrix.

The complexity comparison (see Fig. 3), shows that the C4.5 algorithm generates larger trees than the other tested solutions, despite all being considered interpretable. When the classifier representation is inadequate, the true relationship is only partially captured or with the inclusion of uninformative features, making it difficult to understand and interpret the output model. This is often reflected in the increased size of the generated model, rather than a decrease in accuracy.

Due to the limited space in the paper, the authors only briefly mention that they examined the rules generated by the $RMCT$ and their biological relevance in the TCGA research network [12]. They found that on average, 25% of the features used in the models (especially in the upper parts of the tree) were directly related to the analyzed cancer, and an additional 30–40% were discussed in several papers in the medical literature. An example DT induced by the $RMCT$ for Glioblastoma multi-omic dataset is shown in Fig. 4. However, these are preliminary results, and further work is planned with biologists to better understand the gene-gene relationships generated by the $RMCT$.

4   Conclusions

The presented research explores the use of a decision tree and relative expression analysis for classification of multi-omics data. Three data integration methods, referred to as multi-test, were proposed and validated for determining the split in the tree nodes. Results from initial experiments on both single and multi-omics datasets indicate that the proposed method, RMCT, is able to identify various patterns and improve accuracy compared to other solutions. Future research aims to expand the use of integrated multi-omics analysis on proteomic and metabolomic data for pathway analysis and more specialized classification.

Acknowledgments. This project was funded by the Polish National Science Centre and allocated on the basis of decision 2019/33/B/ST6/02386 from BUT founded by Polish Ministry of Science and Higher Education (first and second author). The third author was supported by the grant WZ/WI-IIT/4/2023 from BUT founded by Polish Ministry of Science and Higher Education.

References

1. Chen, X., Wang, M., Zhang, H.: The use of classification trees in bioinformatics. Wiley Interdisciplinary Reviews: Data Mining and Knowledge 55–63 (2011)
2. Cohen, WW.: Fast effective rule induction. In: ICML95, San Francisco, CA, USA, pp. 115–123. Morgan Kaufmann (1995)
3. Czajkowski, M., Kretowski, M.: Top scoring pair decision tree for gene expression data analysis. Adv. Exp. Med. Biol. **696**, 27–35 (2011)
4. Czajkowski, M., Kretowski, M.: Decision tree underfitting in mining of gene expression data. An evolutionary multi-test tree approach. Expert Syst. Appl. **137**, 392–404 (2019)
5. Czajkowski, M., Jurczuk, K., Kretowski, M.: Relative expression classification tree. A preliminary GPU-based implementation. In: Wyrzykowski, R., Deelman, E., Dongarra, J., Karczewski, K. (eds.) PPAM 2019. LNCS, vol. 12043, pp. 359–369. Springer, Cham (2020). https://doi.org/10.1007/978-3-030-43229-4_31
6. Demsar, J.: Statistical comparisons of classifiers over multiple data sets. J. Mach. Learn. Res. **7**, 1–30 (2006)
7. Eddy, J.A., Sung, J., Geman, D., Price, N.D.: Relative expression analysis for molecular cancer diagnosis and prognosis. Technol. Cancer Res. Treat. **9**(2), 149–159 (2010)
8. Frank, E., Hall, M.A., Witten, I.H.: The WEKA Workbench. Data Mining: Practical Machine Learning Tools and Techniques. Morgan Kaufmann (2016)
9. Geman, D., d'Avignon, C., Naiman, DQ., Winslow, RL.: Classifying gene expression profiles from pairwise mRNA comparisons. Stat. Appl. Genet. Mol. Biol. **3**(19) (2004)
10. Huang, S., Chaudhary, K., Garmire, L.X.: More is better: recent progress in multiomics data integration methods. Front. Genet. **8** (2017)
11. Kotsiantis, S.B.: Decision trees: a recent overview. Artif. Intell. Rev. **39**(4), 261–283 (2013)
12. Multi-Omics Cancer Benchmark TCGA Preprocessed Data repository. http://acgt.cs.tau.ac.il/multiomic_benchmark/download.html
13. Robnik-Sikonja, M., Kononenko, I.: Theoretical and empirical analysis of ReliefF and RReliefF. Mach. Learn. **53**(1–2), 23–69 (2003)
14. Quinlan, R.: C4.5: Programs for Machine Learning. Morgan Kaufmann, San Mateo (1993)
15. Wu, C., Zhou, F., et al.: A selective review of multi-level omics data integration using variable selection. High-Throughput **8**(1), 4 (2019)

Author Index

J. Mikyška et al. (Eds.): ICCS 2023, LNCS 14074, pp. 731–734, 2023.
https://doi.org/10.1007/978-3-031-36021-3

Printed in the United States
by Baker & Taylor Publisher Services